Nitrate Esters Chemistry and Technology

Jiping Liu

Nitrate Esters Chemistry and Technology

Springer

Jiping Liu
College of Materials Science
and Engineering
Beijing Institute of Technology
Beijing, China

ISBN 978-981-13-6645-1 ISBN 978-981-13-6647-5 (eBook)
https://doi.org/10.1007/978-981-13-6647-5

Library of Congress Control Number: 2019931544

This Springer imprint is published by the registered company Springer Nature Singapore Pte Ltd.
The registered company address is: 152 Beach Road, #21-01/04 Gateway East, Singapore 189721, Singapore

Preface

Oxygen-containing nitro compounds, in which carbon and nitrogen atoms are connected through oxygen atom, are nitrate with a characteristic nitroxyl radical $-C-O-NO_2$. This book is the first summary on nitrate and includes 15 chapters, in which nitrate structure, nitration agents and process installation, preparation process of nitrate, and utilization and destruction of nitrate, are discussed and integrated coherently with the author's study. In this book, the nitration is focused on the synthesis process of important nitrogen-containing organic compounds, the nitrate structure and characteristics are particularly focused on the synthesis of important oxygen-containing organic compounds, and the manufacturing process and product performance of nitrates are extensively discussed according to the author's studies on energetic materials.

When nitrate chemistry and technology are discussed, generally, their applications in the defense and war as explosives and propellants are the focusing points. In fact, nitrates used in weapon technology and military field is only a small percentage because their major applications are in aerospace, rockets explosion, satellite communication, transportation, mining, construction, seismic exploration, airbag, waterway development, etc.

In order to use these explosive nitrates as safe as possible, studies on their basic characteristics to optimize their energy levels, combustion, explosion characteristics, stability, processing characteristics and service characteristics, are very important, which is also a way to measure the technology level of a country.

In Chap. 1, the concept of nitrates and their characterization of nitrates were introduced, and some new directions of the nitration theory were also proposed. In Chaps. 2 and 3, we analyzed the nitrating agents and the nitration reaction, and found that the attack of nitryl cation to OH^- was different from the general substitution and replacement, which could not be explained with the classic theory of chemical reactions but related to positive onium ion. Based on the relationship between catalysts and substituents and the relationship between reaction order and reaction conditions, the relationship between technology conditions in nitration process was analyzed from the easy to the complicated.

From Chaps. 4 to 9, we systematically discussed the preparation technologies of unary nitrates, binary nitrates, poly nitrates, and azide nitrates. For example, nitroglycerin is explosive when it is slightly vibrated, so it is not suitable for long-distance transportation. However, with nitration technology today, nitroglycerin can be modified to less sensitive to vibrations and thus suitable for long-distance transportation. Additionally, green nitration process of nitrates, and the properties of nitrates form sugar, starch, lignin and chitinous material, were also reviewed comprehensively and systematically. Through the summary on the recovery and comprehensive utilization of retired nitrates, new ideas to green applications of retired energetic materials were proposed that also offered a peaceful way to end and avoid wars in the world.

This book is an updated summary on nitrates and shows the real level of nitrate manufacturing technology and applications accurately and objectively. Therefore, it should be very helpful for scientists and technicians in the fields of nitrates and other energetic materials to understand the manufacturing processes, formulas, application methods and working performance of energetic materials.

During the writing of this book, apart from tireless work of myself, I would like to sincerely thank Guining Cui, a Senior Engineer, from China Tianchen for his help on drawings. I also would like to thank my colleagues in the Springer for their cooperation in the process of publishing this book.

Although the book is a summary on and a reflection of part of my studies on energetic materials area, there might be still some inevitable mistakes, and I would be very pleased to see it to be corrected by readers.

Beijing, China
March 2017

Jiping Liu

Contents

Chapter 1
Introduction of Nitrates

Nitrate is the earliest discovered class of liquid explosives [1]. It is well known that liquid explosive is nitroglycerin, is also the first explosives for mining and military applications [2], and has played an important role in the World War I and World War II [3, 4]. For nearly half century, every country has developed and produced nitroglycerin substitutes, such as ethylene glycol dinitrate, diethylene glycol dinitrate, triethylene glycol dinitrate, mannitol nitrate, nitrificated diethanolamine, nitro isobutyl glycerol trinitrate, butantriol trinitrate, butantetraol tetranitrate, butanediol dinitrate, neopentanetriol trinitrate, and neopentanetetraol trinitrate, nitrificated sugar, nitrificated starch, etc. [3]. In the literature [5], triethylene glycol dinitrate, butantriol trinitrate, and neopentanetriol trinitrate or their mixtures were mentioned intensively. The technology and performance of these nitrates can have the same level as nitrificated glycol, with breakthrough achieved in propellants of aerospace and missile, other military technologies and natural sciences [6–8].

Nitrates are widely used not only in military, but also in human health and medical technology with many contribution and many saved lives [9, 10]. The n-alkane nitrate played a very important role in energy clean and the environment [11, 12]. The most promising and produced liquid nitrates in the world are around 20 types [13, 14]. In this chapter, the concept of nitrate will be introduced, and manufacturing process, characteristics, and performance of different nitrates will be specifically described in the later chapters.

1.1 Description on the Basic Structure of Nitrate

The characteristic of nitrate is it has nitroxyl radical –C–O–NO_2. In the nitrate, ester radical is connected through oxygen and carbon atoms, which means that nitrates are oxygen-connected nitro compounds (O-nitro compound). This mechanism is confirmed with hydrolysis of nitrate or reduction of nitro group into alcohol [3, 15, 16] as,

J. Liu, *Nitrate Esters Chemistry and Technology*,
https://doi.org/10.1007/978-981-13-6647-5_1

$$R - O - NO_2 \xrightarrow{[H]} R - OH + NH_3$$

This reaction shows that nitrate and nitro compounds are essentially different, and the main difference is that nitro compounds form amine under the same conditions. Although the ester group (–NO_2) in nitrate has the same structure as nitro (C–NO_2) in nitro compound, reactivity of nitrate is high and chemical composition of its hydrolysis product is complex, indicating that nitrate might have a peroxide structure as R–O–O–NO.

Determination of chemical structure of nitrate showed that its ester group does not have peroxide structure [1, 3, 13]. The normal and only structure of nitrate might be $R{-}O{-}NO_2$.

Raman Spectroscopy, infrared spectroscopy, electron diffraction, the dipole moment, x-ray and various kinds of equipment were used to further characterize the structure of nitrate. The results supported that ester group in nitrate did not have peroxide structure, confirming the symmetrical structure of ester group in nitrate

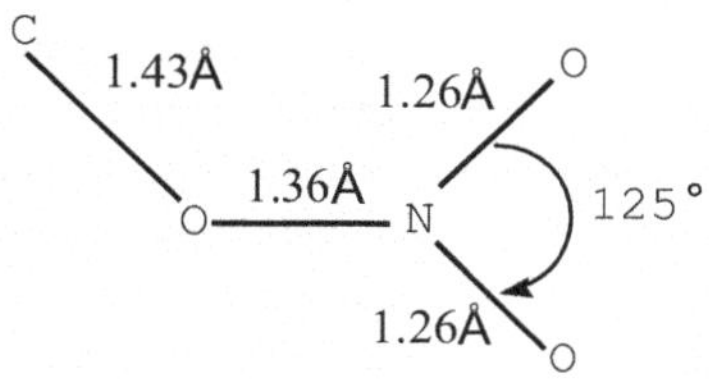

1.2 Physico-Chemical Characteristics of Nitrates

1.2.1 Physical Properties of Nitrates

Nitric acid reacts with alcohols to form esters. Esters from lower alcohols are generally liquid with higher boiling point compared with that of the corresponding alcohol, but the boiling point of the corresponding nitrite ester is much lower than the alcohol. Boiling points of C-nitro compounds are higher than that of the corresponding nitrates. Boiling points of the corresponding primary alcohols and esters were shown in Table 1.1.

Table 1.1 Boiling points of alcohol, ester, and nitro compounds (°C)

Alkyl	Alcohol	Nitrate	Nitrite ester	Nitro
Methyl	65	65	−12	101
Ethyl	78	88	17	114
n-Butyl	96	111	47	131
iso-Propyl	82	102	45	120
n-Butyl	117	136	75	151

Table 1.2 Dielectric constant, density and some other parameters of nitrate

Chemical	ε_{20}	n_{∞}^{20}	d_4^{20}	η_{20}
n-Butyl nitrate	13.10	1.39526	1.0156	0.87
Glycol dinitrate	28.26	1.43235	1.4918	4.61
1,3-Propanediol dinitrate	18.97	1.43476	1.3952	5.8
1,2-Propanediol dinitrate	26.80	1.42720	1.3774	4.65
1,3-Butanediol dinitrate	18.85	1.43259	1.3167	6.00
2,3-Butanediol dinitrate	28.84	1.42754	1.3061	4.7
nitroglycerin	19.25	1.45731	1.5931	37.8
3-Chloro-1,2-propylene glycol dinitrate	17.50	1.45850	1.5323	12.4
1,3-Dichloro-2-propylene glycol dinitrate	13.28	1.46032	1.4630	4.8
Tribromopropane	6.45	1.56190	2.4360	–
Glycerol triacetate	7.19	1.410929	1.1596	–

Alcohols are liquids with hydrogen bonds, which can explain that boiling point difference of alcohols and their esters of nitric acid are very little. Because of this association, viscosity of the alcohol is higher than that of its nitrate. On the other hand, nitro group in nitrate has semipolar bond as nitro compound, leading to its relatively high vapor pressure and its easier volatile than the corresponding alcohols.

Some scientists [3, 17] tested the various physical constants of nitrates (dielectric constant ε_{20}, refractive index n_{∞}^{20}, density d_4^{20} and viscosity η_{20}), and studied their laws. The test results were listed in Table 1.2.

In Table 1.2, the corresponding data of tribromopropane and glycerol triacetate were listed to distinguish a substituent variation.

Differences in viscosity of 1,3-propanediol dinitrate, 1,2-propanediol dinitrate, 1,3-butanediol dinitrate, 2,3-butanediol dinitrate and other similar compounds is due to the optical isomerism. Free rotation of nitrate can generate anti structure with a higher viscosity. This view explains 1,3-propanediol dinitrate and 1.3-Butanediol dinitrate have higher viscosity than other nitrates.

With the gradually deepening understanding and research on nitrate [3, 18], some physical constants nitroglycerin and some glycol dinitrate were determined and the results are shown in Table 1.3.

(1) Dipole moment

Dipole moment is one of the main factors to measure the stability of a material. Usually dipole moments of alkyl nitrates are easily measured. For example, the μ of methyl nitrate is 2.73 Debye, and μ values of alkyl nitrates do not have much relationship to their length of carbon chains. From ε and η_{∞}^{20} values of a number of nitrates, their dipole moments can be calculated out and results are shown in Table 1.4.

Table 1.3 Five constants of nitrate

Ester	Viscosity (η)		Density d_{16}^{16}	Refraction index $n_D^{21.2}$
	Temperature (°C)			
Nitroglycerin	5.1	1.033	1.5985	1.472
	20.0	0.352		
	55.0	0.0875		
Nitro ethylene glycol	7.1	0.0633	1.4918	1.446
	20.0	0.0423		
	54.4	0.0198		
1,3-Propanediol dinitrate	6.3	0.0940	1.4053	1.448
	20.4	0.0550		
	54.4	0.0275		
Diethylene glycol dinitrate	6.0	0.133	1.3890	1.4505
	20.4	0.0727		
	54.4	0.0337		
Triethylene glycol dinitrate	6.0	0.257	1.3291	1.4542
	20.3	0.119		
	54.2	0.015		

Table 1.4 Dipole moment values of some nitrates

Nitrate name	Dipole moment μ (Debye)		Nitrate name	Dipole moment μ (Debye)	
	Pure	Solution in benzene		Pure	Solution in benzene
n-Butyl nitrate	2.90	2.98	Nitroglycerin	3.82	3.16
Ethylene glycol dinitrate	4.00	3.28	3-Chloro-1,2-propanediol dinitrate	3.46	3.05
1,3-Propanediol dinitrate	3.50	3.14	1,3-Dichloro-2-propanediol dinitrate	2.86	2.59
1,2-Propanediol dinitrate	4.24	3.72	Tribromopropane	1.63	1.58
1,3-Butanediol dinitrate	3.74	3.45	Glycerol triacetate	2.50	2.58
2,3-Butanediol dinitrate	4.72	4.12			

According to this theory, the dipole moment difference of pure substance and it in benzene solution is due to a certain amount of trans-rotatory isomers formed in benzene solution.

According to studies on dipole moment [19], pentaerythritol tetranitrate may have some optical isomers. Infrared absorption indicates the general formula of nitrate is $\text{R-CHX-}\underset{\text{ONO}_2}{\text{CH}}\text{-R'}$. In nitrate, when X = ONO_2, NO_2, CN, I, Br, Cl, etc., it has optical isomers.

(2) Spectral characteristics of nitrate

Similar to C-nitro group, the O-nitro group generates an absorption band at about 270 nm in the ultraviolet region. Although the spectrometric studies on ultraviolet absorption of nitrate are limited, it has been determined that the absorption in nitrates without other light groups is lower than that of aliphatic nitro compounds ($\varepsilon = 10 - 20$). UV spectra of some nitrates are shown in Table 1.5.

Some also [20] studied the ultraviolet absorption spectrum data of nitrates. From methyl nitrate to n-butyl nitrate, their infrared spectra had been studied [21]. Some scientists [22] also studied Raman spectra of aliphatic and aromatic nitro compounds of nitrates and inorganic nitrates. According to studies on infrared spectroscopy of about 20 nitrates [23–26], including mononitrate ester, dinitrate ester, nitroglycerine, pentaerythritol diethylene, and triethylene pentaerythritol, it has been found that characteristic stretching frequencies of ester group $-ONO_2$ as

$$\begin{array}{ll} \text{Asymmetric} & 1639 \pm 13\,\text{cm}^{-1} \\ \text{Symmetric} & 1279 \pm 7\,\text{cm}^{-1} \end{array}$$

Stretching frequencies of ethylene glycol dinitrate and pentaerythritol tetranitrate are 1650–1610 cm^{-1} and 1300–1250 cm^{-1}, respectively [27]. These results are consistent to their Raman spectra according to different authors [28, 29], in which measured Raman frequency values of the two characteristic vibrations of NO_2 are close to 1640 and 1290 cm^{-1}. Authors [13, 30], with spectroscopic studies on more than 40 kinds of nitrates, shows that the range of most of the contraction vibration peaks of nitrates as shown in Table 1.6.

When R in $R{-}ONO_2$ molecule is changed, asymmetrical vibration frequencies of NO_2 are susceptible to change. According to the positive and negative inductive effect of R on the frequency of NO_2 ($\nu_{as}(NO_2)$), the primary, secondary and tertiary alkyl nitrates can be differentiated.

When $-ONO_2$ group is closer to other large group, the asymmetric and symmetric vibration bonds of NO_2 are moved to a higher frequency, which can help the identification of the optical isomers. Because of the presence of optical isomers, the

Table 1.5 UV absorption parameters of some nitrates [7, 8]

Chemicals	Solvent	The maximum adsorption	
		Wavelength (10^{-9} m)	ε
$C_2H_5ONO_2$	–	270	10
$C_2H_5ONO_2$	Ethanol	265	14.8
α-Methyl glucose-6-nitrate	Water	265	19.0
2,3,4-trimethyl-α-methyl glucose-6-nitrate	Ethanol	265	21.4

Table 1.6 Range of contraction vibration peak of nitrate

Vibrational peak	Peak (cm^{-1})
Asymmetric stretching $\nu_{as}(NO_2)$	1675–1621
Symmetric stretching $\nu_{a}(NO_2)$	1301–1268
Stretching ν(C–O)	1100–950
Stretching ν(O–N)	880–816
Vibration $\gamma_w(NO_2)$	760–737
Bending $\delta(NO_2)$	710–640
Shaking $\gamma_r(NO_2)$	580

two stretching vibration sections of NO_2 group are split into two distinct distorted peaks as

$$\begin{array}{ll}\text{Asymmetric} & 1-25/\text{cm} \\ \text{Symmetric} & 10-35/\text{cm}\end{array}$$

The internal hydrogen bond between –OH and $–ONO_2$ groups is difficult to characterize with instruments, probably because ONO_2 inhibits the intermolecular hydrogen bonds between itself and OH group 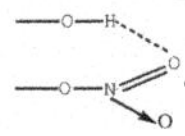.

1.2.2 *Hydrolysis of Nitrate*

Nitrates of monohydroxy alcohol and polyol can be hydrolyzed in weak basic or acidic medium, while its hydrolysis in neutral media is more difficult. The hydrolysis process is as follows:

$$RONO_2 + H_2O \rightarrow ROH + HNO_3$$

When the nitrate is hydrolyzed into alcohol and acid, a number of other compounds, such as aldehyde, ketone, carboxylic acid, unsaturated hydrocarbon, nitrous acid, etc., are formed [31].

In most cases, the original alcohol cannot be generated through the complete hydrolysis of esters, while a series of other compounds are formed.

$$C_3H_4(ONO_2)_2 + 5KOH \rightarrow KNO_2 + 2KNO_3 + CH_3COOK + HCOOK + 3H_2O$$

This equation does not represent the hydrolysis reactions of all esters [32–34]. Products from hydrolysis of different nitrates under the same conditions include phenolic resin, oxalic acid and ammonia. In the basic medium, hydrolysis products do not include glycerol [35].

When the nitroglycerine contains benzyl mercaptan, glycerin id formed from its hydrolysis in basic medium.

Due to the peroxide structure of nitrate, nitric acid can only be generated. Under the influence of agents of hydrolysis, the structure of nitrate can be changed and converted into peroxy compounds, while nitrite is generated.

$$ROONO + H_2O \rightarrow ROOH + HNO_2$$
$$RCH_2ONO_2 + H_2O \rightarrow RCH_2OOH + HNO_3$$
$$RCH_2OOH \rightarrow RCHO + H_2O$$

When nitrate was hydrolyzed in neutral medium, for example, aqueous solution of potassium nitrate was boiled to treat ethanol, ethanol cannot be oxidized [36]. The hydrolysis of tert-butanol nitrate can generate alkene. In this case, in addition to the corresponding alcohol generated, a certain amount of isobutylene is formed.

$$C_4H_9ONO_2 + 2H_2O \rightarrow C_4H_9OH_2 + H_3O^{\oplus} + NO_3^{\ominus}$$
$$C_4H_9ONO_2 + H_2O \rightarrow C_4H_8 + H_3O^{\oplus} + NO_3^{\ominus}$$

When benzyl nitrate is hydrolyzed, exception of benzyl alcohol, a certain amount of benzaldehyde is also formed as

$$C_6H_5CH_2ONO_2 + 2H_2O \rightarrow C_6H_5CH_2OH + H_3O^{\oplus} + NO_3^{\ominus}$$
$$C_6H_5CH_2ONO_2 \xrightarrow{H_2O} C_6H_5CHO + HNO_2$$

According to the number of hydrolysis of nitrates, hydrolysis of nitrate can be concluded [37] and carried out according to the following three equations:

$$RCH_2CH_2ONO_2 + Y^{\ominus} \rightarrow RCH_2CH_2Y + NO_3^{\ominus}$$
$$RCH_2CH_2ONO_2 + Y^{\ominus} \rightarrow RCH = CH_2 + NO_3^{\ominus} + HY$$
$$RCH_2CH_2ONO_2 + Y^{\ominus} \rightarrow RCH_2CHO + NO_2^{\ominus} + HY$$

Y is the hydrolysis agent.

Among these three reactions, the first is a nucleophilic substitution reaction. In the second, a hydrogen atom in the β-position can be removes. In the third reaction, α-hydrogen atom is supposed to remove.

Those experiments can only be achieved in aqueous solution of mononitrate containing aliphatic groups. Especially, methyl nitrate cannot be hydrolyzed as the second reaction, but tert-butyl nitrate cannot follow the third hydrolysis. Attention should be paid to that the third reaction of methyl nitrate is very limited and ignorable, ethyl nitrate follows the first and second reactions with only 4% of conversion, and other nitrates react according to the first hydrolysis. The rates of these three reactions are following 70:7:1.

With sodium hydrosulfide (solution alkalinity is equivalent to pH = 10–11), 93% of the ester is hydrolyzed into nitrite anion, 7% of nitrogen of nitrate is converted into ammonia. When ammonium hydrosulfide is used, concentration of the initially generated nitrite rises to its maximum point, and reduced down to zero due to the reduction of ammonium hydrosulfide. With the change of pH value, the reduction process becomes more significant. Especially, when the pH value is more than 10, it is even more significantly. Reaction will be carried out as follows:

$$C_4H_9ONO_2 + 2NaHS \rightarrow C_4H_9OH + NaNO_2 + NaHS_2$$
$$C_4H_9ONO_2 + 4NH_4HS + 3H_2O \rightarrow C_4H_9OH + 4S + 4NH_4OH$$

Sodium hydrosulfide hydrolyzes nitrocellulose [38] to make the fiber, as a regenerated cellulose, to generate products.

$$C_6H_8O_3(ONO_2)_3 + 2NaHS \rightarrow C_8H_{10}O_5 + 2NaNO_2 + 2S$$
$$2NaHS + 2S \rightarrow Na_2S_3 + H_2S$$

In the presence of hydrogen sulphide, reduction mechanism nitrocellulose is

$$RONO_2 + HO^{\ominus} \rightarrow ROH + NO_3^{\ominus}$$
$$NO_3^{\ominus} + 2HS^{\ominus} \rightarrow NO_2^{\ominus} + HS_2^{\ominus} + HO^{\ominus}$$

After nitrocellulose and primary alcohol nitrate in the presence of dilute hydroxide are hydrolyzed, and all reactions are slow. 0.1 mol pf butyl nitrate and 0.2 mol of NaOH solution in ethanol (60%) were mixed and hydrolyzed at 27 °C for 16 days, only 10% have been converted. Even when the temperature is increased to 60 °C, after 100 h of reaction, only 10% are converted in the reaction. When hydrogen sulfide was added, the hydrolysis is completed only within 4 h.

In the presence of sodium hydrosulfide, nitrate ion (compared with the nitrite ion) can be slowly reduced. Nitrite ion produced from the hydrolysis of nitrate is not the reduced product from nitrate anion. Because the nitrite ion is generated from nitrate hydrolysis, it can only be produced from the bond breaking between oxygen and nitrogen atoms. Therefore, hydrolysis of nitrates with hydrogen sulfide is mainly a reduction reaction, and its reaction mechanism can be expressed as

$$RONO_2 + S_X^{2\ominus} \rightarrow RO^{\ominus} + NO_2S_X^{\ominus}$$
$$NO_2S_X^{\ominus} + HS^{\ominus}(S^{\ominus}) \rightarrow NO_2^{\ominus} + HS_{X+1}^{\ominus}(S_{X+1}^{\ominus})$$
$$HS_{X+1}^{\ominus} + H_2S^{\ominus} \rightleftharpoons HS_X^{\ominus} + HS_2^{\ominus}$$
$$RO^{\ominus} + HS_X^{\ominus} \rightleftharpoons ROH + S_X^{2\ominus}$$
$$HO^{\ominus} + HS_X^{\ominus} \rightleftharpoons H_2O + S_X^{2\ominus}$$
$$RONO_2 + S_X^{2\ominus} \rightarrow ROS_X^{\ominus} + NO_2^{\ominus}$$
$$ROS_X^{\ominus} + HS^{\ominus} \rightarrow RO^{\ominus} + HS_{X+1}^{\ominus}$$
$$HS_{X+1}^{\ominus} + H_2S^{\ominus} \rightleftharpoons HS_X^{\ominus} + HS_2^{\ominus}$$
$$RO^{\ominus} + HS_X^{\ominus} \rightleftharpoons ROH + S_X^{2\ominus}$$
$$HO^{\ominus} + HS_X^{\ominus} \rightleftharpoons H_2O + S_X^{2\ominus}$$

Because of the activity of sulfur ion and sulfide ion, they can lead to O–N bond cleavage to form alkoxyl and sulfur or polysulfide nitrate ion. In various similar chemical reactions, it should be taken account that sulfide has the impact on the reduction of oxygen atom, resulting the generation of nitrite ion which cannot be formed through the reduction of nitrate.

With the presence of hydrazine in nitrate, nitrate mainly works as an antioxidant to be involved in the explosion.

Ethyl nitrate and hydrazine chemical react at high temperatures to generate aniline, ammonium nitrate and nitrogen. When sodium ethoxide is present, even at room temperature, nitrite nitrogen, the hydrolysis can generate nitrite ion, nitrogen, benzene, azide benzene, nitrobenzene, aniline, acetic acid, acetaldehyde, etc.

In the presence of palladium or platinum, hexyl nitrate reacts with hydrazine with products as alcohol, nitrous oxide, and nitrogen.

$$2C_6H_{13}ONO_2 + 2N_2H_4 \rightarrow 2C_6H_{13}OH + N_2O + 3H_2O + 2N_2$$

In the reaction, if methyl hydrazine is used to replace hydrazine, hexyl nitrate cannot generate methyl ether, but hexanol.

In the reaction products of alkyl nitrate and hydrazine, nitrate, nitrite, the corresponding alcohol, alkyl hydrazide, nitrogen oxide, ammonia and trace amounts of formaldehyde can be detected. If the reaction occurs without solvent and with an excess of hydrazine, the principal reaction is reduction [39].

When β-chloroethyl nitrate and 1,3-dichloropropyl nitrate, nitroglycerin and ethylene glycol dinitrate are hydrolyzed, the gas releasing is more violent than that in hydrolysis of non-substituted aliphatic nitrates.

The reaction of nitrate with hydrazine is the reaction between methyl hydrazine and ethyl nitrate, in which methane (not ethane) is formed. These results suggest the existence of monomethyl succinimide as the intermediate as

$$CH_3HNNH_2 + C_2H_5ONO_2 \rightarrow CH_3N = NH + C_2H_5OH + HNO_2$$
$$CH_3N = NH \rightarrow CH_4 + N_2$$

With the reduction over hydrazine, the concentration of nitrite in the products is gradually decreased.

$$HNO_2 - N_2H_4 \rightarrow NH_3 + N_2O + H_2O$$
$$HNO_2 + N_2H_4 \rightarrow NH_3 + 2H_2O$$
$$HNO_2 + HN_3 \rightarrow N_2 + N_2O + H_2O$$

When benzyl nitrate reacts with hydrazine, a series of substitute products can be produced as: $C_6H_5CH_2HNNH_2$ and $(C_6H_5CH_2)_2NNH_2$ and $(C_6H_5CH_2)_2NNHCH_2$ C_6H_5 and $(C_6H_5CH_2)_2NN(CH_2C_6H_5)_2$. With benzyl hydrazine, p-benzyl-dimethyl hydrazine $(C_6H_5CH_2)_2N{-}N{=}CHC_6H_5$ is produced.

Tribenzyl hydrazine may be oxidized, even with the oxygen in the air. When the hydrazine is present, the C–O bond of tert-butyl nitrate is broken to generate isobutene and hydrazine nitrate. The hydrolysis of iso-propyl nitrate in alkaline medium can form iso-propyl alcohol in 70–80% of yield, propylene in 8–14% and acetone in 11–14%. The hydrolysis can be expressed as the following equations:

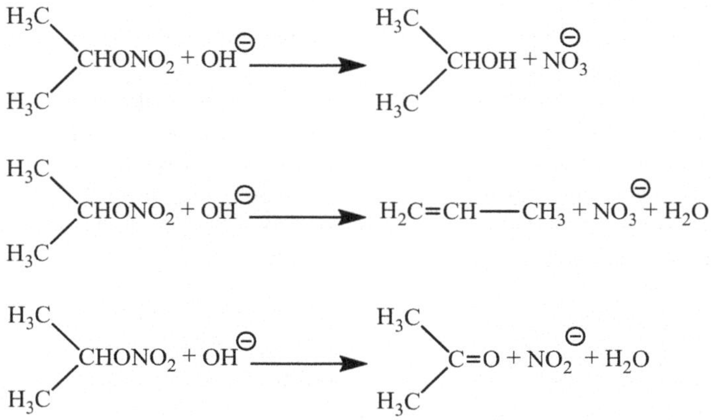

Hydrolysis of iso-propyl nitrate in neutral medium only generates alcohol.

With sodium ethoxide, tert-butyl nitrate reacts with benzyl cyanide to generate butyl ethyl ether and sodium nitrate only. Transesterification reaction does not occur in such cases.

In the mixture with coexisted secondary or primary alcohols and tertiary ester, secondary or primary alcohol esters are generated with further chemical combination at the same time. Because the original ester has a hydrogen atom that can react with α-carbon atom, the generated alcohol cannot be oxidized. Tert-butyl nitrite transesterification can occur because NO is unstable.

1.2.3 Reduction of Nitrate

Nitrate reduction is to replace nitrate groups with hydroxyl groups. There are many reduction methods, and the original alcohol can be formed in most cases. The oldest and most common reduction method is to use reductive hydrolysis of alkali metal sulfide or hydrosulfide (such as sodium or ammonium sulfide or sulfur hydride) [40]. When sulfuric acid is used to hydrolyze nitrate, the transesterification reaction can be caused to form sulfate [41].

The original alcohol can be produced through hydrogenation with palladium or platinum as catalyst under a pressure, reduction with hydrazine as the reducing agent [42]. The yield of reduction can be up to 60–80% and NO, N_2, and H_2O are produced at the same time.

With iron as catalyst, nitrate can be reduced in acetic anhydride by iron and zinc. If Pd is the catalyst, nitrate can be reduced in acetic anhydride only by of zinc.

Electrolysis can reduce nitrate. Nitrate can easily be reduced in the presence of lithium aluminum hydride. This reaction can only be completed in the ether solution and can quantitatively recover the original alcohol. Within reaction products, except of alcohol, there are chemicals with $-NO_2$ and ammonia. In order to complete the reaction, 3.4 mol of $LiAlH_4$ are needed per mole of $-ONO_2$. These methods can be used to reduce sugar acetals nitrates in the boiling solution. However, the reaction is very slow and usually 40–50 h is needed to complete, and product is pure acetal alcohol.

1.2.4 Some Other Reactions of Nitrate

In the presence of sodium ethoxide or potassium ethoxide, nitrate can react with substances containing active methylene group to generate salt of isonitro (–(HO) ON=) compounds. Under this condition, ester can be hydrolyzed.

As nitrites, alkyl nitrate reacts with an excess of Grignard reagent to generate n-dialkyl hydroxylamine.

Nitrate has a notable special property [43], which can catalyze some addition polymerization. When there is nitroglycerine, styrene and methyl methacrylate can self-polymerize. When the amount of nitroglycerin is up to 5–20 equivalents (by weight), equal to about 10% of polymeric material, polymerization rate is increased and polymer molecular weight is less than that from slow polymerization and in situ anionic polymerization.

The polymerization of methyl methacrylate with ethylene glycol dinitrate [44] is a typical reaction, in which their ratio is 0.07–5.9, reaction temperature is 25, 60, 80, and 90 °C, respectively, nitro ethylene glycol is decomposed partially to produce free radicals to induce the polymerization.

1.3 Important Properties of Nitrate

Stability, often chemical stability, can be believed as stability of the pure compound, or product characteristics after stability treatment.

Stability is the primary issue to measure whether the nitrate can be long-term stored or not [45], because their structure has the risk to lead to hydrolytic decomposition, automatic burning, and even explosion. However, esters with a higher purity have less risk of decomposition. Of course, stability depends on nitrate's chemical structure. For example, most of nitrate, including glycol dinitrate, nitroglycerine, ethylene glycol dinitrate, 1,3-propanediol dinitrate, pentaerythritol tetranitrate and nitrocellulose, are in their pure states with very satisfied stability.

The thermal decomposition of nitrate is particularly important in the practical applications [45], because it is related to whether nitrate can be used and long-term stored in weapons and industrial production, and security during its service in army.

The relationship between nitrate decomposition rate and temperature can be expressed as the following formula:

$$V = K_t + C$$

V reaction rate, mg of generated nitrogen per gram of nitrate,
T temperature,
K constant c, where c depends on the purity and nitration of the substance and K depends on the nature of material.

The relationship between alkyl nitrate decomposition and temperature can be well expressed with the following equation as,

$$\log N = a + b \times 0.9932t$$

N: The mg number of produced nitrogen per 2.5 grams of nitrate at certain temperature within 15 min.

$$a, b\text{: constant;}$$

$$t\text{: temperature.}$$

In nitrocellulose with 13.55% of nitrogen, these constants are as follows, a = 8.8842, b = 22.86.

Nitrate can automatically catalytic accelerated decompose [46]. Trace amounts of nitrogen oxides are generated from nitrate decomposition, and these nitrogen oxides combine with trace amounts of water to produce nitric acid and nitrous acid, and these acids accelerate the decomposition of nitrate again.

1.3.1 Determination Methods of Nitrate Stability

Determination method of explosive (especially nitrate) stability is to study on the stability of these substances at increased temperatures first. Normally, stability of nitrate is measured at temperature close to room temperature.

If stability test is conducted at a low temperature, the result cannot reflect the stability and decomposition rate of nitrate, because, at a low temperature, nitrate decomposition cannot be realized. Thus, usually test is carried out at an increased temperature. The test methods are divided into qualitative and quantitative methods.

1.3.1.1 Abel Heat Resistance Test

Abel heat resistance test is one of the most ancient and common qualitative tests, to determine the required heating time for obvious decomposition of compounds or impurities contained in the sample. A longer time in the Abel heat resistance means a higher stability of the tested compound.

Since 1865, heat resistance test method is always used to evaluate the stability of energetic materials in UK. In most countries in the world, this method to evaluate explosives is named as "Abel test", while it is known as "potassium iodide starch test" in US. In this test, potassium iodide starch paper is hanged in the tube after its is wetted with aqueous glycerol solution, nitroglycerine or nitrocellulose samples (dynamite or smokeless explosive) are heated on a constant temperature bath until the test strips turns into light brown. In the original heat resistance test in UK, the heated sample is 2 mL of nitroglycerine with heating temperature is 160 F (71.1 ° C), or 1.3 g of nitrocellulose with heating temperature of 170 F (76.7 °C), or 1.6 g Kodak drug with heating temperature of 180 F (82.2 °C). Their tested periods are 10 min, 15 min, and 30 min, respectively. Test conditions and operation method of the heat resistance test (Abel heat resistance test) at 80 °C are the following:

(1) Reagents

Reagents are glycerin or distilled glycerine, secondary distilled water, and special potassium iodide starch paper (20 × 10 mm).

In the preparation of glycerol aqueous solution, the volume ratio is 1:1.

(2) Instruments and equipment

Test tube: 130–140 mm in length, 16–18 mm in ID, 10–15 mm in wall thickness.

Glass hook: 85–95 mm in length, 3–4 mm in diameter.

Tube rubber stopper: good elasticity with hole in the middle and tightly plugged with glass rods with hook.

Thermometers: –1/5 °C or 1/10 °C of graduation with scale correcting value (scale correction is tested at least once every three months).

Glass rods with holes or paddles: for the hole puncturing on strips.

Puncturer, tweezers, brown bottles for strips, stopwatch (1/10 s), heating thermostats, balance, test tube rack, standard color tubes, funnels.

(3) Operating procedures of determination

Samples are dried according to the 8th article. 1.3 ± 0.01 g of dried sample is weighed and rapidly added into a clean and dry test tube through a funnel, the sample is pressed tightly with a glass rod into the height of 55 mm, and dried rubber plug with an inserted glass hook is stuffed. Then, KI starch strip is taken out from the bottle with clean tweezers, and put on a piece of clean glass with hole (or paddle board). With the puncture, a hole in the middle of one strip end, which is about 2 mm from the edge, is made. Then, the strip is held with tweezers on the end with the hole to make strips erect. A suitable amount of glycerol aqueous solution is taken with a glass rod (about 3–4 mm in diameter) to roll down from the middle of strips (1–1.5 mm lower from the middle) to roll coat the strips. Then, the lower half of strips should be wet with uniform glycerol water, and the discriminating line

Fig. 1.1 The assembled tube

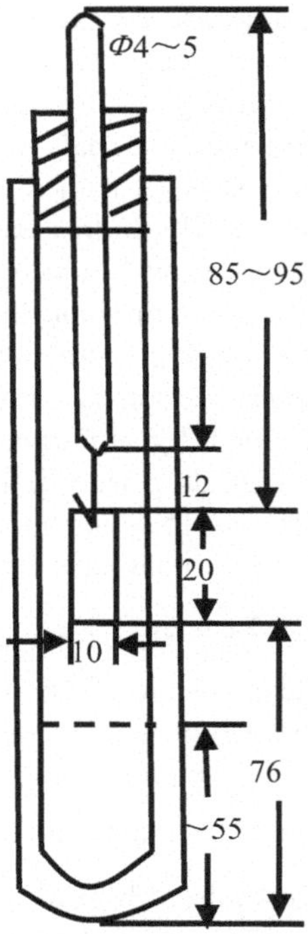

between wet and dry should be just in the middle of strip. The rubber plug is removed and the strip with well-coated aqueous glycerol solution is hanged on the glass hook, then quickly the tube is plugged, where you should make test paper hanged naturally and vertically without touch with the wall, and the bottom edge of strip should be about 76 mm from the bottom. The assembled tube should follow the requirements shown in Fig. 1.1.

The prepared tube is inserted into the constant temperature bath that is heated up to 80 ± 0.5 °C, and record the time. Strip in the tube should remain vertical, and the lower edge of strip should be on the same level as the upper edge of thermostatic bath (the water surface of thermostatic bath should be kept constant with a distance of 6 mm from the edge of thermostatic bath).

When the test is close towards the end, the color is compared with the standard color tube in the thermostatic bath. When boundary lines between wet and dry on strips have the same color depth, the reaction is at its end and the time around 1 min.

Two parallel determination tests are done for every sample, and the results are not averaged.

(4) A Laboratory requirements of bel heat resistance test

Laboratories shall be well ventilated and painted to white for the convenience of regular cleaning. Light should be sufficient in the room, but the direct sunlight should be avoided. Windows should be opened to north with frosted glass installed.

In the heated laboratory, it is not allowed to wash or dry tubes, glass hooks, etc. After the experiment, test tubes are not allowed to open and tested specimens are not allowed to store in the laboratory.

Standards on heat resistance test are different in different country. For example, the test temperature was 75 ± 0.5 °C in the previous Soviet Union, 65.5±1 °C in the United States and 80 °C in China, while more sensitive zinc iodide starch test paper is used in Germany.

Certainly, the heat resistance temperature of the tested material is not necessary to mean the beginning of its thermal decomposition. According to many experiences, it means that the tested material itself, or the existence of more decomposable and unstable impurities (such as residual nitrated acid) or unstable by-products.

The simplicity of this test method indicates the universality of this method to determine the purity of nitrate. When nitrate contains mercuric salts (for example, HgCl), sensitivity of the test will be greatly reduced due to the reaction between the salt of mercury and iodine.

1.3.1.2 Vielle Test (Litmus Test)

In Vielle test method, nitrate (2.5 g) or smoke-free chemical sample is heated in a sealed test tube, and a piece of blue litmus paper is placed in the tube. Tube is heated to 110 °C in the thermostat. Inside the test tube, the temperature is 108.5 °C. In the previous Soviet Union, the thermostat temperature was set at 106.5 °C, and the strips color changing time from blue to red shall not be less than 7 h.

Quantitative sample is placed in a special instrument and heated to decompose at the set temperature, and the stability is the heating time of sample until the color of blue litmus paper turns into red, or brown smoke appears.

(1) Application range: This method is to determine the chemical stability of nitrocellulose and propellant.
(2) Equipment: In this method, the special equipment is Vielle test equipment used as shown in Fig. 1.2.
(3) Test procedures

1. Sample preparation

Nitrocellulose: Nitrocellulose is wrapped a white cloth first. After its water is removed with the manual press, nitrocellulose was sieved through an 8 mm copper sieve. Then, the sample is put in aluminum can, baked in an oven for 2 h at 95–100 °C, removed from the oven and cooled in a desiccant without drying reagent.

(a) (b)

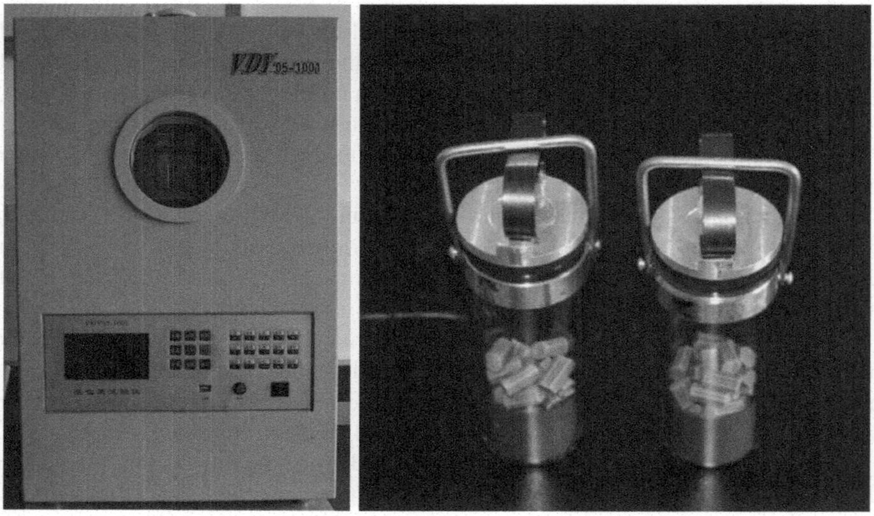

Fig. 1.2 Vielle test device modified in Chinese military. **a** Improved Vielle device, **b** tubes with propellants

Smokeless powder: Propellant grains more than 1 g in weight, and granulated propellants and drug tubes greater than 20 mm in diameter, are cut into powder with a blade and screened through a sieve with 8 mm hole on the top and 5 mm on the bottom. And samples between the top and bottom sieves are taken for the coming test.

Ribbon and flake propellants: Sample is cut into 5–8 mm pieces, in which cracking and burrs should be prevented.

Tubular propellant with burning layer thickness of greater than 1.2 mm is treated as granulated with 1 g in weight. Tubular propellant with burning layer thickness of less than 1.2 mm is cut to remove 5–10 mm at its both ends, then cut into 30 mm smaller tubules.

Single base propellant samples are heated at 95–100 °C for 2 h, and moved into a desiccant without drying reagent to cool for 1 h. Double-base propellants are not necessary to dry.

2. Sample loading

The operator must wear clean gloves (or finger gloves) and masks to remove the litmus paper with tweezers. The paper is rolled into a ring and placed in a beaker with its pattern towards the outside. It is made sure that the paper is very close to the inner wall and bottom of the beaker, and the junction should be overlapped. 2.5 g of NC or 10 g of smokeless powder samples (accurate to 0.1 g) are weighed, loaded into beaker with the installed strips through a funnel, in which sample cannot go between the test strip and the beaker wall. The loaded nitrocellulose can be pressed and flatten gently with a brass rod having a flat head. The baker is covered with the lid having a plastic ring and the brass ring is pulled up. Now, the beaker should be closed tightly until fingers with slight powder can not turn the drum cover), and the overlap of strips should be just below the copper ring item number.

3. Temperature at test

Two temperatures, 106.5 ± 0.5 and 115 ± 0.5 °C, are used in this study. Test at 106.5 ± 0.5 °C is divided again into the common test, accelerated and ten-time repeated test, regular ten-time repeated test, accelerated and 1 h repeated test, and regular 1 h repeated test.

4. Test method

The common test: The beaker with a thermometer is placed in a hole of a thermostat beaker. Power is turned on to rotate the beaker on the beaker rotating rack in the thermostat that is heated. When the temperature is 106.5 ± 0.5 °C in the beaker and is stabilized, the beaker with sample loaded is placed on the other holes of the beaker rack in the thermostat with a hook, and strips overlap is towards outside. After samples are placed into the thermostat, the temperature should be restored within 1 h to 106.5 ± 0.5 °C. During the whole experiment, the beaker rack should be rotating slowly and evenly. Through the observation window, temperature and test strips color change are observed. The color changes gradually from blue into blue purple, purple, purple rose, rose, then to red or brown smoking samples finally. These color-coded symbols are listed in Table 1.7.

5. Conditions of test strips: brittle HO/C—broken after one folding pro and con. Soft and brittle HO/CY—broken after one folding pro and con with a gent pulling. Soft HO/Y—broken after one folding pro and con only with a strong pulling.

6. The end point determination of test strips: When the end point of test strips is difficult to determine, the experiment is continued until the appearance of brown smoke, but heating cannot exceed 7 h. Specimen, that reaches the end point within less than 7 h, is immediately removed from the thermostat without any repeated test again.

7. Data recording: In the record, batch numbers of samples and test strips, number of the thermostat, test paper conditions, color changes, time of brown smoke, temperature of test, tester's name and date should be marked.

Table 1.7 Color-coded symbols of test strips in Vielle test

Color	Symbol	Color	Symbol
Blue	l	Rose	m
Blue purple	lz	Red	h
Purple rose	zm	Brown smoking	zy
Purple	z	Yellow	h

l Blue, *lz* Blue purple, *zm* Purple rose, *z* Purple

Each sample is tested with two parallel experiments, the results are neither used to get an average nor rounded, but connected the measured times with dash line.

(4) Accelerated repeated test (Heated for ten times continuously, or to 1 h)

Different from the operation common test, in accelerated repeated test, test is repeatedly for 10 times or to 1 h, and sample is taken out from thermostat after each test and cooled in a desiccant without drying reagent for 30 min. After the copper cover is opened, sample is poured into the aluminum box, which is placed in ventilation cabinet for 2 h (The sample can be in ventilation cabinet for 5 h if the test strips become red within 7 h of heating). After the ventilation, a new test strip is placed into the baker before new sample is loaded for the next test until ten times or the appearance of red color (or brown smoke) on the strips within less than 1 h. Test results are expressed as the total time of repeated heating, and the final heating time less than 1 h is not included in the results. In the total time, the minute part of time equal to or greater than 20 min is record as half an hour, equal to or greater than 45 min is 1 h. Test results does not take an average, but the measured times are connected with a dash line.

(5) Normal repeated test (Heated for ten times continuously, or to 1 h)

Apart from the difference in ventilation time, other operations in normal repeated test are same as those in accelerated ten-time or 1 h repeated test. The ventilation time of specimen in the fume hood shall be 24 h in the normal repeated test which is the total of heating, cooling and handling time, indicating, at the same time every day, beaker with loaded samples will be placed in a thermostat to heat.

The accelerated (or normal) repeated test of nitrocellulose (or normal) is only repeated for four times, and the whole heating time of sample is used as test result. And the time in the last test will not be included in the result if it is less than 60 min. The result from the common test results will not be rounded and expressed with the measured value.

Test at 115 ± 0.5 °C: It is suitable to stability test of stored nonsmoking propellants. Apart from test temperature and time, others are the same as those in common test at 106.5 ± 0.5 °C. Specimen cannot be heated for more than 150 min. Stability is expressed with the total time from the point when the sample is put into a thermostat to the point when the test paper turns into red or brown smoke appears (accurate to 15 min). If the ending point is not reached even when the sample is heated for 150 min, more than 150 min is written down as the result. The parallel test results will not be used to take an average, but connected with a dash line.

1.3.1.3 Stability Test Under Vacuum

Weight loss test is one of the oldest and simplest method for quantitative determination of stability, and explosive samples are heated at a certain temperature to measure their weight loss. This approach has been widely recognized by every

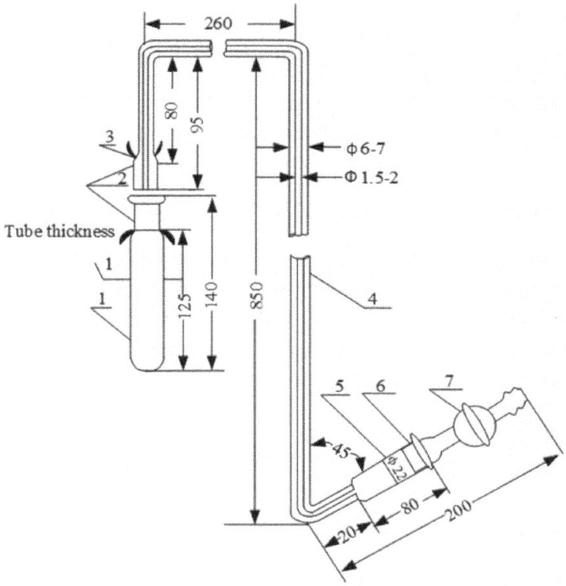

Fig. 1.3 Vacuum stability tester. 1—Heating test tube, 2—frosted connection, 3—ear for hanging, 4—capillary, 5—the mercury cup, 6—frosted connection, 7—vacuum plug

country to measure the stability of explosives. Now, the normal weight loss test is changed into the vacuum stability test, which has become an important index to evaluate the stability of energetic materials. Quantitative samples are heated to decompose under conditions with a constant volume, constant pressure and vacuum, and mercury pressure gauge is used to measure the pressure of the gas evolved over a certain of time before it is converted to volume under standard conditions, which is used to evaluate the stability and compatibility of energetic materials. Test apparatus is shown in Fig. 1.3.

(1) Preparation of test specimens

1. Specimen amount: 5.00 ± 0.01 g, 0.2 ± 0.01 g for dynamite.

2. Temperature of test: 120.0 ± 0.5 °C or 100.0 ± 0.5 °C for explosives, 100.0 ± 0.5 °C for single base propellants, 90.0 ± 0.5 °C for double base propellants, 260.0 ± 0.5 °C for Heat-resistant explosives.

3. Heating time: The energetic material is continuously heated for 48 h, and heat-resistant explosive is continuously heated for 140 min.

(2) Test preparation

1. Volume calibration of heating test tube

Distilled water is added into the heating test tube with a burette to its bottom edge of frosted opening, and the consumption volume of water is the volume of heating test tube.

2. Calibration on volume per unit length of capillary

Mercury (10.00 ± 0.01 g) is weighed and added into the mercury glass, and the mercury is made sure to go into the CD capillary tube. The mercury column should be continuous without air bubbles. On three locations of the capillary, the length of mercury column is measured. The three measurements are averaged to calculate the volume of capillary per unit length as the following Eq. (1.1),

$$V_e = \frac{m}{13.59L} \tag{1.1}$$

where

V_e	capillary volume per unit length, mL/min,
m	quality of mercury, g,
L	the average length of mercury column, mm,
13.59	the density of mercury, g/mL.

(3) Determination of length of AD section of the capillary

With a ruler, length of AB, BC, and CD sections of capillary is measured, separately, then they are added together as the AD length.

(4) The injection of mercury

The marked capillary pressure gauge is washed and dried, and 7 mL of pure mercury are taken with a graduated cylinder and injected into the mercury glass.

(5) Specimen processing

Gunpowder is ground and screened through a chosen sieve, and samples are taken under the sieve. Double base propellant is dried under vacuum at 55 ± 2 °C for 6 h. Single base propellant is used without drying.

Explosives are dried in a vacuum oven at 55 ± 2 °C for 4 h. Column propellant must be crushing and screened through a chosen sieve for use.

(6) Preparation of the thermostat bath

According to the selected test, thermostat bath is prepared. When the temperature meets the requirement and is stable, tests can be carried out.

(3) Test procedure

1. Specimens weighing

According the requirement, samples are weighed and placed in a heating test tube.

2. Connecting the heating test tube and capillary pressure gauge

A relatively thin layer of vacuum grease or silicone rubber is put on the frosted mouth of capillary pressure gauge, the sample-loaded heating test tube is connected and slowly turned to seal. With rubber band or metal wire, heating test tube and the ears of the capillary pressure gauges is tightened well.

3. Vacuum with instrument

The well sealed test instrument is fixed on the vacuum frame, connected with the pumping vacuum system with a vacuum rubber tube, then tilt to a certain angle to make mercury all into the mercury cup without capillary column blocking. The piston of mercury cup is slowly opened to begin vacuum pumping. The vacuum is continued for another 5–10 min after the system pressure is less than 670 Pa. Then, the tilt test instrument is pushed back to its normal position and mercury cup piston is closed to make mercury to go slowly into the capillary. After it is stable, it is observed for 10 min again to confirm that vertical height of capillary mercury has no change before the next step. If the vertical height of mercury column declines, the leakage should be checked and all operations should be repeated.

4. The following data are recorded before the heating:

Mercury column height of capillary CD section H_0, the vertical height from mercury surface in the mercury cup to point D H_0, Atmospheric pressure P_0, Temperature of test chamber T_0.

5. Heating test

The vacuumed heating tube is immersed in the thermostat bath and heated, and the frosted lower edge of heating test tube is kept at 5 mm away from the surface of heated liquid medium, while temperature and heating starting time of the thermostat.

After 48 h of continuous heating, the heating test tube is removed from the thermostat with the time record. After it is cooled at room temperature for 1 h, all heating data are recorded.

During the heating experiment, the change of temperature and height of mercury column in capillary tube and other phenomena should be observed.

In the process of heating, if the height of the mercury column height in the capillary tube is close to that of the mercury cup, heating should be immediately stopped to remove the heating test tube with time recorded.

6. The following data are recorded after the heating:

Vertical mercury column height in capillary CD section H_1, the vertical height from mercury surface in the mercury cup to point D h1, atmospheric pressure P_1, Temperature of the laboratory T_1.

(4) Results calculation

The standard state volume of gas produced from the decomposition of heated specimen is calculated according to the following formula:

$$V_H = 2.69 \times 10^{-3}\left\{[V_0 + V_e(L - H)]\frac{P - P_B}{T_1} - [V_0 + V_e(L - H_0)]\frac{P_0 - P_A}{T_0}\right\} \quad (1.2)$$

In which,

V_H	the standard state volume of gas produced from the decomposition of heated specimen, mL,
V_0	the volume of heating test tube minus the volume of sample (sample volume is obtained from its mass divided by its density), mL,
V_e	capillary volume per unit length, mL/mm,
L	AD section length of capillary, mm,
H_0	the vertical height of mercury column in the capillary before the heating, mm,
H	the vertical height of mercury column in the capillary after the heating, mm,
P_0	the atmosphere pressure before the heating, Pa,
P	the atmosphere pressure after the heating, Pa,
P_A	mercury column pressure in the capillary before the heating, (which is pressure converted from the difference after the mercury column vertical height in the capillary minus the height from the liquid surface of mercury cup to D point before the heating) Pa,
P_B	mercury column pressure in the capillary after the heating, (which is pressure converted from the difference after the mercury column vertical height in the capillary minus the height from the liquid surface of mercury cup to D point after the heating) Pa,
T_0	temperature in the laboratory before the heating, K,
T	temperature in the laboratory before the heating, K,
2.69×10^{-3}	the ratio of temperature over pressure under the standard state conditions.

1.3.1.4 Compatibility Test

Compatibility test is an important parameter to judge whether the energetic materials can be used or not. All countries pay a great attention to it.

(1) Test conditions

1. The amount of sample: the mass of a single specimen is 2.50 ± 0.01 g; the mass of mixed samples is 5.00 ± 0.01 g with a mass ratio of 1:1.

2. Test temperature: temperature of explosive or single base propellant is 100.0 ± 0.5 °C, but tat of double base propellant or three base propellant is 90.0 ± 0.5 °C.

3. Heating time is 40 h with continuous heating.

(2) Contacting material pretreatment and test preparation

1. The metal or plastic components of energetic materials should be crushed fro the convenience of mixing. Textile and paper should be cut into tiny pieces. Coating, paint, adhesive, and other materials containing solvents should be coated on the glass to dry, and the formed film is removed to cut into small pieces.

2. Gunpowder can be filed, ground, and screened through the chosen sieve, and samples are taken under the sieve. Double-base propellant is dried in the vacuum drying oven at 55 ± 2 °C for 6 h. Single base propellant does not need to dry. Explosive is dried in a vacuum drying box at 55 ± 2 °C for 4 h. Grains must be crushing and screened through the chosen sieve and samples are taken under the sieve.

(3) Test procedure

2.50 ± 0.01 g of gunpowder or explosive are weighed and added into the first heating test tube. 2.50 ± 0.01 g of contacting material is weighed and added into the second heating test tube. After 2.50 ± 0.01 g of gunpowder or explosive and 2.50 ± 0.01 g contacting material are mixed evenly, they are added into a third heating test tube.

These three heating test tubes are tested according to the previously mentioned methods. And the continuous heating time is 40 h.

(4) Results calculation

Compatibility is represented with the net increase of outgassing, and calculated according to the following formula:

$$R = V_C - (V_A + V_B) \tag{1.3}$$

where,

R	the net increase of released gas in the reaction, the increased volume of gas released from the mixture compared to that from the single specimen, ml.
V_C	the volume of gas released when the mixed samples from gunpowder or explosive and contacting material is heated to decompose, mL.
V_A	the volume of gas released when gunpowder or explosive is heated to decompose, mL.
V_B	the volume of gas released when contacting material sample is heated to decompose, mL.
V_C, V_A and V_B	are all calculated according to Eq. (1.2).

1.3.2 Determination of Important Physical Properties of Nitrate

In addition to chemical stability of nitrate, nitrogen content, and nitration uniformity of nitrate in its performance characterization are also important parameters, as well as important application basis, to measure nitrate.

1.3.2.1 Determination of Nitration Uniformity of Cellulose Nitrate

The esterification and uniformity of cellulose nitrate are important parameters to judge its applications although there have been many other methods [47–49]. In industrial production, usually the degree of esterification is determined through the nitrogen content determination of cellulose nitrate. However, this approach results can only be used to obtain the mixed nitrogen content of cellulose nitrate macromolecules, and cannot prove esterification uniformity of cellulose. Shao [50] believed that nitrogen content, swelling coefficient and shapes of cellulose nitrate are different in solvent. According to this theory, shape and nitration degree of cellulose nitrate can be identified accurately with a polarized light microscopy [50–52]. The test system of esterification and uniformity of cellulose nitrate is shown in Fig. 1.4.

Operation methods of polarized light microscope to determine nitration uniformity of cellulose are as follows:

(1) Selection of immersion liquid

Due to the rough surface and porous inside of cellulose nitrate fiber, if the fiber is not immersed with appropriate solution, when light passes through fiber there will be scattering and reflection on fiber surface, thus resulting in the loss of light intensity. Theoretical calculations prove that optical losses by about 4–10% after every single interface, therefore, the fiber must be immersed in a suitable liquid

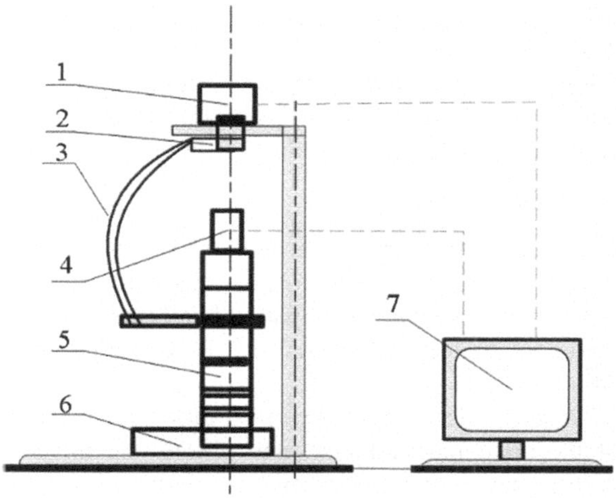

Fig. 1.4 The schematic diagram of an instrument to test esterification and uniformity of cellulose nitrate. 1—Angle sensor, 2—bracket arm, 3—connecting rod, 4—colored video camera, 5—polarized light microscope, 6—bracket pedestal, 7—computer [50]

during the determination of its optical path difference. Meanwhile, reported data also showed that the size of birefringence of cellulose nitrate fiber in the polarized light microscopy depends on the choice of immersion liquid. With different immersion liquids, fiber's birefringence can be increased, disappear or vice versa. Therefore, the correct choice of immersion liquids is the basis and prerequisite to accurately measure nitrogen content and uniformity of cellulose nitrate. The choice of immersion liquid is usually according to the following principles, ① non poisonous, no pungent smell, ② nonvolatile, ③ stable chemical properties, ④ stable refractive index with relatively small temperature coefficient, ⑤ no reaction with the sample, not to dissolve or swell or shrink sample, not to react with sample, ⑥ the viscosity should not be too large, ⑦ clear or tinted transparent is the best, ⑧ with the same refractive index of sample.

After repeated testing and screening, the authors have determined the two-component system of kerosene/α-bromonaphthalene to be the immersion solution. The refractive index of kerosene is 1.4500 and the refractive index of α-bromonaphthalene is 1.6601, thus these two can be mixed in any ratio. Additionally, neither of them can react with fiber. After they are left at room temperature for a month, their refractive index values only have small changes, indicating they are excellent stable.

(2) Preparation of immersion liquid

Calibration on Abbe Refractometer: Abbe Refractometer is an instrument to measure the refractive index of liquid. Before its usage, calibration must be done with distilled water as the base material. The refractive index of distilled water at 20 °C is 1.333.

Preparation of immersion liquid: 5 mL of kerosene are added to the 10 mL reagent bottles, in which α-bromonaphthalene is then added with a dropper drop-by-drop. After they are mixed evenly, refractive index of the solution is measured on the Abbe Refractometer kept at 20 °C, and it is good if the data is within an appropriate range. It should be paid attention that, once you have determined a specific refractive index value, you should try to use this value in the following measurements. The minimum is that the refractive index of immersion liquid should be consistent with the actual measured value at least when the baseline is made.

(3) Sample preparation

Microscope slides and cover glass should be cleaned to remove all organic impurities for the future use. Before the test, nitrocellulose is dried at 105 °C for 2 h. In order to observe the optical path difference of each fiber, every fiber is separated for random sampling, which reflects the esterification degree and distribution of the whole sample.

With stainless steel tweezers, a little fiber is taken from every different location of the dried nitrocellulose, and placed and mixed well on an evaporating dish. A little sample is taken from the well-mixed nitrocellulose and spread on a clean, nickless and scratchless slide, and aggregated nitro cellulose fibers are spread out as much as possible with the pointy tweezers to avoid the overlapping of fibers. Two drops of the prepared immersion solution are dropped on the sample before the cover glass is covered, during which air bubbles between the slide and cover glass are removed. The sample is stored for the future use. The operation system to test esterification and homogeneity of cellulose nitrate is shown in Fig. 1.5.

Before the measurement, the polarization direction and polarizer is orthogonal to the polarization direction of analyzer, and a one-fourth wave plate is inserted. Thus, microscope is under its dark field. The prepared sample is place on the microscope stage, it microscope observing tube is down to its lowest position. Then, you can

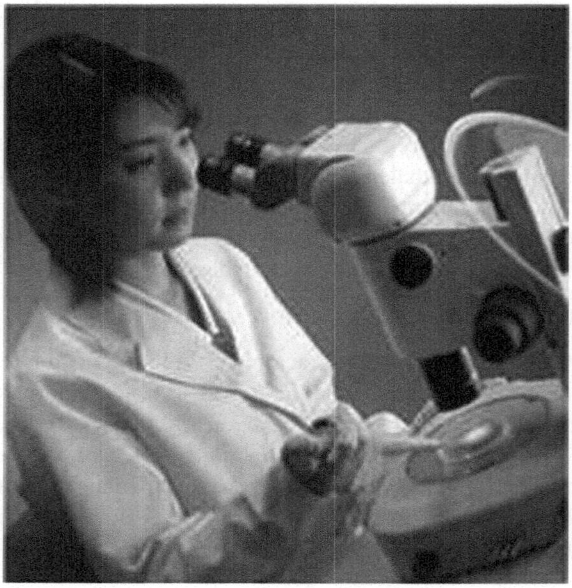

Fig. 1.5 The operation system to test esterification and homogeneity of cellulose nitrate

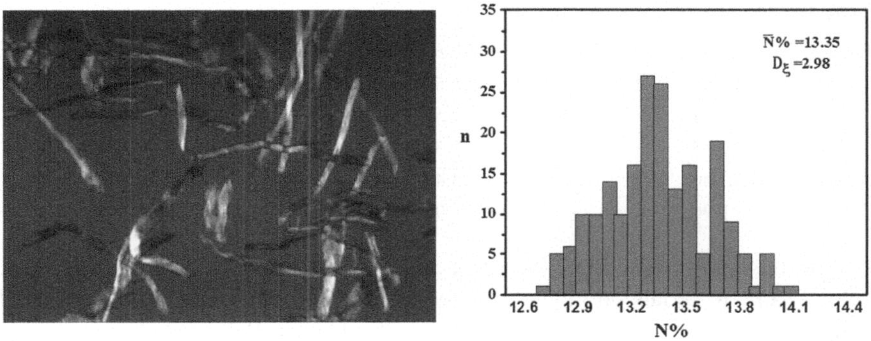

Fig. 1.6 Typical measured results of nitration uniformity of cellulose nitrate [51]

watch the sample while raising the observing tube until the fiber image is the clearest. After a uniform and straight fiber is found, sample stage is turned until the fiber disappears. When the stage is turned by 45° again, the fiber has the clearest image. The analyzer is turned again to make the fiber to disappear. The angle that analyzer turns is the compensation angle of this fiber. This operation is repeated until all required fibers have been measured. Typical measured results of nitration uniformity of cellulose nitrate with different esterification degree are shown in Fig. 1.6.

Characteristics of polarized light microscopy to test the esterification degree and its uniformity of cellulose nitrate are the following:

(1) With this method, we can directly get the average nitrogen content of sample, nitrogen content of every fiber, and column charts of nitrogen distribution and the standard deviations of nitrogen distribution in different zones.
(2) With the help of computer, visual fatigue of operator is reduced and testing efficiency is improved. Tests are not affected by season, temperature and other outside environmental factors, not affected by the viscosity, molecular weight and nitrogen content of cellulose nitrate. The 1#, 2#, and 3# cotton in military, and cotton for civilian use can all be measured with this method.
(3) During the test procedure, it is not necessary to destruct the fiber through burning or other forms. This method is more environmental beneficial compared with other commonly used methods to determine nitrogen content of cellulose nitrate. And fewer samples (a few micrograms) is needed. Therefore, it is particularly suitable for trace samples difficult to obtain.
(4) This method has high precision and speed of measurement. Compared with the interferometer, the absolute deviations of nitrogen content of cellulose nitrate could be controlled within 0.05%. The observed measurement error of nitrogen content is not larger than 0.1%.
(5) Because of the implementation on the automation of data acquisition, transmission and processing, and other special experimental techniques, the time to measure a sample is about 15 min, which greatly shortens the testing cycle, is very convenient and fast. The typical test results are shown in Table 1.8.

Table 1.8 Data of nitrocellulose samples

Sample	Interferometer, N_1 (%)	The invention, N_2 (%)	Standard deviation
1	12.60	12.59	1.32
2	12.59	12.60	2.74
3	13.35	13.39	3.62
4	13.40	13.41	3.52
5	11.96	11.96	1.81

With this test system, Shao, Wang et al. [53–55] studied the influencing law of cellulose raw materials on nitrate nitrogen content and uniformity of the prepared cellulose. And some valuable conclusions were obtained [54–57].

1.3.2.2 Determination of Nitrogen Content of Nitrate (Measuring Nitrogen Content with Devarda Alloys Reduction Method)

Nitrogen-containing materials are saponificated with NaOH in the presence of hydrogen peroxide to form nitrate, which is reduced with Devarda alloy into ammonia. After it is absorbed with boric acid, then titrated with hydrochloric acid standard volumetric solution. Apparatus of Devarda alloy reduction method to measure the nitrogen content is shown in Fig. 1.7.

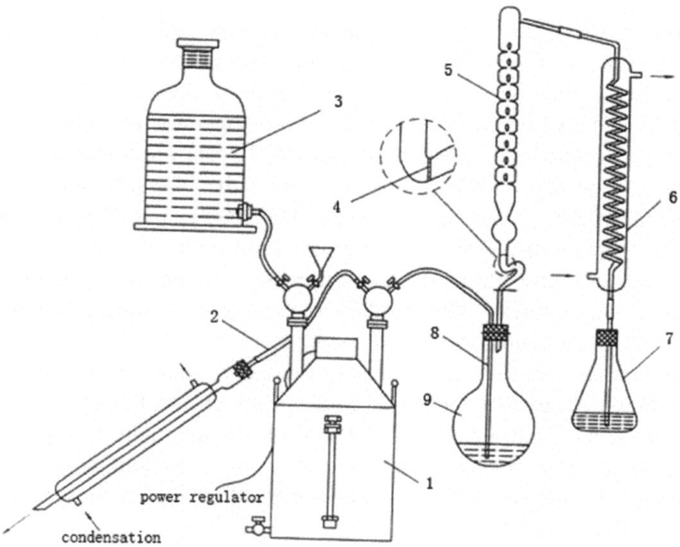

Fig. 1.7 Apparatus of Devarda alloy reduction method to measure the nitrogen content. 1—Electric steam generator, 2—the exhaust pipe, 3—distilled water, 4—filtration plate, 5—fractionating column, 6—the condenser, 7—absorption bottle, 8—intake tube, 9—the reaction bottle

(1) Test materials and instruments

1. Test materials

Boric acid: 4% based on *China National Standards* 628–65 Level 2.
Sodium hydroxide: 20% based on *China National Standards* 628–65 Level 2.
Hydrogen peroxide: 30% based on HGB3109–59 Level 2.
Hydrochloric acid: 0.2 N *standard solution based on China National Standards* 622–65 Level 1.
Ethanol: Distilled or based on *China National Standards* 679–65Level 2.
Potassium nitrate:Standard reagent.
Devarda alloy: HG3–901–76 Level 2.
Its components are Zu 5% + Al 45% + Cu 50%.
Its composition is: < Φ0.25 mm, 1/3.

$$\Phi 0.25 - 0.5\,\text{mm},\ 1/3$$
$$\Phi 0.5 - 1.0\,\text{mm},\ 1/3$$

The mixed indicator is from 0.1% of bromocresol green (HGB3386–60) in ethanol and 0.2% of methyl red in ethanol (HGB3040–59) with a volume ratio of 3:1.

2. Instruments and equipment

Steam generators (including the dashboard),
Electric furnace (adjustable),
Asbestos net, the weighing bottle, dryer,
Reaction flask: 1000 mL flat bottom flask (it is allowed to use 500 mL conical frosted flask)
Erlenmeyer flask: 500 mL
Measuring cup: 50 mL, 5 mL
Pipette: 50 mL, 5 mL, 2 mL
Titration burettes: Special burettes or 50 mL pipette (the used part is corrected every 2 mL)
Water-sealed fractionation column
Condenser: Snake-shape, straight pipe.

(2) Experiment procedure

1. Amount of tested sample

0.4–0.65 g of dry sample are weighed with a weight-known weighing bottle to the accuracy of 0.0002 g, baked in the oven at 110 ± 2 °C for 30 min (or dried with an infrared dryers). The sample is cooled until its weight is constant when its deweighting difference is not greater than 0.005 g.

2. Operating procedures

Before the experiment, the two sets of heaters of the steam generator are opened with the maximum current less than 13 A. During the test intervals, the current is adjusted to maintain a slight boiling and the liquid surface in steam generator should be kept within the specified range.

50 mL of boric acid solution are placed in a 500 mL Erlenmeyer flask, and the lower end of a snake condenser is inserted into the flask to about 5 mm under the liquid surface of an absorption flask.

The weight-constant sample is added into the reaction flask with 5 mL of added ethanol. With 50 mL of distilled water, the weighing flask, and the reaction flask are washed and cleaned. About 2 mL of 30% hydrogen peroxide solution are added before 50 mL of 20% sodium hydroxide solution through a funnel in the reaction flask mouth. After the dissolved is shaken until it is uniform, it is stored for 5–10 min. Saponification is done on a furnace that is pre-warmed to dark red with continuous shaking to avoid the overheating and overheating decomposition of sample. After the solution becomes transparent, it is heated continuously to boiling until the full decomposition of hydrogen peroxide (about 10–15 min). After the saponification is finished, the reaction flask, and the funnel and the reaction flask bottle are rinsed with distilled water having a flushing total volume that is approximately equal to the volume before the saponification. Then, the reaction bottle is cooled down to room temperature.

4 g of Devarda alloy are added into the reaction flask quickly, reaction flask and fractionation column are connected immediately, the plug is twisted tight, the connection is sealed with water, the intake switch is opened, the exhaust switch is closed, reaction time is recorded, and current is adjusted according to the reaction degree (Generally within 3–5 A). After 10 min of reaction, the auxiliary heater is opened, current is adjusted (within 12–13 A), ammonia is purged for 10 min (absorption liquid is controlled within 250–275 mL), absorption flask is removed first, the condenser spout is flushed with a small amount of distilled water, the vent switch is turned on, the reaction flask is removed, and the inlet switch is closed at the end.

0.3–0.5 mL of mixed indicator is added to the absorption flask. The ending point is reached when the color is wine red during the titration with standard hydrochloric acid solution.

3. Blank experiment

Blank experiments are done under the same conditions, but without samples added.

Each specimen is divided into two with the difference no more than 0.8 mL/g, and average is taken and is accurate to 0.1 mL/g.

(3) Calibration of method reliability

Potassium nitrate as the used standard reagent.

With the constant-weight weighing bottle (pre dried at 120 ± 1 °C for 1–2 h), 0.5–0.6 g of potassium nitrate are weighed and be accurate to 0.0001 g, dried in an oven at 120 ± 2 °C for half an hour, cooled until the weight is constant with difference between two less than 0.0002 g. The constant-weight sample is placed in a reaction flask, and the weighing bottle and reaction flask neck are rinsed with 50 mL of distilled water. Then, along the neck, 50 mL of 20% sodium hydroxide solution are added. Determination procedures are done according to the previous experiment.

Under the same conditions, blank test is done. Typically, 50 mL of distilled water and 50 mL of 20% sodium hydroxide solution are taken for experiment according to the previous procedures. The parallel difference of consumption volume of standard hydrochloric acid solution is no more than 0.02 mL.

The purity of potassium nitrate is calculated according to the following formula:

$$X = \frac{101.11N(V - V_0) \cdot 100}{G} \tag{1.4}$$

where

X	percentage of potassium nitrate,
V_0	consumption volume of the standard hydrochloric acid solution in the blank experiment (mL),
V	consumption volume of the standard hydrochloric acid solution of sample (mL),
N	the equivalent concentration of sodium hydroxide solution,
G	sample weight (g),
101.11	molar mass of potassium nitrate

Two experiments are done for each specimen, and each qualified result should be in less than 100.00 ± 0.1%.

1.3.2.3 The Flashpoint Determination of Nitrate

(1) Instrument

Heating oil bath

Rigid test tube: 170 ± 5 mm in length, 16.5 ± 0.5 mm in diameter, 1.0–1.5 mm in wall thickness

Thermometer: 0–200 °C with graduate of 1 °C.

(2) Determination procedures

Sample is dried according to the relevant regulations of energetic materials.

An oil bath with a stirrer and a thermometer is prepared before the experiment. When the temperature is increased to 100 °C, the tube with weighed 0.30 g of dried sample is added into the oil bath with the test tube depth of 100 mm in oil bath. At the central of the exposed part of test tube, an auxiliary thermometer is placed to adjust the heating rate to 3–5 °C/min. Careful attention should be paid to the sample. At the explosion the main and auxiliary temperatures are immediately recorded and accurate to 1 °C.

(3) Results

Flashing point (T) is calculated according to the following formula:

$$T = T_1 + \Delta T = T_1 + [0.00016 \times K(T_1 - T_2)] \quad (1.5)$$

where

T_1	the main thermometer reading, °C,
T_2	the auxiliary thermometer reading, °C,
K	the temperature of the exposed part at the explosion, °C,
0.00016	glass expansion coefficient

Two parallel measurements are done for each sample, and the results are not averaged.

Note: Other verified heaters are allowed.

1.3.2.4 The Acidity Determination of Nitrate

(1) Reagents

Acetone: analytic purity (international reference reagent),

Sodium hydroxide: analytic purity (international reference reagent), 0.01 N of standard solution,

Mixed indicator: 0.1% of bromocresol green in ethanol and 0.2% methyl red in ethanol with a ratio of 3:1.

(2) Determination procedures

Sample is dried according to the relevant regulations of energetic materials.

About 1 g of dry specimen is weighed and accurate to 0.01 g in an Erlenmeyer flask. 100 mL of acetone are added to dissolve the sample completely. With the stirring of magnetic stirrer, 40 mL of new boiled neutral distilled water was dropped inside to make it opaque. Then, a few drops of indicator are added, with 0.01 N of sodium hydroxide standard solution the solution is titrated to its green color as the ending point. The blank test is done under the same conditions.

(3) Result calculation

The acidity of NC is presented with the percentage of sulfuric acid (x), and calculated according to the following formula:

$$X = \frac{(V_1 - V_2) \times N \times 0.0490}{G} \times 100 \tag{1.6}$$

where

V_1	the amount of used standard solution of sodium hydroxide (mL),
V2	the amount of used standard solution of sodium hydroxide in the blank experiment,
N	the equivalent concentration of sodium hydroxide solution,
0.0490	gram per mmol of sulfuric acid,
G	sample weight (g).

Two parallel determinations are done for each sample, and the two are averaged. The calculation precision is up to 0.001%, and the parallel error is less than 0.01%.

1.3.2.5 Determination of Ash in Nitrate

This method is only suitable for cellulose nitrate, starch, sugar nitrates, and nitrates of chitinous material.

(1) Reagents

Nitric acid: AR (international reference reagent)

(2) Measurement procedure

With weight-constant crucible, 0.8–1.0 g of dry specimen are weighed and is accurate to 0.001 g. With the addition of 5–7 mL of concentrated nitric acid, the sample is wet. The sample is carefully heated to decompose on an electric hot plate (or sand bath) until it is dry (spill is not allowed). The sample is further heated to fully carbonized, and the crucible is moved into 700–800 °C high temperature furnace to burn to constant weight.

(3) Calculation of tested results

Percentage of ash (x) is calculated according to the following formula:

$$X = \frac{G_1 - G_2}{G} + 100 \tag{1.7}$$

where

G_1	weight of crucible and residue after burning, g,
G_2	weight of crucible, g,
G	weight of sample, g.

1.4 The Toxicity of Nitrate

Nitrate is a toxic substance. No matter what type of nitrate a chemical is, it is toxic with different degree. Typically, the toxicity of aromatic nitrate is larger than aliphatic nitrate, and toxicity of aliphatic nitrate is increased with the increase of nitrate groups. All nitrates have the function to expand blood vessels, especially, the blood vessel expanding effect of aliphatic nitrates is very significant with very outstanding reducing blood pressure capability. If a large amount of nitroglycerine vapor is inhaled, poisoning can be caused through the skin or the mucous membranes. The sensitivity of nitroglycerin poisoning effect varies to a great extent with different persons [58].

The main symptom of nitroglycerin poisoning is severe and squeezed headache. Poisoning of alcohol addiction is shown by the body that quickly becomes accustomed to nitroglycerin, and this adaptation can be formed in a few days. However, when the contact with nitroglycerin is interrupted, this adaptability is lost. When the contact is resumed, the body system must adapt to it again. Lots of workers on double base propellant production line knead and screen dynamites with bare hands, and they are used to contact nitroglycerin without any hurt feelings. Nitroglycerin does not produce chronic poisoning symptoms, and the substance should not be seen as industrial poison. But, it is found that very few people may have allergic phenomena, and are not accustomed to contact nitroglycerine [59, 60]. The accidental poisoning symptoms include headache, vomiting, cyanosis skin, visual disturbance, itching and swelling of limbs, etc. Up to present, fatal poisoning cases due to nitroglycerin has not been found. The headache due to nitroglycerin poisoning can be detoxificated through muscular injection of caffeine and sodium benzoate, and taking sulfuric acid n-phenyl propane orally [61].

References

1. Liu JP (1987) Basic chemical of propellants and explosives. GuiZhou Press, Guiyang
2. Bao DM, Liu JP (2011) The development and research situation of liquid explosive. Hazard Mater Secur Emerg Technol 12:114–121
3. Urbanski T (1976) Chemistry and technology of explosives (II). National defense industry Press, Beijing
4. Sun YB (1992) Military mixed explosives. National Defense Industry Press, Beijing
5. Wu YC (1980) Nitroglycerine chemical technology. National Defense Industry Press, Beijing
6. Austruy H (1993) Double-base propellants. In: Technology Solid Rocket Propulsion (ed) A. Davenas), Pergamon Press, Oxford, UK, pp 404–405
7. Agrawal JP, Singh H (1993) Qualitative assessment of nitroglycerin migration from double-base and composite modified doublebase rocket propellants: concepts and methods of prevention, *propellants*. Explos Pyrotech 18:106–110
8. Powell J (2005) Reduced Vulnerability Minimum Smoke Propellants for Tactical Rocket Motors. In: 41st AIAA/ASME/SAE/ASEE joint propulsion conference & exhibit, Tucson, AZ, USA, July 10–13, AIAA, pp 2005–3615

9. Maruhashi T, Nakashima A, Matsumoto T et al (2014) Relationship between nitroglycerine-induced vasodilation and clinical severity of peripheral artery disease. Atherosclerosis 235:65–70
10. Reinmutha N, Meyerb A, Hartwigsenc D et al (2014) Randomized, double-blind phase II study to compare nitroglycerin plus oral vinorelbine plus cisplatin with oral vinorelbine plus cisplatin alone in patients with stage IIIB/IV non-small cell lung cancer (NSCLC). Lung Cancer 83:363–368
11. Hu YX, Lv JZ, Liu X et al (1997) Synthesis of a cetane number improvers in diesel fuel (I)—Isooctyl Nitrate. J Petrochem Univ 10(4):16–18
12. Wang CP, Li HY (2005) The new process of a cetane number improvers in diesel fuel. Chem Eng 11:58–59
13. Yang RH (2010) Preparation of typical nitrate esters. Beijing Institute of Technology
14. Chavez DE, Hiskey MA, Naud DL, Parrish D (2008) Synthesis of an Energetic Nitrate Ester. Angew Chem Int Ed 47:8307–8309
15. Matsushima Y (1950) Coll Papers Fac Sci Osaka Univ Chem 11:1
16. Crigee C, Schnorrenberg W (1949) Ann 563:93
17. Zeng XL, Chen WH, Liu JC (2007) J Acta Phys Chim Sin 23(1):192
18. Gong XD, Xiao HM (1997) J Acta Phys Chim Sin 13(1):36
19. Mortimer CT, Spedding H, Springall HD (1957) J Chem Soc 188
20. Dalmon R, Bellin G (1942) Compt Rend 215:316
21. Plyler EK, Steele PJ (1929) Plys Revs 34:604
22. Danieu A, Jele F, Kohlrausch KWP (1931) Monatsh 58:428
23. Kumler WD (1953) J Am Chem Soc 75:4346
24. Mccallum KS, Emmons WD (1956) J Org Chem 21:377
25. Namba K, Yamashita T, Tntaka S (1954) J Ind Explos Soc Jpn 15:282
26. Jr Brown J F (1955) J Am Chem Soc 77:6432
27. Bellamy LJ (1958) The infra-red spectra of complex molecules. Methuen, London
28. Liu ZQ, Liu JP, Wang YJ, Zhang L (2011) Laser-raman spectrum for ethyl nitrate. Hazard Mater Secur Emerg Technol 12:497–498
29. Nielsen JE, Smith DC (1943) Ind Eng Chem (Anal Ed) 15:609
30. Urbanski T, Witanowski M (1963) Trans Faraday Soc 59:1039–1046–1510
31. Berthelot M (1990) Compt Rend 131: 519;Aubertein P (1948) Mem Pourdres 30:7
32. Vignon L, Bay J (1903) Bull. Soc. Chim. France [3], 29: 26
33. Silberrad O, Farmer RC (1906) J Chem Soc 89:1759
34. Berl E, Delpy IM (1910) Ber. 43:1421
35. Carlson T (1907) Ber 40:4192
36. Farmer RC (1920) J Chem Soc 117:806
37. Baker JW, Easty D M (1950) Nature 166:156; (1952) J Chem Soc 1193:1208
38. Nadai A, Physik Z (1928) Chem 136:289
39. Cristol SJ, Franzus B, Shadan A (1955) J Am Chem Soc 77:2512
40. Pujo AM, Boileau J, Frejacques C (1955) Bull Soc Chim France 974
41. Bonner TG, Friesel DE (1959) J Chem Soc 3902
42. Oldham JWH (1925) J Chem Soc 127:2840
43. Urbanski T (1933) Roczniki Chem 13:399; (1934) Roczniki Chem 14:925; (1935) Roczniki Chem 15:191; (1936) Roczniki Chem 16:359; (1937) Roczniki Chem 17:585
44. Hicks JA (1956) Trans Faraday Soc 52:1526
45. Bi XL, Liu JP (2011) Research progress in synthesis and nature pyrolysis of nitrate ester. In: Hazard Mater Secur Emerg Technol 35–38
46. Roginskii SZ (1932) Fiz Zh 1:640; (1932) Physik Chem (B) 18:364
47. Kamide K, Okada T, Terakawa T et al (1978) Characterization of cellulose nitrate by thin-layer chromatography. Polym J 10(5):547–556
48. Clark D, Stephenson P, Heatley F (1981) Partial degrees of substitution in cellulose nitrates determined by means of C–13 magnetic-resonance studies. Polymer 22(8):1112–1117

49. Clark D, Fowler A, Stephenson P (1983) Application of modern analytical techniques to the investigation of cellulose nitrates. J. Macromol. Sci Rev Macromol Chem Phys C 23(2):217–246
50. Shao ZQ, Wang WJ (2009) Structure and properties of nitrocellulose. National Defence Industry and Technology, Beijing
51. Shao ZQ (2002) Processing and equipment of nitrocellulose. Beijing Institute of Technology Press, Beijing
52. Zhang YH, Shao ZQ et al (2013) Effect of nitrification system and refined cotton drying process on the nitrogen distribution uniformity of nitrocellulose. Chin J Explos Propellants 36 (1):68–72
53. Zhang YH, Shao ZQ et al (2013) Effect of nitrification system and refined cotton drying process on the nitrogen distribution uniformity of nitrocellulose, pp 1007–7812
54. Wang WJ, Shao ZQ et al (2011) Effect of nitrification condition and cellulose pretreatment on nitrogen content and its distribution uniformity of nitrocellulose. Chin J Explos Propellants 34 (3):72–75
55. Zhang YH et al (2011) Effect of Nitration System and Temperature on Nitration Uniformity of Nitrocellulose Fibers. Acta Armamentarii 30(11):1750–1755
56. Wang WJ et al (2010) Effect of cotton linter and nitrification conditions on nitrogen content and its distribution uniformity of nitrocellulose. Acta Armamentarii 31(10):1363–1371
57. Wang WJ, Shao ZQ et al (2007) One method for determining nitrogen content and uniformity of nitrocellulose: CN200710119614.1
58. Ellis HV et al (1984) Subacute and Chronic Toxicity Studies of Trinitroglycerin in dogs, rats and mice. Appl Toxicol 4(2):248
59. Carmichael P, Lieben J (1963) Sudden Peath in Explosives workers. J Arch Environ Health Int 7(4):424–439
60. Feng YZ et al (1991) Study on Toxicity of Nitroglycerine. Ind Health Occup Dis 17(4):207–210
61. Gong XD, Xiao HM (2000) J. Mol Struct (Theochem) 498:181

Chapter 2
Characteristics of Nitrating Reagents

The nitration is a very important reaction to add nitro group to organic compounds, in which nitrating reagent is particularly important Bayat and Mokhtari (Def Sci J 61 (2):171–173, 2011, [1]). In most cases, people only know that nitric acid is a very strong acid. According to the acid–base theory, nitric acid is the offensive reagent as a base to form nitroxyl ion, and then to participate in the chemical reactions Bayat et al. (J Energ Mater 30:124–134, 2012, [2]), Poret Jay and Shaw Anthony (J Propllants Explos Pyrotech 38:35–40, 2013, [3]). Glaser and Lewis (Org Lett 1(2):273–276, 1999, [4]) believed that, under the action of different strong and weak acids, attacking of protons formed from nitric acid on the corresponding position of different structural organisms was different. According to different characteristics of these actions, different nitrating agents were introduced and analyzed in this chapter.

2.1 Overview

According to the characteristics of nitrates, each compound has an ester group on its structure. When it is produced, an ester group ($-ONO_2$) needs to be introduced into an organic molecule to carry out the nitration reaction [5]. Up to present, the so-called nitration reaction is a reaction to introduce nitrate group or nitro group into organic compounds.

$$ROH + HONO_2 \rightleftharpoons RONO_2 + H_2O$$

$$ROH + HNO_3 \rightarrow RONO_2 + H_2O$$

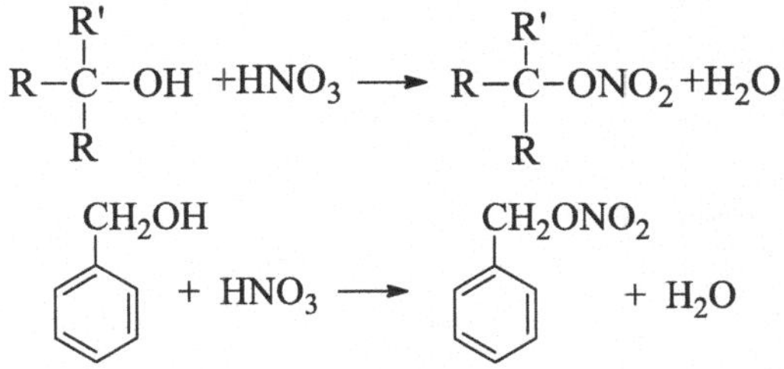

where in, R represents an alkyl group and R′ represents an aromatic hydrocarbon group.

J. Liu, *Nitrate Esters Chemistry and Technology*,
https://doi.org/10.1007/978-981-13-6647-5_2

Through other reactions, nitrate group can also be introduced into the molecular, such as oxidation, addition, etc.

$$CH_2 = CH_2 \xrightarrow{HNO_3} HO - CH_2 - CH_2 - NO_2 + O_2N - CH_2 - CH_2 - ONO_2$$

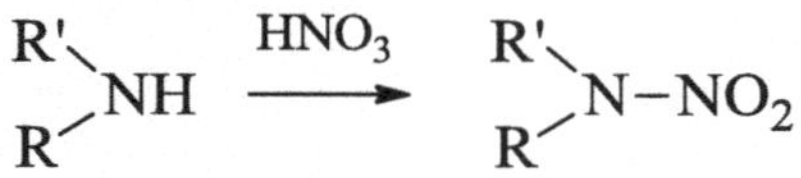

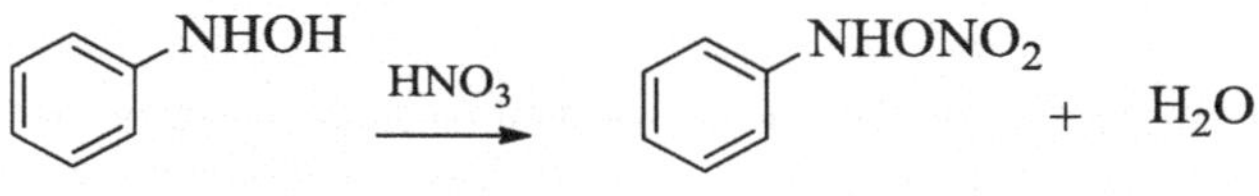

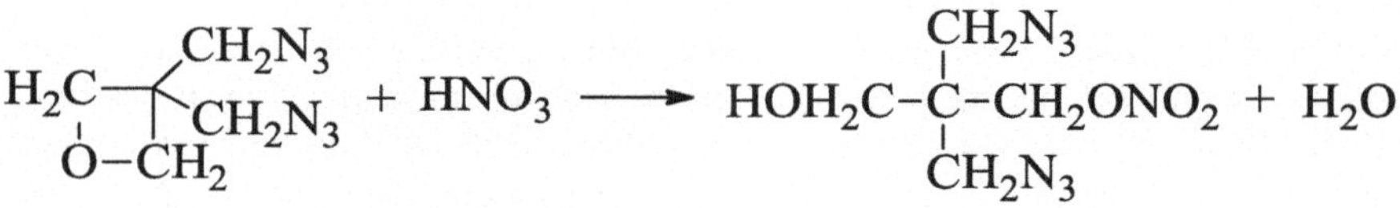

Reaction with direct introduction of nitro groups into organic molecules through substitution and addition mechanisms is nitration. Nitration reaction includes the following reaction types (not including the name reaction):

$$RO\ H + HNO_3(\text{or other material}) \rightarrow R - ONO_2 + H_2O$$

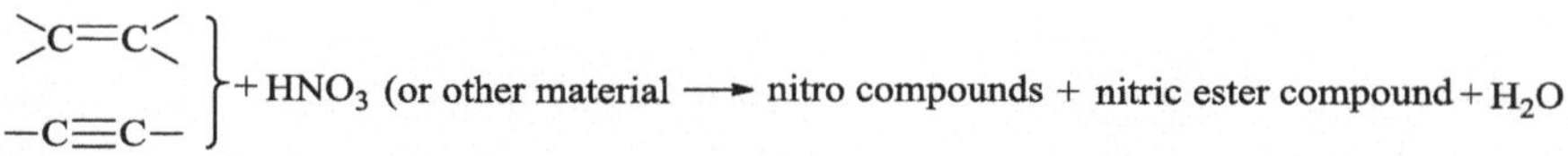

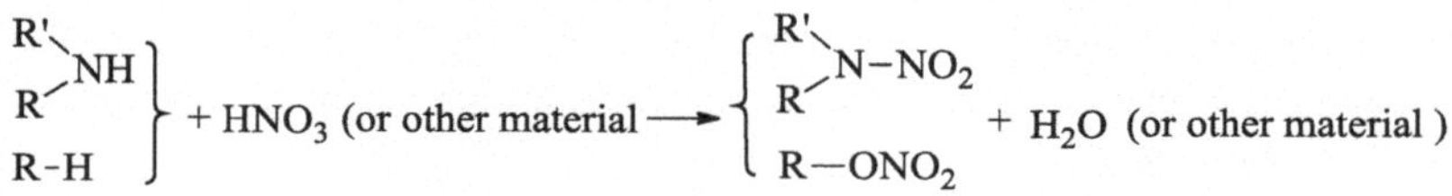

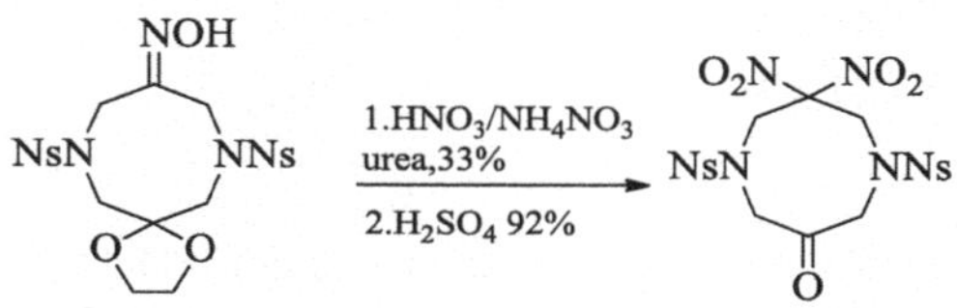

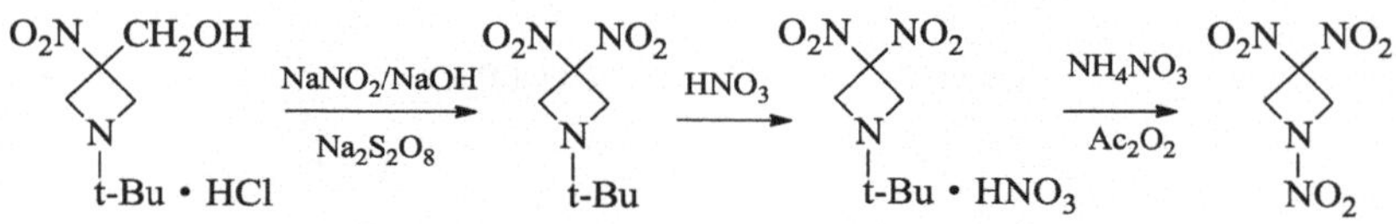

The reaction of nitric acid with aromatic hydrocarbon is governed by the charge on the aromatic ring. Usually, because of the influence of charge density, nitro group always replaces hydrogen on the meta-position of the group with large charge density, followed by the one on the para-position. After three nitro groups are on an aromatic ring, the further nitration is difficult because of the increase of electrons and steric hindrance. For example, the first nitration on cubane is easy, and the nitration difficulty is increased with the increase of nitro group number [6].

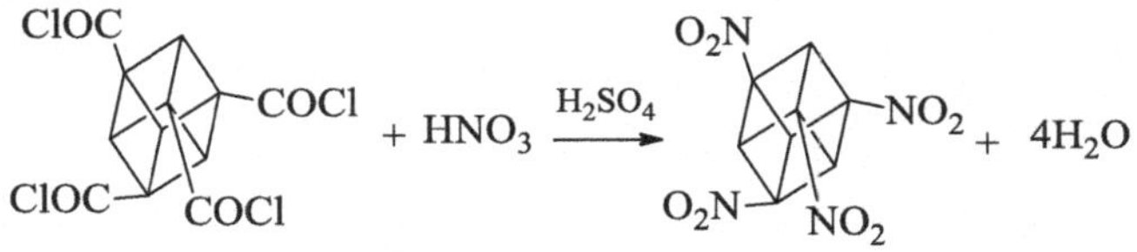

2.2 Acidity Function

Numerous nitration reactions are acid–base reactions. In order to compare and obtain the strength order of acid and alkali in the concentrated solution, it is necessary to introduce the concept of acidity function.

2.2.1 Acidity Function

In 1932, Hammett [7] proposed that, when the activity of hydrogen ion is meaningless, its ability to convert the charge-neutral base into a conjugate acid is its acidity function.

When the reaction is

$$BH_3{}^+ \overset{KBH^+}{\rightleftharpoons} H^+ + B$$

HB^+ and B represent conjugated acid and neutral base, respectively.

Then, the equilibrium constant K_{BH}^+ can be expressed as

$$K_{BH^+} = \frac{\alpha_{H^+}\alpha_B}{\alpha_{HB^+}}$$

$$pK_{BH^+} = -\log K_{BH^+} = -\log\frac{\alpha_H + \alpha_B}{\alpha_{HB^+}} = -\log\frac{\alpha_H + c_B\gamma_B}{c_{HB^+}\gamma_{HB^+}} = -\log\frac{\alpha_H + \gamma_B}{\gamma_{HB^+}} - \log\frac{C_B}{C_{HB^+}} \tag{2.1}$$

where

α activity,
c concentration, and
r activity factor.

Then, the following can be obtained:

$$H_0 = -\log\frac{\alpha_{H^+}\gamma_B}{\gamma_{HB^+}} \quad (2.2)$$

so

$$pK_{BH^+} = H_0 - \log\frac{c_B}{c_{HB^+}}$$

or

$$H_0 = pK_{BH^+} + \log\frac{c_B}{c_{HB^+}} \quad (2.3)$$

According to experiments, the above formula only shows that the property $\frac{\gamma_B}{\gamma_{HB^+}}$ of medium in its diluted solution is independent of the type of neutral base, and the ratio $\frac{\alpha_{H^+}\gamma_B}{\gamma_{HB^+}}$ is independent of the type of neutral base. Therefore, H_0 only represents a function of the nature of medium. If pK_{BH}^+ of the neutral base is known and the $\frac{c_B}{c_{HB^+}}$ value of the solution is measured, H_0 can be obtained according to the formula (2.3). And, H_0 values of some common acids are listed in Tables 2.1, 2.2, and 2.3.

Basing on the ionization constants and activities of nitric acid and sulfuric acid, Miller [8] calculated out the H_0 of the sulfuric acid–nitric acid–water system. The results are shown in Fig. 2.1.

In the medium, H_0 and pH are two different concepts. The pH value represents the hydrogen ion activity in the solution and can be expressed as

Table 2.1 H_0 values of several solutions (25 °C) [9]

Concentration of acid(mol/L)	HNO_3	HCl	$HClO_4$	H_2SO_4	H_3PO_4	HF
0.1	+0.98	+0.98	–	+0.83	+1.45	–
0.5	+0.21	+0.20	–	+−0.13	+0.97	–
1.0	−0.18	−0.20	−0.22	−0.26	+0.63	+1.20
2.0	−0.67	−0.69	−0.78	−0.84	+0.24	+0.91
3.0	−1.02	−1.05	−1.23	−1.38	−0.08	+0.60
4.0	−1.32	−1.40	−1.72	−1.85	−0.37	+0.40
5.0	−1.57	−1.76	−2.23	−2.23	−0.69	+0.28
6.0	−1.79	−2.12	−2.84	−2.76	−1.04	+0.15
7.0	−1.99	−2.56	−3.61	−3.32	−1.45	+0.02
8.0		−3.00	−4.33	−3.87	−1.85	−0.11
9.0		−3.39	−5.05	−4.40	−2.22	−0.24
10.0		−3.68	−5.79	−4.89	−2.59	−0.36

Measured at 19°C

Table 2.2 H_0 values of some strong acids (25 °C)

Acid	H_2SO_3F	H_2SO_3Cl	$H_2S_2O_7$	BF_3H_2O	H_2SO_4	HF	CH_3OSO_3H	HNO_3	HCO_2H
H_0	−12.8	−12.8	−12.2	−11.4	−11.0	−10.2	−7.86	-6.3	-2.2

Table 2.3 H_0 of fume sulfuric acid (25 °C)

Sulfuric acid concentration (%)	100	101	102	103	104	105	106	107
SO_3 concentration (%)	0.00	4.44	8.9	13.3	17.8	22.2	26.6	31.1
H_0	−11.01	−11.82	−12.06	−12.28	−12.47	−12.62	−12.74	−12.87

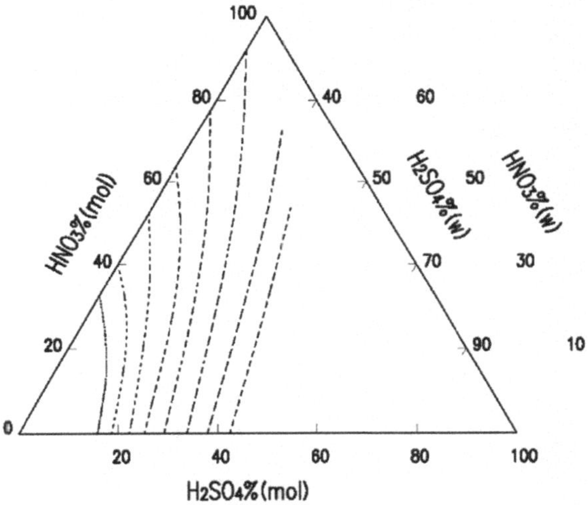

Fig. 2.1 H_0 of H_2SO_4–HNO_3–H_2O system (25 °C)

$$pH = \log \alpha_{H^+} \tag{2.4}$$

While, H_0 is not only related to α_H^+, but also related to $\frac{\gamma_B}{\gamma_{HB^+}}$. Only when the hydrogen ion activity of the solution is not high and $\frac{\gamma_B}{\gamma_{HB^+}} = 1$, H_0 is numerically equal to pH. When the hydrogen ion concentration of the solution is increased, H_0 decreases faster than the pH. Thus, in the concentrated acid solution, α_H^+ value is small, but the ability of medium to convert the neutral base into a conjugate acid is very strong at this time. For example, in 100% sulfuric acid, the α_H^+ value produced from self-ionization of sulfuric acid is small (self-ionization pKa of sulfuric acid is 3.47), but its ability to convert neutral base into conjugate acid is very strong. A smaller H_0 value (or a larger H_0) means a stronger ability to convert the neutral base to the conjugate acid. From (2.4), the following can be obtained:

$$\log \frac{C_B}{C_{HB^+}} = H_O - pK_{BH^+} \tag{2.5}$$

For a given neutral base, pK_{BH^+} is a certain value, and a smaller H_0 of the medium means a smaller value on the right side of the equation, and the $\frac{C_B}{C_{HB^+}}$ value is also smaller, which indicates that the proportion of neutral base converted to conjugate acid is larger.

When a salt having the same ionic effect is added to the acid solution, its H_0 will be reduced. And, the addition of ammonium sulfate into concentrated sulfuric acid is a typical example.

Although H_0 is related to the concentration of acid, it is not related to the neutral base used during the determination theoretically. In fact, the pK_{BH^+} value of neutral base is slightly different with different mediums used during the determination, and measured H_0 is also different with different neutral bases.

Similar to the definition of H_0, some researchers have proposed and defined H_- and H_+ as

$$H_- = -log \frac{\alpha_{H^+} \gamma_{B^-}}{\gamma_{HB}} = pK_{BH} + log \frac{C_{B^-}}{C_{HB}} \tag{2.6}$$

$$H_+ = -log \frac{\alpha_{H^+} \gamma_{HB^+}}{\gamma_{H_2B^{2+}}} = pK_{BH_2^{2+}} + log \frac{C_{HB^+}}{C_{H_2B^{2+}}} \tag{2.7}$$

In the following reactions:

$$HB \rightleftharpoons H^+ + B^-$$
$$H_2B^{2+} \rightleftharpoons HB^+ + H^+$$

H_- and H_+ indicate the ability of medium to convert anionic or cationic bases into a conjugate acid, which have a significant difference from H_0 in values.

2.2.2 Meanings of Acidity Function J_0 and C_{0+}

Alcohols, nitric acid, nitrous acid, hypochlorous acid, and other typical XOH-type neutral bases can have the following reactions in strong acid medium:

$$XOH + H^+ \rightleftharpoons XOH_2^+$$
$$XOH_2^+ \rightleftharpoons X^+ + H_2O$$

Westheimer and Kharasch [10] proposed another method to represent the acidity function J_0 of this type of neutral base.

$$X^+ + H_2O \overset{K_{X^+}}{\rightleftharpoons} XOH + H^+$$

then,

$$K_{X^+} = \frac{\alpha_{H^+}\alpha_{XOH}}{\alpha_{X^+}\alpha_{H_2O}}$$

$$pK_{X^+} = -\log\frac{\alpha_{H^+}\alpha_{XOH}}{\alpha_{X^+}\alpha_{H_2O}} = -\log\frac{\alpha_{H^+}\alpha_{XOH}}{\alpha_{X^+}\alpha_{H_2O}} \times \frac{\gamma_{XOH_2^+}}{\gamma_{XOH_2^+}}$$

$$= -\log\frac{\alpha_{H^+}\gamma_{XOH}\gamma_{XOH_2^+}c_{XOH}}{c_{X^+}\gamma_{X^+}\alpha_{H_2O}\gamma_{XOH_2^+}} = -\log\frac{\alpha_{H^+}\gamma_{XOH}}{\gamma_{XOH_2^+}} - \log\frac{1}{\alpha_{H_2O}}$$

$$-\log\frac{c_{XOH}}{c_{X^+}} - \log\frac{\gamma_{XOH_2^+}}{\gamma_{X^+}} = H_0 + \log\alpha_{H_2O} - \log\frac{c_{XOH}}{c_{X^+}} - \log\frac{\gamma_{XOH_2^+}}{\gamma_{X^+}}$$

In the sulfuric acid medium with the concentration of 80–95%, $\frac{\gamma_{XOH_2^+}}{\gamma_{X^+}}$ of the ionized neutral base is a constant, and the acidity function J_0 is proposed as

$$J_0 = -\log\frac{\alpha_{H^+}\gamma_{XOH}}{\gamma_{XOH_2^+}} + \log\alpha_{H_2O} \tag{2.8}$$

then,

$$J_0 = pK_{X^+} + \log\frac{c_{XOH}}{c_{X^+}} + \log\frac{\gamma_{XOH_2^+}}{\gamma_{X^+}} = pK_{X^+} + \log\frac{c_{XOH}}{c_{X^+}} + \text{constant}$$

This reaction was confirmed through the measurement of J_0 of the sulfuric acid–water system with triphenylmethanol compound as the neutral base. In the further studies, it has been found that there is a linear relationship between nitration rate constant k_2 of aromatic hydrocarbon and J_0 in the 80–90% sulfuric acid medium, because, when the nitric acid is a neutral base, the corresponding reaction of basic reaction can be written as

$$H^+ + HONO_2 \rightleftharpoons H_2O + NO_2{}^+$$

Deno et al. [11] found $\frac{\gamma_{XOH_2{}^+}}{\gamma_{X^+}}$ was constant only in 80–95% sulfuric acid. If sulfuric acid concentration was beyond this range, $\frac{\gamma_{XOH_2{}^+}}{\gamma_{X^+}}$ was not constant. Therefore, another acidity function C_0 (or H_R) was proposed as

$$C_0 = -\log\frac{\alpha_{H^+}\gamma_{XOH}}{\gamma_{XOH_2{}^+}} + \log\alpha_{H_2O} - \log\frac{\gamma_{XOH_2{}^+}}{\gamma_{X^+}} = J_0 + \log\frac{\gamma_{X^+}}{\gamma_{XOH_2{}^+}} \tag{2.9}$$

Table 2.4 C_0 and J_0 of aqueous solution of sulfuric acid

Concentration of sulfuric acid (wt%)	C_0	$-\frac{dC_0}{d\%H_2SO_4}$	J_0	$-\frac{dJ_0}{d\%H_2SO_4}$	$\log\frac{\gamma_{X^+}}{\gamma_{XOH_2^+}}$
1.0	0.92	–	0.97	–	0.05
5.0	−0.07	0.17	0.23	–	0.30
10	−0.72	0.12	−0.18	0.08	0.54
20	−1.92	0.12	−0.95	0.07	0.97
30	−3.22	0.14	−1.66	0.08	1.56
40	−4.80	0.16	−2.52	0.10	2.28
50	−6.60	0.20	−3.67	0.13	2.93
60	−8.92	0.26	−5.11	0.16	3.81
70	−11.52	0.26	−6.91	0.20	4.61
80	−14.12	0.26	−9.11	0.25	5.01
90	−16.72	0.26	−11.80	0.26	4.92
92	−17.24	0.26	−12.32	0.26	4.92
95	−18.08	0.33	−13.16	0.33	
98	−19.64		−14.72		

or

$$C_0 = pK_{X^+} + \log\frac{c_{XOH}}{c_{X^+}} \tag{2.10}$$

C_0 and J_0 values of aqueous solutions of sulfuric acid, nitric acid, and perchloric acid were determined by Deno et al. and shown in Tables 2.4 and 2.5.

As listed in Tables 2.4 and 2.5, the meaning of C_0 is the tendency of the medium to convert HOX-type neutral base into X^+. The smaller C_0 value means the greater ability of the medium to convert HOX to X^+. Therefore, C_0 decreases faster than H_0 and J_0 with the increase of sulfuric acid concentration.

Table 2.5 Acidity functions of nitric acid and perchloric acid (25 °C) [12]

Concentration (wt%)	$HClO_4$				HNO_3			
	C_0	$-\frac{dC_0}{d\%HClO_4}$	H_0	$-\frac{dH_0}{d\%HClO_4}$	C_0	$-\frac{dC_0}{d\%HNO_3}$	H_0	$-\frac{dH_0}{d\%HNO_3}$
1.0	0.97	–			0.74		0.80	
10	−1.04	0.14	−0.16	0.09	−1.30	0.4	−0.39	0.07
20	−2.43	0.13	−0.47	0.06	−2.63	0.12	−0.95	0.05
30	−3.79	0.15	−1.12	0.07	−3.83	0.12	−1.42	0.05
40	−5.54	0.20	−1.88	0.10	−5.07	0.13	−1.77	0.04
50	−7.86	0.29	−3.12	0.16	−6.40	0.15		
58					−7.18	0.18		
60	−11.14	0.42	−4.78	0.17				

Table 2.6 H_0 comparison of HF and fluorine-containing Lewis acids

Composition	H_0
HF + 0.02 M NdF_5(wt%:0.4)	−12.5
HF + 0.36 M NdF_5(wt%:6.7)	−13.5
HF + 0.36 M SdF_5(wt%:6)	−14.3
HF + 3.0 M NdF_5(wt%:60)	−15.2

2.2.3 *The Acidity Function of Lewis Acid*

Lewis acid can also be measured through the determination of proton acid acidity function with neutral base to compare the strength of Lewis acid. In the benzene or chlorobenzene solution and with a neutral base as an indicator, the relative strength of some Lewis acids was qualitatively compared and the following strength order was obtained:

$AlBr_3 > AlCl_3 > FeCl_3 > H_2SO_4$(fuming)
96%H_2SO_4 >70% $HClO_4 > SnCl_4 > SnCl_5 > SnCl_3$
$BiCl_3 > PCl_5 > PCl_3$
$HCl > AsCl_3 > ZnCl_2$
$CdCl_2 > CuCl_2 > HgCl_2 > AgNO_3$

Although the order of Lewis acids is given here, the strength of relationship between different groups is not given.

Lewis acid mixture is a very strong acid to hydrides, and Mackor [13] measured H_0 of boron trifluoride (7wt%) in anhydrous hydrofluoric acid as −16.8, which was 5 less than H_0 of 100% sulfuric acid. The acidity of $AlCl_3$–HCl is close to that of $AlBr_3$–HBr and HF–BF_3, thus it is also very strong acid. The mixture of anhydrous hydrofluoric acid and other fluorine-containing Lewis acid is also very strong. The value comparison is listed in Table 2.6.

The balance of HF and fluorine-containing Lewis acids in Table 2.6 can be established as follows:

$$2HF + SbF_3 \rightleftharpoons H_2F^+ SbF_6^-$$

Therefore, the chemical reaction in the acid is acid-catalyzed reaction, and its catalytic effect is very strong.

2.3 Mixture of Nitric Acid and Its Protonic Acid

Nitric acid and the mixture of nitric acid and proton acid are the common nitrating agents to prepare nitrates that are usually nitriles, which are also most widely used ones in all industrial and laboratory applications.

2.3.1 Nitric Acid

Pure nitric acid is a colorless transparent liquid with the molecular formula of HNO_3, relative molecular mass of 63.0087, and the relative density of 1.52 g/mL. Its structural formula is

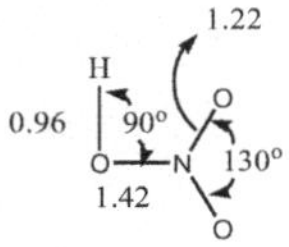

Since the nitrogen and oxygen atoms are in the same plane and there are hydrogen bonds between molecules, the association of the nitric acid structure has been described in the literature as follows:

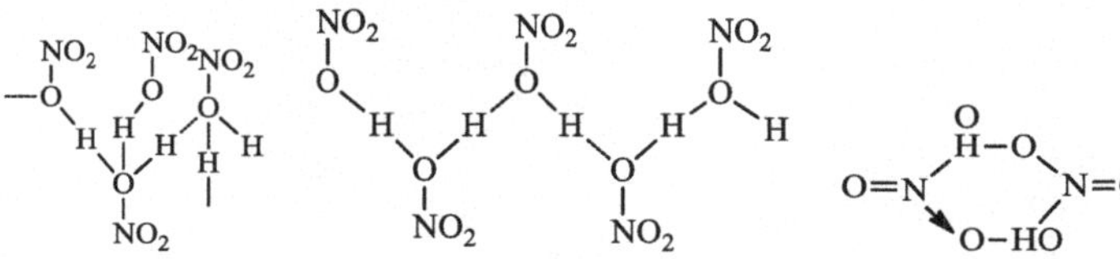

Tetrahedron Chain Ring

Nitric acid is highly soluble in water and is present as an association way. The association phase diagram of nitric acid aqueous solution is shown in Fig. 2.2.

In Fig. 2.2, points A and B correspond to the molecular complexes $HNO_3{\cdot}3H_2O$ and $HNO_3{\cdot}H_2O$, and their melting points are −18.53 and −37.68 °C, respectively. When the temperature is increased, complexes are decomposed into nitric acid and water. At room temperature and slightly higher temperature, nitric acid aqueous solution has the complex $HNO_3{\cdot}2H_2O$, meaning that nitric acid aqueous solution

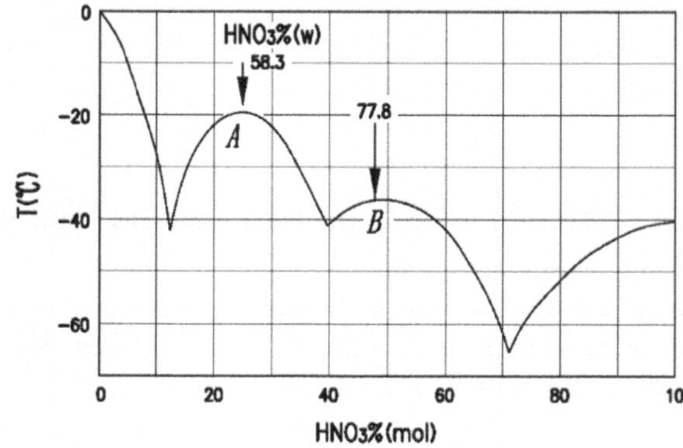

Fig. 2.2 The phase diagram of nitric acid–water system

Table 2.7 Constants of Pure nitric acid

Item	Result
Density d_4^{25}	1.503
Specific heat capacity (300 K)	0.416 cal g^{-1} K^{-1}
Melting point	−41.59 °C
Boiling point	86 °C
Enthalpy of formation	−174.31 kJ mol^{-1}
Evaporation enthalpy (20 °C)	38495.78 J mol^{-1}

with this composition has the smallest molar volume and the maximum viscosity. The physical constants of nitric acid are listed in Table 2.7.

Nitric acid (100%) is actually a mixture with dissolved nitrogen oxides, including a small amount of dinitrogen pentoxide. And N_2O_5 automatically decomposes and generates nitrous oxide and oxygen according to the following process:

$$2HNO_3 \rightleftharpoons N_2O_5 + H_2O$$
$$N_2O_5 \rightarrow N_2O_4 + 1/2O_2$$

This decomposition of nitric acid shows that pure nitric acid is extremely unstable. However, pure nitric acid in solid state can be stored for a long time. Pure liquid nitric acid will automatically cause the decomposition, even when it is stored in the dark. And ultraviolet rays and sunlight can accelerate its decomposition reaction. After the decomposition of nitric acid, the color of it is from yellow to reddish brown due to the presence of dinitrogen tetroxide and then its color begins to change and is gradually deepened after 30–60 min at 20 °C. The degree of decomposition was increased with the increase of temperature. At different temperatures, nitric acid decomposes according to the following decomposition degree. And the decomposition percentage is listed in Table 2.8.

$$4HNO_3 \rightleftharpoons 4N_2O + 2H_2O + O_2$$

With a lower concentration, the stability of the better is higher, and 70% nitric acid can be long-term preserved and very stable. Industrial concentrated nitric acid has different degrees of yellow, and AR grade fuming nitric acid (concentration is greater than 90%) also is light yellow.

Nitric acid is a strong acid, and it can be all ionized in its diluted aqueous solution as

Table 2.8 Decomposition of nitric acid at different temperatures

Temperature (°C)	85	100	130	160	190	220	250	255
Decomposition percentage (%)	9.53	11.77	18.78	28.96	49.34	72.07	93.03	100

$$2HNO_3 \overset{K_a}{\rightleftharpoons} NO_2^+ + NO_3^- + H_2O$$

The pKa value is approximately −1.4. According to the formula, with the increase of nitric acid concentration, the relative amount of ionization is gradually decreased, and the nitric acid hydrate is gradually increased. And most of the concentrated nitric acid is in the form of nitric acid molecule. Therefore, when the concentration of nitric acid aqueous solution is 96.12%, its conductivity is extremely small. When nitric acid concentration is increased, its conductivity is gradually increased again because of the self-ionization of nitric acid.

$$HNO_3 + HNO_3 \rightleftharpoons H_2NO_3^+ + NO_3^-$$
$$H_2NO_3^+ \rightleftharpoons H_2O + NO_2^+$$
$$H_2O + HNO_3 \rightleftharpoons H_3O^+ + NO_3^-$$

The total reaction is

$$3HNO_3 \rightleftharpoons NO_2^+ + H_3O^+ + 2NO_3^-$$

The percentages of molecules and ions in nitric acid with different concentrations are shown in Fig. 2.3.

In the Raman spectra of anhydrous nitric acid, in addition to the lines produced by the nitric acid molecules, there are two weak lines at 1400 and 1050 cm^{-1}, which are from nitryl ions and nitrate ions. With the increase of water in the acid, the spectral line of 1400 cm^{-1} is weakened, while the line at 1050 cm^{-1} is

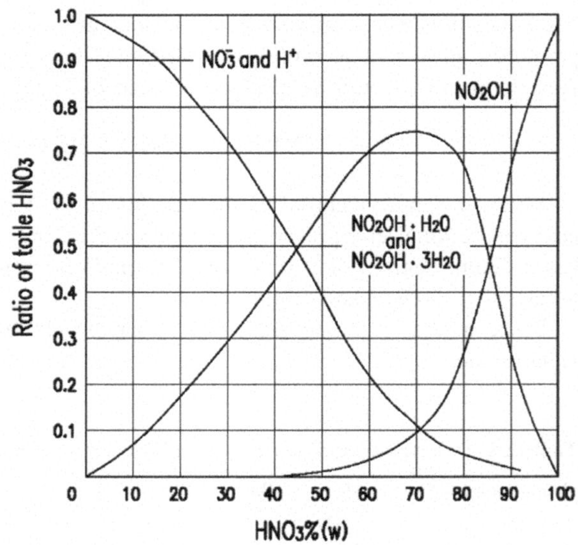

Fig. 2.3 Molecules and ions of nitric acid aqueous solution

enhanced. When the nitric acid concentration is 94–95%, the 1400 cm^{-1} line disappears completely. A similar situation occurs in the infrared spectroscopy determination of nitric acid. Nitryl ions produce a 2360 cm^{-1} line in the infrared spectrum, which is gradually weakened and eventually disappears if water is added to anhydrous nitric acid. These results suggested that nitric acid with high concentration could produce nitryl ions through the self-ionization.

The ionization of nitric acid can be written as

$$2HNO_3 \overset{K_a}{\Leftrightarrow} NO_2^+ + NO_3^- + H_2O$$
$$K = \frac{\alpha_{NO_2^+}\alpha_{NO_3^-}\alpha_{H_2O}}{\alpha^2_{HNO_3}}$$

Lee et al. [14] measured the equilibrium constant of self-ionization of ammonium nitrate with the method to determine freezing point and conductivity. The results are listed in Table 2.9.

According to these equilibrium constants, the calculated concentration of nitroxylate is not high even in anhydrous nitric acid. According to the results of freezing point measurement, about 3.4% of anhydrous nitric acid is ionized at −40 °C to produce 1.2% of nitroxyl, 1.7% of nitrate, and 0.5% of water. With the decrease of nitric acid concentration, the concentration of nitryl ions is decreased rapidly until the nitric acid concentration was 95%, where the spectrum of nitryl ions cannot be detected. However, it does not mean that there are no nitroxyl ions in nitric acid with this concentration. The isotope exchange experiment of $H_2^{18}O$-diluted nitric acid shows that oxygen isotopic exchange in concentrated nitric acid is very rapid, which is the strong evidence of the presence of nitryl ions.

$$HNO_3 + HNO_3 \rightleftharpoons H_2NO_3^+ + NO_3^-$$
$$H_2NO_3^+ \rightleftharpoons H_2O + NO_2^+$$
$$NO_2^+ + H_2{}^{18}O \rightleftharpoons NO_2{}^{18}OH_2^+ \rightleftharpoons NO_2{}^{18}OH + H^+$$

When nitric acid concentration is less than 40% (mole, equivalent to about 70 wt %), the oxygen isotopic exchange reaction occurs only when nitrous acid is present, indicating probably no nitryl ion there. Nitric acid catalyzes the isotopic exchange reaction as

$$HNO_3 + HNO_2 \rightleftharpoons HNO_2^+ + NO_3^-$$
$$HNO_2^+ \rightleftharpoons H_2O + NO^+$$
$$NO^+ + H_2{}^{18}O \rightleftharpoons NO{}^{18}OH_2$$
$$NO{}^{18}OH_2 \rightleftharpoons H{}^{18}O \cdot NO + H^+$$

Table 2.9 The equilibrium constant of nitric acid ionization

Temperature (°C)	−4	−20	−10	25
10^5 K	2.44	1.602	0.930	0.280

This indicates that, if nitrous acid is present in diluted nitric acid, it is possible to produce another electrophilic reagent, nitrite ion NO^+.

When the nitric acid concentration exceeds 70 wt%, the rate constant of oxygen exchange reaction is substantially equal to that of the nitration of activated aromatic hydrocarbon, which indicates that the attacking agent of nitration reaction is still the nitryl ion with such a concentration of nitric acid.

Generally, nitric acid always contains a certain amount of nitrous acid. Nitryl ions are present only in high concentrations of nitric acid, and both nitryl ions and nitrosyl ions are electrophilic offensive agents. In organic solvent with low dielectric constant, nitration with anhydrous nitric acid can generate ion pair of nitryl ions.

$$5HNO_3 \rightleftharpoons NO_2{}^+NO_3{}^- \cdot HNO_3 + H_3O^+NO_3{}^- \cdot HNO_3$$

The nitrogen dioxide molecule has properties of free radical that can initiate free radical reactions, especially in solvents with low dielectric constant. When the nitration temperature is increased, the decomposition rate of nitric acid and the content of nitrogen dioxide are increased, but the balance constant of self-ionization of nitric acid into nitryl ion and nitryl ion concentration are decreased. Therefore, when nitric acid is used to nitrate alkyl hydrocarbon, with the changes in nitric acid concentration and reaction temperature, nitration mechanism is gradually transformed from the ion mechanism into free radical one. And nitrated compounds with different structures give different nitration products with different carbon atom positions to connect nitro groups.

(1) Nitration of alcohols

Nitric acid with medium concentration (60%) can nitrate the primary alcohols to nitrates. The nitration of secondary alcohols needs nitric acid with concentration of more than 80%. Nitration of tertiary alcohols produces tert-nitroalkane, but the yield is very low as only about 10% without practical value.

$$(CH_3)_3 \cdot C - OH \xrightarrow{HNO_3} (CH_3)_3 \cdot C - NO_2$$

(2) Nitration of saturated and unsaturated carbon atoms with nitric acid.

Nitric acid nitration of aliphatic saturated carbon atoms, especially the nitration of cycloalkanes, mostly follows free radical substitution mechanism [15]. Nitration of unsaturated atoms with nitric acid is the electrophilic addition first. At relatively high temperature, nitration of unsaturated hydrocarbons with diluted nitric acid may be the addition of free radicals (nitrogen dioxide) to double bonds first.

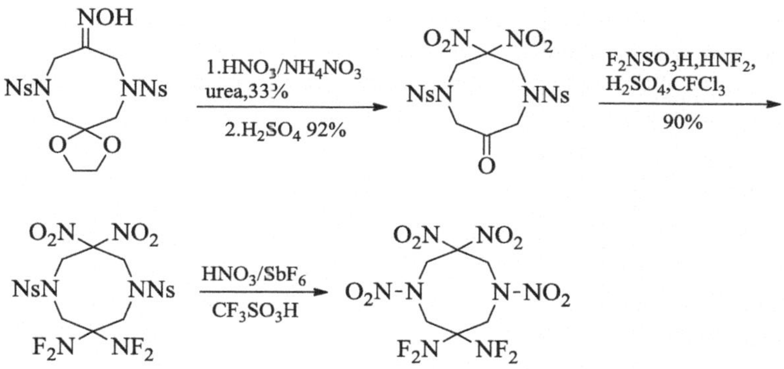

(3) Nitration of aromatic compounds

Nitric acid, nitric acid aqueous solution, and nitric acid in an organic solvent can nitrate various aromatic compounds. The nitration is generally carried out in a liquid phase. Depending on the nitric acid concentration and the activity of nitrated product, one, two, or three nitro groups can be introduced onto the benzene ring,

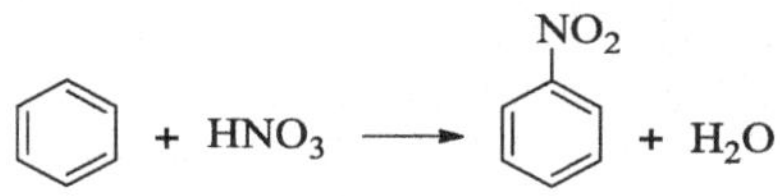

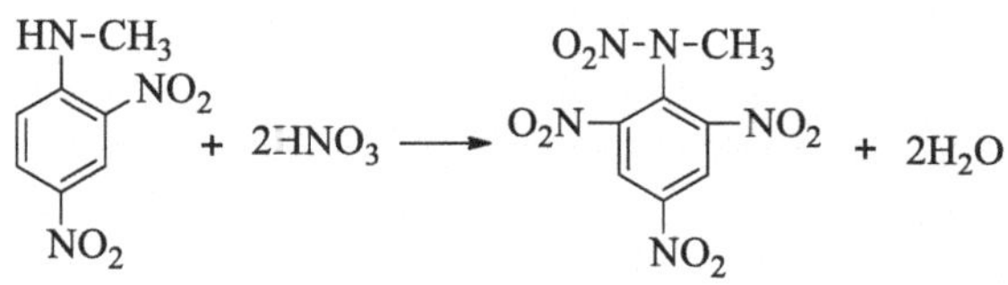

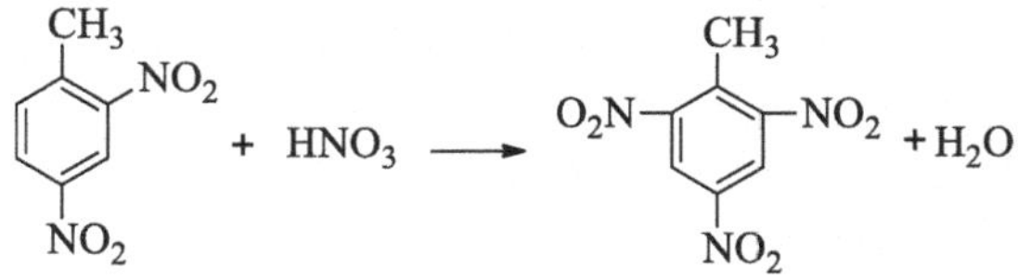

In these reactions, nitric acid is both a nitrating reagent and a strong oxidizing agent. When the concentration of acid is higher, the nitration ability is stronger and the oxidation capacity is weaker. With a lower concentration, the nitration capacity of nitric acid is weaker and its oxidation capacity is stronger. When temperature is increased during the nitration reaction, nitration reaction rate is speeded up, but side reactions due to oxidation are more significant with higher temperature.

In general, the nitration of aromatic hydrocarbon with nitric acid is an electrophilic substitution reaction, and the reagent that attacks the benzene ring is nitryl ion or its carrier. However, in some special cases, it may be another reagent. The nitration of aromatic compounds those are electrophilically active with diluted nitric may be done through the nitroxyl ion or its carrier. For example, with dilute nitric acid, phenol or aromatic amine compounds can be nitrated into nitro derivatives.

According to the above description and the specific situations (the nature of nitrated compounds, reaction conditions, etc.), nitration of aromatic compounds with nitric acid can follow different mechanisms. However, the products from free radical mechanism are relatively complicated, and electrophilic substitution reaction is mainly used to prepare a variety of aromatic nitro compounds in the industry.

It should be ensured that there are sufficient nitro ions (including their carriers) in the reaction environment to nitrate aromatic compounds through electrophilic substitution, which requires a certain concentration of nitric acid. If the concentration is too low, the oxidative side reactions will be increased, even without nitration reaction. And this minimum concentration depends mainly on the activity of the nitrated compound although the reaction temperature also has a certain impact. In order to maintain this concentration, an excess of nitric acid is required. For example, in the nitration of toluene into 1-nitrotoluene with nitric acid, the final concentration of nitric acid cannot be less than 70%. If the concentration of nitric acid is 98%, the amount of nitric acid required to nitrate 1 mol (92 g) of toluene is the following:

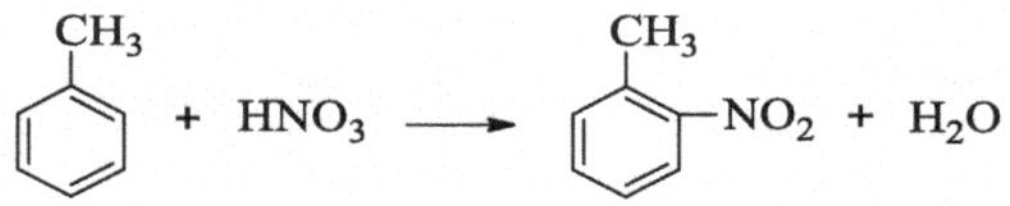

Assuming x g of 98% nitric acid is needed, we can get

$$\frac{0.98x - 63}{x - 63 + 18} = 0.7$$

so

X = 112.5 g
in which the amount of nitric acid is
112.5 × 0.98 = 110.25 g
that is more than theoretical amount as
110.25 − 63 = 47.25 g

In fact, due to nitric acid loss through side reactions and evaporation, the amount of nitric acid is more than required. Therefore, when only nitric acid is used as nitrating reagent, the amount of nitric acid used is relatively large, and the nitration product can be separated sufficiently after the reaction mixture is diluted with water after the reaction, which again increases the difficulty of nitric acid recovery.

Nitric acid is very corrosive to metal, and therefore the equipment to treat nitric acid requires different acid-resistant materials according to the different needs, such as high ferrosilicon, special stainless steel, or pure aluminum, which is a shortcoming of nitric acid alone as nitrating reagent.

However, under some conditions, nitric acid as a nitrating reagent has its own special advantages. Therefore, pure nitric acid is still widely used in industry as one of the nitrating reagents.

(4) Nitration of amines

Diluted nitric acid and amines form nitrate salts of amines. Concentrated nitric acid can nitrate secondary amine and tertiary amine into nitramine. Because most of nitramines decompose in the concentrated sulfuric acid, during the manufacture of nitramines, nitric acid is commonly used to nitrate amines.

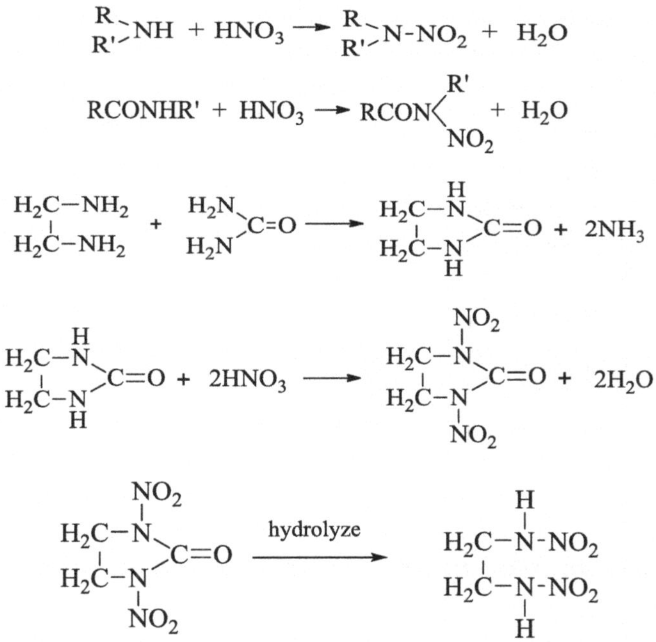

(5) Nitration of fused heterocycles with nitric acid

There have been very classic examples about the nitration of high energy density materials with nitric acid as the attacking reagent. For example, there have been large production cases of nitric acid nitration to produce RDX and HMX. The fused ring nitration is the same [16, 17].

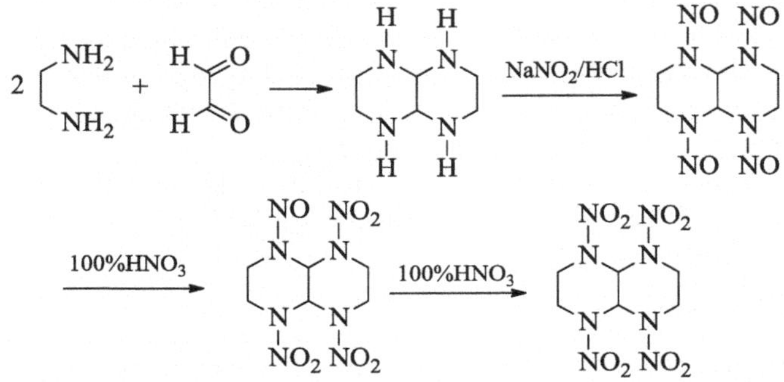

(6) Nitration of adamantane

The nitration of adamantane is different from that of general alkanes [18, 19]. If nitrating reagent of general alkanes is used to nitrate adamantane, the target product cannot be obtained but only multistep nitration can be carried out and the nitration reaction can be carried out under special nitration conditions.

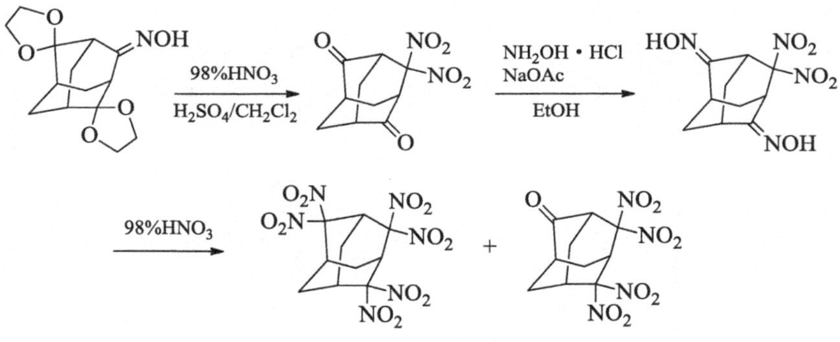

2.3.2 Nitrating Reagent from Nitric Acid and Sulfuric Acid

According to the acidity function value of either H_0 or H_R, capacity of sulfuric acid to convert the neutral alkali HOX into its conjugate acid H_2OX^+ or X^+ is stronger

Table 2.10 Relationship between $\frac{C_{NO_3^-}}{C_{HNO_3}}$ and sulfuric acid concentration

Sulfuric acid concentration(%)	30	34	38	42	46	50	54	58
$\log \frac{C_{NO_3^-}}{C_{HNO_3}}$	0.76	0.49	0.31	0.12	−0.09	−0.32	−0.54	−0.81

Table 2.11 The relationship between the concentration of sulfuric acid and the dissociation of nitric acid into nitryl ions

Sulfuric acid concentration (%)	100	97.5	95	92.5	90
Quality of nitryl ion (%)	100	100	100	70	60
Measured sulfuric acid concentration ①(%)	98.5	96.0	93.6	91.3	89.2

than that of nitric acid. In the mixture of nitric acid and sulfuric acid, sulfuric acid plays the role of acid and nitric acid plays the role of base. With the appropriate proportion, the following equations can be obtained:

$$H_2SO_4 + HNO_3 \rightleftharpoons H_2NO_3{}^+ + HSO_4{}^-$$
$$H_2NO_3{}^+ \rightleftharpoons NO_2{}^+ + H_2O$$
$$H_2O + H_2SO_4 \rightleftharpoons H_3O^+ + SO_4{}^-$$

The total reaction is

$$HNO_3 + 2H_2SO_4 \rightarrow NO_2{}^+ + H_3O^+ + 2HSO_4{}^-$$

Sulfuric acid helps nitric acid to get converted into nitryl ions. The above reaction can be proved by the following data.

(1) Spectral analysis

When nitric acid was added into pure sulfuric acid, a Raman spectrum line at 1400 cm^{-1} appears. When sulfuric acid is added into pure nitric acid, the 1400 cm^{-1} line is significantly enhanced. Because the Raman spectrum of the nitryl salt solution also shows a 1400 cm^{-1} line, the 1400 cm^{-1} line is due to the generated nitryl ions.

(2) Raoult's law

The decrease of freezing point of the diluted solution is proportional to the number of proton points (i) produced from the solute in solvent. When nitric acid is added into sulfuric acid, the decreased value of freezing point (i) is equal to 4 (the measured value is around 4).

Table 2.12 Ionization proportion of nitric acid into nitryl ions in the anhydrous mixed acid

Components of the mixed acid (wt%)	HNO_3	5	10	15	20	40	60	80	90	100
	H_2SO_4	95	90	85	80	60	40	20	10	0
Proportion of nitric acid into $NO_2{}^+$ (%)	100	100	80	62.5	28.8	16.7	9.8	5.9	1	

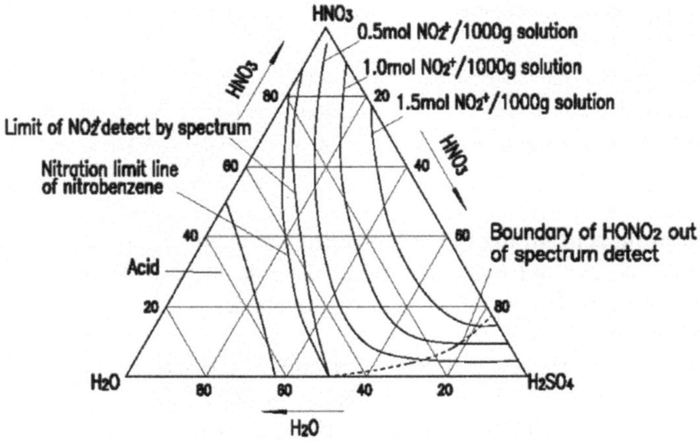

Fig. 2.4 Molar concentration of nitryl ion (c%) in sulfuric acid–nitric acid–water system [20]

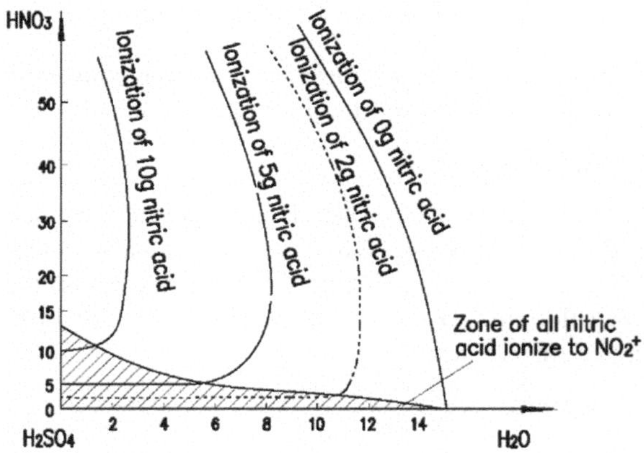

Fig. 2.5 Mass concentration of nitryl ions (%) in nitric acid–sulfuric acid–water system [20]

Some people have studied the presence of nitric acid with various concentrations of sulfuric acid [21], and found that, when nitric acid concentration is 10^{-2}–10^{-4} mol/L, there are the following possibilities:

(1) When the sulfuric acid concentration is less than 20%, all nitric acid is ionized into nitrate ions,

$$HNO_3 + H_2O \rightarrow H_3O^+ + NO_3^-$$

(2) When the sulfuric acid concentration is 20–60%, nitric acid is present as a mixture of nitric acid molecule and nitrate ion, and the relationship between concentrations of these two substances and sulfuric acid concentration is shown in Table 2.10.

When sulfuric acid concentration is 44%, the concentration of nitrate ions is equal to the concentration of nitric acid molecules.

(3) When sulfuric acid concentration is 65–82%, most of the nitric acid is in its molecular state.

(4) When the concentration of sulfuric acid is increased, nitric acid begins to dissociate into nitryl ions. When nitric acid concentration is 5% in the solution, relationship of the percentage of nitric acid that is dissociated into nitryl ion and sulfuric acid concentration is listed in Table 2.11.

$$H_2SO_4 + HNO_3 \rightleftharpoons H_2O + NO_2{}^+ + HSO_4{}^-$$

Sulfuric acid concentration is calculated after the produced water is included.

(5) When the sulfuric acid concentration is 90%, the extinction coefficient of absorption band at 220–270 nm is the maximum value due to the formation of nitryl sulfuric acid.

$$NO_2{}^+ + HSO_4{}^- \rightleftharpoons NO_2HSO_4$$

When nitric acid content in the anhydrous mixed acid is over a certain limit, the ionization ratio of nitric acid into nitryl ions will decrease with the increase of nitric acid content. The proportion of nitryl ions in the anhydrous mixed acid is shown in Table 2.12.

Equilibrium constant K of the reaction is expressed as

$$K = \frac{[NO_2{}^+][H_3O^+][HSO_4{}^-]^2}{[HNO_3][H_2SO_4]^2} \tag{2.11}$$

At 33.9 °C, K is equal to 33.6. The above is an exothermic reaction, and the balance will shift to the left when the temperature is increased. The relationship between nitryl ion concentration and mixed acid composition is shown in Fig. 2.4.

As shown in Fig. 2.4, mole percent is used to represent nitryl ion concentration. The amount of nitric acid (c%) that is ionized into nitryl ions in 100 g of mixed acid is shown in Fig. 2.5.

In Fig. 2.5, mass percentage is used and only a portion of the entire triangular graph is taken into account. As shown in Fig. 2.4, the concentration of nitryl ions is increased when sulfuric acid is added into nitric acid. When the content of nitric acid in the mixed acid is reduced to about 18% (molar percentage, equivalent to 12.4% of mass percentage), almost all nitric acid is ionized into nitryl ions. When the molar ratio of nitric acid to sulfuric acid in the mixed acid is kept constant, the nitryl ion concentration is decreased with the increase of water content. Because water is stronger base than nitric acid, the spectrum of nitryl ion disappears when water content in the mixed acid exceeds a certain limit. When the water content in the mixed acid exceeds 50(mol)%, the spectrum of nitryl ions cannot be seen regardless of the molar ratio of nitric acid to sulfuric acid. However, this result is

Table 2.13 The change rates of k_2 and C_0 with the change of concentration

Sulfuric acid concentration (%)	$-\frac{dC_0 1}{d\%H_2SO_4}$	$-\frac{d \log k_2}{d\%H_2SO_4}$ in the nitration of Ar-Y, when Y is								
		$-OCH_3$	H	F	Cl	Br	CN	NH_3^+	NO_2^+	$N(CH_3)_3^+$
40	0.16	0.16								
42.5	0.17	0.17								
45	0.18	0.18								
47.5	0.19	0.19								
50	0.20	0.20	0.21							
52.5	0.21	0.21	0.22							
55	0.23	0.24	0.24	0.23						
57.5	0.24	0.25	0.24	0.23	0.23	0.24				
60	0.26	0.28	0.26	0.26	0.27	0.26				
62.5	0.26	0.31	0.29	0.29	0.32	0.31				
65	0.26		0.36	0.36	0.35	0.36				
67.5	0.26		0.36	0.36	0.35	0.36				
70	0.26				0.35	0.36				
72.5	0.26				0.35	0.36				0.40
75	0.26									0.40
77.5	0.26									0.40
80	0.26						0.39		0.39	0.40
82.5	0.26						0.39	0.37	0.39	
85	0.26						0.39	0.37	0.39	
87.5	0.26							0.37	0.39	

not equal to say that there is no nitryl ion in the mixed acid in other zones, where the nitryl ion spectrum is not detected because its concentration is very low. Kinetic methods have proved that the attacking agent may still be nitryl ion in the nitration of anisole in the mixture of sulfuric acid (sulfuric acid concentration is 11%) and nitric acid with 60 wt% of water. The homogeneous nitration of aromatic compound with mixed acids is a secondary reaction, and the reaction rate can be written as

$$\upsilon = k_2[HNO_3][ArH] \tag{2.12}$$

in which

υ	Nitration rate,
k_2	Rate constant of secondary reaction, and
$[HNO_3]$, $[ArH]$	are the concentrations of nitric acid and aromatic compound, respectively.

Nitric acid, as a positively charged reagent, in the nitration system, might have the following two forms:

$$H^+ + HNO_3 \rightleftharpoons H_2NO_3{}^+$$
$$H_2NO_3{}^+ \rightleftharpoons NO_2{}^+ + H_2O$$

If the attacking agent is a nitric acid–proton combined ion ($H_2NO_3^+$), the change in k_2 should be parallel to the H_0 of the medium as

$$\frac{d\log k_2}{d\%H_2SO_4} = -\frac{dH_0}{d\%H_2SO_4} \tag{2.11}$$

If the nitryl ion is an attacking agent, the change in k_2 should be parallel to C_0 (or H_R) as

$$\frac{d\log k_2}{d\%H_2SO_4} = -\frac{dC_0}{d\%H_2SO_4} \tag{2.12}$$

Experiment results show that the rate of k_2 change is equal to that of -C_0 with the change of sulfuric acid concentration when the aromatic compound is nitrated in the mixed acid at 25 °C. The effect of sulfuric acid concentration on the rate is shown in Table 2.13.

As listed in Table 2.13, when nitration is in sulfuric acid with concentration of less than 60%, $\frac{d\log k_2}{d\%H_2SO_4}$ and $-\frac{dC_0}{d\%H_2SO_4}$ are substantially equal. With relatively high concentration of sulfuric acid, their difference is large, but their changes are almost parallel to each other, and this deviation is caused by the solvent effect.

The concentration of nitryl ion in low-concentration sulfuric acid is very low. When the molar ratio of H_2SO_4 and HNO_3 is 16.7 and 5%, the concentration of nitryl ion is about 10^{-9} mol/L.

Sulfuric acid can promote the conversion of nitric acid into nitryl ions, which inevitably leads to enhanced nitration ability of nitric acid electrophilic substitution reaction. The nitration ability of the electrophilic substitution reaction in the mixed acid is related to the concentration of nitryl ion. Prior to this, many people in nitration have mistaken that sulfuric acid has a strong water absorption capability and only acts as dehydrating reagent in the mixed acid to form hydrates, and thus to cause the highest concentration of nitric acid. Therefore, sulfuric acid is usually called dehydrating agent in the mixed acids. As shown in Fig. 2.4, after sulfuric acid is added into nitric acid, the concentration of nitryl ion in the solution is much higher than that in anhydrous nitric acid.

The concentration and the nitration ability of nitryl ion in the mixed acid also change with the change of the acid component. Nitration capacity of mixed acid with low water content or anhydrous mixed acid is much stronger than that of nitric acid, especially the introduction of one, two, or three nitro groups onto the benzene ring, and sometimes even the introduction of four nitro groups.

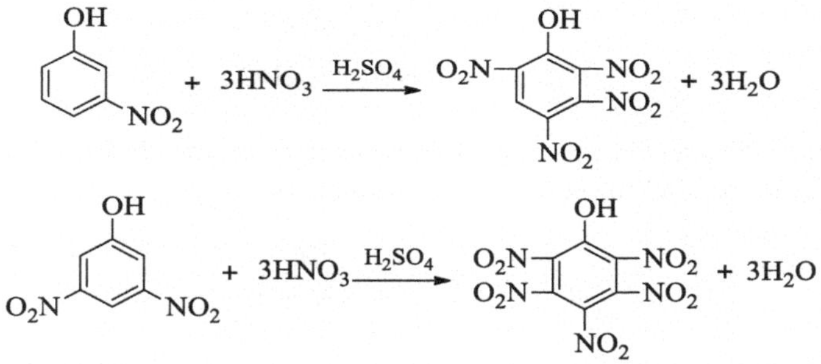

When nitric acid is used alone as a nitrating reagent, the above reaction cannot be carried out. With mixed acids, it can be fully effective to play the role of nitric acid, while nitric acid content in the wasted acid can be decreased to relatively low.

In order to enhance the nitration capacity of mixed acids, or to prevent the decomposition of certain nitrated products in the aqueous acid medium, the fuming sulfuric acid or sulfur trioxide can be mixed with the high concentration of nitric

Table 2.14 Melting points of nitryl salt formed from SO_3

Formed nitryl salt	Melting point (°C)
$HNO_3 + 2SO_3 \rightarrow (NO_2^+)(HS_2O_7^-)$	100–105
$N_2O_5 + 2SO_3 \rightarrow (NO_2^+)_2(S_2O_7^{2-})$	100–150
$N_2O_5 + 3SO_3 \rightarrow (NO_2^+)_2(S_3O_{10}^{2-})$	150

acid to form the mixed acid containing free sulfur trioxide. In fuming sulfuric acid, there is the following balance:

$$H_2SO_4 + SO_3 \rightleftharpoons H_2SO_7$$

According to the comparison of numerical values of acidity function, the acidity of pyrosulfuric acid is stronger than that of sulfuric acid because free sulfur dioxide can bind with water molecule from nitration reaction to form sulfuric acid.

$$SO_3 + H_2O \rightleftharpoons H_2SO_4$$

In the nitrous oxide–sulfur trioxide–water system, melting points of formed nitryl salts are shown in Table 2.14.

When the sulfur trioxide content is limited, the following can be obtained:

$$HNO_3 + SO_3 \rightleftharpoons NO_2^+ + HSO_4^-$$

When the sulfur trioxide content is sufficient, all nitric acid molecules are in the form of nitryl salts, and their Raman spectroscopy shows a strong line at 1400 cm^{-1}.

Thus, the attacking agents in N_2O_5–SO_3–H_2O (including HNO_3–H_2SO_4–H_2O and HNO_3–H_2SO_4–SO_3) systems, in most cases, are nitryl ions (including ion pairs and their carriers). The nitration of different compounds with mixed acids is

(1) Nitration of alcohols to form nitrates

Nitration of alcohol with α-hydroxyl group can usually be done with the mixed acids from nitric acid with sulfuric acid. However, when most of the monohydric alcohols are nitrated with nitric–sulfuric mixed acid, the corresponding nitrates are

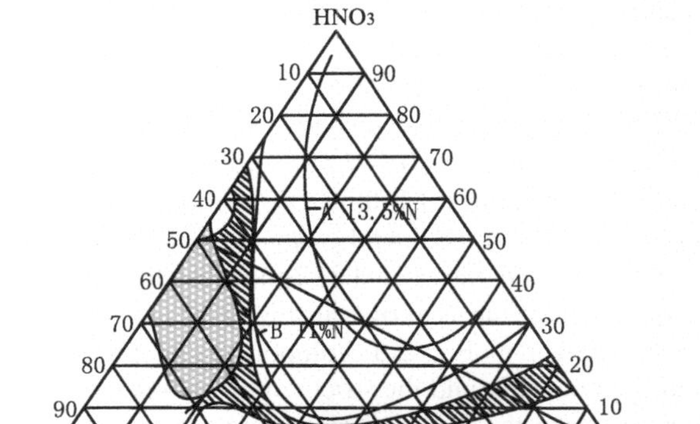

Fig. 2.6 Phase diagram of mixed acids during the cellulose nitration

obtained, and the risk of nitration may be increased due to oxidation and the accumulation of nitrogen oxides.

1. Nitration of monomer alcohol

$$CH_3CH_2OH \overset{\text{protectant}}{\rightarrow} CH_3CH_2ONO_2 + H_2O$$

The monohydric alcohol can be directly nitrated to the corresponding ester without the need for sulfuric acid. For nitrogens of diols, triols, and polyols, they need to be carried out in mixed acid. But this nitration reaction can sometimes be completed in 0.5–1.0 s.

$$HOCH_2CH_2OH \overset{HNO_3 + H_2SO_4}{\rightarrow} O_2NOCH_2CH_2NO_2$$

2. Nitration of unsaturated carbon atoms

The first reaction between carbon–carbon double bonds and mixed acids is addition reaction. For example, ethylene is fed with a mixture from 1 equivalent of nitric acid (its density is 1.4) and 2 equivalents of sulfuric acid containing 20% of sulfur trioxide, and in addition to the oxidation product, ethylene glycol dinitrate and β–nitrate ethyl alcohol nitrate are generated.

$$CH_2 = CH_2 \overset{HNO_3 + H_2SO_4}{\rightarrow} O_2NO - CH_2 - CH_2 - ONO_2 \text{ - } O_2N - CH_2 - CH_2 - ONO_2$$

3. Nitration of cellulose

The nitration of cellulose can be carried out directly with nitric acid, nitric acid vapor, or nitric anhydride (N_2O_6) without any catalyst.

The addition of reagents other than nitric acid into the nitration mixture is technical and economic reasons only although there are various types of composition used in the mixed acids. Typically, nitric acid concentration is 18–25%, and sulfuric acid concentration is 60–70%. When the nitration degree is more than 212.5 mL/g during the preparation of mixed acids, the formula tolerance of the moisture content is ±0.3–0.5%, and water formula tolerance in the mixed acids is controlled within ±0.1–0.3% in the other nitrogen contents (NC). During the production of mixed acid with low nitrogen content NC, water is controlled to ±0.12% and nitric acid is controlled to ±0.5%. The phase diagram of mixed acids for nitration is shown in Fig. 2.6.

As shown in Fig. 2.6, components of mixed acids are closely related to the nitrogen content, and the strict control of the composition of mixed acid is the basic guarantee on the stable nitrogen content.

(2) Nitration of aliphatic saturated carbon atoms

Aliphatic nitro compounds with α—hydrogen are unstable in sulfuric acid, so nitro-sulfuric acid is limited only to nitrate aliphatic compounds without α—hydrogen to produce nitro compounds and polynitro compounds without α—hydrogen.

(3) Aromatic compounds

The most majority of aromatic compounds (including monocyclic, polycyclic, and fused aromatic hydrocarbons and their derivatives, heterocyclic aromatic hydrocarbons) can be nitrated with mixed acid, and industrial nitration of aromatic compounds is done with mixed acid in all cases, because the mixed acid component can be adjusted to nitrate different active substances. The strong acid action of sulfuric acid can also greatly reduce the oxidation reaction, improve the utilization rate of nitric acid, and reduce the loss of nitric acid. When the total acidity of mixed acid is more than 80%, the corrosion of carbon steel is not significant and carbon steel can be used to make equipment. In addition, due to the use of a large number of sulfuric acid, the heat capacity of nitration system is increased, and the reaction temperature is relatively easy to control.

If compound is difficult to be nitrated, anhydrous mixed acids from anhydrous nitric acid and fuming sulfuric acid with high concentration can be used.

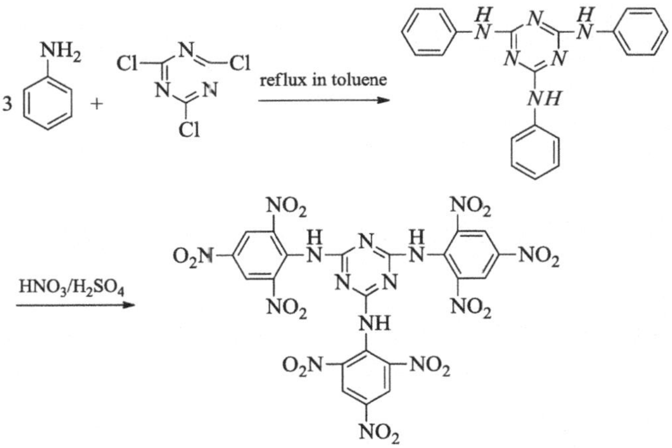

(4) Nitration of aliphatic secondary and tertiary amines

The nitration product of secondary and tertiary amines is secondary nitramine. Some secondary nitramines are stable in anhydrous sulfuric acid, and thus secondary nitramines can be produced from the corresponding amines with anhydrous mixed acids.

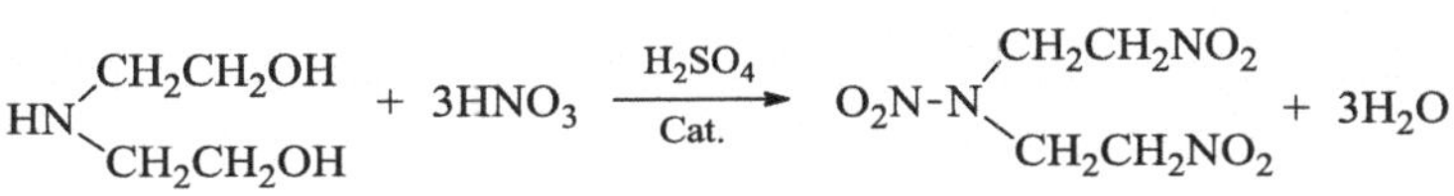

This reaction is the classic aza-nitration reaction. The main reaction on the nitrogen heterocycle is

$$NH_3 + SO_3 \xrightarrow{H_2SO_4} HSO_2NH_2$$

$$NH_3 + HSO_2NH_2 \rightarrow NH_4SO_3NH_2$$

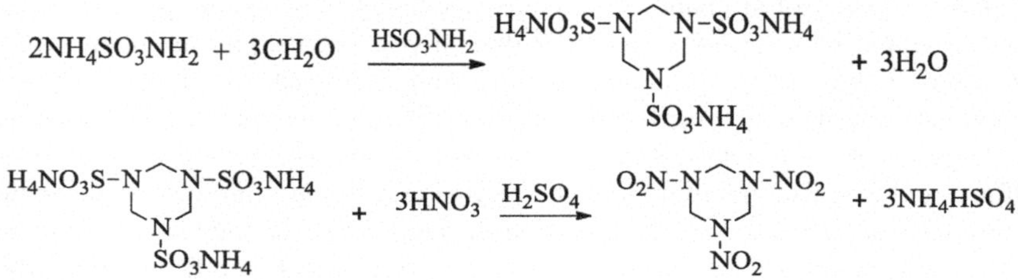

(5) Nitration of N-contained heteroalkanes with mixed acids

The nitration of nitrogen-containing heteroalkane with the mixed acids is relatively stable [22, 23].

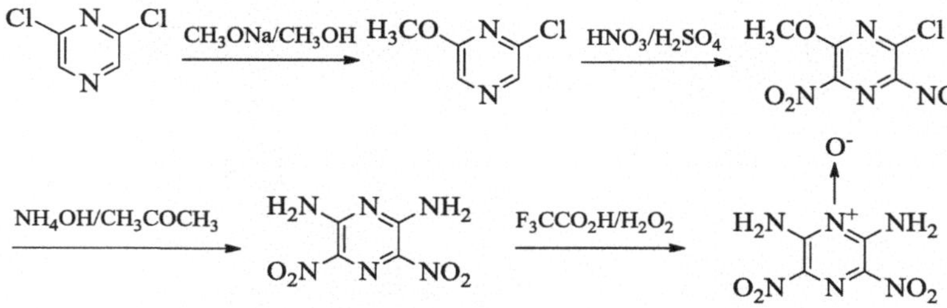

The role of mixed acids is shown in the nitration of another compound. For example, it shows good performance in the following olefin synthesis and nitration [24–26].

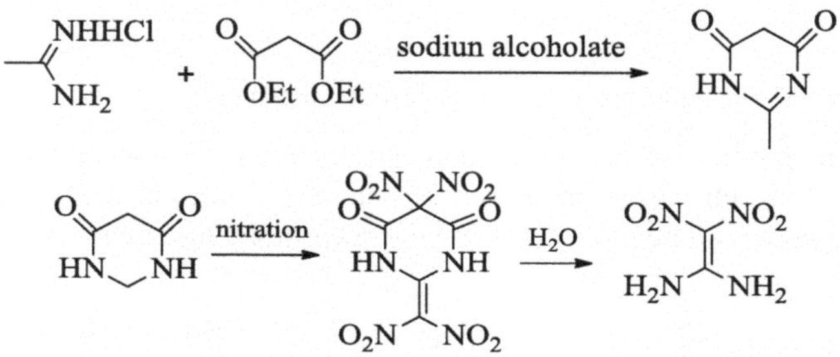

2.3.3 Nitration Agent Composed with Nitric Acid and Other Strong Acids

The acidity of perchloric acid is stronger than that of nitric acid, and nitric acid is also dissociated into nitryl ions in perchloric acid with the Raman spectroscopy line at 1400 cm^{-1}.

$$HNO_3 + 2HClO_4 \rightarrow NO_2^+ + H_3O^+ + 2ClO_4^-$$

This mixture also has a strong nitration capacity. However, perchloric acid is a strong oxidant and can form an explosive mixture with the chemical (hydrocarbon).

The Raman spectrum of the mixed nitric acid and selenic acid also exhibits a strong 1400 cm^{-1} line. The reaction between nitric acid and selenic acid to produce nitryl ions is similar to that of nitric acid–sulfuric acid.

$$HNO_3 + 2H_2SeO_4 \rightarrow NO_2^+ + 2HSeO_4^- + H_3O^+$$

Phosphoric acid is relatively weak acid, but it has a strong hygroscopicity. Especially, the phosphorus pentoxide has a very strong absorption capacity, which can keep nitric acid in a high concentration state. Thus, nitric acid–phosphorus pentoxide mixture also showed a 1400 cm^{-1} line, and the dissociation reaction of nitric acid into nitryl ions is present.

$$P_2O_5 + 6HNO_3 \rightleftharpoons 3N_2O_5 + 2H_3PO_4$$
$$3N_2O_5 \rightleftharpoons 3NO_2^+ + 3NO_3^-$$

The mixed acid from nitric acid and phosphoric acid can nitrate the aromatic compounds smoothly. Due to the different dissociation forms of acid, various isomers of the product can be obtained.

2.3.4 Nitration Agent from Nitric Acid, Acetic Acid, and Acetic Anhydride

Acetic acid is a weak acid with a pKa of 4.7 at room temperature that is smaller than the dielectric constant of water, and thus the ionization of nitric acid in acetic acid is not complete.

$$HNO_3 \rightleftharpoons H^+ + NO_3^-$$

Because the acidity of acetic acid is much weaker than that of nitric acid, acetic acid in the mixed acids plays the role of alkali and nitric acid is acid.

According to the above ionization equation, the addition of acetic acid reduces both the concentration of nitroxyl ions in nitric acid and the nitration capability of nitric acid in electrophilic substitution. In acetic acid–nitric acid mixture, there is not any spectrum line of nitryl ion. When the concentrated nitric acid is gradually diluted with acetic acid, nitration capacity of the latter on the benzene ring is decreased, but the declining degree is slower than that when nitric acid is diluted with water. However, it makes the nitration capacity on the carbon atoms in the side chain of the branched aromatic hydrocarbon molecules to show gradually.

In concentrated nitric acid, there are nitryl ions and nitrogen dioxide, and the former is a nitrating reagent of electrophilic substitution and the later is a nitrating reagent of free radical reaction. Because the electrophilic substitution reaction of toluene is easy to occur, its nitration with nitric acid is mainly electrophilic substitution reaction to form nitrotoluene. When nitric acid is diluted with acetic acid, the nitryl ion concentration is decreased significantly, which weakens the nitrating ability of nitric acid through electrophilic substitution, but it has little effect on the ability of free radical substitution. However, under the conditions suitable for free radical reactions (high temperature is conducive to the decomposition of nitric acid into nitrogen dioxide), the free radical reaction is changed to be significant.

Acetic acid is a good solvent for a variety of organic compounds. The mixture of acetic acid and nitric acid usually allows the reaction to proceed under homogeneous conditions. Therefore, this mixture is sometimes used in nitration and kinetics studies.

Pictet [27] believed that acetic acid and concentrated nitric acid (d = 1.52) are mixed to produce diacetyl nitric acid.

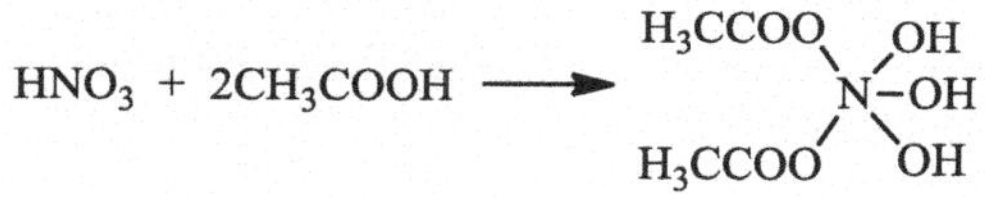

The boiling point of this compound is 127.7 ° C. The acetic anhydride is mixed with nitric acid with a specific gravity of 1.4 to produce this compound. It is terminated to study that this compound is in the acetic acid–nitric acid mixture because there is insufficient evidence. MaJIKOBa [28] suggested that the molecular compound $CH_3COOH \cdot HNO_3$ is present in the mixture of glacial acetic acid and anhydrous nitric acid.

Table 2.15 k_1, k_2, and K of the reaction of acetic anhydride with nitric acid

Temperature (°C)	K	10^5k_1 (L·mol^{-1}·s^{-1})	10^5k_2 (L·mol^{-1}·s^{-1})
−10	–	3.81	1.26
0	2.46	16.9	1.68
5	2.00	21.4	10.7
10	1.67	–	
15	1.40	32.4	23.1
20	1.18	41.2	39.4
25	1.00	51.0	51.0

Acetic acid is a weak acid with a pKa of 4.7 at room temperature, which is smaller than that of water; therefore, the ionization of nitric acid in acetic acid is not complete.

$$HNO_3 \rightleftharpoons H^+ + NO_3^-$$

Because the acidity of acetic acid is much weaker than that of nitric acid, acetic acid in the mixed acids plays the role of alkali and nitric acid plays the role of acid.

$$HNO_3 + CH_3COOH \rightleftharpoons NO_3^- + CH_3COOH_2^+$$
$$CH_3COOH_2^+ \rightleftharpoons CH_3CO^+ + H_2O$$

According to the above ionization equation, the addition of acetic acid reduces both the concentration of nitroxyl ions in nitric acid and the nitration capability of nitric acid in electrophilic substitution. In acetic acid–nitric acid mixture, there is not any spectrum line of nitryl ion. When the concentrated nitric acid is gradually diluted with acetic acid, nitration capacity of the latter on the benzene ring is decreased, but the declining degree is slower than that when nitric acid is diluted with water. However, it makes the nitration capacity on the carbon atoms in the side chain of the branched aromatic hydrocarbon molecules to show gradually.

In the mixture of nitric acid and acetic anhydride, acetyl nitrates are first formed as the nitrating reagent in nitration of amine to prepare nitramine, which has a strong application value.

Acetic anhydride can be hydrolyzed into acetic acid with water. When the acid is present, the rate of hydrolysis is increased.

$$(CH_3CO)_2O + H_2O \rightarrow 2CH_3COOH$$

In the reaction between nitric acid and acetic anhydride, acetyl nitrates are generated first.

$$(CH_3CO)_2 + HNO_3 \underset{k_2}{\overset{k_1}{\rightleftharpoons}} CH_3COONO_2 + CH_3COOH$$

The reaction rate constants k_1, k_2 and the equilibrium constant K of the above reactions are listed in Table 2.15.

Acetyl nitrite is a liquid, with a specific gravity of 1.24 at 15 °C, boiling point of 22 °C under 70 ml Hg pressure. It can form smoke in air to show a strong hygroscopicity and decompose into nitric acid and acetic acid in water.

$$CH_3COONO_2 + H_2O \rightarrow CH_3COOH + HNO_3$$

The acetonitrile can decompose into nitrogen oxides, tetranitromethane, and yellow oil where it is heated to 60 °C. When acetyl nitrates are rapidly heated or in contact with active metal oxides, such as alumina, violent explosion can happen. Acetyl nitrates are generally prepared according to the following methods:

$$P_2O_5 + (CH_3CO)_2O \rightleftharpoons 2CH_3COONO_2$$

Additionally, anhydrous nitric acid can be added into slightly excessive acetic anhydride. Through distillation under reduced pressure, acetyl nitrate can be obtained. In the nitration reaction, in order to increase the performance of nitrating reagent, it is not necessary to produce pure acetyl nitrate. When concentrated nitric acid is added into acetic acid or acetic anhydride, or 70% of nitric acid is added into acetic anhydride, a solution of acetyl nitrate in acetic acid–acetic anhydride can be prepared. Acetyl nitrite is a nitrating reagent, and can nitrate alcohols, amines, active aromatic compounds, etc., and this nitrating reagent should be prepared just before use without storage because the concentrated nitric acid can nitrate acetic anhydride into tetranitromethane.

(1) Nitration mechanism of acetic anhydride–nitric acid nitration system

Someone has studied the conductivity and infrared spectra of the acetic anhydride–nitric acid system, and others have studied its Raman spectra and infrared spectra. They have obtained almost identical conclusions [20, 29]. Overall, according to the molar ratio of these two substances, the mixture can be divided into the following four states:

1. When the nitric acid content is increased from 0 to 50%, the conductivity is gradually increased in this region, indicating that nitric acid reacts with acetic anhydride to produce a mixture with a small ionization degree. When the nitric acid content reaches 50% (i.e., the molar ratio of acetic anhydride to nitric acid is 1:1), the absorption band of acetic anhydride in IR spectrum is very weak and

Table 2.16 Effect of additives on 2360 cm^{-1} absorption rate (A)

Added reagent	H_2SO_4		$NaNO_3$		$(CH_3CO)_2O$		H_2O	
Length of sample cell (μm)	20.2		18.4		31.1		20.1	
Nitric acid concentration (%)	99.8		99.6		99.1		100	
	H_2SO_4 content (mol%)	A	$NaNO_3$ content (mol%)	A	$(CH_3CO)_2O$ content (mol %)	A	H_2O content (mol%)	A
	0	0.083	0	0.076	0	0.056	0	0.08
	0.6	0.093	1.05	0.052	8	0.252	0.9	0.048
	3.6	0.154	2.1	0.041	18	0.152	1.7	0.035
	7.1	0.264	4.9	0.028	31	0.00	2.7	0.014
	13.0	0.357	about 7.5	0.022	47	0.01 ①	3.2	0
					81	0.00 ①	4.4	0

the absorption band of acetic acid is very strong. In addition, an absorption band corresponding to esters of organic acids (1370 and 830 cm^{-1}) and $-\overset{O}{\overset{\|}{C}}-ONO_2$ (1240–1250 cm^{-1}) were also present, indicating that the main reaction in this region is

$$(CH_3CO_2)O - HNO_3 \rightleftharpoons CH_3COONO_2 + CH_3COOH$$

(2) After the nitric acid content exceeds 50%, the conductivity is increased rapidly. When the nitric acid concentration reaches 78.5%, the conductivity of the mixture reaches the maximum. This result indicates that a compound having a significant conductivity is formed. When the nitric acid concentration is 69%, the absorption band of nitroxide (2360 cm^{-1}) starts to appear in the infrared spectrum. Thus, the reaction within this region is the protonation of acetyl nitrates and the starting of ionization of protonated acetyl nitrates into nitryl ions.

$$CH_3COONO_2 + HNO_3 \rightleftharpoons (CH_3COONO_2H) + NO_3^-$$
$$(CH_3COONO_2H)^+NO_3^- \rightleftharpoons NO_3^- + NO_2^+ + CH_3COOH$$

(3) When the concentration of nitric acid is increased continuously, the conductivity of the mixture is decreased and the absorption band of the nitryl ion is increased. When the nitric acid concentration in the mixture reaches 82.5%, the

conductivity reaches the minimum value, while the nitryl ion absorption band reaches the maximum at the concentration of 92%. Due to the excess of nitric acid, the reaction in this region is

$$(CH_3CO)_2O + HNO_3 \rightleftharpoons CH_3COONO_2 + CH_3COOH$$
$$CH_3COONO_2 + HNO_3 \rightleftharpoons (CH_3COONO_2H)^+NO_3^-$$
$$(CH_3COONO_2H) + NO_3^- \rightleftharpoons CH_3COOH + N_2O_5$$
$$\text{or } (CH_3CO_2)O + 2HNO_3 \rightleftharpoons N_2O_5 + 2CH_3COOH$$
$$N_2O_5 \rightleftharpoons NO_3^- + NO_2^+$$

When the nitric acid content increased continuously, nitryl ion concentration is decreased and the conductivity is gradually increased and is gradually approaching the conductivity of concentrated nitric acid. In addition to the formation of nitrous oxide in this region, there are reactions as follows:

$$HNO_3 + CH_3COOH \rightleftharpoons NO_3^- + CH3COOH_2^+$$
$$CH_3COOH_2^+ \rightleftharpoons H_2O + CH_3CO^+$$
$$H_2O + HNO_3 \rightarrow H_3O^+ + NO_3^-$$

Fresco and Marcus measured the effects of various additives on the absorbance of nitryl ion at 2360 cm^{-1}. The results are shown in Table 2.16.

As described above, in the compounds of acetic anhydride and nitric acid with different ratios, there are acetyl nitrates and protonated acetyl nitrates, nitrogen pentoxide, nitrogen dioxide, nitryl ion, and nitric acid. The choice of attacking reagent should be based on specific conditions with specific analysis, which cannot be generalized for all cases.

In other words, in the nitric acid–acetic anhydride operation method, the ratio between nitric acid and acetic anhydride should be kept at a certain value, and the nitration capacity of mixed acids has a direct relationship with their preparation method. A typical example is that after the mixture of nitric acid–acetic anhydride is placed for a period of time, its composition changes. When nitric acid is mixed with acetic anhydride, acetic anhydride is nitrated into tetranitromethane. Therefore, mixed acids should be prepared just before use. When preparation temperature is different, nitration capacity of mixed acids is very different.

① When the nitric acid with a concentration of 70% is added to acetic anhydride and the preparation temperature is −10 °C, there is no heat release during the preparation.

② When the nitric acid with a concentration of 70% is added to acetic anhydride, the preparation temperature is 20–25 °C, a lot of heat is released during the preparation.

Preparation ratios of the two materials are exactly the same. The molar ratio of nitric acid over acetic anhydride is 1:6.

The nitration capability of mixed acids from method ② is good. The yield of ethylene glycol-N-nitramine dinitrate was about 96.8% after 10 min of nitration at −10 °C. At −15 to −20 °C, it can make polyvinyl alcohol to nitrate.

The nitration capacity of the mixture prepared with method ① is weak. It cannot nitrate polyvinyl alcohol into nitrate at −15 to −20 °C. When the ethylene glycol amine is nitrated at −10 °C, the yield of nitrate of ethylene glycol-N-nitramine is very low.

When urea is added into the two mixed acids to precipitate the free nitric acid in the form of urea nitrate, it has been found that only 30–35% of nitric acid from method ② can be recovered from the mixed acid, and the recovered nitric acid is decreased to 22% after 1 h of storage. However, with mixed acid prepared with method ①, all nitric acid is basically precipitated with urea. These results showed that nitric acid reacts with acetic anhydride first to form acetyl nitrates, while nitric acid solution coexists with acetic anhydride–acetic acid mixed solution with weak nitration capability.

If a small amount of sulfuric acid is added into acetic anhydride–nitric acid mixture, its nitration capacity can be increased dramatically. If a small amount of sulfuric acid is added into the mixed acid from method ②, nitration products of polyvinyl alcohol not only include polyvinyl alcohol nitrate, but also have ethylene glycol dinitrate and ethanol nitrate.

The mixed acid from nitric acid–acetic anhydride is not only a nitrating reagent, but also an acylating agent. Especially, when the nitric acid content is low and acetic anhydride content is high, its acylation performance is very significant, mainly because of the presence of acetyl cations in the mixed acid, leading to the acetyl oxidation of nitrated compound first (the introduction of $H_3C\text{-}\overset{O}{\overset{\|}{C}}\text{-}O$ group). The nitrating reagent from nitric acid–acetic anhydride can follow four typical nitration reactions. The nitration of atoms with different structures with nitrating reagent from acetic anhydride and nitric acid is as follows:

(1) Nitration of aliphatic carbon atoms

The mixture of acetic anhydride–nitric acid can nitrate the organic carboxylic acid ester and ethyl acetoacetate, and nitration performance is better than that with nitric–sulfuric mixed acid.

Table 2.17 Comparison of nitration capability of different nitrating reagents to toluene

Nitrating reagent	Temperature (°C)	Percentage of isomers (%)		
		o-	m-	p-
Nitric acid with specific gravity of 1.47	30	57	3	40
Formyl nitrates in acetonitrile	0	59	5	36
Acetonitrile nitrate solution of carbon tetrachloride		88	–	12

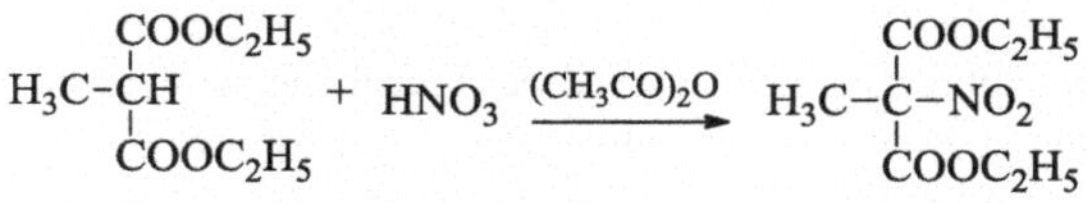

The yield of 48%

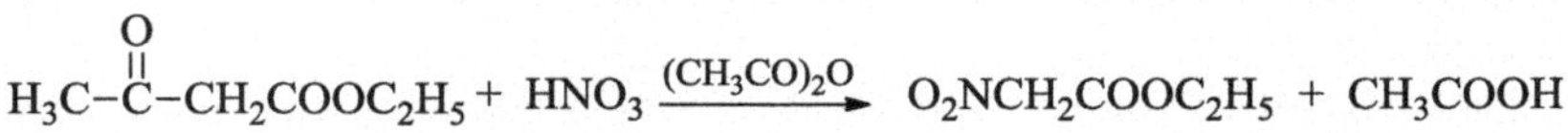

The yield of 32%

(2) Nitration of olefins

From −15 to −20 °C, when olefin is added to the acetic anhydride–nitric acid mixture, the nitration is completed after 5 min and the product and yield are [30]

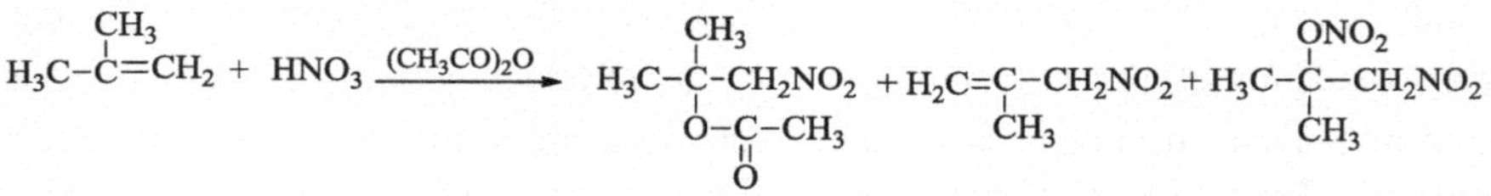

The yield is 65%. The yield is 20%. The yield is 15%

This reaction is of general significance and practical value to the preparation of nitrocellulose acetate.

(3) Nitration of alcoholic hydroxyl groups

At −10 to 10 °C, the nitration of ethylene glycol amine in the nitrating reagent of mixed nitric acid and oxalic anhydride produces unstable product acetyl diethanolamine dinitrate.

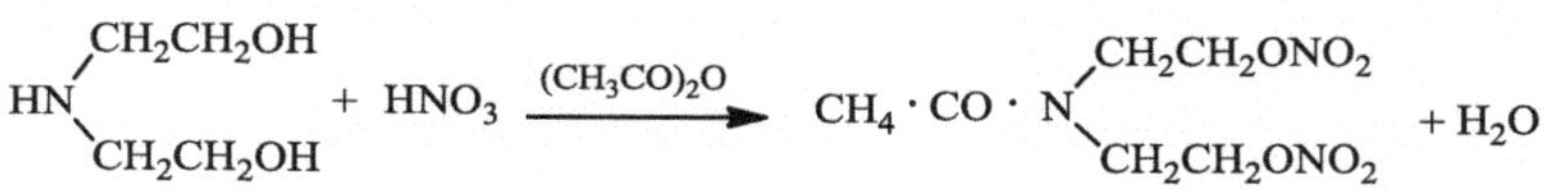

Since the hydrogen at the end of amine group has been replaced by $CH3CO^+$, the above product is still not ethylene glycol amine dinitrate, but an intermediate. It is necessary to have the catalysis of HCl to get the final product.

(4) Secondary amine and tertiary amine

Acetic anhydride–nitric acid mixture is mainly used to nitrate amines to nitramines in industry.

The nitration of fats, olefins, and alcoholic hydroxyl groups has been discussed, and it is the same in the nitration of benzene ring. In the nitration of benzene, the mixed acid of nitric acid–acetic anhydride can generate benzoyl nitric anhydride in benzene, and benzoyl nitric anhydride can also be used as a nitrating reagent. The benzoyl chloride which excess 20% is dissolved in carbon tetrachloride, cooled to −15 °C, and then fine powder of silver nitrate is added. After the reaction, silver salt was removed with sand filtration, and then benzoyl nitrate solution in carbon tetrachloride contains 15–20% of benzoic anhydride.

Benzoyl nitrate is very sensitive to water, and can cause explosive decomposition even when it is filtered with wet filtration flask. The nitration of benzoyl nitrate is similar to that of acetonitrile nitrate, which can nitrate benzene, xylene, trimethylbenzene, and other aromatic hydrocarbons. It can also nitrate phenol and phenolic ether compounds. When it nitrifies tetra-, penta-, and hexamethylbenzene, nitration of side chain occurs. The isomer products from nitration of toluene with different nitration agents are shown in Table 2.17.

Because this nitrating reagent has no special advantage but expensive, it is not practical.

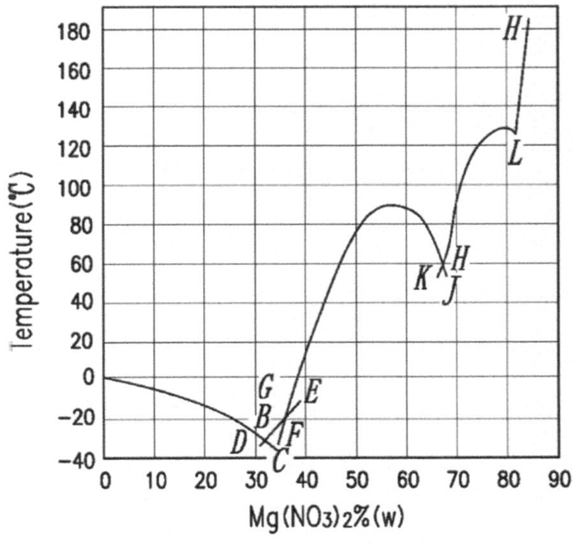

Fig. 2.7 Phase diagram of magnesium nitrate–water system

Nitric acid–acetic anhydride mixture is a widely used nitrating reagent, the generated water in the reaction reacts with acetic anhydride to form acetic acid, and thus the side reaction of oxidation is relatively limited. The disadvantages are that acetic anhydride is relatively expensive, the equipment to recover acetic anhydride from acetic acid is relatively complex, and the nitric acid–acetic anhydride mixture and the produced waste acid are relatively strong corrosive to metals.

2.4 Other Nitrating Reagents

2.4.1 The Nitrating Reagent from Nitric Acid and Lewis Acid

Many researchers have studied on the nitration of aromatic compound with the nitric acid–boron trifluoride systematically [1, 3, 31]. 18 g of BF_3 is mixed with 0.25 mol (31 g) of nitrobenzene. After the temperature has been raised to 80 °C, 20 g of nitric acid (d = 1.5) is added dropwise. Then, the temperature is raised to 100 °C. After 30 min at 100 °C, the reaction mixture is poured into 500 mL of cold water and cooled down with stirring. Dinitrobenzene is filtered out and recrystallized with ethanol to give 36.5 g of dinitrobenzene with a yield of 87.5%. A similar method also is used to nitrate benzoic acid, p-toluenesulfonic acid, methyl benzoate, phthalic anhydride, etc., and the yield of nitration reaction is 77–92%.

$$ArH + HNO_3 + BF_3 \rightarrow Ar - NO_2 + BF_3 \cdot H_2O$$

After the reaction, the added water converts $BF_3{\cdot}H_2O$ into $BF_3{\cdot}2H_2O$, and BF_3 can be distilled out under vacuum.

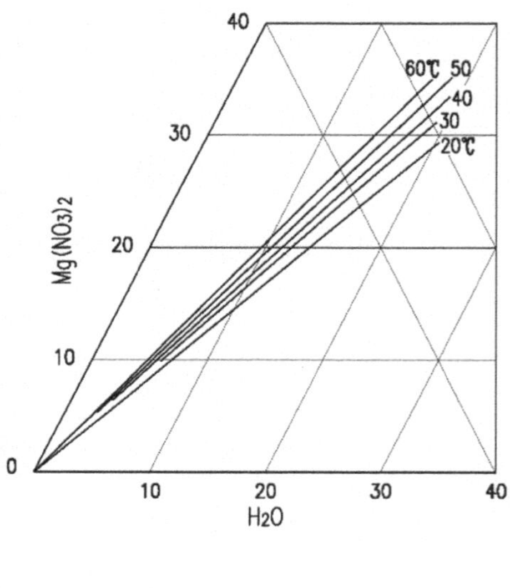

Fig. 2.8 Solubility of magnesium nitrate in aqueous nitric acid [32]

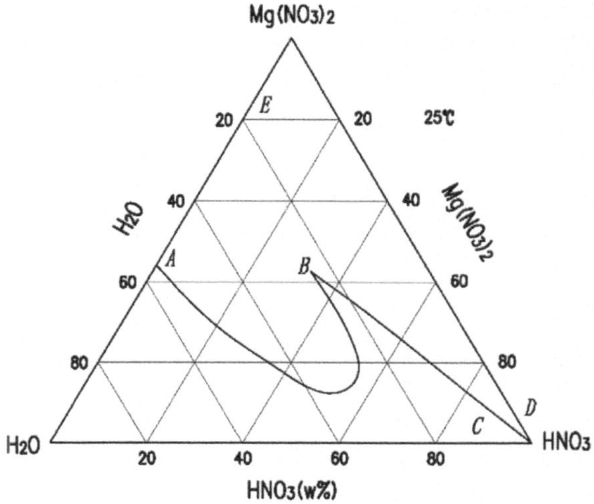

Fig. 2.9 Phase equilibrium of magnesium nitrate–nitric acid–water system [33]

$$BF_3 \cdot 2H_2O + CaF_2 \rightarrow Ca(BF_4)_2 + 4H_2O$$
$$Ca(BF_4)_2 \rightarrow 2BF_3 + CaF_2$$

Boron trifluoride is added into nitric acid to form a viscous liquid. When the composition of the mixture is close to $HNO_3{\cdot}2BF_3$, crystals are beginning to form after a few days and finally solidified to crystal $HNO_3{\cdot}2BF_3$. Melting point of the crystal is 53 °C.

The nitric acid–boron trifluoride mixture has a good nitration capacity. Spectroscopic studies prove that $HNO_3{\cdot}BF_3$ is nitryl salt [34] as

$$HNO_3 + BF_3 \rightarrow NO + 2BF_3OH^-$$

$HNO_3 \cdot \frac{1}{2}BF_3$ is essentially a mixture of nitric acid and NO_2BF_3OH. The attacking reagent in the nitration with a mixture of nitric acid–boron trifluoride is nitroxyl, which can also be used to nitrate alcohols and amines.

2.4.2 Nitrating Reagent from Nitric Acid and Magnesium Nitrate

Magnesium nitrate has three stable crystalline hydrates as ninehydrate, hexahydrate, and dihydrate. Based on analysis, unstable tetrahydrate and monohydrate have been also detected. The phase diagram of magnesium nitrate with crystal water is shown in Fig. 2.7.

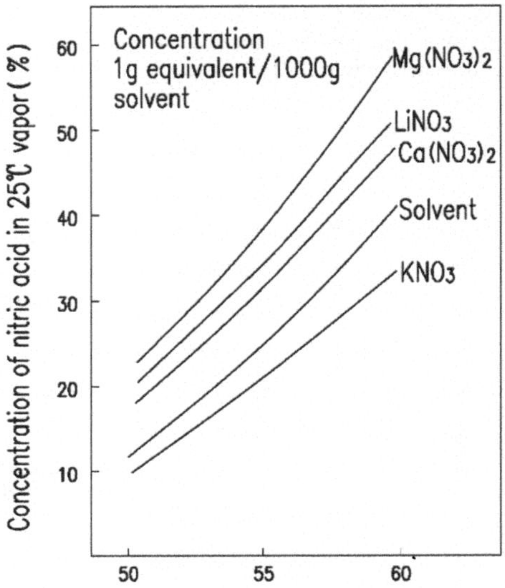

Fig. 2.10 Effects of metal ions on the vapor composition of nitric acid aqueous solution

In Fig. 2.7, the OC line indicates that the solid phase is ice in which the BC line is the metastable zone. The DE line is magnesium nitrate ninehydrate, in which the BD and GE segments are metastable zones. The FJ line is magnesium nitrate hexahydrate, in which GF and HJ lines are metastable zones. The KL line indicates that the solid phase is magnesium nitrate dihydrate, where HK is a metastable zone. The LH line is anhydrous magnesium nitrate, where the point L is the coexistence of magnesium nitrate dihydrate with anhydrous magnesium nitrate.

31.5–17.1 °C is the stable zone of magnesium nitrate ninehydrate. The converting temperature (G point) of magnesium nitrate hexahydrate into magnesium nitrate ninehydrate is from −18.0 to −14.7 °C. The stable zone of magnesium nitrate hexahydrate is from 18 °C (or 14.7 °C) to its melting point of 89.3 °C, that of magnesium nitrate dihydrate is from about 55 to 130 °C, and that of anhydrous magnesium nitrate is 130 °C or more. In addition, there are three eutectic mixtures, corresponding to the B, H, and L points in Fig. 2.7, in which B point is the eutectic of water and magnesium nitrate hexahydrate with a eutectic point of ~31.5 °C, H point is the eutectic of magnesium nitrate hexahydrate and magnesium nitrate dehydrate with a eutectic point of about 55 °C and the corresponding composition of $Mg(NO_3)_2{\cdot}3.98H_2O$, and L point is the eutectic of magnesium nitrate dihydrate and anhydrous magnesium nitrate with a eutectic point of about 130 °C. Usually, the magnesium nitrate used in experiment is the mentioned product.

The solubility of magnesium nitrate in nitric acid–water system is shown in Figs. 2.8 and 2.9.

Figure 2.8 is the solubility of magnesium nitrate in the nitric acid–water system between 20 and 60 °C. And the solid phase equilibrating with the liquid phase is magnesium nitrate or magnesium nitrate dihydrate. As shown in this figure, with the increase of temperature, the solubility of magnesium nitrate in nitric acid aqueous solution is increased, the nitric acid content in nitric acid–water system is increased, and the solubility of magnesium nitrate is decreased.

Figure 2.9 is the solubility of magnesium nitrate in 0–100% nitric acid aqueous solution at 25 °C. A and E points are equivalent to magnesium nitrate hexahydrate and magnesium nitrate dihydrate. The solid phase equilibrating with AB line is magnesium nitrate hexahydrate. The solid phase equilibrating with BC line is magnesium nitrate dihydrate and that equilibrating with CD line is anhydrous magnesium nitrate. Thus, with the increase of the moisture content in the system, the solid phase equilibrating with the liquid-phase anhydrous magnesium nitrate is converted to magnesium nitrate hexahydrate through magnesium nitrate dihydrate.

According to the definition of Lewis acid, metal cations are acids. Although the method to compare acid strengths of magnesium ion and nitric acid has not been found, magnesium ion will be hydrated in the aqueous medium through its empty electron orbitals. In the magnesium nitrate–nitric acid–water system, the degree of hydration of magnesium ion will depend on the relative value of the partial vapor pressure of magnesium nitrate–water and nitric acid–water system. If the partial vapor pressure of the former is larger than that of the latter, magnesium hydrate will be dehydrated and nitric acid will be hydrated. Conversely, the magnesium ions are hydrated and the nitric acid hydrate will lose water.

Flatt et al. [35] determined the effect of addition of metal ions into 50–60% nitric acid aqueous solution on the vapor pressure of the solution. The results are shown in Fig. 2.10.

As shown in Fig. 2.10, in addition to potassium nitrate, the addition of metal nitrate increases the concentration of nitric acid in the vapor phase. This indicates that the addition of metal salts increases the activity of nitric acid in the liquid phase. According to these experiment results, Flatt extrapolated that, with this concentration of nitric acid, the hydration number of magnesium ion is about 11–12, that of

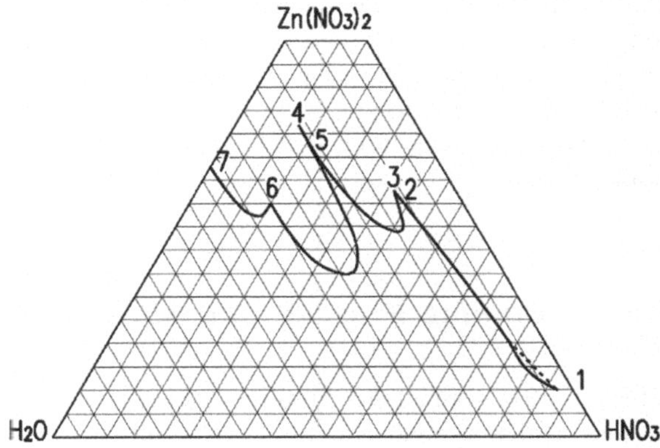

Fig. 2.11 Phase diagram of zinc nitrate–nitric acid–water system (25 °C) [20]

calcium ion is 8, that of lithium-ion is 4–5, and that of sodium ion is 2. Of course, when the concentration of nitric acid changes, the hydration number will change.

In United States, the preparation of nitrocellulose with nitric acid–magnesium nitrate–water mixed nitrating agent has been built in the 80s of last century to produce single-base nitrocellulose to replace the traditional nitrocellulose made with nitrating reagent from nitric acid--sulfuric acid mixture.

In the magnesium nitrate–nitric acid–water system, magnesium ions play the role of dehydrating reagent, which enhances the activity of nitric acid. However, it can be speculated that, with nitric acid–nitric acid–water system, the degree of hydration of magnesium ions will be increased during the nitration, and the waste acid after the nitration can be sent to recycle as concentrating nitric acid with magnesium nitrate as the dehydrating reagent. It is also possible to separate magnesium nitrate in the form of hydrates from the waste acid, reducing the amount of water in the waste acid, and then a portion of the waste acid is used to form nitrated acid with concentrated nitric acid and dehydrated magnesium nitrate, and the remaining waste acid is concentrated again and recovered, leading to the recycling of part of the waste acid.

The advantage of magnesium nitrate–nitric acid–water system to replace nitric acid–sulfuric acid as a nitrating reagent is that it is not necessary to concentrate sulfuric acid from the waste acid, and thus to eliminate the sulfuric acid mist pollution to the atmosphere [36].

The dehydrating capacity of magnesium ions is relatively strong among all of the two-valent metal ions, which is presumably only weaker than beryllium ion, and thus its dehydrating capacity is weaker than the concentrated sulfuric acid. Therefore, magnesium nitrate–nitric acid–water system, as nitrating reagent, cannot nitrate compounds that are difficult to nitrate, and it cannot replace nitric acid–sulfuric acid in all fields. In order to improve the nitration capacity of the salt–nitric acid–water system, other metal ions can be found as dehydrating agents. Both theory and experiment have proved that metal ion with larger ionic charge and smaller ionic radius has the greater enthalpy change in hydration and the stronger hydration ability. Therefore, dehydration capacity of the aluminum ion, iron ion, chromium ion (Cr^{3+}), gallium ion, and others is stronger than that of magnesium ion. These issues are beyond the scope of the book and can learn from books in physical inorganic chemistry. Of course, during the new dehydrating reagent searching, the dehydration capacity is an important factor, but we should comprehensively measure both its advantages and disadvantages from the engineering point of view.

2.4.3 Nitrating Reagent from Nitric Acid and Zinc Nitrate

Cloetens [37] studied the nitration capacity of the zinc nitrate–nitric acid–water system. Under normal conditions, zinc nitrate exists in the form of ninehydrate compound. And, although its dehydration capacity is weaker than that of magnesium nitrate, its solubility in the nitric acid aqueous solution is relatively large. Especially, its solubility in concentrated nitric acid is larger than that of magnesium nitrate, which is beneficial to the improvement of the system's nitration capacity. The phase diagram of the zinc nitrate–nitric acid–water system is shown in Fig. 2.11.

As shown in Fig. 2.11, line segment 1–2 represents the solid phase equilibrating with the liquid phase is zinc nitrate monohydrate, line segment 2–3 indicates that the solid phase equilibrating with the liquid phase is zinc nitrate dihydrate, line segment 5–6 represents the liquid phase equilibrating with the solid phase is zinc nitrate tetrahydrate, and line segment 6–7 represents the solid phase equilibrating with the liquid phase is zinc nitrate hexahydrate. Because the solubility of zinc nitrate in nitric acid aqueous solution is relatively large, the nitrogen content of product from the nitration of cellulose is high up to 11.2–13.85%, while with magnesium nitrate–nitric acid–water system as nitration reagent, the nitrogen content of product is 10.4–13.65%.

2.4.4 *Nitryl Halide*

(1) Nitryl fluoride

Nitryl fluoride [CNO_2F] is a colorless gas and is not strong corrosive to glass at room temperature. Its melting point is −166 °C, boiling point is −72.5 °C, and density is 1.571 g/cm^{-1} at −101 °C. The nitryl fluoride was prepared through the following reaction:

$$N_2O_5 + NaF \rightarrow NO_2F + NaNO_3$$

or

$$NaNO_2 + F_2 \rightarrow NaF + NO_2F$$

Or nitrous oxide and fluorine are mixed together to generate nitryl fluoride. And, infrared, nuclear magnetic resonance, and Raman spectroscopy have shown that nitryl fluoride has a planar structure as

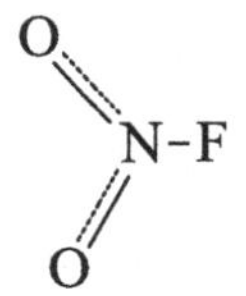

Hetherington and Robinson [38] used the pure nitryl fluoride to nitrate aromatic compounds. When nitryl fluoride is added into aromatic compounds (or its solution), the following results can be obtained: ① the nitration of inactive aromatic

compounds (such as nitrobenzene, etc.) does not occur, ② the nitrated products of benzene, toluene, bromobenzene, salicylic acid, naphthalene, and anthracene are mononitro-compounds, and ③ active aromatic compounds, such as m-cresol, anisole, diphenyl ether, aniline, furan, quinoline, etc. become resin during their nitration.

The nitration capacity of nitryl fluoride is that it can dissociate into nitryl ions. Nitryl fluoride and sulfuric acid can occur as the following:

$$NO_2F + H_2SO_4 \rightleftharpoons NO_2^+ + HF + HSO_4^-$$
$$HF + 2H_2SO_4 \rightleftharpoons HSO_3F + HSO_4^- + H_3O^+$$

Therefore, the mixed solution of nitryl fluoride and sulfuric acid also has nitration capacity.

The Lewis acid with a formula of MFn (M represents metal, n is its valence), such as boron trifluoride, phosphorus pentafluoride, or antimony pentafluoride, can enhance the nitration capacity of nitryl fluoride. With these substances as catalysts, nitryl fluoride can nitrate dihalogenated benzene and polyhalogenated benzene with the yield of more than 80%, and can reduce side reactions caused by the hemolysis of nitryl fluoride into free radicals.

There are also some nitration reactions under the protective conditions. For example, during the synthesis of 3,3,7,7-tetrakis(difluoroamino)-1,5-diazacyclooctane (HNFX), the first step is to protect p-nitroamine with p-nitrobenzenesulfonyl group in order to get the fluoroamination done, and then the protected groups are deprotected to form ammonium nitrate [39].

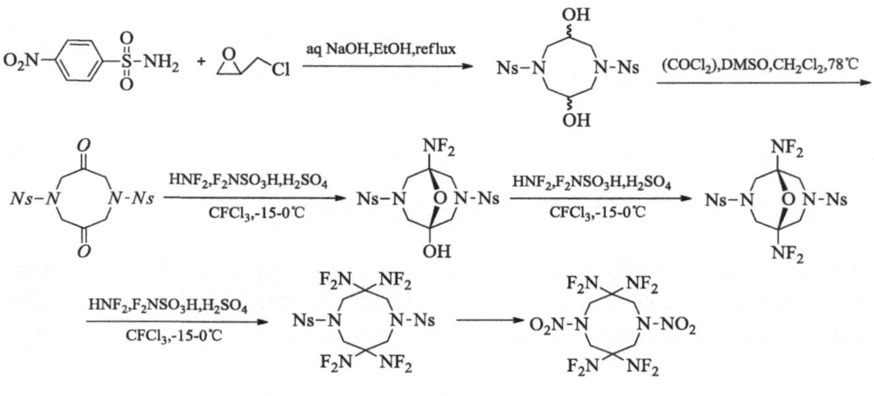

(2) Nitryl chloride (NO_2C1)

After fuming sulfuric acid is used to dehydrate nitric acid, chlorosulfonic acid is added at 5–10 °C to prepare nitryl chloride as

$$HNO_3 + ClSO_3H \rightarrow NO_2Cl + H_2SO_4$$

or

$$PCl_5 + N_2O_5 \rightarrow 2NO_2Cl + POCl_3$$

Nitryl chloride has a melting point of −145 °C, a boiling point of −15.9 °C, and a density of 1.50 g/cm^3 at −80 °C. Its spectrum proves that nitryl chloride has a planar structure as

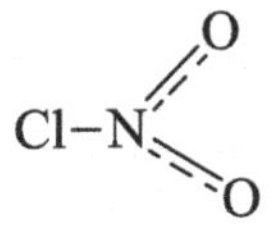

Without any catalyst, the solution of nitryl chloride with a polar solvent is yellow and chlorinates the aromatic compound to form chlorinated aromatic hydrocarbon. Nitryl chloride decomposition can produce chlorine as

$$2NO_2Cl \rightleftharpoons N_2O_4 + Cl_2$$

Nitryl chloride solution with a nonpolar solvent is colorless and cannot play the role of chlorinating reagent. Therefore, according to different conditions, nitryl chloride can be used as nitrating reagent or chlorinating reagent. Its chlorinating activity order is the following:

$$C1^{+} > H_2OC1^{+} > C1_2 > NO_2Cl > CH_3COCl > HOCI$$

At low temperature, the product of nitryl chloride with alkylbenzene is chlorinated derivative, those with naphthalene are 1-chloronaphthalene and 1-nitronaphthalene, while that with phenol is o-nitrophenol at low temperature but 4,6-dichloro-2-nitrophenol at room temperature.

The nitration capacity of nitrochloride can be improved with hydrofluoric acid or Lewis acid as catalyst. With hydrofluoric acid, aluminum trichloride or titanium tetrachloride as catalyst, nitryl chloride can nitrate naphthalene, benzene, toluene, ethylbenzene, chlorobenzene, bromobenzene, p-bromotoluene, dichlorobenzene, benzyl chloride and salicylic acid, and other substances to produce nitro compounds, but the yield is not too high. In the case of aromatic compounds with high nitration activity, not only compounds with nitro group cannot be obtained, but also the benzene ring is destroyed due to its oxidation [40, 41].

Nitration capacity of nitryl chloride is weaker than that of nitryl fluoride, but its halogenation tendency is stronger than that of the latter.

The nitration capacity of nitryl bromide is weaker than that of nitryl chloride, and its halogenation tendency is stronger than that of nitryl chloride. Therefore, orders of their nitration capacity and halogenation tendency of the three nitro halides are as follows:

$$\text{Nitration capacity}: \quad NO_2F > NO_2Cl > NO_2Br$$
$$\text{Halogenation tendency}: \quad NO_2Br > NO_2Cl > NO_2F$$

At low temperature and with dry inert solvent, addition reaction between nitryl chloride and carbon–carbon double bond occurs to form chloronitrosoalkane, and it also generates chlorinated nitrite and dichloroethane with B4, which is a free radical addition reaction as follows:

$$NO_2Cl \rightarrow \cdot NO_2 + Cl\cdot$$

$$CH_2 = CH_2 \xrightarrow{NO_2Cl} \underset{\text{The yield of 45\%}}{Cl - CH_2 - CH_2NO_2} + \underset{\text{The yield of 35\%}}{Cl - CH_2 - CH_2 - ONO} + Cl - CH_2 - CH_2 - Cl + N_2O_4$$

This reaction is of general interest and can be used to make chloronitrosoalkanes.

In the addition reaction of nitroxyl chloride and carbon–carbon triple bond, the solvent has a great influence. The most suitable solvent is dichloromethane because nitryl chloride decomposes into chlorine and nitrous oxide in it. The addition product of nitryl chloride and carbon–carbon triple bond is chloronitrone.

$$C_2H_5 - C \equiv C - C_2H_5 + NO_2Cl \rightarrow C_2H_5 - ClC = CNO_2 - C_2H_5$$

The yield of this reaction is not high.

(3) Nitroxyl chloride (NOCl)

The addition reaction of nitroxyl chloride and the carbon–carbon double bond first generates unstable β-chloro-α-nitrosoalkane, which is then oxidized by the excessive nitryl chloride into nitro compound. The free chlorine and unreacted olefin react to produce a by-product of dichloroalkane. This reaction is electrophilic addition, and nitro entering position is in accordance with the Markovnikov rule.

$$NOCl \rightarrow NO^+ + Cl^-$$

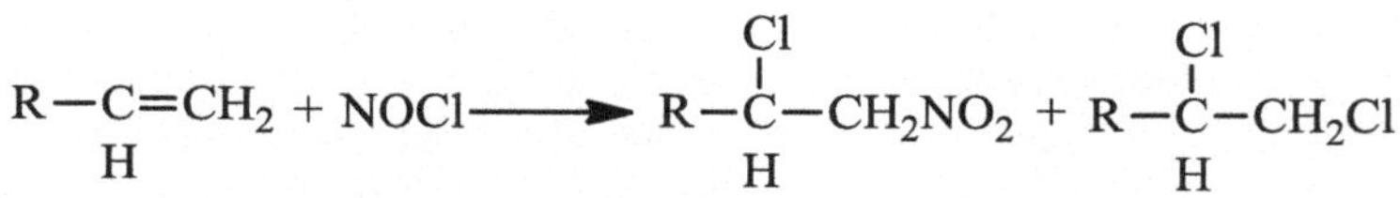

In the reaction, oxygen and nitrous oxide can also be used as oxidizing agents to oxidize the intermediate nitroso compounds.

The yield of this reaction with olefins, nitrones, carbonylenes, or alkoxyenes is not high. But the yield of reaction with halogenated olefins is relatively good.

Nitrosyl chloride can also be added to alkynes to form chloronitrone.

$$C_6H_5{-}C{\equiv}CH + NOCl \xrightarrow{\text{oxidation}} C_6H_5{-}\overset{\overset{\displaystyle Cl}{|}}{C}{=}\underset{H}{C}{-}NO_2$$

2.4.5 Metal Nitrate

The metal nitrate and its mixture with protonic acid or Lewis acid can nitrate aromatic compounds and aliphatic saturated carbon atoms.

(1) The mixture of metal nitrate and water

Only the mixture of easy hydrolyzed metal nitrate and water can be used as a nitrating reagent, and these nitrates are bismuth nitrate, mercury nitrate, aluminum nitrate, silver nitrate, manganese nitrate, cobalt nitrate, nickel nitrate, copper nitrate, etc.

1. Nitration of aliphatic compounds when metal nitrate nitrifies aliphatic compound, the amount of water added is to hydrolyze the nitrate to form 12% nitric acid (specific gravity is 1.075), in which the effects of bismuth salt and aluminum salt are relatively good. Alkali metal nitrates and ammonium nitrate do not function in the nitration.

10 g of cyclopentane, 24 g of aluminum nitrate, and an appropriate amount of water are charged in a tube, sealed, heated at 100–115 °C for 6 h, and cooled to separate the organic phase. Unreacted cyclopentane is distilled out, the distillation residue is treated with potassium hydroxide solution, the insoluble oil is separated and removed, and carbon dioxide is purged through the potassium hydroxide solution to separate nitro cyclopentane, which is finally purified with a boiling point of 90–91 °C through distillation under 40 mm Hg pressure.

Similarly, the yield of mononitrocyclohexane is 57%. That of nitrocyclohexane is 58%.

2. Metal nitrate can only nitrate easily nitrated aromatic compounds, and due to the side reaction oxidation, better results can be produced only in the nitration of phenol and cresol.

At 250–360 °C, an aqueous solution of copper nitrate or silver nitrate can be used to nitrate benzene or benzoic acid to nitrobenzene. The reaction is carried out in an autoclave. As the temperature is high, benzoic acid is accompanied by a carboxyl loss reaction.

(2) A mixture of nitrate and protonic acid (or anhydride).

(1) A mixture of nitrate and sulfuric acid before the industrialization of ammonia oxidation method to produce nitric acid, sodium nitrate, and sulfuric acid mixture has been used in the nitration of aromatic compounds. During the World War I, France had used the mixture of sulfuric acid and sodium nitrate to nitrate phenol tinto di- and trinitrophenol. Sulfuric acid and sodium nitrate generate free nitric acid according to the following reaction:

$$NaNO_3 + H_2SO_4 \rightarrow NaHSO_4 + HNO_3$$

$$NaNO_3 + H_2SO_4 \overset{\text{heating}}{\rightarrow} Na_2SO_4 + HNO_3$$

Since the acidity of sulfate is weaker than that of nitric acid, the second reaction can be smoothly moved to the right only if the nitric acid is continuously removed; otherwise, they only can establish an equilibrium.

According to the reactions above, the reaction between the formed nitric acid and the excessive sulfuric acid is the same as that in the nitric acid–sulfuric acid mixture. Therefore, the nitration of nitrate–sulfuric acid mixture is not much different from that of nitric acid–sulfur acid mixture, and the proportion of isomers in the product is also roughly the same. But their nitration rate is greater than that of the corresponding mixed acid, which may be due to the salt effect. The order of effect of cations on nitration rate is as follows:

$$Li > Ag > Na > K > NH_4 > H$$

(2) A mixture of nitrate and acetic anhydride or acetic acid The mixture of nitrate and acetic anhydride has a moderate nitration capacity, which can nitrate benzene and aromatic compounds containing o-, p-positional groups (in electrophilic reactions), such as aniline, chlorobenzene, toluene, naphthalene, etc., with the product of nitro compound. The nitration is as follows: 5 g of to-be-nitrated compound are dissolved in acetic anhydride, and the solution is added dropwise to a mixture of 15 g of ferric nitrate and 25 g of acetic anhydride. And, the nitration temperature is determined according to the nitrated compound. It is 80 °C for benzene, 40–45 °C for chlorobenzene, and 40 °C for toluene. The phenol is nitrated to trinitrophenol at 75–80 °C. The following balance exists between metal nitrate and acetic anhydride:

$$MNO_3 + (CH_3CO)_2O \rightleftharpoons CH_3COONO_2 + CH_3COOM$$

That shows the formation of acetyl nitrate.

The mixture of metal nitrate and glacial acetic acid can nitrate the active aromatic compound, but its nitration ability is weaker than that of metal nitrate and acetic anhydride mixture. For example, with a mixture of metal nitrate and glacial acetic acid, the nitration temperature of naphthalene or acetylaniline into mononitro derivative is 105 °C or 75 °C, respectively, and it is only 30 °C with the mixture of metal nitrate and acetic anhydride. Metal nitrate and glacial acetic acid have the following balance:

$$MNO_3 + CH_3COOH \rightleftharpoons CH_3COOM + + HNO_3$$

It is also possible to nitrate an aliphatic saturated carbon atom having an electronegative substituent with a mixture of nitrate and acetic anhydride as

$$CH_3COOC_2H_5 \xrightarrow[<50^{\circ}C]{Cu(NO_3)_2-(CH_3CO)_2O} NO_2 - CH_2 - COOH$$

The yield of 50%

(3) The mixture of metal nitrate and Lewis acid TonqNeB systematically studied the nitration of metal nitrate–Lewis acid mixture. The nitrates used were nitrates of silver, barium, sodium, potassium, ammonium, and lead,

The Lewis acids used were boron trifluoride, silicon tetrachloride, ferric chloride, aluminum trichloride, etc. The effects of various catalyst activities, reaction temperature, reaction time, and catalyst dosage were studied. In the nitration method, the nitrated compound and nitrate are added into a three-mouth bottle first, and the temperature can be increased to 30–40 °C without external cooling. Among the Lewis acids used, the catalytic abilities of aluminum trichloride and boron trifluoride are relatively strong. Activity order of various nitrates in the nitration of benzene is

$$AgNO_3 > KNO_3 > NaNO_3 > NH_4NO_3 > Pb(NO_3)_2 > Ba(NO_3)_2$$

The mechanism is

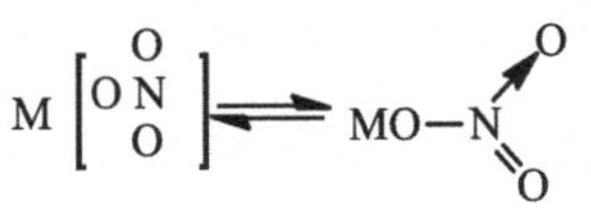

$$C_6H_6 + AlCl_3 \rightarrow (C_6H_5AlCl_3)H$$

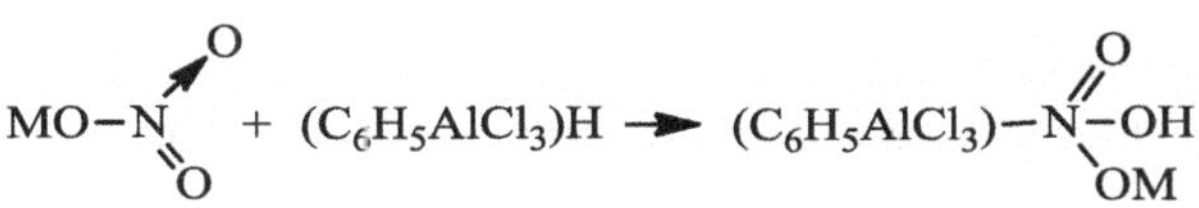

Table 2.18 Reaction parameters and yield of CL-20 with different nitrating reagents

Nitrating reagents	Temperature (°C)	Time (h)	Yield (%)
HNO_3(98%)	85	3	76
H_2SO_4	80	1	96
CF_3COOH	85	2	93
N_2O_4	80	2	97
NH_3NO_3	85	2	91
Silicon sulfate	85	3	94
AC_2O	85	3	89
KNO_3	85	2	85
$NaNO_3$	85	2	84
Ammonium cerium nitrate	85	3	87

The catalytic mechanism of boron trifluoride is

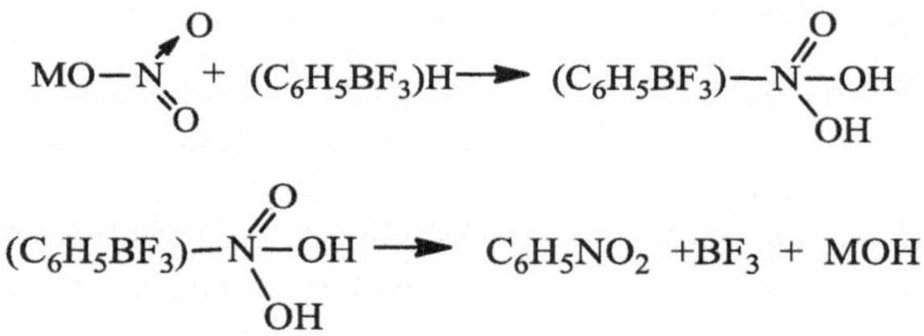

This means that the catalyst first forms a complex with the to-be-nitrated compound, which is activated through the nitration to form a three-component complex with nitrate, and finally the complex decomposes into nitro compound. But, this mechanism is not consistent with modern nitration theory. According to the modern nitration theory, the catalytic effect of Lewis acid is to react with nitrate to form nitryl salts, in which nitryl ion nitrifies the aromatic compounds.

$$\mathrm{KN}O_3 + \mathrm{A1C1}_3 \rightarrow \mathrm{N}O_2^+\,\mathrm{A1C1}_3K^-$$

or

$$3\mathrm{KN}O_3 + 8\mathrm{A1C1}_3 \rightarrow 3\mathrm{N}O_2{}^+\mathrm{A1C}l_4{}^- + 3K^+\mathrm{A1C}l_4{}^- + \mathrm{A1}_2O_3$$

Another possibility is that nitrate reacts with Lewis acid first to form nitrogen pentoxide, which reacts with Lewis acid to form nitryl salt. The formation of

$N_2O_5{\cdot}2BF_3$ has been confirmed through the reaction between potassium nitrate and boron trifluoride.

$$6KNO_3 + 14BF_3 \longrightarrow 3NO_2^+\left[ON\begin{smallmatrix}\nearrow OBF_3\\ \searrow OBF_3\end{smallmatrix}\right]^- + B_2O_3 + 6KBF_4$$

The side reactions of aromatic compound nitration with metal nitrate and Lewis acid are limited, and there is not any oxidative side reaction and the purity of nitrated product is high.

2.4.6 The Nitrating Reagent from Nitric Acid and Nitrate

In the first few parts of this chapter, green nitrating reagents, such as magnesium nitrate, zinc nitrate, and other metal salts, have been mentioned more or less.

Since Nielsen [42] synthesized hexanitrohexaazamethane (Cl-20) first in 1987, studies on the industrialization green nitrating reagents of high-energy explosives have been carried out in every country. The initial synthesis method is to form tetraacetyl dibenzyl hexaazaisowurtzitane (TADB) intermediate from benzylamine and glyoxal through condensation and hydrogenation, then to use tetrafluoroboronitrite ($NOBF_4$) to generate tetraacetyl dinitroso-hexaazaisowurtzitane (TADN), and finally to use tetrafluoroboronitride (NO_2BF_4) to produce Cl-20.

Every country has done series of modifications on nitrating reagents and conditions to remove the nitrosation agent $NOBF_4$ and nitrating reagent NO_2F_4 in the original method with the replacement of dinitrogen tetraoxide (N_2O_4), nitric acid, sulfuric acid, or nitric acid/sulfuric acid. The Gunpowder and Explosive Group used the improved technology with N_2O_4 as nitrosation reagent and nitric acid/sulfuric acid mixture as nitrating reagent to achieve the CL-20 pilot production first, and the product purity is >99%.

In order to further understand the effect of nitric acid on the yield of CL-20, Iranian researchers, with 2,2,8,12-tetraacetyl-2,4,6,8,10,12-hexaazaisowurtzitane (TAIW) as raw materials, studied effects of different nitrating reagents on the temperature, reaction time, and yield. The results are shown in Table 2.18. The highest yield of product is 97% with N_2O_4 as the nitrating reagent due to the increase of nitronium ion concentration (NO_2^+) with the presence of N_2O_4 [1, 43].

According to the latest disclosure by Iranian researchers, with nitric acid/heteropoly acid as nitrating reagent in the environmental synthesis technology of CL-20, the effect of nitric acid concentration on product purity and yield was studied. It was found that the purity and yield of the product are decreased with the decrease of nitric acid concentration, because the increase of water content in nitric acid solution leads to the decrease of nitronium ion concentration and solubility of some nitration product. When the nitric acid concentration is 98%, the yield and purity of CL-20 are the best [3]. TADF is nitrated with nitric acid at 125 °C to prepare CL-20.

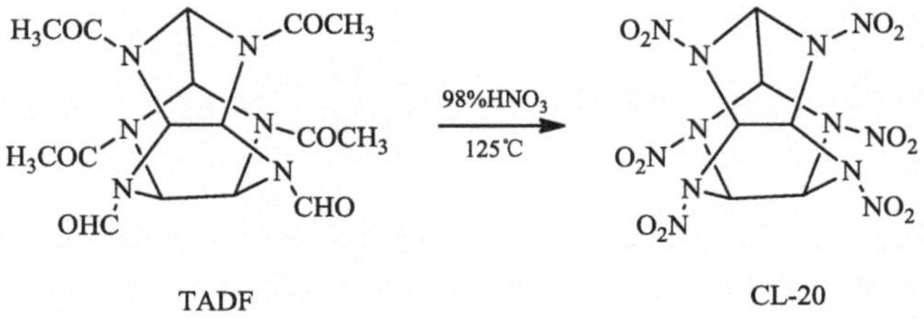

In 2008, the researchers at the US Navy Air Force Center announced another new “no benzylamine, no heavy metal” CL-20 synthesis process [6], in which 1-propenylhexaazaisowurtzitane (HPIW) was produced through the base-catalyzed isomerization of the starting material hexaallylhexaazaisowurtzitane (HAllYlIW), which was synthesized according to method from Herve et al. in French Gunpowder and Explosive Group. Then, through the oxidation with single bond oxygen (produced from oxygen photolysis), C = C bond of acrylamide was broken to form formamide, and then generated hexacyl formyl hexaazaisowurtzitane, and CL-20 was produced through nitration with the yield of 47%. HPIW can also be directly nitrated to produce CL-20, but the yield (11.6%) is not as high as that of photooxidation. The CL-20 produced through the direct nitration of HPIW can be achieved as follows:

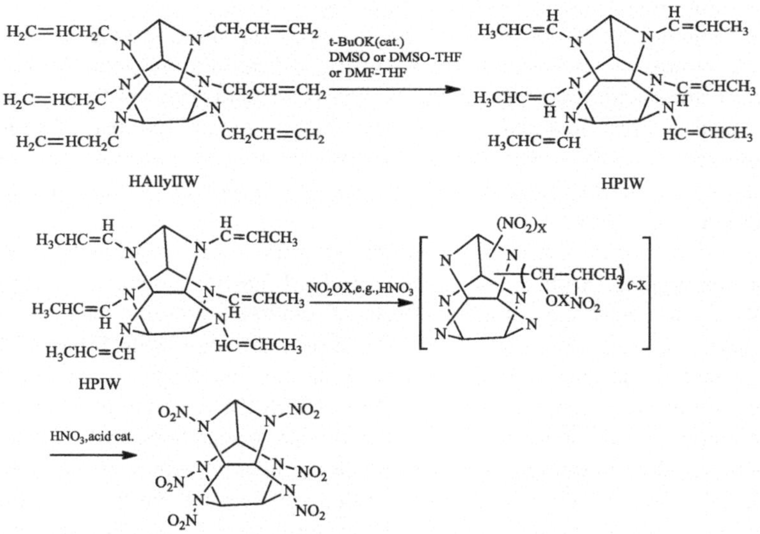

The mechanism of reaction between allylamine with glyoxal to synthesize cage-like HAllylIW from Herve et al. in French Gunpowder and Explosive Group is the same as that of HBIW preparation from Nielsen et al.

Iran has carried out a special study on compositions of nitrating mixed acid, tested the nitration reaction in different nitration systems, such as sodium bisulfate monohydrate, trichloroisocyanuric acid, phosphorus pentoxide–silica composite acid, lanthanum trichloride, silicon sulfate, melamine trihydrosulfate, sodium nitrate, and other reagents. In the preparation of CL-20, TADBIW is the starting material, solid acid melamine trihydrosulfate is reactant, and nitric acid is the nitrating reagent. And, features of the process are simple, efficient, green, low cost, and high yield.

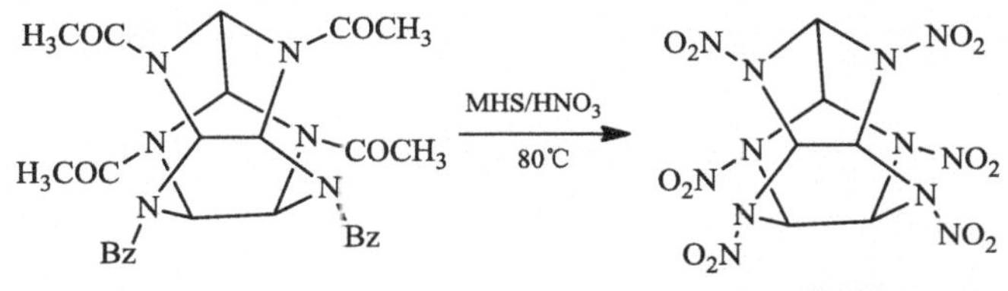

Synthesis of the other compound 3,3-bis (difluoroamino)-1,5,7,7-tetranitro-1,5-diazacyclooctane (TNFX) [15, 44] is shown as follows:

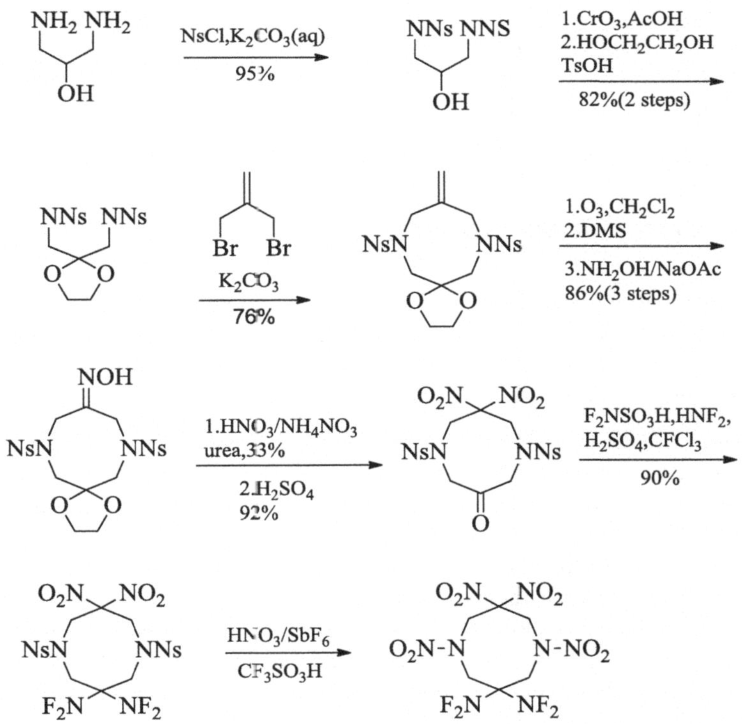

Some compounds cannot be directly nitrated, only with nitrite as a replacement, which is a representative nitration reaction.

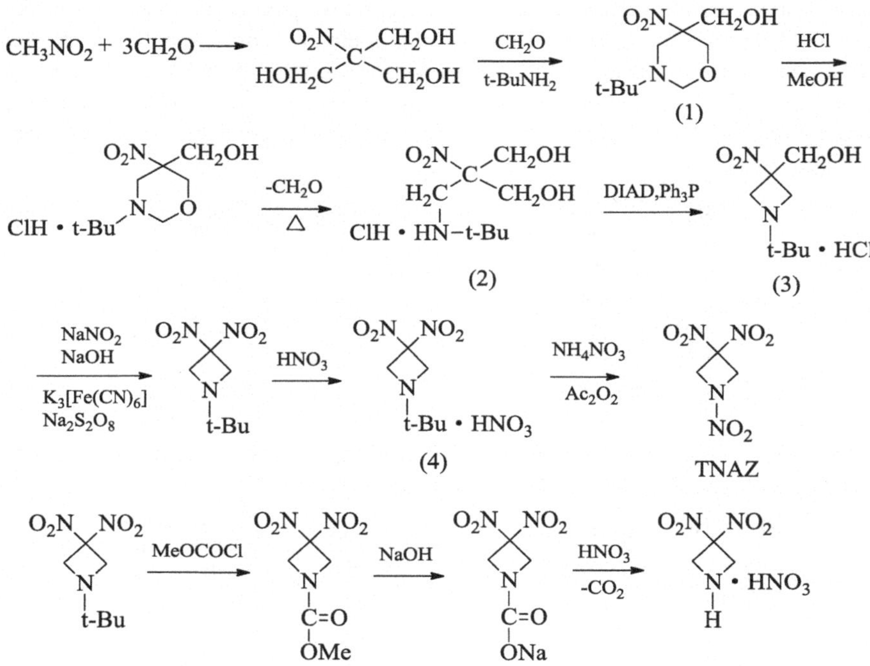

2.4.7 *Nitrate Nitrating Reagent*

Nitrate has no nitration capacity. However, when a protonic acid, a Lewis acid, or an alkaline substance is present, it can nitrate different organic compounds. Therefore, when the nitrate is used as the nitrating reagent, the reaction may be carried out in an acidic medium or in a basic medium.

(1) Mixture of nitrate and sulfuric acid. In the presence of sulfuric acid, many nitrates can nitrate aromatic compounds. Nitrates can be ethyl nitrate, butyl nitrate, amyl nitrate, β-chloro ethyl nitrate, nitroglycerin, nitrocellulose, pentaerythrite tetranitrate, nitromannitol, etc. For example, when toluene and ethyl nitrate are mixed together, concentrated sulfuric acid is gradually added with stirring and the temperature is maintained at about 40 °C, and the yield of mononitrotoluene is close to the theoretical value. Similarly, benzothiazole ($C_6H_4N:CHS$) and its derivatives can be nitrated to give mononitro derivatives of benzene ring. A mixture of concentrated sulfuric acid and nitroglycerin can nitrate nitrobenzene to m-nitrobenzene.

According to the freezing point study, 1 mol of alkyl mononitrate was dissolved in concentrated sulfuric acid to produce five proton points, while the spectrum of sulfuric acid and nitric acid solution showed nitryl ion line. Kuhn [44] suggested that nitryl ions are present in this solution.

$$R\text{-}ONO_2 + H_2SO_4 \longrightarrow \left[R\text{-}\overset{\overset{\displaystyle H}{|}}{O}\text{-}NO_2\right]^+ + HSO_4^-$$

$$\left[R\text{-}\overset{\overset{\displaystyle H}{|}}{O}\text{-}NO_2\right]^+ \longrightarrow ROH + NO_2^+$$

$$ROH + 2H_2SO_4 \rightarrow R - O - SO_3H + H_3O^+ + HSO_4^-$$

$$ROH + 2H_2SO_4 \rightarrow R - O - SO_3H + H_3O^+ + HSO_4^-$$

The total reaction is

$$R - O - NO_2 + 3H_2SO_4 \rightleftharpoons R - O - SO3H + NO_2^+ + H_3O^+ + 2HSO_4^-$$

(2) A mixture of nitrate and Lewis acid

Lewis acid has the catalytic effect on the nitration of benzene with ethyl nitrate, and the used Lewis acids include boron trichloride, tin tetrachloride, phosphorus trichloride, aluminum trichloride, silicon tetrachloride, antimony pentachloride, and trichloride iron. Among them, aluminum trichloride, tin tetrachloride, antimony pentachloride, and ferric chloride have a certain degree of catalytic effect. The reaction mechanism is

$$C_2H_5\text{-}O\text{-}\overset{\overset{\displaystyle O}{\uparrow}}{N}\text{=}O + AlCl_3 \longrightarrow C_2H_5\text{-}O\text{-}\overset{\overset{\displaystyle O}{\uparrow}}{N}\text{=}O\cdots\cdots AlCl_3$$

$$C_2H_5\text{-}O\text{-}\overset{\overset{\displaystyle O}{\uparrow}}{N}\text{=}O\text{-}AlCl_3 + C_6H_6 \longrightarrow C_2H_5OH + C_6H_5NO_2 + AlCl_3$$

In modern terms, nitrate and Lewis acid form nitryl salt, which are attacking reagent in the nitration reaction.

$$C_2H_5ONO_2 + AlCl_3 \rightarrow Al(OC_2H_5)Cl_3^-$$

(3) Nitration with nitrate in alkaline media. In an alkaline medium, nitrate can nitrate amines and aliphatic compounds containing active methylene groups.

1. Good yields can be obtained with nitrate as nitrating reagent in the nitration of aliphatic compound containing active methylene groups nitrate contains in alkaline medium.
2. Nitration of amines α-Cyano-α-methyl ethanol nitrate is particularly suitable to nitrate amines. Since the reaction is carried out in alkaline medium, the

Table 2.19 Results of alcohol nitration with NO_2BF_4

R in ROH	CH_3	C_2H_5	N-C_3H_7	N-C_4H_9	N-C_8H_{17}	C_2H_4F	C_2H_4Cl	C_2H_4Br	CH_3CH_2
Yield of nitrate (%)	87	92.5	87	91	85.5	88	85	85.5	72

oxidation side reaction is small, and the yield of nitrated primary amine is about 60% and that of nitrated secondary amine is 55–80%. The side effects to cause this declining is

$$R_2NH + HCN + (CH_3)_2C{=}O \longrightarrow R_2N\text{-}\overset{\displaystyle CH_3}{\underset{\displaystyle CH_3}{\overset{|}{\underset{|}{C}}}}\text{-}CN$$

2.4.8 *Nitryl Salt as Nitrating Reagent*

The salt of nitrating reagent $NO_2^+\,A^-$ is called nitryl salt, some of which are extremely unstable and some of which have very good stability. The nitryl salt can be made from nitric acid with nitric oxide and protonic acid or anhydride, or Lewis acid.

$$N_2O_5 + HClO_4 \rightarrow NO_2^+ClO_4^- + HNO_3$$
$$HNO3 + 2SO3 \rightarrow NO_2{}^+HS_2O_7{}^-$$
$$HNO_3 + HF + BF_3 \rightarrow NO_2{}^+BF_4{}^- + HNO_3$$
$$HNO_3 + HF \rightarrow H_2NO_3{}^+F^- \overset{PF_5}{\rightarrow} H_2NO_3{}^+PF_6{}^- \overset{PF_5}{\rightarrow} NO_2{}^+PF_6{}^- + 2HF + POF_3$$

A series of nitryl salts can be prepared through the reaction between nitryl halites and Lewis acids or nonmetallic fluorides.

$$2NO_2Cl + SnCl_4 \rightarrow (NO_2)_2SnCl_6$$
$$NO_2F + BF_3 \rightarrow NO_2BF_4$$
$$2NO_2F + SiF_4 \rightarrow (NO_2)_2SiF_6$$
$$2NO_2F + GeF_4 \rightarrow (NO_2)_2GeF_6$$
$$NO_2F + PF_5 \rightarrow NO_2PF_6$$
$$NO_2F + AsF_5 \rightarrow NO_2AsF_6$$
$$NO_2F + SbF_5 \rightarrow NO_2SbF_6$$
$$NO_2F + SeF_4 \rightarrow NO_2SeF_5$$
$$NO_2F + IF_5 \rightarrow NO_2IF_6$$

A variety of nitryl slats have been made and will no longer be listed one by one here. All nitryl slats have strong hygroscopicity. Generally speaking, fluorine-containing nitrate has good thermal stability. For example, NO_2BF_4 decomposes only at a temperature over 170 °C into nitryl fluoride and boron trifluoride, and can be long-term stored at room temperature. Their solubility in organic solvents are not large.

Nitromethane can dissolve 0.3 wt% of tetrafluoroboronitronate or 0.5% of nitric acid perchlorate at 25 °C, but NO_2SbF_6, NO_2PF_6, and NO_2AsF_6 have relatively large solubility in nitromethane and can make 50% solution at 25 °C. Nitryl salt has a large solubility in tetramethylsulfoxide, and the solubility of NO_2BF_4 is 7%. At room temperature, acetonitrile can dissolve 10% NO_2BF_4. The studies on freezing point decrease show that nitryl salts are present in these solvents essentially with the form of non-ionized ion pairs. For example, the amount of ionized NO_2C1O_4 is less than 6% in nitromethane, but the Raman spectra and infrared spectra of these solutions show strong bands of nitryl ions.

In the study on mechanism and kinetics of electrophilic substitution nitration, the organic solvent solution of nitryl salt is widely used as nitrating reagent.

Because, in this solution, the attacking agents certainly are and only can be nitryl ions or their ion pairs without any other electrophilic attack reagents, and we can rule out many other factors of interference. In special cases, it can also be used as a laboratory preparation reagent. For example, it can be used to nitrate compounds that are easily hydrolyzed and oxidized in acidic media.

(1) Aromatic compounds

In the nitration of aromatic compounds, generally tetrasodium sulfone is used as solvent, because tetramethylsulfone has a good ability to dissolve both aromatic compounds and nitroxides, and tetramethylsulfone can be miscible with water, and the product can be separated after dilution with water at the end of reaction. In order to produce pure nitro compounds, excess to-be-nitrated compounds are usually used to avoid the production of polynitro compounds. This reaction needs to be carried out under anhydrous conditions.

At room temperature, mononitro-compounds can be formed with high yields except of compounds with strong passivation substituents [45, 46]. The two and three nitrations should be carried out at relatively high temperature.

$$ArH + NO_2BF_4 \rightarrow ArNO_2 + HF + BF_3$$

Nitration yield of aromatics with basic heteroatoms (oxygen, sulfur, and nitrogen) is low at 5%, and usually with the formation of nitro salts of heterocyclic bases.

$$C_5H_5N + NO_2ClO_4 \rightarrow [C_5H_5NNO_2]^+ClO_4^-$$

where only thiophene can be nitrated to nitrothiophene with NO_2BF_4 with the yield of 91%. In nitration of furan, the yield of mononitro-compounds was only 14%.

(2) Alcohols can react with nitroxides to form nitrates [47] as shown below:

$$ROH + NO_2BF_4 \rightarrow RONO_2 + BF_3 + HF$$

Table 2.19 lists the yields of alcohols reacting with nitrates.

(3) Amines in tetramethylsulfoxide or sulfur dioxide solution, NO:BF nitrifies the primary, secondary amine, and dialkyl amide into the corresponding nitramine as

$$2R_2NH + NO_2BF_4 \rightarrow R_2N - NO_2 + R_2NH \cdot HBF_4$$
$$R_2NCOR' + NO_2BF_4 \rightarrow R_2N - NO_2 + R'COF + BF_3$$

(4) Olefins

NO_2BF_4 reacts with olefins to form polymers [48]. The low molecular weight polymer was formed by NO_2BF_4 and butene-1 was fractionated to give two fractions. One fraction is free of nitro and the other is a more viscous liquid containing nitro. It is clear that the nitryl ion initiates the polymerization reaction to produce a nitro group-containing polymer, and the proton released when the polymerization chain is interrupted causes the polymerization to produce another polymer without nitro groups. The reaction mechanism is proposed as follows:

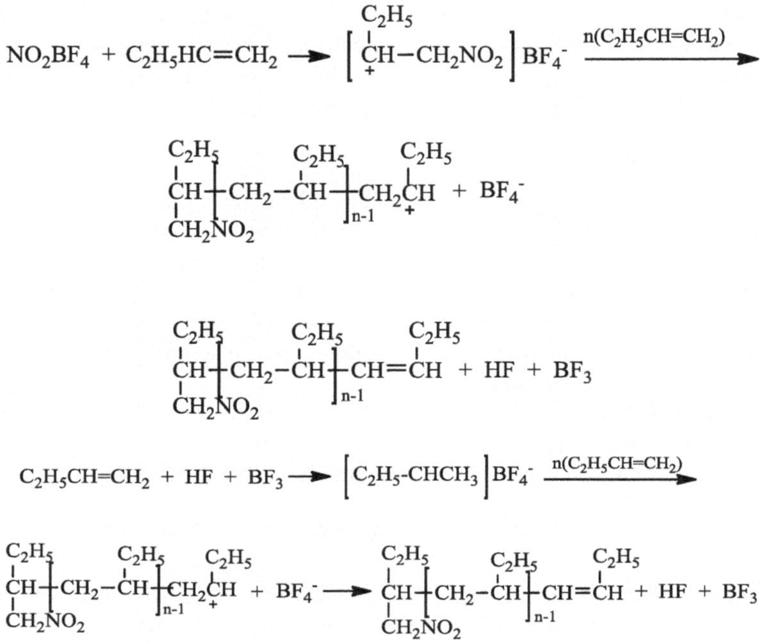

Non-nitro polymer

(5) Alkane

Olah and Lin [49] nitrated alkanes (including cycloalkanes) with nitryl slats (NO_2PF_6, NO_2SbF_6, and NO_2BF_4) in methylene chloride-tetramethylsulfone solution at 25 °C to obtain a small amount of nitroalkane. But the amount of reacted alkane is very small, and the nitration of methane can only get 0.1% nitromethane. If the solvent is changed to hydrofluoric acid and fluorosulfonic acid, the yield can be increased by 10 times. The nitration yields of high alkanes are higher as from 2 to 5%. And that of adamantane is 10%.

Obviously, this reaction could only be electrophilic substitution, and the authors proposed the reaction mechanism as follows:

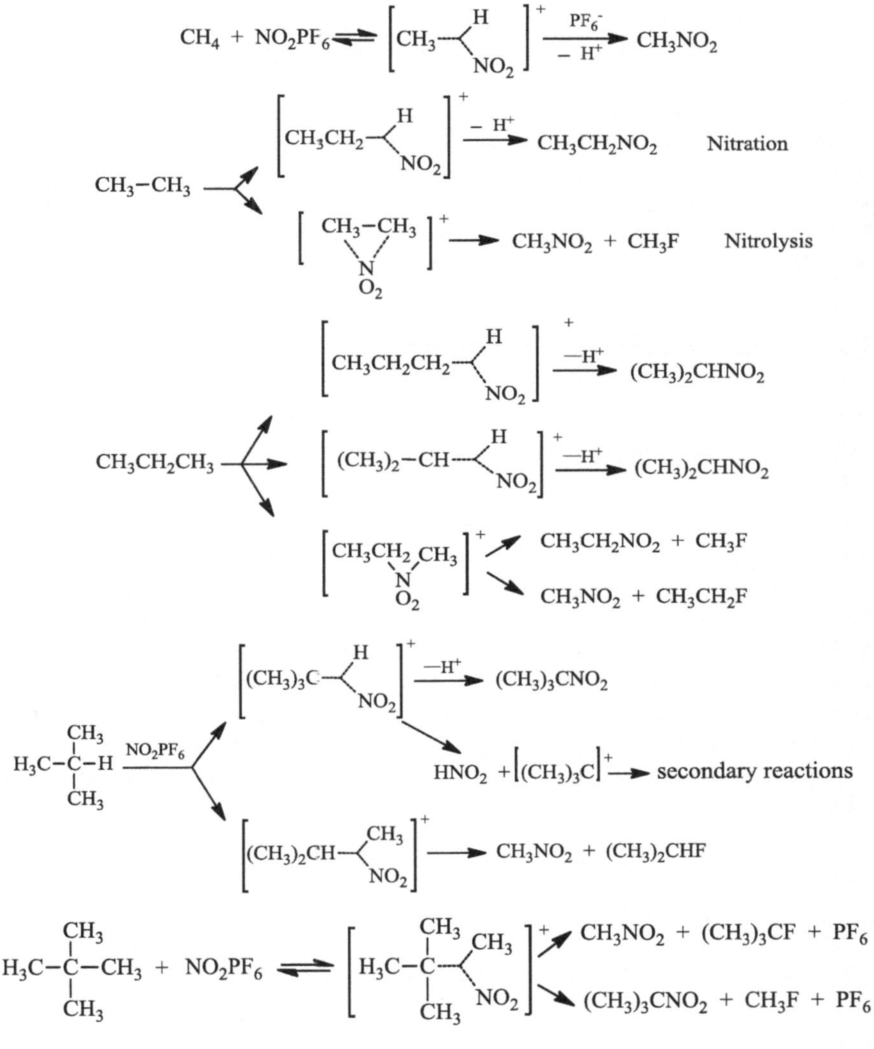

The hydrogen on the tertiary carbon is the most active. Carbon–carbon bonds are more active than hydrogen atoms on secondary or primary carbons, leading to the first nitration of n-alkanes.

Although the yield of this reaction is low without practical value, it has certainly been shown that the mononitroalkane can be produced through electrophilic substitution of alkane. It is different from the electrophilic substitution nitration of aromatic compounds. In the electrophilic substitution, nitration of aliphatic compounds is affected by proton to have denitration reaction, which is more important for the tertiary, secondary carbon nitro compounds. Even at −80 °C,

2-nitro-2-methylpropane also undergoes denitration reaction in the $FSO_3H–SbF_5$, $HF–SbF_5$, or $HF–PF_6$ solution to form t-butyl cations and nitrous acid instead of nitryl ion.

$$(CH_3)_3 - C - NO_2 \overset{H^+}{\rightleftharpoons} (CH_3)_3C - NO_2H^+ \rightarrow (CH_3)_3C^+ + HNO_2$$

If isobutane is nitrated in nitric acid–sulfuric acid at 50 °C, the formed nitrobutane will react with strong acid to denitrify into t-butyl cation, which can undergo further reaction, such as isomerization and degradation. Therefore, composition of the product is the following, 28% of nitromethane, 4.9% of nitroethane, 0.5% of 2-nitropropane, and 66.6% of tert-nitrobutane. If the temperature is too high, the substitution with a free radical mechanism will occur.

2.4.9 *Nitric Oxide as Nitrating Reagent*

The formula of nitrogen monoxide is NO as a colorless gas with a melting point of −163.6 °C and boiling point of −151.8 °C under 760 mm Hg pressure. Its electronic structure is

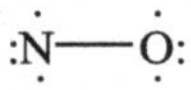

The nitrogen and oxygen are connected with a σ bond, a π bond, and a three-electron π bond. The distance between atoms is 1.14 Å, which is between the double bond (1.18) and the triple bond (1.06). This three-electron bond makes nitric oxide stable, not easy to form dimer, not active as other compounds containing sole electron. It is not prone to free radical reactions like nitric oxide, but it is easily oxidized by oxygen or oxygen in the air to nitrogen dioxide.

Nitric oxide reacts with olefins to form dinitroso compounds. If the dinitroso compound has a α hydrogen atom in its molecule, it will get converted into oxime.

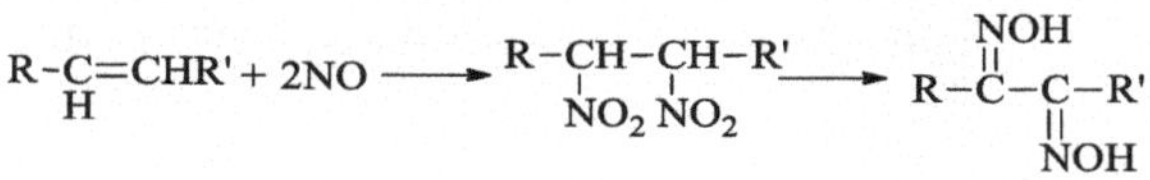

Oxime or nitroso compounds can be oxidized to nitro compounds.

2.4.10 *Nitrous Oxide as Nitrating Reagent*

Nitrous oxide [N_2O_3] is a gas with a boiling point of 3.5 °C and melting point of −102 °C. It decomposes at boiling. The gaseous nitrous oxide is almost completely dissociated into nitric oxide and nitrogen dioxide. Solid nitrous oxide is gray blue. It is dark blue when it is liquid due to its partial decomposition into nitric oxide and nitrogen dioxide, and it becomes green when temperature is close to its boiling point. The structure of nitrous oxide is not yet fully defined, but there may be two structural formulas as follows:

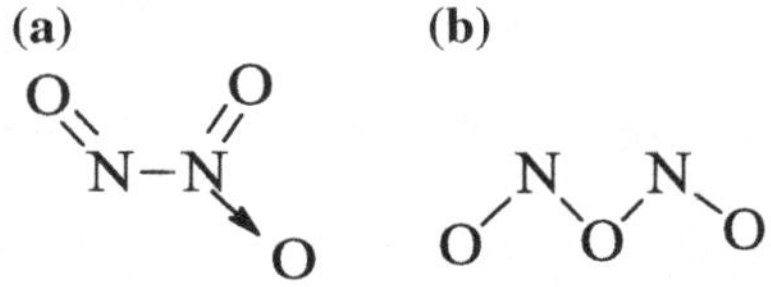

Compound b has been caught at very low temperatures. The structural formula a can be split into two free radicals as NO and NO_2, or it can be split into two ions as nitrosyl ion and nitrite. Structural formula b can only be split into nitrosyl ion and nitrite. Nitric oxide is dissolved in water to form nitrous acid.

$$N_2O_3 + H_2O \rightarrow 2HNO_2$$

(1) Action on aromatic hydrocarbons

Nitrous oxide solution in sulfuric acid does not nitrate aromatic hydrocarbons. Nitric oxide is ionized in sulfuric acid to become nitrosyl ion.

$$N_2O_3 + 3H_2SO_4 \rightleftharpoons NO_2{}^+ + H_2SO_4{}^- + H_3O^+$$

The electrophility of nitrosyl ion is much weaker than that of nitryl ion. With the same concentration, the amount of water required to hydrolyze nitryl ions into nitric acid is much less than that required for nitrosyl ion to nitrite, proving the weak electrophility of nitrosyl ion. Therefore, it is not reactive to medium active aromatic compounds, such as benzene and alkylbenzene. Another reason is that the nitroso compound is unstable and readily further reacts with excess nitrosyl ions.

Nitrosyl ion can react easily with highly active aromatic compounds to produce nitroso compounds, which reacts with low-concentration nitric acid to produce nitro species. Its reaction with resorcinol is a typical example.

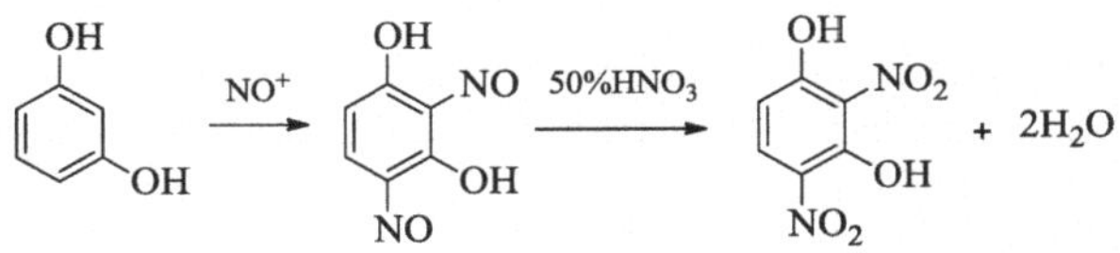

The reaction to 1,3,5-tricarboxylic acid follows the above reaction rule.

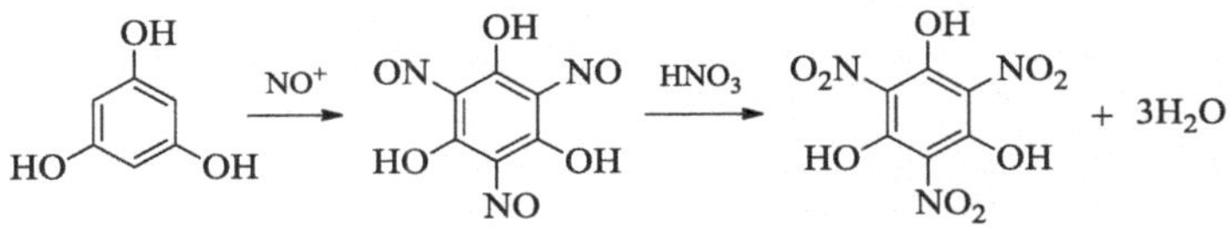

This method has a certain practical value. Each hydroxyl group on the benzene ring can be replaced to introduce one nitroso group with this method, which is then converted to nitro through oxidation. With this method, the nitrophenol compounds with the same number of nitro groups and hydroxyl groups can be prepared.

Dinitro-resorcinol is a practical detonating chemical. If the conventional method with sulfonation first and then nitration, only trinitroresorcinol can be prepared.

(1) Nitrosation

First, the resorcinol is dissolved in water to prepare 2.75% aqueous solution of resorcinol. Sulfuric acid is added into this solution. After it is cooled to −2 °C, 10% aqueous solution of sodium nitrite is added with stirring, and the molar ratio of resorcinol:sulfuric acid:sodium nitrite is equal to 1:2:2. The yellow dinitroso-resorcinol crystal is washed with water and the yield is close to its theoretical value.

(2) Oxidation

The dinitroso-resorcinol was oxidized with 50% nitric acid at a temperature below 0 °C. The reactor is charged with nitric acid cooled to −5 °C and then filled with wet dinitroso-resorcinol. The nitric acid concentration should be taken into account when the nitric acid was calculated. When dinitroso-resorcinol is added, concentrated nitric acid is added to the reactor at the same time to maintain the concentration of acid in the reactor. After the reaction, the reaction mixture was dropped into ice water to separate dinitroso-resorcinol, washed with a small amount of cold water and recrystallized with hot water, and the melting point of the product is not lower than 147 °C.

Nitrous acid can have a variety of reactions with aromatic amine in inorganic acid solution. Normally, diazotization of primary amine occurs. If the resulting diazonium salt is hydrolyzed, the product is phenol. However, under the appropriate conditions, nitration of benzene ring can be done. The concentrated sulfuric acid solution of nitrosyl sulfuric acid can nitrosate α-nitramine.

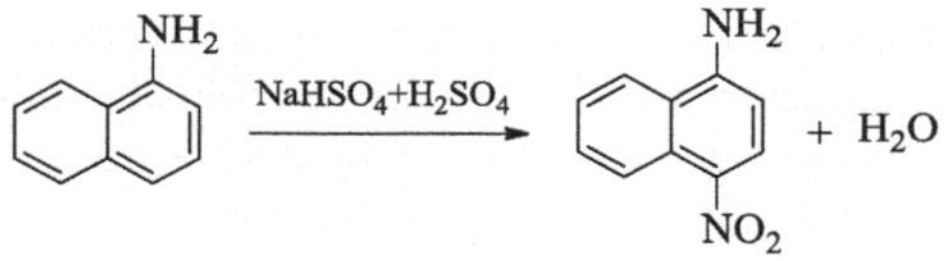

Aryl secondary amine reacts with nitrous acid to produce N-nitrosamine, and this reaction is reversible. With hydrogen chloride or hydrogen bromide, N-nitrosoamine rearranges into nitroso-aromatic amine, and this isomerization is called Fischer–Hepp reaction.

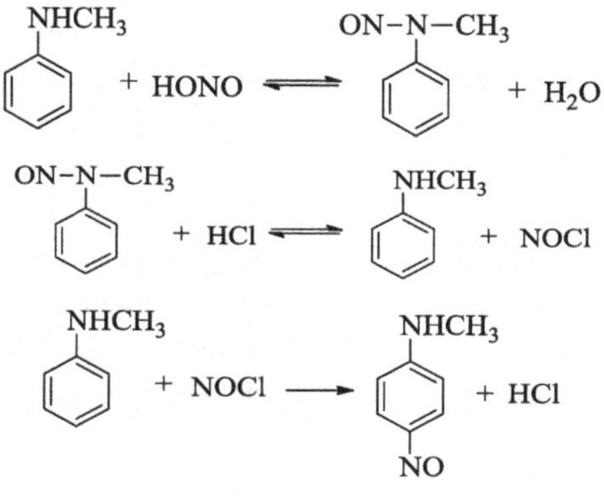

If the aromatic secondary amine has an ortho-substituent, 4-nitroso-aryl secondary amine is directly formed due to the space barrier. Nitrous acid reacts with an aryl tertiary amine to form 4-nitroso-aryl tertiary amine.

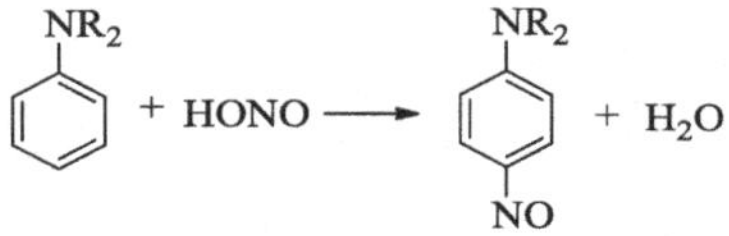

(3) Reaction catalyzed by Lewis acid

When the molar ratio of nitrous oxide to boron trifluoride is 1:1, a white crystal is formed. And, this white crystal can be sublimated at room temperature, but it cannot be melted when heated up to <300 °C in a closed tube. The infrared spectroscopy shows the absorption band of nitrite and tetrafluoroboride ions, indicating this crystal might be the complex $N_2O_3{\cdot}BF_3$, which is exactly the same as $NOBF_4$. Thus, the possible reaction between nitrous oxide and boron trifluoride to form nitrosyl tetrafluoroborate is proposed as

$$3N_2O_3 + 8BF_3 \rightarrow' 6NOBF_4 + B_2O_3$$

Bachman et al. [50] used this white crystal to treat the aromatic hydrocarbons and found that it was a good nitrosating reagent with a weak nitrating capacity. Only after a very long time, it was possible to introduce nitro groups onto the benzene ring. Therefore, isomer ratio in the product is completely different from that with nitric acid as nitrating reagent. The yield of nitronaphthalene in the air is much higher than that in nitrogen. The route to form nitro compounds in this reaction may have the following two ways:

1. Nitrosofication occurs first and the produced nitroso compound is oxidized into nitro compound again. (2) $N_2O_3{\cdot}BF_3$ is oxidized to $N_2O_4{\cdot}BF_3$, and the latter is a nitrating reagent.
2. Alcohol: Nitrous acid can convert aliphatic alcohols into nitrites. First, alcohol and aqueous solutions of sodium nitrite are mixed, and then the mixture is cooled and acidified with hydrochloric acid or sulfuric acid under stirring to form nitrite ester. The yield is up to about 90%.

$$ROH + HONO \rightarrow RONO + H_2O$$

3. Amine: Aliphatic tertiary amines react with nitrous acid to form nitrosamines. For example, urotropine reacts with nitrous acid to form trimethylenetriylenetetramine.

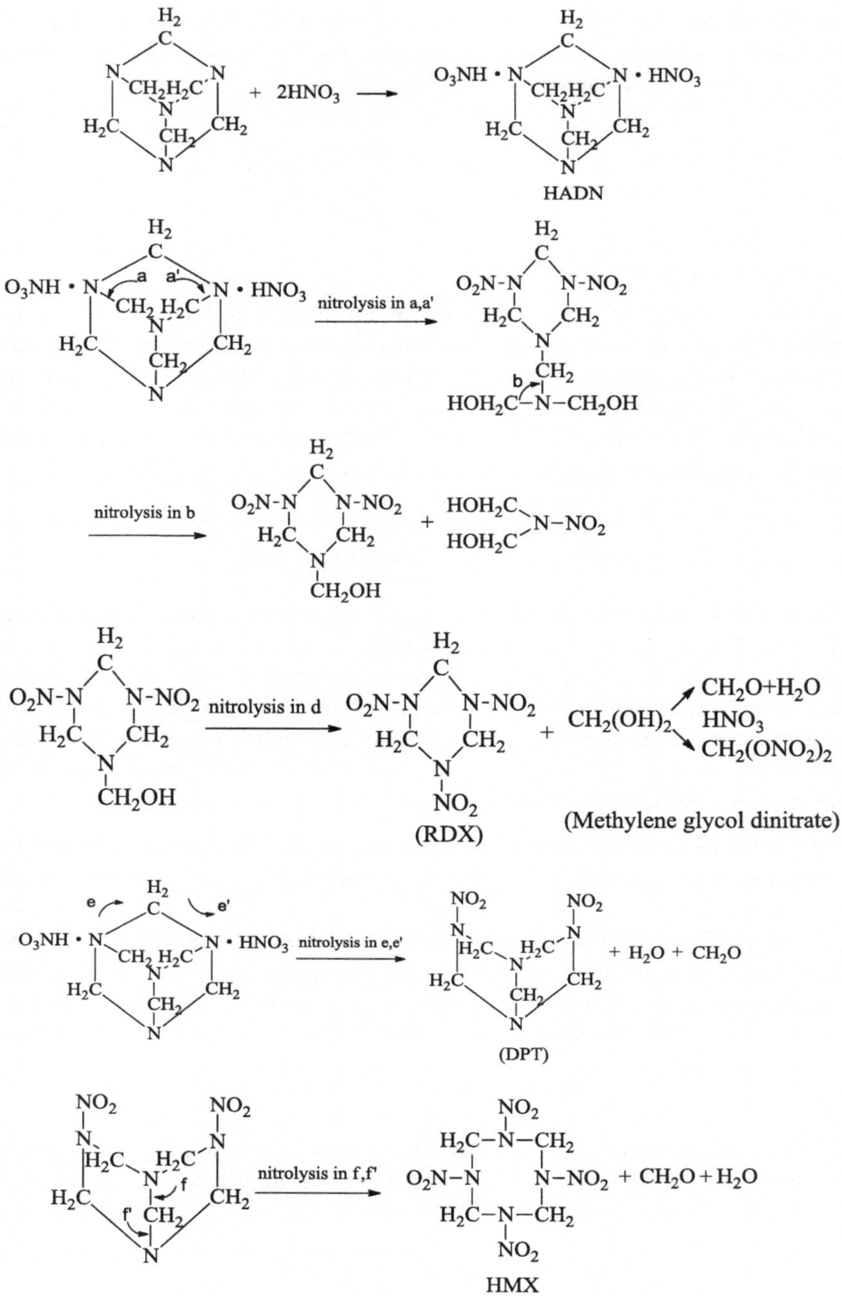
+ 2HNO3
HADN
nitrolysis in a,a'
nitrolysis in b
nitrolysis in d
(RDX)
(Methylene glycol dinitrate)
nitrolysis in e,e'
+ H2O + CH2O
(DPT)
nitrolysis in f,f'
+ CH2O + H2O
HMX

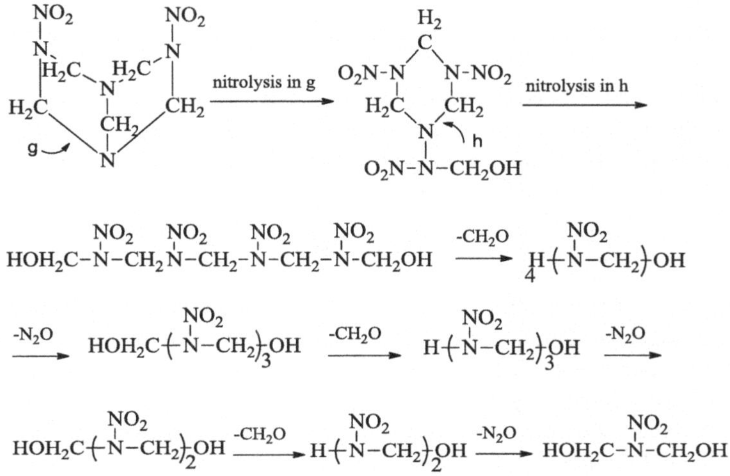

4. Olefin: Generally, nitrous oxide reacts with olefins to produce nitroso-hydroxamates.

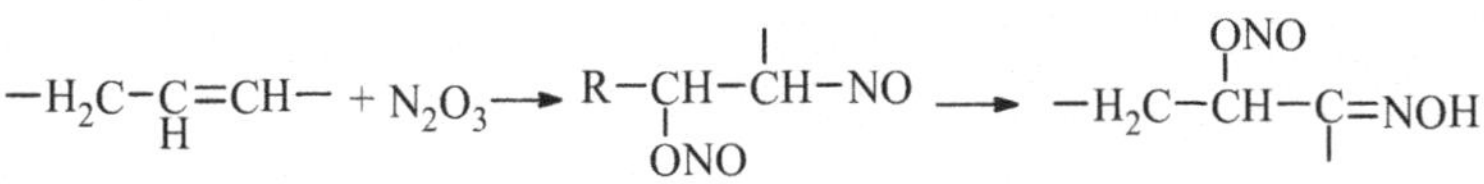

Under the action of base, the latter loses nitrous acid to become unsaturated oxime.

The α, β—unsaturated ketone reacts with nitrous oxide to produce nitro–nitroso compound first, which is finally converted to the derivative of isoxazole.

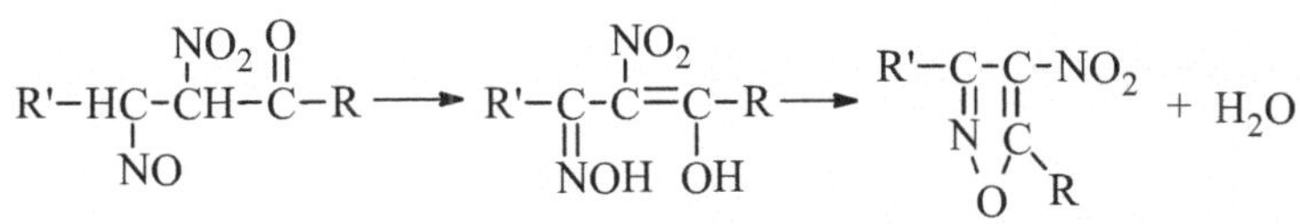

2.4.11 Nitrating Reagent from Nitrogen Dioxide and Nitrous Oxide

The melting point of nitrous oxide is −9.3 °C, its boiling point is 21.3 °C, and it decomposes at boiling. Pure solid nitrogen tetroxide is a colorless crystal. At room temperature, it is a yellow liquid, which contains about 1% of the brown nitrogen dioxide. When it is heated to 100 °C, about 90% of it decomposes into nitrogen dioxide.

The molecular structure of nitrogen dioxide is

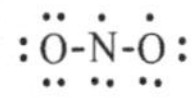

In addition to the σ bond between the nitrogen and oxygen atoms, three-electron π bond including three atoms is also formed. Thus, it is more active than nitric oxide, and also easy to dimize into nitrous oxide. There is a balance between nitrogen dioxide and nitrous oxide as

$$N_2O_4 \rightleftharpoons 2NO_2$$

The decrease of temperature or the increase of pressure moves the balance to the left. The nitrogen dioxide molecule has a planar structure with an O–N–O bond angle of 132° and a bond length of 1.20 ± 0.02 angstroms, which is close to the bond length of a double bond.

In solid, liquid, and gaseous nitrous oxides, the location and intensity of the main absorption band are not significantly different, and thus the phase structure has not changed after the phase change. The molecular structure of nitrous oxide can be believed as a planar given as,

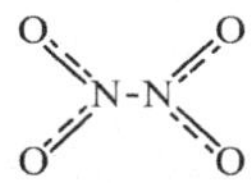

But, at very low temperature, its structure may be

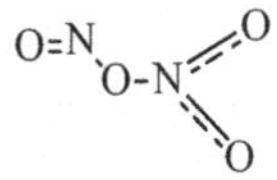

The dissociation of nitrous oxide may have the following three ways:

$$N_2O_4 \rightleftharpoons NO^+ + NO_3^-$$
$$N_2O_4 \rightleftharpoons NO_2^+ + NO_2^-$$
$$N_2O_4 \rightleftharpoons 2NO_2$$

The first two reactants are the cracking of nitrous oxide into two ions, the last one is its homolysis into two molecules with free radical properties. High dielectric constant solvent and low temperature are conducive to different crackings. On the contrary, the hemolysis is favorite.

(1) Nitration of aromatic hydrocarbons,

1. Nitration with nitrogen dioxide or nitrous oxide.

In the nineteenth century, the nitration of nitrogen dioxide was found on aromatic hydrocarbons. At room temperature, nitrogen dioxide acts on benzene. After

7 days, in addition to non-reacted benzene, oxalic acid, butylbenzene, and nitro compounds of phenol were found as the product.

Benzene and nitrous oxide in the tube are sealed up and heated to 80 °C. In addition to oxidation products, the trinitrobenzene is obtained. When nitrobenzene is used to replace benzene, there is no reaction.

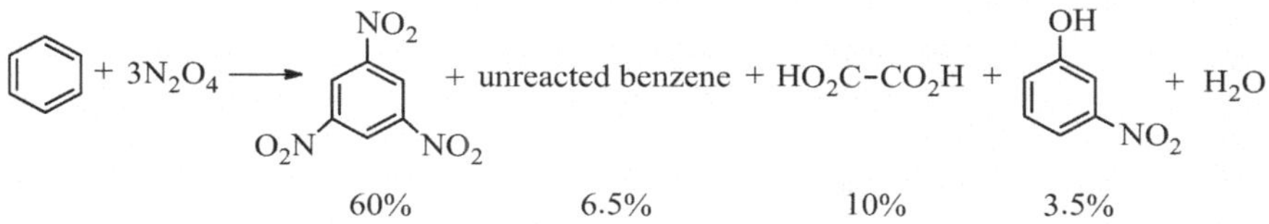

When nitrous oxide reacts on anthracene in chloroform solution, 9,10-dinitrodihydrocene anthracene is obtained.

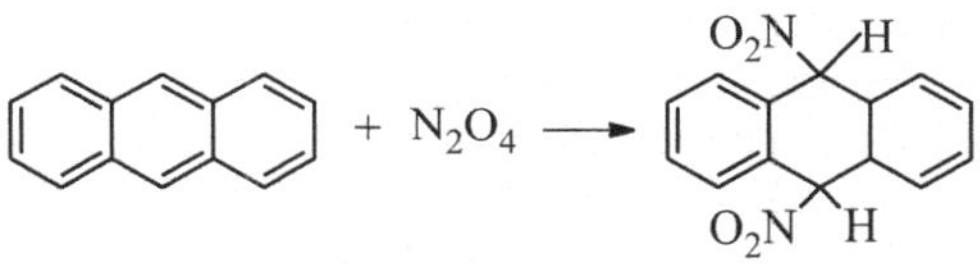

These reactions have significant properties of free radicals.

Nitrous oxide can nitrate aromatics, phenols, aromatic amines, and fused aromatic hydrocarbons in liquid or gaseous conditions. When the gas nitrous oxide nitrifies these chemicals, ultraviolet irradiation is used to compare the effect of ultraviolet light on gas nitration.

① The main nitration products of benzene, naphthalene, phenanthrene, and other aromatic products are nitro compounds. Among them, the nitration of naphthalene is the best at 18–20 °C with a nitration yield of 96%, and the product does not contain polynitronaphthalene. When nitration temperature is higher than 60 °C, a small amount of polynitronaphthalene is generated. Benzene is the most difficult to nitrated, with a nitration yield of 32% at 55–60 °C.

② The nitration product of anthracene is 9,10-dinitroanthracene and the best yield is 80%.

③ Ultraviolet radiation has no effect on nitration results of aromatic hydrocarbons with symmetrical structures (such as benzene and naphthalene).

④ The nitration of aromatic hydrocarbons with side chains can occur either on the benzene ring or on the side chain. Ultraviolet radiation can increase the yield. In the case of ultraviolet irradiation, the optimum yield of nitrotoluene was 46%, and that of phenyl nitromethane was 9% (nitration temperature was 140–150 °C). The optimum nitration yield to nitroethylbenzene was the best as 40%, with that to methyl phenyl nitromethane of 24% (nitration temperature was 78–80 °C).

⑤ During the nitration of aromatic hydrocarbons with asymmetric structures, the use of ultraviolet radiation can improve the yield. For example, during the nitration of xylene, with ultraviolet radiation, the yield of mononitro-compounds was from 81 to 99%

⑥ Phenols are easier to nitrate to generate bi- or trinitrophenol. For example, the product of nitrated phenol is 2,4-dinitrophenol with the yield of 76%, the nitration products of cresol are 2,4,6-trinitro-m-cresol and 2,4-dinitrophenol with the total yield of 41%, and the nitration product of β-naphthol was 1,6-nitro-naphthol with a yield of 81%.

⑦ The reaction direction of nitrous oxide and arylamine depends on whether the amino group has a substituent. The main nitration products of aniline are 2,4-dinitrophenol and a small amount of p-nitroaniline, and thus the main reaction mechanism may include the occurrence of diazotization first, and then nitration.
The yield of dimethylaniline nitration was p-nitro-dimethylaniline at 72 °C and the yield of acetanilide nitration was 60% of p-nitroaniline and 30% of o-nitroaniline at 60–70 °C.

⑧ Heterocyclic compounds are difficult to nitrate. At 95–100 °C, the yield of nitroquinoline was only 11% in the quinoline nitration. When the nitration temperature was increased to 155–160 °C, the yield of dinitroquinoline was 10–12%. The yields of pyridine and 8-nitro-pyridine were 6% at 115–120 °C, and most of the pyridine could be recovered.

⑨ Close to nitration naphthalene results with mixed nitric oxide in industrial production of nitric acid, naphthalene nitration with nitrous oxide at 18–20 °C has a nitronaphthalene yield of 96%.

2. Nitration with nitrous oxide and sulfuric acid mixture

The mixture of nitrous oxide and sulfuric acid can nitrate a variety of aromatic compounds. In order to obtain good yields, sulfuric acid must be excessive, and there is a suitable sulfuric acid concentration for every compound. If sulfuric acid concentration is too low, nitration will stop. When benzene is nitrated, the concentration is 61%.

The nitrating attack agent in the nitrous oxide solution is nitryl ion, which is produced as a result of the following reaction:

$$N_2O_4 + 3H_2SO_4 \rightleftharpoons NO^+ + NO_2{}^+ + H_3O^+ + 3HSO_4{}^-$$

The Raman spectra of the solution show the line of nitryl ions, because when water was added to the solution and the molar ratio of $\frac{H_2O}{H_2SO_4}$ was 0.53, the spectrum of nitryl ions disappeared. At this point, the concentration of sulfuric acid in the solution was 91% according to the chemical analysis.

(2) Nitration under Lewis acid catalysis

Schaarschmidt first studied and used aluminum trichloride or ferric chloride and nitrous oxide mixture to nitrate benzene. When the liquid or gaseous nitrogen dioxide is added to the mixture of excess aromatic hydrocarbons and catalyst, aluminum chloride is gradually dissolved to form a red solution. When water is added or it is heated, nitrobenzene is formed. The catalytic capacity of ferric chloride is weaker than that of aluminum trichloride. With this way, benzene, toluene, and halogenated benzene are nitrated. Nitration of toluene also produces oxidative by-products. Nitration of halobenzene, the nitro halobenzene isomer ratio is not the same as that with the general nitrating reagent, in which para-isomer is relatively large. Aluminum trichloride, carbon tetrachloride, phosphorus trichloride, phosphorus pentachloride, chromium trichloride, antimony pentachloride, tin dichloride, thallium chloride, and magnesium bromide have no catalytic effect.

In the reaction, if the tetramethylsulfoxide is used as the solvent, the reaction can be carried out smoothly under the homogeneous conditions with titanium tetrachloride, boron trifluoride, boron trichloride, phosphorus pentafluoride, or arsenic pentafluoride as the catalyst. The yield of nitrobenzene is 32–67%, and the yield of nitrochlorobenzene is 28–76% The activity order of catalysts is

$$AsF_5 > PF_5 > BF_3 > TiCl_4 > BCl_3$$

It is worth mentioning that the important phenomenon is, when bromide is used as catalyst, the bromination on the benzene ring, bromobenzene or its derivatives are formed. With chloride as catalyst, the main reaction is nitration, but also a small amount of chlorination, and nitrobenzene is the main product if aluminum trichloride is the catalyst, but with a small amount of chlorobenzene. Only when the fluoride is the catalyst, halogenation reaction cannot occur.

With Lewis acid as catalyst, when nitrous oxide nitrifies aromatic compounds, nitroxyl salt or nitryl halide is formed.

$$3N_2O_4 + BF_3 \rightarrow 3NOBF_4 + 3NO_2BF_4 + B_2O_3$$
$$2N_2O_4 + TiCl_4 \rightarrow 2NO_2Cl + 2NOCl + TiO_2$$
$$3N_2O_4 + 2BBr_3 \rightarrow NO_2Br + 3NOBr + B_2O_3$$

The Raman spectra of N_2O_4–BF_3 were identical to that of $NOBF_4$–NO_2BF_4, confirming that tetrafluoroboronitronate and tetrafluoroboronitrite were formed in the system of nitrous oxide and boron trifluoride. Nitryl chloride generated from the second reaction is nitrating reagent, but also has a certain chlorinating capacity.

$$NO_2Cl \rightarrow NO_2^+ + Cl^-$$
$$2NO_2Cl \rightarrow 2NO_2 + Cl_2$$

The main tendency of nitryl bromide is homolysis into nitrogen dioxide and bromine, and thus when bromide is a catalyst, the bromination reaction occurs.

Nitrogen dioxide and Lewis acids can also form salts with ionic structures, and the reactions carried out are similar to those described above.

(3) Nitration of alkanes and olefins

Diphenylacetylene, as a mixture of two isomers, is nitrated into diphenyldyenitroethylene by N_2O_4,

m.p. 186~187°C b.p. 105~107°C

Nitration of phenylacetylene forms phenyl dinitroethylene, which is an unstable compound and decompose according to the following reaction into benzonitrile, carbon dioxide, and nitrous acid.

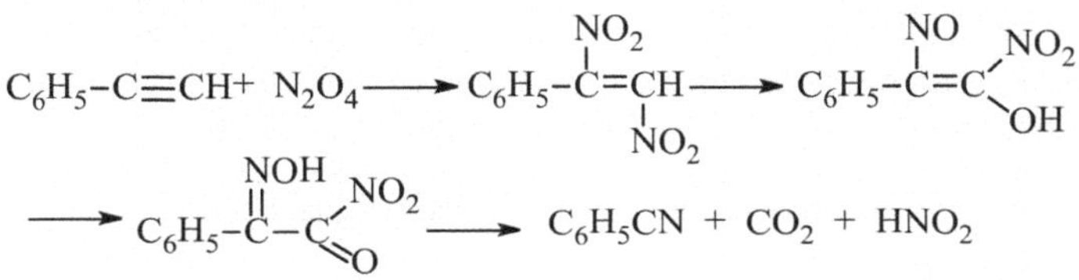

(4) Deep nitration of the clathrate compound

Nitration of clathrate nitro compound is difficult to complete. For example, the nitration of octanitrocubane indicates that the nitrating reagent is not only acid, but can also be composed of salt. The initial can be carried out smoothly nitration in the mixed acid, once five or six nitro groups are on the cubane structure, the mixed acid cannot nitrate it, but causes its decomposition. Although N_2O_4 and NO_2Cl can indeed nitrate certain cubic alkane anions, none of them reacts with nitro-cubic alkanes to produce octanitrocubane [51, 52].

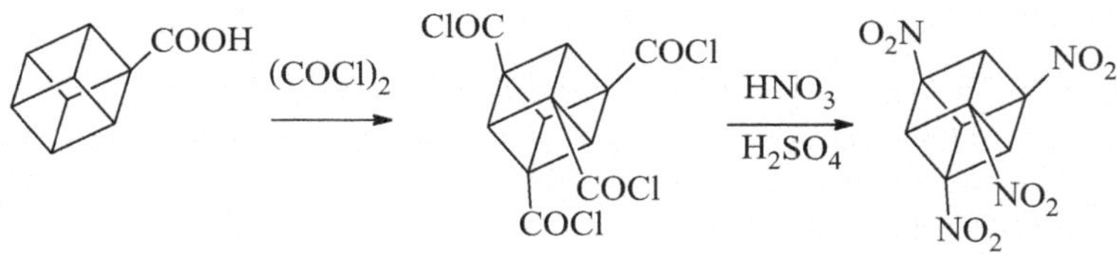

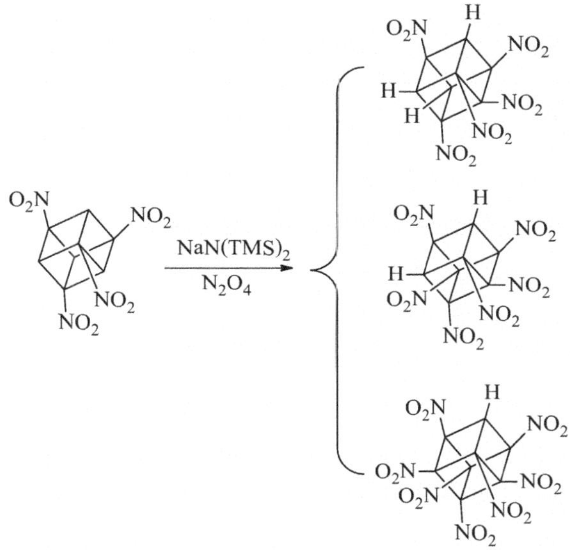

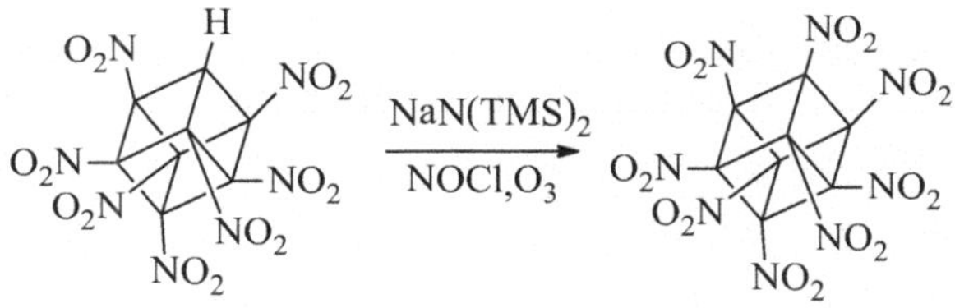

In this synthesis, there is actually a cyclization polymerization intermediate process as

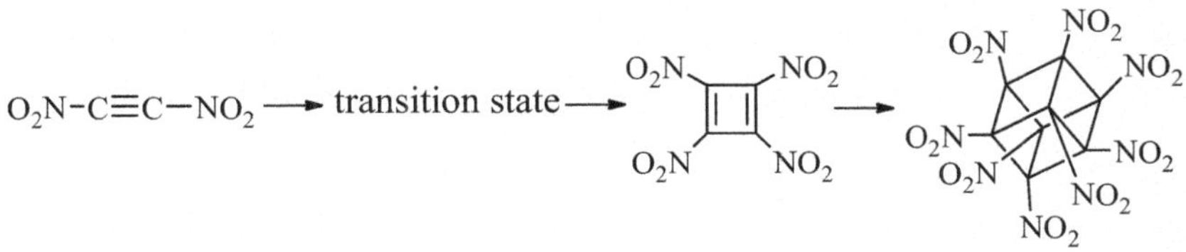

2.4.12 Nitrous Oxide as Nitrating Reagent

Nitrous oxide [N_2O_5] is a colorless crystal and is stable when it is exposed to daylight at the temperature below 8 °C, but it decomposes when heated or exposed to the sun. The melting point of nitrous oxide is 30 °C with the boiling point of 47 ° C. It decomposes during its melting and the decomposition products are nitrogen dioxide, nitric oxide, and oxygen.

The crystalline nitrous oxide has an ionic structure $NO_2^+NO_3^-$. But, at very low temperature, at gaseous state, and in nonpolar solvents, its structure is covalent as

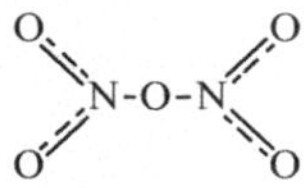

Therefore, nitrous oxide can be dissociated in two forms. In polar solvent, it is

$$N_2O_5 \rightarrow NO_2^+ + NO_3^-$$

In a nonpolar solvent or in a gaseous state, it is

$$N_2O_5 \rightarrow \cdot NO_2 + NO_3 \cdot$$

Nitrous oxide is nitric acid anhydride, and it is soluble in water to form two molecules of nitric acid, which is given as

$$N_2O_5 + H_2O \rightarrow HNO_3$$

Nitric oxide can be formed from nitric acid dehydration with phosphorus pentaoxide. Nitric oxide is a strong oxidant and can oxidize iodine into iodium acid anhydride I_2O_5.

(1) Nitration of aromatic compounds

1. The sulfuric acid solution of nitrous oxide is a strong nitrating agent, and its nitration is very similar to that of nitric acid–sulfuric acid. But, when they have the same molar concentration in the sulfuric acid solution, the nitrating capacity of nitrous oxide is about the double of that of nitric acid, because the nitrous oxide in sulfuric acid solution decomposes as the following:

$$N_2O_5 + 3H_2SO_4 \rightleftharpoons 2NO_2^+ + 3HSO_4^- + H_3O^+$$

where the nitration capacity of a molecule of nitrous oxide is equivalent to that of two nitric acid molecules, and the water produced is less than half of that with nitric acid. Therefore, nitrous oxide solution of sulfuric acid has a strong nitration capacity.

2. Lewis acid catalysis N_2O_5 and BF_3 form a stable complex $N_2O_5{\cdot}BF_3$, which is insoluble in aromatic hydrocarbons. If this complex is prepared in nitroalkane, a white solid is formed. This complex can nitrate benzene, benzoic acid, trifluorobenzene, chlorobenzene, benzonitrile, and other substances. It is also possible to use titanium tetrachloride, tin tetrachloride, phosphorus pentafluoride, and others as catalyst. A more convenient method of nitration is with the mixture of nitrous oxide and Lewis acid. A nitrous oxide–Lewis acid solution in a molar ratio of 1:1 in tetramethylsulfone was formed. The solution is added to the solution of nitritrified compound in tetramethylsulfoxide at room temperature and under stirring (such as toluene, ethylbenzene, benzene, butylbenzene, xylene, and trimethylbenzene). After it is heated to room temperature which was followed by another 15 min of stirring, the yield is 87–95%.

   The infrared spectrum of N_2O_5–BF_3 is completely similar to that of NO_2BF_4, indicating that the nitryl salt is formed in this system. The reaction equation is

$$N_2O_5 + 8BF_3 \rightarrow 6NO_2BF_4 + B_2O_3$$

3. Solution of nitrous oxide in inactive solvent. In the aprotic solvent, such as nitromethane, acetonitrile, chloroform, and carbon tetrachloride, nitrous oxide may be present in a covalent state. This solution has a strong nitration capability, and most of nitration products are ortho-isomers, meaning a higher o/p-isomer ratio in the product compared with that with general nitrating reagent. Therefore, some researchers believed that, in this solvent, the nitrous oxide molecules are the attacking reagent of nitration reaction.

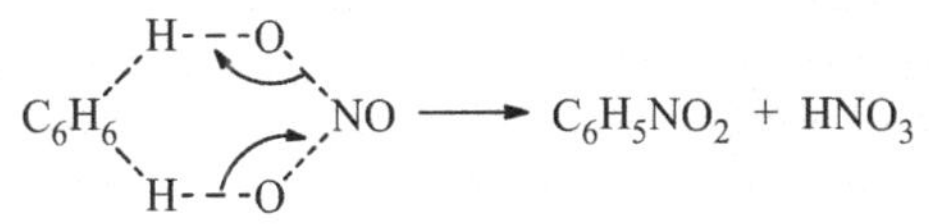

The produced nitric acid has a catalytic effect on the nitration reaction.

In the aminocarbamate method to prepare dinitramide, N-nitrocarbamate ammonium salt is synthesized first, further nitration is done with HNO_3/H_2SO_4, and then secondary nitration is done with N_2O_5 to achieve both safe and guaranteed yield [53, 54].

$$NH_2COOEt \xrightarrow{HNO_3/H_2SO_4} O_2NHNCOOEt \xrightarrow{NH_3}$$

$$NH_4 \cdot N(NO_2)COOEt \xrightarrow{N_2O_5} (O_2N)_2NNH_4$$

Many experiments show that $NO_2^+BF_4^-$ is the nitrating reagent during the second nitration, the reaction is required to occur at −15 °C and under anhydrous conditions, and the cost of the process is relatively high. Experiment with N_2O_5 as a nitrating reagent has a high yield and equipments also meet industrial production.

(2) Nitration of alkanes

Nitration of alkane with nitrous oxide belongs to radial reaction mechanism.

2.4.13 Tetranitromethane Nitrating Reagent

It was found that tetranitromethane [$C(NO_2)_4$] can nitrate organic compounds. The properties of tetranitromethane as a nitrating reagent are the requirement of a base. The reaction is carried out in an alkaline medium. It has been now known that the nitrifiable compounds with tetranitromethane in alkaline media are aromatic amines and phenols, such as aniline and phenols. It can nitrate nitroalkanes into di- and trinitroalkanes, arylene into nitroalkenes or aryltetranitroalkane via addition reaction, some acid-sensitive substances such as heterocyclic compounds, tyrosine, protein, etc. These substances are typically the general nitro compounds and have no prospect of any application in the manufacture of explosives.

In the nitration of aromatic compounds, pyridine is generally used as a basic material, tetranitromethane is converted into nitromethane with the formation of nitration products, and the latter and pyridine form nitro-pyridine salt. However, in the nitration of anilines, because of the alkaline of the nitrated compound, it is not necessary to add alkaline.

There is no direct evidence for the attacking reagent in this nitrating reagent, thus there are many different proposals. Soklovsky et al. [55] suggested that the following dissociation occurred in tetranitromethane:

$$NO_2C(NO_2)_3 \rightleftharpoons NO_2{}^+C(NO_2)_3^-$$

Therefore, the attacking reagent is nitryl ion. Mare [56] and others believed that the first nitrosofication reaction was followed by the oxidization to form nitro compounds.

With nitroso methane nitration phenol is considered free radical mechanism reaction, and the phenolic nitration process is as follows:

$$\text{XPhOH} + \text{OH}^- \rightarrow \text{XPhO} + \text{H}_2\text{O}$$

The phenoxy anion forms a charge transfer complex with tetranitromethane, which is given as

$$\text{XPhO}^- + \text{C(NO}_2)_4 \rightleftharpoons \left[\text{XPhO} \rightarrow \underset{\text{O}}{\overset{\text{O}}{\text{N}}}\text{-C(NO}_2)_3\right]^-$$

Electron transfer occurs within the charge transfer complex, which is given as

$$\left[\text{XPhO} \rightarrow \underset{\text{O}}{\overset{\text{O}}{\text{N}}}\text{-C(NO}_2)_3\right]^- \xrightarrow{\text{electron transfer}} \left[\text{XPhO}^\bullet + {}^\bullet\text{NO}_2 + \text{C(NO}_2)_3^-\right]$$

Redecomposition of electron transfer products is

$$\left[\text{XPhO}\cdot + \cdot\text{NO:} + \text{C}(NO_2)_3^-\right] \rightarrow \cdot\text{NO}_2 - \text{XPhO}^- + \text{C}(NO_2)_3^-$$
$$\left[\text{XPhO}\cdot + \cdot\text{NO:} + \text{C}(NO_2)_3^-\right] \rightarrow \text{XPhO}\cdot + \cdot\text{NO}_2\cdot + \text{C}(NO_2)_3^-$$
$$\text{XPhO}\cdot + \cdot\text{NO:} \rightarrow \text{XPhO}\cdot + \cdot NO_2{}^-$$
$$\text{nPhO}\cdot \rightarrow \text{Coupling product}$$

In these formulas, X represents other substituents and Ph represents phenylene.

The nitration of hexanitroethane is similar to that of tetranitromethane, but its nitration rate is lower than that of tetranitromethane. And, the nitration production of hexanitroethane is difficult, and it is no practical value.

In the presence of alkoxides, tetranitromethane can nitrate alkane with a nitro group at the end into polynitrosoalkane.

$$\text{RCH}_2\text{NO}_2 \xrightarrow{\text{C(NO}_2)_4 + \text{KOR}} \text{RCH(NO}_2)_2 \xrightarrow{\text{C(NO}_2)_4 + \text{KOR}} \text{RC(NO}_2)_3$$

With the method from tetranitromethane into nitroform, the following nitroalkanes can be produced:

$$\text{CH}_3\text{CH}_2\text{CH(NO}_2)_2 \rightarrow \text{CH}_3\text{CH}_2C\text{(NO}_2)_3 \quad \text{The yield is 32\%}$$
$$C_6H_5\text{CH}_2\text{CH}_2\text{CH}_2\text{NO}_2 \rightarrow C_6H_5\text{CH}_2\text{CH}_2\text{CH(NO}_2)_2 \quad \text{The yield is 31.6\%}$$
$$C_6H_5\text{CH}_2\text{CH}_2\text{CH(NO}_2)_2 \rightarrow C_6H_5\text{CH}_2\text{CH}_2C\text{(NO}_2)_3 \quad \text{The yield is 32.6\%}$$

In some special cases, tetranitromethane can also be used as a nitrating reagent. It reacts with the compound ArCH = CH–CH_3 first through addition and then loses one molecule of trinitromethane to form nitroalkene. The reaction is usually carried out at 0 °C in a mixed solvent of acetone–pyridine [57].

Sometimes the intermediate product is stable, and tetranitro derivative is formed. Tetranitromethane reacts with 1-(4-methoxyphenyl)propylene-1 in diethyl ether solution to give l,1,1,3-tetranitro-2-(4-methoxyphenyl)butane [58].

2.4.14 *Nitroguanidine Nitrating Reagent*

Aniline, phenols, and other easily oxidized compounds can be nitrated by the newly prepared nitroguanidine sulfuric acid solution.

Nitroguanidine is decomposed into nitramine and aminonitrile in concentrated sulfuric acid.

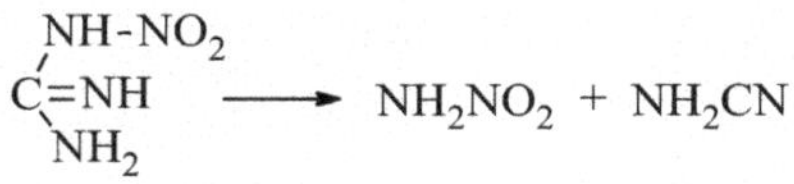

They can be further decomposed as

$$NH_2NO_2 \rightarrow N_2O + H_2O$$
$$NH_2CN + H_2O \rightarrow 2NH_3 + HNO_3$$

Nitroguanidine sulfuric acid solution can turn diphenylamine into blue, and this reaction seems to contain nitric acid, which is likely to be nitric acid hydrolyzed from nitramine when there is some compound to react with nitric acid.

$$NH_2NO_2 + H_2O \rightleftharpoons NH_3 + HNO_3$$

Nitric acid is converted to nitryl ions by sulfuric acid, resulting in electrophilic substitution of nitration.

Another possibility is that nitroguanidine is dissolved in concentrated sulfuric acid, and the following balance is established:

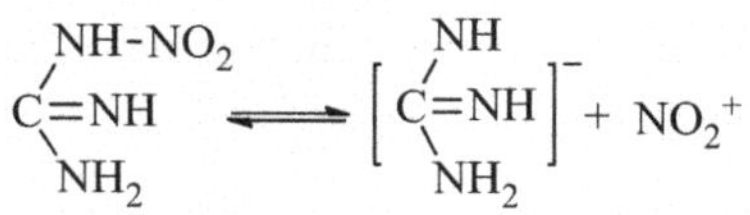

2.4.15 A Mixture of Nitric Acid with a Sulfonic Acid or Sulfonic Acid Ion Exchange Resin

Some researchers [59] have studied the nitration of toluene with nitric acid–sulfonic acid polystyrene and believed that this sulfonic acid ion exchange resin also promotes the dissociation of nitric acid into nitroxides

$$HNO_3 + 2R_{es}SO_3H \rightleftharpoons NO_2{}^+ + H_3O^+ + 2R_{es}SO_3{}^-$$

where Res represents a resin group.

This ion pair is formed from an electrophilic attack reagent and nitryl ion. This nitrating agent is characterized by a relatively high content of p-nitrotoluene in the nitration product. When toluene is nitrated with nitric acid–sulfuric acid mixture, the o-/p-nitrotoluene ratio in the product is 1.57 ± 0.1, while the ratio can be low to 0.68 with nitric acid–sulfonic acid polystyrene.

2.5 The Introduction of Other Groups by Nitrating Reagent

When nitro group is introduced into organic molecules with some nitrating reagents, other groups are introduced too. Of course, other groups introduced here do not include those from side reactions, but nitrating reagents that have these properties. With this property, two kinds of groups can be introduced simultaneously into the nitrated compound.

Through the nitration of addition reaction, other groups are introduced with the nitro group during the nitration.

$$CH_2 = CH_2 + HNO_3 \rightarrow HOCH_2 - CH_2NO_2 \xrightarrow{HNO_3} O_2N - O - CH_2 - NO_2$$
$$CH_2 = CH_2 + NO_2Cl \rightarrow Cl - CH_2 - CH_2 - NO_2$$

As well as, it includes the addition of diazoxide to the double bond.

This suggests that, when some nitrating reagents react with the substitution mechanism, only nitro groups are introduced in organic molecules. While, when the reaction is carried out through the addition mechanism, two kinds of groups are introduced into the nitrated compound, which is the impact of nitrated compound.

2.5.1 Nitrate Containing Catalyst

Researchers worldwide have carried out in-depth research on various new nitration processes, and the development of nitrating reagents and catalysts has become the focus of research. Clean catalytic nitration system cannot only reduce pollution and energy consumption, achieve energy-saving emission reduction purposes, but also improve resource utilization and achieve the purpose of increasing economic efficiency.

Zeolite catalyst (especially zeolite beta) is used in combination with nitric acid/acetic anhydride nitrating reagent, and their heterogeneous catalytic nitration performance is excellent, with not only the high yield of nitro products, but also the high selectivity of isomers. If the problem of catalytic deactivation can be solved, it will be a very practical nitration catalytic reaction system.

Catalytic nitration of nitrogen oxides is the nitration of N_2O_5, N_2O_4/O_2, N_2O_4/air, or N_2O_4/O_3 as a nitrating reagent. Currently, the widely studied catalysts are acetal acetone/zeolite molecular sieve catalyst.

Catalytic nitration of nitric acid is the nitration process, in which NO^{+2} from the nitric acid dissociation on catalyst surface acid sites, as an electrophilic substitution proton, attacks benzene ring. Currently, the rapidly developed nitration catalysts mainly are zeolite/ionic liquid catalyst and solid superacid.

In order to selectively nitrate nitrotoluene to produce the widely used and important fine chemical intermediate p-nitrotoluene in the field of energy-containing materials, many chemical workers worldwide have done a lot of research work and actively looked for appropriate nitrating reagents and catalysts to the establishment of new nitration reaction systems to improve the selectivity in toluene nitration reaction and to improve the nitration reaction environment, which is the hot spot in the field of nitration.

The results from Ban et al. [60] showed that the catalyst $TiSiW_{12}O_{40}/TiO_2$ had a good catalytic activity in the nitration of glycerol. When the amount of catalyst is increasing, the nitration selectivity is also enhanced. When the amount of catalyst is 10 g/L, glycerol can be quantitatively nitrated into nitroglycerin.

2.5.2 *Nitric Acid/Mercury Nitrate Nitrating Reagent*

When the nitric acid or nitrous oxide containing mercury nitrate is heated with benzene, the resulting product varies depending on the concentration of nitric acid. With concentrated nitric acid, the product is nitrobenzene through the normal nitration. If the concentration of nitric acid is low, nitro group is introduced with hydroxyl group and the product is a mixture of nitrophenols. For example, when 50% of nitric acid with 6% of mercury nitrate, the nitration product is a mixture of 2,4-dinitrophenol (90%) and 2,4,6-trinitrophenol (10%) with a total yield of 51%. This method is known as the former oxidative nitration. In order to avoid the confusion with the Kaplan-Shekertel reaction, the above reaction is named as hydroxyl nitration reaction here.

If 65% nitric acid with mercury nitrate is used to nitrate chlorobenzene or naphene, the products are nitrochlorobenzene and trinitro-m-chlorophenol, or 2-nitro-1-naphthol and 2,4-dinitro-1-naphthol. However, during the nitration of inactive materials, such as nitrobenzene, the introduction of hydroxyl reaction cannot occur.

In the nitration of toluene, xylene, and other side-chain aromatic hydrocarbons, in addition to the introduction of hydroxyl groups on the benzene ring, the side-chain oxidation occurs. The nitration products of toluene are nitrotoluene, trinitro-m-cresol, and p-nitrobenzoic acid. The nitration products of m-xylene are 4-nitro-3-methyl benzoic acid. When 50% nitric acid with 6% mercury nitrate is used to nitrate p-xylene at 28 °C, oxidation and nitration occur on the side chain. Only at 2–4 °C, its nitration can generate 4,6-dinitro-2,5-dimethylphenol.

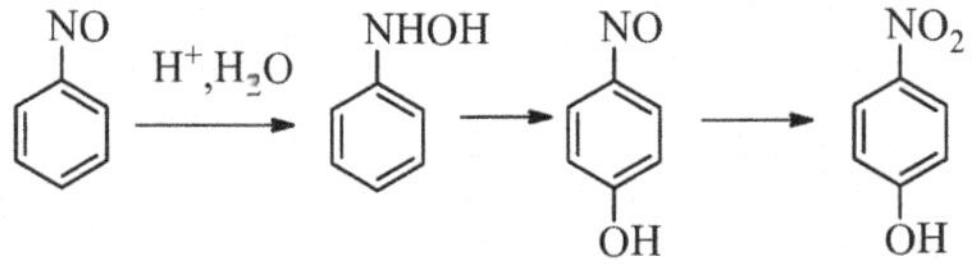

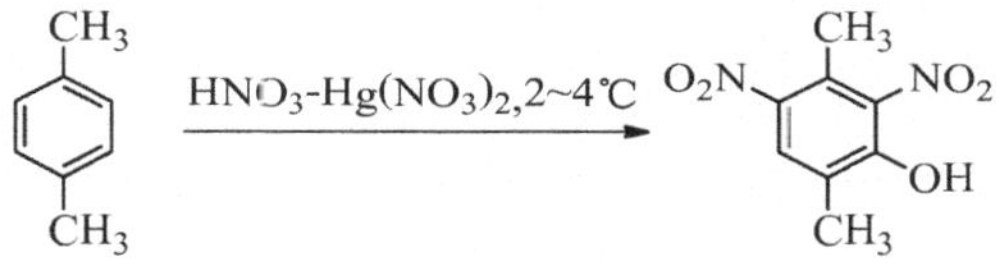

In the hydroxylnitration, nitric oxide must be present in the nitric acid. Otherwise, the organic mercury species cannot be converted into nitroso compounds, but decompose reversibly.

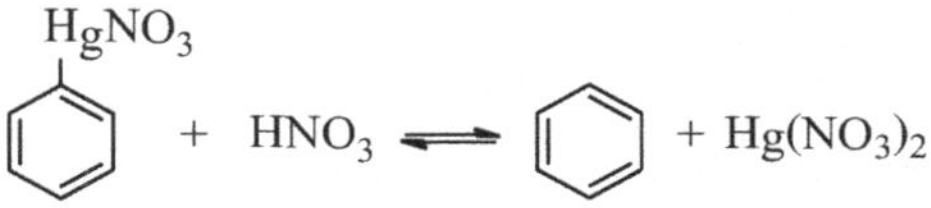

When the nitric acid concentration is sufficient, the organic mercury species can be nitrated directly by nitric acid, which is given as

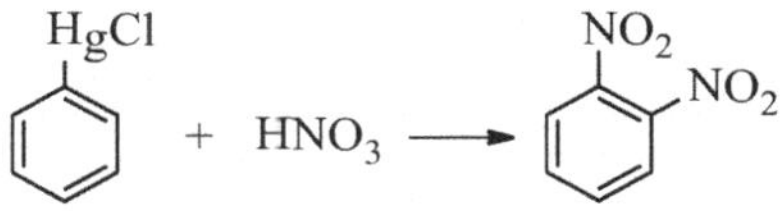

The conversion rate is 91%.

In the hydroxylnitration, oxidation side reaction occurs vigorously and its waste acid contains a large number of oxalic acid and other oxidation products. When mercury nitrate is added with about 2% of aluminum and 5% of manganese, the catalytic capability can be enhanced. Aluminum can accelerate the reaction rate. Manganese can completely oxidize the oxalic acid in the waste acid and improve the purity of the product.

2.5.3 Peronitrous Acid Nitrating Reagent

Peronitrous acid is peroxyacid of nitrous acid, and its structural formula is

$$HOONO$$

When hydrogen peroxide and nitrous acid are combined in the aqueous solution, nitrous acid is oxidized into nitric acid and peronitrous acid is formed in the middle.

$$HNO_2 + H_2O_2 \rightarrow HO_2NO + H_2O$$
$$HO_2NO \rightarrow HO\cdot + \cdot NO_2 \rightarrow HNO_3$$

Halfpenny studied the nitration of aromatic hydrocarbons in aqueous nitrous acid solution, in which nitrous acid aqueous solution was treated with hydrochloric acid to form nitrous acid, and then nitration reagent was formed through the oxidation with the added hydrogen peroxide. Then, the prepared nitrating reagent was added dropwise into the nitrated compound, and in addition to the produced nitro compound, there are hydroxyl nitration product and conjugated hydroxyl nitration products. Due to the low concentration of nitric acid, this reaction is not possible to have the reaction mechanism to produce nitryl ions. And, the authors consider the mechanism is free radical reaction.

$$HO_2NO \rightarrow' HO\cdot + \cdot NO_2$$

Hydroxyl radical and nitrated compound form the following free radicals:

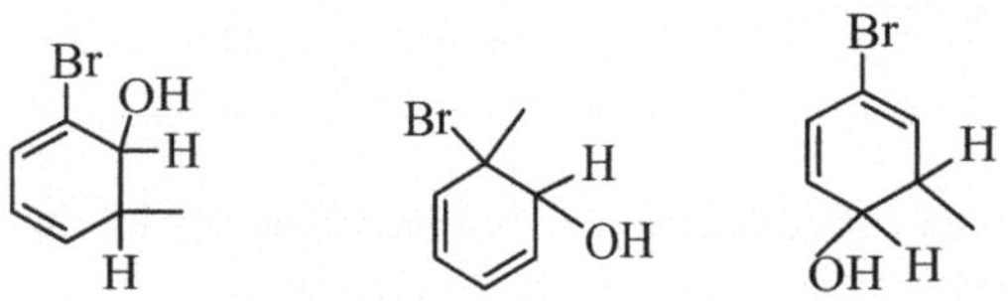

The resulted peronitrous acid reacts with NO_2 to form an intermediate as

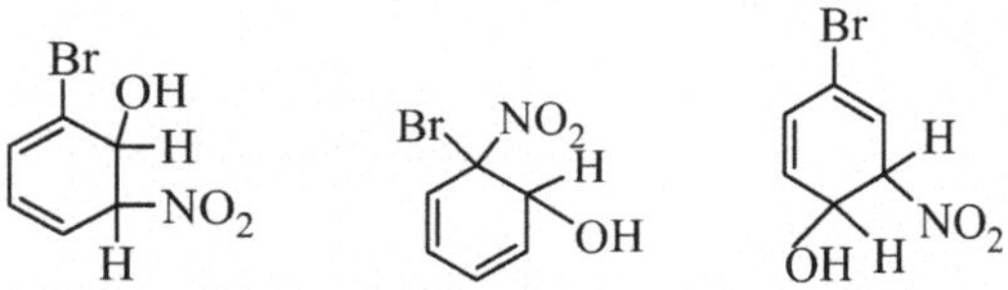

The above three intermediates then generate nitro compounds, phenols, and nitrophenols according to the following reaction equations:

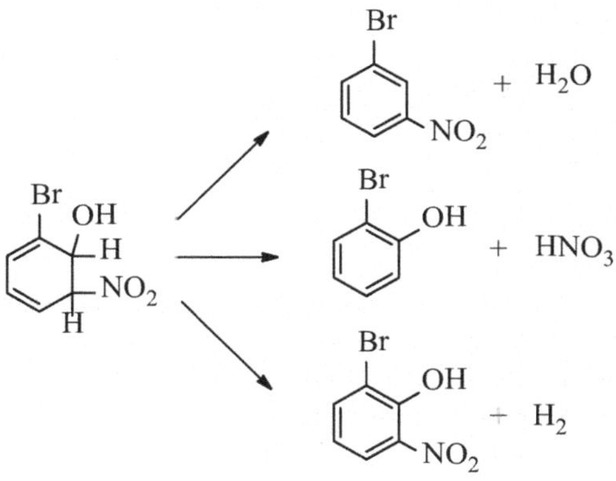

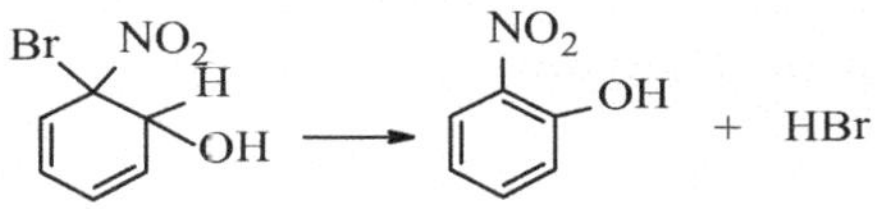

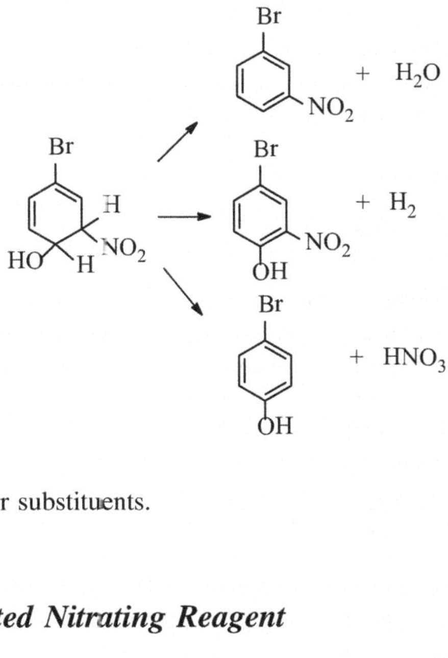

R represents other substituents.

2.5.4 *Chlorinated Nitrating Reagent*

In the case of nitration of active aromatic compound with a mixture of metal nitrate and chloroacetyl, chlorine and nitro groups are introduced into the benzene ring at the same time. If the copper nitrate and chloroacetyl mixture are used to nitrate phenol, 4,6-dichloro-2-nitrophenol is formed.

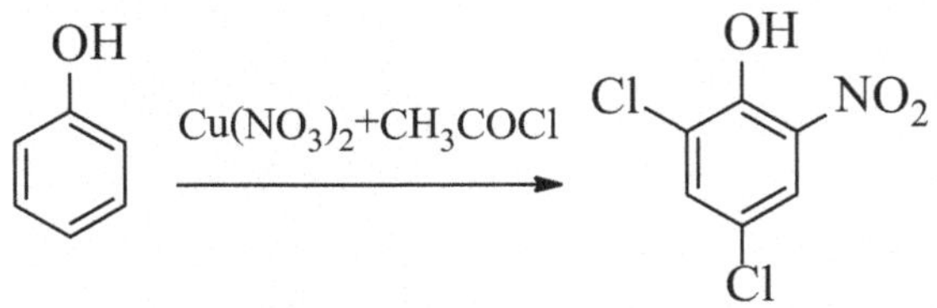

In methylene chloride, 4,6-dichloro-2-nitrophenol, and 3,6-dichloro-2-nitrophenol are produced when phenol is treated with nitryl chloride at −50 °C. If anisole is treated, a mixture of 4-chloro- and 2-chloroanisole is obtained in addition to the small amount of nitrophenol.

2.6 Green Nitrating Reagent

The nitration of various attacking reagents is known worldwide. In order to fully and effectively utilize the nitrating reagent and improve the yield of the product, the mechanism of some physical effects on the nitration reaction has been gradually studied and improved and also adopted in the industry, which is of great significance to improve the nitration environment and protect the resources.

2.6.1 Electrolytic Nitration

In some reported electrolytic nitration methods, when 52% of nitric acid is used to electrolytically nitrate nitronaphthalene, l-nitronaphthalene without dinitronaphthalene was obtained in the anode area with relatively high yield.

The green nitration technology with N_2O_5 as nitrating reagent has the characteristics of mild reaction conditions, fast reaction rate, non-acid reaction medium, good selectivity, high yield, and limited waste. This method not only can overcome various shortcomings of the traditional nitration technology, but also has limited side effects and can nitrate acid-sensitive, water-sensitive, and easy-oxidize chemicals. Nitric oxide can be generated through the electrolysis of nitric acid, which has reached the scale of industrial production in foreign countries. The excess nitrous oxide after the nitration generates nitric acid after treatment, which can be recycled and reused. Therefore, as a green nitrating reagent, the application prospect of N_2O_5 is very wide, which is recognized as the most likely green nitrating reagent to achieve industrial production. In fact, Russia, the United States, and the United Kingdom have used nitrous oxide in the industrialization of nitration. The nitration capacity of nitrogen pentoxide with some acid catalysts is very strong, and it can achieve the nitration activity of nitric acid–sulfuric acid mixture, but it can avoid the use of difficult-dealt sulfuric acid with serious pollution. Therefore, this process has strong social and environmental benefits.

The electrolytic oxidation of preparation N_2O_4–HNO_3 generates N_2O_5. In this method, nitric acid solution with dissolved N_2O_4 is filled in the anode chamber, the N_2O_5 nitric acid solution is obtained through electrolytic oxidation. After the cathode chamber is filled with concentrated nitric acid, nitric acid is reduced to N_2O_4 with current flow. When the cathodic fluid reaches a certain concentration, it enters into the anode chamber after going through the distilled water to replenish the nitric acid solution.

The nitrating agent used in the electrolytic nitration is generally nitric acid or nitric–sulfuric mixed acid. Nitric acid is concentrated in the anode region, resulting in nitration. Because of the product of this nitration method has no difference from that of the general nitration method, and there are relatively large numbers of oxidation side reactions in the anode area, the loss of raw materials is increased and product purity is decreased. Therefore, it has no practical value.

2.6.2 Radiation Nitration

The work in this field involves irradiating the reaction mixture with a certain amount of radiation (including ultraviolet, X-ray, γ-ray, fast electron beam, etc.) in liquid- or vapor-phase nitration. Due to the different experimental conditions of the various researchers, the results are inconsistent.

To make a chemical system react, usually it is needed to supply a certain amount of energy to initialize the reaction, and heat is the most commonly used energy. But, you can also consider the use of other energy to achieve a special purpose. In other forms of energy, radiant energy is one. In the chlorination of toluene with chlorine, if the energy is supplied in the form of ultraviolet light, the chlorination occurs on the side chain selectively. When the radiation passes through the object, the energy is transferred to the medium, and thus the reaction can be promoted to start sometimes.

In general, the mechanism of nitration under high-intensity radiation may be different from that of general nitration. The former often has the characteristics of free radical reaction only with theoretical significance, and this nitration method has no practical value to present.

(1) Nitration with the effect of ultraviolet irradiation

Effect of ultraviolet irradiation (no wavelength range) on the nitration of aromatic compound with N_2O_4 has been studied. Foote [61] studied the reaction between 70% nitric acid and nitrobenzene, which was irradiated with ultraviolet light (wavelength of 3660 angstroms produced with a medium pressure mercury arc lamp of 550 W and filtered) with nitrogen purging through the reaction solution to take away the nitrogen oxides. If the reaction mixture was placed in the dark, no reaction occurs. With UV irradiation, the reaction product is dinitrobenzene and picric acid, but the reproducibility

Table 2.20 Nitration results of toluene under microwave effects

Reaction conditions t/min		W (Isomer of nitration product)			Δ	Yield/%
		o-	m-	p-		
a.t.	30	0	0	0		0
a.t.	60	47.6	2.1	50.3	1.06	93
a.t.	90	46.9	2.5	50.6	1.08	100
MW	4	44.1	2.2	53.7	1.22	54.7
MW	6	43.0	2.1	54.9	1.28	89.0
MW	8	39.7	1.8	58.5	1.47	100

of the experiment is poor. The reproducibility of the experiment was improved after urea, potassium sulfamate, etc. were added to remove the nitrous acid. When the nitrogen oxides were present in the reaction mixture, the amount of m-dinitrobenzene was increased. However, in all cases, the increased ratios of para-isomer in dinitrobenzene were different from that of general nitration.

Nitrobenzene is placed in $D_2SO_4 - D_2O$, the hydrogen isotope exchange does not happen. Therefore, this nitration is not an electrophilic substitution mechanism, but may be a free radical reaction mechanism.

(2) Nitration under X-ray

The nitrate of potassium, sodium, barium, or copper is suspended in benzene, and then irradiated with X-rays with stirring. The yield of nitrobenzene reaches to 95%, and the value of G (the number of reaction molecules with every 100 eV of radiation) is up to 39, but the yield is significantly decreased without stirring.

(3) Nitration with γ-radiation

Generally, Co-60 is used as the radiation source to study the effect of γ-ray on nitration under various conditions.

Benzene is placed in aqueous calcium nitrate solution and irradiated with γ-radiation, the yield of nitrobenzene reached to 45%, and a certain amount of picric acid was also produced.

Sugioto [62] studied the nitration of benzene with nitric acid in the presence of γ-radiation, in which the concentration of benzene in the solution was 0.016–0.022 mol/L, the nitric acid concentration was 1.7–0.23 N, the radiation dose rate was $2.9 \times 10^3 - 5.6 \times 10^4$ dela h^{-1}, nitration temperature was 15–25 °C, and the main nitration products were p-nitrophenol with a small amount of o-nitrophenol, but the G value was not high as 0.28–0.96. The author proposed the nitration mechanism was that the dissociation of nitric acid into the free radicals was first and then it was followed by the nitration reaction as

$$HNO_3 \rightarrow HNO_3^{※} \rightarrow {}'HO\cdot + \cdot NO_2$$

(Activated nitric acid molecule)

$$C_6H_6 + \cdot OH \rightarrow C_6H_5 + H_2O$$
$$C_6H_5 \cdot + \cdot NO_2 \rightarrow \cdot C_6H_5NO_2$$

But there is no phenol in the reaction product because the phenol is nitrated to nitrophenol by nitric acid.

$$C_6H_5OH + \cdot OH \rightarrow \cdot C_6H_4OH + H_2O$$
$$C_6H_5OH + \cdot NO_2 \rightarrow O_2N - C_6H_4 - OH$$

Smenata [63] studied the effect of Co-60-produced radiation on the nitrous oxide-benzene system at 20 °C. The main product was mononitrobenzene with a small amount of dinitrobenzene at a dose of 0.5 mega dela h^{-1} radiation. The yield of the product was proportional to the total radiation dose, but the G value was low. The G of p-dinitrobenzene (DNB), that of the 1,2-dinitrobenzene (p-DNB), and that of (GN) (P-DNB) and m-dinitrobenzene (m-DNB) were 0.138, 0.01, and 0.033, respectively, when the amount of benzene was excessive. When nitrous oxide was excessive, G(GNB), G(p-DNB), and G(m-DNB) were 1.18, 0.012, and 0.076, respectively. In the product, the amount of meta-isomer was about 77%, that of para-isomer was about 23%, and that of ortho-isomer was very small, which was different from that in nitration product with mixed acids (m-:p-:o- ratio was equal to 93:0.3:64). Therefore, it is considered that the main reaction mechanism is the nitration mechanism of nitryl ions under the influence of γ-ray, but free radical substitution mechanism occurred, resulting in the change of the ratio of o- and p-isomer in dinitrobenzene. In the benzene-nitrous oxide system, free radical substitution is very much possible.

$$NO_2 + O \rightleftharpoons HNO_3$$
$$NO_3 + NO_2 \rightleftharpoons N_2O_5$$
$$N_2O_5 \rightleftharpoons NO_2{}^+ + NO_3{}^-$$

As Castorina et al. [64]. found that a product of N_2O_4 under the γ-ray radiation was N_2O_5.

Under the radiation of γ-ray, the nitration product of toluene contains at least 10 kinds of ingredients, and the main reaction is the side-chain nitration.

γ-ray radiation has no significant effect on the gas-phase nitration of ethane.

(4) Nitration under the effect of fast electrons

Under the action of fast electrons, when toluene is nitrated with the mixture of N_2O_4 and oxygen, 95% of the obtained nitrotoluene was p-nitrotoluene.

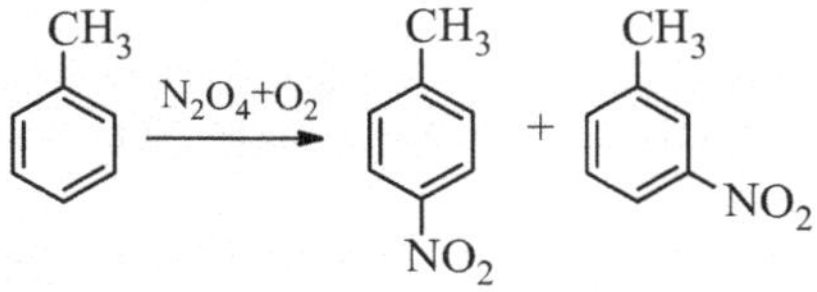

The fine powder of alumina is suspended in 1 mol of toluene to prepare a suspension containing 0.7% of alumina, and a mixture of N_2O_4 and oxygen, which is washed with nitric acid aqueous solution, and is introduced into the mixture. The gas-sucked suspension is irradiated with fast electrons at 50 °C for 2 s until the absorbed amount reaches to 1.3×10^6 rep (radiation dose unit lrep = 0.93 den = 93×10^{-7} J g^{-1}). The reaction mixture is incubated at 50 °C until the reaction is completed. After a few minutes, it is irradiated again until the total absorption reaches to 3.9×10^6 rep. When it is cooled down, crystals with 95% 4-nitrotoluene and 5% m-nitrotoluene is precipitated. The yield of 4-nitrotoluene is 70%.

2.6.3 *Microwave Nitration*

In 1986, Gedye et al. [65] found that the use of microwave heating could promote organic chemical reactions, and the reaction rate was even more than a thousand times of that with the traditional heating technology, which was a sign of the beginning of microwave organic synthesis chemistry. Microwave radiation technology is a new type of green chemistry method, which has the characteristics of quick reaction speed, simple operation, limited by-products, high yield, easy purification, and environmental friendliness in organic synthesis. Thus, it has attracted increasing attention, and gradually developed into a promising new field for MORE chemistry, namely, Microwave Promotion of Organic Chemistry or Microwave Induced Catalytic Organic Reaction Chemistry.

The use of microwave technology can overcome the shortcomings of traditional heating technology, to greatly reduce the reaction time. Microwave heating is an internal heating process, different from the traditional heating method to heat from the outside of the material to the inside, acts directly on the media molecules at the same time to heat the entire material at the same time, and is a direct heating. Therefore, the microwave has a rapid heating, uniform of heated system, high thermal efficiency, easy controlling, and high-quality products. Microwave heating will be a new technology that replaces many of the current traditional methods that consume energy, wastes time, and causes environmental pollution, and it can be said to be a forward-looking green technology that is environmentally friendly for environmentally sustainable development.

Lu et al. [66] studied the nitration of toluene under microwave action and found that the o-/p- ratio and yield had changed significantly, the reaction rate could be increased by more than 10 times, and the contents of isomers and other data were listed in Table 2.20.

The results in Table 2.20 showed that, under the action of microwave, the same yield could be achieved in a very short time, and the o-/p-ratio was obviously improved.

2.6.4 *Ultrasonic Nitration Technology*

The effect of ultrasonic wave on the chemical reaction is due to the cavitation phenomenon and the free radical reaction mechanism caused by the ultrasonic wave in the solution. The cavitation phenomenon is that the tiny bubble nuclei in the liquid are intensified under the action of ultrasonic wave, which is manifested as a series of mechanical processes, such as turbulence, growth, contraction, and collapse. In the very short period of collapse of the cavitation bubble, hot spots appear within the space of the cavitation bubble and its surrounding minimal space and result in large amount of energy. Under these extreme conditions, supercritical reaction of substances entering into the cavitation nuclear happens at the high temperature and under high pressure. The acoustic chemical reaction is the sound cavitation mechanism and the sound-causing free radical mechanism, with the combined advantages of wet catalyst and photocatalysis at supercritical point, therefore, ultrasonic catalyst is a new green and efficient chemical reaction. Ultrasonic is used in the nitration due to the reaction rate acceleration and the reaction yield improvement [67, 68].

When the RDX was synthesized from TRAT, the RDX yield is extremely low with 95 wt% HNO_3, 100% HNO_3, HNO_3/H_2SO_4 systems, and the yield of RDX was only 40% with the P_2O_5/HNO_3 system, which can be increased to 57.3% under ultrasonic conditions. The yield of HMX synthesized from DAPT under ultrasonic action can be increased from 9.6 to 66.8% [69].

References

1. Bayat Y, Mokhtari J (2011) Preparetion of 2,4,6,8,10,12-hexanitro-2,4,6,8,10,12-hexaazaisowurtzutane from 2,6,8,12-tetraacety1 2,4,6,8,12-hexaazaisowurtzutane using Various Nitrating reagent s. Def Sci J 61(2):171–173
2. Bayat Y et al (2012) Hwteropolyacids: an efficient Catalyst for Synthesis of CL-20. J Energ Mater 30:124–134
3. Poret Jay C, Shaw Anthony P et al (2013) Development and performence of the W/Sb_2O_3/krO4/Lubricant Pyrotechnis Delay in the US Army Hand-Held Sigal. J Propllants Explos Pyrotech 38:35–40
4. Glaser R, Lewis M (1999) Single- and double-proton-transfer in the aggregate between cytosine and guaninediazonium ion. Org Lett 1(2):273–276
5. Liu J (2015) Liquid explosives. Springer

6. Balasubramanian K (2007) Rovibronic symmetry and nuclear spin analysis of octanitrocubane. Chem Phys Lett 441(1–3):29–38
7. Hammett LP, Deyrup AJ (1932) J Am Chem Soc 54:2721
8. Miller RC, Noyce DS, Vermulen T (1964) Ind Eng Chem Soc 56(6):43
9. Paul MA, Long FA (1957) Chem Rev 57:1
10. Westheimer FH, Kharasch MS (1946) J Am Chem Soc 68:187
11. Deno NC, Jaruzelski JJ (1955) J Am Chem Soc 77:3044
12. Deno NC et al (1959) J Am Chem Soc 81:2344
13. Mackor EL et al (1958) Trans Farad Soc 54(66):186
14. Lee WM, Millen DJ (1956) J Chem Soc 4463
15. Klapötke, TM (2011) Chemistry of high-energy materials. DE GRUYTER
16. Ou Y (2014) Explosives. Beijing Institute of Technology Press
17. Cao XF, Li BD, Wang M (2014) An efficient method to synthesize TNAD by the nitration of 1,4,5,8-tetraazabicyclo-[4,4,0]-decane with N_2O_5 and acidic ionic liquids. Chin Chem Lett 25(3):423–426
18. Lungare S, Bowen J, Badhan Raj (2016) Development and evaluation of a novel intranasal spray for the delivery of Amantadine. J Pharm Sci 105(3):1209–1220
19. Jacob A, Sood R, Chanu KV, Bhatia S (2016) Amantadine resistance among highly pathogenic avian influenza viruses (H5N1) isolated from India. Microb Pathog 91:35–40
20. Sun RK et al (1981) Chemistry and technology of high explosive. National Defense Industry Press
21. Deno NC, Peterson HJ, Sacher E (1961) J Phys Chem 65:199
22. Mansfield LP, Billett E, Olsen E, Forsythe SJ (xxxx) Variation in salmonella core lipopolysaccharide as detected by the monoclonal antibody M105. Lett Appl Microbiol 23(2):104–106
23. Mansfield LP, Forsythe SJ Demonstration of the Rb-1 lipopolysaccharide core structure in Salmonella strains with the monoclonal antibody M105. J Med Microbiol 50(4): 339–344
24. Karlsson S, Qstmark H, Eldsater C et al (2002) Detonation and sensitivity properties of FOX-7 and formulations eontaining FOX-7. In: 12th detonation (Int.) symposium, San Diego, CA
25. Astratev AA, Dashko DV (2001) Some specific features if acid nitration of 2-substituted 4,6-dihydroxpyrimidines. Nucleiohilic Cleavage id the Nitration Products. Russ J Org Chem 37(5):729–733
26. Latypov NV, Johansson M (2007) On the synthesis of 1,1-diamino-2,2-dinotroethene (FOX-7) by nitration of 4,6-dihydroxy-2-methylpyrimidine. J Org Process Res Decelopment 11:51–59
27. Pictet A, Genequand P (1902) Ber. 35:2526
28. ManssoBa TB (1954) K, 06 m. XEM 24:1157
29. Marcus RA, Fresco JM (1967) J Chem Phys 27:564
30. Bordwell FG, Garbisch EW Jr (1960) J Am Chem Soc 82:3588
31. Thomas RJ, Anzilloti WF, Hennion GF (1940) Ind Eng Chem 32:908
32. Gunter B et al (1967) Chem Tech 19:765
33. Handbuch G der anorganische. Chemie Bd 27:94
34. Gerding H (1956) Ghent Weekblat 52:204
35. Flatt R, Benguerel F (1962) Helv China Acta 45:765
36. Ren TS (1994) Chemical and technology of nitramine and nitrate explosives. Weapon Industry Press
37. Cloetens G (1966) Explosifs 58:90–117
38. Hetherington JA, Robinson PL (1954) J Chem Soc 3512
39. Chapman RD, Gilardi RD et al (1999) Nitrolysis of a Highly Deactivated Amide by Protonitronium. Synthesis and Structure of HNFX. J Org Chem 64:960–965
40. Kuhn SJ (1961) J Am Chem Soc 83:4564
41. Olab GA, Noszko L, Pavlathe A (1957) Natare 179:146
42. Blangey L (1938) Helv Chim Acta 21:1579

43. Geick JL, Bumpus JA (2014) Characterization of non-cage structural analogs of octanitrocubane and CL-20. In: 247th national Spring meeting of the American-Chemical-Society (ACS), Dallas, TX. 637-CHED
44. Chapman RD (2002) Theodore Axenrod etal. Genimal-dinitro-1-5 diazocine derivatives. US Patent 6417355
45. Olab GA, Kuhn SJ (1956) Chem Ind 98
46. Olab GA, Kuhn SJ, Mlinko A (1956) J ChemSoc 4257
47. Olab GA, Kuhn SJ, Szelke M (1956) Ber 89:2374
48. Olab GA, Qinn HW, Kuhn SJ (1960) J Am Chem Soc 82:426
49. Olab GA, Lin HC (1971) J Am Chem Soc 93:1259
50. Bachman GB, Hokamo T (1957) J Am Chem Soc 79:4370
51. Cheng KF, Liu MH, Ho PH (2015) Comparative Theoretical Synthesis of High-Energy Density Materials 2,4,6-Trinitro-1,3,5-Triazine. Int J Quantum Chem 114(21):1457–1465
52. Yang Z, He YH (1999) Pyrolysis of Octanitrocubane via molecular dynamics simulations. Acta Phys Chim Sin 32(4):921–928
53. Lan YH, Zhai JX, Li DH, Yang RJ (2015) The influence of solution chemistry on the morphology of ammonium dinitramide crystals. J Mater Sci 50(14):4933–4939
54. Matsunaga H, Izato Y, Habu H, Miyake A (2015) Thermal decomposition characteristics of mixtures of ammonium dinitramide and copper(II) oxide. J Therm Anal Calorim 121(9):319–326
55. Soklovsky M et al (1966) Biochem. 5:3582
56. de la Mare PBD, Ridd JH (1955) Aromatic substitution, nitration and halogenation p 55
57. Shnlgin AT (1966) J Med Chem Soc 9:445
58. Altuchov KV et a1 (1969) 0pr. XJIM 5:2246
59. Wright OL, Teipel J, Thoennes D (1965) J Org Chem 30:1301
60. Ban D M, Liu J P (2011) $TiSiW_{12}O_{40}/TiO_2$ catalytic synthesis nitroglycerin. In: Proceedings of the national conference on hazardous materials and safety emergency technology, pp 75–77
61. Foote CS, Engel P, Delpeseo TW (1965) Tetrahydron Lett 2669
62. Sugimoto K et al (1963) Bl Chem Soc Jpn 36:124
63. Smetana AF, Castorina TC (1973) Explosivstoffe 21:146
64. Castorina TC, Alien AO (1965) J Phys Chem 69:3547
65. Xue YQ (2003) Modern organic synthesis methods and techniques. Chemical Industry Press, Beijing, pp 277–286
66. Lǜ CX (2011) Study progress on green nitration. Chin J Explos Propellants 34(1):165–173
67. Hawket DL, Steigman J (1954) Analy Chem Soc 26:1989
68. Tonaaea AB (1938) Hampo Banae yraeBoxopoIIoB II apyrHe 44. L Blangey. Helv Chim Acta 21:1579
69. Qian H, Ye Z, Lǜ C (2008) Ultrasonically promoted nitro ly sis of dapt to HMX in ionic liquid. J Ultrason Sonochemistry 15(4):326–329

Chapter 3
Nitration Reaction Reactor and Operation Technology

Nitration is a unit process in the organic synthesis industry and a process link in the manufacture of certain nitrogen-containing organic compounds, which plays an important role in the synthesis of explosives and nitrates. Indeed, nitration technology in the industrial application has more than a hundred years of history. As early as 1830, chemists began to investigate the nitration of glycerol. In 1847, Ascanio Sobrero [1] prepared nitroglycerin by nitration reaction using a mixed nitric and sulfuric acid at the low temperature. Later on, nitrobenzene was prepared by the nitration of benzene, followed by the reduction of nitrobenzene to aniline. For more than 100 years, the application scope of nitration has been expanded dramatically, and the number of nitrogen products has been continuously increased. At the same time, the production technology has been improved continuously. Specifically, the reaction mode has been changed from batch to continuous style processing, and from manually operated unit operation to automatic continuous operation. In the case of nitrate and nitro compound production, a continuous nitration reaction has been implemented almost in every chemical plant worldwide. In addition to the use of nitro-sulfuric acid or nitric acid as a nitrating agent, there are other widely used nitrating agents, such as mixed nitric acid and acetic anhydride, mixed nitric acid and magnesium nitrate, as well as mixed nitric acid and phosphoric acid. With an increasing variety of nitrate compounds, almost all nitrates or nitro compounds can be prepared by the following process units and operation technologies, for example, Nitroalkane or other nitro compounds by gas nitration method, nitrocellulose by U-tube method and four-chamber method, nitroglycerin and multinitrate by bubbling or injection method, aromatic hydrocarbon nitro compounds, hexanitrohexaaza-isowurtzitane (HNIW, CL-20) and adamantine nitrate by piping or other continuous nitration methods [2–4]. From the mode of operation, the nitration operation has continuous style and batch methods. According to the type of reactor, continuous nitration can be further divided into several types. According to the type of nitrating agent, the nitration reactions can be classified as nitric acid nitration, mixed nitric acid nitration, and other nitration processes. According to the physical state of reactants, the nitration reactions include liquid–solid-phase nitration, gas-phase nitration,

J. Liu, *Nitrate Esters Chemistry and Technology*,
https://doi.org/10.1007/978-981-13-6647-5_3

gas–liquid nitration, and liquid nitration. According to the aggregate state of reactants, there are homogeneous and heterogeneous nitration reactions [5].

3.1 Nitration Reactor

The nitration reactor is closely related to the chemical reaction process [6, 7] while nitration involves chemical reaction conditions, product type, manufacturing process and product quality, and other issues. For many chemical reactions, the reactor determines the process, which in turn ensures product quality and production costs. Therefore, the chemical reaction device is particularly important.

(1) Batch nitration reactor

Batch nitration reactor is a nonmechanical operation of the nitration equipment with high safety, which is one of the most widely used equipment. The nitration reactor is made of acid-resistant stainless steel. The equipment is mainly composed of main reactor, inlet, outlet, mixed acid, and waste acid, which is suitable for liquid–liquid nitration. The structure of the reactor is shown in Fig. 3.1.

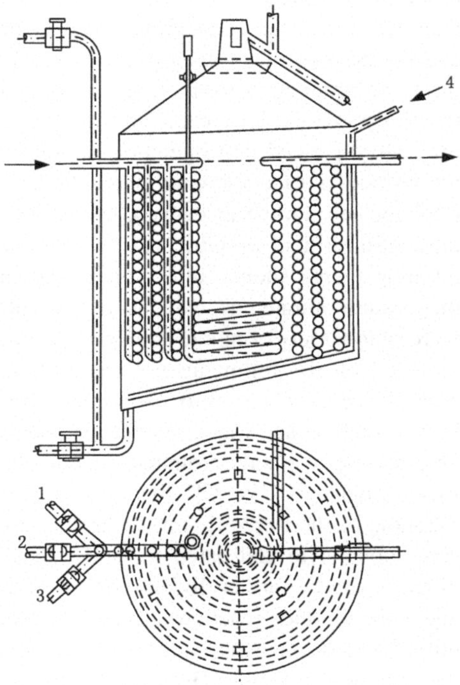

Fig. 3.1 Batch nitration reactor: 1—waste acid discharge port, 2—discharge port, 3—mixed acid replacement port, and 4—inlet

The batch nitration reactor has the following characteristics: the raw material for nitration is dispersed by compressed air without mechanical operation; heat conduction is also controlled by compressed air stirring, and cooling is regulated by a serpentine cooled brine system; temperature is controlled by an electrical contact remote control system connected to a safety switch; and nitrating mixed acid is injected in one batch, and the amount the raw material for nitration is added by the manual control.

On the other hand, the drawbacks of batch nitration reactor include a large amount of raw material, long contact time of acid and waste acid in the equipment, incompatible for rapid discharge in the event of an emergency, low production efficiency, and low conversion rate.

(2) Mechanical stirring continuous nitration reactor

The mechanical stirring continuous nitration reactor can be divided into a propeller type and a worm type. The propeller type reaction has a brine tube cooling system (Fig. 3.2) while the worm type is a multilayer coil cooling. This reactor is suitable

Fig. 3.2 Mechanical stirring continuous nitration reactor (the unit of number is mm)

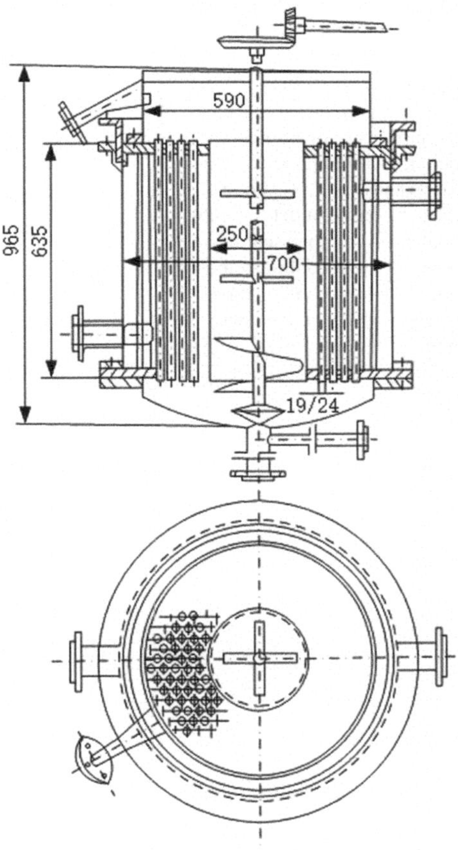

for liquid–liquid nitration processes, such as nitration of glycerol, glycerol replacement, and aromatics.

Characteristics of mechanically agitated continuous nitration reactor feature the full advantage of mechanical stirring. Therefore, glycerol is highly dispersed in the mixed acid, which leads to nitroglycerin and mixed acid to form an insensitive emulsion, thus ensuring the production safety. The continuous and quantitative feeding of glycerol and mixed acid, a continuous outflowing of product, heat transfer, and mass transfer are controlled by the automatic agitator. This type of reactor typically has a high degree of automation and large production capacity. On the other hand, it has some limitations, such as a large amount of loss, high energy consumption, and complicated equipment.

(3) Air stirring continuous nitration reactor

As shown in Fig. 3.3, air stirring continuous nitration reactor consists of stainless steel with a diameter of 300 mm, a height of 1900 mm, and heat transfer area of 7.6 m^2. There are 14 pairs of tube coolers and an external cooling jacket in the interior and exterior of the reactor, respectively. The reactor is stirred by the compressed air. The reaction material is sprayed from a tube inserted into the bottom of

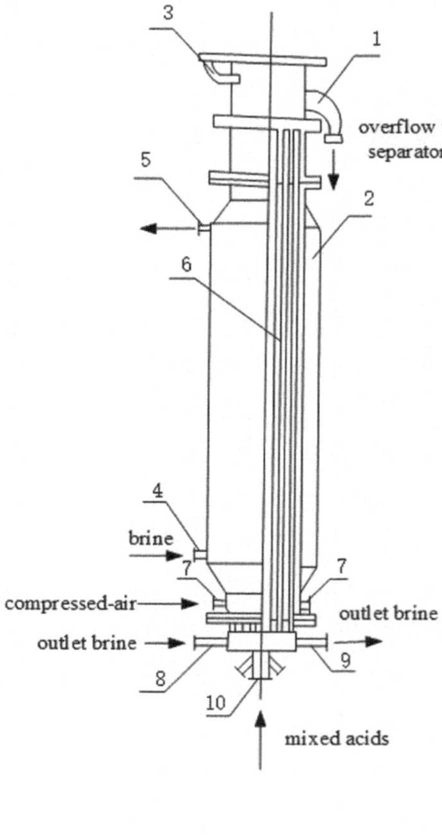

Fig. 3.3 Air stirring continuous nitration reactor: 1—nitroglycerin emulsion outlet, 2—cooling jacket, 3—exhaust port, 4—brine inlet, 5—brine outlet, 6—tube cooler, 7—compressed air inlet, 8—tube cooling brine inlet, 9—tube cooling brine outlet, and 10—mixed acid inlet

the reactor, and the mixed acid is injected from the bottom of the reactor. The mixture after the reaction is discharged and recovered from the top of the reactor.

The feeding of raw materials and discharging of products are remotely controlled in the air stirring continuous nitration reactor, in which the nitration temperature is controlled and adjusted by the feeding rate of raw materials. The reaction temperature of air stirring continuous nitration reactor is also monitored remotely for the temperature measurement. The production capacity is typically 300–400 kg/h. The characteristics of this type of reactor are as follows. The nitration equipment is simple and compact because the stirring method is the utilization of the compressed air. The amount of product loss is less than that of the mechanical stirring method. However, uneven reaction temperature control is the shortcoming of air stirring continuous nitration reactor.

(4) U-tube nitration reactor

U-tube nitration reactor is made of stainless steel (Fig. 3.4) and is suitable for the production of nitrocellulose. Each nitration reactor is composed of four to six

Fig. 3.4 U-tube nitration reactor

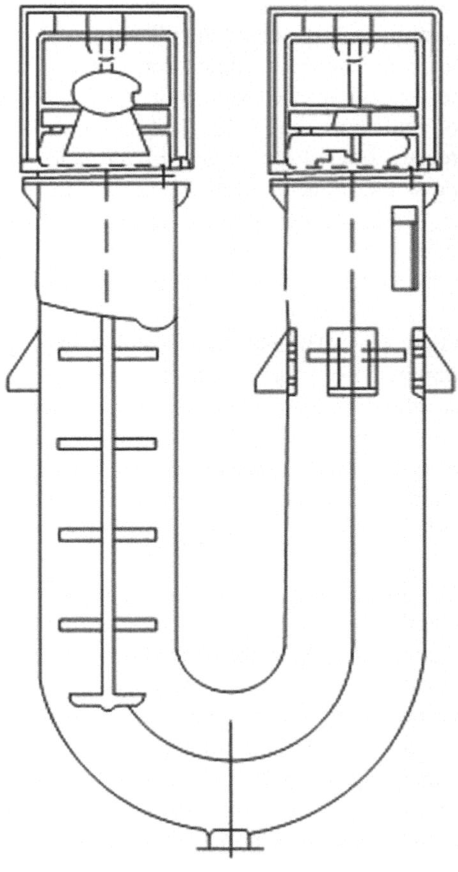

U-shaped tubes in tandem. The U-shaped tubes are connected in a tangential direction. There are two ways to connect U-tubes, namely, linear connection and S-shaped connection. Between each U-shaped tube, there is a certain height difference from front to back to facilitate the flow of cellulose. Dried and refined cellulose and mixed acid are introduced from the first U-tube upper inlet and flow along the U-shaped tubes to the discharge port on the top of the last U-tube. In the straight pipe part of each U-shaped tube, there are two paddle stirrers with opposite rotation directions. The front agitator presses down the cellulose material, and the rear agitator lifts the cellulose upward. The last agitator has a slow stirring speed to ensure a sufficient nitration reaction time.

In addition, reactors suitable for the nitration of cellulose include batch nitration reactor and multichamber nitration reactor. The low part of batch stirring nitration reactor is a cone shape, which is convenient to discharge and drain out. The cover is equipped with oval-shaped feeding port (covered) and an exhaust port (connected with suction duct). There are two paddle stirrers inside the tank. The position of these two paddle stirrers is staggered each other with 90° angle, the opposite direction of rotation, and a speed of 30–60 r/min. The blade has a certain degree of inclination to facilitate the introduction of cellulose into the mixed acid. There are prominent and welded the reflective walls on both sides of the middle part of reactor to avoid the dead angles during the stirring. The sidewall of the upper nitration reactor is equipped with an acid inlet, connected with ring acid spray tube. The end of acid spray tube is blocked, and the bottom of the spray wall to the center of the reactor has a lot of holes. Mixed acid is ejected through these holes to soak and press the refined cellulose into the mixed acid.

Multichamber nitration reactor is made of stainless steel, which is divided by three baffles into four reaction chambers. Dried cellulose and mixed acid from the upper inlet are introduced into the first chamber. The reaction mixture is flowed from the bottom of the first chamber into the second chamber, and then from the upper part of the second chamber into the third chamber. Finally, the product is pumped from the fourth chamber into a pulsating discharge centrifuge for acid separation and recovery. Each chamber is equipped with two turbo agitators and one paddle stir wings on the same axis. Each agitator shaft is individually driven by a motor and a reducer (or belt drive). Four chambers, six chambers, or eight chambers can be used for extension of nitration time.

(5) Continuous nitration reactor

Continuous nitration reactor is made of acid-resistant stainless steel and can be classified into the tank and tubular types (Fig. 3.5). Typically, the raw material and the mixed acid are introduced into the reactor simultaneously. Specifically, the tank reactor is suitable for liquid–liquid reaction, liquid–solid reaction, as well as the liquid-phase reaction generating solid product, such as nitration of cellulose and pentaerythritol, and production of trinitrotoluene. This type of reactor is also equipped with a feeding tank in the front with a manual discharge. The tubular reactor is only suitable for the nitration of liquid–liquid reaction, which employs compressed air to proceed the reaction and to discharge the products.

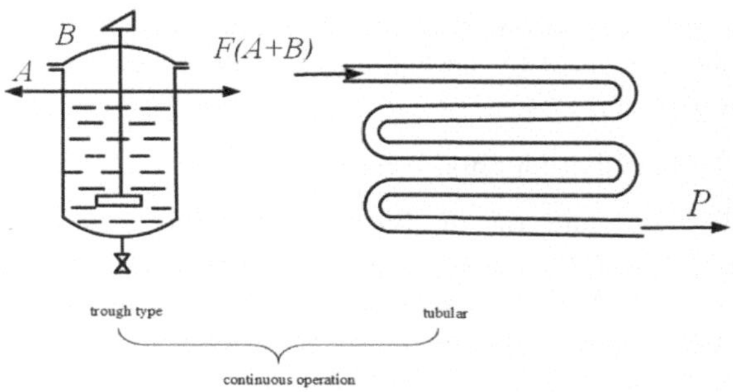

Fig. 3.5 Continuous nitration reactor

The characteristic of continuous nitration reactor is that the feeding system is controlled by the metering device without manual feeding or field control. The reaction temperature is also controlled by the remote temperature control system to ensure the safety of production. Reaction efficiency, the particle size of nitrate product, and intermittent product separation are drawbacks of this type of reactor.

(6) Bubbling nitration reactor

The bubbling nitration reactor is a conventional reactor for the preparation of nitrates, which consists of the raw material feed port, mixed acid inlet, and bubbler (Fig. 3.6). The lower part of the reactor is equipped with an electromagnetic self-discharge device.

Bubbling nitration reactor cannot achieve continuous nitration. After finishing one batch of nitration reaction and displace the reaction substrate, the second batch of nitration reaction can be conducted. The most prominent characteristic of the bubbling nitration reactor is the electromagnetic block opening valve. When the

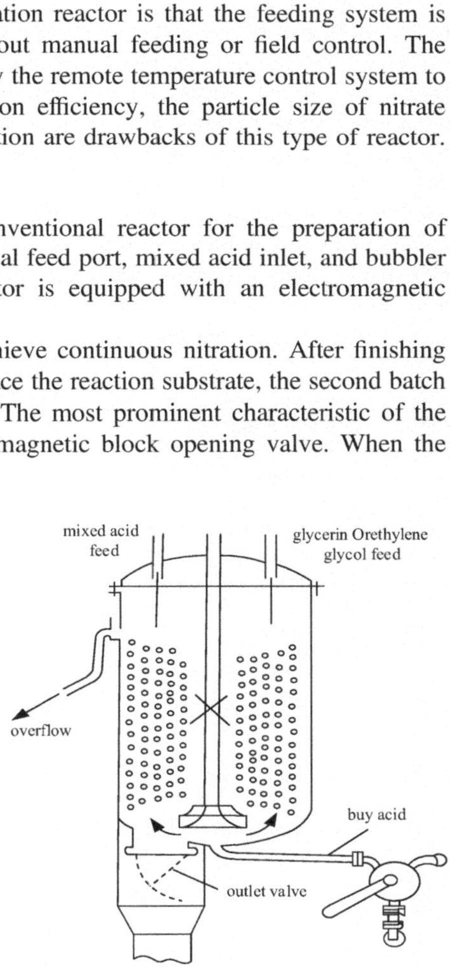

Fig. 3.6 Bubbling nitration reactor

nitration reaction temperature control fails, it can use the electromagnetic block opening valve to discharge the acidic nitrate into the water, which is a safe and reliable process.

(7) Injection nitration reactor

The structure of the injection nitration reactor (Fig. 3.7) consists mainly of nozzles, throat, diffuser, a suction chamber, and so on. The injector utilizes the working fluid (also known as working medium), that is the conversion between static pressure and kinetic energy of nitrating mixed acid flow to suck and discharge of liquid. When the nitrating mixed acid with a high-pressure flow through the nozzle, it is ejected with a high speed from the nozzle. The static pressure of nitrating mixed acid can be converted into kinetic energy, resulting in low pressure (a certain degree of vacuum) and suction of glycerol. The nitrating mixed acid and glycerol flow through the throat and mix with each other by the high turbulence. During the mixing process, the nitration reaction takes place rapidly. Subsequently, the reaction mixture flows into the expansion tube (i.e., diffuser). As the diameter of the diffusion tube gradually increases, the flow speed of reaction mixture gradually decreases. The kinetic energy is gradually converted to static pressure, thus resulting in the discharge of the mixture. The nitration reaction of glycerol is very fast and can be completed in 0.5–1 s, and the shortest reaction time is 0.2 s when glycerol is highly dispersed. Since the residence time of the mixture in the injector is about 0.5 s, the nitration reaction is basically completed after the reaction mixture is ejected out of the ejector.

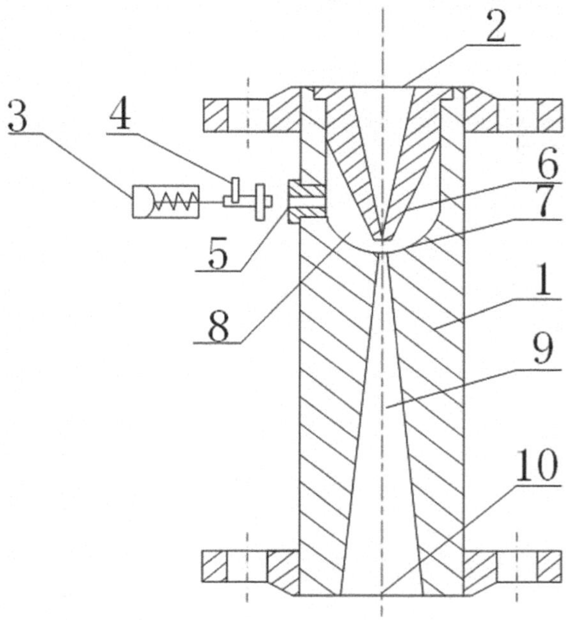

Fig. 3.7 Injection nitration reactor: 1—injector body, 2—nitrocellulose inlet, 3—safety switch, 5—air conditioning port (control vacuum), 6—nozzle, 7—throat (mixing), 8—diffusion tube, and 10—nitrate outlet

The characteristics of injection nitration are as follows. Since the external cycle of a reaction waste acid cooling is employed to absorb the reaction heat, the nitration process temperature is directly determined by the feeding ratio of glycerol to nitric acid and regulated by the use of the linear relationship between nitration coefficient and temperature, which is a simple and reliable control protocol; the reaction mixture through the injector is homogeneous, and the glycerol is highly dispersed, which are beneficial to the nitration reaction. This method fully utilizes the fast reaction of glycerin nitration reaction, thus considerably reducing the residence time and eliminating the local overheating. Therefore, injection nitration affords short residence time, high production efficiency, a dramatically reduced product loss. For example, for the nitration system with a production capacity of 800 kg/h, the product loss is only about 15 kg; because the injection nitration is relied on a high-speed flow of nitric acid to produce negative pressure to suck the raw material into the ejector, the risk caused by the destruction of nitration coefficient is significantly reduced. However, improper operation or control failure still can destroy nitration coefficient and cause an accident, so caution is required during the nitration process; other advantages of this injection process include equipment simplification, investment reduction, easy processing, and small site area.

3.2 Mass Transfer in the Nitration Process

The contact between the reactants is a necessary condition for the reaction. In particular, when the mutual solubility between the reactants is not significant, and the reaction mixture is heterogeneous, the impact of the mass transfer on the reaction is very critical.

Numerous industrial nitration reactions are two-phase reactions, which are typically two liquid phases. One phase is a saturated solution of organic matter (raw material and nitration products) in the acid phase, and another is acid (nitric acid, sulfuric acid, and water) saturated solution in the organic phase. For the sake of convenience, the former is referred to as the acid phase (or inorganic phase), and the latter is referred to as the oil phase (or organic phase). The physical process of the two-phase reaction is the added reactants to dissolve each other. After the reaction takes place in the phase or boundary layer, the reaction product is transferred out, and the new reactant is dissolved again. The reaction is continued until the reaction is complete. In the case of nitration of dinitrotoluene to make trinitrotoluene, after the dinitrotoluene is mixed with the mixed acid, a part of dinitrotoluene is dissolved in the mixed acid, and a part of the mixed acid (whose composition is different from the acid phase) is dissolved in the organic phase, thus forming two solutions, in which each solution represents a phase. Dinitrotoluene is nitrated to trinitrotoluene in the acid phase. When the amount of trinitrotoluene in the acid phase exceeds the solubility, trinitrotoluene is transferred to the organic phase. As a result of the reaction, the concentration of dinitrotoluene in the acid phase is reduced, and the dinitrotoluene in the organic phase is transferred to the acid phase. The situation in the organic phase is similar, but the reaction product is different because the oxidative side reaction predominates in the organic phase.

Hougen [8] believed that the dynamic expression of the two-phase nitration should include two reversible reactions in two phases, where the reaction rate constant is k, and the equilibrium constant is K.

$$A + B \rightleftharpoons R + S \tag{3.1}$$

The reaction rates in the phases A and b are as follows:

$$r_a = k_a\left(C_A C_B - \frac{1}{K} C_R C_S\right)_a$$

$$r_b = k_b\left(C_A C_B - \frac{1}{K} C_R C_S\right)_b$$

where, r represents reaction rate;

C represents reactant concentration;

Subscripts a and b represent two phases. A, B, R, and S represent the reactants and corresponding products.

When the physical process between the two phases reaches equilibrium, m is employed to represent the partition coefficient. The equilibrium is expressed as follows:

$$C_{Ab} = m_A C_{Aa};$$

$$C_{Bb} = m_B C_{Ba};$$

$$C_{Rb} = m_R C_{Ra};$$

$$C_{Sb} = m_S C_{Sa}$$

The volumes of the two phases are expressed as V_a and V_b, respectively. The mole number of reaction per unit time is R:

$$R = r_a V_a + r_b V_b$$

$$R = k_a\left(C_{Aa} C_{Ba} - \frac{1}{K_a} C_{Ra} C_{Sa}\right) V_a + k_b\left(C_{Ab} C_{Bb} - \frac{1}{K_b} C_{Rb} C_{Sb}\right) V_b \tag{3.2}$$

If the reaction proceeds only in one phase, one of the two terms on the right side of the above equation is equal to zero. The single phase nitration includes the nitration reactions of pentaerythritol, cellulose, aromatic compounds, and so on, in which nitration reaction takes place only in the acid phase, whereas oxidation and other side reactions occur in the organic phase. The nitration reaction of aromatic compounds has a very large equilibrium constant, which in fact can be considered as an irreversible reaction, namely, $\frac{1}{K} \cong 0$. Therefore, the above equation can be rewritten as

$$R = k_a C_{Aa} C_{Ba} V_a$$

Using the partition coefficient and assuming that the mass balance is reached, the concentration of a substance in a phase can be expressed by the concentration of the substance in the other phase:

$$R = k_a C_{Aa} \left(\frac{C_{Bb}}{m_B} \right) V_a$$

or

$$r_a = \frac{R}{V_a} = \frac{k_a}{m_B} C_{Aa} C_{Bb} \quad (3.3)$$

Strictly speaking, the concentration C in the above equations should be changed to the degree of activity a.

The distribution of nitric acid, sulfuric acid, and water in these two phases was determined by Hetherington and Masson [9]. The results are shown in Fig. 3.8.

To determine the effect of various factors on the equilibrium of distribution, Schiefferle et al. [10] investigated the solubility of p-nitrobenzene in relatively dilute mixed acid. The results are shown in Figs. 3.9, 3.10, 3.11 and 3.12.

Figure 3.9 illustrates the effect of acid phase on the solubility of benzene. Specifically, when sulfuric acid content increased from 15% to 20%, the solubility of benzene increased; when the sulfuric acid content remained constant, and the

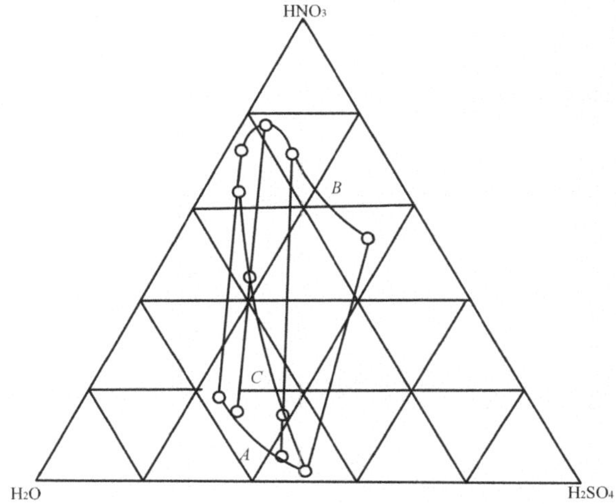

Fig. 3.8 Distribution of nitric acid, sulfuric acid, and water in two phases in the nitration of benzene [17]

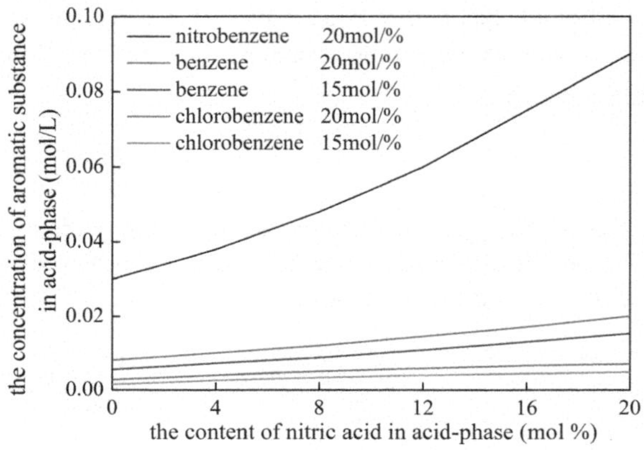

Fig. 3.9 Solubility of aromatic compounds in the mixed acid (5 °C)

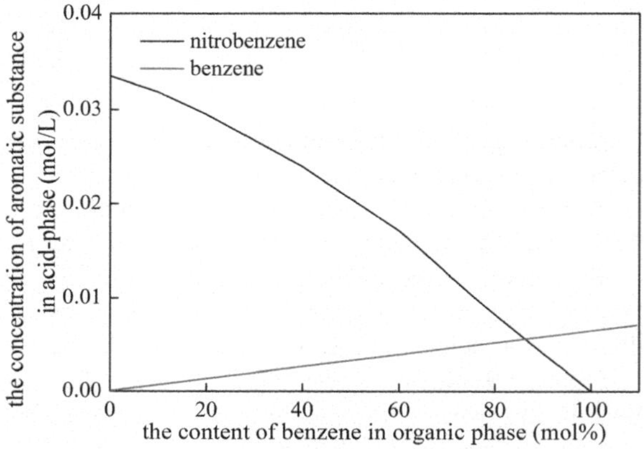

Fig. 3.10 The effect of organic phase composition on the distribution of benzene and nitrobenzene in the two phases (33 °C)

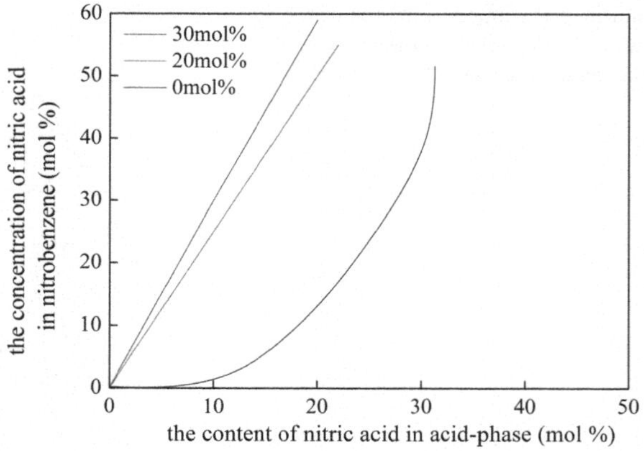

Fig. 3.11 Distribution of nitric acid in mixed acid and nitrobenzene (26 °C)

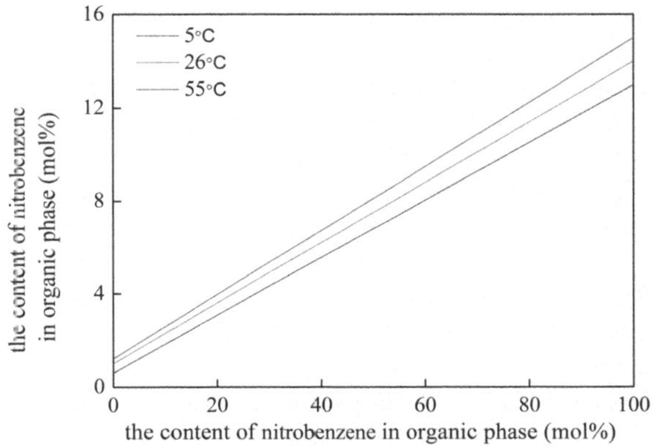

Fig. 3.12 Distribution of nitric acid between two phases (20% H_2SO_4, 10% HNO_3, mol concentration)

nitric acid content increased (water content correspondingly reduced), the solubility of benzene slightly increased. Their solubility order in the mixed acid was: Nitrobenzene > Benzene > Chlorobenzene.

Figure 3.10 shows the effect of the organic phase composition on the distribution of benzene in the organic phase and sulfuric acid aqueous solution. It is evident that benzene content in the organic phase was higher, so was benzene concentration in the acid phase.

Figure 3.11 shows the effect of acid-phase composition on the distribution of nitric acid in nitrobenzene and acid phase. The nitric acid content dissolved in the nitrobenzene was relatively high, which increased with increasing content of sulfuric acid in the acid phase.

Figure 3.12 illustrates the effect of nitrobenzene content in the organic phase on the nitric acid content in the organic phase. With the increase of nitrobenzene content, the solubility of nitric acid in the organic phase also increased.

As can be seen from Figs. 3.8, 3.9, 3.10, 3.11 and 3.12, the solubility of the aromatic compound in the acid increased slightly with increasing temperature.

(1) The concentration of sulfuric acid in the organic phase is relatively low while the concentration of nitric acid is relatively high so that only the oxidative side reaction occurs in the organic phase rather than nitration reaction. The increase in the ratio of acid phase in the reaction mixture is beneficial to the nitration reaction and is not conducive to side effects.

(2) The distribution coefficient is not a constant.

In addition to the temperature, the composition of the two phases also affects the partition coefficient. In the case of nitration of aromatic compounds, the raw material and nitration products are often miscible. The higher the content of nitration products in the organic phase, the lower the concentration of raw material in the acid phase and the higher the concentration of nitric acid in the organic phase. For aromatics and mononitroaromatics, the latter have good solubility in the acid phase. Therefore, some aromatics that can be nitrated readily do not tend to stop in the mononitration stage. As a result, the mononitrated compounds are often contaminated with a certain amount of dinitrated compounds.

According to the determination of homogeneous nitration, the nitration rate of toluene is about 20 times that of benzene under the same conditions. However, their nitration rates in industrial nitration reactions are almost the same. Except for the slight difference in the composition of mixed acid, the solubility of benzene in the mixed acid is about four times that of the toluene.

To reach an equilibrium between the two phases, it is essential to use stirring in the tank reactor. For the two-phase reaction, there are two types of factors determining the reaction rate, namely, chemical factors and physical factors. Chemical factors are determined by the reaction temperature and the property and composition of the reactants. Physical factors refer to mass transfer, which depends on the temperature, the physical and chemical properties of the two phases, and their mixing degree. For a certain reaction system, the chemical reaction rate depends on the temperature while the mass transfer rate is determined by the temperature and the stirring intensity. Since the mass transfer velocity is related to the diffusion coefficient and the boundary area of the two phases, the diffusion coefficient increases with increasing temperature. The small diameter of the droplet in the dispersed phase leads to a great contact surface of the two phases, thus promoting the mass transfer.

$$\frac{C_{Ab}}{C_{Aa}} \ll m_A \tag{3.4}$$

The reaction rate depends on the mass transfer rate, which is called the mass transfer controlled reaction.

In the case of

$$\frac{C_{Ab}}{C_{Aa}} \text{ is close to or equal to } m_A$$

The reaction rate depends on the chemical factors, which is called chemical controlled reaction. In the transitional state, the mass transfer rate and the chemical factor have a certain influence on the reaction rate [11, 12].

The mass transfer requirements for different chemical reactions are different. If the chemical reaction is very slow, the reaction rate depends on the chemical factor. For example, the nitration of dinitrotoluene into trinitrotoluene belongs to this situation. If the chemical reaction rate is very fast, it requires a high speed of mass transfer. Otherwise, it is a mass transfer controlled reaction or both chemical factor and mass transfer rate affect the reaction rate. The industrial nitration of aromatic compounds, such as toluene and xylene, belongs to this category. Many researchers have studied the effect of agitation intensity on the reaction rate of nitration of toluene with mixed acid. It was found that reaction rate increased with increasing stirring speed. When the stirring intensity exceeded a certain limit, the effect of stirring intensity on the reaction rate was less significant. Therefore, it was regarded as chemical factor controlled reaction rate and put forwarded the empirical formula for calculating the reaction rate [13].

Stirring intensity not only affects the reaction rate but also influences the heat transfer rate. Since the reactivity of the reaction mixture is related to its Reynolds number in the tank reactor, the enhanced stirring can increase the heat supply coefficient. When nitrocellulose is nitrated to form nitrate, the degree of nitration is related to the stirring speed.

Reasonable design and reliable operation of the stirring device has a great impact on the product quality and production safety. If the stirring is not uniform, the temperature and material composition in the reactor is not homogeneous, even resulting in local overheat to cause hazard [14]. In the production practices, the cessation of stirring repeatedly causes nitrate explosion because of stirring device failure or power failure. The mechanical stirrer is widely used in the tank reactor. Besides mechanical stirrer, compressed air stirrer is also frequently used. In the past, compressed air stirring has been widely used in the production of nitroglycerin. However, the equipment of compressed air stirring is complicated. In addition, a large number of compressed air will take away part of the material, thus resulting in a significant material loss.

The utilization of compressed air stirring process becomes more and more scarce, which has been replaced by ultrasonic emulsification technology. Ultrasonic emulsification can accelerate the nitration rate of m-xylene [15]. However, the energy consumption of this method is too large. The utilization of this method in industrial production is rare due to the consideration of production costs.

3.3 Heat and Heat Balance of Nitration Reactor

3.3.1 Enthalpy Change and Equilibrium Constant for Nitration

The enthalpy change of the chemical reaction can be calculated according to the law of thermodynamics.

$$aA + bB \rightarrow cC + dD$$

According to the first law of thermodynamics, the enthalpy change of the above reaction equation can be calculated according to the enthalpy of formation of the four substances A, B, C, and D.

$$\begin{aligned}\Delta H^0 &= c\Delta H_{f(c)} + d\Delta H_{f(d)} - b\Delta H_{f(b)} - a\Delta H_{f(a)} \\ &= \sum \Delta H_{f(product)} - \sum \Delta H_{f(reactant)}\end{aligned} \tag{3.5}$$

Since the enthalpy of formation in the general thermodynamic data table is the standard enthalpy of formation, the enthalpy change is calculated in the standard state.

The enthalpy change of nitration benzene under standard conditions (1 atm, 25 °C):

$$C_6H_6 + HNO_3 \rightarrow C_6H_5NO_2 + H_2O$$

$$\Delta H_f \; 48.95 \quad -173.22 \quad 22.18 \quad -285.85 \text{ kJ/mol}$$

$$\Delta H^0 = (-285.85 + 22.18) - (48.95 - 173.22) = -139.40 \text{ kJ/mol}$$

The enthalpy change of nitration of 1 mol benzene (78 g) is −139.40 kJ (exothermic), which does not include the enthalpy change caused by the change of the mixed acid composition.

If the reaction temperature is not 25 °C, the effect of the reaction temperature on the enthalpy change can be calculated by the Kirchhoff equation.

The temperature in Eq. (3.5) is differentiated to give

$$\begin{aligned}\left[\frac{\partial(\Delta H^0)}{\partial T}\right] &= \left[c\left(\frac{\partial(\Delta H_{f(c)})}{\partial T}\right)_P + d\left(\frac{\partial(\Delta H_{f(D)})}{\partial T}\right)_P\right] - \left[a\left(\frac{\partial(\Delta H_{f(A)})}{\partial T}\right)_P + b\left(\frac{\partial(\Delta H_{f(B)})}{\partial T}\right)_P\right] \\ &= cC_{P(C)} + dC_{P(D)} - aC_{P(A)} - bC_{P(B)} \\ &= \sum C_{P(product)} - \sum C_{P(reactant)} \Delta C_P\end{aligned}$$

The temperature coefficient of constant pressure reaction is equal to the difference between the heat capacities at a constant pressure of the product and the reactant.

The above equation is integrated over T to yield

$$\Delta H = \Delta H^0 + \int_{T_0}^{T} \Delta C_P dT$$

If the reactants or products undergo phase transitions in the $T_0 \sim T_1$, phase change enthalpy should be added to the above equation.

$$\Delta H = \Delta H^0 + \int_{T_0}^{T} \Delta C_P dT + \sum \Delta H_{\text{phase transition}}$$

$$\sum \Delta H_{\text{phase transition}} = \sum \left(\Delta H_{\text{phase transition}}\right)_{\text{product}} - \sum \left(\Delta H_{\text{phase transition}}\right)_{\text{reactant}}$$

Since the temperature of liquid nitration reaction is generally around 25 °C, the effect of reaction temperature on the reaction enthalpy change is often negligible in engineering calculations. However, the phase change enthalpy should be taken into account when the physical state at the reaction temperature is different from that of the standard state (i.e., a phase change occurs), especially if only the reactant or product undergoes a phase change.

The enthalpy change of the isothermal–isobaric chemical reaction occurring at a pressure different than 1 atm can be calculated by the following method.

Equation (3.5) is differentiated by P to provide

$$\left[\frac{\partial(\Delta H^0)}{\partial P}\right] = c\left(\frac{\partial(\Delta H_{f(c)})}{\partial P}\right)_T + d\left(\frac{\partial(\Delta H_{f(D)})}{\partial P}\right)_T - a\left(\frac{\partial(\Delta H_{f(A)})}{\partial P}\right)_T - b\left(\frac{\partial(\Delta H_{f(B)})}{\partial P}\right)_T$$

Since the effect of pressure on the enthalpy of formation of the solid and liquid phases is very slight, the right end of the above equation can be approximated as zero. Therefore, the effect of the pressure on the enthalpy change of the solid- or liquid-phase reactions is negligible.

If the pressure is not too high, but the temperature is high (such as alkane gas nitration at the temperature of 400–450 °C under the pressure of 0.8–1.2 MPa), the gas-phase reaction can still be regarded as an ideal gas. Pressure does not affect the enthalpy of ideal gas, so the effect of pressure on the reaction enthalpy change can also be ignored.

Due to the steric hindrance of cubane, the equilibrium constant K of nitration reaction of cubane is as great as that of aromatic compounds, which can be regarded as an irreversible reaction. The equilibrium constant under the standard conditions can be obtained from standard enthalpies of formation and the standard entropies of the reactants.

$$\Delta G^0 = \Delta H^0 - T\Delta S^0 \tag{3.6}$$

where

ΔG^0 the standard Gibbs function change of the reaction;

ΔS^0 the standard entropy change of the reaction can be obtained by the standard entropy

$$\Delta G^0 = -RT\ln K$$

The followings are the calculation of the equilibrium constant of nitration of 2,4-dinitrotoluene to 2,4,6-trinitrotoluene under the standard conditions. The reaction equation is as follows:

$$2,4\text{–}C_6H_3(NO_2)_2CH_3(s) - HNO_3(l) \rightleftharpoons 2,4,6\text{–}C_6H_2(NO_2)_3CH_3(s) + H_2O(l)$$

The standard enthalpy of formation and the standard entropy of each compound are shown in Table 3.1:

Table 3.1 ΔH_f and S^0 value of each compound in example 8

Term	Compound			
	2, 4, 6-TNT(solid)	2, 4-DNT (solid)	Nitrate (liquid)	H_2O (liquid)
ΔH_f (kJ/mol K)	−80.75	−64.43	−173.22	−285.77
S^0 (J/mol K)	285.35	243.93	155.64	69.87

$$\Delta H^0 = (-285.77 - 80.75) - (-173.22 - 64.43) = -128.87\,\text{kJ mol}^{-1}$$

$$S^0 = (69.87 + 285.32) - (243.93 + 155.64) = -71.35\,\text{J mol}^{-1}$$

$$\Delta G^0 = \Delta H^0 - T\Delta S^0 = -101.87 - 298\left(-71.35 \times 10^{-3}\right) = -88.65\,\text{kJ/mol}$$

$$\text{lnK} = -\frac{\Delta G^0}{RT} = \frac{27.64 \times 10^{-3}}{1.987 \times 298} = 32.5$$

$$K = 1.3 \times 10^{14}$$

The equilibrium constant is very large at the actual nitration temperature so that it can be considered as an irreversible reaction.

3.3.2 *Heat Balance of the Reactor*

Nitration reaction should be controlled at a certain temperature. The heat balance is reached when the nitration reactor is under stable operation. If the heat balance is broken, the reaction temperature will fluctuate [16].

When the heat balance is reached, the amount of heat input is equal to the heat output per unit time in the reactor.

The main heat inputs include:

Reaction heat Q_1 is equal to the negative value of the reaction enthalpy change.

The thermal effect Q_2 of composition change of mixed acid is equal to the negative value of the enthalpy change.

The heat input Q_3 refers to the heat brought by the nitration material, in which 0 °C is used as the reference temperature.

$$Q_3 = \sum G_1 C_{P1} t_1$$

In the equation, G_1, C_{P1}, and t_1 represent the mass of the various reactants, the specific heat capacity under the constant pressure, and the Celsius temperature, respectively.

Heat outputs include:

Q_4 refers to the heat carried away by the nitration material;
Q_5 refers to the heat loss of nitration reactor; and
Q_6 represents the heat carried away by the heat transfer medium.

$$Q_4 = \sum G_2 C_{P2} t_2$$

where G_2, C_{P2}, t_2 represent the mass of the various reactants, the specific heat capacity under the constant pressure, and the Celsius temperature of various materials leaving the nitration reactor, respectively.

The condition of heat balance is

$$Q_1 + Q_2 + Q_3 = Q_4 + Q_5 + Q_6$$

In addition to the heat mentioned above, there is heat generated by side reactions, as well as the dissolution heat of materials, etc. These types of heat are difficult to be calculated theoretically, and the calculated result Q_6 by the above method is very close to the actual. The correct nitration heat (the heat generated by nitration of a unit mass of raw material) can only be determined experimentally. For a given nitration reaction, the heat of nitration is not a constant, which varies with the composition and reaction conditions of the nitrating agent.

The heat balance is very important both in practical operation and in the design of nitration plants [12, 14, 15]. Nitration reaction is an exothermic process. In addition, nitration mixture and nitration products are mostly explosive or flammable. When the nitration temperature is out of control, the increase in temperature promotes nitration reaction rate and side reaction rate, thus dramatically increasing heat output. Simultaneously, both the reaction temperature and the reaction rate increases. If the reaction is not controlled timely, it will lead to an explosion.

For the continuous nitration reactor under the steady operation state, the operating variables do not change with time, and the amount of heat transfer per unit time also remains stable.

The operating variables of the batch nitration reactor vary with time. Even if the amount of feeding remains unchanged in a certain period of time, the heat transfer is not fixed. In the nitration reaction, the mixed acid is added to the nitration reactor first, followed by the addition of benzene under the stirring and cooling conditions. Even if the addition rate of benzene is constant, the change in the composition of the acid and the change of the thermal effect vary with the change of the acid composition. As a result, the heat transfer also changes constantly. In the initial stage of nitration reaction, the concentration of acid is relatively concentrated, and the dilution heat is relatively large. Therefore, the feeding speed of the batch operation is not fixed. In the initial stage of the reaction, the feeding speed is slow but tends to fast in the middle of reaction. Occasionally, it is appropriate to add the raw materials slowly.

3.4 Properties and Composition Design Calculation of Mixed Nitro-sulfuric Acid

The mixed nitric and sulfuric acid is one of the most widely used nitrating agents. It is very important to study and understand the properties of mixed acid for the preparation and production of nitrates [5, 17, 18].

3.4.1 The Calculation of Mixed Acid

Nitrating mixed acid is mainly composed of sulfuric acid, nitric acid, and water. The main free components of this composition include sulfur trioxide, nitrous oxide or nitrous oxide. Because of their small content, their effect on the nitration reaction can be ignored.

(1) The conversion of mass percent and mole percent

The mixed acid component can be expressed as a percentage by mass or by mole percent. In industrial nitration, mixed acid is usually expressed in mass percentage but can also be expressed in mole percent. For the convenience of use, the composition of mixed acid is shown in Table 3.2.

The conversion relation of the two representations can be expressed as

$$s = \frac{\frac{S}{98}}{\frac{S}{98} + \frac{N}{63} + \frac{W}{18}} = \frac{9S}{9S + 14N + 49W} \tag{3.7}$$

$$n = \frac{\frac{N}{63}}{\frac{S}{98} + \frac{N}{63} + \frac{W}{18}} = \frac{14N}{9S + 14N + 49W} \tag{3.8}$$

$$w = \frac{\frac{W}{18}}{\frac{S}{98} + \frac{N}{63} + \frac{W}{18}} = \frac{49W}{9S + 14N + 49W} \tag{3.9}$$

Table 3.2 The composition of mixed acid

Name	Wt%	Mol%
Sulfuric acid	S	s
Nitrate	N	n
H_2O	W	w

The mol composition of known mixed acid can be expressed as a percentage of mass

$$S = \frac{98s}{98s + 63n + 18w} \times 100\%$$

$$N = \frac{63n}{98s + 63n + 18w} \times 100\%$$

$$W = \frac{18w}{98s + 63n + 18w} \times 100\%$$

(2) The design of formula of nitrating mixed acid

According to the requirements of nitration reaction and given the starting acid and mixed acid composition, the amount of raw acid required for the preparation of a certain amount of mixed acid can be calculated by the following two methods.

1. Algebraic method

The formula needs the following raw materials:

Nitric acid, mass percentage N_1.

Sulfuric acid, mass percentage S_1.

The waste acid containing mass percentage N_2 of nitric acid and mass percentage S_2 of sulfuric acid to formula a mixed acid contains nitric acid N%, sulfuric acid S%, water W% in mass percentage.

The unit mass of mixed acid is used as a calculation basis, and the required amount of raw material nitric acid, sulfuric acid, and the amount of waste acid are set as x, y, and z kilograms, respectively. According to the material balance calculation, the following material balance equation is obtained:

$$\left.\begin{array}{r} x + y + z = 1 \\ N_1 x + N_2 z = N \\ S_1 y + S_2 z = S \end{array}\right\} \tag{3.10}$$

After solving Eq. (3.10), the equations for x, y, z material balance are obtained.

$$\left.\begin{array}{r} x = \dfrac{N(S_1 - S_2) + N_2(S - S_1)}{S_1(N_1 - N_2) - N_1 S_2} \\ y = \dfrac{S(N_1 - N_2) + S_2(N - N_2)}{S_1(N_1 - N_2) - N_1 S_2} \\ z = \dfrac{N_1(S_1 - S) - NS_1}{S_1(N_1 - N_2) - N_1 S_2} \end{array}\right\} \tag{3.11}$$

By the matrix equations,

$$y_1S_1 + y_2S_2 = (y_1 + y_2)S$$

$$D_x = \begin{vmatrix} 1 & 1 & 1 \\ N & 0 & N_2 \\ S & S_1 & S_2 \end{vmatrix}$$

$$D_y = \begin{vmatrix} 1 & 1 & 1 \\ N_1 & N & N_2 \\ 0 & S & S_2 \end{vmatrix}$$

$$D_z = \begin{vmatrix} 1 & 1 & 1 \\ N_1 & 0 & N \\ 0 & S_1 & S \end{vmatrix}$$

As a result,

$$x = \frac{D_x}{D};$$

$$y = \frac{D_y}{D};$$

$$z = \frac{D_z}{D}$$

The starting acid:

$$\begin{array}{ll} \text{Sulfuric acid } 95\% & S_1 = 0.95 \\ \text{Nitric acid } 98\% & N_1 = 0.98 \end{array}$$

Waste acid containing sulfuric acid 70%, nitric acid 2% $S_2 = 0.70$ $N_2 = 0.02$. The composition of waste acid:

$$\text{Nitric acid } 10\%\ N = 0.10$$

$$\text{Sulfuric acid } 80\%\ S = 0.80$$

$$\text{Water } 10\%\ W = 0.1$$

The material balance of each starting acid is calculated to make 5000 kg of mixed acid.

The unit mass of mixed acid is used as a calculation basis and substituting the above values into Eq. (3.11). The obtained x, y, and z are the amount of various raw materials for preparing the mixed acid.

$$
\begin{aligned}
x &= \frac{N(S_1 - S_2) + N_2(S - S_1)}{S_1(N_1 - N_2) - N_1 S_2} \\
&= \frac{0.1(0.95 - 0.70) + 0.02(0.8 - 0.95)}{0.95(0.98 - 0.02) - 0.98 \times 0.70} \\
&= 0.0973
\end{aligned}
$$

$$
\begin{aligned}
y &= \frac{S(N_1 - N_2) + S_2(N - N_2)}{S_1(N_1 - N_2) - N_1 S_2} \\
&= \frac{0.8(0.98 - 0.02) + 0.70(0.1 - 0.98)}{0.95(0.98 - 0.02) - 0.98 \times 0.70} \\
&= 0.6726
\end{aligned}
$$

$$
\begin{aligned}
z &= \frac{N_1(S_1 - S) - N_1 S_1}{S_1(N_1 - N_2) - N_1 S_2} \\
&= \frac{0.98(0.95 - 0.8) - 0.1 \times 0.95}{0.95(0.98 - 0.02) - 0.98 \times 0.70} \\
&= 0.2301
\end{aligned}
$$

By solving the matrix equations, the results are obtained similarly.

$$
D = \begin{vmatrix} 1 & 1 & 1 \\ 0.98 & 0 & 0.02 \\ 0 & 0.95 & 0.7 \end{vmatrix} = 0.226
$$

$$
D_x = \begin{vmatrix} 1 & 1 & 1 \\ 0.1 & 0 & 0.02 \\ 0.8 & 0.95 & 0.7 \end{vmatrix} = 0.022
$$

$$
D_y = \begin{vmatrix} 1 & 1 & 1 \\ 0.98 & 0.1 & 0.02 \\ 0 & 0.8 & 0.7 \end{vmatrix} = 0.152
$$

$$
D_z = \begin{vmatrix} 1 & 1 & 1 \\ 0.98 & 0 & 0.1 \\ 0 & 0.95 & 0.8 \end{vmatrix} = 0.052
$$

$$
x = \frac{D_x}{D} = \frac{0.022}{0.226} = 0.0973\,\text{kg}
$$

$$
y = \frac{D_y}{D} = \frac{0.152}{0.226} = 0.6726\,\text{kg}
$$

$$z = \frac{D_z}{D} = \frac{0.052}{0.226} = 0.2301\,\text{kg}$$

From the calculation of Eq. (3.11), the material balance parameters of the various acids are shown in Table 3.3 when preparing 5000 kg of mixed acid.

Taking into account the nitric acid loss by evaporation and mechanical loss during the preparation, the actual amount of raw materials used in the reaction is slightly more than the calculated amount.

When using fuming sulfuric acid for the preparation of mixed acid, the amount of free sulfur trioxide per hundred grams of fuming sulfuric acid is regarded as g grams, then:

$$S_1 = 100 + 0.225\,g$$

If fuming sulfuric acid contains 20% sulfur dioxide, then g = 20:

$$S_1 = 100 + 0.225 \times 20 = 104.5\%$$

The calculation steps are the same as those described above.

When the process requires the use of two different concentrations of nitric acid (or sulfuric acid) to prepare the third acid, calculation by the crossing method is more convenient.

$$\text{S1} = 100 + 0.225\,g$$

Such as sulfur dioxide containing 20% of fuming sulfuric acid, then g = 20:

$$\text{S1} = 100 + 0.225 \times 20 = 104.5\%$$

When the process requires the use of two different concentrations of nitric acid (or sulfuric acid) to prepare the third acid, calculation by the cross method is more convenient.

The sulphuric acid of G kg (equal to y_1+y_2) is prepared using Y_1 kg sulfuric acid at a concentration of S_1 and Y_2 kg at a concentration of S_2, and the material balance equation can be expressed as

$$y_1S_1 + y_2S_2 = (y_1 + y_2)S$$

Table 3.3 Material balance parameters when preparing 5000 kg of mixed acid

The name or content of the raw material	Match quality (kg)
98% Nitric acid	0.0973 × 5000 = 486.5
95% Nitric acid	0.6726 × 5000 = 3363
Waste acid	0.2301 × 5000 = 1150.5
Total	5000

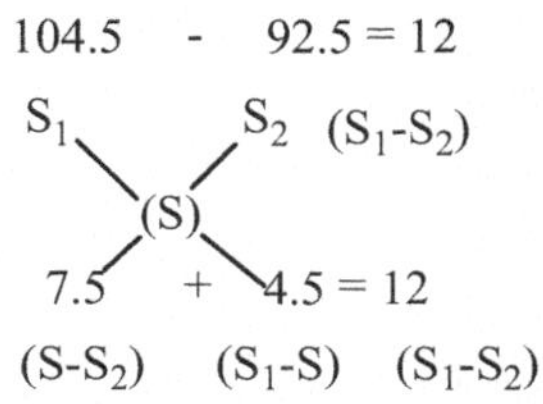

Fig. 3.13 The diagram of crossing method for calculation of mixed acid

The equations are sorted to give the following expression:

$$\frac{y_1}{y_1+y_2}=\frac{S-S_2}{S_1-S_2}\quad y_1=(y_1+y_2)\frac{S-S_2}{S_1-S_2}=G\frac{S-S_2}{S_1-S_2}$$

$$\frac{y_2}{y_1+y_2}=\frac{S_1-S}{S_1-S_2}\quad y_2=(y_1+y_2)\frac{S_1-S}{S_1-S_2}=G\frac{S_1-S}{S_1-S_2}$$

In the preparation of nitrating mixed acid, if the use of 104.5% sulfuric acid and 92.5% sulfuric acid to prepare 100 kg of 100% sulfuric acid, the crossing method can be used to calculate the content of each component quickly. The calculation method is shown in Fig. 3.13.

The amount of S_1 and S_2

$$y_1=G\frac{S-S_2}{S_1-S_2}=100\times\frac{7.5}{12}=62.5\,\text{kg}$$

$$y_2=G\frac{S_1-S}{S_1-S_2}=100\times\frac{4.5}{12}=37.5\,\text{kg}$$

2. Graphic method

The graphic method is the use of triangular diagram to represent the composition of mixed acid, which can be used to calculate the formula of mixed acid.

There are two mixtures R and S consisting of three substances A, B, and C, and the third mixture M is formulated with the two mixtures. The composition of point M must be on the line of R and S. The G_R and G_S corresponding to relative amounts of R and S, respectively, should be subjected to the following relationship:

$$\frac{G_R}{G_S}=\frac{RM}{MS} \tag{3.12}$$

which is equal to the ratio of line segments RM and MS. This relationship can be confirmed by the following method:

W_{AR}, W_{AS}, and W_{AM} represent the mass percentages of substance A in the ternary mixtures R, S, and M, respectively.

W_{CR}, W_{CS}, and W_{CM} represent the mass percentages of substance C in the ternary mixtures R, S, and M, respectively.

GR, GS, and GM are the masses of the ternary mixtures R, S, and M, respectively.

Three material balance equations are listed as follows:

$$\left.\begin{aligned} G_M &= G_R + G_S \\ G_M W_{AM} &= G_R W_{AR} + G_S W_{AS} \\ G_M W_{CM} &= G_R W_{CR} + G_S W_{CS} \end{aligned}\right\} \tag{3.13}$$

After solving these equations, the results are as follows:

$$\frac{G_S}{G_R} = \frac{W_{AR} - W_{AM}}{W_{AM} - W_{AS}}$$

$$\frac{G_S}{G_R} = \frac{W_{CR} - W_{CM}}{W_{CM} - W_{CS}}$$

As a result,

$$\frac{W_{AR} - W_{AM}}{W_{AM} - W_{AS}} = \frac{W_{CR} - W_{CM}}{W_{CM} - W_{CS}} \tag{3.14}$$

In Eq. (3.14), when the M point is on the RS line, the triangles RPM and MQS are similar, so the above proportion formula is validated. If the M point is not on the RS line but located at the position of M′, the triangle RP′M′ is not similar to M′Q′S′, and the above scale cannot be satisfied. Therefore, M must be on the RS line. The compositional relationship between the two ternary mixtures is obtained by Eq. (3.14), as shown in Fig. 3.14.

As can be seen from Fig. 3.14,

$$\frac{RM}{MS} = \frac{W_{AR} - W_{AM}}{W_{AM} - W_{AS}} = \frac{W_{CR} - W_{CM}}{W_{CM} - W_{CS}} \tag{3.15}$$

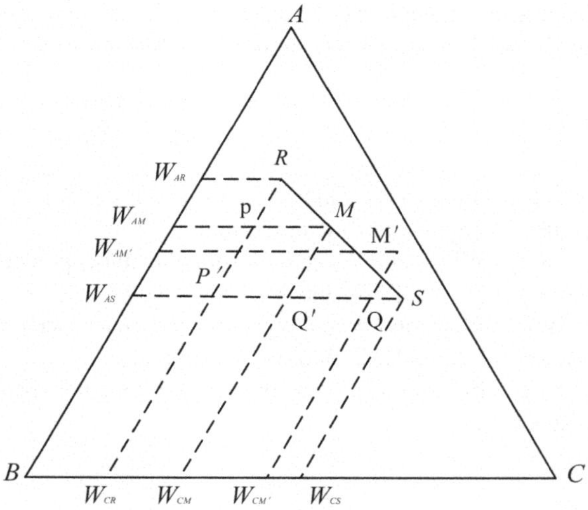

Fig. 3.14 Compositional relationship of two ternary mixtures when mixed

The following result can be obtained

$$\frac{G_S}{G_R} = \frac{RM}{MS}$$

Therefore,

$$G_S = \frac{RM}{RS} G_M$$

$$G_R = \frac{SM}{RS} G_M$$

When preparing 1000 kg of mixed acid containing 89% nitric acid, 7.5% sulfuric acid, and 3.5% water by using 92.5% sulfuric acid and 98% nitric acid, a triangle diagram is drawn for the above raw materials, as shown in Fig. 3.15.

In Fig. 3.15, A, B, and O represent sulfuric acid, nitric acid, and to be prepared mixed acid, respectively.

$$W_M = 1000\,\text{kg}$$

$$G_A = 1000\frac{OB}{AB} = 81.8\,\text{kg}$$
$$G_B = 1000\frac{AO}{AB} = 918.2\,\text{kg}$$

The same method can be used to prove that the fourth mixture M that prepared from the three mixtures of A, B, and C must be located inside the triangle ABC, as shown in Fig. 3.16.

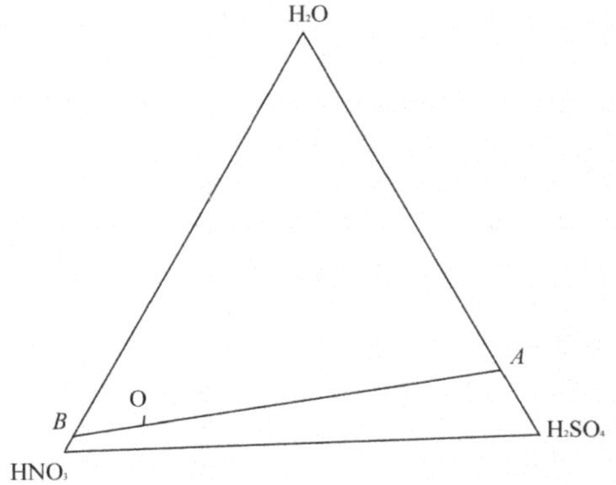

Fig. 3.15 Calculation diagram for the mixed acid [17]

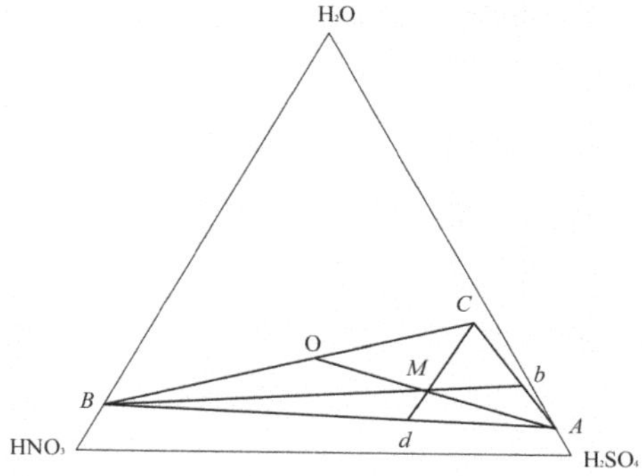

Fig. 3.16 Diagram for the calculation of three different mixed acids by using graphics method [17]

In Fig. 3.16, the amount of A acid = the amount of mixed acid $\times \frac{MO}{AO}$; the amount of B acid = mixed acid amount $\times \frac{Mb}{Bb}$; and the amount of C acid = mixed acid amount $\times \frac{Md}{Cd}$.

As the length of the line cannot be measured accurately, the results obtained by the graphical method are not very accurate. However, the advantages of the graphic method are simple and clear. It can be quickly judged whether the raw materials used can prepare the desired mixed acid.

3. The expression of mixed acid nitration capability

The mixed acid has various components. An efficient method for the expression and comparison of the nitration capacities of different mixed acids is a topic needs an urgent solution.

As a nitrating agent, the mixed nitric and sulfuric acid is mainly used for the nitration reactions with electrophilic substitution mechanism. According to this reaction mechanism, the major and the most active reagent is nitronium ion in the nitration reaction. For the sake of convenience, most scholars use the concentration of nitronium ion in the mixed acid to denote the nitration capacity of the mixed acid, which is considered to be accurate but not all true because some of the nitrates are beyond the spectroscopic detection limit. If the nitronium ion of mixed acid is beyond the spectroscopic detection limit, the concentration of nitronium ion cannot be used to compare its nitration ability.

Miller et al. [19] calculated the H_0 of mixed acid based on the H_0 of sulfuric acid and nitric acid. Subsequently, according to the H_0 of mixed acid and the nitric acid content of mixed acid, the nitronium ion of mixed acid is obtained from the measured nitration rate of meta-bis-(trichloromethyl)benzene. However, when the water content in the mixed acid is high, the calculated concentration of nitronium ion is very low, which is difficult to be confirmed by the experimental method.

From the perspective of the production process, the content of the mixed acid is changing in the batch process, in which the content of nitric acid is gradually consumed to generate water. Therefore, the nitration capacity of mixed acid is declining. At the end of nitration reaction, the mixed acid becomes waste acid, and

Table 3.4 The composition of waste acid

Name	Content (%)
H_2SO_4	72
HNO_3	1
N_2O_3	2
H_2O	24
Nitro compound	1

the nitration capacity is reduced to the minimum. Similarly, the nitration capacity of waste acid is also the lowest in the continuous nitration process. For this reason, the nitration capacity of the waste acid is very important, which should be able to ensure the completion of nitration reaction.

(1) Dehydrating value of sulfuric acid (DVS), which refers to the mass concentration ratio of sulfuric acid to water in the waste acid.

$$\mathrm{DVS} = \frac{S_f}{W_f}$$

where S_f and W_f represent the mass percentages of sulfuric acid and water in the waste acid, respectively. The content of waste acid is shown in Table 3.4.

The dehydrating value of sulfuric acid:

$$\mathrm{DVS} = \frac{72}{24} = 3$$

The method for calculation of dehydrating value of sulfuric acid by using the composition of waste acid:

$$\mathrm{DVS} = \frac{S}{W'} = \frac{S}{W + W_r}$$

where

S and W	represent the mass percentages of sulfuric acid and water in the waste acid, respectively;
W'	refers to the mass of water per 100 parts of mixed acid after the nitration; and
W_r	refers to the mass of generated water per 100 parts of mixed acid after the nitration.

If all the nitric acid in the mixed acid is consumed in the nitration reaction, the result is as follows:

$$W_r = \frac{72}{24}N = 0.286N$$

where

N refers to the mass percentage of nitric acid in the mixed acid.

If the excessive equivalent of nitric acid is $\rho\%$, the equation would be as follows:

$$W_r = \frac{18N}{63(1 + 0.01\rho)} = \frac{0.286N}{1 + 0.01\rho}$$

If benzene is nitrated with the mixed acid having the following composition and with $\rho = 0$:

H_2SO_4	60%
HNO_3	32%
H_2O	8%

After the calculation,

$$W_r = 0.286N = 0.286 \times 32 = 9.14\,\text{kg}$$

$$\text{DVS} = \frac{S}{W + W_r} = \frac{60}{8 + 9.14} = 3.50$$

If the excessive equivalent of nitric acid is 5%,
The result is the following:

$$W_r = \frac{0.286N}{1 + 0.01\rho} = \frac{0.286 \times 32}{1 + 5\%} = 8.72\,\text{kg}$$

$$\text{DVS} = \frac{60}{8 + 8.72} = 3.59$$

(2) Effective concentration of sulfuric acid (Φ value)

In the nitration study, the Russian chemists usually use the Φ value to represent the nitration capacity of the mixed acid, which can be expressed as a percentage of sulfuric acid to the sum of sulfuric acid and water in the waste acid.

$$\Phi = \frac{S_f}{S_f + W_f} \times 100\%$$

Taking into account the reaction between nitrous oxide and sulfuric acid in the mixed acid:

$$2H_2SO_4 + N_2O_2 = 2HNSO_5 + H_2O$$

$$\text{Molecular weight} \quad 2 \times 98 \quad 76 \quad 2 \times 127 \quad 18$$

The mass per unit of nitrogen dioxide consumed by sulfuric acid and the amount of generated water are

$$\text{The amount of consumed sulfuric acid} = \frac{2 \times 98}{76} = 2.58\,\text{parts}$$

$$\text{The amount of generated water} = \frac{18}{76} = 0.24\,\text{parts}$$

$$\text{The amount of generated } HNO_3 = \frac{2 \times 127}{76} = 3.34\,\text{parts}$$

$$\Phi = \frac{S_f - 2.58[N_2O_3]_f}{S_f - 2.58[N_2O_3]_f + 0.24[N_2O_3]_f + W_f} \times 100\% = \frac{S_f - 2.58[N_2O_3]_f}{S_f - 2.34[N_2O_3]_f + W_f} \times 100\%$$

where $[N_2O_3]_f$ is the mass percentage of nitrous oxide in the waste acid.

If the composition of waste acid is as follows: H_2SO_4 72%′; HNO_3 1%; N_2O_3 2%; H_2O 24%; and the content of nitrate is 1%, then the Φ value is

$$\Phi = \frac{S_f - 2.58[N_2O_3]_f}{S_f - 2.34[N_2O_3]_f + W_f} \times 100\% = \frac{72 - 2.58 \times 2}{72 - 2.34 \times 2 + 24} \times 100\% = 73.2\%$$

The method of estimating the value of Φ from the mixed acid composition:

When the amount of nitric acid is zero without consideration of the presence of nitrous oxide

$$\Phi = \frac{S}{S + W + W_r} \times 100\% = \frac{S}{100 - N + W_r} \times 100\%$$

When the amount of nitric acid is $\rho\%$, the concentration of nitrous oxide $[N_2O_3]$ in the mixed acid is

$$\Phi = \frac{S - 2.58[N_2O_3]}{100 - 3.34[N_2O_3]_f - N + 0.286(\frac{N}{1 + 0.01\rho})} \times 100\%$$

As the nitration reaction accompanied by the oxidative side reaction and the decomposition of nitric acid, the amount of nitrous oxide in the process of nitration increases gradually. Therefore, the calculated median value of waste acid is more accurate.

The Φ and ρ values required for the nitration of certain aromatic compounds are listed in Table 3.5. The values listed in the table vary with the nitration conditions.

Since the effect of nitrous oxide is considered when calculating the Φ value, it is more reasonable to use the median value to evaluate the nitration capability of waste acid for nitration reaction in which the catalysis of nitrous acid does not positively catalyze the nitration reaction.

The above calculation of the effect of nitrogen dioxide on the Φ value can only be applied to mixed acid. If the water content is high, nitrosyl sulfuric acid will be hydrolyzed. The addition of nitrosyl sulfuric acid to this mixed acid will lead to an increase of nitration rate of toluene [20].

(4) The choice of mixed acid composition

Reasonable choice of mixed acid composition significantly impacts the economy of the production process. Even if the values of Φ and ρ are known, the composition of waste acid cannot be determined because various mixed acids have the same Φ value. In theory, there are an infinite number of mixed acids to have the same Φ value. Figure 3.17 is the mixed acid triangle diagram regarding mole percent.

Figure 3.17 is utilized to analyze the nitration process, in which the mole fraction of sulfuric acid in the mixed acid remains constant in the nitration process. The consumption of one mole of nitric acid generates one mole of water, and the total moles of the acid are unchanged. Mixed acid components move along a straight line parallel to the HNO_3–H_2O axis towards the direction of increasing the molar percentage of water. If the acid composition is A, the composition of acid shifts to the direction of A′ along the line of AA′ without the consideration of side reactions. In Fig. 3.17, lines AA′, BB′, and CC′ represent three different with lines with same Φ values that expressed in mole percent. Therefore, the fixed Φ value does not mean the composition of mixed acid is determined.

In general, the required Φ value ρ for the nitration of a particular compound is known. Alternatively, the composition range of the mixed acid can be determined experimentally. The mixed acid within this composition range can afford a better yield and product purity. Finally, a reasonable mixed acid composition can be selected from the investigated composition range.

In the design of mixed acid composition, the choice of mixed acid components should also consider the following factors: ① to make full use of nitric acid; ② fully exert the role of sulfuric acid; ③ the composition of the mixed acid components should be in the range that starting acids can provide; and ④ the mixed acid and waste acid are less corrosive to the metal equipment.

Table 3.5 Φ and ρ values in the nitration of aromatic compounds

Nitration reaction	Φ (%)	ρ (%)	Nitration reaction	Φ (%)	ρ (%)
Benzene → Nitrobenzene	70	3–5	Naphthalene → Mononitronaphthalene	61	3–5
Mononitrotoluene → Dinitrotoluene	88	10–15	Mononitromethylnaphthalene → Dinitromethylnaphthalene	72	30–40
Toluene → Mononitrotoluene	70	3–5	Dinitroxylene → Trinitroxylene	90	50–60
Dinitrotoluene → Trinitrotoluene three-segment method	93	80–100	Mononitrotoluene → Dinitrotoluene	82	10–15
Two-segment method	87	70–80	Chlorobenzene → Mononitrochlorobenzene	81.5	–
Xylene → Dinitroxylene	72	10–15	Mononitrochlorobenzene → Dinitrochlorobenzene	88.2	–

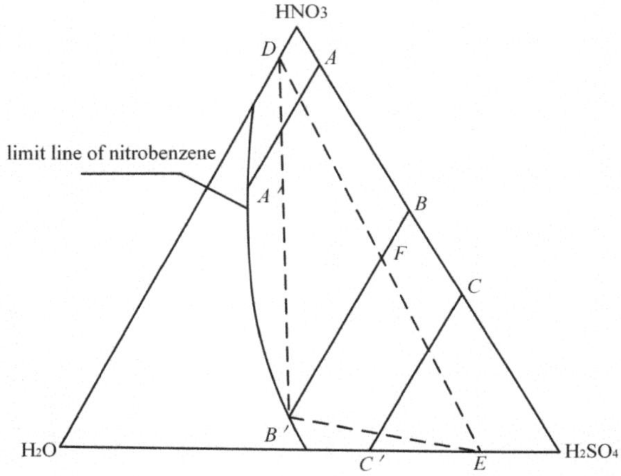

Fig. 3.17 The change of mixed acid composition in the process of nitration

The principle of mixed acid design is further illustrated in Fig. 3.17. In the case of preparation of nitrobenzene by nitration, all the mixed acids in the right of the limit line can nitrate benzene into dinitrobenzene. If the mixed acids above the AA′ line are selected, the composition of waste acid is A′, which contains a large number of nonreactive nitric acid. As a result, nitric acid is not fully used. If the mixed acids above the CC′ line are used, the nitric acid is completely consumed and converted into waste acid C′, but the role of sulfuric acid is not fully exerted. Since C′ is located to the right of the limit line, the addition of nitric acid to C′ can still nitrate benzene. Therefore, the mixed acid compositions above BB′ line are more reasonable, which not only can make full use of nitric acid but also can fully exert the role of sulfuric acid. However, there are various mixed acid compositions above BB′ line. The selection of mixed acid composition depends on the starting acid and other factors. If the mixed acid composition is close to B′, the mixed acid composition does not change significantly after the nitration reaction. As a result, a large amount of mixed acid is required. When the mixed acid composition is B′, the required amount of mixed acid is infinite in theory, which is definitely very uneconomical and impossible. When the mixed acid composition is C′, the amount of mixed acid is the minimum, but it is anhydrous mixed acid, which cannot be formulated with starting acid (concentrated sulfuric acid) and D (concentrated nitric acid). The mixed acid composition made of acids E and D is above the ED line, the mixed acid at the intersection of ED and BB′ is utilized to prepare nitrobenzene to generate waste acid composition B′. Therefore, if E, D, and B′ three acids are used to prepare the mixed raw acid, the acid composition should be on the FB′ line. When the composition of mixed acid is closer to F, the prepared mixed acid per mass unit contains more E and D acids, which can nitrate more benzene per unit mass of mixed acid. In contrast, the mixed acid composition is closer to B′, the less amount of E and D acids are used in the preparation of the mixed acid, and more amount of B′ acid is used. As a result, less amount of benzene can be nitrated per unit mass of mixed acid.

3.4.2 The Property of Mixed Acid

(1) Specific heat capacity

The correlation between the specific heat capacity of the mixed acid and its composition at 20 °C is shown in Figs. 3.18 and 3.19. The mixed acid with a composition of H_2SO_4 40%, HNO_3 40%, and H_2O 20% at room temperature has the following specific heat capacity.

As can be seen from Fig. 3.18, the total acidity of the mixed acid is 40 + 40 = 80; the content of nitric acid in the total acidity is $\frac{40}{40+40} \times 100\% = 50\%$.

In Fig. 3.19,starting from a point on the abscissa axis with a total acidity of 80, a vertical line is drawn upward and intersects with a curve named as "50". The ordinate of the intersection is the specific heat capacity of the mixed acid, which is equal to 2.125 J/g k^{-1}. The specific heat capacity of mixed acid is determined to be 2.125 J/g k^{-1} by using Fig. 3.19 directly.

(2) The relative enthalpy of the mixed acid

Any changes in the composition of the acid will have a thermal effect, such as dilution of sulfuric acid or nitric acid with water, mixing of sulfuric acid with nitric acid, and the mixing of two different concentrations of acid. Since the composition of mixed acid is constantly changing in the nitration process, the calculation of the thermal effect associated with variation in the mixed acid composition is an important part to assess the heat balance of nitration reactor.

There are several commonly used methods to calculate the thermal effect associated with the change in the acid composition, such as dilution enthalpy method [21, 22], the heat transfer method [23], the enthalpy change method [8], and the relative enthalpy of the acid method [24, 25].

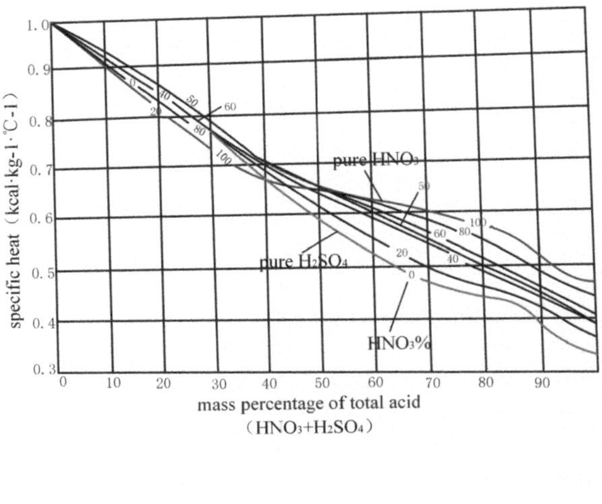

Fig. 3.18 The specific heat capacity of mixed acid at 20 °C (w%) [24]

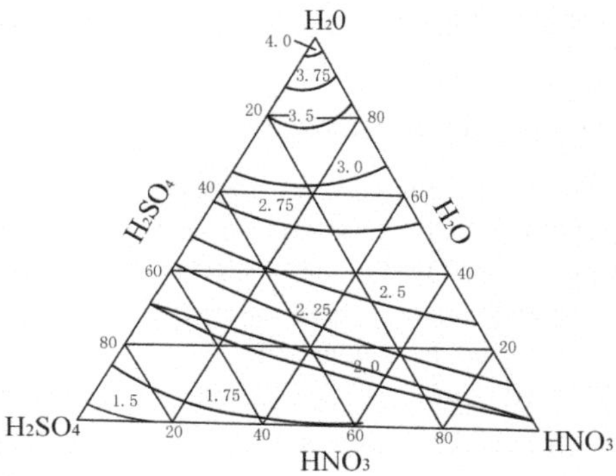

Fig. 3.19 The specific heat capacity of mixed acid at 20 °C (mixed acid composition in mol%, specific heat capacity J/g k $^{-1}$) [36]

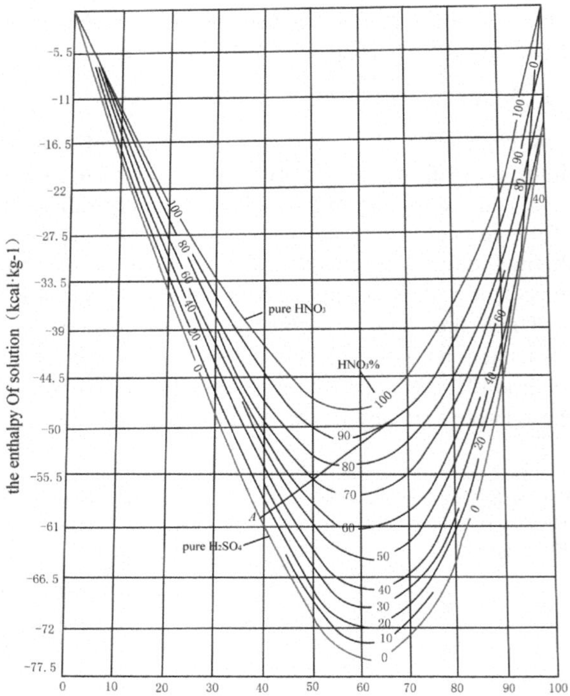

Fig. 3.20 The mass percentage of total mixed acid

If the enthalpy of a pure substance (water, sulfuric acid or nitric acid) at 0 °C and 1 atmospheric pressure is zero, the relative enthalpies of the mixture at various concentrations are shown in Fig. 3.20. The thermal effect associated with the change of acid composition at any temperatures can be calculated by using Figs. 3.18, 3.19, and 3.20 (not much difference from room temperature without phase change).

Table 3.6 The energy parameters vary with the change of mixed acid composition

Mass	G_N	G_S	G_W'	G_M	g
Relative enthalpy	H_N	H_S	0	H_M	J/g
Total enthalpy	$G_N H_N$	$G_S H_S$	0	$G_M H_M$	J

For example, when the composition of a mixed acid is A with a composition of HNO_3 30%, H_2SO_4 20%, and H_2O 50%, how to calculate the relative enthalpy of the mixed acid at 32 °C?

From the previous theory, the total amount of acid in the mixed acid is 20 + 30 = 50; the total nitric acid content in the acid is $\frac{30}{50} \times 100\% = 60\%$. The mass percentage of total mixed acid is shown in Fig. 3.20.

In the abscissa axis of Fig. 3.20, starting from a point corresponding to the total amount of acid value of 50, a vertical line is drawn upward to intersect with the curve representing anhydrous mixture with 60% nitric acid. The ordinate axis of the intersection point is the relative enthalpy of mixed acid A at 0 °C, which is equal to –246 J/g. In Fig. 3.18, the specific heat capacity of the mixed acid A is determined to be 2.76 J/g k^{-1}, and the relative enthalpy of the mixed acid A at 32 °C is

$$H_{32} = -246 + (32 - 0) \times 2.76 = -157.68\ J/g$$

The principle of the thermal effect associated with the change in the acid composition is shown in Fig. 3.20. It was assumed that the temperatures before and after the change in the acid composition are 0 °C, the enthalpy change of mixing G_N kg nitric acid with concentration N_1, G_S kg of sulfur acid with a concentration of S_1, and G'_W kg of water to prepare G_M kg of mixed acid is shown in Table 3.6.

$$N_1 + S_1 + \text{Water} \rightarrow M$$

As a result:

$$G_M = G_N + G_S + G_W$$

$$\begin{aligned} \Delta H &= \boldsymbol{G_M H_M - G_N H_N - G_S H_S} \\ &= \text{enthalpy of the product acid} - \text{enthalpy of the starting acid} \end{aligned} \tag{3.16}$$

The above calculation method is suitable for calculating the thermal effect of changes in acid composition during the nitration reaction (Fig. 3.21). If the mixed acid M with a relative enthalpy G_{Mg} of HM J/g is used for the nitration reaction, it becomes the waste acid W with an enthalpy G_{Wg} of H_W J/g after the nitration.

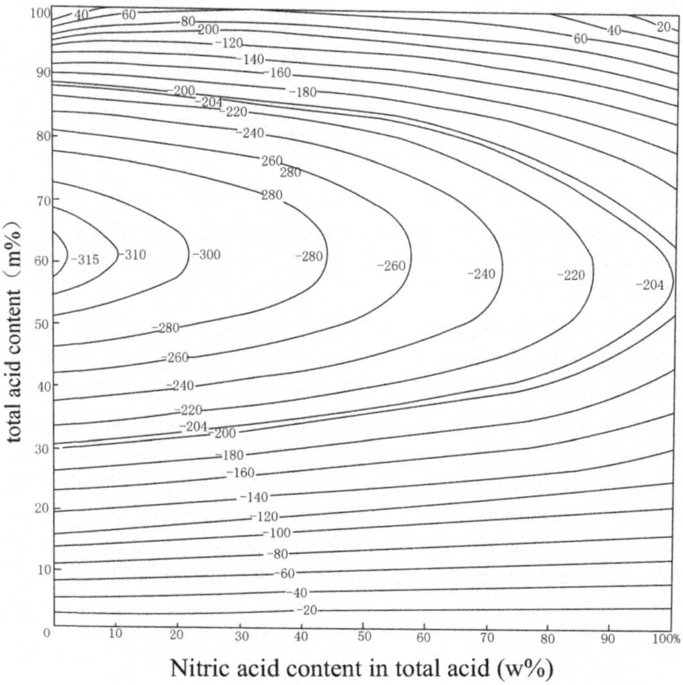

Fig. 3.21 Relative enthalpy of H_2SO_4–HNO_3–H_2O system (0 °C, 760 mm Hg) [17]

That is,

$$\begin{array}{lcc} & M \xrightarrow{\Delta H} & W \\ \text{Mass} & G_M & G_W \\ \text{Enthalpy} & H_M & H_W \end{array}$$

According to the first law of thermodynamics, the thermal effect of this process is equal to the sum of the thermal effects of the following two processes:

(1) The separation of pure nitric acid G'_N from G_M g mixed acid M affords the waste acid W_1 with G_{W_1} of g. The thermal effect of this process is ΔH_1

$$\begin{array}{lcccc} & M \xrightarrow{\Delta H_1} & W_1 + & HNO_3 & \\ \text{Mass} & G_M & G_{W_1} & G'_N & \text{g} \\ \text{Enthalpy} & H_M & H_{W_1} & 0 & \text{J/g} \end{array}$$

$$G'_N + G_{W_1} = G_M$$

$$\Delta H_1 = G_{W_1}H_{W_1} - G_M H_M$$

(2) G'_Ng nitric acid generates G'_Wg water via nitration reaction, which is mixed with waste acid W_1 to form G_{Wg} waste acid W.

$$W_1 + H_2O \xrightarrow{\Delta H_2} M$$

Mass	G_{W_1}	G'_W	G_W	g
Enthalpy	H_{W_1}	0	H_W	J/g

$$G_W = G'_W + G_{W_1}$$

$$\Delta H_2 = G_W H_W - G_{W_1}H_{W_1}$$

$$\Delta H = \Delta H_1 + \Delta H_2 = G_{W_1}H_{W_1} - G_W H_W + G_W H_W - G_{W_1}H_{W_1} = G_W H_W - G_M H_M$$
$$= \text{enthalpy of product acid} - \text{enthalpy of starting acid}$$

In the process of continuous nitration, the starting acid is constantly added to the nitration reactor, and waste acid is generated simultaneously. The waste acid is going through the following changes during the nitration process:

$$N_1 + S_1 \xrightarrow{\Delta H} W$$

Mass	G_N	G_S	G_W	g
Enthalpy	H_N	H_S	H_W	J/g

From this process, we can see that the conversion of the acids mentioned above and mixed acid into waste acid consists of two processes. The thermal effect is equal to the sum of the two processes.

$$\Delta H = G_M H_M - G_N H_N - G_S H_S + G_W H_W - G_M H_M = G_W H_W - G_N H_N - G_S H_S$$
$$= \text{enthalpy of product acid} - \text{enthalpy of starting acid}$$

If the temperatures before and after the process are not 0 °C, the effect of temperature in Fig. 3.19 on the calculation using the above equation should be considered.

Table 3.7 The calculation of the relative enthalpy in the preparation process

Acid	Specific heat capacity($J\ g^{-1}\ k^{-1}$)	Enthalpy (J/g)		
		0 °C	24 °C	35 °C
Mixed acid	1.966	229.28	–	−160.67
70% HNO_3	2.552	−187.86	−126.78	–
95% H_2SO	1.506	−102.09	−66.11	–
H_2O	4.184	0	+100.42	–

At the temperature of 24 °C, 100 kg of mixed acid with the following composition is prepared with 30 kg of 70% HNO_3, 65.3 kg of 95% sulfuric acid, and 4.7 kg of H_2O.

$$\begin{array}{ll} HNO_3 & 21\% \\ H_2SO_4 & 62\% \\ H_2O & 17\% \end{array}$$

The temperature of prepared mixed acid is 35 °C. The thermal effect of the preparation process is calculated.

The specific heat capacity and the relative enthalpy of the mixed acid at 0 °C are first determined from Figs. 3.18, 3.20 and 3.21, and then the relative enthalpy at the corresponding temperature is calculated. The results are shown in Table 3.7.

The relative enthalpy of the preparation process is calculated.

The total enthalpy of the raw materials at 24 °C is

$$\begin{array}{ll} HNO_3 & -126.78 \times 30 = -3803.26\,\text{kJ} \\ H_2SO_4 & -66.11 \times 65.3 = -4316.63\,\text{kJ} \\ \text{Water} & 100.42 \times 4.7 = 471.96\,\text{kJ} \\ \text{Total} & -7702.41\,\text{kJ} \end{array}$$

The relative enthalpy of 100 kg mixed acid at 35 °C:

$$-160.67 \times 100 = -16067\,\text{kJ}$$

The thermal effect is

$$-16067 - (-7702.41) = -8364.59\,\text{kJ}$$

or

$$83.65\ \text{J/g}$$

Table 3.8 The mass and composition of the waste acid

Name	Mass (kg)	Content (%)
Nitric acid	2.5	70.8
Sulfuric acid	1.44	12.5
H_2O	0.59	16.7
Total	3.53	100

5000 g of glycerol is nitrated with 5000 g of mixed acid with the following composition:

$$H_2SO_4 \quad 50\%$$
$$HNO_3 \quad 50\%$$

When the acid temperatures before and after the nitration are both 15 °C, the thermal effect of mixed acid composition change in the process of nitration 1000 g glycerol can be calculated by the following method.

According to the preceding statement, the waste acid and the composition are determined first.

The chemical reaction is

$$C_3H_5(OH)_3 + 3HNO_3 \rightarrow C_3H_5(ONO_2)_3 + 3H_2O$$

$$\text{Molecular weight} \quad 92 \quad 3 \times 63 \quad 3 \times 18$$

$$\text{The consumed amount of } HNO_3 \ \frac{3 \times 63}{92} = 2.06\,\text{kg/kg Glycerol}$$

$$\text{The generated amount of water } \frac{3 \times 18}{92} = 0.59\,\text{kg/kg Glycerol}$$

The mass and composition of the waste acid are shown in Table 3.8.

The enthalpy at 15 °C is calculated using Figs. 3.18 and 3.21, and the results are shown in Table 3.9.

The thermal effect of the change in acid composition: −729.69 + 188.28 = −541.4 kJ/kg Glycerol^{-1} (or −541.4 J/g glycerol^{-1}).

When the temperatures of the starting acid and the waste acid are still 15 °C, and the mixed acid in the calculation is changed to 2.5 kg of pure sulfuric acid, and 2.5 kg of pure nitric acid, the thermal effect of the acid composition change in the process of nitration of 1 kg glycerol is shown in Table 3.10.

The thermal effect of acid composition change: −729.69 − (52.72 + 65.69)= −848.10 kJ/kg glycerol^{-1}.

Table 3.9 The enthalpy for the nitration of glycerin 1000 g

Acid	Specific heat capacity (kJ kg^{-1} k^{-}1)	Enthalpy (kJ kg^{-1})		Total enthalpy (kJ)
		0 °C	15 °C	
Mixed acid	0.4	−62.76	−37.66	−188.28
Waste acid	0.44	−56	−49.4	−729.69

Table 3.10 Thermal effect changes in Table 3.9 caused by changing the composition of mixed acid

Acid	Specific heat capacity (kJ kg^{-1} k^{-1})	Enthalpy (kJ kg^{-1})		Total enthalpy (kJ)
		0 °C	15 °C	
H_2SO_4	1.41	–	21.09	52.72
HNO_3	1.75	–	28.12	65.69
Waste acid	1.84	−234.30	−206.69	−729.69

According to the above calculation, the micro-changes in the mixed acid composition in Tables 3.9 and 3.10 show that the thermal effect of preparation of 5000 g mixed acid from 2500 g sulfuric acid and 2500 g nitric acid is as follows:

$$-188.28-(52.72+65.69)=-306.69\,\text{kJ}$$

It is very coincidental that the above results are combined with the thermal effects of Table 3.9 for −541.4 kJ/kg glycerol^{-1}, which is exactly equal to 848.10 kJ/kg glycerol^{-1} and identical to the thermal effect in Table 3.10.

If the fuming sulfuric acid is used in the process of preparation of waste acid and nitration, the enthalpy change of dilution of fuming sulfuric acid to 100% sulfuric acid can be added by the results in Figs. 3.20 and 3.21, as shown in Table 3.11.

1000 g of sulfuric acid containing 20 g sulfur trioxide is diluted to 2000 g of diluted sulfuric acid at the temperature of 15 °C. The thermal effect of this process can be calculated as follows.

Table 3.11 The enthalpy changes of diluting fuming sulfuric acid into 100% sulfuric acid

T (°C)	Free sulfur trioxide content in fuming sulfuric acid (%)						
	3.2	10	20	30	40	50	60
	kJ/mol^{-1}						
15	−88.70	−90.79	−93.72	−97.07	−100.42	−108.78	−119.24
20	−89.54	−92.05	−94.98	−98.32	−102.09	−110.46	−121.34
40	−93.72	−96.23	−99.58	−103.76	−107.95	−117.57	−130.12
60	−97.91	−100.42	−104.60	−109.20	−113.80	−124.68	−140.58
80	−101.67	−105.02	−109.62	−114.64	−119.66	−126.36	−134.72
100	−105.86	−109.20	−114.22	−119.66	−125.52	−132.21	−138.49
	kJ/kgH_2O^{-1}						
15	−4937.12	−5041.72	−5188.11	−5397.36	−5606.56	−6024.96	−6610.72
20	−4978.96	−5104.48	−5271.84	−5439.20	−5732.08	−6108.64	−6736.24
40	−5188.16	−5355.52	−5522.88	−5773.92	−6024.96	−6527.04	−7238.32
60	−5439.20	−5564.72	−5815.76	−6066.80	−6359.68	−6945.44	−7531.20
80	−5648.40	−5857.60	−6108.64	−6359.68	−6652.56	−7238.32	−7824.08
100	−5899.44	−6066.80	−6359.68	−6652.56	−6987.28	−7565.80	−8158.80

Sulfur trioxide reacts with water to form sulfuric acid

$$SO_3 + H_2O \rightarrow H_2SO_4$$

$$\text{Molecular weight} \quad 80 \quad 18$$
$$200 \quad \chi$$

$$\chi = \frac{200}{80} \times 18 = 45\text{g}$$

It can be seen from Table 3.11 that the enthalpy of dilution of 20% fuming sulfuric acid to 100% sulfuric acid is −5188.16 J/g.

Therefore, the enthalpy change of dilution of 1000 g 20% fuming sulfuric acid to 1045 g pure sulfuric acid is

$$-5188.16 \times 0.045 = -233.5\,\text{kJ}$$

The concentration of diluted sulfuric acid:

$$\frac{104.5\%}{2} = 52.25\%$$

The enthalpy change of the reaction of sulfur trioxide with water and dilution to H_2SO_4 is shown in Table 3.12.

The enthalpy change of dilution of 1045 g pure sulfuric acid into 2000 g 52.25% sulfuric acid:

$$-529 - 21 - 60 = -610\,\text{kJ}$$

The total enthalpy change:

$$\Delta H = -610 - 233.5 = -843.5\,\text{kJ}$$

The above calculation ignores the effect of specific heat capacity on enthalpy change when sulfuric acid is diluted into pure sulfuric acid.

Table 3.12 The enthalpy change of the reaction of sulfur trioxide with water to form sulfuric acid

Name	Specific heat capacity (J/g k^{-1})	Enthalpy (J/g 1)		Total enthalpy (kJ)
		0 °C	15 °C	
100% H_2SO_4	5.86	0	87.86	87.86
52.25% H_2SO_4	10.46	−1263.57	−1106.67	−2213.34
H_2O	17.49	0	262.34	251.04

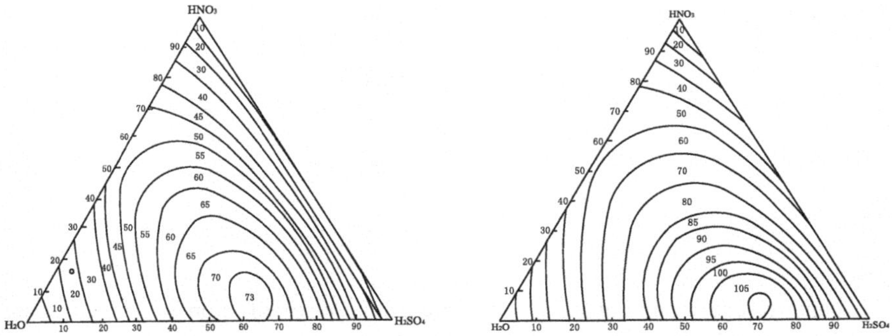

a, the mixing pure sulfuric acid with pure nitric acid and water
b, the mixing 20% fuming sulfuric acid with pure nitric acid and water

Fig. 3.22 Enthalpy change in the preparation of mixed acid (−ΔHJ/g mixed acid) and calculation of the thermal effect of acid composition changes

(3) Mixed enthalpy change method

McDavid measured the enthalpy change of mixing pure sulfuric acid with pure nitric acid and water or mixing 20% fuming sulfuric acid with pure nitric acid and water. The results are shown in Fig. 3.22.

The enthalpy change is calculated when 96% sulfuric acid and 65% nitric acid are used to prepare the mixed acid with the following composition.

H_2SO_4	64%
HNO_3	21.7%
H_2O	14.3%

The preparation of 100 g of mixed acid needs:

96% H_2SO_4	66.7 g
65% HNO_3	33.3 g

As can be seen from Fig. 3.22:

The enthalpy change of mixing pure sulfuric acid, pure nitric acid, and water to the mixed acid is −238.49 J g mixed acid^{-1}.

The enthalpy change of mixing pure sulfuric acid and water into 96% sulfuric acid is −83.68 J g 96% sulfuric acid^{-1}.

The enthalpy change of mixing pure nitric acid and water into 65% nitric acid is −198.74 J g 65% nitric acid^{-1}.

Therefore,

$$\Delta H = -238.49 \times 100 - [-83.68 \times 66.7 - 198.74 \times 33.3] = -11649.50\,\text{J}$$

If 20% fuming sulfuric acid and 90% nitric acid are used for the preparation of 100 g mixed acid containing 58% sulfuric acid, 40% nitric acid, and 2% water, it needs the following amount of starting acids:

20% fuming sulfuric acid 55.5 g
90% HNO_3 44.5 g

As can be seen from Fig. 3.22:

The enthalpy change in the preparation of the mixed acid with the above composition from 20% fuming sulfuric acid, water, and nitric acid is −205 J g mixed acid^{-1}.

The enthalpy change of mixing 100% nitric acid with water to prepare 90% nitric acid is −97.49 J g 90% nitric acid^{-1}.

Therefore,

$$\Delta H = 100 \times (-205) - 44.5 \times (-97.49) = -20500 + 4338.31 = -16161.69\,\text{J}$$

The effect of temperature is not taken into account in the above calculations, but the temperature has a significant effect on the enthalpy change of the process. As a result, using Fig. 3.22 will lead to an error.

(4) Dilution enthalpy change method

The thermal effect of the acid composition change can also be calculated by dilution enthalpy change.

ΔHN, ΔHS, and ΔHM represent the infinite dilution of enthalpy changes of nitric acid, sulfuric acid, and mixed acid, respectively. Their enthalpy changes can be calculated using the following empirical equation:

$$\Delta H_N = -111 + \frac{312W}{W + 98.5}$$

$$\Delta H_S = -183 + \frac{324W}{W + 49}$$

$$\Delta H_M = \frac{-\Delta H_N \Delta H_S}{\Delta H_S - \Delta H \Delta\chi + \Delta H_N \Delta\chi} \quad \chi = \frac{S}{S + N} \tag{3.17}$$

where S, N, and W represent the mass percent of sulfuric acid, nitric acid, and water in the acid, respectively. Figure 3.23 is a nomogram of Eq. (3.17).

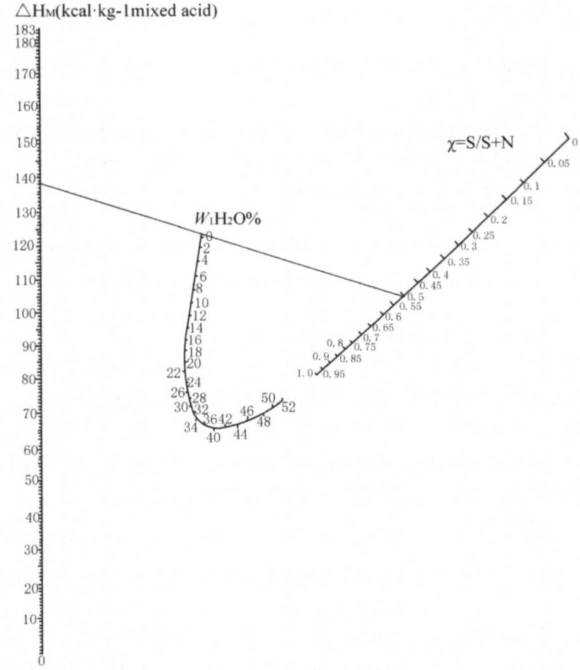

Fig. 3.23 Dilution enthalpy change of mixed acid [17]

The thermal effect of mixing acid and acid composition changes during the nitration process can be calculated by using the formula (3.17) or Fig. 3.23. However, since reference conditions of these equations are not provided, it is inconvenient to use.

(5) Other properties

1. Viscosity. Hackel et al. [26] measured the viscosity of the mixed acid, and the results are shown in Fig. 3.24.
2. Specific gravity (or relative density).

Figure 3.25 shows the relative density of the mixed acid at room temperature with water as the reference substance.

The relative density of the mixed acid is calculated by experimental method [27]:

$$D_4^{15} = bS + a \tag{3.18}$$

where

S is the mass percentage of sulfuric acid in the mixed acid;

a and b are constants, and their relationship with the nitric acid content in the mixed acid is shown in Table 3.13.

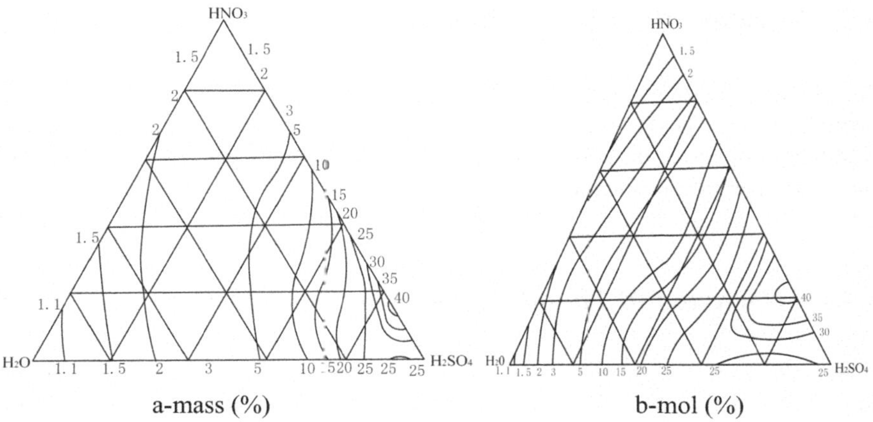

Fig. 3.24 The viscosity of the mixed acid (centipoise) [17]

Fig. 3.25 The density of the mixed acid (mass%) [17]

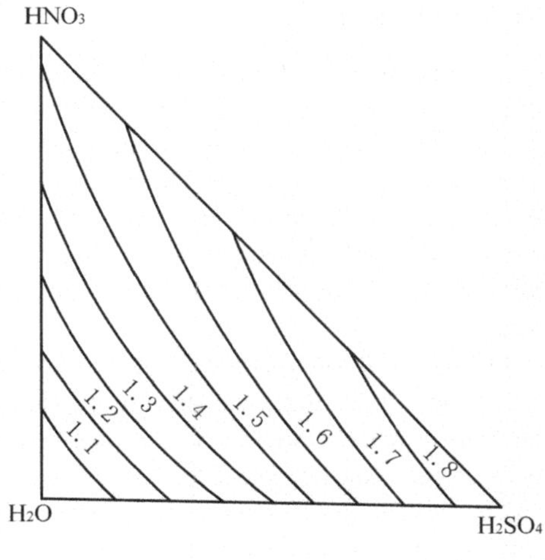

Table 3.13 The values of a and b in Eq. (3.18)

Nitric acid content (%)	5.0	10	15	20	30	40	50
a	1.0266	1.0555	1.0872	1.1186	1.1844	1.2534	1.3148
$b \times 10^3$	8.587	8.791	8.844	8.959	8.997	8.636	8.868

If the content of nitric acid in the mixed acid is known, the sulfuric acid content can be calculated according to its specific gravity. Alternatively, the specific gravity is determined according to the sulfuric acid and nitric acid contents in the mixed acid.

3.5 Nitration Reaction Technology

This section focuses on the effects of physicochemical factors of mixed acid (including nitric acid) as a nitrating agent on the nitration reactions, including liquid–liquid-phase nitration, liquid–solid-phase nitration, gas nitration reaction, and their operation techniques [28].

3.5.1 Operation of Liquid–Liquid Nitration Reaction

Compared with the gas-phase reaction, the liquid-phase reaction has various advantages such as a small volume of reactants and relatively easy to control the reaction, which is widely used in the production of explosives. According to the classification of the reactor, the following three methods are used in the production of explosives:

(1) Semi-batch process

This mode of operation uses a tank reactor. First, one reactant is added to the reactor, followed by the gradual addition of another reactant. After the completion of addition, according to the requirement of the reaction, the temperature can be controlled at another range. The reaction temperature is remained for a certain time to complete the reaction, and then the separation and subsequent processing are conducted to afford the product. After the nitration product is unloaded, the next batch reaction can be started. This type of operation mode is called discontinuous production.

In this mode of operation, the reactant concentration and the reaction rate vary with time. Under the ideal mixing conditions, the concentration of the material in the entire reactor should be homogeneous, so the material concentration and reaction rate can be considered to be the same in each part of the reactor.

The advantage of semi-batch process is convenient to start and stop the operation, which is suitable for small batch production. The drawbacks of this process include low production capacity, heterogeneous quality of the product, tedious operation process, and incompatible with automation and centralized control. Therefore, mass production generally uses a continuous process. Semi-batch nitration can generally be carried out in the following two ways:

1. The addition of raw material to the acid

The acid is first placed in a nitration reactor, and then the substrate for nitration is gradually added to the acid under stirring and cooling. As the acid volume is often greater than the volume of the substrate for nitration, the cooling surface of the equipment is fully utilized in this method. However, at the beginning of the nitration reaction, the concentration of mixed acid is very high and excessive, and the nitration capability is strong. As a result, the first added substrate tends to be over nitrated. For example, when toluene is nitrated to mononitrotoluene, the product often contains a large amount of dinitrotoluene. The more reactive the substrate for nitration, the more obvious the degree of nitration. In contrast, for the substrate that is hard to be nitrated, such as nitration of dinitrotoluene into trinitrotoluene, there is no problem of over nitration. When preparing the nitro compound, as long as the number of nitrate group does not exceed the required nitrate group of the product, the entire process is feasible. Therefore, in the process of batch production of explosives, most of the methods employ the addition of nitrate to the acid.

2. The addition of acid to the substrate for nitration

The substrate for nitration is placed in the nitration reactor, followed by the gradual addition of acid with constant stirring and cooling. The advantages and disadvantages of this method are just the opposite of the above method. The disadvantage is the insufficient utilization of cooling surface. Thus the reaction requires longer reaction time. In the case of the same raw material composition and quantity, the reaction time is significantly longer than that of the first method. As the acid is added to the substrate for nitration, the substrate is excessive in the initial stage of the reaction, and it is not prone to over nitration. In the dye industry, the manufacture of mononitrotoluene utilizes this method and batch production to avoid the formation of dinitrotoluene, which could block the tower during the distillation and even cause an explosion.

Since the average concentration of acid in the nitration process is low, and the organic phase is all put into the reactor at the beginning, the volume fraction of the organic phase in the whole reaction process is large. Therefore, there are more oxidative side reactions when nitrating the aromatic compounds.

(2) Continuous tank process

In the steady stage of operation, the acid and substrate for nitration are continuously added to the reactor in a certain proportion. The reactor is equipped with stirring and heat transfer devices, which can rapidly mix the reactants and maintain them at a certain reaction temperature. The reaction mixture continuously flows from the reactor and enters the separation device. The liquid level in the reactor remains constant as the amount of material leaving the reactor per unit time is equal to the amount of material entering the reactor.

Under the ideal mixing conditions, the added material is rapidly mixed with the material in the reactor, and the composition of the material in the reactor is homogeneous and is the same as the composition of the effluent material. Therefore, under the steady-state operation and the ideal mixing conditions, the composition of the material and the reaction rate are independent of the reaction time and location. In addition, the reaction temperature is constant. Overall, all the operating variables do not change with time, but they change slightly in the real situation. Therefore, if the quality of raw materials is stable, the quality of the product is more uniform, which is favorable for achieving automation and centralized control.

Another advantage of this process is that the variables of the reactor, such as waste acid composition and temperature, can be controlled under optimum conditions, which can improve the product quality and reduce the consumption of raw materials.

The volume of the tank reactor is generally 1–10 m. Because of the mechanical mixing device, it is particularly suitable for heterogeneous and slow chemical reaction. When a reactor cannot ensure the desired reaction time, several reactors can be used in tandem.

Since the composition of the material in the reactor is the same as that of the effluent material, the concentration of the reactants in the reactor is low, thus reducing the reaction rate and reducing the production capacity of the reactor. This is a drawback of a single tank continuous reactor. To overcome this drawback, multiple tank reactors can be connected in series. After the connected in tandem, the flow of material between the reactors has the following two methods:

(1) Concurrent flow

The main connection of the concurrent flow reactor group is shown in Fig. 3.26.

The feedstock is continuously fed to the reactor r_1, and the overflowed material is passed through the subsequent reactor. If necessary, the raw material can be added to the reactor in the rear, and the effluent from the last reactor enters the separator unit S. The concurrent flow reactor group is suitable for the nitration reaction with relatively fast rate and homogeneous nitration reaction.

1. Counterflow

The main connection of the reactor group in the counterflow reaction is shown in Fig. 3.27.

The raw materials A and B are continuously fed into the reactor r at both ends of the reaction system. Since the miscibility of the two materials is not significant, two phases are formed in the reaction system. The reaction mixture is well mixed by the stirring device and divided into two phases in the separator. These two liquid phases flow in opposite directions. Therefore, the concentration of raw material A in r_1 is

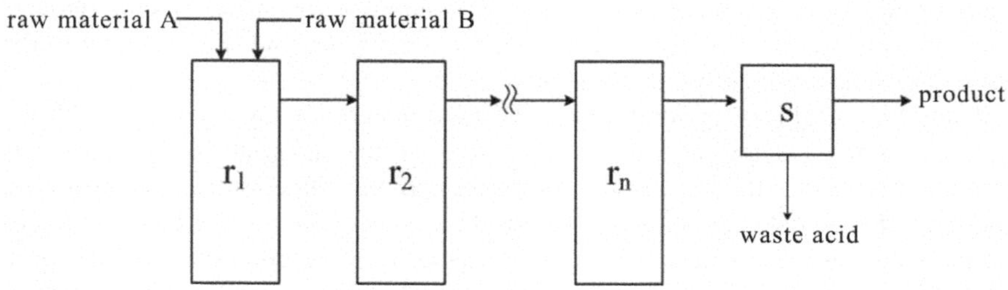

Fig. 3.26 The main connection of the concurrent flow reactor group: R—reactor and S—separator

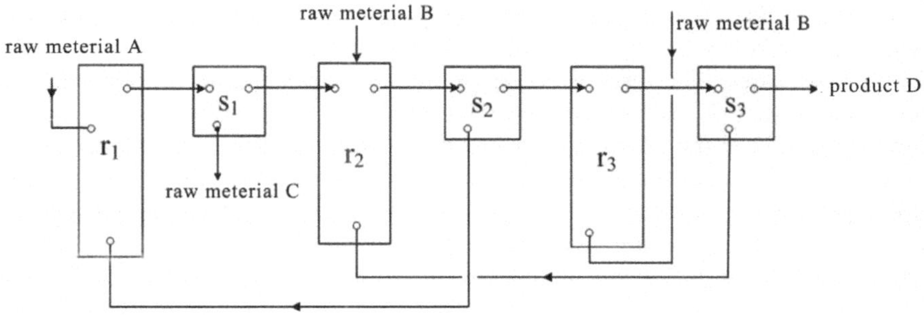

Fig. 3.27 The connection of the reactor group in the counterflow reaction: R—reactor and S—separator

the highest, and then gradually decreased because of the progress of the reaction. The order of concentration of the raw material A in the reactor is

$$r_1 > r_2 > r_3 \ldots$$

The flow direction of the raw material B is opposite to that of A, so that the order of concentration of the raw material B in each reactor is

$$r_n > r_{n-1} > r_{n-2} \ldots > r_1$$

If necessary, the raw material can be added to a middle reactor to maintain a suitable concentration gradient.

This reaction system is suitable for the reaction with a low reaction rate and two-phase. The advantage of this system is a certain concentration of raw material decreases while another concentration of raw material increases, thus maintaining a certain reaction rate. However, the system needs more equipment, and each nitration reactor should be equipped with a separator.

In addition to the above two typical connection methods, the hybrid connection can also be used, in which some are concurrent flow reactors, and some are counterflow reactors in a group of reactors. The ultimate purpose is to achieve reasonable economic and technological results.

(3) Tubular continuous nitration reactor

The tubular reactor is also a continuously operated reactor. The feedstock is fed from one end of the tube. The raw material flows in the tube with simultaneous nitration reaction. The reaction is complete when it flows out of the tube. The degree of reaction is proportional to the length of the tube through which the material flows. Steady-state operation and the plug flow, the material composition of each point in the tube is a function of the length of the material flowing through the tube rather than time.

To ensure the reactants are mixed evenly, the raw material passes through a mixing device before entering the tubular reactor. In the glycerin nitration and TNT washing processes, the ejector is used as a mixing device, and the nitrating acid or the washing water is used as the conveying fluid. The negative pressure is sucked into the glycerin or the molten TNT when the fluid flows through the nozzle. After mixed in the mixing chamber, the raw material is passed through the tubular reactor to complete the nitration reaction or washing. For the heterogeneous reaction, to prevent the reaction mixture stratified in the tube and affect the mass transfer, the raw material should maintain a high degree of turbulence in the tube. Therefore, for the heterogeneous reaction, the tubular reactor can only be applied to the reactions with high rate. Otherwise, the tube is too long, which will lead to a dramatic increase in friction loss of fluid flow.

For the intense exothermic reaction (nitration reaction is a typical example), when the reaction rate is fast, the material in the mixing device takes place a vigorous reaction and releases a lot of heat. If appropriate measures are not taken, the reaction temperature will be too high or even causes explosion accident. The amount of transferred liquid should be increased and precooled to a low temperature to absorb the reaction heat to control the reaction temperature in a certain range. Therefore, in the nitration of glycerol, a part of the waste acid can be recycled to save the amount of mixed acid.

The advantages of the tubular reactor include continuous operation, easy to achieve automation and centralized control, high production capacity per unit volume of the reactor, and relatively large tube surface area. Compared with the tank reactor, the tubular reactor has a large heat transfer surface per unit of reaction volume and a high heat transfer capacity.

Although nitrated product contacts with the mixed acid having a strong nitration ability from the beginning, the difference between the tubular reactor process and semi-batch operation is that the mixed acid and the nitration substrate maintain a certain ratio. In particular, the acid in the tubular reactor is not very excessive and the residence time of the material in the tube is very short. Therefore, over nitration reaction does not occur.

For the production of very dangerous explosives, the outstanding advantages of tubular reactors include high production capacity and a small volume of the reactor, which can dramatically reduce the quantity of product in the production process, thereby increasing the safety of the production and reducing the safety distance and site area.

3.5.2 Operation of Injection Nitration

The working principle of the ejector is the use the conversion of static pressure and kinetic energy of the mixed nitrating acid flow of to suck and discharge liquid. When the liquid is ejected from the nozzle at high speed, the static pressure can be converted into kinetic energy, resulting in low pressure (a certain degree of vacuum) and suction of glycerol. When passing the throat, nitrating acid and glycerol mix with each other under a high degree of turbulence. During the mixing process, rapid nitration reaction takes place, and nitration reaction can be completed in 0.5–1 s. The residence time of nitration mixture in the rejecter is about 0.5 s, so the reaction is basically complete.

The main parameter and characteristic curve.

The main parameter of the injector is the injection coefficient, which is the weight ratio of the liquid (glycerol) to the working liquid (nitrating mixed acid)

$$\mu = G_{\text{glycerol}}/G_{\text{nitrating mixed acid}} \tag{3.19}$$

It is related to the efficiency of the injector and the nitration coefficient (the coefficient of injection is the reciprocal of the nitration coefficient). Therefore, injection coefficient has an important effect on the production safety. Each injector has its characteristics. According to the experimental result, the injector has a parabolic characteristic curve. Figure 3.28 shows the characteristic curve of the nitrating mixed acid pumped by the injector.

It can be seen from Fig. 3.28 that the μ value increases with increasing mass flow rate G (kg/min) of the nitrating mixed acid. When G increases to a certain extent, μ reaches a maximum value, which shows the highest point on the curve. If G continues to increase, μ value is reduced. At the initial stage of nitration reaction the amount of nitrating mixed acid increases, which causes the increases in the flow rate of the ejector, vacuum degree of the suction chamber, and the amount of glycerol intake, thus increasing μ value. When G increases to a certain extent, the throat resistance also increases, which hinders the suction of glycerol, thus resulting in a slow increase in glycerol intake and then reduced intake amount.

Fig. 3.28 Characteristic curve of the injector [29]

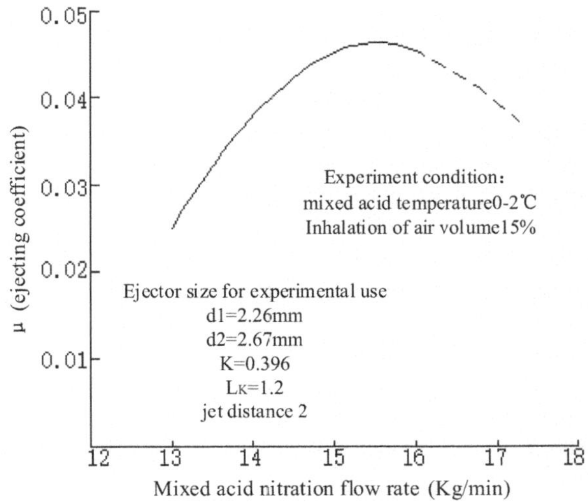

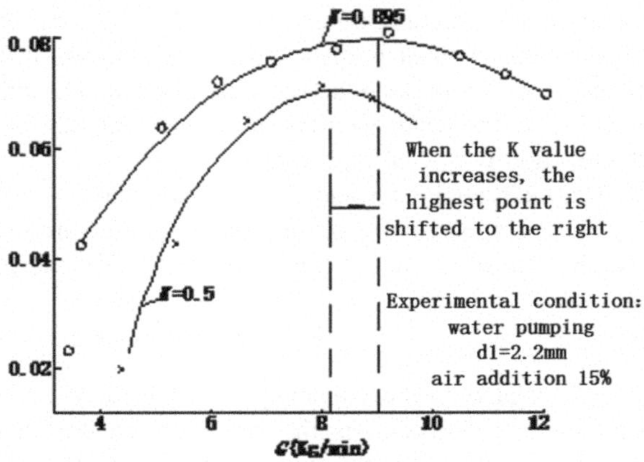

Fig. 3.29 Characteristics of injectors with different K values

It is well known from production practice that the performance of the ejector depends mainly on the K value, namely, the ratio of the annular cross section of the throat to the nozzle cross section. As can be seen from the following formula:

$$K = \frac{F_2 - F_1}{F_1} = \frac{F_2}{F_1} - 1 = \frac{0.785 d_2^2}{0.785 d_1^2} - 1 = \left(\frac{d_2}{d_1}\right)^2 - 1 \quad (3.20)$$

where

F_2 throat cross section (mm^2);
d_2 throat diameter (mm);
F_1 nozzle cross section (mm^2); and
d_1 throat diameter (mm).

From the experiment, K value and μ value have the following relationship, as shown in Fig. 3.29.

It can be seen from the μ-G curve of different K values, the injector with a small value of K has a small maximum value of μ with an early appearance of the maximum point while the injector with a large value of K has a large maximum value of μ with a late appearance of the maximum point. According to practical experience, the optimal μ value is in the range of 0.067–0.083 (about nitration coefficient range of 12–15). In particular, a sudden increase in μ value should be avoided to prevent an accident. The K value of the injector is typically in the range of 0.4–0.7. If the μ value is small, the K value of the injector should also be correspondingly small.

When selecting the K value, in addition to considering the μ value, the nozzle outlet flow rate W should also be considered. When W is large, K value should be slightly small. Conversely, when W is small, K should be slightly large.

The nitration capacity is determined by the injector, whereas the injector depends on the following main parameters:

(1) Nozzle

The size of the nozzle determines its production capacity. The flow rate of nitrating mixed acid at the nozzle outlet is related to the amount of sucked glycerol and the mixing degree of glycerol with mixed acid. Therefore, the flow rate at the nozzle outlet must be fast enough to ensure sufficient kinetic energy, so that the suction chamber can generate a certain degree of vacuum. The nozzle size is closely related to the flow rate. According to experience, the nozzle diameter and the flow volume and flow rate of nitrating mixed acid have the following relationship:

$$d_1 = 3.52\sqrt{G/W}$$

where

d_1	diameter of nozzle outlet (mm);
G	nitrating mixed acid flow volume (kg/min); and
W	linear speed of nitrating mixed acid at the outlet (m/s) (usually in the 12–20 m/s range, the typical speed is about 15 m/s).
3.52	Conversion Constant

To reduce the resistance, the angle of nozzle inlet tube is typical 20°–26°, and the length of the nozzle is 0.5–1 mm.

(2) Throat

The throat size should be appropriate, in which the nitrating mixed acid mixes with glycerol. If the throat diameter is slightly small, it can produce strong turbulence, which is conducive to mixing and reaction. If the diameter of the throat is too large, the nitrating mixed acid cannot cover the throat completely, which causes a part of the air to flow into the suction chamber, thus reducing the degree of vacuum and reducing the amount of sucked glycerol or even failing the suction of glycerol. In particular, when the glycerol flow is small, this negative effect is more prominent. On the other hand, if the diameter of the throat is too small, the resistance increases accordingly, especially when the flow of the nitrating mixed acid flow is very fast, which will significantly increase the throat resistance. As a result, the sucked glycerol is difficult to pass through the throat, thus reducing the flow rate.

After the nozzle diameter and K value are determined, the pipe diameter can be calculated using the following equation:

$$d_2 = d_1\sqrt{K+1}$$

where

d_2 pipe diameter (mm);
d_1 throat diameter (mm); and
K the ratio of throat annular section to nozzle cross section (generally in the range of 0.4–0.7).

The distance between the nozzle and the throat should be as short as possible and is generally in the range of 0.5–2 d_1. Thus, the operation is stable, and the acid is not easy to backward flow.

(3) Diffuser angle

To reduce the power loss, the diffuser angle is generally in the range of 6°–8°. The outlet diameter of the diffuser is about three times that of the tube.

To enhance the nitration reaction rate and to allow the reaction mixture to be rapidly removed and cooled, and the safety of the delivery, the nitration mixture is required to be in the form of an emulsion. It is therefore desirable that the mixture should be in a turbulent state to ensure that the Reynolds number of the turbulence mixture is no less than 2100. Reynolds number can be calculated as follows:

$$Re = DV\rho/\eta \tag{3.21}$$

where

Re Reynolds number;
D the diameter of tube (cm);
V the flow rate of mixture (cm/s);
ρ the density of mixture (g/cm^3); and
η the viscosity of mixture (centipoise).

(4) Examples of main performance parameters of the injection nitration reactor

d_1—5.3 (mm),
d_2—6.8 (mm), and
K—0.64

Table 3.14 shows the injection parameters of the injection nitration reactor.

Table 3.14 The injection parameters of the injection nitration reactor

Injection parameter	Value
The pressure of nitrating mixed acid (MPa)	0.35
The linear speed of nitrating mixed acid (W, m/s)	17.24
The degree of vacuum in production (mmHg)	300
The quantity of nitrating mixed acid (G, kg/h)	2350
The flow volume of glycerol (kg/h)	220
Injection coefficient (μ)	0.074
Nitration coefficient	13.8
Production capacity (kg/h)	550

3.5.3 Operation of Liquid–Solid-Phase Nitration

The reaction system consists of a multiphase reaction process of liquid and solid phases. This kind of reaction process includes: (1) the reactant dissolves in the liquid phase, and the solid phase only catalyzes the liquid–solid-phase reaction, such as the trimerization reaction of formaldehyde (aqueous solution) in the ion exchange resin catalyst; (2) both solid and liquid phases are involved liquid–solid-phase non-catalytic reaction, such as a variety of acidic hydrolysis reaction of ore and causticization reaction of soda. Since the process of liquid–solid-phase reaction is similar to that of gas–solid-phase reaction, some operation principles of gas–solid-phase reaction can also be used for liquid–solid-phase reaction. However, most of the gas–solid-phase reactions are the diffusion of the reaction components in the gas phase to the solid phase, where the chemical reactions occur. On the other hand, the liquid–solid-phase reaction may be carried out either in the solid phase or the liquid phase. As the liquid–solid-phase reaction is mostly ion reaction, and the reaction rate is usually fast, the liquid mass transfer coefficient is much lower than the gas mass transfer coefficient. Therefore, the effect of liquid–solid-phase reaction process on the apparent reaction rate is more significant, resulting in a lower apparent reaction rate and a longer reaction time. For liquid–solid-phase reaction, in addition to the use of fixed bed reactor, moving bed reactor, fluidized bed reactor, batch or continuous operation of kettle-type reactor is also widely used reactor, which relies on violent turbulence caused by stirring and the use of fine particles to increase the contact surface area, thus improving the mass transfer rate per unit volume.

The nitration of cellulose is a typical liquid–solid-phase nitration reaction [29, 30]. Currently, cellulose nitration process includes the batch and continuous nitration processes. Early nitration of cellulose is completely manually operated batch kettle nitration. First, nitrating mixed acid is added to the reactor, and then the refined cellulose, stirring, discharge of product, and extrusion of excess waste acid are all operated manually. Finally, the product is washed with water to yield the nitrocellulose. Later on, this process is changed to mechanically operated centrifugal nitration. The batch nitration is an early process for the manufacture of NC, which has the characteristics of simple operation and equipment. The main drawbacks include poor nitration uniformity, low yield, significant acid consumption, intense labor, and poor operating conditions. At present, this process has been replaced by stirring nitration process, which is a kind of nitration method with a highly stable product quality among various batch nitration methods. This process has various advantages such as simple equipment, high production efficiency, less acid consumption, and uniform quality, which is still utilized in plants with small production capacity. NC continuous nitration processes include four-chamber, multichamber, U-tube, and multiple stirring nitration reactions connected in tandem. All these type reactors are industrial automation equipment for liquid–solid-phase nitration reaction. These processes are characterized by continuous operation, high yield, highly uniform quality, low labor intensity, and safe production.

Table 3.15 Composition of mixed acid used for different nitrogen contents

Nitration degree (ml/g)	Composition of mixed acid	
	HNO_3 (%)	Water content (%)
>214.0	20–23	8–10
213–214	20–22	9–10.5
212–213	19–21	10–12
211–212	18–21	12–13.5
193–198	18–20	16–17.5
188–193	18–20	16.5–18.0
180–188	20–24	17.5–19.0
172–180	20–24	18.0–20.0

Table 3.16 The process conditions for batch stirring nitration

Composition of mixed acid (%)			Loading amount (kg) (dry weight)	Nitration coefficient	Nitration temperature (°C)	Nitration time (min)	
H_2O	HNO_3	H_2SO_4				Addition	Nitration
16–18	18–22	60–66	24–27	36–42	28–38	2–6	25–35

Table 3.17 The process conditions for continuous nitration

Composition of mixed acid (%)			Loading amount (kg) (dry weight)	Nitration coefficient	Nitration temperature (°C)	Nitration time (min)
16–18	18–22	60–66	12–18	40–50	30–40	15

Table 3.18 The process conditions for four-chamber continuous nitration

Composition of mixed acid (%)			Loading amount (kg) (dry weight)	Nitration coefficient	Nitration temperature (°C)	Nitration time (min)
16–18	18–22	60–66	3–7	40–50	28–38	25

In the nitration of cellulose, the preparation of the mixed acid with different nitrogen content is shown in Table 3.15, in which the operation process is slightly different when the nitration equipment is different. Tables 3.16, 3.17, 3.18 and 3.19 show the preparation of nitrocellulose with nitration degree of 188–193 ml/g under the conditions of batch stirring nitration, continuous nitration, four-chamber continuous nitration, and U-tube nitration.

The stirring rate of the first, second, and third chambers is 150–200 revolutions per minute, and the stirring rate of the fourth chamber is 70–120 revolutions per minute. The height difference between the chambers and the chamber is less than 40 mm.

Table 3.19 The process conditions of U-tube nitration

Number item	Composition of mixed acid (%)			Loading amount (kg) (dry weight)	Nitration coefficient	Nitration temperature (°C)	Nitration time (min)
1	8–13	18–22	65–74	7–10	50–60	20–30	<20
2	10–18	18–22	60–66	7–10	50–65	20–30	<25
3	16–18	18–22	60–66	7–15	50–65	30–35	25

Stirring speed is 100–150 r/min, and the last U-shaped tube is 80–100 r/min

3.5.4 Operation of Gas-Phase Nitration

Since acid-phase nitration has brought various adverse effects, a large number of new nitration systems have been explored to reduce or avoid the above adverse effects. In recent years, researchers have developed many green reaction systems, such as solid acid catalytic liquid nitration reaction, lanthanum lanthanide metal salt catalytic nitration system, and nitrogen oxides as nitration agent gas-phase nitration system. One of the most notable is the green nitration process utilizing nitroxide as nitration agent.

A process for the production of C1−C3 lower nitroalkanes from the corresponding hydrocarbons, such as methanol, was proposed by using a catalyst reported in a US patent in 1984 [31]. The catalyst was made of a Group II metal oxide or salt. Methanol and nitric acid were fed into the respective evaporators by the pump. After mixing, the mixed gas was fed into a fixed bed catalytic reactor. After the reactor was heated to 270 °C, the nitrogen was introduced to dilute and maintain the system pressure of 0.15 MPa with a preheating temperature of 180 °C and nitrogen consumption of 4000 mL/min. The nitric acid and methanol mixture was then fed to the reactor at a rate of 0.0076 mol/min with a material ratio of methanol: nitric acid: nitrogen of 4: 1: 24. Analysis of the condensate at the outlet of the reactor showed that the methanol conversion was 17% and the nitromethane selectivity was 60%. In 1985, Wang and Sherwin [32] used C1–C10 fatty alcohols to prepare the corresponding nitro compounds using NO_2 nitration reagent in the presence of O_2 and/or water under 0.5–2 MPa pressure, the temperature of 100–500 °C, and the contact time of 0.5–20 s. For example, the gas-phase nitration reaction of ethanol with NO_2 yielded nitromethane in the presence of water and oxygen at 300 °C and 1 MPa pressure. The ratios of oxygen to NO_2 and NO_2 to alcohol were 0.05–1.0 and 0.3–3, respectively. The contact time was 1–12 s. In addition to nitromethane, the products also contained CO, CO_2, ethyl formate, ethyl acetate, acetal, and acetaldehyde. The unreacted raw materials were easily separated and recycled.

US patent 4476336 describes the process of nitration of methane, ethane, propane, and their mixtures by using NO_2 and O_2. The gas flow consisting of ethane 3825 mml/h (34%), propane 2928 mml/h (26%), nitrogen dioxide 1372 mml/h (12.3%), oxygen 462 mml/h (4%), and nitrogen 2626 mml/h (23.4%) afforded a product mixture containing nitromethane 5.8%, nitroethane 7.7%, 1-nitropropane 5.4%, and 2-nitropropane 29% at the temperature of 300 °C, the pressure of 10 bars, and the reaction time of 6.8 s.

Nitric acid or nitrogen oxides were utilized to organic nitrate compounds to the corresponding nitroalkanes in US patent 4260838, in which nitroxides were used as nitrating agents at the temperature range of 280–350 °C. In the nitration reaction of NO_2 with propane, the experimental results showed that the content of nitromethane increased dramatically and the contents of 1-nitropropane and nitroethane increased slightly while the content of 2-nitropropane decreased with increasing temperature.

Li et al. [33] studied the gas-phase nitration of cyclohexane. The result indicated that cyclohexane could be nitrated to nitrocyclohexane using NO_2 with a selectivity of 89.31% at 240 °C in the absence of any catalyst. However, Si/Al zeolite catalyst has a certain catalytic effect on the reaction. In particularly, βzeolite exhibited the most potent catalytic activity.

With special zeolite catalyst, NO_2, N_2O_3, or N_2O_4 was used as nitrating agent to carry out gas-phase continuous nitration of benzene, toluene, and chlorobenzene, such as nitration of benzene to nitrobenzene by NO_2 with molecular sieve treatment, nitration of chlorobenzene to nitrochlorobenzene by NO_2 in the presence of palladium chloride, and non-catalytic gas-phase nitration of cyclohexane.

These methods have various advantages such as reduction of the corrosion and pollution caused by traditional mixed nitration method, avoiding the use of liquid sulfuric acid catalytic nitration, effectively inhibiting the formation of phenolic byproducts, and reduction on the dinitrated byproducts.

A nitro compound can be selectively formed by homogeneous gas-phase reaction of NO_2 with one of C2 to C10 carboxylic acids, aldehydes, or ketones. A nitroalkane with one less carbon (n − 1) atom can be obtained with high selectivity if an aldehyde, acid or alcohol with several carbon atoms other than ketone is selected for gas-phase nitration. In 1984, a United States patent reported [34] a process employing the homogeneous gas nitration of C2–C10 (preferably C3–C5) carboxylic acid with nitric acid or NO_2 in the presence of O_2 and water to produce nitroalkanes. The acetic acid conversion and nitromethane yield were as high as 43.5% and 44%, respectively, under the reaction temperature of 300 °C, pressure 1 MPa, acetic acid to the NO_2 ratio of 0.73, nitrogen to the NO_2 ratio of 5.2. Furthermore, a United States patent published in 1985 [32] reported the gas-phase nitration of aldehydes to prepare nitro compounds. In the case of acetaldehyde gas-phase nitration reaction, acetaldehyde conversion rate and nitromethane yield were 23.8–40.5% and 21.0–25.5%, respectively, under the reaction temperature of 150–250 °C, pressure 0.5–1.2 MPa, NO_2 and acetaldehyde ratio of 0.51–0.66. In 1985, a United States patent [35] proposed gas nitration of ketones to prepare nitro compounds. In the case of acetone gas-phase nitration reaction, nitromethane was obtained in 9% yield at the reaction temperature of 350 °C, pressure 1.0 MPa, and NO_2 and acetone ratio of 0.54.

References

1. Marsh AN (2000) Short history of nitroglycerine and nitric oxide in pharmacology and physiology. Clin Exp Pharmacol Physiol 27(4):313–319
2. Dietmar K, Linke J, Teichmann L et al (2016) Centrohexaindane: six benzene rings mutually fixed in three dimensions—solid-state structure and six-fold nitration. Phys Chem Chem Phys 18(17):11722–11737
3. Ji W, Li X, Wang J (2015) Preparation and characterization of CL-20/EPDM by a crystal refinement and spray drying method. Cent Eur J Energ Mater 12(4):831–840
4. Shokova EA, Kovalev VV (2012) Adamantane functionalization. Synthesis of polyfunctional derivatives with various substituents in bridgehead positions. Russ J Org Chem 48:1007
5. Ren CY (2010) Study on the vapor-phase nitration form cyclohexane to nitro-cyclohexane. Xiang Tan University
6. Campora J, Giambastiani G (2011) Olefin upgrading catalysis by nitrogen-based metal complexes II. Berlin: Springer
7. Kühn O, Ludger Wöste L (2007) Analysis and control of ultrafast photoinduced reactions. Springer, Berlin
8. McDavid JW (1922) J Soc Chem Ind 41:246T
9. Hetherington JA, Masson I (1933) J Chem Soc 105
10. Schiefferle DF, Hanson C, Albright LF (1976) ACS Symposium. Set 22:176
11. EL-Kabeir SMM, Modather M, Rashad AM (2015) Heat and mass transfer by unsteady natural convection over a moving vertical plate embedded in a saturated porous medium with chemical reaction, soret and dufour effects. J Appl Fluid Mech 8(3):453–463
12. Fagbade AI, Omowaye AJ (2016) Influence of thermal radiation on free convective heat and mass transfer past an isothermal vertical oscillating porous plate in the presence of chemical reaction and heat generation-absorption. Bound Value Probl 2016:97
13. Chamkha AJ, Rashad AM, Al-Mudhaf HF (2012) Heat and mass transfer from truncated cones with variable wall temperature and concentration in the presence of chemical reaction effects. Int J Numer Meth Heat Fluid Flow 22(3):357–376
14. Mabood F, Shateyi S, Rashidi MM et al (2016) MHD stagnation point flow heat and mass transfer of nanofluids in porous medium with radiation, viscous dissipation and chemical reaction. Adv Powder Technol 27:742–749
15. Miyagawa J (1949) J Soc Org Synth Chem Jpn 7:167
16. Satyanarayana V, HarishBabu D (2016) Numerical study of MHD heat and mass transfer of a Jeffrey fluid over a stretching sheet with chemical reaction and thermal radiation. J Taiwan Inst Chem Eng 59:18–25
17. Sun RK, Ren TS et al (1981) Chemistry and technology of high explosive. International Industry Press
18. Liu JP (2015) Liquid explosives. Springer, Berlin
19. Miller DC, Noyce DS, Vermeulen T. Ind Eng Chem 5
20. Kobe KA, Lakemayer JL (1958) Ind Eng Chem 50:1691
21. Rhodes FE, Nelson CC (1938) Ind Eng Chem 30:648
22. Prang AH, Andrianof JK (1955) Organic intermediates and dyes industry special equipment. Higher Education Press, Lingxian Sun translation
23. Yoshida T, Iwai Y, Yamamoto J (1960) Ind Powder 21:26
24. Mckinley C, Brown GG (1942) Chem Met Eng 49:142
25. McCurdy JL, Mckinley C (1942) Ind Eng Chem 34:1002
26. Hackel J et al (1900) Chemia Stosowana 4:44
27. Yoshida Y (1954) Ind Powder 12:43
28. Maestri F, Dionigi LR, Rota R (2006) Safe and productive operation of homogeneous semibatch reactors. the nitration of n-(2-phenoxyphenyl) methane sulfonamide. Am Chem Soc 45(24):8014–8023
29. Ren TS (1982) Process of nitric acid ester. National Defense Industry Press

30. Liu JP, Qian HJ (1993) Production and manufacture of nitrocellulose. Air Force Air Ammunition Research Institute
31. William VH (1984) Catalytic preparation of nitroalkanes, U.S. 4431842
32. Wang SP, Sherwin MB (1985) Preparation of nitro compounds by vapor phase nitration of organic alcohols. US 4517392
33. Li GH, Yin DL et al (2007) Vapor phase nitration process of cyclohexane to nitrocyclohexane. J Nat Sci Hunan Norm Univ 30(2):80–82
34. Wang SP, Sherwin MB (1985) Preparation of nitro compounds by vapor phase nitration of carboxylic acids. U.S. 4517394
35. Wang SP, Sherwin MB (1985) Preparation of nitro compounds by vapor phase nitration of ketones. U.S. 4517393
36. Pascal et al (1920) Bull Soc Chim France 27:8

Chapter 4
Mononitrate Ester

A monohydric alcohol nitrate typically refers to a nitric acid ester derived from an aliphatic or aromatic monohydric alcohol. This chapter covers the preparation and properties of mononitrate ester and focuses on several mononitrate ester having significant applications in national defense industries and agricultural production, such as methyl nitrate, ethyl nitrate, n-butyl nitrate, isopentyl nitrate, and isopropyl nitrate.

4.1 Methyl Nitrate

Methyl nitrate CH_3ONO_2 is not only one of the simplest nitrates but also is one of the most powerful liquid explosives [1, 2], which is made by the nitration of methanol. The primary reaction is:

$$CH_3OH + HNO_3 \rightarrow CH_3ONO_2 + H_2O$$

Methyl nitrate has high volatility and low boiling point. Mixed methyl nitrate with H_2O_2 or nitrogen tetroxide is used as a liquid explosive or liquid propellant. Methyl nitrate can be used as a warming agent for gasoline. When 1–2% amount of methyl nitrate is added to gasoline, the calorific value, or total work capacity of gasoline can be increased by 6.2%. In addition, it is a gas-generating agent in an airbag.

J. Liu, *Nitrate Esters Chemistry and Technology*,
https://doi.org/10.1007/978-981-13-6647-5_4

4.1.1 Physical and Chemical Properties of Methyl Nitrate

Molecular formula: CH_3NO_3
Chemical structure:

```
   H
   |
H- C -ONO2
   |
   H
```

Molecular weight: 77.02
Oxygen balance: −10.4% (based on the generation amount of CO_2)
Nitrogen content: 18.18%

Methyl nitrate is a slightly fragrant and colorless oily liquid with a specific gravity of 1.21 and a boiling point of 65–66 °C. The flow ability of methyl nitrate is good, but its viscosity is lower than water. Methyl nitrate can be miscible with any alcohols. At the temperature of 25 °C, 100 ml of water can dissolve 3.85 g of methyl nitrate, which has a fragrant and pleasant odor. Methyl nitrate has a strong ability to dissolve cotton and high nitrogen nitrocellulose. The vapor of methyl nitrate is flammable and explodes when heated to 150 °C. Methyl nitrate under sealed condition heated to 250–300 °C takes place gas phase explosion with a spark or ultraviolet radiation.

4.1.2 Explosive Properties and Thermal Decomposition Properties of Methyl Nitrate

The explosion heat of methyl nitrate is 6732.1–6744.15 J/g, which is 1.02% higher than that of nitroglycerin. The explosion of methyl nitrate is initiated by dropping 2 kg weight from 40 mm height, which is less sensitive than nitroglycerin. The burning velocity of methyl nitrate is four times faster than that of nitroglycerin. The burning velocity in the glass tube is 0.12 cm/s, and it can continue to burn slowly under the pressurized condition (0.15 MPa). When the pressure is higher than a certain value, it can reach pulsating combustion. Only when the pressure reaches 0.4 MPa, it will escalate from combustion to detonation.

The decomposition of methyl nitrate can trigger a spontaneous explosion [3] because of the exothermal property of the reaction.

$$2CH_3ONO_2 \rightarrow CH_3OH + CH_2O + 2NO_2$$

At the temperature of 25 °C, the gas explosion of methyl nitrate induced by spark [4] is different from the explosion reaction induced by heat. The explosion decomposition mechanism initiated by the spark is shown as below

Induced reaction

$$CH_3ONO_2 \rightarrow CO + NO + H_2O + \frac{1}{2}H_2$$

Explosion reaction

$$CH_3ONO_2 \rightarrow \frac{1}{2}CO + \frac{3}{2}H_2O + \frac{1}{2}CO_2 + \frac{1}{2}N_2$$

A series of oxidation and reduction reactions alternately proceed among the generated intermediates, resulting in the formation of nitrous oxide, carbon monoxide, and other gases. When the methyl nitrate is heated to 300 °C, the oxidation reaction takes place in the form of spontaneous decomposition.

$$CO + H_2O \rightarrow CO_2 + H_2$$
$$2NO + 2CO \rightarrow 2CO_2 + N_2$$
$$2NO + 2H_2 \rightarrow 2H_2O + N_2$$

The overall reaction can be expressed as

$$CH_3ONO_2 \xrightarrow{Q} 0.95NO + 0.025N_2 + 0.75CO + 0.25CO_2 + 0.7H_2 + 0.8H_2O$$

In the case of a spark-induced reaction

$$CH_3ONO_2 \xrightarrow{Spark} 0.175NO + 0.41N_2 + 0.49CO + 0.51CO_2 + 0.175H_2 + 1.31H_2O$$

The decomposition of methyl nitrate under UV irradiation is different from those induced by other methods.

$$CH_3ONO_2 \xrightarrow{h\gamma} 0.30NO + 0.35N_2 + 0.50CO + 0.50CO_2 + 0.30H_2 + 1.2H_2O$$

From the results of above experiments, it is evident that all the decomposition reactions of methyl nitrate proceed through free radical decomposition pathway.

$$2CH_3ONO_2 \rightarrow CO_2 + CO + 3H_2O + N_2$$

or

$$2CH_3ONO_2 \rightarrow 2CO_2 + H_2 + 2H_2O + N_2$$

In the initial stages of thermal decomposition, the reactions of methyl nitrate are same as those of other simple nitrates [5]. The reactions are as follows:

$$RCH_2ONO_2 \rightleftharpoons RCH_2O^{\ominus} + NO_2$$

$$RCH_2O^{\ominus} + RCH_2ONO_2 \rightarrow RCH_2OH + R\overset{\ominus}{C}HONO_2$$

$$R\overset{\ominus}{C} + HONO_2 \rightarrow RCHO + NO_2$$

$$RCH_2O^{\ominus} \rightarrow \frac{1}{2}RCHO + \frac{1}{2}RCH_2OH$$

In a tube with a diameter of 30–40 mm, the detonation velocity is about 8,000 m/s. The detonation velocity is approximately 1500 m/s in a tube with a small diameter, whereas the detonation velocity exceeds 6000 m/s in a large tube.

4.1.3 *The Preparation of Methyl Nitrate*

The early [1, 6, 7] process for the preparation of methyl nitrate is the nitration of methanol with nitric acid containing urea nitrate or a mixed nitric and sulfuric acid (40:60) at 18 °C for 20–40 min [8]. The density of employed nitric acid is 1.4 g/cm^3 without containing nitrogen dioxide. After nitration, the crude product is washed with water to neutral and distilled at 40 °C or 18 °C carefully to afford methyl nitrate. The yield of nitration products ranges from 45 to 80%. This process not only has a long production cycle but also has a low yield.

The exchange reaction of nitrate salt with sodium methoxide has been reported for the preparation of methyl nitrate in a yield of 85% [9]. In addition, a mixed acid consisting of acetic acid and nitric acid was also employed [10] as a nitrating agent to prepare methyl nitrate in a yield of 72–81% by nitration of methanol at 5 °C. However, this process requires a high concentration of mixed acid, which causes a vigorous exothermic reaction and significant hydrolysis in the water washing step, thus leading to severe environmental pollution.

Those preparation methods are very complicated, and their maximum yields are not greater than 81.5%. In addition, the reaction temperature of nitration process must be accurately controlled. Otherwise, an explosion might occur if the reaction temperature is out of control.

Because of the low boiling point, high volatility, and low viscosity, methyl nitrate cannot be washed by hot water in the production process but can be washed with cold water and aqueous solution. Considering these factors [2, 4], the author developed the following process for the preparation of methyl nitrate. The basic process is shown by the following equation:

$$CH_3OH + HNO_3 \xrightarrow{CON_2H_4} CH_3ONO_2 + H_2O$$

In the industrial production process, in order to obtain a highly pure product, it is required to avoid an excessive amount of methanol to reduce the impurities that

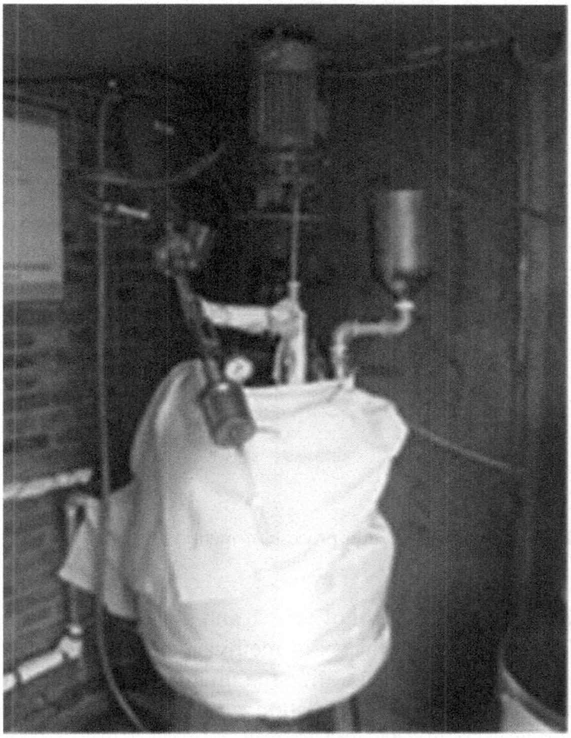

Fig. 4.1 Schematic diagram of continuous nitration for methyl nitrate

might be dissolved in methyl nitrate, thus ensuring the complete conversion of the nitric acid to methyl nitrate. The schematic diagram of methyl nitrate production by continuous nitration is shown in Fig. 4.1.

Two regular funnels filled with a methanol solution containing water and urea are separately equipped at two sides of the reactor. In the middle of the reactor is equipped with a distillation head, a temperature controller, a condenser, and a receiver. The procedure of the continuous nitration for methyl nitrate is described as follows. HNO_3 and distilled water are slowly added to urea in the reactor to control the reaction temperature around 110 °C. The temperature at the distillation head is controlled at the range of 80–85 °C. The methanol solution and nitric acid are added dropwise at a rate of 80–160 drop/min. The collected fraction in the receiver gradually separates into two layers. In the meantime, the temperature of the distillation head decreases to 65 °C, and the reaction temperature decreases to 95 °C. After the completion of the addition, the reaction mixture is heated at 95 °C for another 30 min to ensure the complete distillation. The product at the lower layer of the receiver is methyl nitrate.

The acidic crude product is washed with water to neutral and then washed with sodium carbonate solution to pH = 7.2–7.5 range. The crude methyl nitrate product

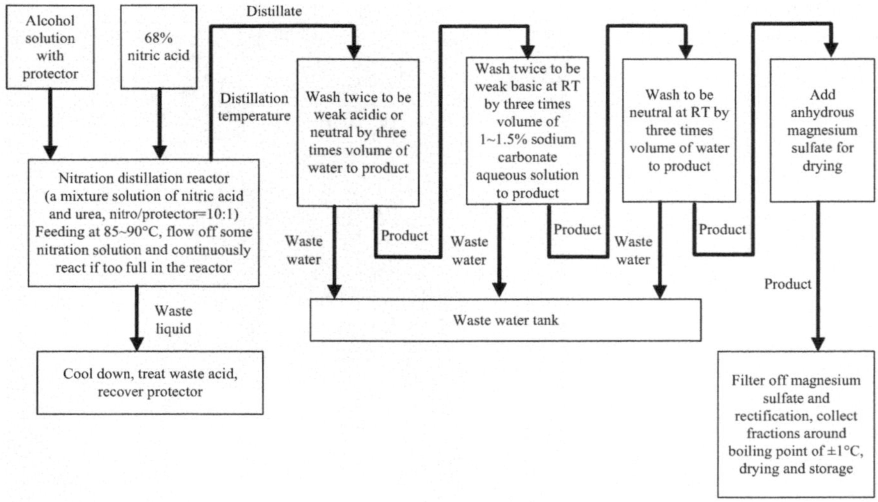

Fig. 4.2 The production process of methyl nitrate

is obtained at a yield of 98.7% and stored with the desiccant of magnesium sulfate. The production process of methyl nitrate is shown in Fig. 4.2.

As shown in Fig. 4.2, the main steps of methyl nitrate production process include nitration, distillation, washing, drying, rectification, and treatment of wastewater, waste acid and other waste generated in the production process. The crude methyl nitrate is transferred into the distillation unit for distillation, and the fraction with a boiling range of 64.5–65 °C is collected as the pure methyl nitrate. The infrared spectrum of methyl nitrite is shown in Fig. 4.3.

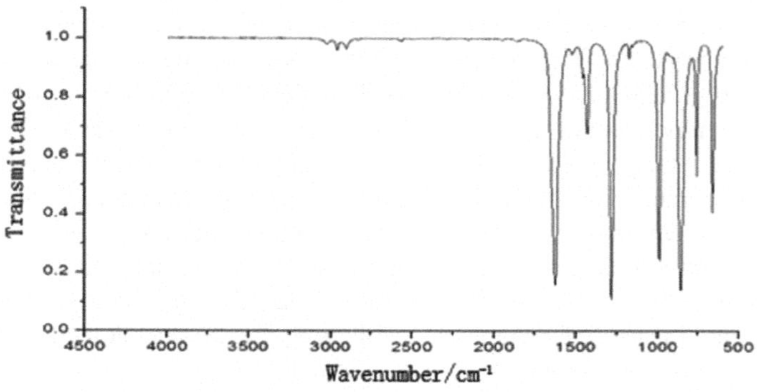

Fig. 4.3 IR spectrum of methyl nitrate

It can be seen from Fig. 4.3 that the bands around 1625 and 1282 cm^{-1} correspond to stretching vibration $-ONO_2$ group of methyl nitrate. A series of strong and medium-intensity bands are observed near 855 and 760 cm^{-1}.

The following correlations are investigated in the above-mentioned production process and reactor, such as the correlations of nitrate product yield with different nitration catalysts, the nitrate yield with feeding rate in the nitration of 1000 g CH_3OH in a 5 L reactor, the yield of qualified crude nitrate with the washing times of 1000 g distilled product. The results of this study are shown in Tables 4.1, 4.2 and 4.3.

It can be seen from Tables 4.1, 4.2 and 4.3 that the preparation of methyl nitrate is related to the reactor. More importantly, the correlations of product yield with the reaction conditions and catalyst selection are more obvious.

Table 4.1 The nitrate yield by using different catalysts in the nitration of 100 g methanol

No.	Catalyst	Yield of nitrate (%)	Remark
1	0	9–15	The reaction produces large amounts of brown smoke
2	Sodium nitrate	73	Reaction mixture turns to light yellow
3	Magnesium nitrate	86	Vigorous reaction with lots of bubbles
4	Ammonium nitrate	89	Smooth reaction with lots of bubbles
5	50% Magnesium nitrate + 50% Ammonium nitrate	93	Vigorous reaction with transparent reaction solution
6	Urea	100	Transparent reaction solution without odor
7	Urea formaldehyde	98.5	Transparent reaction solution with strong odor
8	Ammonium urate	100	Transparent reaction solution with odor

Table 4.2 The correlation between the addition rate and the nitrate yield in the nitration of 1000 g of methanol

Addition rate(ml/min)	Temperature of distillation head (°C)	Weight of refined methyl nitrate (g)
5	83	215.2
10	84	230.5
20	86	235.3
30	80	197.8
50	84	226.7

Table 4.3 The correlation between the washing number of 1000 g recovered product and the yield of qualified crude product

Number of washing	The ratio of crude product to water	pH value	Yield (%)	Water temperature (°C)
1	1:6	<1	95.9	20
2	1:6	2.5	92.3	20
3	1:6	4	88.7	20
4	1:6	6	85.1	20
5	1:6	6.5	82.2	20
6	1:6	9	80.3	20

By using the above-mentioned preparation process, the crude methyl nitrate product is obtained after nitration reaction and water washing, which is subjected to rectification to afford refined methyl nitrate with distillation temperature of 64.5–66 °C. By this preparation method, 100 parts of methanol can produce 234.98 parts of methyl nitrate by weight, which equals to the theoretical yield of 97.5%.

4.1.4 Toxicity of Methyl Nitrate

Methyl nitrate is a compound with moderate toxicity. Due to the low viscosity and high volatility of methyl nitrate, it is typically poisonous through respiratory exposure. The inhalation of methyl acetate vapor through the respiratory tract can cause a headache. This physiological effect is similar to the effect of inhaled nitroglycerin, except that the symptom of methyl nitrate poisoning comes and goes faster than nitroglycerin.

4.2 Ethyl Nitrate

4.2.1 Physical and Chemical Properties of Ethyl Nitrate

Ethyl nitrate, also known as ethanol nitrate, is a transparent oily liquid with slightly aromatic odor. It is slightly soluble in water but miscible with alcohols and most organic solvents. It also can effectively dissolve nitrocellulose [11].

Molecular formula: $C_2H_5NO_3$
Structure:

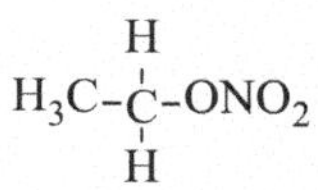

Molecular weight: 91.07
Oxygen balance: −61.5% (based on the generation amount of CO_2)
Nitrogen content: 15.38%
Boiling point: 87 °C
Density: 1.12 g/cm^3
Combustion heat: 1347.25 kJ/mol

The nitration heat of ethanol is 25.94 kJ/mol. According to Thomson's calculation, the combustion heat $-\Delta H$ and detonation heat (liquid water) of ethyl nitrate were determined to be 1355.78 kJ/mol and 3568.95 J/g, respectively. Although the explosive properties of ethyl nitrate have long been well known, few attentions have been paid to this compound for an extended period, except that it is used as a component of liquid rocket fuel and mixed with other oxidants to make mixed explosives for engineering blasting.

4.2.2 Explosive Properties of Ethyl Nitrate

The water-filled lead block expansion value of ethyl nitrate was determined to be 345 cm. The detonation velocity of ethyl nitrate in a wide tube is 6000–7000 m/s, and a steady detonation can be formed in a thick-walled metal tube filled with ethyl nitrate. The detonation velocities in steel tubes with diameters of 27 and 60 mm are 5800 and 6020 m/s, respectively. In a steel tube with a diameter of 10 mm, the detonation cannot occur. The lead block expansion value of ethyl nitrate is greater than that of picric acid approximately by 20%. However, its shock sensitivity is low. In the drop weight test, 2,000 g weight dropping from 50 cm height causes 50% explosion of ethyl nitrate.

4.2.3 Preparation of Ethyl Nitrate

The preparation of ethyl nitrate is carried out by dissolving cold ethanol in nitric acid (density: 1.41 g/cm^3) [1, 6] without nitrogen oxide, followed by distilling the product [12]. When nitrogen oxide is present in the nitric acid, ethanol is readily oxidized to acetaldehyde, which can interfere with the nitration reaction if the nitric acid contains a significant amount of nitrogen oxide.

Recently, the authors [11] invented a continuous nitration process for the preparation of ethyl nitrate. Ethanol and the nitric acid solution are added dropwise to the reactor containing the carbonyl group protective agent. In the process of nitration reaction, the nitrated product is continuously removed from the reactor by distillation. The removal of nitrate product from the reaction mixture not only shifts the reaction equilibrium to the right side but also increases the yield of nitration reaction. More importantly, this process reduces the heating time of nitrate in the reactor, thus reducing the generation of nitrite and reaction intermediates and improving the safety of the entire production process. The reaction process can be expressed by the following equation:

$$CH_3CH_2OH + HNO_3 \rightarrow CH_3CH_2ONO_2 + H_2O$$

The left and right sides of the reactor are separately equipped with two regular funnels filled with a mixed alcohol solution (containing a certain amount of protective agent alcohol solution) and the corresponding mass of 68% nitric acid solution. In the middle of the reactor is equipped with a distillation head, a thermometer, a condenser, and a receiver. Zeolites are added to the reactor.

The overall composition of nitrating agent includes HNO_3 75–90%, H_2O 5–15%, and protective agent 5–10%. Among them, the protective agent consists of urea, dicyandiamide, and melamine.

A certain amount of protective agent is dissolved in distilled water and added to the reactor, followed by the slow addition of a corresponding amount of HNO_3 to the reactor. After the temperature of the reactor is raised to the appropriate nitration temperature, the addition of reactants is started. The temperature of the oil bath is controlled at about 130 °C, and the temperature of the distillation head is controlled within a certain range. When the protective agent is dissolved, and the solution is refluxed at a stable temperature of set point, the vapor of nitric acid is full of the reaction. Anhydrous ethanol and nitric acid solution are added to the reactor slowly and at a controlled rate. Oil droplets can be observed in the condenser, and the solution in the receiver starts to be layered. In this process, the reaction temperature and the addition rate of reactants are carefully controlled to ensure a smooth reaction.

After the addition of reactants, the oil bath temperature is maintained for additional 20 min to ensure the remaining product in the reactor can be completely distilled out. In the meantime, a clear layer separation can be observed, in which the lower layer is the nitrated product. The lower layer is separated, washed with aqueous sodium carbonate solution and water to pH 7.2–7.5, and dried and stored with magnesium sulfate granule. The resulting crude product could be transferred into a rectification system with zeolites to obtain highly pure nitrate.

In order to study the relationship between the nitration reaction conditions and the yield of ethyl nitrate, the author conducted a more comprehensive study on the effects of the equivalents of urea, dicyandiamide and melamine, the concentration of nitric acid, and the addition time of ethanol on the yield of ethyl nitrate. The effects of the feeding device and the feeding method on the yield of ethyl nitrate were also examined in the nitration process of 15 g ethanol in a four-necked flask. The effect of the amount of melamine on the product yield was studied by carrying out the reactions under the conditions (100 ± 2 °C, 4% dicyandiamide, 15% ethanol, 68% nitric acid 100 ml, addition time of 30 min in dropwise) with various amount of melamine. The yields of ethyl nitrate are shown in Table 4.4.

It can be seen from Table 4.4 that the yield is the highest when the amount of melamine is 6.0%. When the amount of melamine is 5.5%, and the other conditions

Table 4.4 The effect of amount of melamine on the yield of ethyl nitrate

Melamine/g	4.0	5.0	5.5	6.0	6.5	7.0
Ethyl nitrate/g	18.33	19.59	22.59	24.97	19.02	16.18
Yield/%	58	62	71.5	79.1	60.2	51.2

Table 4.5 The effect of nitric acid concentration on the yield of ethyl nitrate

Ethyl nitrate concentration/%	55.0	70.0	73.0	75.0	77.0	80.0
Ethyl nitrate/g	20.00	26.29	29.10	23.41	20.73	13.33
Yield/%	53.3	83.2	90.2	74.1	65.6	42.2

Table 4.6 The effect of the amount of dicyandiamide on the yield of ethyl nitrate

Dicyandiamide/g	7.0	9.0	11.0	13.0	15.0
Ethyl nitrate/g	18.52	20.19	27.71	26.95	25.21
Yield/%	58.6	63.9	87.7	85.3	79.8

Table 4.7 The effect of the addition time of ethanol on the yield of ethyl nitrate

Addition time/min	8	10	20	30
Ethyl nitrate/g	26.38	30.55	26.51	21.33
Yield/%	83.5	90.67	83.9	67.5

remain same, the effect of the variation in the concentration of nitric acid on the yield of ethyl nitrate is studied. The yields of ethyl nitrate are shown in Table 4.5.

As can be seen from Table 4.5, when the nitric acid concentration is 73%, the yield of ethyl nitrate is 90.2% When the other conditions are fixed, the variation in the amount of dicyandiamide results in the different yield of ethyl nitrate. The yields of ethyl nitrate are shown in Table 4.6.

As can be seen from Table 4.6, when the amount of dicyandiamide is 11%, the yield of ethyl nitrate is the highest. The addition time of ethanol has a significant effect on the yield of ethyl acetate. The yields of ethyl nitrate are shown in Table 4.7.

It can be seen from Table 4.7 that the addition time of ethanol has a significant effect on the yield of ethyl nitrate. An extended addition time causes more oxidation of ethanol, thus decreasing the yield of ethyl nitrate.

The above process has significantly modified the traditional nitration method for the preparation of ethyl nitrate. With the protection of the carbonyl group, the vapor of nitrate product is effectively employed to remove the product from the reactor quickly. Moreover, the orderly control of the feed rate and product formation rate significantly reduce the residence time of the product in the heating environment, thus suppressing the formation of the nitrite product and improving the safety of the production process.

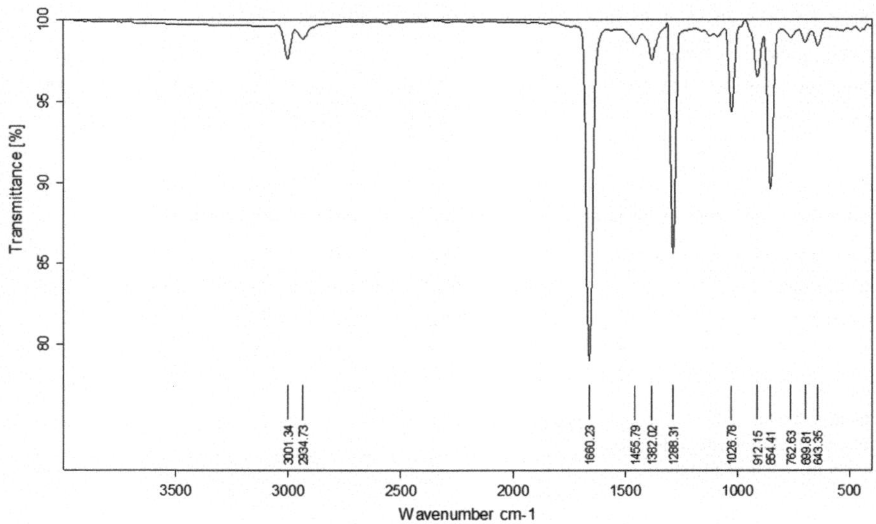

Fig. 4.4 IR spectrum of ethyl nitrate

The resulting crude product is transferred to a distillation flask for rectification to afford the colorless liquid product with a boiling point range of 88–91 °C as refined ethyl nitrate. The infrared spectrum of ethyl nitrate is shown in Fig. 4.4.

As shown in Fig. 4.4, the bands around 3,000 cm^{-1} correspond to C–H stretching and bending vibrations with frequency multiplication. The bands around 1660 and 1288 cm^{-1} are assigned to the asymmetric and symmetrical stretching vibrations of NO_2, respectively. The bands around 1455 and 1382 cm^{-1} correspond to the methyl and methylene CH bending vibrations, respectively. The bands around 1026 and 864 cm^{-1} are attributed to the C–O and N–O stretching vibrations, respectively. Furthermore, the ^{1}H-NMR spectrum of ethyl nitrate is shown in Fig. 4.5.

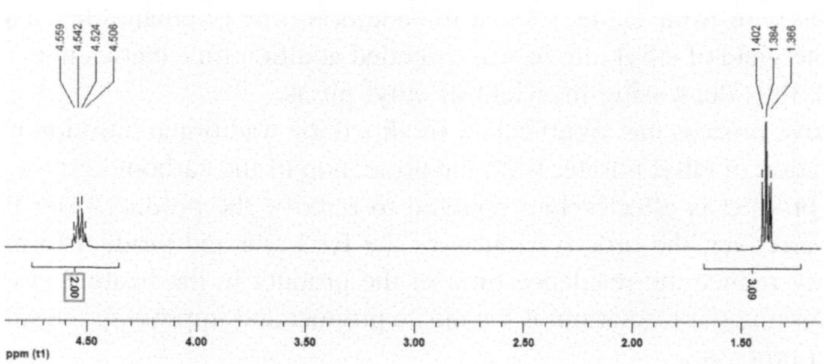

Fig. 4.5 H-NMR spectrum of ethyl nitrate

The ^{1}H-NMR spectrum of ethyl nitrate is described as follows. δH: 4.506–4.559 (2H, m), 1.366–1.402 (3H, t). Based on the IR and H-NMR spectra, the product can be unambiguously identified as ethyl nitrate with a yield of 94.5%.

4.3 n-Propyl Nitrate

n-Propyl nitrate is also known as n-propanol nitrate.

4.3.1 Physical and Chemical Properties of n-Propyl Nitrate

Molecular formula: $C_3H_7NO_3$
Chemical structure:

$$\begin{array}{c} \qquad\qquad\quad H \\ H_3C\text{–}H_2C\text{–}\overset{|}{\underset{|}{C}}\text{–}ONO_2 \\ \qquad\qquad\quad H \end{array}$$

Molecular weight: 105.09
Oxygen balance: −99.0% (based on the generation amount of CO_2)
Nitrogen content: 13.33%

n-Propyl nitrate is a flammable liquid with a boiling point of 110.5 °C and a specific gravity of 1.063. Since the explosive property of n-propyl nitrate is not vigorous, it is more difficult to be detonated. The calculated explosion heat is 2297 J/g (water vapor condition). The lead block expansion value after rectification of the detonator is only 15 cm. The vapor can form an explosive mixture with air, with an explosion limit of 2–100%, a flash point of 20 °C, and an ignition point of 176.7 °C in the air. The vapor is heavier than air and can be ignited by distant fire. The ignition flame could be returned to the n-propyl nitrate source. Being heated, n-propyl nitrate would be exploded. When mixed with diatomite at a ratio of 71.50/28.50, significant detonation could occur with a lead block expansion value of 230 cm. n-Propyl nitrate can be used as a sensitizer of cloud detonation, rocket propellant, and a promoter in enhancing cetane numbers for diesel engine fuels.

4.3.2 Preparation of n-Propyl Nitrate

Decades ago, Hinkamp et al. [13] prepared n-propyl nitrate in a yield of 66.5% by nitration of n-propanol using a mixed acid consisting of 20% HNO_3, 68% H_2SO_4, 12% H_2O (wt%). This process not only affords a low yield but also generates an acid mist that results in environmental pollution during the treatment of waste acid involving the concentration of dilute nitric acid with sulfuric acid. Later on, the nitrating agent and nitration yield were improved for the nitration of n-propanol [14], which leads to significant improvement of the n-propyl nitrate yield and reduction of waste acid on the environment pollution.

Recently, the author developed the following process for the production of n-propyl nitrate [6, 15]. A mixture of 68% nitric acid and urea is added to a reactor and heated to a vapor temperature of 98 °C, in which the liquid surface temperature is 112 °C, and the oil bath temperature is 142 °C. Subsequently, n-propyl alcohol solution containing water and 5% urea, along with 68% nitric acid, is added to the reactor in dropwise within 90 min. The temperature of the reaction mixture is kept for additional 15 min after the addition. At the beginning of the addition, the temperature of the distillates is 98–106 °C. When the reaction is stable, the vapor temperature reaches 102–104 °C with a solution surface temperature of 112 °C and an oil bath temperature of 142 °C. The crude product with a yield of 89% is washed with water, an alkaline solution (1.5% sodium carbonate), and water successively to pH 7. The separated oil layer is dried over anhydrous magnesium sulfate for 24 h, filtered, and rectified. The fraction at 109.5–110.5 °C is collected as the final product with a yield of 86.3% and a purity of 99.62%.

4.4 Isopropyl Nitrate

Isopropyl nitrate, also known as isopropyl alcohol nitrate, is an isomer of n-propyl nitrate and one of the most widely used monohydric alcohol nitrates.

4.4.1 Physical and Chemical Properties of Isopropyl Nitrate

Molecular formula: $C_3H_7NO_3$
Chemical structure:

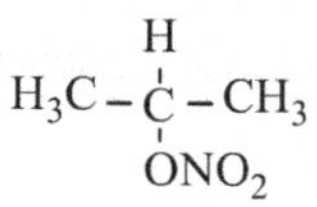

Molecular weight: 105.09
Oxygen balance: −99.0% (based on the generation amount of CO_2)
Nitrogen content: 13.33%
Boiling point: 101–102 °C.

4.4.2 Preparation of Isopropyl Nitrate

The physicochemical properties and manufacturing methods of isopropyl nitrate are different than those of n-propyl nitrate. The boiling point of isopropyl nitrate is 101–102 °C, and the relative density is 1.036, which are different than those of n-propyl nitrate. Moreover, the direct nitration of isopropyl alcohol is more difficult because of the steric hindrance of the secondary hydroxyl group of isopropyl alcohol. Vast literature reported that isopropyl nitrate could only be prepared by the reaction of 2-iodopropane with silver nitrate as isopropyl nitrate is difficult to be directly prepared because of the oxidation of secondary carbon atom in secondary alcohol [16]. However, a British Chemistry Patent [13] described the continuous nitration of isopropanol in the presence of urea. At the temperature below the boiling point of nitric acid, isopropanol and urea were added to nitric acid (above 40%). The unstable product was removed from the reaction mixture by a flowing gas. The yield of this method was about 80% of the theoretical yield [17]. In general, the preparation methods of isopropyl nitrate are as follows: (1) reaction of silver nitrate with halogenated alkane; (2) nitration using a nitro-sulfur mixed acid; (3) inert adsorption nitration [18, 19]. Based on the above-mentioned preparation methods, the author [6, 20–22] invented a binary inert complementary adsorption nitration method by using nitric acid with a concentration of more than 50% at the nitration temperature of 108–110 °C. This method has been used for the industrial production of isopropyl nitrate. In this nitration process, a certain amount of urea and ammonium nitrate is added to nitric acid as a nitration substrate to absorb the generated nitrogen oxides. The addition of urea and ammonium nitrate can inhibit the undesired oxidation and catalytic hydrolysis of isopropyl nitrate by the nitrogen oxides that generated in the high nitration temperature. The reaction mechanism of isopropyl nitrate prepared by inert adsorption nitration method is as follows:

$$(CH_3)CHOH + HNO_3 \xrightarrow[100-105\,^{\circ}C]{(NH_2)_2CO + NH_4NO_3} (CH_3)_2CHONO_2 + H_2O$$

The inert adsorption nitration process is suitable for the industrial production of isopropyl nitrate because it has various advantages, such as simple process, cheap raw material, low preparation cost, and easy treatment of waste acid.

70% nitric acid is added to a stainless steel reactor, followed by the addition of urea and ammonium nitrate in proportion. The mixture is slowly heated to 70–80 °C with stirring until the urea and ammonium nitrate are completely dissolved. The temperature is increased to 95–100 °C, and then isopropanol is added in dropwise using a dropping funnel with the lower end of funnel below the surface of nitration solution. In the meantime, the cooling water and vacuum system are turned on to pump out the generated nitrogen oxide gas from the reaction mixture. The addition time of isopropyl alcohol is in the range of 20–50 min at the temperature of 100 °C. After the completion of the addition, the temperature of the reaction mixture is kept at 100–105 °C for additional 10–20 min until no reflux and condensation products can be observed. After turn on the agitation, vacuum, and condensation systems, the condensate is separated into two layers by using a separation unit. The lower layer is waste acid and discarded, and the nitrate layer is washed with a small amount of 3% aqueous Na_2CO_3 solution and deionized water until pH = 7. The lower layer is collected, dried, and rectified under atmospheric pressure. The pure product with a yield of 97.6% is collected at the boiling point of 101–102 °C. The product is a colorless liquid with a relative density of 1.036 and a purity of 99.6% by gas chromatography.

In order to study the relationship between the nitration reaction conditions and the yield, the authors [21, 23] studied the effects of the amount of urea and ammonium nitrate dosage, the concentration of nitric acid, and the addition time of isopropyl alcohol on the yield of isopropyl nitrate. In addition, the author also investigated the influence of the feeding device and the feeding method on the yield of isopropyl nitrate. The effects of the amount of urea on the yield of isopropyl nitrate are summarized in Table 4.8 under the reaction conditions of 100 ± 2 °C, 13% ammonium nitrate, 12% isopropyl alcohol, 70% nitric acid 50 ml, and addition time of 30 min.

As can be seen from Table 4.8, when the amount of urea is 5.5%, the yield is the highest. When the urea is 5.5% with other conditions remaining the same, the effect of nitric acid concentration on the yield of isopropyl nitrate is described in Table 4.9.

The results in Table 4.9 show that the yield of isopropyl nitrate is 88.9% when the nitric acid concentration is 70%. When the other conditions remain the same as above, the amount of ammonium nitrate is changed, and the yields of isopropyl nitrate are shown in Table 4.10.

Table 4.8 The effect of the amount of urea on the yield of isopropyl nitrate

Urea/g	4.0	5.0	5.5	6.0	6.5	7.0
Isopropyl nitrate/g	13.7	14.8	16.6	15.0	13.1	11.5
Yield/%	63.7	68.9	77.3	69.7	60.9	53.6

Table 4.9 The effect of nitric acid concentration on the yield of isopropyl nitrate

Nitric acid concentration/%	65.0	70.0	73.0	75.0	77.0	80.0
Isopropyl nitrate/g	13.9	18.6	18.7	15.4	14.1	8.9
Yield/%	66.2	86.3	88.9	71.7	65.6	41.3

Table 4.10 The effect of ammonium nitrate amount on the yield of isopropyl nitrate

Ammonium nitrate/g	9.0	11.0	13.0	15.0	17.0
Isopropyl nitrate/g	12.8	15.0	18.9	16.2	13.7
Yield/%	59.7	69.9	87.9	75.3	63.8

Table 4.11 The effect of the addition time of isopropanol on the yield of isopropyl nitrate

Addition time/min	35	20	12	7
Isopropyl nitrate/g	14.8	17.8	21.0	18.2
Yield/%	68.7	82.6	97.7	84.5

As shown in Table 4.10, when the amount of ammonium nitrate is 13%, the yield of isopropyl nitrate is the highest. It is also found that the addition time of isopropanol has a significant effect on the yield of isopropyl nitrate. The yields of isopropyl nitrate with different addition time are shown in Table 4.11.

The results in Table 4.11 show that the maximum yield of isopropyl nitrate is achieved in the reaction time of 7–15 min, which is attributed to an accelerated nitration rate of isopropanol at high temperature and suppressed oxidation reaction with short reaction time.

The nitration of isopropanol is carried out at the temperature above the boiling point of isopropanol (82.5 °C). If the isopropyl alcohol is added to the surface of nitration liquid, it will be evaporated before being nitrated. However, when the end of the addition funnel is inserted into below the surface of nitration solution, the isopropanol is rapidly nitrated to isopropyl nitrate (boiling point: 102 °C), which is vaporized at 100 °C and escapes the reaction solution. In the meantime, the unreacted isopropyl is refluxed and returned to the reactor. Isopropyl nitrate is collected by condensation to suppress the oxidation reaction at the high temperature, thus improving the yield. The modified process exhibits high production efficiency and stable yield. The industrial production device of isopropyl nitrate is shown in Fig. 4.6.

The optimal conditions for the production of isopropyl nitrate are as follows: the nitration temperature of 106 °C, the holding temperature of 100–105 °C, the addition time of about 15 min, and the holding time of 5–10 min. The mass ratio of

Fig. 4.6 Industrial production device of isopropyl nitrate

reactants is isopropanol (99.5%): nitric acid (70%): urea (99%): ammonium nitrate (98%) = 1: 6: 0.43: 0.96. Under the optimal conditions, the yield of isopropyl nitrate is up to 97.6% with a purity of 99.6%.

4.5 n-Butyl Nitrate

4.5.1 Physical and Chemical Properties of n-Butyl Nitrate

Molecular formula: $C_4H_9NO_3$
Chemical structure:

$$\begin{array}{c} \qquad\qquad\qquad\quad H \\ H_3C-H_2C-H_2C-\overset{|}{\underset{|}{C}}-ONO_2 \\ \qquad\qquad\qquad\quad H \end{array}$$

Molecular weight: 119.12
Oxygen balance: −127.6% (based on the generation amount of CO_2)
Nitrogen content: 10.52%.

4.5.2 Preparation of n-Butyl Nitrate

The traditional preparation process for n-butyl nitrate is the nitration of n-butanol [24, 25] using a mixed nitric and sulfuric acid. This process not only affords a low yield but also has various safety risks [26]. The author prepared n-butyl nitrate by using continuous nitration with dilute nitric acid and protective agent. The route is shown in Fig. 4.7.

It can be seen from Fig. 4.7 that the preparation route of n-butyl nitrate is different from that of other monohydric nitrates. The main differences are as follows: (1) the nitration process is different. Methyl nitrate and other nitrates are prepared by using continuous nitration and distillation process, whereas the preparation of n-butyl nitrate is the use of water segregator equipment for the separation of reflux water. (2) the washing process is different. Since the density of n-butyl nitrate is close to that of water, the product is washed with a saturated sodium chloride aqueous solution rather than water in the washing process.

The key equipment for the preparation of n-butyl nitrate is the water segregator, and the water segregator is rare process equipment in the chemical equipment. According to the actual requirements of the n-butyl nitrate process and the actual situations of the process equipment, water segregator for the preparation of n-butyl

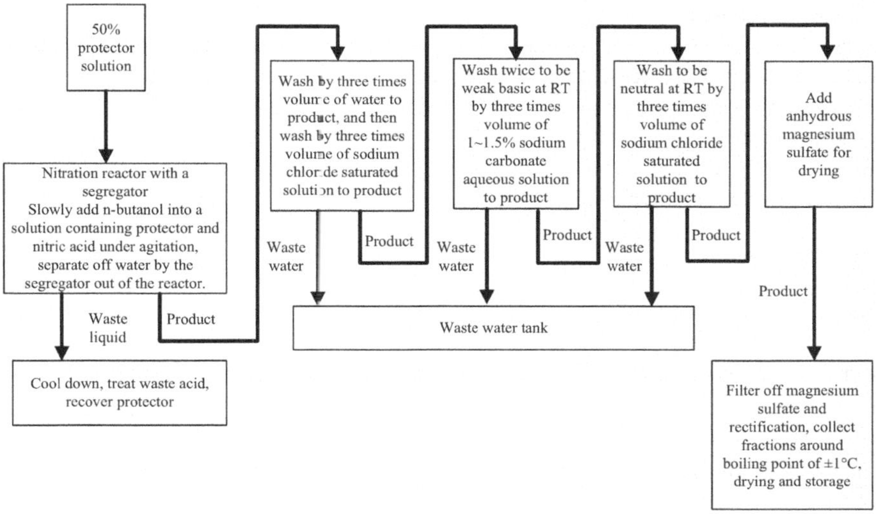

Fig. 4.7 Preparation process of n-butyl nitrate

Fig. 4.8 Facilities for producing n-butyl nitrate. Left: Nitration reactor, right: Segregator with a glass mirror

nitrate is designed and manufactured. The production facilities for n-butyl nitrate are shown in Fig. 4.8.

The volume of nitration reactor for the preparation of n-butyl nitrate is 110 L. The reactor is equipped with explosion-proof motor, cycloid reducer, reactor seal, anchor stirrer, temperature sensor, pressure gauge, a feed opening with three different diameters, water segregator with a glass mirror, 1000 mm condenser, heat oil jacket, and other devices. The facilities also include a thermal oil heating cycle temperature controller, an agitation controller, a washing vessel, and other devices to meet the requirements of n-butyl nitrate preparation process. The yield of n-butyl nitrate produced by the above-mentioned production facilities is up to 95.7%, and the purity after rectification is 98.8%.

4.6 n-Amyl Nitrate

4.6.1 Physical and Chemical Properties of n-Amyl Nitrate

Chemical formula: $C_5H_{11}NO_3$
Chemical structure:

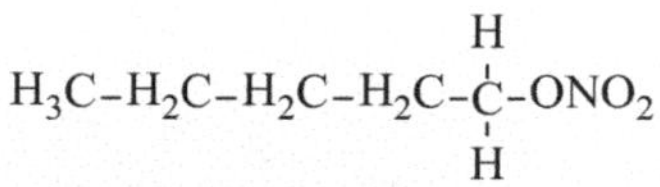

Molecular weight: 133.15
Oxygen balance: −150.2% (based on the generation amount of CO_2)
Nitrogen content: 10.52%.

4.6.2 Preparation of n-Amyl Nitrate

A certain amount of nitric acid mixture is added to the reactor [24]. The mass composition of the nitration mixture is HNO_3 60%, H_2O 30%, and the protective agent 10%. The protective agent is composed of urea 80% and ammonium nitrate 20% by mass. When the nitration mixture is heated to 82 °C, n-amyl alcohol and 15% aqueous solution of urea are added in dropwise. If the addition rate of n-amyl alcohol is too fast, there may be a risk of overflowing, so it is important to ensure slow addition of n-amyl alcohol to the three bottles. After the temperature is increased to 95 °C, the water in the system will be separated. The oil bath is heated at about 160 °C, and distillation head thermometer temperature is controlled in a certain range. When the vapor temperature is increased to close to 100 °C, water reflux in the condenser is observed. With a continuing heat of the reactor, two liquid layers are formed in the water segregator, in which the upper and lower layers are the nitrate and water, respectively. Water is separated continuously from the water segregator, and the amount of separated water is calculated. When the vapor temperature in the reactor is increased to about 150 °C, and the separated water reaches about 75%, the reaction is stopped. It can be seen that the upper layer of the reactor is light yellow, which is a nitrated layer, and the lower layer is the water layer. Under the conditions of natural cooling with stirring, until the temperature of reaction mixture drops to about 30 °C, the solution is poured out. After filter out the precipitates, the filtrate has a clear layer separation, of which the upper layer is nitrate product. The crude product is obtained after washing and drying of the separated upper organic layer. Since the nitrate product contains dissolved waste acid, it requires separating the waste acid from the nitrate. The procedure for the separation of waste acid from the nitrate is to wash the nitrate first by distilled water with five times volume of the nitrate. Subsequently, the nitrate layer is washed with a saturated $NaCl–Na_2CO_3$ solution with pH = 10 in five times volume of the nitrate to alkalescence. After washing, the nitrate solution has a pH value in the range of 7.2–7.5, and then magnesium sulfate particles are added for drying and storage. The crude product n-amyl nitrate is obtained in a yield of 97.0%. The crude n-amyl nitrate is fed to a rectification equipment and added with zeolite. The fraction with a boiling point of 154.5–155.5 °C is collected as highly pure n-amyl nitrate in a final yield of 92.4%.

4.7 Other Monohydric Nitrates

4.7.1 Properties and Preparation of Chloroglycerol Mononitrate

Chloroglycerol mononitrate is a homolog of nitroglycerin, in which a hydroxyl group is substituted by nitrate and a hydrogen on the methylene position is replaced by chlorine. Chloroglycerol mononitrate has a very weak explosive property.

Chloroglycerol mononitrate has the following three different isomers with different substitution positions:

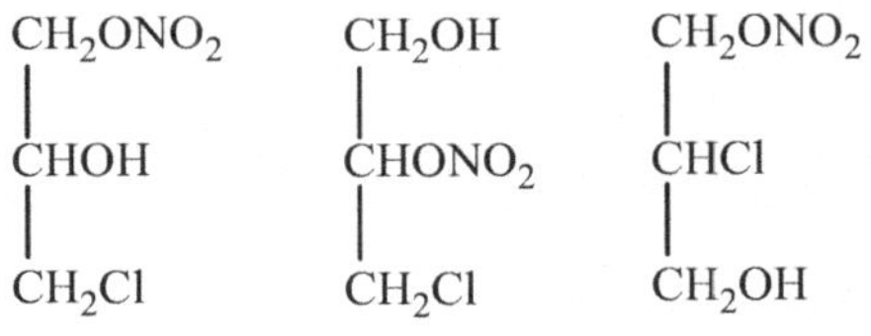

Molecular weight: 155.58
Oxygen balance: −51.42% (based on the generation amount of CO_2)
Nitrogen content: 9.00%

Chloroglycerol mononitrate is a kind of liquid explosive with weak explosive characteristics and a weak shock sensitivity. It only detonates under the action of booster charges.

Table 4.12 Properties of glycerol mononitrates with different configurations [3]

Configuration	Density (g/cm^3)	Melting point (°C)	Boiling point (°C)
α	1.53	58.5–59.0	159 (15 mmHg)
β	1.52	54	155 (15 mmHg)

4.7.2 Properties and Preparation of Glycerol Nitrate

Glycerol nitrate (or nitroglycerin) has α and β isomers [27]. Their melting points and boiling points have slight differences, which are shown in Table 4.12.

Molecular formula: $C_3H_7NO_5$
Chemical structure:

Molecular weight: 137.08
Oxygen balance: −52.52% (based on the generation amount of CO_2)
Nitrogen content: 10.22%

Glycerol mononitrate has α and β two different isomers. The properties of glycerol mononitrate are shown in Table 4.12.

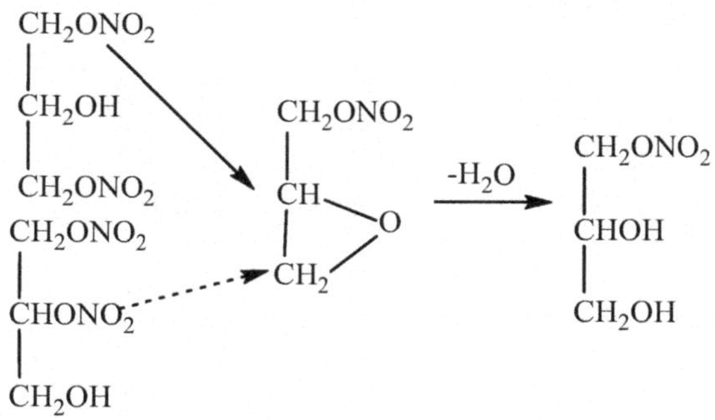

Glycerol mononitrate is a very stable material. When heated to above 100 °C, it does not decompose. It only explodes at about 170 °C in a sealed tube.

Glycerol mononitrate is a weak explosive with a significant negative oxygen balance. The explosive heat of glycerol mononitrate (water vapor condition) is only

1815.9 J/g. Impact sensitivity is so low that it can be regarded as an insensitive agent. The shock sensitivity of glycerol mononitrate to detonation is related to its physical state. For example, the lead block expansion value of glycerol mononitrate is only 75 cm^3 while crystal glycerol mononitrate can be easily detonated with a lead block expansion value of 245 cm^3.

4.7.3 Properties and Preparation of Glycidyl Nitrate

Molecular formula: $C_3H_5NO_4$
Chemical structure:

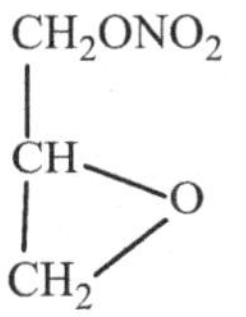

Molecular weight: 119.07
Oxygen balance: −60.46% (based on the generation amount of CO_2)
Nitrogen content: 11.76%

The glycidyl nitrate (nitrated glycidyl) boils and decomposes at 174–175 °C under atmospheric pressure and boils at 94 °C under 20 mm Hg pressure. The density of glycidyl nitrate is 1.332 at 20 °C. The glycidyl nitrate is slightly soluble in cold water with a solubility of 5 grams in 100 ml of water at 20 °C. Glycidyl nitrate is hydrolyzed to α-glycerol nitrate in boiling water. The solubility of collodion in glycidyl nitrate is very high [27].

The explosive heat of glycidyl nitrate is 3447.62 J/g. It is susceptible to detonate under shock. The average height of weight drop to cause detonation is 10–20 cm. It explodes when continuously heated to 195–200 °C with a water-filled lead block expansion value of 430 cm (2 kg drop hammer).

The glycidyl nitrate is formed by the hydrolysis of two isomers of glyceryl dinitrates. The glycerol dinitrates can be hydrolyzed at room temperature in 30% KOH solution to generate oily glycidyl nitrate in a yield of 95%. The final compound is obtained after separation, stability treatment, washing with water to neutral, and drying over a desiccant.

4.7.4 Properties and Preparation of Dichloroglycerol Nitrate

Dichloroglycerol nitrate does not have explosive property, which is formed by replacing two hydroxyl groups in glycerol with two chlorine groups [27].

Molecular formula: $C_3H_5NO_3Cl_2$
Chemical structure:

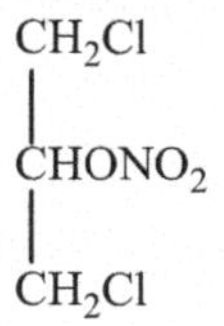

βisomer

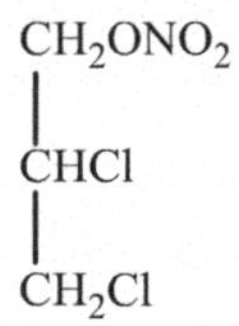

αisomer

Molecular weight: 174.07
Oxygen balance: −50.55% (based on the generation amount of CO_2)
Nitrogen content: 8.04%

Dichloroglycerol nitrate has two isomers, namely α- and β-isomers, which have boiling points of 183 and 174 °C, respectively. The mixture of such isomers can be added to nitroglycerin to produce dynamite with a low freezing point. Many of the properties of dichloroglycerol nitrate are inconsistent with those of explosives. Dichloroglycerol nitrate is more volatile than chloroglycerol dinitrate and is a poor solvent for collodion. The explosive power of dichloroglycerol nitrate is negligible. The explosive heat of dichloroglycerol nitrate is 1464.4 J/g, and the lead block expansion value is 75 cm^3 with booster charges. Dichloroglycerol nitrate is readily prepared by nitration of dichloroglycerol with anhydrous acid in a yield of 94.5%. In other words, 100 parts of dichloroglycerol can yield 127.5 parts of dichloroglycerol nitrate by weight.

References

1. Liu JP (1987) Basic chemical explosives. Guizhou Press, Guiyang
2. Liu JP, Yang R, Wang H et al (2011) The preparation of methyl nitrate. China Patent CN 201110456374
3. Jones RN, Thorn GD (1931) Can J Res B 27:282
4. Cowley EG, Partington JR (1933) J Chem Soc 1252
5. Bi XL, Liu JP (2011) Research progress in synthesis and nature pyrolysis of nitrate ester. In: Hazardous materials and security emergency technology
6. Yang RH (2010) Preparation of typical nitrate esters. Beijing Institute of Technology
7. Xu B, JP Liu (2011) Research on method for preparation of methyl nitrate. In: Hazardous materials and security emergency technology
8. Lea C (1862) Jahresber f Chemise 387
9. Desseigne G (1948) Mem poudres 30:59
10. Technical Report P.B.925, Explosives Plants D.A.G.., Krümmel, Düneberg, Christianstadt; U. S. Dept of Commerce, Washington (1945)
11. Liu JP, Yang RH, Fan BZ (2014) Method for preparing ethyl nitrate by continuous nitration of nitric acid steam. China Patent, ZL 201110456374.0, Mar 2014

12. Naoúm P (1924) Nitroglycerin u. Nitroglycerinsprengstoffe. Springer, Berlin
13. Hinkamp JB (1956) U.S. 2734910
14. Xi MH (1995) Improvement of preparation method of propyl nitrate. Chin J Energ Mater 3 (3):34–39
15. Tian Q, Liu J, Zhang X (2011) The preparation in laboratory and characterization of isopropyl nitrate. In: Hazardous materials and security emergency technology
16. Allen WG, Tobin TJ (To Imperial Chemical Industries Ltd.), Brit. Pat. 696489 (1953); See also Brit. Pats. 749734 and 749844 (1956)
17. No land WF (1955) Chem Rev. 55:137
18. Xi MH (1998) Study on preparation technology of isopropyl nitrate. Chin J Energ Mater 6 (2):70–72
19. Xie F, Zhang QZ, Zhang SS (1998) Progress in synthesis of nitrate and nitrite. J Henan Norm Univ (Nat Sci Ed) 26(2):42–50
20. Zhang HD (1998) Preparation of fuel air explosive. Beijing Institute of Technology
21. Zhang P (1996) Preparation and characterization of isopropyl nitrate. Beijing Institute of Technology
22. Liu JP (2002) Study on explosion characteristics of isopropyl nitrate. In: D 04 proceedings of the workshop. Beijing Institute of Technology
23. Wang X, Xu B, Liu J (2011) Research on the thermal properties of isopropyl nitrate. In: Hazardous materials and security emergency technology
24. Liu JP, Yang RH (2011) A method for preparing C4–C10 alkyl straight chain monohydric alcohol nitrate ester. China Patent. ZL 201110452889.3, 2011
25. Wang H, Liu JP, Wang S et al (2011) Research on method for synthesis of n-butyl nitrate. In: Hazardous materials and security emergency technology
26. Wang CP, Li HY (2005) A new process of diesel sixteen alkane improver. Chem Eng 11:58–59
27. Liu JP (2015) Liquid explosives. Springer, Berlin

Chapter 5
Dinitrate Ester

With the rapid development of explosive and propellant industries, the performance shortcomings of nitroglycerin as raw material and application product are becoming more severe. As a dinitrate, dihydric alcohol nitrate can complement some shortcomings of nitroglycerin [1]. In the book "Selection and Identification of Solid Propellants" published by NASA in 1971, triethylene glycol dinitrate and neopentyl triol trinitrate have been included in the catalog of propellants [2]. According to the relevant information, triethylene glycol dinitrate is included in the high-energy, cross-linked, and double-base propellant for intercontinental ballistic missile "Trident" that is deployed in the recently developed United States submarines. In recent years, the United States is still expanding the applications of dihydric alcohol nitrates, which are included in many formulations for practical applications.

To address the drawbacks of nitroglycerin, such as high freezing point and high shock sensitivity, researchers have prepared and investigated glycerol mononitrate and dinitrate, ethylene glycol nitrate, chloroglycerin nitrate, and other nitrates mentioned in the previous chapter. However, dinitrates cannot completely replace the nitroglycerin [2] because of their lower total energies, poor availability of raw materials, or some deteriorated properties compared with nitroglycerin. For example, some dinitrates are difficult to process because they are solids at room temperature.

The production process and the equipment used for the production of dinitrate are basically the same as those used to produce nitroglycerin. In many plants, the same production line is used to complete the production tasks of various liquid nitrates with a slight modification in the process conditions. Therefore, the properties and the primary process conditions of nitrates are briefly described without repeating the production theory in the following sections.

Dinitrate is the most widely used nitrate, which is highly explosive and has a high risk during production. The focus of this chapter is to describe the conventional processes for the preparation of more than ten types of dinitrates, such as ethylene glycol dinitrate, diethylene glycol dinitrate, triethylene glycol dinitrate, ethylene glycol-N-nitramine dinitrate, with an emphasis on the new methods of injection nitration and vapor phase nitration.

J. Liu, *Nitrate Esters Chemistry and Technology*,
https://doi.org/10.1007/978-981-13-6647-5_5

5.1 Ethylene Glycol Dinitrate

Ethylene glycol dinitrate, also known as nitrated ethylene glycol, code-named EGDN or GDN. Ethylene glycol dinitrate is a transparent liquid and is the only liquid explosive with a zero oxygen balance [3], which is a very important explosive in the energetic materials.

5.1.1 Properties of Ethylene Glycol Dinitrate

5.1.1.1 General Properties of Ethylene Glycol Dinitrate

Molecular formula: $C_2H_4N_2O_6$
Chemical structure:

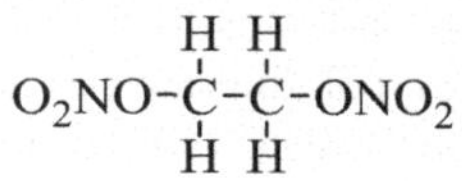

Molecular weight: 152.06
Oxygen balance: 0% (based on the generation amount of CO_2)
Nitrogen content: 18.42%
Freezing point: −22.3 to −22.80 °C
Boiling point: 197.5 °C
Density: 1.489 g/ml (2 °C)
The latent heat of melting of the solidified product: 125.52 J/g.

Ethylene glycol dinitrate is nonhygroscopic. Its solubility in water is slight but increases with increasing temperature. The solubility of ethylene glycol dinitrate in 100 ml of water is listed in Table 5.1.

Ethylene glycol dinitrate is soluble in most organic solvents but slightly soluble in carbon tetrachloride and gasoline. The solubility of collodion in ethylene glycol dinitrate is higher than that of nitroglycerin.

Table 5.1 Solubility of glycol dinitrate in 100 mL water

Temperature (°C)	Solubility (g)
0	0.57
15	0.62
20	0.68
50	0.92

Table 5.2 The volatilization loss of ethylene glycol dinitrate

Time (h)	Volatilization loss		Remark
	20 °C	35 °C	
24	0.14	3.6	15 g sample heated on a glass-surface vessel with a diameter of 56 mm and height of 40 mm
48	0.37	8.1	
168	1.63	21.7	
288	2.06	46.6	
720	6.91	68.5	
4320	47.0	100.0	

The volatility of ethylene glycol dinitrate is 20 times more than that of nitroglycerin. Its volatility dramatically increases with increasing temperature. At the temperature of 60 °C, the volatility of ethylene glycol dinitrate is 2.2 mg/cm^2/h. The volatilization loss of ethylene glycol dinitrate is shown in Table 5.2.

The vapor pressure of ethylene glycol dinitrate increases with increasing temperature, as shown in Table 5.3.

The viscosities of ethylene glycol dinitrate are 0.08056 and 0.02831 poise at the temperatures of 15 and 40 °C, respectively, which are about 10 times smaller than that of nitroglycerin. The correlation between viscosity and temperature is shown in Table 5.4.

The refractive index of ethylene glycol dinitrate varies slightly with increasing temperature. The refractive index of ethylene glycol dinitrate at different temperatures is shown in Table 5.5.

Table 5.3 Vapor pressure of ethylene glycol dinitrate [4, 5]

Temperature (°C)	Vapor pressure (mmHg)
0	0.0044
20	0.046
30	0.125
40	0.289
50	0.735
60	1.302
70	3.986
80	5.976
90	13.972
100	22.062

Table 5.4 The viscosity of ethylene glycol dinitrate at different temperatures

Temperature (°C)	15	20	23.6	40
Viscosity (centipoise)	8.056	4.2	3.63	2.831

Table 5.5 The relationship between the refractive index of glycol dinitrate and temperature [6]

Temperature (°C)	0	15	25	35
N	1.4546	1.4491	1.4454	1.4417

Ethylene glycol dinitrate can be hydrolyzed when heated in an aqueous solution of sodium hydroxide or potassium hydroxide. When contacted with an ethanol solution of potassium hydroxide, the reaction is very vigorous and produces potassium nitrite and potassium glycolate.

Ethylene glycol dinitrate can be hydrolyzed in hot water. Ethylene glycol dinitrate of 0.008% can be hydrolyzed in water at 60 °C for 120 h. The ethylene glycol dinitrate can decompose after contact with concentrated acid for a long time.

The solubility of ethylene glycol dinitrate in water is greater than in nitroglycerin. At the temperature of 20 °C, the solubility of ethylene glycol dinitrate in water is four times than in nitroglycerin, which is noteworthy during the production process of ethylene glycol dinitrate. The solubility of ethylene glycol dinitrate in water is shown in Table 5.6.

(1) The solubility of ethylene glycol dinitrate in nitric acid aqueous solution is shown in Table 5.7.

The solubility of ethylene glycol dinitrate in nitric acid aqueous solution is shown in Fig. 5.1.

(2) The solubility of ethylene glycol dinitrate increases with increasing alkali concentration in an alkaline solution. The solubility of ethylene glycol dinitrate in aqueous sodium carbonate solution is listed in Table 5.8.
(3) Ethylene glycol dinitrate can be well dissolved in organic solvents, such as methanol, ethanol, ether, chloroform, acetone, benzene, nitrobenzene, and toluene, but is hardly dissolved in carbon tetrachloride. At room temperature, it is readily dissolved in alcohol, ether, and nitroglycerin.

Ethylene glycol dinitrate can dissolve nitrocellulose well (nitrogen content of 11.8–12.2%). Even at low temperatures, ethylene glycol dinitrate is a good solvent of weak cotton. Ethylene glycol dinitrate can plasticize nitrocellulose at room

Table 5.6 The aqueous solubility of ethylene glycol dinitrate at different temperatures

Temperature (°C)	0	15	20	50	60
Solubility (g/100 g of water)	0.57	0.62	0.68	0.92	0.85

Table 5.7 The solubility of ethylene glycol dinitrate in nitric acid aqueous solution

The content of nitric acid (%)		0	4.93	9.99	19.67	29.58
The weight of ethylene glycol dinitrate dissolved (%)	0 °C	0.57	0.77	1.02	1.52	2.03
	20 °C	0.631	0.81	1.10	1.61	2.24

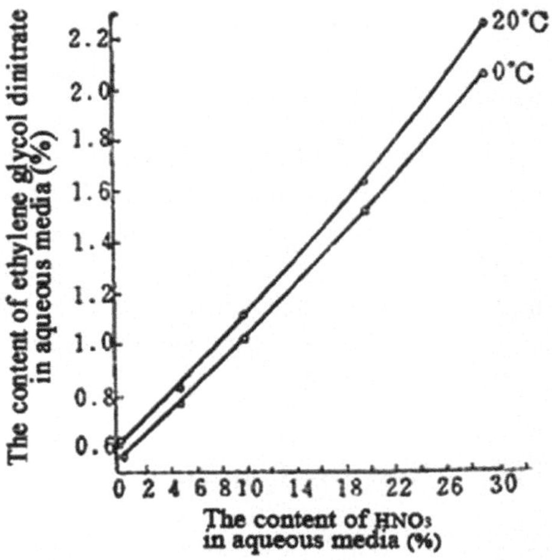

Fig. 5.1 The solubility of ethylene glycol dinitrate in nitric acid aqueous solution

Table 5.8 The solubility of ethylene glycol dinitrate in aqueous sodium carbonate solution at 20 °C

The content of Na_2CO_3 (%)	The solubility of ethylene glycol dinitrate (wt%)
1.39	0.48
12.5	1.14

temperature. The performance of dissolved nitrocellulose in ethylene glycol dinitrate is better than that of nitroglycerin. However, it cannot be used in gunpowder because of high volatility.

5.1.1.2 Thermal Properties and Chemical Stability of Ethylene Glycol Dinitrate

The 5 s delayed explosion temperature of ethylene glycol dinitrate is 257 °C.
The formation heat of ethylene glycol dinitrate: 1529.3 J/g (constant volume); 229.3 kJ/mol (constant volume).
The combustion heat of ethylene glycol dinitrate: 7380.6 J/g (constant volume); 1121 kJ/mol (constant volume); 7334.6 J/g (constant pressure).
The latent heat of the solidification product of ethylene glycol dinitrate is 125.5 J/g.
The thermal coefficient hi: 73.43 kJ/g.
Explosion coefficient βi: 74.25 kJ/g.
Specific volume coefficient ωi: 0.007368 L/g.
The reaction heat of ethylene glycol dinitrate:

$$C_2H_2(OH)_2 + 2HNO_2 \rightarrow C_2H_4(ONO_2) + 2H_2O + \Delta H$$
$$\Delta H = 35.6\,kJ/nmolethylene\,glycol$$

Ethylene glycol dinitrate has high thermal stability and chemical stability [7]. It does not decompose when heated at 75 °C for 260 h, indicating its stability is better than that of nitroglycerine. In addition, ethylene glycol dinitrate has high volatility. In Abel test, the heat resistance of glycol dinitrate is 30 min. Ethylene glycol dinitrate explodes when heated to 195–200 °C when the explosion occurred. The activation energy of ethylene glycol dinitrate during thermal decomposition is E = 149.37 kJ/mol.

5.1.1.3 Explosive Properties

Ethylene glycol dinitrate is a powerful liquid explosive with zero oxygen balance. The theoretical explosion reaction equation for complete detonation is

$$\begin{array}{l} CH_2ONO_2 \\ | \\ CH_2ONO_2 \end{array} \longrightarrow 2CO_2 + 2H_2O + N_2$$

As all the carbon, hydrogen, and oxygen are fully utilized in the explosion reaction, the explosion heat of ethylene glycol dinitrate is 6610.72 J/g (water vapor condition) or 7133.72 J/g (water liquid condition). The formation heat of ethylene glycol dinitrate is $-\Delta H_f$ 283.26 kJ/mol. It is relatively insensitive to shock with a characteristic height of 20–25 cm (2000 g weight dropping hammer). At the temperature of 10–11 °C and the charge of 0.1 g, the characteristic height is 7 cm (5 kg weight dropping hammer) with 100% detonation.

Ethylene glycol dinitrate is very sensitive to friction. It detonates when ground in a porcelain mortar. All samples detonate in 10 friction pendulum tests.

The authors performed the composition analysis for the explosion product of ethylene glycol dinitrate. The result of composition analysis is illustrated in Table 5.9.

According to the results of gas composition analysis, ethylene glycol dinitrate explosion reaction is expressed as

$$C_2H_4(ONO_2)_2 \rightarrow 1.34CO_2 + 0.66CO + 1.96H_2O + 0.04H_2 + N_2 + 0.35O_2$$

Ethylene glycol dinitrate is very sensitive to flames [8, 9] and can be stably combusted at atmospheric pressure in contact with flames, which is slightly

Table 5.9 The composition of ethylene glycol dinitrate explosion product

Explosion product	Weight (%)	Volume (%)
CO_2	57.9	40
H_2O	23.7	40
N_2	18.4	20

different from nitroglycerin. The combustion velocity of ethylene glycol dinitrate is linearly proportional to the pressure in a glass tube with a diameter of 4 mm at the temperature of 14–15 °C under a pressure range of 300–770 mmHg.

$$U = A + BP$$

where

U combustion velocity (cm/s);
P pressure (mmHg);
A, B constants.

Ethylene glycol dinitrate cannot be burnt under the pressure of 230 mmHg but is easily detonated. This linear relationship between the combustion velocity and the pressure remains until the pressure reaches 10 kg/cm. If the pressure is increased further, the combustion velocity will be increased dramatically until an explosion occurs. The local overheating or heating a large amount of ethylene glycol dinitrate can lead to detonation.

The shock sensitivity of ethylene glycol dinitrate is greater than that of nitroglycerin. It can be detonated easily by a detonator, and a mixture containing ethylene glycol dinitrate also has the similar property. Ethylene glycol dinitrate has a detonation velocity of 7300 m/s (1.485 g/cm^3), a detonation temperature of 4553 K, a specific volume of 737.2 ml/g, and the explosive force of 12,32,000 kg dm/kg.

The water-filled lead block expansion value of ethylene glycol dinitrate is 650 cm, which is 110% of NG. The brisance values of ethylene glycol dinitrate are determined to be 62.7 g in sand testing and 22–24 mm compression value by lead cylinder method. The detonation velocity varies with the charge condition. The relationship between the detonation velocity of ethylene glycol dinitrate and the charging conditions is shown in Table 5.10.

Table 5.10 Correlation between the detonation velocity of glycol dinitrate and the loading condition

Loading condition/method	Materials of container	Initiation approach	Detonation velocity (m/s)
Loading diameter φ36 mm	Seamless steel tube (wall thickness 2.5 mm)	Directly by 8# detonator	7780
Loading diameter φ38 mm	Seamless steel tube (wall thickness 2.5 mm)	80 g mandelic acid as booster charge, and directly by 8# detonator	7960
Loading diameter φ45 mm	Seamless steel tube (wall thickness 2.5 mm)	80 g mandelic acid as booster charge, and directly by 8# detonator	8100
φ60 mm Porcelain crucible	Open	Directly by 8# detonator	1830
φ60 mm Porcelain crucible	Cool down to −70 °C	Directly by 8# detonator	7980

5.1.2 Preparation of Ethylene Glycol Dinitrate

5.1.2.1 Preparation Process of Ethylene Glycol Dinitrate

(1) Early preparation methods of ethylene glycol dinitrate

The early preparation method of ethylene glycol dinitrate involves the nitration of ethylene [10]. A mixture of β-nitroethanol and ethylene glycol dinitrate was obtained by using this method.

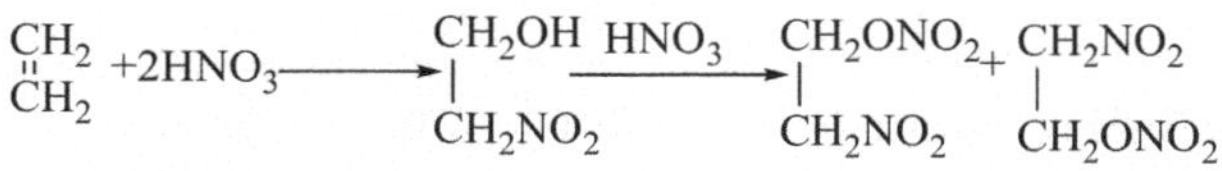

Later on, it was found that ethylene glycol dinitrate could be directly obtained by dehydration in the process of nitration of ethylene.

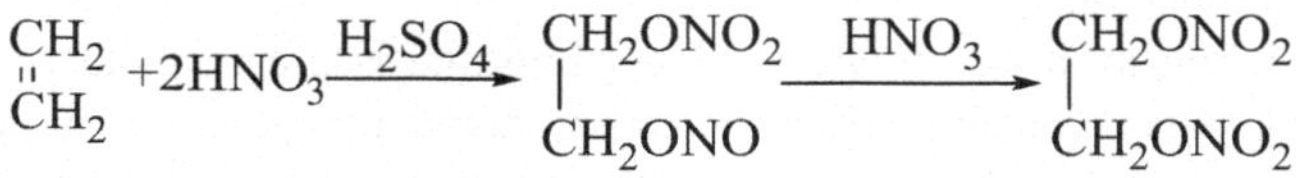

In the above reactions, 50–60% of ethylene is converted via the first reaction, and 40–50% of ethylene is converted via the second reaction.

(2) Production of ethylene glycol dinitrate by electrolytic ethylene

Electrolytic ethylene method [11] can be used to produce ethylene glycol dinitrate and diethylene glycol dinitrate. The platinum wire mesh immersed in the calcium nitrate acetone solution is used as the anode, and the ethylene in the gaseous state continuously flows through the solution. The cathode space behind the aluminum cathode is filled with calcium nitrate in nitric acid solution. On the anode, the nitrate ions are converted into free radicals $\dot{N}O_3$, part of which react with ethylene to form ethylene glycol dinitrate.

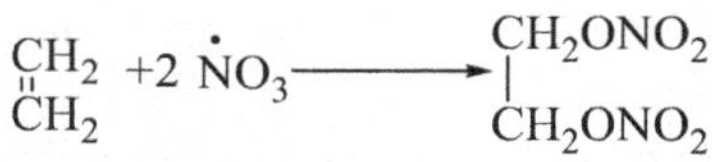

The interreactions among nitric acid-free radicals generate oxygen-free radicals and $\dot{N}O_3$ free radicals, which further react with ethylene to form diethylene glycol dinitrate.

$$2NO_3 \rightarrow N_2O_5 + [O] \xrightarrow{+H_2O} 2HNO_3 \rightarrow 2NO_3$$

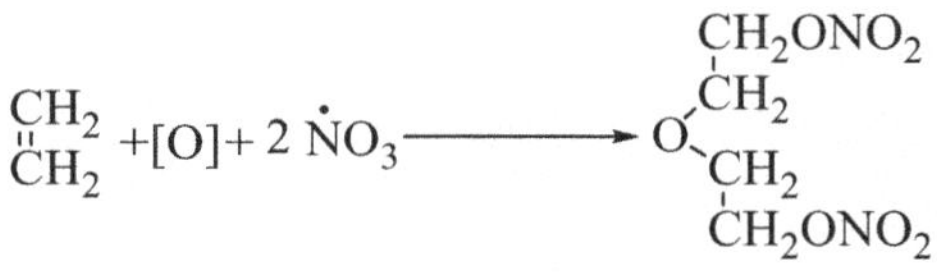

After the neutralization of anionic liquid with calcium hydroxide and removal of acetone by distillation, the resulting products are mainly calcium nitrate and ethylene glycol dinitrate. The calcium nitrate is separated from ethylene glycol dinitrate by filtration. The filtrate is subjected to water wash, stability treatment, and drying to afford ethylene glycol dinitrate product.

(3) Modern manufacturing technologies of ethylene glycol dinitrate [10, 12, 13]

It is well known that the nitration of ethylene glycol in a mixed nitric and sulfuric acid yields ethylene glycol dinitrate. The optimal conditions of nitration are as follows, a mixed acid composed of H_2O 11.5–20%, HNO_3 18–40%, and H_2SO_4 45–68.5%, a nitration temperature in the range of 5–15 °C, and the nitration coefficient range of 12–24. The conversion rate of nitration by this method is typically up to 92–96%.

The nitration of ethylene glycol is an exothermic reaction. The reaction heat is $\Delta H = 355.64$ J/mol per mol of ethylene glycol.

Because of the accumulation of nitration heat, the reaction heat in industrial nitration production typically exceeds that of the laboratory scale. The primary reaction mechanism of ethylene glycol nitration is described as follows:

$$C_2H_4(OH)_2 + 2HNO_3 \xrightarrow{H_2SO_4} C_2H_4(ONO_2)_2 + H_2O + 355.64\,\text{J/mol}$$

The reaction heat of nitration reaction of ethylene glycol in the industrial production is several times of that small-scale nitration reaction. As a matter of fact, when the glycol and mixed acid are fully mixed, the nitration reaction rate is so fast that the nitration reaction only takes 1–1.5 s to complete.

The authors [14] studied the effects of different nitration conditions, mixed acid spray nitration process, and air agitation nitration process [15]. These processes have been successfully applied to the industrial scale production of ethylene glycol dinitrate. Ethylene glycol used in the preparation of ethylene glycol dinitrate should meet the specifications shown in Table 5.11.

Table 5.11 Specification of glycol for nitration reaction [28]

Parameter	Value
Boiling point (°C)	196–200
Density (g/cm3)	1.1130–1.1135
Acidity (H_2SO_4-based %)	<0.1
Saponification value (Na_2O %)	<1.0

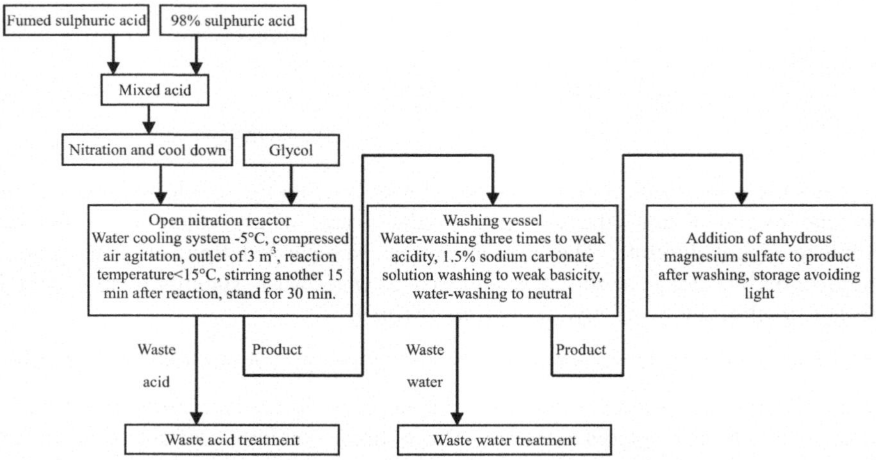

Fig. 5.2 Production process of ethylene glycol dinitrate by the batch method

The nitration processes of ethylene glycol dinitrate and nitroglycerin have some differences because of slightly different properties of raw materials and products. The nitration of ethylene glycol does not need preheating, and the optimal nitration temperature is 10–15 °C as a high nitration temperature will reduce the product yield. The production process of ethylene glycol dinitrate by the batch method is shown in Fig. 5.2.

The nitration of 100 kg of ethylene glycol at the 10–12 °C can produce ethylene glycol dinitrate 229.075 kg, which equals to a theoretical yield of 93.5%.

5.1.2.2 Ethylene Glycol Dinitrate Production Process

(1) Process conditions

Reaction for the preparation of ethylene glycol dinitrate

$$\begin{array}{ccccc} C_2H_4(OH)_2 + & 2HNO_3 \rightleftharpoons & C_2H_4(ONO_2)_2 + & 2H_2O + & Q \\ 62 & 126 & 152 & 36 & 35.56\,\text{kJ/mol} \\ 100 & 203.2 & 245.2 & 58.1 & \end{array}$$

In the ethylene glycol dinitrate production process, the main nitration reaction equipment and nitration process are container nitrator and continuous nitration, respectively. The production process of ethylene glycol dinitrate by continuous nitration is shown in Fig. 5.3.

The main process conditions for the production of ethylene glycol dinitrate by continuous nitration are shown in Table 5.12.

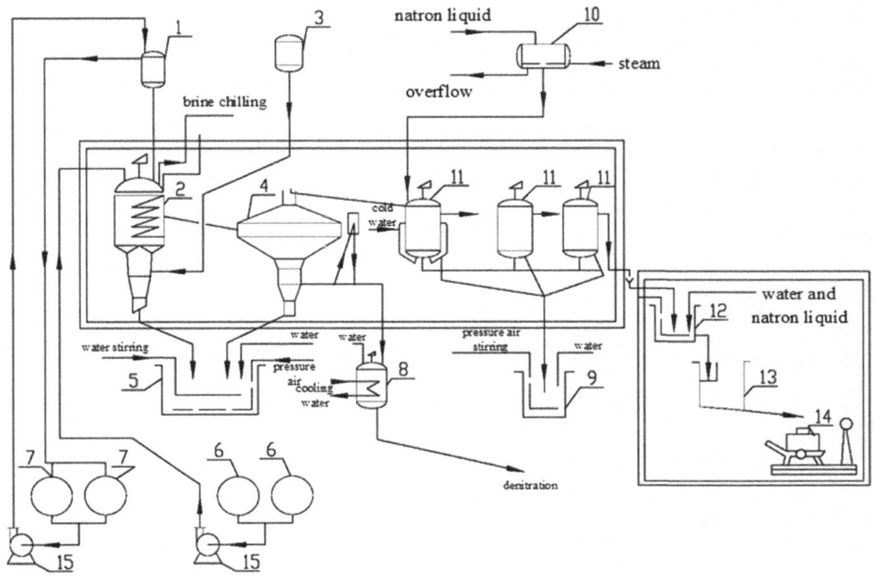

Fig. 5.3 The production process of ethylene glycol dinitrate by continuous nitration

Table 5.12 The production process of ethylene glycol dinitrate

Nitration equipment	Ethylene glycol is nitrated in a container nitrator alone
Composition of mixed acid	HNO_3 41%; H_2SO_4 58%; H_2O 1.0%
Nitration coefficient	5.9
Nitration temperature	10–15 °C

Table 5.12 shows the main process conditions for the production of ethylene glycol dinitrate.

(2) Raw material specification

The raw material quality requirements for the manufacture of ethylene glycol dinitrate are shown in Table 5.13.

Compared with the nitration of glycerol, the nitration reaction of ethylene glycol has a faster reaction rate, smaller reaction heat, and lower freezing point of the product. Therefore, the production of ethylene glycol dinitrate is suitable for rapid nitration at a low temperature. The effects of the water content of a mixed acid on the exothermic reaction and the heat of nitration reaction are shown in Fig. 5.4.

It can be seen from Fig. 5.4 that the reaction is basically completed at the time of 0.6 s when the water content in the nitric acid is low. Even with 15% of water content, the reaction is completed only in 1–1.2 s.

Table 5.13 Technical specifications for ethylene glycol

Specification	Criteria
Appearance	Colorless or light yellow transparent liquid without visible impurities
Density (d_4^{20})	1.1130–1.1140
Content of ethylene glycol	$\nless$97%
Boiling range (by volume)	196–199 °C distillate $\geq$97%
Refractive index n_D^{20}	1.4312–1.4320
Reaction	Neutral
Solubility in water	Transparent when mixed with water
Chloride (as $C1^-$)	$\ngtr$0.001%
Sulfate (as SO_4)	$\ngtr$0.001%
Iron (as Fe+++)	$\ngtr$0.0025%
Ash content	$\leq$0.003%
Saponification value	The consumption of KOH by 1 g Ethylene glycol is less than 0.45 mg
Reduced product	None

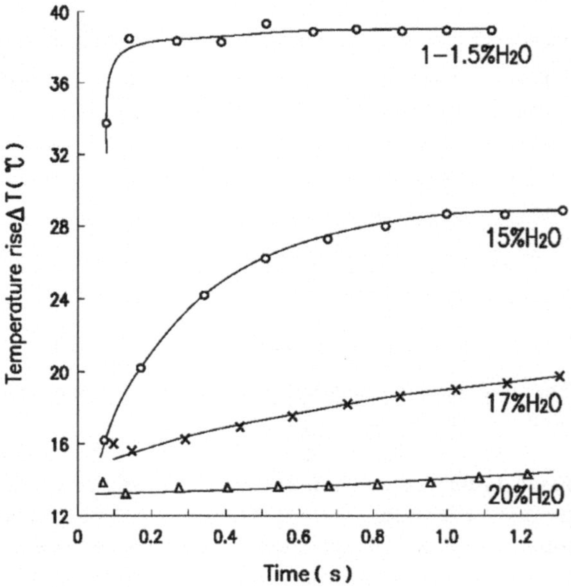

Fig. 5.4 The heat quantity of the nitration of ethylene glycol with different water contents in the mixed acid

Since the viscosity of ethylene glycol is low, the control of air bubble pressure at 0.02–0.05 MPa is sufficient to complete the nitration of ethylene glycol by using a mixed acid addition method. The nitration of ethylene glycol can be completed fastly because of the low freezing point of nitrate product. For the same reason, the reaction can be enhanced by cooling without risk of freezing. The optimum nitration temperature of ethylene glycol is 10–15 °C, and the yield is decreased when the temperature is higher than this range.

The aqueous solubility of ethylene glycol dinitrate is high, so as little as possible water should be used during the washing and the temperature of the water should be in the range of 25–45 °C. Otherwise, it will cause the loss of volatile product and the reduction of yield. Because of the low viscosity of ethylene glycol dinitrate, during the washing and stability treatment, the stirring speed should be slower than nitroglycerin to reduce the loss caused by evaporation.

The theoretical yield of ethylene glycol dinitrate is 245.2%, and the actual yield by continuous nitration is 231.2%, which is equivalent to 94.29% of the theoretical yield.

The author studied the dynamics of ethylene glycol nitration reaction in detail during the investigation of injection nitration reaction. After numerous experiments, it was found that the composition of the mixed acid under different nitration conditions should be different. When the injection nitration process is used, the composition of mixed acid should be controlled in the range of nitric acid 18–40% HNO_3, 45–69% H_2SO_4, 11.0–20% H_2O. Nitration coefficient is in the range of 12–25 under this condition.

5.1.3 Problems in the Production of Ethylene Glycol Dinitrate

5.1.3.1 Equilibrium in the Ethylene Glycol Dinitrate Production

The theoretical basis of the production of ethylene glycol dinitrate is basically the same as that discussed in the previous chapters. In this chapter, only the following factors are discussed to guide the production, such as the reaction equilibrium and phase equilibrium in the nitration of ethylene glycol, the phase equilibrium and distribution in the washing process, and the relationship between product loss and stability treatment.

(1) Equilibrium of ethylene glycol nitration

When the water content is greater than the hydrate $H_2SO_4{\cdot}H_2O$, the equilibrium of waste acid is expressed as

$$C_2H_4(ONO_2)_2 + 2H_2O \rightleftharpoons C_2H_4(OH)(ONO_2) + HNO_3$$

When the water content is less than the hydrate $H_2SO_4{\cdot}H_2O$, the equilibrium of waste acid is described as

$$C_2H_4(ONO_2)_2 + H_2SO_4 \rightleftharpoons C_2H_4(ONO_2)(OSO_3H) + HNO_3$$

The solubility of ethylene glycol dinitrate in waste acid is closely related to the water content in waste acid. When the water content of waste acid is higher than that of hydrate, the content of ethylene glycol dinitrate in waste acid increases with increasing content of water in waste acid, which is attributed to the influence of hydrolysis. The content of ethylene glycol dinitrate in waste acid is shown in Figs. 5.5 and 5.6.

As shown in Figs. 5.5 and 5.6, the nitration equilibrium constants of the nitroglycerin system and the ethylene glycol dinitrate system are significantly different. When the water content in the waste acid of the ethylene glycol dinitrate nitration system is more than that of the hydrate point, the content of nitrate increases slowly. In contrast, the increase in total oil content is negligible.

Fig. 5.5 The correlation between the content of ethylene glycol dinitrate with the water content in waste acid

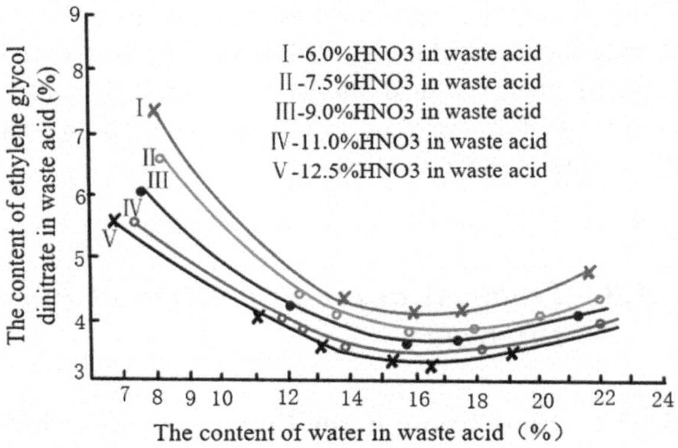

Fig. 5.6 The correlation between the content of ethylene glycol dinitrate and nitroglycerin with the water content in waste acid

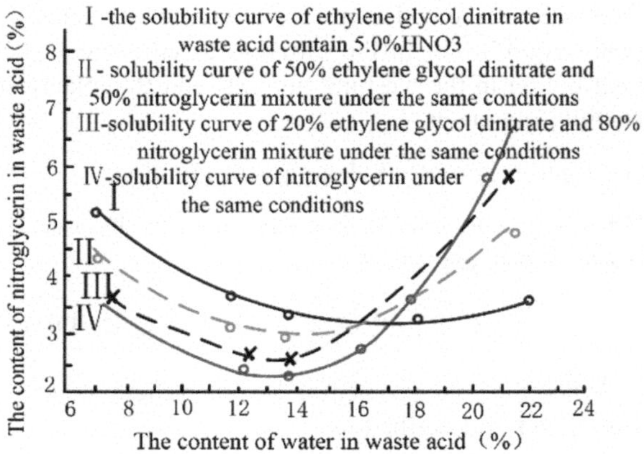

The composition of the equilibrated nitrate phase in the nitrated ethylene glycol phase mainly contains ethylene glycol mononitrate, nitric acid and water, and a small amount of sulfuric acid. The composition of these components and the composition of waste acid restrict each other. The relationship between the composition of the nitrated ethylene glycol phase and the water content in waste acid is shown in Fig. 5.7.

The relationship between the amount of nitric acid and the composition of the waste acid in the oil phase consisting of the mixture of nitrated ethylene glycol and nitroglycerin is shown in Fig. 5.8.

As shown in Fig. 5.8, the curves of the mixtures of ethylene glycol dinitrate and nitroglycerine with different ratios are between the curves of these two compounds. When the water content is low, curves of these two mixtures are parallel; when the water content increases, these two curves are close to each other and intersects with the water content of 22%. The hydrolysis of nitroglycerin is more vigorous (about 15 times) than that of ethylene glycol dinitrate, probably because the water content exceeds the hydrate critical point. At this time, the chemical equilibrium in the oil phase is described as

$$\begin{array}{l} CH_2ONO_2 \\ | \\ CH_2OH \end{array} + HNO_3 \rightleftharpoons \begin{array}{l} CH_2ONO_2 \\ | \\ CH_2ONO_2 \end{array} + H_2O$$

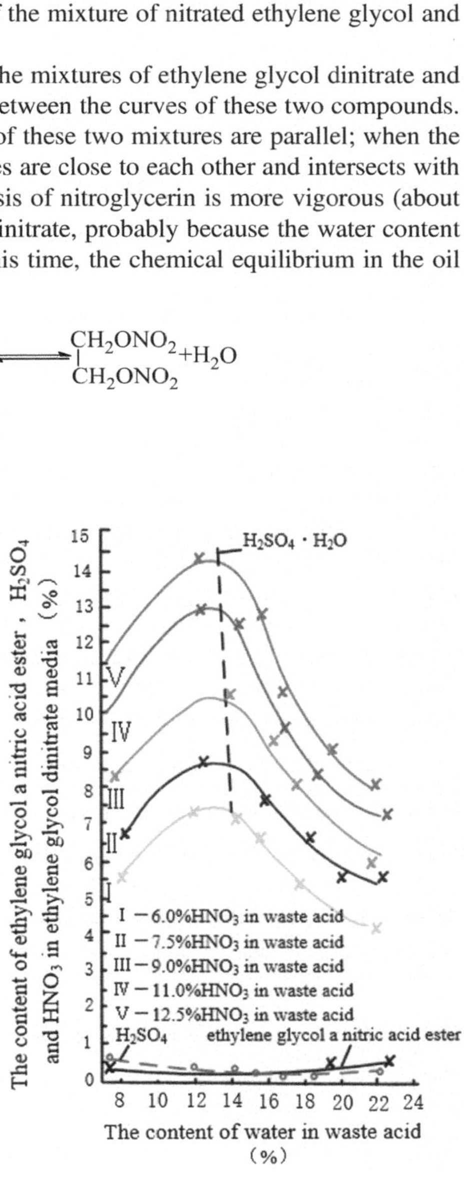

Fig. 5.7 Relationship between composition of nitrated ethylene glycol phase and water content in waste acid

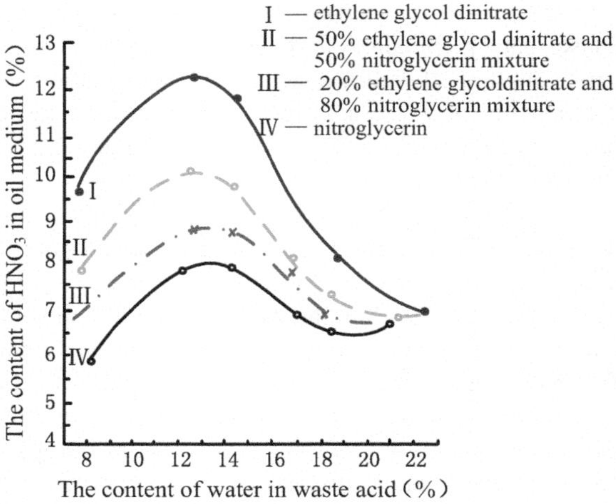

Fig. 5.8 Relationships between nitric acid content and water content in waste acid

The percentages of composition at the reaction equilibrium are shown in Table 5.14.

The equilibrium factor can be expressed as

Equilibrium factor = a · b/c · d

As shown in Fig. 5.9, the content of ethylene glycol mononitrate and equilibrium factor of equilibrated oil phase are related to the water content in the oil phase.

In the nitroglycerin manufacturing system, the association of glycerol dinitrate molecules during nitration reduces the water content in the nitration system, and the effect of nitric acid dissociation on the nitration is much less significant. The difference between the two systems is mainly due to the association effect in the nitroglycerine system, which reduces the equilibrium factor in the nitroglycerine system. At the time of nitration, the equilibrium factor of the ethylene glycol nitration system is about 15 times greater than that of nitroglycerin system when the water content of the glycerol phase is 0.3%.

(2) The equilibrium of ethylene glycol nitrate in the washing phase

1. The distribution of the glycol-glycerol mixed phase and the aqueous phase.

In general, since a mixture of nitroglycerin and nitrated ethylene glycol is used for practical applications, the semifinished product containing the mixed acid after nitration is fed into the washing system simultaneously. This distribution system can be considered as two parallel systems, namely, distribution of ethylene glycol

Table 5.14 Percentages of composition at reaction equilibrium

Number	Composition	Number	Composition (%)
a	Ethylene glycol dinitrate	c	Nitric acid
b	H_2O%	d	Ethylene glycol moninitrate

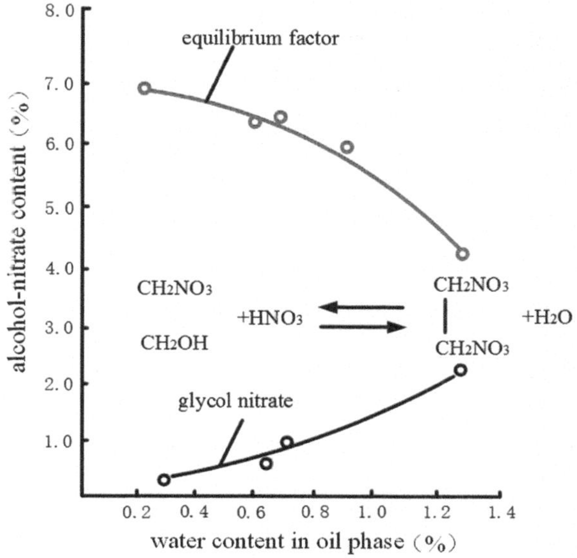

Fig. 5.9 The content of ethylene glycol mononitrate and equilibrium constant

dinitrate between nitroglycerin and distribution of nitroglycerin between ethylene glycol dinitrate and water. Both systems do not have an ideal distribution constant that conforms to the Nairnst's law, which is mainly attributed to the effect of small amount of associated products in the production process. The distribution relationship between nitrate mixture system and water is shown in Fig. 5.10.

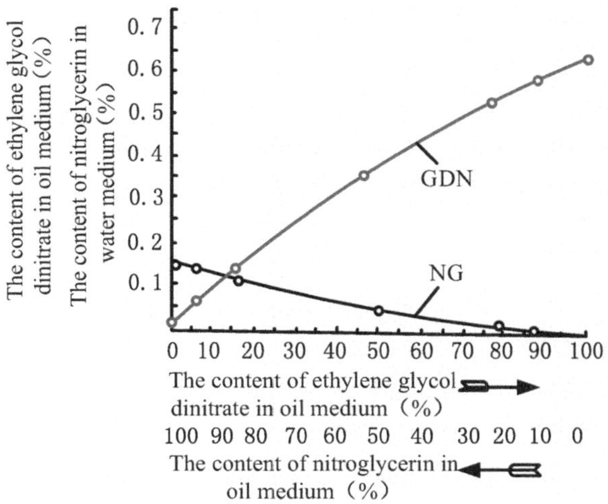

Fig. 5.10 The distribution relationship between the mixture of ethylene glycol dinitrate and nitroglycerin and water

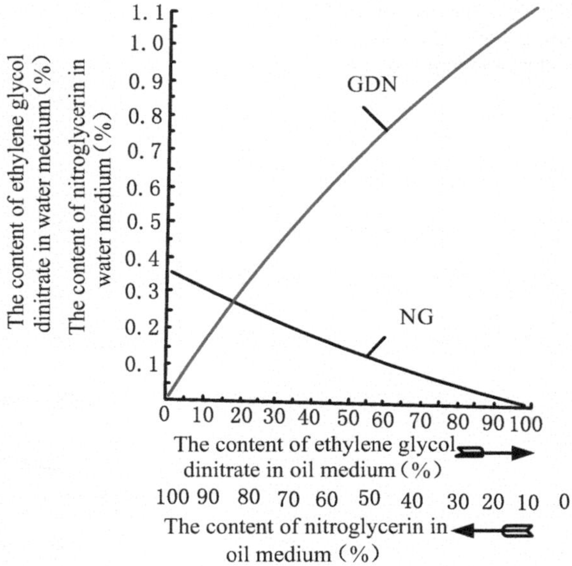

Fig. 5.11 Distribution relationship between ethylene glycol dinitrate-nitroglycerin mixed phase and 10% nitric acid phase

2. The distribution relationship between the mixed phase of ethylene glycol nitrate-nitroglycerin and the 10% nitric acid phase. The distribution relationship between the mixed nitrate phase and 10% nitric acid aqueous phase is shown in Fig. 5.11.

3. The partition coefficients of ethylene glycol dinitrate in acidic and neutral water.

C1 and C represent the percentages of ethylene glycol dinitrate in the oil phase and ethylene glycol dinitrate in the aqueous phase, respectively. The distribution factor can be expressed as C/C1. The relationship between the content of ethylene glycol dinitrate in the oil phase and the partition coefficient is shown in Fig. 5.12.

4. The distribution of nitric acid in ethylene glycol dinitrate and water.

The distribution relationship of nitric acid in ethylene glycol dinitrate and water is similar to that of nitroglycerin system. Due to the dissociation of nitric acid, the partition coefficient C_O/C_W is not a constant. In this case, distribution relationship can be described by $\sqrt{C_O}/C_W$, whose value is close to a constant, as shown in Table 5.15 and Fig. 5.13.

(3) The loss of ethylene glycol dinitrate during washing in the presence of mixed acid

1. The relationship between the washing loss of ethylene glycol dinitrate and the amount of washing water.

The content of ethylene glycol dinitrate and mononitrate in the aqueous phase of the washing system can be calculated using the relationship of the partition coefficients.

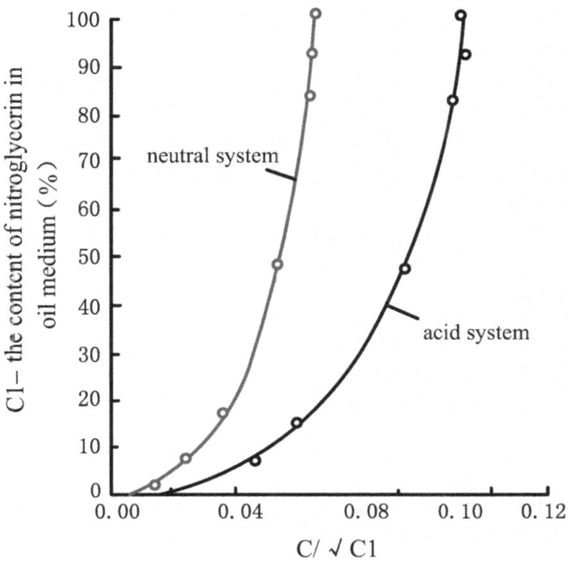

Fig. 5.12 The content of ethylene glycol dinitrate in the oil phase and the partition coefficient

Table 5.15 Partition coefficient of nitric acid in ethylene glycol nitrate and water phases at the temperature of 20 °C

The content of HNO_3 in water phase (%) (C_W)	The content of HNO_3 in oil phase (%) (C_O)	Partition coefficient (C_O/C_W)	Partition coefficient ($\sqrt{C_O}/C_W$)
4.93	0.004	0.0008	0.013
9.99	0.007	0.0007	0.008
19.67	0.037	0.002	0.010
29.58	0.154	0.005	0.013

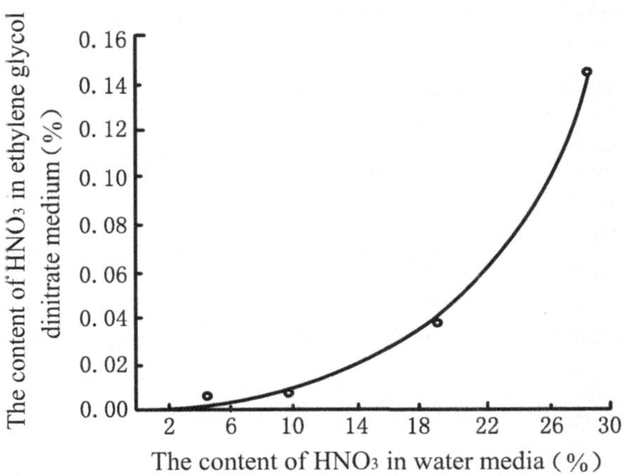

Fig. 5.13 The distribution relationship of 0.2% ethylene glycol dinitrate in nitric acid and water

It is also feasible to calculate the functional relationship between the wash loss of nitrate and the amount of wash water in acidic and neutral systems. As shown in Tables 5.16 and 5.17 and Fig. 5.14, the product loss in alkali washing used for safety treatment can be considered to be the same as that in neutral washing.

The nitrated product contains 9% nitric acid, 0.2% ethylene glycol mononitrate, and the rest is ethylene glycol dinitrate, which is the most common outcome in the production.

The two solid lines in Fig. 5.14 are expressed as: $Y = 0.0152\ a + 1.55$ (acidic system); $Y = 0.0162\ a + 0.07$ (neutral system).

Where

a the amount of water (kg) for washing 100 kg of oil;

Y the loss amount of nitrated ethylene glycol (kg) for washing 100 kg of raw material ethylene glycol.

Table 5.16 The washing loss of ethylene glycol dinitrate with 9% nitric acid and 0.2% ethylene glycol mononitrate

The weight ratio of unwashed oil to washing water	The ratio of ethylene glycol to nitrate in oil phase (%)	Water phase				Washing loss (kg)	
		Nitric acid (%)	Ethylene glycol mononitrate (%)	Ethylene glycol dinitrate (%)	Total amount of oil (%)	Based on 100 kg ethylene glycol	Based on 100 kg nitrated ethylene glycol
100/20	0.13	31.0	0.30	2.34	2.64	1.88	0.77
100/50	0.09	15.3	0.21	1.37	1.58	2.29	0.93
100/100	0.06	8.3	0.14	1.01	1.15	3.07	1.25
100/130	0.05	6.5	0.11	0.92	1.03	3.51	1.43
100/200	0.03	4.3	0.08	0.82	0.90	4.61	1.88

Table 5.17 Washing of neutral nitrated ethylene glycol containing 0.1% ethylene glycol mononitrate

The weight ratio of unwashed oil to washing water	The ratio of ethylene glycol to nitrate in oil phase (%)	Water phase			Washing loss (kg)	
		Mononitrate (%)	Ethylene glycol mononitrate (%)	Total amount of oil (%)	Based on 100 kg ethylene glycol	Based on 100 kg nitrated ethylene glycol
100/20	0.07	0.16	0.63	0.79	0.39	0.16
100/50	0.05	0.11	0.63	0.74	0.91	0.37
100/100	0.02	0.07	0.63	0.70	1.72	0.70
100/130	0.02	0.06	0.63	0.69	2.20	0.90
100/200	0.02	0.04	0.63	0.67	3.29	1.34

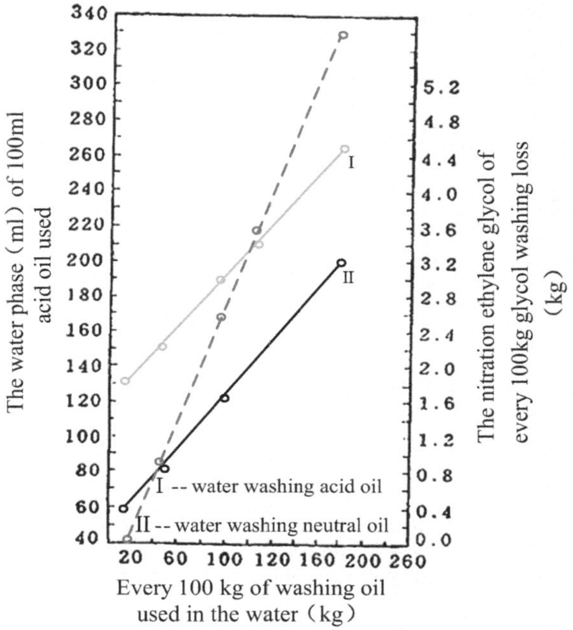

Fig. 5.14 The relationship between the washing loss and the amount of washing water (the content of ethylene glycol mononitrate is 0.2%)

Figure 5.14 shows the relationship between the mass losses of the ethylene glycol dinitrate per 100 kg of raw material ethylene glycol and the amount of wash water used. The amount of washing water is the number of kg of water for washing 100 kg of oil. Since the ratio between the water phase and the oil phase is controlled by measuring their volumes in the continuous production, Fig. 5.14 also shows a dashed line that illustrates the relationship between the weight and the volume in the acidic system. In a neutral system, to convert water volume into weight, the division of the water volume used for 100 cm^3 of ethylene glycol dinitrate by 1.5 gives the weight of the aqueous phase used for the nitration of 100 g of ethylene glycol.

2. The relationship between the washing loss and the washing times of the nitrated ethylene glycol-nitroglycerin mixture

According to the partition relationship, the mixture contains 20–30% nitrated ethylene glycol, and the rest is nitroglycerin mixture. The content of each product after each washing is shown in Tables 5.18, 5.19, and Fig. 5.15.

(4) The production yields of nitrated ethylene glycol and nitroglycerin

The production yields of nitrated ethylene glycol and nitroglycerin can be shown in Fig. 5.16.

Table 5.18 Washing of ethylene glycol dinitrate-nitroglycerin (20:80) mixture

Washing time	The loss of nitroglycerin per 100 kg mixed nitrate (kg)	The loss of ethylene glycol dinitrate per 100 kg mixed nitrate (kg)	The content of ethylene glycol dinitrate in mixed oil phase (%)
1	0.25	0.28	19.83
2	0.36	0.45	19.71
3	0.47	0.62	19.59
4	0.58	0.79	19.48
5	0.69	0.96	19.36
6	0.80	1.13	19.24
7	0.91	1.30	19.12
8	1.02	1.47	19.00

Table 5.19 The washing of nitrated ethylene glycol-nitroglycerin (30:70) mixture

Washing time	The loss of nitroglycerin per 100 kg mixed nitrate (kg)	The loss of nitrated ethylene glycol per 100 kg mixed nitrate (kg)	The content of nitrated ethylene glycol in mixed oil phase (%)
1	0.21	0.41	29.77
2	0.30	0.66	29.62
3	0.39	0.91	29.47
4	0.48	1.16	29.32
5	0.57	1.41	29.17
6	0.66	1.66	29.01
7	0.75	1.91	28.86
8	0.84	2.16	28.70

As shown in Fig. 5.16, the comparison of ethylene glycol dinitrate and nitroglycerin systems reveals that the difference in their yield curves is attributed to their different nitration equilibriums. The yield of nitroglycerin is rapidly decreasing when the amount of nitrating acid is reduced to about 580 kg per 100 kg of glycerol because the nitration coefficient of nitrating acid is reduced. In other words, the content of water in waste acid increases after the nitration reaction, which enhances the hydrolysis reaction and increases the contents of glycerol dinitrate in both waste acid and the oil phases, thus decreasing the product yield. In contrast, the yield increases with decreasing amount of nitrating acid in the nitrated ethylene glycol system (compared with the curve of 48% nitric acid in the nitrating acid), which is attributed to the fact that the nitration degree of nitrated ethylene glycol is less than that of nitroglycerin. When the acid concentration is high with less hydroxyl group, nitrated ethylene glycol has a less tendency to be hydrolyzed than nitroglycerin under the effective acid. With increasing nitration coefficient of mixed acid, the

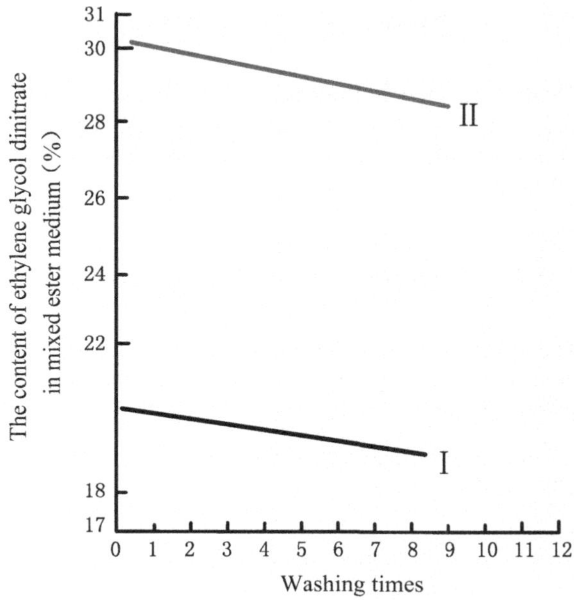

Fig. 5.15 The relationship between the content of ethylene glycol dinitrate and the washing times, 20% ethylene glycol dinitrate and 80% nitroglycerin in the I-mixed ester; 30% ethylene glycol dinitrate in the II-mixed ester and 70% nitroglycerin I-20% ethylene glycol dinitrate and 80% nitroglycerin in the mixed ester;II-30% ethylene glycol dinitrate and 70% nitroglycerin in the mixed ester

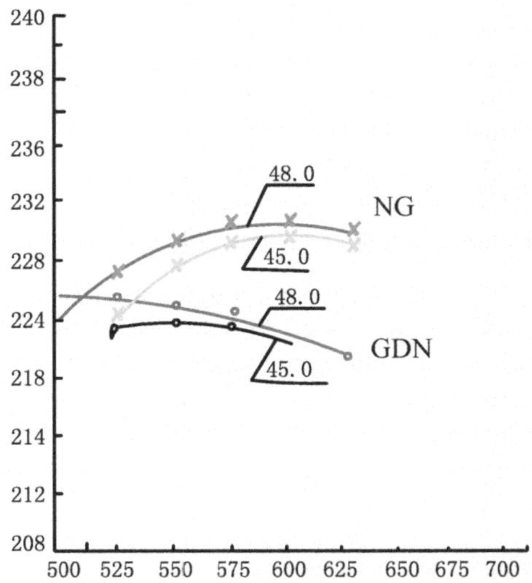

Fig. 5.16 System yield of ethylene glycol nitrate and nitroglycerin

amount of dissolved nitrate in the waste acid also increases, which means the residence time of dinitrate in the mixed acid increases. Therefore, the tendency of dinitrate toward oxidation increases, and the hydrolysis rate and amount of dinitrate also increase, which results in a reduced yield.

5.2 Diethylene Glycol Dinitrate

Diethylene glycol dinitrate, also known as diethyl glycol dinitrate, bis(ethyl-2 alcohol) ether dinitrate, 2,2-dihydroxy diethyl ether dinitrate. The commercial name is dinitrate diethylene glycol or nitrated diethylene glycol with a code of DEGN or DEGDN.

5.2.1 Properties of Diethylene Glycol Dinitrate

Diethylene glycol dinitrate is a colorless or pale yellow oily liquid. With very high energy and strong explosive properties, it is widely used in propellants and propellant liquid explosives [16].

Chemical formula: $C_4H_8N_2O_7$
Chemical structure:

```
          H H   H H
          | |   | |
O2N-O-C-C-O-C-C-O-NO2
          | |   | |
          H H   H H
```

Molecular weight: 196.12
Oxygen balance: −40.8% (based on the generation amount of CO_2)
Nitrogen content: 14.28%
Density: 1.385 g/cm^3
Boiling point: 160 °C
Decomposition temperature: 160 °C
Refractive index: 1.4517 (20 °C).

Diethylene glycol dinitrate has two kinds of crystallines. The stable form has a melting point of +20 °C, and unstable form has a melting point of −10.9 °C. It is an oily liquid with a freezing point of −11.3 °C (or −12 °C). The vapor pressure of the diethylene glycol dinitrate increases with increasing temperature. The vapor pressures at 20 and 60 °C are 0.0032 and 0.130 mmHg, respectively. The change in vapor pressure of diethylene glycol dinitrate with temperature is shown in Table 5.20.

Table 5.20 Vapor pressure of DEGN

Temperature (°C)	Vapor pressure (mmHg)	
25	0.0036	
	60	0.130
	70	0.875
	80	1.216
	90	1.412
	100	1.836

The volatility of diethylene glycol dinitrate at 60 °C is 0.19 mg/cm^3 h, and the volatilization amount at 60 °C is 6.0%.

The diethylene glycol dinitrate is miscible with the aromatic nitrate compound to form a eutectic as shown in Table 5.21.

Diethylene glycol dinitrate is soluble in nitroglycerin, ethylene glycol dinitrate, and other nitrates, and it also dissolves in the ether, methanol, and other organic solvents. However, it does not readily dissolve in ethanol, carbon tetrachloride, carbon disulfide, and other solvents. Diethylene glycol dinitrate has a very low hygroscopicity with a hygroscopicity of 0.19% at the relative humidity of 65%. It also has very low water solubility. The water of solubility diethylene glycol dinitrate are 0.40 and 0.46 g in 100 g of water at 25 and 60 °C, respectively. Its water solubility increases with increasing temperature. The dissolving ability of diethylene glycol dinitrate in 15% nitric acid aqueous solution is 1.06%, which increases with increasing concentration of nitric acid. The solubility of diethylene glycol dinitrate in water is shown in Table 5.22.

Diethylene glycol dinitrate decomposes readily when reacts with the acid, which is mainly attributed to the presence of ether oxygen bond (–COC–) in the molecular structure of diethylene glycol dinitrate. The catalytic decomposition rate increases with increasing acid concentration. Diethylene glycol dinitrate is unstable and decomposes significantly in the waste acid.

The hydrolysis effect of diethylene glycol dinitrate in water and the alkaline solution is similar to that of nitroglycerine. When heated in water, the hydrolysis rate of diethylene glycol dinitrate is slower than that of nitroglycerin. Even if it is heated continuously for 5 days in water at 60 °C, the acid value from the hydrolysis is only 0.003%. In the presence of acid or alkali, the hydrolysis diethylene glycol dinitrate is also difficult.

Table 5.21 The list of eutectic mixture formed by diethylene glycol dinitrate and nitro aromatics

Nitro aromatics	Eutectic mixture			
	Stable		Unstable	
	Fraction of nitro aromatics (%)	Melting point (°C)	Fraction of nitro aromatics (%)	Melting point (°C)
Dinitrobenzene	8.0	−13.2	11.5	−1.4
Trinitrobenzene	17.0	−15.5	20.5	−3.0
Dinitrobenzene naphthalene	19.0	−11.2	25.0	−1.5

Table 5.22 The aqueous solubility of DEGN

Temperature (°C)	Solubility (%)
25	0.4
60	0.46
80	0.63
100	0.92
120	1.37

With a viscosity of 8.1 centipoises at 20 °C, the viscosity of diethylene glycol dinitrate is lower than that of nitroglycerin. Since diethylene glycol dinitrate has a high dissolving ability for collodion, it is an excellent energetic solvent and energy source for double-base propellants.

5.2.2 Stability and Thermal Decomposition of Diethylene Glycol Dinitrate

The heat of formation of diethylene glycol dinitrate is about 2077.36 J/g or 407.52 kJ/mol, or 428.02 kJ/mol under constant pressure. The combustion heat under constant pressure is 11681.73 J/g, and the combustion heat under constant volume is 11706.8 J/g or 2297.02 kJ/mol. The latent heat of vaporization is 93.3 kJ/mol.

The chemical stability of diethylene glycol dinitrate is equivalent to that of nitroglycerin. Compared with nitroglycerin, diethylene glycol dinitrate is insensitive to the flame and is not easy to burn. It is even difficult to be ignited by using the direct flame. For example, diethylene glycol dinitrate does not burn at 15 °C in the glass tube with the diameter of 5–6 mm. When the diameter of the glass tube is increased to 8 mm or the initial temperature is increased to 40 °C, it starts to burn. The linear velocity of combustion of diethylene glycol dinitrate is only 1/2 of that of nitrated ethylene glycol. The burning rate increases with increasing pressure.

The combustion rate can be expressed as

$$U = 0.0162 + 0.0122P$$

where

U combustion rate (mm/s);
P pressure (mm Hg).

Under the pressure of 55 kg/cm^2, the combustion rate increases rapidly with pulsating characteristic, which is similar to the properties of most nitrates.

When heated in water, the hydrolysis progress of diethylene glycol dinitrate is very slow. When the water mixture is heated at 60 °C for 120 h, only 0.003% of diethylene glycol dinitrate is hydrolyzed. However, it decomposes readily in the presence of waste acid. The stability of acidic diethylene glycol dinitrate is similar to that of nitroglycerin. Acidic diethylene glycol dinitrate is very unstable and can start to decompose within a few hours. Therefore, the separated product should be washed and finished with stability treatment as soon as possible in the nitration production.

The denitration rate of diethylene glycol dinitrate in mixed acid containing 7.5–10.5% HNO_3 and 0.5–1.1% H_2O is shown in Fig. 5.17.

The chemical stability of diethylene glycol dinitrate is related to the nitration temperature. When the nitration temperature is slightly higher, the content of residual acid is less. As a result, the stability of nitrate is better than that of the

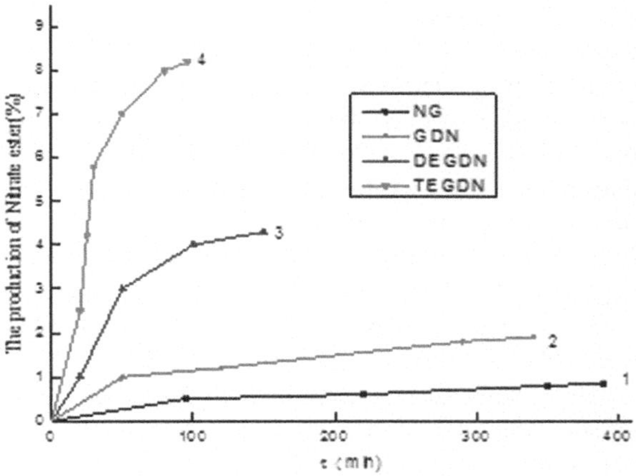

Fig. 5.17 The denitration rate of several nitrates in a mixed acid

Table 5.23 The relationship between nitration temperature and the chemical stability of diethylene glycol dinitrate

Nitration temperature (°C)	Composition of residue acid (%)			Noise reduction test (min)
	HNO_3	H_2SO_4	NO_2	
0	0.217	0.0053	0.0015	12
5	0.191	0.0047	0.0012	16
15	0.138	0.0025	0.0009	22
25	0.0069	0.0015	0.0004	30

product prepared at the low nitration temperature. The relationship of nitration temperature with residual acid content and product stability is shown in Table 5.23.

According to these characteristics of diethylene glycol dinitrate, it is suitable for the manufacture of smoke-free, non-ablative, flameless propellant, as well as liquid mixed explosives.

The stable product of diethylene glycol dinitrate made in the United States and the United Kingdom must be subjected to Abel's heat resistance test with no less than 15 min.

5.2.3 Explosive Properties of Diethylene Glycol Dinitrate

Diethylene glycol dinitrate is a highly energetic and explosive liquid explosive. The explosion equation is expressed as

$$2O(C_2H_4ONO_2)_2 \rightarrow 3CO_2 + 5CO + 3H_2O + 5H_2 + 2N_2$$

Diethylene glycol dinitrate has an explosion heat of 4476.96 J/g with explosion temperature of 3070 K, heat coefficient h of 44.20 J/g, and the specific volume CO_2 of 10.28 ml/g. The explosive force of diethylene glycol dinitrate is 11,80,000 (kg dm/kg). However, the impact sensitivity of diethylene glycol dinitrate is lower than that of nitroglycerin. The dropping weight of 500 g from a height of 110 cm does not cause the explosion of diethylene glycol dinitrate, whereas the drop weight of 2 kg from a height of 175–180 cm causes 100% explosion. Among 10 times of friction sensitivity testing, 7 testing cause an explosion. Diethylene glycol dinitrate explodes when heated to 200 °C, but direct ignition is difficult by flame. Different than nitroglycerin, it would not explode after ignition. In the diameter of 5–6 mm glass tube, it would not be burnt at 15 °C.

The detonation velocity of diethylene glycol dinitrate varies with the characteristic conditions of the charge. The detonation velocity is up to 6800 m/s after the optimization of charge conditions. However, the detonation velocity can only reach 1800–2300 m/s under some charge conditions. The detonation ability of diethylene glycol dinitrate is lower than that of nitroglycerin. The lead block expansion value of diethylene glycol dinitrate is 425 cm, which is 70% of nitroglycerin. Ballistic pendulum test value for the diethylene glycol dinitrate is 127, only 82% of the power of nitroglycerin. A mixture of diethylene glycol dinitrate with nitroglycerin and ethylene glycol dinitrate composes a liquid mixed explosive with an explosion strength of nearly two times of TNT.

5.2.4 Production of Diethylene Glycol Dinitrate

Diethylene glycol dinitrate is prepared by the condensation reaction of saturated ethylene oxide solution with ethylene glycol at 100 °C [14], followed by the nitration of diethylene glycol in mixed nitric and sulfuric acid. After washing and stability treatment, diethylene glycol dinitrate is obtained. The composition of the mixed nitric and sulfuric acid is

HNO_3	50%
H_2SO_4	45%
H_2O	5%

Usually, the nitration coefficient is controlled in the range of 3–6, and the nitration reaction temperature is in the range of 5–25 °C. Under this condition, 100 kg of diethylene glycol can produce 168.7 kg of nitrate product with a theoretical yield of 91.3%.

The first step for the preparation of diethylene glycol dinitrate is the conversion of ethylene glycol and ethylene oxide into diethylene glycol.

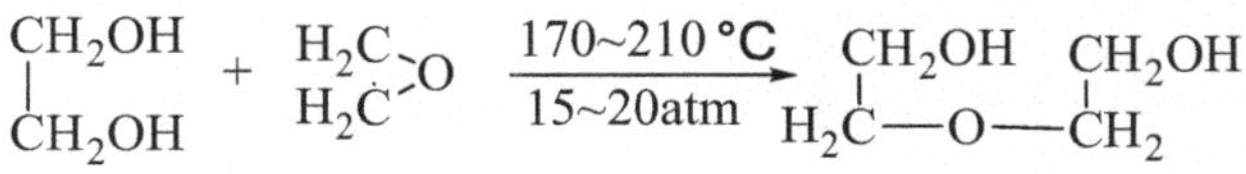

The quality indexes of diethylene glycol for nitration reaction are shown in Table 5.24.

The composition of the mixed nitric and sulfuric acid for the nitration of diethylene glycol is

$$HNO_3 \quad 65\%$$
$$H_2SO_4 \quad 35\%$$

Although diethylene glycol nitration process is mainly performed in an ejecting continuous nitration reactor, bubbling continuous nitration process is also utilized in some plants. In this chapter, we only describe the nitration in an open reactor, which is composed of a compressed air agitator, a circulating water jacket, a temperature sensor, an explosion-proof quick-opening solenoid valve, a safety sink, and other devices. This system is also equipped with a corresponding ice brine circulation cooling system, a constant temperature water circulation system, an air compressor, a washing kettle, a hot kettle, and other ancillary facilities. The nitration reactor is shown in Fig. 5.18.

During the nitration process, the nitration temperature should be controlled at no more than 25 °C. The nitration reaction is complete after 25 min, and then the nitration mixture in the reactor is cooled to below 15 °C for the separation. The preparation process of diethylene glycol dinitrate is shown in Fig. 5.19.

As shown in Fig. 5.19, the addition rate of diethylene glycol to the mixed acid should not be too fast. The reaction temperature should be strictly controlled throughout the reaction process to ensure the temperature of no more than 35 °C, and the addition rate is proportional to the reaction temperature. After the separation

Table 5.24 The quality indexes of diethylene glycol for nitration reaction

Specification	Value
Color/appearance	Colorless, odorless, transparent liquid
Density (20 °C)	1.1157–1.1165
H_2O (%)	<0.5
Acidity (%, H_2SO_4-based)	<0.02°
Saponification value (%, Na_2O-based)	<0.05
Chloride	Trace
Reducing substance	0.00
Glycol (%)	<2.00
Distillate (%)	99% in 241.0–246.5 °C, 1.0% in 247–249 °C

Fig. 5.18 Facilities and control systems of preparation of diethylene glycol dinitrate, 1-Open nitration reactor, 2-Stability treatment vessel, 3-Temperature-controlled meter, 4-Frequency variator

of nitration products, the crude product containing waste acid is obtained. The crude product is processed with water washing and stability treatment, and the waste acid is treated for denitration. The content of waste acid after separation is shown in Table 5.25.

The separated diethylene glycol dinitrate is washed twice with 300 L of water each time, and the composition of the acidic washing water is shown in Table 5.26.

The diethylene glycol dinitrate is washed with cold water several times, and then washed with hot water (60 °C), finally washed and neutralized with 5% sodium carbonate solution, followed by washing with water to give the product. The product is subjected to Abel's heat resistance test with no less than 15 min at 82 °C.

By the method of intermittent nitration, each feedstock is 450 kg with 1500 kg of mixed acid. After the stability treatment, diethylene glycol dinitrate 778 kg can be obtained at a yield of 93.5%.

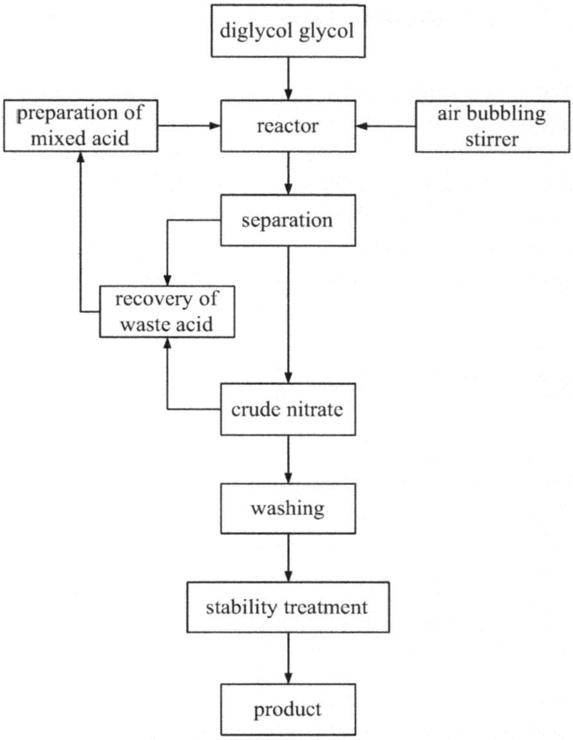

Fig. 5.19 The preparation process of diethylene glycol dinitrate

Table 5.25 Composition of the separated waste acid after nitration reaction for diethylene glycol dinitrate

Compound	Fraction (%)
HNO_3	28–29
H_2SO_4	44–46
H_2O	20–24
Diethylene glycol dinitrate	4–5

Table 5.26 Composition of the waste washing water for the separated diethylene glycol dinitrate

Compound	Fraction (%)
HNO_3	3–4
H_2SO_4	5–6.5
Diethylene glycol dinitrate	2–3

5.3 Triethylene Glycol Dinitrate

Triethylene glycol dinitrate is also known as dinitro-triethylene glycol, nitrated triethylene glycol, which is abbreviated as TEGDN. Triethylene glycol dinitrate is the homolog of ethylene glycol dinitrate and diethylene glycol dinitrate but has lower volatility than them. To date, it is the compound possessing the best plasticizing ability towards nitrocellulose among polyhydric alcohols. As one of the most widely used nitrates in the world, the other properties of triethylene glycol dinitrate are similar to diethylene glycol dinitrate.

5.3.1 The Property of Triethylene Glycol Dinitrate

Molecular formula: $C_6H_{12}O_8N_2$
Chemical structure:

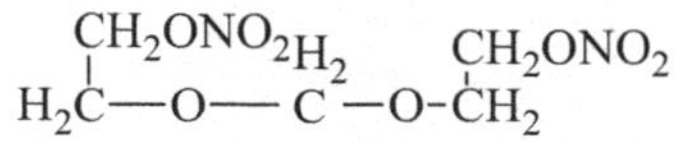

Molecular weight: 240.17
Oxygen balance: −66.6% (based on the generation amount of CO_2)
Nitrogen content: 11.66%
Density: 1.335 g/cm^3
Freezing point (°C): −40
Viscosity: 2.5 times of nitroglycerin at 20 °C
Hygroscopicity: 0.144 at a relative humidity of 65, similar to diethylene glycol dinitrate
Volatility: about 1/5–1/6 value of nitroglycerin.

Triethylene glycol dinitrate is an oily liquid with similar chemical properties to diethylene glycol dinitrate. Its thermal stability is higher than nitroglycerin with more than 40 min in Abel's heat resistance test at 72 °C. At the temperature of 20 °C, 100 g of water can dissolve 0.7 g of triethylene glycol dinitrate, and the solubility increases with decreasing pH value. Triethylene glycol dinitrate is soluble in diethyl ether, acetone, acetic acid, but insoluble in alcohol. It has a high plastic solubility for nitrocellulose, which is one of the polybasic alcohol nitrates that have strongest ability to dissolve plastics.

5.3.2 Explosive Properties of Triethylene Glycol Dinitrate

The explosion heat of the triethylene glycol dinitrate is 3138 J/g with the explosion heat at constant volume of 2510.4 J/g, the formation heat of 2629.2 J/g or 613.37 kJ/mol, and the specific volume of 1214 ml/g (water liquid condition). The explosion temperature and the explosion point are 2,100 K and 245 °C, respectively, and the lead block expansion value is 320 mL. The thermal coefficient and the coefficient of the specific volume are 44.18 J/g and 12.13 ml/g, respectively. Triethylene glycol dinitrate is insensitive to shock. Its shock sensitivity is similar to dinitrobenzene but lower than the ethylene glycol dinitrate. The characteristic height is 100 cm for 2000 g weight dropping hammer.

5.3.3 Preparation of Triethylene Glycol Dinitrate

The preparation principle, production equipment, and production method of triethylene glycol dinitrate are exactly identical to those of other nitrates, such as nitroglycerin and diethylene glycol dinitrate [14]. In the early time, triethylene glycol dinitrate is produced by batch production equipment, and later on, it is changed to air stirring and blowing type continuous nitration. The global production of triethylene glycol dinitrate has been largely changed to ejecting continuous nitration with remote control production.

The main reaction formula for the production of triethylene glycol dinitrate is

$$
\begin{array}{cccc}
C_6H_{12}O_2(OH)_2 + & 2HNO_3 \Leftrightarrow & C_6H_{12}O_2(OON_2)_2 + & 2H_2O \\
150 & 126 & 240.7 & 36 \\
100 & 160 & &
\end{array}
$$

The theoretical yield is 160%, but the actual production yield is about 89.2% of the theoretical yield of with the maximum yield of 91.55%. The main process conditions are shown below. The composition of mixed acid for nitration is as below:

HNO_3—70%;
H_2SO_4—30%;
Nitration coefficient: 2.5.

In a container-type nitration reactor, the nitration of 500 kg of triethylene glycol can be completed within 30 min. The preparation process of triethylene glycol dinitrate is shown in Fig. 5.20.

The preparation process of triethylene glycol dinitrate is well developed. After the nitration reaction, a large quantity of waste acid is removed by coarse separation. The separated triethylene glycol dinitrate semifinished product that contains waste acid needs immediate treatment.

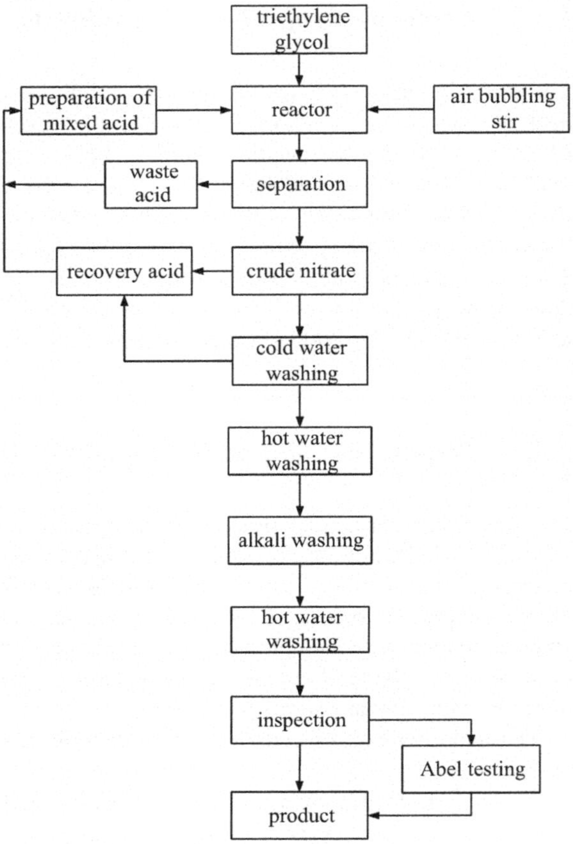

Fig. 5.20 The preparation process of triethylene glycol dinitrate

Triethylene glycol dinitrate has a high solubility in the waste acid, up to 8–9%, so the waste acid decomposes readily. The treatment of waste acid is also dangerous. After the separation of the waste acid, water must be added to prevent the precipitation of triethylene glycol dinitrate. The diluted waste acid should be treated promptly.

The separation of triethylene glycol dinitrate from waste acid separation is slow. For example, the separation takes more than 30 min by using gravity method. Thus, it increases the risk of separation.

The separated acidic triethylene glycol dinitrate is washed twice with water and then with aqueous Na_2CO_3, and finally washed with water to slightly alkaline. After washing, the product is subjected to with more than 40 min in Abel's heat resistance test. The resistance time should be no less than 20 min at 82 °C or no less than 40 min at 72 °C.

The quality requirements of triethylene glycol for the production of triethylene glycol dinitrate are shown in Table 5.27.

Table 5.27 Specifications of triethylene glycol

Specifications	Value
Color/appearance	Transparent yellow liquid with slightly odorous
Density (20 °C)	1.1217
Refractive index (20 °C)	1.4552
H_2O (%)	<0.1
Acidity (H_2SO_4-based)	<0.005
Saponification value (Na_2O-based)	<0.630
Chloride	Trace
Reducing substance	0.00
Glycol (%)	0.00
Distillate in 278.8–310 °C (%)	<95.2
Residues	<0.001

The separated oil is washed twice with water, sodium carbonate solution, and finally washed with water. After washing and neutralization of the triethylene glycol dinitrate product, the heat resistance test at 82 °C should be no less than 20 min. In general, 500 kg of triethylene glycol can produce 714.11 kg of triethylene glycol dinitrate.

5.4 Ethylene Glycol-N-Nitramine Dinitrate

Ethylene glycol-N-nitramine dinitrate, also known as nitrated ethylene glycol amine and Gina with an abbreviation of DINA. Ethylene glycol-N-nitramine dinitrate is a powerful explosive with a light yellow crystalline solid. G. F. Wright [17] prepared ethylene glycol-N-nitramine dinitrate for the first time in 1942. In 1944, the United States began small batch production, and the Soviet Union established the first 100 tons/year production line in 1945. In China, the research and development of ethylene glycol-N-nitramine dinitrate were initiated in 1958. After the small batch trial production was conducted in 1978, the production of ethylene glycol-N-nitramine dinitrate has been put into small batch production. The explosion performance of DINA is similar to RDX with an energy range similar to NG, but the specific volume of ethylene glycol-N-nitramine dinitrate is large.

5.4.1 Physicochemical Properties of Ethylene Glycol-N-Nitramine Dinitrate

Ethylene glycol-N-nitramine dinitrate is a light yellow crystalline compound with α and β crystal forms. It is an explosive compound and mostly used as the propellant.

Molecular formula: $C_4H_8N_4O_8$
Chemical structure:

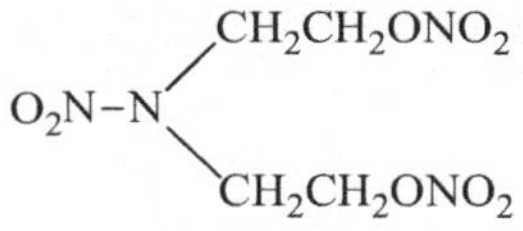

Molecular weight: 240.06
Oxygen balance: −26.67% (based on the generation amount of CO_2)
Oxygen content: 17.72%
Density (g/cm^3): 1.67–1.70
False density (g/cm^3): 0.8–1. 0
Melting point (°C): 49.5–51.5
Specific heat (cal/g °C): 0.38
The heat of fusion (J/g): 98.32
The heat of formation (J/g): 1182.82

Hygroscopicity: It does not absorb moisture at room temperature. The compound absorbs water 0.00375% in the relative humidity of 95% at the temperature of 40 °C for 8 h.

Solubility: Ethylene glycol-N-nitramine dinitrate is soluble in most organic solvents, such as acetone, ethyl acetate, ethyl ether, benzene, toluene, and dichloromethane, and is hardly soluble in CCl_4, CS_2, or petroleum ether. It is insoluble in water. However, it turns to slightly soluble with increasing temperature [18]. The miscibility of ethylene glycol-N-nitramine dinitrate with other nitrates is very good, such as nitroglycerin and diethylene glycol dinitrate. Ethylene glycol-N-nitramine dinitrate has a certain solubility in different concentrations of acetone solution, and its dissolution capacity at 25 °C is shown in Table 5.28.

DINA is miscible in 98% nitric acid but insoluble in hydrochloric acid. The solubility of acetic acid varies with concentration and temperature, as shown in Tables 5.29 and 5.30.

DINA is insoluble in water, but the solubility increases with increasing temperature. At 100 °C, 1.0 g of DINA can be dissolved in 100 ml of water. The solubility of DINA in water is shown in Table 5.31.

DINA has a strong plasticization ability on nitrocellulose with a variety of nitrogen contents, especially the plasticization ability of DINA for high nitrogen content nitrocellulose is far greater than nitroglycerin. However, there is crystallization as the solubility changes with the temperature changes, which is the shortcomings of DINA application.

Table 5.28 The solubility of DINA in different concentrations of acetone at 25 °C

Acetone concentration (%)	0	10	20	30	40	50	60	70	80
Dissolved grams of DINA in 100 g of acetone solution	0.053	0.095	0.260	0.590	1.226	~1.7	3.53	5.87	9.39

Table 5.29 Solubility of DINA in acetic acid at different temperatures

Temperature	20	25	30	40
Dissolved grams of DINA in 100 g of 30% acetic acid solution	0.230	0.353	0.444	0.943

Table 5.30 The relation between DINA solubility and acetic acid concentration

Acetic acid concentration (%)	10	20	40	50	60
Dissolved grams of DINA in 100 g of acetic acid solution at 25 °C	0.099	0.178	0.729	1.43	2.66

Table 5.31 The solubility of DINA in water at different temperatures

Temperature	20	25	30	40	60	100
Dissolved grams of DINA in 100 g of water	0.042–0.05	0.053	0.068	0.129–0.14	0.27	1.0

Volatility: DINA volatility is so low that the volatile loss is 0.7% at 60 °C for 60 h.

DINA chemical structure contains three liable functional groups, namely, ammonium nitrate, $-ONO_2$, and N−C bond, which will break under the strong external influence.

The reaction with alkali: 10% of the alkali solution can cause the saponification and slow decomposition of DINA. When heated in NaOH ethanol solution, DINA is completely decomposed via saponification to yield brown precipitate and NH_3, which can turn red litmus test paper into blue. The saponification decomposition products of nitroglycerin and diethylene glycol dinitrate are brown red and rose precipitation, respectively.

The reaction with concentrated sulfuric acid: concentrated sulfuric acid decomposes DINA completely even at 0 °C. DINA is liable to acids, and the reaction of DIAN with concentrated sulfuric acid is shown below.

$$O_2N\text{-}N\text{-}(CH_2CH_2ONO_2)_2 + H_2SO_4 \longrightarrow 2SO_2\begin{matrix}\diagup OH\\ \diagdown ONO\end{matrix} + 2CH_3CHO + 2NO + 2H_2O$$

(1) The reaction of DINA with the reducing agent is as follows:

$$NO_2-N-(CH_2CH_2ONO_2)_2+FeCl_2+HCl\xrightarrow{120-130^\circ C} NO+\text{other products}$$
$$\text{HCl concentration } 10-20\%$$

In the presence of dilute acid and weak reducing agent, DINA is usually a relatively stable compound. Typically, reducing agent does not react with DINA in a neutral environment. However, reducing agent reacts with DINA in an acidic medium to generate the corresponding aldehyde. In the case of a vigorous reaction, the products decompose.

Under the condition of concentrated acid and strong reducing agent, the reaction is as follows:

$$NO_2-N-(CH_2CH_2ONO_2)_2+FeSO_4+H_2SO_4\xrightarrow{220-250^\circ C}NO+N_2+NH_3+\text{other products}$$

70% concentration H_2SO_4 30% concentration $FeSO_4$

5.4.2 The Explosive Properties of Ethylene Glycol-N-Nitramine Dinitrate

The explosive properties of DINA are similar to that of RDX. The explosion reaction is as follows:

$$O_2N-N-(CH_2CH_2ONO_2)_2\rightarrow 3.1CO+0.9CO_2+0.9H_2+3.1H_2O+2N_2$$

When the reaction is performed under pressure condition, it will not fully follow the above equation.

Explosion heat: Q_{v2} (water liquid condition) = 5230 J/g
Specific volume: W_1 = 930 ml/g
Explosion temperature: TV = 3450 K
Explosive force: f_v = 1326000 kg dm/kg
Explosion strength: Lead block expansion value = 480 ml
Brisance value: Lead column compression = 22 mm (H/d = 60/40 mm)
Copper column compression = 4.7 mm
Detonation velocity: D = 7800 m/s (Δ = 1.6 g/cm^3)
7350 m/s (Δ = 1.47 g/cm3)
Sensitivity: shock sensitivity and friction sensitivity are lower than PETN, but greater than Hexogen.

(1) Shock sensitivity: The dropping of 10 kg hammer from 25 cm height causes the explosion rate of 84%. 2 kg of drop hammer requires the minimum height of 44 cm for detonation.
(2) Friction sensitivity: The explosion rate is 10% with the swing angle of 90° and pressure of 39.85 kg/cm^2.
(3) Thermal sensitivity: The explosion point is 230 °C with 5 s delay period.
(4) Shooting sensitivity: Explosion rate of 100%.

5.4.3 *Stability of Ethylene Glycol-N-Nitramine Dinitrate*

The thermal stability of dry and pure DINA is very good. It begins to decompose significantly at 160 °C and turns to decompose vigorously at 180 °C. The ignition point is 240 °C. In Abel's heat resistance test, the resistance time is up to 60 min at 72 °C.

When the diethanolamine contains triethanolamine impurity, ethylene glycol-N-nitramine dinitrate is unstable.

Probably due to the influence of the nitroamine group $>N{-}NO_2$ on the nitrate, DINA is very inactive. As a result, the thermal decomposition of DINA is much slower than that of nitroglycerine. Nitroamine is generally more stable than nitrate because the free electron pair on the amine group shifts significantly, resulting in the N–N bond with double bond characteristics. Therefore, it is not easy to undergo thermal decomposition.

The thermal decomposition capacity of DINA is 188.28 kJ/mol, but the presence of water or acid accelerates the decomposition.

The hydrolysis of DINA in water is much slower than that of nitroglycerin. The decomposition rate is about 3% in 80 °C heating water for 2 h, whereas decomposition rate increases to 6% in 100 °C heating water for 6 h. The decomposition product contains NO_2, which can acidify reaction solution.

The formation heat of DINA is Qv = 1180.3 J/g, or Q′v = 283.25 kJ/mol, Q′p = 307.5 kJ/mol.

The combustion heat Q = 10079.3 J/g.

5.4.4 *Preparation of Ethylene Glycol-N-Nitramine Dinitrate [15, 17, 19]*

5.4.4.1 Preparation of Ethylene Glycol-N-Nitramine Dinitrate

There are two kinds of reaction groups, amine (–NH–), and hydroxyl (–OH), in the molecular structure of diethanolamine $HN{-}C(CH_2CH_2OH)_2$. The secondary amino group has a strong alkaline, which reacts with inorganic acid to form salt typically. The hydroxyl group reacts with nitric acid via esterification reaction, as shown in the following formula:

$$HN\begin{matrix}\diagup CH_2CH_2OH\\ \diagdown CH_2CH_2OH\end{matrix} + 3HONO_2 \rightleftharpoons HNO_3{\cdot}NH\begin{matrix}\diagup CH_2CH_2ONO_2\\ \diagdown CH_2CH_2ONO_2\end{matrix} + 2H_2O$$

The amino terminal of diethanolamine has a strong ability to accept protons because of its free electron pair on the nitrogen atom, which reacts with HNO_3 to form a complex covalent salt.

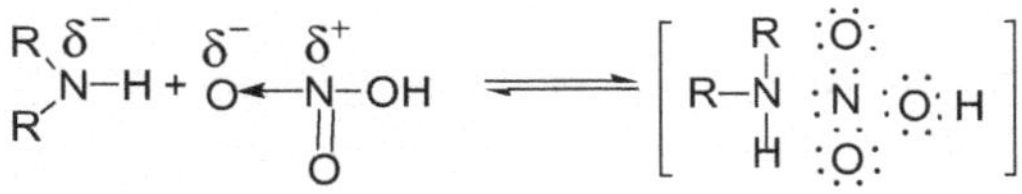

Nitrate of diethanolamine dinitrate is a crystalline compound with a melting point of 123.8 °C. It is very unstable and tends to eliminate nitric acid. The pure compound does not absorb moisture with a low shock sensitivity. For example, the explosion rate is 34% when 10 kg weight of hammer drops from 1 m height. The denotation velocity is 5560 m/s when the density is 1.0 g/cm^3, and the denotation velocity reaches 5560 m/s when the density is 1.4 g/cm^3. The lead block expansion value of nitrate of diethanolamine dinitrate is about 30% greater than that of picric acid. The stability of this compound is very poor. For example, it deflagrates when heated for 8 min at 130 °C, and it detonates and decomposes within 5 h when heated at 85 °C. This property should be taken into account in the preparation of DINA.

Since the amino group not only has a strong alkaline but also can be oxidized readily, the content of nitric oxide in the nitric acid raw material should be strictly controlled.

A secondary amino group has a strong ability to accept protons, which is easy to react with NO^+ to generate nitroso compound

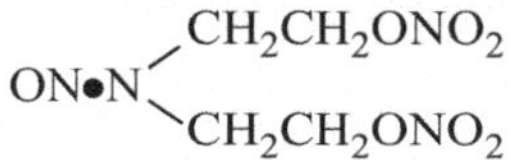

Ethylene glycol-N-nitrosamine dinitrate is unstable and tends to decompose at a melting point of 43.5–45.5 °C.

From the above analysis, in combination with nitroglycerin preparation principle, it is evident that the excessive high concentrations of nitric acid and dehydrating agent should be used in the preparation of DINA.

The amino group in the diethanolamine has characteristics of alkaline and easy oxidation. To change this property of the amino group, the protection of the amino group is beneficial for nitration.

By the above mechanism, it can be predicted that using different catalysts and dehydration methods will lead to different results in the preparation of DINA.

(1) Acetic anhydride method

As the sulfuric acid can decompose and damage DINA, other dehydrating agents should be used in the preparation of DINA rather than sulfuric acid. Normally, acetic acid is used as dehydration agent for the amine nitration in the preparation of explosives. The reaction of acetic anhydride with water produces acetic acid, and the reaction is as follows:

$$(CH_3CO)_2O + H_2O \longrightarrow 2CH_3COOH$$

Acetic anhydride and nitric acid constitute nitration agent. In addition to dehydration, acetic anhydride also has a catalytic role. The reaction of acetic anhydride and nitric acid first generates acetyl nitrate, which has a very strong electrophilic and ability to bind to electron pair, thus further improving the nitration rate of ethylene glycol amine. The main reaction mechanism is as follows:

$$2HNO_3 + CH_3CO - O - OCCH_3 \rightleftharpoons N_2O_5 + 2CH_3COOH$$
$$N_2O_5 + CH_3CO - O - OCCH_3 \rightleftharpoons 2CH_3COONO_2$$
$$2CH_3COONO_2 \rightarrow 2NO_2^+ + 2CH_3COO^-$$

In addition, acetic anhydride can also react with diethanolamine via acetylation reaction, thus changing the property of the amino group.

$$(CH_3CO)_2O + RR'NH \longrightarrow CH_3CO \cdot NRR' + CH_3COOH$$

After the acylation of the amino group, π electron in carbonyl double bond and a pair of non-shared electron pair in the amino nitrogen atom form a p-π conjugated system.

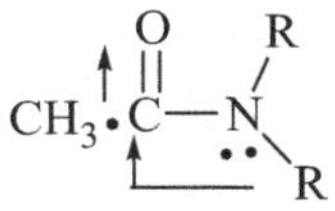

Because of the conjugation effect, the electron cloud density of the nitrogen atom is decreased significantly, thus weakening the alkalinity. Therefore, the acetylation reduces the ability of the amine to accept protons and protects the amino group.

In medium consisting of acetic acid and acetic anhydride, diethanolamine converts into acetyl diethanolamine dinitrate.

$$CH_3 \cdot CO \cdot N(CH_2CH_2ONO_2)_2$$

In the preparation reaction of diethanolamine dinitrate, the addition of HCl to the medium of nitric acid and acetic anhydride results in the corresponding chloramine in a yield of about 30%, in which HCl acts as a catalyst.

The addition of trace amounts of $ZnCl_2$ to the reaction mixture is beneficial to the overall reaction and can increase the reaction efficiency by 60%. $ZnCl_2$ produces about 39% of the positively charged chlorine in the nitric acid and acetic anhydride reaction medium.

Promoted by the positive charge chlorine, the amino group in the diethanolamine generates chlorinated amine, which eliminates the alkaline of the secondary amino group, thus preventing the formation of nitrate salt. Chloroamine further reacts with nitric acid to form nitroamine. The chemical reactions for the preparation of DINA from diethanolamine in the nitric acid and acetic anhydride medium are mainly composed of the following two steps.

First step.

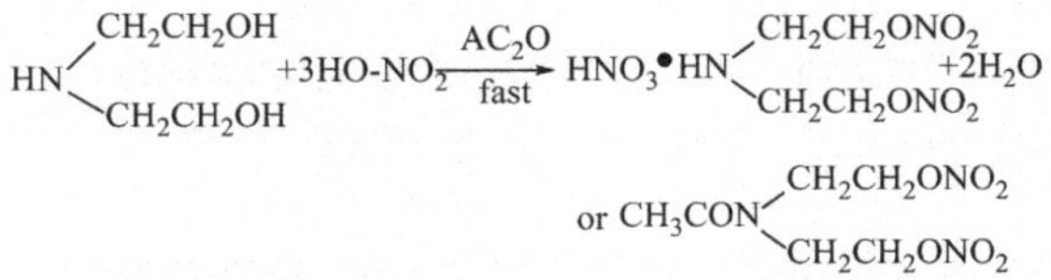

Second step.

$$HNO_3 \cdot HN(CH_2CH_2ONO_2)_2 \xrightarrow[\text{HCl} \atop \text{fast}]{HNO_3 \atop AC_2O} ONO_2 \cdot HN(CH_2CH_2ONO_2)_2 + H_2O$$

As mentioned above, the generated nitrate salt in the first step is an unstable product. In the absence of a suitable catalyst, amine hydrogen cannot be replaced by NO'_2. Therefore, to carry out the second step reaction, it is necessary to add a catalyst. In the meantime, an appropriate increase in the reaction temperature and extension of the reaction time can significantly promote the nitration reaction. The rate and yield of the overall chemical reaction depend on the second step reaction. When the mixture of nitric acid and acetic anhydride is used as the nitrating agent with the presence of catalyst HCl, the chemical reaction pathway for the preparation of DINA is as follows:

$$\text{(a) } 2HCl + 2HNO_3 + 3(CH_3CO)_2O \rightarrow 2CH_3COOCl + N_2O_3 + 4CH_3COOH$$
$$\text{(b) } CH_3COOCl + HN(CH_2CH_2OH)_2 \rightarrow ClN(CH_2CH_2OH)_2 + CH_3COOH$$
$$\text{(c) } ClN(CH_2CH_2OH)_2 + HNO_3 \rightarrow O_2N - N(CH_2CH_2OH)_2 + HOCl$$
$$\text{(d) } O_2N - N(CH_2CH_2OH)_2 + 2HNO_2 \rightarrow O_2N - N(CH_2CH_2ONO_2)_2 + 2H_2O$$
$$\text{(e) } HOCl + (CH_3CO)_2O \rightarrow CH_3COOCl + CH_3COOH$$
$$\text{(f) } (CH_3CO)_2O + H_2O \rightarrow 2CH_3COOH$$

In the Eq. (a), the generation of CH_3COOCl from HCl can be proceeded by the following reactions:

$$2HNO_3 + 4HCl \rightarrow 2Cl_2 + 3H_2O + N_2O_3$$
$$2HNO_3 + 6HCl \rightarrow 2Cl_2 + 4H_2O + 2NO$$
$$Cl_2 + H_2O \rightarrow HCl + HClO$$

The first step is the oxidation of HCl (or another chloride) in the nitric acid medium to produce free chlorine, which reacts with water under certain conditions via a self-dual oxidation–reduction reaction. Since Cl_2 and H_2O are regenerated by the reactions of HCl and HOC1 that are produced by the reaction of chlorine and water, the reaction of water with chlorine is an equilibrium. Studies have shown that the reaction of water and chlorine reaches equilibrium when about 1/3 of the dissolved chlorine reacts with water. In other words, in this process, only 1/3 of the chlorine converts into HOCl with positively charged chlorine.

To determine the reasonable reaction conditions and to utilize the catalytic effect of HCl, it is necessary to take into account the hydrolytic nature of chlorine and instability of hypochlorous acid. Because the catalytic effect of HCl is conditional, the positive charge of chlorine is not easy to produce in the absence of suitable conditions. Thus, it cannot effectively play a catalytic role in the process. Therefore, the timing of the introduction of the catalyst is extremely important in the design of production process.

(2) Magnesium nitrate method

As the recovery process of acetic acid is complicated and the price of acetic anhydride is high, the cost of acetic anhydride is high. Moreover, the adverse and toxic effect of acetic anhydride on the human is severe. In addition, acetic anhydride can lead to some adverse reactions. For example, acetic anhydride reacts with concentrated nitric acid to produce tetranitromethane:

$$4(CH_3CO)_2O + 4HNO_3 \rightarrow C(NO_2)_4 + 7CH_3COOH + CO_2$$

Tetranitromethane can be easily decomposed.

$$C(NO_2)_4 \rightarrow CO_2 + 2N_2 + 3O_2$$

When acetic anhydride is used as a dehydrating agent for the nitration reaction, the acid mixture can only be prepared immediately before the reaction. The preparation of nitration agent requires low temperature and a slight fluctuation to avoid impact on the industrial production. Therefore, seeking other alternative dehydrating agents is beneficial to the environment protection and production costs.

Magnesium oxide (MgO) is a basic metal oxide, which reacts with concentrated nitric acid to produce magnesium nitrate.

$$MgO + 2HNO_3 \rightarrow Mg(NO_3)_2 \cdot H_2O$$

Magnesium nitrate can be combined with six crystal water, so it has a certain ability to initiate.

$$Mg(NO_3)_2 \cdot H_2O + 5H_2O \Leftrightarrow Mg(NO_3)_2 \cdot 6H_2O$$

The production cost can be significantly reduced by using HNO_3–$Mg(NO_3)_2$ nitrating agent that is prepared from MgO and HNO_3. In the preparation of DINA, the acid preparation and nitration are usually carried out at higher temperatures without the requirement of a large cooling system. The cheap sodium chloride can be used as a catalyst to replace HCl.

The process of converting NaCl into a positive charge in HNO_3–$Mg(NO_3)_2$ solution is shown by the following formula:

$$HNO_3 + NaCl \Leftrightarrow NaNO_3 + HCl$$
$$HNO_3 + HCl \rightarrow NOCl + Cl_2 + 2H_2O$$
$$Cl_2 + 2H_2O \Leftrightarrow HOCl + HCl$$

In the nitration of a secondary amine with strong alkaline, ammonium nitrate salt and chlorine with the positive charge are produced first, which is a necessary condition for the catalytic reaction. The intermediate product is a perchloroamine. Therefore, the nitration reaction process of DINA can be described as

$$(a)\ MgO + 2HNO_3 \rightarrow Mg(NO_3)_2 \cdot H_2O;$$
$$(b)\ HN(CH_2CH_2OH)_2 + 3HONO_2 \Leftrightarrow HNO_3 \cdot HN(CH_2CH_2ONO_2)_2 + 2H_2O;$$
$$(c)\ HNO_3 \cdot HN(CH_2CH_2ONO_2)_2 \rightarrow NH(CH_2CH_2ONO_2)_2 + HNO_3;$$
$$(d)\ 3HNO_3 + 2NaCl \rightarrow 2NaNO_3 + NOCl + HOCl + H_2O;$$
$$(e)\ HNO_3 \cdot HN(CH_2CH_2ONO_2)_2 + HOCl \rightarrow HNO_3 \cdot ClN(CH_2CH_2ONO_2)_2 + H_2O;'$$
$$(f)\ NH(CH_2CH_2ONO_2)_2 + HOCl \rightarrow ClN(CH_2CH_2ONO_2)_2 + H_2O;$$
$$(e)\ ClN(CH_2CH_2OH)_2 + HOCl;$$
$$(h)\ Mg(NO_3)_2 \cdot H_2O + 5H_2O \Leftrightarrow Mg(NO_3)_2 \cdot 6H_2O$$

5.4.4.2 Refinement of Ethylene Glycol-N-Nitramine Dinitrate

The separation of nitration product DINA with waste acid is achieved by dilution to obtain crude DINA. In addition to water, the crude product contains at least ten kinds of impurities. Typically, these impurities include HNO_3, NaCl, nitration intermediates $HNO_3 \cdot HN(CH_2CH_2ONO_2)_2$, dehydration agent $Mg(NO_3)_2$, nitrated ethanolamine and the corresponding nitrate salt, hydrolysis product, partial nitration product $O_2N{\cdot}N(CH_2CH_2OH)_2$ or its mononitrate product 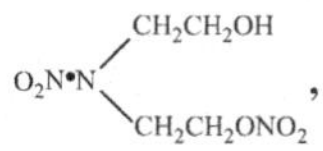, $ON{\cdot}N(CH_2CH_2OH)_2$, nitrated triethanol and nitric acid salt, and nitrosamine products. DINA is very unstable in the acidic medium, and the stability of DINA containing nitric acid triethanolamine and other impurities does not change significantly. Therefore, these impurities must be removed.

Water-soluble impurities in the crude DINA nitration, such as ethanolamine, Mg $(NO_3)_2$, NaCl, and HNO_3 can be quickly removed by washing with hot water. However, the removal of $HNO_3{\cdot}HN(CH_2CH_2ONO_2)_2$ needs washing with an alkali solution. The refining process of DINA includes several repeating cycles of acid washing, alkali washing, and crystallization, which significantly improves the product quality. In addition, the refluxing time in alkali and acid solution should be sufficient, (refluxing in an alkali solution for 20 min and acid solution for 10 min). After the alkali and acid treatment, nitrated triethanolamine can be removed completely. Finally, the final hydrolysis products can be completely removed by repeated recrystallization in acetone.

Normally, $N(CH_2CH_2OH)_3$ cannot be removed without acid reflux. After the preparation, refluxing $N(CH_2CH_2OH)_3$ in an acidic solution (pH 4–5) for 5 min can remove the nitrated triethanolamine. In contrast, after treatment with refluxing alkali (pH = 9–10) for 10 min, nitrated triethanolamine remains in DINA. However, the process of the first acid refluxing, followed by alkali refluxing, cannot remove nitrated triethanolamine. Therefore, in the refinement process of DINA, acid refluxing must be carried out after alkali refluxing. First, nitrated triethanolamine exists in the form of $HNO_3{\cdot}HN(CH_2CH_2ONO_2)_3$ salt, which is more stable under acidic conditions than under alkaline washing conditions. Through alkali refluxing, first, the nitric acid is neutralized to release free $HN(CH_2CH_2ONO_2)_3$, which can be removed by acid reflux. Finally, alkali washing removes the residual acid caused by acid refluxing to ensure DINA is a weak alkaline.

In general, refinement by the prewash (hot water), refluxing in an alkali solution (or alkaline washing), refluxing in an acid solution (or acid washing), alkali washing, washing with water and other processes, DINA with approximately 98% purity and certain thermal stability can be obtained.

After the prewash during the production to remove the water-soluble impurities, the product dissolved in the water continues to emit bubbles, indicating DINA, nitrated triethanolamine, and nitrated ethanolamine in the acidic water still occur hydrolysis and other reactions. Without prewash, the subsequent alkali washing can also remove nitrated ethanolamine, nitric acid, and other impurities.

A large number of oily bubbles are suspended between the DINA and the alkali layer interface, which are CO_2 bubbles produced by the reaction of $NaCO_3$ with HNO_3 that is freed from $HNO_3{\cdot}HN(CH_2CH_2ONO_2)_2$ and $HNO_3{\cdot}N(CH_2CH_2ONO_2)_3$ in the presence of alkaline solution. After the decomposition of impurities, under the action of an alkali, some of the impurities will be saponified into a viscous liquid, which is blown into the water along with DINA, thus forming oily viscous foams. Since DINA will be partially saponified when treated with the alkaline refluxing, the alkaline concentration for the refine process of DINA should not be too strong, which is typically controlled at pH 8–9.

$N(CH_2CH_2OH)_3$ in the refluxing acid water decomposes rapidly, whereas DINA does not decompose. As a result, brown smoke is constantly generated in the process of acid water refluxing. However, the acidity of acid refluxing should not be too strong. Typically, the control of the pH value of 4–5 is sufficient to destroy nitrated triethanolamine. To reduce the washing loss, washing liquid should not be too much, which is controlled as the volume ration of DINA to washing liquid 1.1–1.2.

To facilitate the subsequent refinement of DINA, the preparation of DINA, the content of raw material triethanolamine should be strictly controlled to ensure the complete removal of nitrated triethanolamine in DINA, which is the key to obtain DINA with higher stability. Production record shows that contamination of DINA with nitrated triethanolamine causes a series of issues, such as the liquification in the storage process, increased dark color, significant acidity, increased melting point range, decreased melting point, thus significantly reducing the stability of DINA.

According to the DINA application and long-term storage of loaded ammunition, it is found that propellant containing nitrated triethanolamine is very unstable. The surface of propellant begins to precipitate at room temperature after two years. The longer the storage, the more significant precipitation. Precipitated frost or fog is soluble in deionized water and exhibits strong acidity, which is attributed to the cohesion of the acidic groups $-O-\overset{\delta^+}{N}(=O)\rightarrow O$ and the basic groups $\overset{\delta^-}{N}R_3$ in the nitrated triethanolamine molecule.

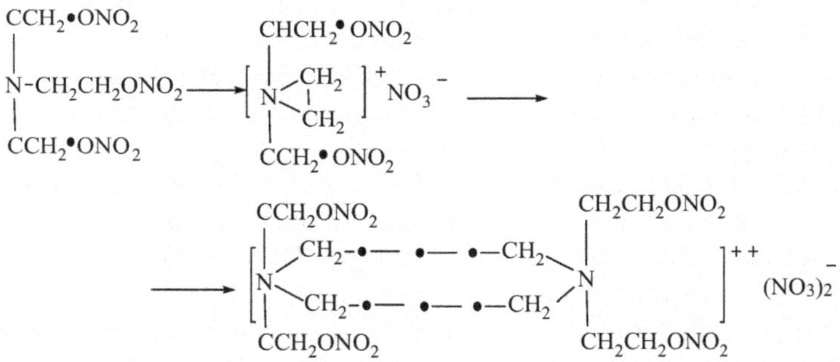

Because of abovementioned reason, cohesion mainly results from the nitrated triethanolamine dissolved in DINA. The cohesive liquid exudes from DINA, which is volatile, unstable, easily decomposed to release HNO_3 or NO_2, further resulting in the acceleration of DINA decomposition and significantly decreased stability. The melting point of DINA is lower than the standard, indicating it contains impurities that are more difficult to remove, such as diethanol-N-nitrosamine dinitrate and possible incomplete nitration product of nitrated triethanolamine, which are important factors affecting the safety of DINA.

5.4.4.3 Raw Material and Process Control Conditions for the Production of Ethylene Glycol-N-Nitramine Dinitrate

(1) The quality of raw material for the production of ethylene glycol-N-nitramine dinitrate

Raw material diethanolamine is prepared by the condensation reaction of ethylene oxide with ammonia, which produces three products, namely ethanolamine, diethanolamine, and triethanolamine at 60 °C and under high pressure. The resulting mixture is purified by fractionated distillation under reduced pressure to yield pure diethanolamine.

$$(CH_2)O + NH_3 \rightarrow \left\{ \begin{array}{l} NH(CH_2CH_2OH)_2 \\ NH_2 \cdot CH_2CH_2OH \\ N(CH_2CH_2OH)3 \end{array} \right. \xrightarrow{\text{fractionated distillation}} NH(CH_2CH_2OH)_2$$

The quality of diethanolamine must be strictly controlled to have a melting point of 28 °C and specific gravity of 1.098. If diethanolamine contains a small amount of monoethanolamine impurity, it reacts with harmful substances nitrite in the nitric acid. However, high content of monoethanolamine affects the reaction yield and stability of the product. The contamination of diethanolamine with triethanolamine is the worst situation because the triethanolamine trinitrate is very unstable and extremely difficult to remove, which affects the product stability considerably. Therefore, the content of triethanolamine must be strictly controlled. The quality of diethanolamine should meet the set requirements in Table 5.32.

The quality requirements of nitric acid for nitration of diethanolamine are same as the nitroglycerin production. The content of nitrogen oxides should be strictly controlled. Specifically, N_2O_4 content should be $\leq 0.4\%$, and nitric acid content should be $\geq 97\%$.

Light magnesium oxide used in the manufacture of DINA (catalyst) should meet the requirements of China Chemical Industry Standard HG1-324-77. The quality indexes are shown in Table 5.33.

The quality of alkali used for stability treatment should meet the technical criteria in Table 5.34.

The catalyst is refined and dried table salt without other special requirements.

Table 5.32 Quality specifications for the production of diethanolamine for DINA

Specification	Criteria
Appearance	Pale yellow liquid or colorless crystal
Monoethanolamine content (%)	$\not> 1.5$
Diethanolamine content (%)	$\not< 97$
Triethanolamine content (%)	$\not> 1.0$
Isopropanolamine content (%)	To be determined
Other impurities (%)	$\not> 0.5$

Table 5.33 Quality indexes of magnesium oxide for manufacture of DINA

Specification	Criteria
Appearance	White powder
Magnesium oxide content (%)	≥92
Calcination loss (%)	≤2
Calcium content (%)	≤2
Sulfate content (%)	<0.4
Iron salt content (%)	<0.1
Sieve residue (%)	≤0.5 (40 Pore/cm)
Apparent specific volume (mL/g)	≥5

Table 5.34 The technical criteria of alkali for the preparation of DINA

Specification	Criteria
Total amount of alkali (%)	≮98.0
Iron content (%)	≯0.02
Sodium chloride content (%)	≯1.2
The content of water-insoluble substance (%)	≯0.20
Sodium sulfate content (%)	≯0.10
Calcination loss (%)	≯0.7

(2) Preparation process of DINA

The preparation of ethylene glycol amine-N-nitramine dinitrate mainly includes an acetic acid method and magnesium nitrate method.

1. Acetic anhydride method

According to the different order of feeding and the mode, as well as the different orders of crystallization and refining, acetic anhydride method also has different processes. The basic technological process for the preparation of ethylene glycol amine-N-nitramine dinitrate by acetic anhydride nitration is shown in Fig. 5.21.

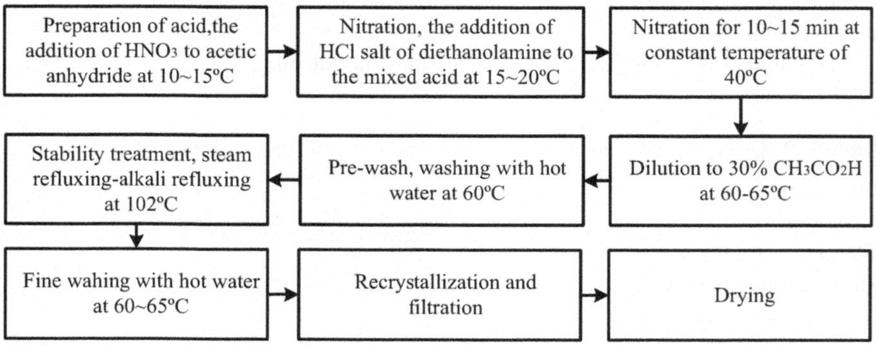

Fig. 5.21 Preparation process of DINA by acetic anhydride method

The ratio of raw materials for nitration is diethanolamine: nitric acid: acetic anhydride: hydrochloric acid = 1: 3.2: 3.4: 0.05 (molar ratio).

2. Magnesium nitrate method

Magnesium oxide is evenly added to the concentrated nitric acid with stirring mixing and cooling to make a mixed acid of magnesium nitrate and nitric acid. After the preparation of mixed acid, diethanolamine is added to the mixed acid. The mixture is added to nitration reactor in a quantitative ratio. After the addition, the reaction mixture is heated with a water bath, followed by the addition of the catalyst sodium chloride. The nitration reaction is complete at a predetermined temperature for a certain period.

The nitration reaction mixture is transferred into the separator, in which the hydrolysis reaction is conducted. Subsequently, nitrated diethanolamine is transferred into a scrubber for washing with water to remove residual acid, followed by refluxing with slightly alkaline and slightly acidic aqueous solutions successively to remove unstable impurities. Finally, the product is washed with a slightly alkaline aqueous solution.

After washing, the oily nitrated diethanolamine is filtered and flows into the crystallization vessel, which crystallizes in the agitated cold water. After drying and packaging, the finished product is obtained.

The nitrated diethanolamine in the separated waste acid is recovered by the cooling and precipitation. The residual waste acid in the nitrated diethanolamine is removed.

The wastewater after washing should also be cooled and precipitated to recover the nitrated diethanolamine. The production process is shown in Fig. 5.22.

Under the process conditions of the magnesium nitrate method, the ratio of ingredients (weight ratio) is diethanolamine: nitric acid: magnesium oxide: sodium chloride = 1: 4.6: 0.52: 0.042. The nitration process conditions are shown in Table 5.35.

Stability treatment conditions

Washing: washing 1–2 times with the water to the material ratio of 2, the temperature of 70–80 °C, stirring for 3–5 min, discharge of wastewater.

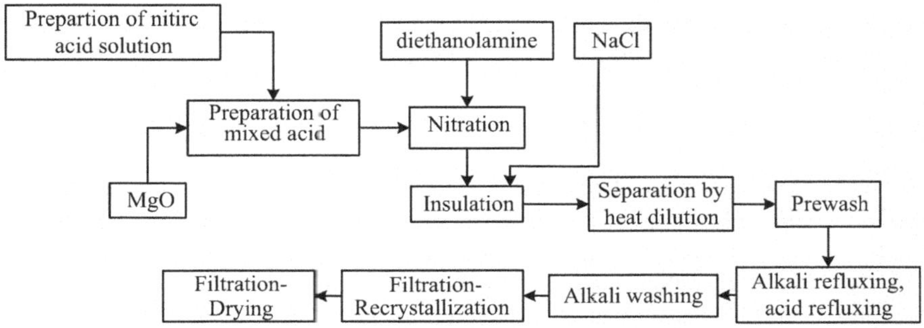

Fig. 5.22 Preparation process of DINA by magnesium nitrate method

Table 5.35 The magnesium nitrate process conditions for preparation of DINA

Process name	Process conditions
Temperature for preparation of mixed acid (addition of MgO)	58 ± 2 °C
Nitration temperature (addition of diethanolamine)	55 ± 2 °C
Temperature for addition of catalyst (addition of NaCl)	60 ± 1 °C
Storage temperature	65 ± 2 °C
Storage period	30 min

Alkali refluxing: refluxing 1 time with the water to the material ratio of 2 at the boiling temperature. $NaCO_3$ 10% aqueous solution is added to adjust the pH to 10–11. The mixture is refluxing for 15–25 min and standing for 5–10 min. The wastewater is discharged.

Acid refluxing: refluxing 1 time with the water to the material ratio of 2 at the boiling temperature. 50% dilute nitric acid is added to adjust the pH $\leq$ 2. The resulting mixture is refluxing for 15–25 min and standing for 5–10 min. The wastewater is discharged.

Alkali washing: washing 1 time with the water to the material ratio of 2 at the temperature of 70–80 °C for 3–5 min. Alkali 10% aqueous solution is added to adjust the pH to 9–10. After washing, the solution is transferred to the crystallization process.

3. Recrystallization process conditions

After alkali washing, the product solution is efficiently stirred, which crystallizes evenly with stirring.

The drying process is performed at 30–40 °C to dry until reaches the required moisture.

Magnesium nitrate production process has the advantages of simple operation and low cost. However, the magnesium nitrate method has certain disadvantages. For example, the yield is lower than that of acetic anhydride, and it is 80% or less of the theoretical yield. The theoretical yield is 228.5%, but the actual yield is 163%, which is about 72% of the theoretical yield. For acetic anhydride method, the raw material acetic anhydride is expensive, but the yield is higher than that of the magnesium nitrate method, which is higher than the theoretical yield 93%. Also, the quality of the product made by acetic anhydride method is better than the magnesium nitrate method.

4. Toxicity

DINA is a toxic substance. The physiological effect of DINA on the human body is similar to that of nitroglycerin. Inhalation of DINA gas can cause dizziness, headache, and fast heartbeat. Some people have the symptom of red eyes.

DINA is different than nitroglycerin, to which human have adaptability. The exposure to DINA always causes the symptoms mentioned above.

At room temperature, the smell of DINA is not obvious. During the refluxing procedure of the stability treatment, DINA has a strong irritation to human. Prevention and treatment for DINA poisons are the same as those for nitroglycerin.

5.5 The Property and Preparation of Glyceryl Dinitrate

Glyceryl dinitrate, also known as dinitro glycerol, is a viscous and slightly yellow liquid with less explosive properties than nitroglycerin [14, 20]. Glyceryl dinitrate has two isomers.

5.5.1 The Physicochemical Properties of Glyceryl Dinitrate

Molecular formula: $C_3H_6N_2O_7$
Chemical structure:

CH_2ONO_2	CH_2ONO_2
|	|
$CHOH$	$CHONO_2$
|	|
CH_2ONO_2	CH_2OH
α isomer;	β isomer

Molecular formula: 182.06
Oxygen balance: −17.58% (based on the generation amount of CO_2)
Nitrogen content: 15.38%.

Glyceryl dinitrate has two isomeric industrial products with a density of 1.47 (15 °C) and freezing point of 40 °C. Under a pressure of 15 mmHg, the distillate is hardly decomposed at 146 °C. It is readily soluble in most organic solvents and is easy to dissolve collodion, but insoluble in carbon tetrachloride and gasoline. The solubility of glyceryl dinitrate in water is 8.6% in 20 °C [21].

The heat resistance of glyceryl dinitrate is better than that of nitroglycerin. In Abel's experiment, signs of decomposition are observed only after heated for 16 days. When heated to 150 °C, glyceryl dinitrate releases the nitrogen oxide as brown smoke. The explosion and decomposition temperatures of glyceryl dinitrate are greater than 170 °C.

5.5.2 The Explosive Properties of Glyceryl Dinitrate

Glyceryl dinitrate can be exploded in the following manner.

$$C_3H_5(OH)(ONO_2)_2 \rightarrow 2CO_2 + CO + 2H_2O + N_2 + H_2 + 5024.98\,J/g$$

The shock sensitivity of glyceryl dinitrate is only lower than that of nitroglycerin. The characteristic height of dropping weight is 7–10 cm (2 kg drop hammer). Specifically, the characteristic height of the α-glycerol dinitrate hydrate crystal is 90–100 cm, while the characteristic height of the liquid β-isomer is only 30–40 cm.

The lead block expansion value of glyceryl dinitrate is 500 cm^3, which is 82% of nitroglycerin. The detonation velocity of 92% glyceryl dinitrate gel is 3320 m/s while the corresponding value of nitroglycerin is 7400 m/s. The detonation velocity of 75% diatomaceous dynamite prepared with glycerol dinitrate is industrial production of glyceryl dinitrate 4300 m/s.

5.5.3 Industrial Production of Glycerol Dinitrate

There are four different methods for the industrial production of glyceryl dinitrate. The first method is nitration of glycerol by using a mixed nitric and sulfuric acid that contains less sulfuric acid than the commonly used mixed acid to produce a mixture of glyceryl dinitrate and nitroglycerin with any ratio. The composition of the mixed acid is shown in Table 5.36.

The resulting oily product has a ratio of dinitrate to trinitrate of about 70:30. The subsequent dilution of the waste acid with water separates the nitrate. The resulting precipitation yields an additional amount of nitrate, which contains glyceryl dinitrate in a ratio of 90%.

The second method is the nitration with dilute acid (e.g., 9% water) to produce glycerol dinitrate, followed by the dilution with water to separate glyceryl dinitrate. This method is very simple but not economical. The third method is the production of glyceryl dinitrate by using DuPont's ammonium method. According to DuPont's protocol [22], ammonia is added to neutralize the dilute nitric acid in the glyceryl dinitrate solution to recover ammonium nitrate. A glyceryl dinitrate containing relatively zero concentration of trinitrate is prepared by addition of 1.0 part of glycerol to 3.3 parts of nitric acid (specific gravity 1.5) at 15 °C or lower. After the reaction is complete, the solution is diluted with 1 part of water and neutralized with calcium carbonate. The product is precipitated as oil, and a small amount of the product in the aqueous solution is extracted with ether.

Table 5.36 Composition of the mixed acid with low content of sulfuric acid

Composition	Content
HNO_3	60
H_2SO_4	30
H_2O	10

In addition, glyceryl dinitrate can also be prepared by denitration of nitroglycerin with sulfuric acid. Simply dissolve of nitroglycerin in concentrated sulfuric acid without cooling results in denitration. The solution is dissolved in water, and the glycerol dinitrate is extracted with ether.

Based on numerous previous studies, the author [14] summarizes a new production process of glyceryl dinitrate. At the temperature below 15–20 °C, glycerol is added in dropwise to 3–4 parts of 68% nitric acid with stirring. After the addition, the stirring is continued for additional 30 min, and the reaction mixture is allowed to stand for 2–6 h. Subsequently, the reaction mixture is diluted with equal volume of ice water and neutralized with calcium carbonate to yield a concentrated calcium nitrate solution (1.51 g/ml). Glyceryl dinitrate is insoluble in the reaction mixture and forms an oil layer on the surface of the solution. The oil layer is separated and washed with a concentrated sodium chloride solution containing sodium carbonate or ammonia. Sodium chloride is added to reduce the solubility of glycerol dinitrate so that the loss of the product is negligible even at 40–50 °C (100 parts of water dissolve 8 parts of glycerol dinitrate at 20 °C while 100 parts of 30% Sodium chloride solution dissolves only about 2 parts of glycerol dinitrate). Finally, the glyceryl dinitrate is washed with sodium chloride solution to neutral. The yield of glyceryl dinitrate obtained in the production process of Fig. 5.23 is shown in Table 5.37. The raw material consumption for the production of 100 kg of glycerol dinitrate is as follows:

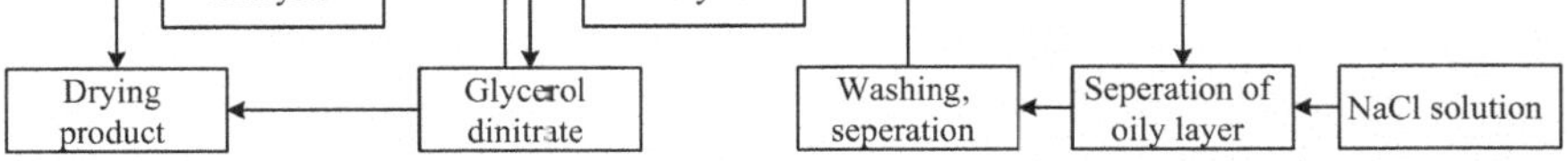

Fig. 5.23 Glyceryl dinitrate production process

Table 5.37 The raw material consumption for the production of 100 kg of glyceryl dinitrate

Raw material name	Consumption amount (kg)
Glycerin	70.4
Nitric acid	240
Ice	80
Calcium carbonate	120
Ammonium sulfate	165
Ammonia	1.5
H_2O	3500
NaCl	5

Table 5.38 Content of nitroglycerin in glycerol dinitrate

The ratio of nitric acid for nitration	Yield of nitrate (KG)			Content of composition (%)	
	Total amount of nitrate	Glycerol dinitrate	Nitroglycerin	Glycerol dinitrate	Nitroglycerin
100:250	115.8	108.2	7.6	93.44	6.56
100:350	151.7	143.5	8.2	94.59	5.41
100:450	179.5	150.4	29.1	88.35	11.65
100:500	184.4	146.9	37.5	79.66	20.34

Production of 100 kg of glycerol dinitrate yields the following by-products:
Ammonium nitrate 200 kg
Calcium sulfate 165 kg

By using the abovementioned raw materials, the production cost of glycerol dinitrate is slightly lower than that of nitroglycerin. The content of nitroglycerin in the produced glycerol dinitrate is shown in Table 5.38.

5.6 Other Dihydric Alcohol Nitrates

5.6.1 The Property and Preparation of Methylene Dinitrate [23, 24]

Molecular formula: $CH_2N_2O_6$
Chemical structure:

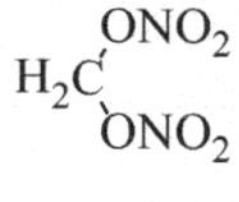

Molecular weight: 138.04
Oxygen balance: 34.8% (based on the generation amount of CO_2)
Nitrogen content: 20.29%.

Methylene dinitrate or "nitrated methylene glycol" is an oily liquid having a boiling point of 75–77 °C under 20 mmHg. It is a powerful liquid explosive that can dissolve in collodion and easy to be hydrolyzed. In early time, methylene dinitrate is regarded as no practical value because of the poor stability and strong explosive property of methylene dinitrate.

Methylene dinitrate is prepared by dissolving formaldehyde in sulfuric acid, followed by slow addition to the mixed nitric and sulfuric acid at 2–5 °C. The resulting oil layer is separated, washed with water, and subjected to a stability treatment to afford product in 35% yield.

5.6.2 Property and Preparation of 1,3-Propanediol Dinitrate

Molecular formula: $C_3H_6N_2O_6$
Structure:

$$O_2NO - CH_2 - CH_2 - CH_2 - ONO_2$$

Molecular weight: 166.09
Oxygen balance: −28.9% (based on the generation amount of CO_2)
Nitrogen content: 16.87%.

1,3-Propanediol dinitrate is an oily liquid with a boiling point of 180 °C under 10 mmHg and a specific gravity of 1.393 at 20 °C. 1,3-Propanediol dinitrate has higher volatility and a lower viscosity than nitroglycerin. The miscible solvents of 1,3-propanediol dinitrate are same as those of nitroglycerin. It can dissolve collodion very well. It would not be decomposed when heated at 75 °C for 600 h.

1,3-Propanediol dinitrate is a strong liquid explosive with explosion heat of 4763.5 J/g (water vapor condition) and an explosion point of 225 °C. The shock sensitivity is very low. For example, the characteristic height of 100 cm (2 kg drop hammer) does not cause an explosion. The water-filled lead block expansion value is 540 cm, which is equivalent to 90% of the expansion value of nitroglycerin. The 93% blasting gelatin prepared by 1,3-propanediol dinitrate has a lead block expansion value of 470 cm, which equals to 80% of nitroglycerine blasting gelatin.

The nitration temperature of 1,3-propylene glycol should be controlled between 0 and 10 °C. The separation of nitrated products from waste acid is also easy to carry out at the temperature below 10 °C. Under the optimized conditions, the conversion rate of 1,3-propanediol into 1,3-propanediol dinitrate is up to 96%. The authors have prepared and separated 210.3 kg 1,3-propanediol dinitrate from 100 kg raw material of 1,3-propanediol.

5.6.3 Properties and Preparation of 1,2-Propanediol Dinitrate

1,2-Propanediol dinitrate, also known as methyl ethylene glycol dinitrate, is important dihydric alcohol nitrate because it has been extensively used as liquid propellants [25]. In addition, it can be used directly as liquid explosive and mixed explosive with urea perchlorate and hydrazine.

Molecular formula: $C_3H_6N_2O_6$
Structure:

$$\begin{array}{l} CH_3\text{-}\underset{|}{C}HONO_2 \\ \quad\;\; CH_2ONO_2 \end{array}$$

Molecular weight: 166.09
Oxygen balance: −28.9% (based on the generation amount of CO_2)
Nitrogen content: 1 6.87%.

1,2-Propanediol dinitrate is an oily liquid with a boiling point of 92 °C (10 mmHg), a specific gravity of 1.368 (20 °C), and a freezing point of below −20 °C. 1,3-Propanediol dinitrate is highly volatile. The explosion heat and water-filled lead block expansion value are 4644 J/g (water vapor condition) and 540 cm, respectively.

According to vast literature [26], the mixed acid is composed of 40% of HNO_3 and 60% of H_2SO_4. The nitration of 100 parts of 1,2-propanediol at 20 °C yields 187 parts of 1,2-propanediol dinitrate, which equals to 86% of the theoretical yield.

A mixed acid composed of HNO_3 47.5%, H_2SO_4 45.5%, and H_2O 7%, combined with a 10% excessive HNO_3, is used to nitrate 1,2-propanediol at 10 °C to afford 1,2-propanediol dinitrate in a theoretical yield of 91–93%.

5.6.4 Properties and Preparation of 1,3-Butanediol Dinitrate

Molecular formula: $C_4H_8O_6N_2$
Structure:

$$\begin{array}{l} CH_3\text{-}\underset{|}{C}H\text{-}CH_2\text{-}CH_2\text{-}O\text{-}NO_2 \\ \qquad\;\; ONO_2 \end{array}$$

Molecular weight: 180.12
Oxygen balance: −53.2% (based on the generation amount of CO_2)
Nitrogen content: 15.55%.

1,3-Butanediol dinitrate is an oily liquid with excellent chemical stability, a freezing point of −20 °C, a specific gravity of 1.32, and a lead block expansion value of 460. Its explosion strength is about 75% of nitroglycerine. It is a highly volatile and good solvent for collodion.

The 1,3-butanediol dinitrate is prepared by the nitration of 1,3-butanediol at a higher conversion rate. In general, 100 kg 1,3-butanediol is nitrated to yield 187 kg 1,3-butanediol dinitrate with a yield of 93.7%. In the author's study, 1,3-butanediol as raw material, 100 kg of raw material 1,3-butanediol is nitrated directly by using nitric acid with an addition of 2–5% amount of ammonium nitrate at the nitration temperature of 17 °C. After separation, washing, stability treatment, and other procedures, 1,3-butanediol dinitrate 193 kg is obtained in a yield of 96.71%.

5.6.5 Properties and Preparation of Dimethylol Nitroethane Dinitrate [27]

Molecular formula: $C_4H_7O_9N_4$
Structure:

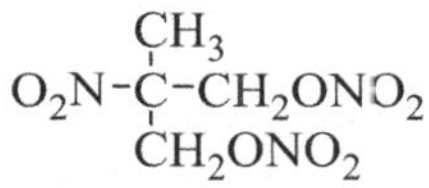

Molecular weight: 255.10
Oxygen balance: 15.68% (based on the generation amount of CO_2)
Nitrogen content: 21.96%.

Dimethylol nitroethane dinitrate (2-nitro-2-methylpropanediol 1-3-dinitrate) is solid with a melting point of 38 °C and low volatility. It dissolves in nitrocellulose readily.

The explosion strength of dimethylol nitroethane dinitrate is slightly weaker than that of nitroglycerin but slightly stronger than that of Hexogen. Its explosion heat is about 5230 J/g. Its shock sensitivity is significantly lower than nitroglycerin but slightly higher than tetryl. Compared with Hexogen, their shock sensitivities are very similar.

5.6.6 The Properties and Preparation of Dimethylol Nitropropane Dinitrate

Molecular formula: $C_5H_9O_8N_3$
Structure:

Molecular weight: 239.13
Oxygen balance: 36.8% (based on the generation amount of CO_2)
Nitrogen content: 17.57%.

Dimethylol nitropropane dinitrate (2-ethyl-2-nitropropyleneglycol-1,3-dinitrate) is a liquid that does not dissolve nitrocellulose. It is a relatively weak liquid explosive. The explosive performance of dimethylol nitropropane dinitrate is slightly stronger than that of picric acid. The lead block expansion value of dimethylol nitropropane dinitrate is 25% greater than that of picric acid, but its shock sensitivity is lower than that of picric acid.

5.6.7 The Properties and Preparation of Nitromethoxy Methanol Nitrate

Nitromethoxy methanol nitrate, also known as methyl ether nitro nitrate, is a colorless oily liquid with slightly higher explosive performance than ethylene glycol dinitrate. It is a liquid explosive with a great application prospect.

(1) The properties of nitromethoxy methanol nitrate

Molecular formula: $C_2H_4O_6N_2$
Structure:

CH_2ONO_2
O
CH_2NO_2

Molecular weight: 151.96
Oxygen balance: 0% (based on the generation amount of CO_2)
Nitrogen content: 18.43%
Boiling point: 48 °C (15 mmHg)
Density: 1.50 (20 °C)
Ignition point: 150 °C (220 °C for ethylene glycol dinitrate).

The shock sensitivity of nitromethoxy methanol nitrate is higher than that of ethylene glycol dinitrate. Studies have shown that 10% explosion probability of ethylene glycol dinitrate requires shock energy of 0.91 kg m/cm^2, while nitromethoxy methanol nitrate only requires shock energy of 0.11 kg m/cm^2. Nitromethoxy methanol nitrate dissolves collodion readily and is more volatile than ethylene glycol dinitrate. The chemical stability and explosion property of nitromethoxy methanol nitrate and its nitrocellulose solution are not very stable.

Nitromethoxy methanol nitrate has a lead block expansion value of 420 cm^3, whereas, in the same group of experiments, the net lead block expansion value of ethylene glycol dinitrate is 450 cm^3. Nitromethoxy methanol nitrate has similar explosive properties and low chemical stability. It has a high impact sensitivity and a lead block expansion value of 440 cm^3.

(2) Preparation method [28]

Nitromethoxy methanol nitrate can be prepared by the nitration of symmetrical bischloromethyl ether. Bischlorodimethyl ether is formed by the reaction of 40% formaldehyde solution and gaseous hydrogen chloride.

$$2CH_2O+2HCl \longrightarrow \begin{array}{c} CH_2Cl \\ | \\ O \\ | \\ CH_2Cl \end{array} + H_2O$$

$$\begin{array}{c} CH_2Cl \\ | \\ O \\ | \\ CH_2Cl \end{array} \xrightarrow[H_2SO_4]{HNO_3} \begin{array}{c} CH_2ONO_2 \\ | \\ O \\ | \\ CH_2NO_2 \end{array}$$

The reaction of formaldehyde with hydrogen chloride also produces a byproduct, chlorinated ether $ClCH_2OCH_2OCH_2Cl$, which converts into another nitrate O_2NOCH_2–O–$CH_2OCH_2NO_2$ with a boiling point of 88 °C (9 mmHg) in the subsequent nitration reaction. Nitromethoxy methanol nitrate is more volatile than ethylene glycol dinitrate but less volatile than other nitrates.

5.6.8 Properties and Preparation of Chloroglycerine Dinitrate

Chloroglycerine dinitrate is a liquid explosive with explosion strength only less than nitroglycerin [29]. Such explosive nature of the chloroglycerin dinitrate has attracted considerable attention. As early as 1904, chloroglycerine dinitrate was used to mix with nitroglycerine to manufacture cold-tolerant dynamite. The commercial product of chloroglycerin dinitrate is a mixture of two isomers, predominantly alpha-isomer. Because of its excellent antifreeze performance characteristics, chloroglycerine dinitrate remains as a mobile solution even at −20 °C.

(1) The properties of chloroglycerine dinitrate

Molecular formula: $C_3H_5N_2O_6Cl$
Structure:
Two isomers of chloroglycerine dinitrate

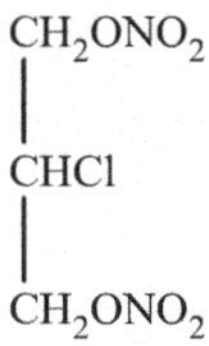

β-isomer

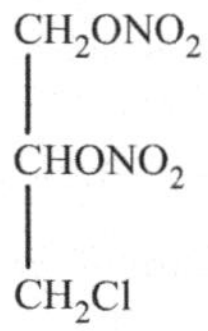

α-isomer

Molecular: 200.57
Oxygen balance: −19.94% (based on the generation amount of CO_2)
Nitrogen content: 13.96%.

The melting point of α-chloroglycerin dinitrate is 5 °C while the melting point of β-chloroglycerin dinitrate is 16.5 °C, which are significantly different. The specific gravity of α-chloroglycerin dinitrate is 1.541 (15 °C), and the heat of explosion is 4769.76 J/g. It slightly decomposes at 117.5 °C with a false boiling (10 mmHg). The commercial chloroglycerin dinitrate boils and rapidly decomposes at 190 °C. Its volatility is slightly higher than that of nitroglycerine. For example, the weight losses of chloroglycerin dinitrate and nitroglycerin are 3.1% and 0.35%, respectively, at a temperature of 75 °C in a 50 mm diameter open vessel for standing 24 h. Chloroglycerin dinitrate is not hygroscopic, and its solubility in water is slight (2.3 g of chloroglycerin dinitrate dissolves in 1 L of water at about 15 °C). Chloroglycerin dinitrate is soluble in most organic solvents other than carbon disulfide and gasoline. It is miscible with nitroglycerin to form a mixture with a very low freezing point. The mixture of 75% nitroglycerin and 25% chloroglycerin dinitrate is nonfreezing in general applications. Under ambient temperature conditions, collodion is difficult to be dissolved in pure chloroglycerin dinitrate but can be dissolved rapidly at 40 °C. When cooled down, chloroglycerin dinitrate is precipitated from the "gel". The mixture of chloroglycerin dinitrate and nitroglycerin has a very strong capability to dissolve nitrocellulose. Chloroglycerin dinitrate has a very low viscosity, about 1.55 times of water. Its chemical stability is higher than that of nitroglycerin with a resistance time of 30 min at 72 °C in Abel's heat test. When continuously heated at 75 °C for 10–12 days, no trace of decomposition is found. When a small amount of chloroglycerin dinitrate is heated to 170 °C or higher in the test tube, decomposition occurs to release nitric oxide. It can completely decompose at 180 °C without explosion. When heated to 190 °C, a weak explosion decomposition occurs. The explosive decomposition reaction is as follows:

$$C_3H_5Cl(ONO_2)_2 \rightarrow CO_2 + 2CO + 2H_2O + HCl + N_2 + 4184\,J/g$$

If the oxygen is sufficient, the explosion reaction is

$$C_3H_5Cl(ONO_2)_2 \rightarrow 2CO_2 + CO + H_2O + HCl + N_2 + 4393\ J/g$$

Chloroglycerin dinitrate is difficult to be ignited or exploded. When it is placed in a tank and thrown into the fire, chloroglycerol dinitrate in the open system can still maintain a calm burning. The shock sensitivity is lower than nitroglycerin and glycerol dinitrate [18]. The characteristic height is about 10–20 cm for 2 kg dropping hammer.

When detonated with No. 8 and No. 1 detonators, the water-filled lead block expansion values are 475 and 380 cm, respectively. The lead block expansion value of blasting gelatin made with 42% chloroglycerin dinitrate is 440 cm^3. The explosive strength of chloroglycerin dinitrate is about 80% of nitroglycerin. As a component of mining explosives, chloroglycerin dinitrate in explosion process would produce hydrogen chloride gas, which will cause environmental pollution. Thus, it is unsuitable for the explosion work under pit operation.

(2) Preparation of chloroglycerin dinitrate [29]

The preparation of the chloroglycerin dinitrate can be made by the nitration of monochloro glycerol with mixed nitric and sulfuric acid in similar method and equipment to the nitration of glycerol. The heat of nitration reaction is less than that of glycerol nitration. The viscosity of chloroglycerin dinitrate is lower than that of nitroglycerin. The nitration time of chloroglycerin dinitrate is shorter than that of nitroglycerin. The nitration coefficient is generally greater than 4.5. The composition of mixed nitric and sulfuric acid is shown in Table 5.39.

At the reaction temperature of 8–12 °C, after water washing and stability treatment, 100 kg of chloroglycerol is nitrated to afford 172.5 kg of chloroglycerin dinitrate (two isomers), which equals to a yield of 93.35%.

According to the nitration coefficient of 4.75, the composition of the waste acid after the nitration is shown in Table 5.40.

Table 5.39 The composition of mixed nitric and sulfuric acid for the nitration of monochloro glycerol

Acid	Content by weight (%)
HNO_3	40
H_2SO_4	60

Table 5.40 Composition of the waste acid after the nitration of 100 kg of chloroglycerol based on the nitration factor of 4.75

Composition	Content (%)
HNO_3	116.9
H_2SO_4	75.0
H_2O	7.9
NO	0.2

5.6.9 The Properties and Preparation of Formoxyl Glycerine Dinitrate

Formoxyl glycerine dinitrate is a liquid explosive with strong explosion strength.

(1) The properties of formoxyl glycerine dinitrate

Molecular formula: $C_4H_6N_2O_8$
Structure:

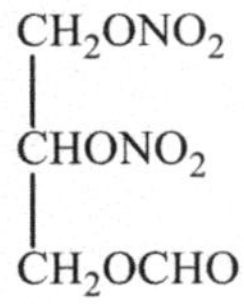

Molecular weight: 210.01
Oxygen balance: −22.86% (based on the generation amount of CO_2)
Nitrogen content: 13.34%
Explosion heat: 4284 J/g.

(2) The properties and preparation of formoxyl glycerine dinitrate [29]

Formoxyl glycerine dinitrate is prepared by the following process.

The mixture consisting of 2 mol glycerol and 1 mol anhydrous oxalic acid is heated at 140–150 °C for 20 h to form formyl glycerol, which is existed as formyl glycerol in glycerol solution as a part of the glycerol remains unreacted. The main reaction equation is as follows:

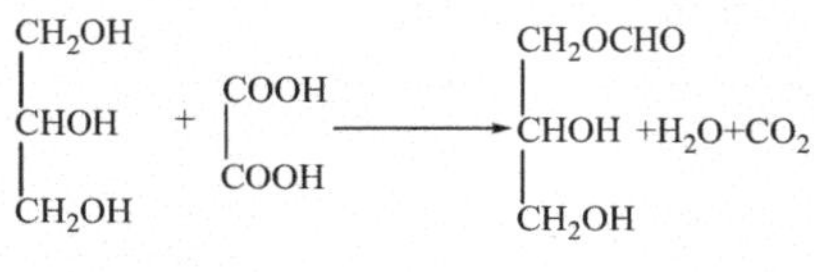

Subsequently, the solution is nitrated with a mixed nitric and sulfuric acid to form a mixture containing 70% nitroglycerin and 30% formylglycerol dinitrate.

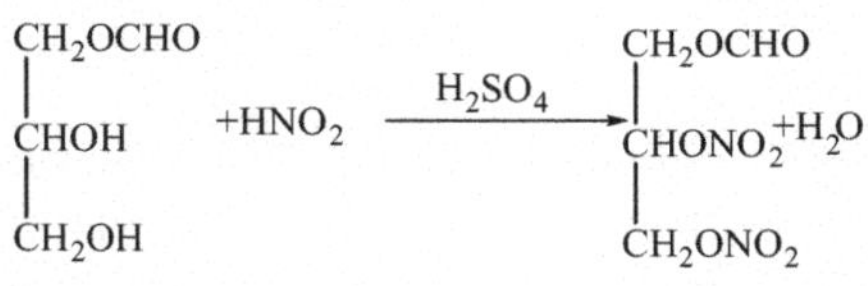

The obtained nitrate mixture, as an additive to nitroglycerin, is used to manufacture a low freezing point of Dynamite, which has 90% explosion strength of the nitroglycerin.

5.6.10 *The Properties and Preparation of Acetyl Glycerine Dinitrate*

(1) The properties of acetyl glycerine dinitrate

Acetyl glycerine dinitrate is a liquid explosive with a low freezing point and 92% explosive strength of nitroglycerin.
Chemical formula: $C_5H_8N_2O_8$
Structure:
Acetyl glycerine dinitrate has two isomers

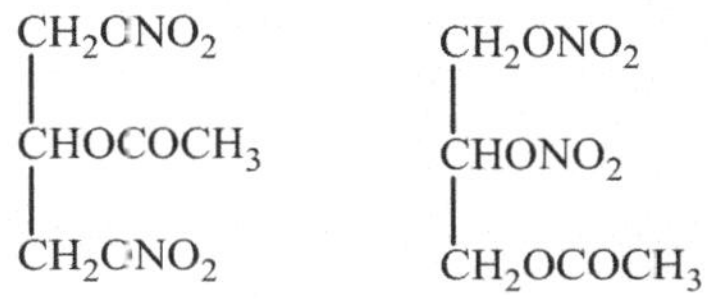

Molecular weight: 169.99
Oxygen balance: −56.47% (based on the generation amount of CO_2)
Nitrogen content: 16.47%.

The acetyl glycerine dinitrate has two isomers with a freezing point below −20 °C, a boiling point of 147 °C under 15 mmHg, a density of 1.42 (15 °C), an explosion heat of 2761.4 J/g. The volatility of commercial acetyl glycerine dinitrate is higher than that of nitroglycerin but is lower than that of chloroglycerine dinitrate. It is miscible with water and dissolves nitrocellulose readily. Its stability is lower than nitroglycerin. For example, it decomposes significantly after continuously heated at 75 °C for 24 h, accompanied by the release of nitrogen oxide. It begins to decompose when heated at 160 °C, and it decomposes vigorously at 170–180 °C.

Acetyl glycerine dinitrate is very insensitive to shock. Detonated with 8 # detonator, the lead block expansion value of acetyl glycerine dinitrate is 200 cm^3. The lead block expansion of blasting gel made of 92% acetyl glycerine dinitrate is 145 cm^3.

(2) The preparation of acetyl glycerine dinitrate

The equipment for the preparation of acetyl glycerine dinitrate is same as the equipment [29] used for the nitration of glycerol. The nitrated monoacetyl glycerine (glycerol monoacetate) is prepared with anhydrous mixed acid at 10–15 °C. After water washing and stability treatment, the refined product is obtained. 100 parts of glycerol monoacetate (by weight) affords 159 parts of acetyl glyceryl dinitrate, which equals to 95% of the theoretical value.

(3) The application of acetyl glycerine dinitrate

When acetylglycerol dinitrate is used as liquid explosives, a mixture of 80% nitroglycerin and 20% acetylglycerol dinitrate is often used. The explosive strength of this mixture is equivalent to 90% of nitroglycerin.

References

1. Edward P et al (1981) Evaluation of the neurophysiologic effects of 1,2-propylene glycol dinitrate by quantitative ataxia and oculomotor function tests. Am J Ind Med 2:365–378
2. Ju XH, Fan XW et al (2007) Intermolecular interactions between triethylene glycol dinitrate and polymer binder in mixed system. Chem Propellants Polym Mater 05:44–47
3. Yang SL, Song HC et al (1998) Thermal cracking mechanism of nitrated ethylene glycol. J Nanjing Univ Sci Technol 02:62–68
4. Department of defense test method standard, safety and performance tests for the qualification of explosives, MIL-STD-1751A (2001)
5. Rinkenbach WH (1951) Encyclopedia of chemical technology (ed Kirk, Othmer), vol 6. Interscience, New York
6. Lindner V (2000) Propellants, Kirk-Othmer encyclopedia of chemical technology, p 22
7. Bi XL, Liu JP (2011) Research progress in synthesis and nature pyrolysis of nitrate ester. In: Hazardous materials & security emergency technology
8. Andreyev KK (1957) Termicheskoye raslozheniye I goreniye vzryvchatykh veshchestv, Gosensrgoizdat, Moskva-Leningrad
9. Altukhova TF, Andreyev KK according to K. K. Andreyev, ref. [17], p. 125
10. Chavez DE, Hiskey MA, Naud DL, Parrish D (2008) Synthesis of an energetic nitrate ester. Angew Chem Int Ed 47:8307–8309
11. Oehman V (1936) Elektrochem Z 42:862
12. Fedoroff T, Sheffield OE (1972) Encyclopedia of explosives and related items, vol 8. Picatinny Arsenal, P86–P121
13. Wang QF, Shi F et al (2009) Synthesis of nitroglycerin with N_2O_5. Chin J Energ Mater 17 (3):304–306
14. Yang RH (2010) Preparation of typical nitrate esters. Beijing Institute of Technology
15. Wu CY (1980) Nitroglycerin chemical process. National Defense Industry Press, Beijing
16. Liu JP (1987) Basic chemical of propellants and explosives. GuiZhou Press, Guiyang
17. Chen Z et al (1982) Technological design of continuous manufacture of DINA. J East China Inst Technol 2:97–108
18. Xing X et al Dissolution properties of DINA in DMSO. Chin J Explos Propellants 34(3):25–27
19. Ma Y (1995) Effect of DINA on the performance of double base propellant. Energ Mater 3 (2):31–36
20. Urbanski T (1976) Chemistry and technology of explosives (II). National Defense Industry Press, Beijing
21. Mccallum KS, Emmons WD (1956) J Org Chem 21:377
22. Reinmutha N, Meyerb A, Hartwigsenc D et al (2014) Randomized, double-blind phase II study to compare nitroglycerin plus oral vinorelbine plus cisplatin with oral vinorelbine plus cisplatin alone in patients with stage IIIB/IV non-small cell lung cancer (NSCLC). Lung Cancer 83:363–368
23. Travagli GO (1935) Italy Patent 338080
24. Travagli GO (1938) Gazz Chim Ital 68, 718; (1938) Chimica Ed Ind 20, 389

25. Shi F, Wang QF, et al (2007) 1, 2-Green synthesis of propylene glycol dinitrate. Chin J Explos Propellants 30(2):75–77
26. Edward P, Horvath MD, Richard MPH, Ilka A et al (1981) Evaluation of the neurophysiologic effects of 1, p=propylene glycol dinitrate by quantitative ataxia and oculomotor function tests. Am J Ind Med 2:365–378
27. Wu Q, Xie HT et al (2004) Research progress of green nitration of nitrate. Chem Propellants Polym Mater 2(3):5–10
28. Wang QF, Shi F et al (2007) Nitrate green synthesis. Chin J Energ Mater 15(4):416–420
29. Liu JP (2015) Liquid explosives. Springer, Berlin

Chapter 6
Nitroglycerin

Nitroglycerin (NG), also known as glyceryl trinitrate, glycerol trinitrate, and trinitroglycerin, is produced by the reaction of glycerol with nitric acid. It is an extremely dangerous and powerful explosive sensitive to impacts, frictions, and vibrations, yet widely used in various fields. So far, NG is still a major component of propellants, especially energetic propellants, dual-based propellants, and explosives.

NG was first synthesized by Ascanio Sobrero in 1847 [1]. He nitrated glycerol with a mixture of nitric acid and carbonic acid as the nitration agent at low a temperature and called it "pyroglycerine" due to its highly explosive properties. It attracted considerable attentions back then because of the explosive properties. However, the high sensitivity of NG brought great dangers and thus was once considered impossible to handle.

In 1861, Alfred Nobel successfully prepared diatomaceous earth explosives with NG and started the large-scale NG production for mining in 1862. Later in 1875, he invented gelignite by absorbing NG in nitrocellulose, with which and the addition of ammonium nitrate or sodium nitrate, he achieved dynamite. In 1888, he applied NG to double-based propellants [2].

The continuous and safe production of NG was a mystery, and the world had been striving to improve its manufacturing safety and broaden its application. In 1908, Nathan, Thomson, and Rintoul et al. systematically studied the factors affecting the nitration of glycerol, such as conditions, yield, and so on, in detail. In 1927, Schmid, for the first time, realized the continuous production of NG by the container-type mechanical agitation [3]. In 1935, Biazzi built a continuous, programmed, and remote-controlled NG production line [4]. In 1956, Gyttorp proposed a new continuous NG production process with injection nitration and high-speed centrifugation separation [5]. The complete equipment for NG production was not commercialized until 1961. Germany and Sweden led the NG production equipment in the last century. In 1990s, their products were all over the world. China also imported NG production lines from Sweden. Based on these

J. Liu, *Nitrate Esters Chemistry and Technology*,
https://doi.org/10.1007/978-981-13-6647-5_6

works discussed above, the authors were able to improve the stability of NG by a catalytic nitration method and optimized stabilization conditions [6]. The produced NG can well tolerate long-distance transportation and minor impacts.

6.1 Properties of NG

Chemical formula: $C_3H_7N_3O_9$;
Chemical structure:

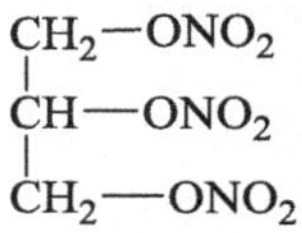

Molecular weight: 229.10;
Oxygen balance: 3.5% (calculated with the produced CO_2);
Nitrogen content: 18.34%.

NG is a colorless oily liquid. The industrial NG is usually light yellow (olive drab) or light brown, depending the purity of raw material and production conditions. It emits "sweet" smells as heated to 50 °C and is tasted bitter sweet and spicy.

6.1.1 Physical Properties of NG

NG has two forms with different freezing points and crystalline forms [7, 8]. The NG with low melting point is unstable and can be transformed into the stable form with a higher melting point. The melting points reported by different researchers are listed in Table 6.1. Table 6.2 lists the solidification rates of NG into the stable form at different temperatures.

Table 6.1 Melting points of different NG forms (°C)

Form	Ref. [9]	Ref. [9]	Ref. [9]	Ref. [9]
Unstable	–	2.1–2.2	1.9	1.9
Stable	12.4	13.2	13.0	13.0

Table 6.2 Solidification rates of NG into the stable form at different temperatures

Temperature(°C)	Rate (mm/min)
+5	0.145
0	1.183
−5	0.267
−17	0.125

Table 6.3 Solubility of NG in 1000 mL water

Temperature (°C)	Solubility (g)
15	1.6
20	1.8
50	2.5

NG is slightly soluble in water. Table 6.3 lists its solubility in water at different temperatures. The solubility of NG in sodium carbonate aqueous solutions (1–10%) is even lower. For example, the solubility of NG in 8% sodium carbonate solution is only 1.0 g/1000 mL. NG itself is an excellent solvent. It is soluble in most organic solvents, even miscible with many solvents, such as methanol, ethyl acetate, and so on, at room temperature. It is also miscible with nitrate ester, azido nitrate ester, and dichloroethylene. However, 100 parts of trichlorethylene can only dissolve 20 parts of NG, and NG is almost insoluble in tetrachlorethylene and pentachloroethane. The solubility of NG in ethanol is affected by temperature and the water content. In anhydrous ethanol, the solubility of NG increases with the increase of temperature and NG is miscible with 96% or anhydrous ethanol at 50 °C. Table 6.4 lists the solubility of NG in ethanol under various conditions.

NG is slightly soluble in polyols. It is partially soluble in propanol, isopropanol, and pentanol at room temperature and become miscible at 90–100 °C. One hundred grams of ethylene glycol can dissolve 20 and 12 g of NG at 200 °C and 80 °C, respectively. The solubility of NG in glycerol is very low. It is soluble in 65% acetic acid, but insoluble in carbon disulfide and slightly soluble in gasoline, crude oil, kerosene, paraffin oil, lubricants, petrolatum oil, and other aliphatic hydrocarbons. Increasing temperature can slightly increase its solubility. It is miscible with vegetable oils. Table 6.5 lists the solubility of NG in various aliphatic hydrocarbons.

NG is slightly soluble in ammonia, sodium hydroxide, and potassium hydroxide aqueous solutions, but highly soluble in concentrated sulfuric acid and miscible with anhydrous nitric acid. Table 6.6 lists the solubility of NG in H_2SO_4: H_2O = 5.8, and H_2SO_4: HNO_3 = 10.4.

Table 6.7 lists the solubility of NG in the sulfuric acid/nitric acid aqueous solutions with the similar ratios of sulfuric acid to nitric acid to that of the mixture M_3 in Table 6.6. NG can be partially hydrolyzed in these acid solutions. In general, NG is insoluble in concentrated hydrochloric acid at room temperature and decomposed as heated to form dark yellow nitrosyl chloride.

Table 6.4 Solubility of NG in ethanol

Ethanol	Temperature (°C)	Solubility (g/100 mL)
Anhydrous ethanol	0	~30
Anhydrous ethanol	20	~43
96%	20	31.6
50%	20	1.8
25%	20	0.7

Table 6.5 Solubility of NG in various aliphatic hydrocarbons

Solvent	Temperature (°C)	Solubility (g/100 g)
Crude oil	20	~1.5
	80	~6.0
Gasoline	20	~2
	80	~6
Refined oil	20	~4
	80	~9
Oliver oil	20	20
	100	25
Castor oil	20	26
	100	30

Table 6.6 Solubility of NG in the mixtures of sulfuric acid and nitric acid

Mixture	Content (%)			Solubility (%)
	HNO_3	H_2SO_4	H_2O	
M_1	24	50	26	13.5
M_2	38	20	42	1.6
M_3	13	75	12	2.5

Table 6.7 Solubility of NG in the mixed sulfuric acid and nitric acid solutions

Content (%)	I	II	III	IV	V	VI
HNO_3	10	10	10	15	15	15
H_2SO_4	70	75	80	80	75	75
H_2O	20	15	10	5	10	15
Solubility (g/100 g)	6.00	3.55	3.33	4.37	2.60	2.36

The freezing point of NG can be significantly decreased by mixing with dinitroglycerin and dinitrates, such as ethylene glycol dinitrate, which is favorable to the preparation of a variety of liquid explosives and colloidal explosives. NG is compatible with nitro compounds and can effectively increase their energy. For example, highly energetic liquid explosives with low sensitivity can be prepared by adding NG in nitroglycol, chloroglyceryl dinitrate, or diglyceride tetranitrate that is then mixed with pentaerythritol tetranitrate or RDX. Table 6.8 lists the low melting point inclusions prepared with NG and aromatic compounds.

NG can dissolve the nitrocellulose with low nitrogen contents. The consistency of the obtained solution is affected by the percentage of nitrocellulose in NG and viscosity of the nitrocellulose. The mechanical properties of the propellant are mainly affected by the nitrocellulose content.

The stability of NG is affected by the trace impurities, especially the trace metals and alkali metals introduced by the nitration and water washing, and nitration temperature [33]. The NGs formed at high temperatures contain less acid residues

Table 6.8 Low melting point inclusions of NG and aromatic compounds

Compounds	Stable form		Unstable form	
	In inclusion (%)	Melting point (°C)	In inclusion (%)	Melting point (°C)
Nitrobenzene	45.5	−15.2	57.5	−22.9
1, 3-dinitrobenzene	82.5	5.0	88.0	−5.4
2,4-dinitrobenzene	72.7	6.1	89.0	−4.1
2,4,6-trinitrobenzene	82.9	6.3	90.0	−4.0
Tetryl	90.0	9.8	94.0	−0.6
Hexogen	99.5	12.3	99.7	1.2
Nitroglycol	~20	~40	~40	~−40
Pentaerythritol tetranitrate	98.5	12.3	98.9	1.3
Benzoyl urethane	90.0	9.6	93.0	−0.8
Phthalide	76.5	0.8	8.10	−8.2
Methyl centralite	96.5	10.6	98.0	0.8
Diphenylamine	93.0	9.8	96.0	−0.4
HMX	99.4	11.8	99.8	6.2
CL-20	98.6	13.6	99.3	5.4

before the stabilization, e.g., water and sodium carbonate washings. The relationship between nitration temperature and stability of NG is shown in Table 6.9.

NG boils and decomposes at temperatures above 180 °C, along with the release of nitrogen oxides and water vapor. Dilute nitric acid and small amounts of NG can be found in the distillate. The boiling point of NG is 180 °C under the pressure of 50 mm Hg, and 125 °C under 2 mm Hg pressure. The vapor pressures of NG at different temperatures are listed in Table 6.10.

The enthalpy of vaporization (ΔH_{vap}) of NG is barely affected by temperature, as shown in Table 6.11.

Based on the vapor pressure of diethyleneglycol dinitrate, researchers calculated the vapor pressures of the mixtures of NG and diethyleneglycol dinitrate at different temperatures, as shown in Table 6.12.

NG is slightly volatile. Table 6.13 lists the volatilization of 20 g NG in a glass tube of 70 mm diameter heated at different temperatures for 24 h. The volatilization of NG is affected by both temperature and contact area. Its volatilization at 60 °C is 0.11 mg/cm^2/h.

Table 6.9 Relationship between nitration temperature and stability of NG

Nitration temperature (°C)	Residue acids		Abel heat test time (min)
	HNO_3 (%)	H_2SO_4 (%)	
15	0.0132	0.0024	15
25	0.0057	0.0017	24
40	0.0052	0.0008	30

Table 6.10 Vapor pressures of NG at different temperatures

Temperature (°C)	Vapor pressure (mm/Hg)
20	0.00025
30	0.00038
35	0.0020
40	0.0024
45	0.0072
50	0.0081
60	0.0188
70	0.043
80	0.095
93	0.29

Table 6.11 ΔH_{vap} of NG at different temperatures

Temperature (K)	Enthalpy of vaporization ΔH (kJ/mol)
100	86.36
140	85.27
180	84.14

Table 6.12 Vapor pressures of the mixtures of NG and diethyleneglycol dinitrate

Diethyleneglycol dinitrate (%)	Total vapor pressure (mm Hg)			
	10 °C	30 °C	40 °C	50 °C
10	0.0027	0.0186	0.0433	0.1002
20	0.0051	0.0345	0.1159	0.1841
40	0.0093	0.0624	0.1475	0.3314
60	0.0137	0.0874	0.2417	0.6316

Table 6.13 Volatilization of 20 g NG at different temperatures

Temperature (°C)	Volatile (g)	Percentage of volatile (%)
50	0.04	0.2
75	0.32	1.6
100	2.00	10.0

The specific gravity of NG slightly decreases with the increase of temperature, as shown in Table 6.14. The volume of NG decreases 8.3% as it is solidified. Its specific gravity at 15 °C is 1.5995 g/cm^3. The specific gravity of its stable form at 100 °C is 1.735 g/cm^3.

The refractive index of NG $n_D^{20} = 1.4732$. Its dielectric constant $\varepsilon 20 = 19.25$ and molecular dipole moments in different solutions are shown in Table 6.15.

Table 6.14 Specific gravities of NG at different temperatures

Temperature (°C)	Specific gravity (g/cm^3)
4	1.614
15	1.600
25	1.590

Table 6.15 Molecular dipole moments of NG in different solutions

Solution	Dipole moment (Debye)
Pure NG	3.82
NG in hexane	2.56
NG in tetrachloromethane	2.88
NG in benzene	3.16

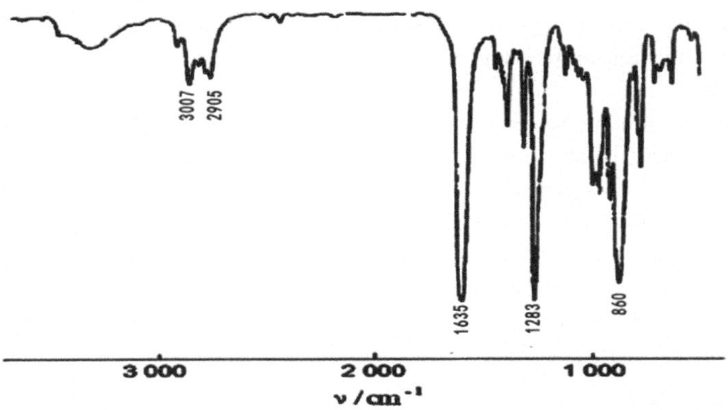

Fig. 6.1 IR spectrum of NG

Figure 6.1 shows the IR spectrum of NG. The absorption peaks at 1675, 1654, and 1645 cm^{-1} are caused by the asymmetric stretching vibration of NO_2. The symmetric stretching vibration of NO results in the absorption peaks at 1293 and 1276 cm^{-1}. The multiple absorption frequencies are attributed to the different enantiomers of NG.

6.1.2 Chemical Properties of NG

(a) Reactions of NG with acids

NG can react with concentrated sulfuric acid by transesterification to produce mixed ester and sulfuric acid ester.

$$C_3H_5(ONO_2)_3 + H_2SO_4 \rightleftharpoons C_3H_5(ONO_2)_2(OSO_3H) + HNO_3$$
$$C_3H_5(ONO_2)_2(OSO_3H) + H_2SO_4 \rightleftharpoons C_3H_5(ONO_2)(OSO_3H)_2 + HNO_3$$
$$C_3H_5(ONO_2)(OSO_3H)_2 + H_2SO_4 \rightleftharpoons C_3H_5(OSO_3H)_3 + HNO_3$$

Hydroiodic acid can decompose NG into glycerol and hydroxides as shown below.

$$C_3H_5(ONO_2)_3 + HI \rightarrow C_3H_5(OH)_3 + NO_2 + NO + I_2 + H_2O$$

Some inorganic acids can diffuse into NG. NG can dissolve 6–9% nitrates and 0.3% sulfuric acid. NG is insoluble in cold hydrochloric acid, but can be degraded into nitroxyl chloride as heated in hydrochloric acid.

(b) Reactions of NG with bases

NG is barely soluble in the cold aqueous solutions of sodium hydroxide, potassium hydroxide, and ammonia, and can be saponified in these solutions at high temperatures.

NG can be subject to hydrolysis, redox, and other chemical reactions in the aqueous or ethanol solutions of sodium hydroxide and potassium hydroxide as heated, which breaks the nitrate structure of NG and produces corresponding inorganic salts, inorganic nitrates, oxalic acid, ammonia, etc.

Generally, NG cannot be saponified to produce glycerol in aqueous alkali solutions. However, its saponification can occur in sodium hydroxide solutions containing benzene sulfuric acid, where NG is saponified to produce glycerol and benzene sulfuric acid is converted into diphenyl sulfide.

(c) Reduction of NG

In general, NG is highly resistant to oxidization, but is easily reduced by many reducing agents, such as hydrochloric acid, zinc and acetic acid, alkali metal sulfides, and hydrosulfides, to produce glycerol and ammonia. NG can be reduced to nitric oxide by ferrous chloride in hydrochloric acid as shown below.

$$C_3H_5(ONO_2)_3 + 5FeCl + HCl \rightarrow CH_3COOH + HCOOH + 5FeCl_3 + 2H_2O + 3NO$$

(d) Hydrolysis of NG

Like other esters, NG can be hydrolyzed. However, heating NG in an aqueous solution or alcohol solution of sodium hydroxide or potassium hydroxide cannot hydrolyze it, but causes redox reactions to produce organic acids, nitrates, and nitrites. It is reduced to glycerol in the presence of easily oxidized substances. NG can be hydrolyzed by sodium hydroxide to glycerol in the presence of phenothioate, along with the formation of p-phenylene sulfide from phenothioate. Strong reducing acids can rapidly decompose NG into glycerol and nitrogen oxide.

6.1.3 Stability of NG

(1) Thermochemical properties of NG

(a) Specific heat and the heat of crystallization of NG

The specific heat of NG is 1.49 J/g, and those of its crystal (stable form) and liquid NG are 1.317 J/g and 1.318 J/g, respectively. Table 6.16 lists the heat of crystallization (coagulation) of NG.

(b) Heats of combustion and formation of NG

The heat of combustion of NG in a fixed capacity (ΔH) is 1541.39 kJ, equivalent to 6791 J/g. Its enthalpy of formation (−ΔHf) is +356.9 kJ, equivalent to +1564.82 J/g. The heat of the nitration for the industrial production of NG with glycerol is 502.08–711.28 J/g glycerol.

(2) Chemical stability of NG

NG is decomposed to acidic compounds in 3–4 day as heated at 75 °C and produce nitric acid at day 6. The decomposition of NG can be very slow if the volatile decomposition products are rapidly removed. The decomposition of a small amount of NG in an open system is not dangerous. However, in a seal space, NG is decomposed to produce NOx that can catalyze and accelerate the decomposition reaction. The accumulated high pressure of the gas products and rapid chain reaction cause explosions. For example, a mixture of 0.3 g of NG and 5% nitric acid sealed in a glass tube at 42 °C can explode in 6 h. Figure 6.2 shows the temperature change during the catalyzed decomposition of NG by nitric acid.

The decomposition of NG is actually a catalyzed reaction of NG by the diluted acid of NO_2 in water formed during the decomposition. The decomposition products of N_2, NO, CO, and CO_2 can accelerate the decomposition. NG starts to decompose as the water content is over 0.01% in 30 h. As the water content increased to 1.5%, NG starts to decompose in 2 h. The thermogravimetric analysis of NG at 90–135 °C indicates that the volatile decomposition products can be removed with a CO_2 flow.

The pH of NG solution increases with the prolongation of heating time. Table 6.17 lists the pHs of the water extracts of NG heated at 110 °C for different periods of time. Figure 6.3 shows the relationship between the pH of NG solution and heating time.

Table 6.16 Heats of crystallization of NGs in different forms

Phase	Heat of crystallization (J/g)
Stable NG	138.91
Unstable NG	21.76
Stable NG transformed from unstable NG	117.20

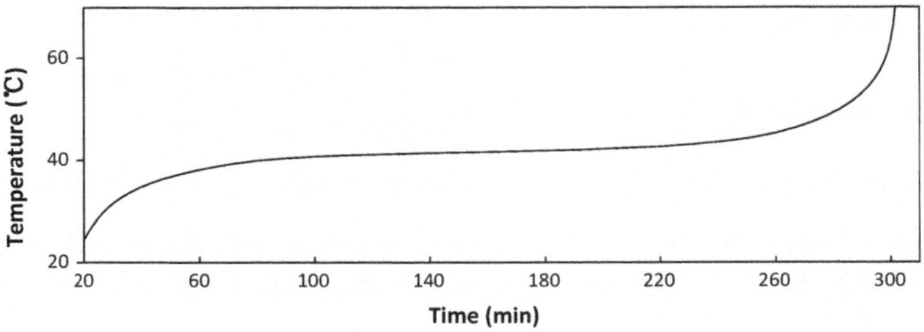

Fig. 6.2 Temperature change during the 5% nitric acid-catalyzed decomposition of NG [10]

Table 6.17 pHs of the water extracts of NG heated at 110 °C for different periods of time

Mass of NG (g)	pH after heated at 110 °C for different periods of time			
	1 h	2 h	3 h	4 h
5	5.88	5.40	4.90	4.31
10	11.82	10.90	9.36	7.63

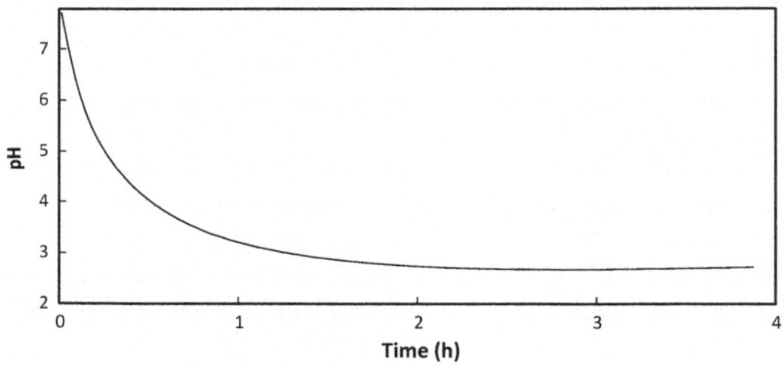

Fig. 6.3 pH change of a water extract of NG with heating time at 110 °C

The decomposition mechanism of NG at 135 °C is significantly different from that at lower temperatures. NG becomes red due to the adsorption of the decomposition products. As heated to 145 °C, the decomposition becomes more vigorous, along with the release of nitrogen oxide gas products and extremely vigorous as temperature increased to 165 °C. NG is then decomposed into dilute nitric acid that distilled with glycerol nitrates. The residue mainly contains glycerol, mononitrates, dinitrates, and trace amounts of unknown impurities. NG becomes very vicious as heated to 180–185 °C and eventually explodes as the temperature increased to 215–218 °C. This procedure has been well recognized. As can be seen, unlike the false impression, NG can be handled safely.

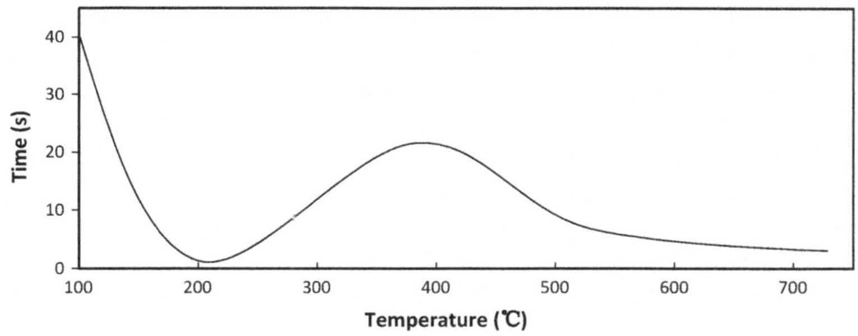

Fig. 6.4 Decomposition of NG on a heated metal plate [10]

NG can easily explode at extremely high temperatures. For example, a drop of NG on a metal plate of 215–250 °C can form a stable explosion, causing a significant deformation of the plate. NG dropped on the metal plate of 420 °C or higher temperatures can form mild combustion balls. Figure 6.4 shows the explosion behaviors of NG on the heated metal plate.

The activation energy (E) required for the thermal decomposition of NG can be calculated from the decomposition process of NG on the heated metal plate shown in Fig. 6.4 as E = 156.48 kJ/mol at 215–250 °C and 20.82 kJ/mol at temperatures over 470 °C.

The activation energies required for the decomposition of NG at low temperatures are E = 209.2 kJ/mol at 150–190 °C, E = 188.28 kJ/mol at 125–150 °C, and E = 178.24 kJ/mol at 20–125 °C.

NG starts to decompose under the irritation of UV light in 1–2 min and stops decomposing in 2–3 days. For example, the pH of 3 g NG irradiated under incandescent light of 3200–4100   at 15 °C for 60 min is 6.86, and becomes 6.12 after 6 h irradiation. NG can be instantly detonated as irradiated under 900 J of UV light at 100 °C. NG is very sensitive to γ-rays, and can release N_2, nitrogen oxides, H_2, CO, and CO_2 under γ-ray irradiation.

6.2 Explosive Properties of NG

The explosion reaction of NG under normal conditions can be expressed as

$$C_3H_5(ONO_2)_3 + 5FeCl + HCl \rightarrow CH_3COOH + HCOOH + 5FeCl_3 + 2H_2O + 3NO$$

Based on the reaction equation, the gas product is composed of 58.15% CO_2, 19.82% H_2O, 18.50% N_2, and 3.53% O_2.

The explosion heat of NG under normal condition is 6213 J/g with the specific volume V_0 = 715.7 mL/g. The heat released from NG can raise the temperature to 4250 °C with the explosive power f_0 = 9292 m.

The sensitivity of NG is significantly affected by its stability treatment. After conducting a large number of experiments, the authors have found that the NG stabilized with filtered water can explode under the vibration of 200 Hz, while that prepared under same conditions but stabilized with deionized water can tolerate vibrations of frequencies up to 800 Hz.

NG is very sensitive to shock. The NG absorbed by a solid possesses an enhanced impacted force. Liquid NG or its shock wave has the minimum impact force. Table 6.18 lists the shock sensitivities of different NG phases.

The shock sensitivity of a mixture of liquid and crystalline NG is higher than either liquid or crystalline NG, due to the friction between the sensitive liquid and the crystallites. For example, probability of NG explosion of gun shooting is 100%. The lead block expansion value of NG is governed by the intensity of detonation. Table 6.19 lists the lead block expansion values of NG detonated by different detonators.

The detonating intensity of NG is different from those of ethylene glycol dinitrate and methyl nitrate that can be detonated and form a complete explosion even by a weak detonation. The detonation of NG is affected by its charge configuration. The currently available charge structures suggest that NG is easily and steadily detonated in a 5-mm-thick steel vessel by No. 1 detonator. Table 6.20 lists the lead block expansion values of NG, and Fig. 6.5 shows the pictures of the resulted lead block expansion.

Table 6.18 Shock sensitivities of NG in different forms

Phase	Impact force (kg m/cm^2)	Explosion rate (%)
Liquid	0.08	10
	0.11	50
	0.40	100
Stable crystallite	0.51	10
	0.65	50
	0.82	100
Unstable crystallite	0.63	10
	0.78	50
	0.93	100

Table 6.19 Lead block expansion values of NG detonated by different detonators

Detonator	Lead block expansion value (cm)
1	190
2	225
6	460
8	590

Table 6.20 Lead block expansion values of NG detonation

NG phase	Pure NG (cm^2)	Sand filling (cm^2)	Water filling (cm^2)
Stable crystallite	–	380	540
Unstable NG crystallite	–	390	525
Liquid	–	390	530

Fig. 6.5 Pictures of the lead block explosion caused by the NG detonation

NG prepared with deionized water is less sensitive and can tolerate the friction on the coarse surface of mortar [6].

The detonation velocity of liquid NG ranges from 1000 to 8000 m/s, depending on the charge diameter and charge configuration. The authors, after a thorough investigation of the relationship between the charge configuration of NG and detonation velocity, found that the detonation velocity of NG increased with the increase of charge diameter. The detonation velocity can only reach 920 m/s with the charge diameter less than 5 mm. The detonation velocities at various charge diameters are shown in Table 6.21.

The emulsions of NG can exhibit certain detonation velocities as shown in Table 6.22 for the detonation velocities of NG/water emulsions.

The NG emulsions with water contents more than 2:1 cannot explode. The explosion reaction of the low detonation velocities follows the equation below:

$$C_3H_5(ONO_2)_3 \rightarrow 2CO + CO_2 + 2\frac{1}{2}H_2 + 1\frac{1}{2}N_2O + 1\frac{3}{4}O_2 + 87.86\ \text{KJ}$$

Table 6.21 Relationships between detonation velocity and charge configuration and detonator

Charge medium	Charge diameter (mm)	Detonation condition	Detonation velocity (m/s)
Lead tube	30	No. 8 detonator	1560
Aluminum tube	3.0	No. 8 detonator	915
Aluminum tube	9.0	No. 8 detonator	1130
30CrMnSi tube	30	No. 8 detonator	3780
Glass tube (wall thickness, 3.0 mm)	22	No. 8 detonator	1165
Organic glass tube	28	No. 8 detonator, 15 g Tetryl booster	7800
30CrMnSi tube (wall thickness, 1.0 mm)	37.5	No. 8 detonator, 2 g Mercury (II) fulminate booster	8520
Organic glass tube	25	No. 8 detonator, 50 g Tetryl booster	7410
30CrMnSi tube	22	No. 8 detonator, 75 g Tetryl booster	7760
#12 corrosion-resistant aluminum tube	40	No. 8 detonator, 20 g Tetryl booster	8560
#12 corrosion-resistant aluminum tube	35	No. 8 detonator, 20 g Tetryl booster	7830
#12 corrosion-resistant aluminum tube	30	No. 8 detonator, 20 g Tetryl booster	6970
#12 corrosion-resistant aluminum tube	25	No. 8 detonator, 20 g Tetryl booster	6100
#12 corrosion-resistant aluminum tube	20	No. 8 detonator, 20 g Tetryl booster	5870

Table 6.22 Detonation velocities of NG/water emulsions

H_2O:NG	Detonation condition	Detonation velocity (m/s)
0.75:1	No. 8 detonator, 15 g Tetryl booster	6370
1:1	No. 8 detonator, 15 g Tetryl booster	5520
1.5:1	No. 8 detonator, 15 g Tetryl booster	5150
2:1	No. 8 detonator, 15 g Tetryl booster	4300

The explosion reaction of NG in a sealed setup detonated by a booster can be expressed as

$$C_3H_5(ONO_2)_3 \rightarrow 3CO_2 + 2\frac{1}{2}H_2O + 1\frac{1}{2}N_2 + 1\frac{3}{4}O_2 + 1334.7\ \text{KJ}$$

The final products are carbon dioxide and water vapor, which can result in a high detonation velocity.

To assess the explosion effects of NG, the authors designed a special thin-walled glass steel ball filled with 3000 g NG that was suspended in the air 800 mm from the ground. It was detonated No. 8 detonator with 20 g blunt black aluminum explosives as the booster to determine the shock wave, fireball, and the impact on ground. The experimental results indicate that the overpressure at the explosion center is 1.52 MPa, and that 2000 mm away from the explosion center is 0.67 MPa. The explosion caused a fireball with a diameter of 6268 mm and loosed the highly solid ground. The effects of NG explosion are shown in Fig. 6.6.

The explosion of NG can form bigger fireballs than those of other nitrates, and cause stronger impacts on the ground than the same amount of TNT.

NG is considered dangerous and scaring due to its explosive power and high sensitivity [11]. In fact, NG is easy to be combusted, but difficult to get fired. Ignition of NG with a gas flame or hot wire in a glass tube with a diameter of 20 mm can only partially combust and decompose the NG and no explosion occurs. However, igniting the pre-heated or burned spot of NG can cause a rapid explosion, without the precombustion occurs. When the NG flame is to be extinguished, the critical pressure varies between 234 mm Hg and 375 mm Hg, depending on the experimental conditions. The combustion velocity of NG is usually ~0.075 cm/s and can reach 0.13 cm/s at 98 °C under the critical pressure of 320 mmHg. NG can be combusted constantly under the pressure of 400 mm Hg and the combustion velocity varies with pressure as shown in Fig. 6.7.

The constant combustion of NG fits Eq. 6.1.

$$v = 0.0067 + 0.216P^{0.831} \quad (6.1)$$

Based on this formula, the combustion velocity of NG under the standard atmospheric pressure is determined to be 0.23 cm/s. The nitrocellulose absorbed NG is easier to be burned than NG alone.

(a) Propellant (b) Fireball (c) Impacted ground

Fig. 6.6 Explosion effects of 3000 g NG hung 800 mm from ground. **a** Propellant, **b** fireball, **c** impacted ground

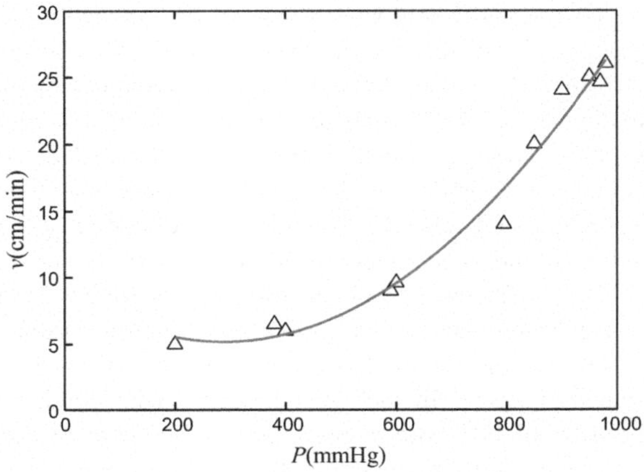

Fig. 6.7 Relationship between the combustion velocity of NG and pressure

6.3 Production Process of NG

The history of NG production can be traced back over 100 years ago, during which the production process has been developed from the discontinuous process to the continuous process and from manual operation to automatic and programmed operations. The direct production process has been developed into the isolated and remote-controlled processes. However, no matter how the NG production process has evolved, the fundamental process is similar. In the early development stage, NG was prepared by directly esterifying glycerol with nitric acid as shown below.

$$\begin{array}{l} CH_2OH \\ | \\ CHOH \\ | \\ CH_2OH \end{array} + 3HNO_3 \longrightarrow \begin{array}{l} CH_2ONO_2 \\ | \\ CHONO_2 \\ | \\ CH_2ONO_2 \end{array} + 3H_2O$$

Later, it was found that the reaction rate of such direct esterification was low and intermediate products stayed in the oxidation environment for a long period of time, leading to more side reactions and increasing the safety risk of the production. Therefore, nitro-sulfuric acid is used as the nitrating agent and sulfuric acid acts as the dehydrating agent in modern NG production processes.

It is well known that NG can be achieved by the nitration of glycerol in the mixture of nitric acid and sulfuric acid. The typical composition of the mixed acid is of 47–50% HNO_3, 49–52% H_2SO_4, and <1.0% H_2O. The optimal nitration temperature is 5–15 °C. The nitration coefficient is usually in the range of 5–8, and the conversion rate of such reactions can reach up to 94–97%.

The nitration of glycerol is an exothermal reaction with the exothermic ΔH = 439.5 J/mol glycerol. However, the reaction heat of the industrial nitration of glycerol is higher than that of bench-scale reaction due to the nitration heat accumulation. The reaction mechanism of glycerol nitration can be expressed as

$$C_3H_5(OH)_3 + 3HNO_3 \xrightarrow{H_2SO_4} C_3H_5(ONO_2)_3 + 3H_2O + 439.5 \text{ mol}$$

The large-scale industrial production of NG can produce more, even double, heat than the small-scale reaction, which significantly accelerates the reaction. The nitration can be finished in only 1–1.5 s after the glycerol is fully mixed with the nitrating agent.

The authors have studied reaction mechanism under various nitration conditions, and the factors affecting the nitration and the difference between the mixed acid injection nitration and the air-agitated nitration [6, 11–13] were studied. These processes have been successfully applied in the industrial manufacture of NG.

6.3.1 The Production Flow Chart of NG

Glycerol contains both α-hydroxy and β-hydroxy group. The primary hydroxyl has a stronger electron donating ability, e.g., more electronegative, and thus can be easily nitrated. The secondary hydroxyl (γ hydroxyl) has a higher steric hindrance and its nitration is slow. The nitration of glycerol occurs through the following process.

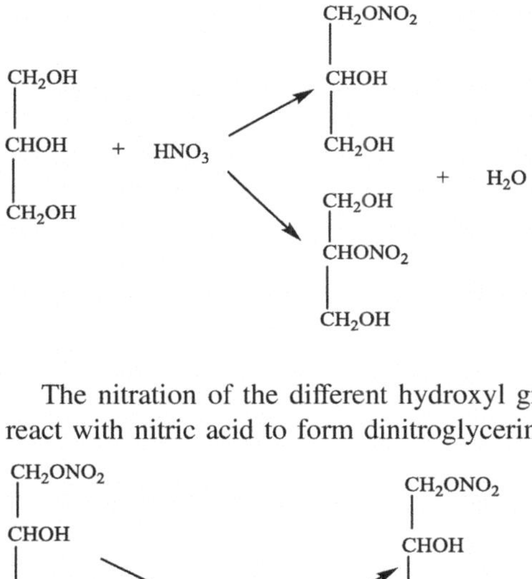

α-glycerol-mononitrate

β-glycerol-mononitrate (3.5~4.0%)

The nitration of the different hydroxyl groups resulted in isoforms that further react with nitric acid to form dinitroglycerin.

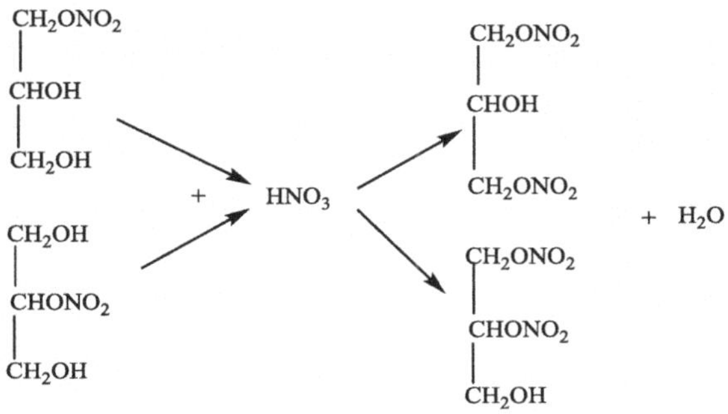

α'-glycerol-dinitrate (70%)

β'-glycerol-dinitrate (30%)

The dinitroglycerin reacts with nitric acid to form glycerol trinitrate.

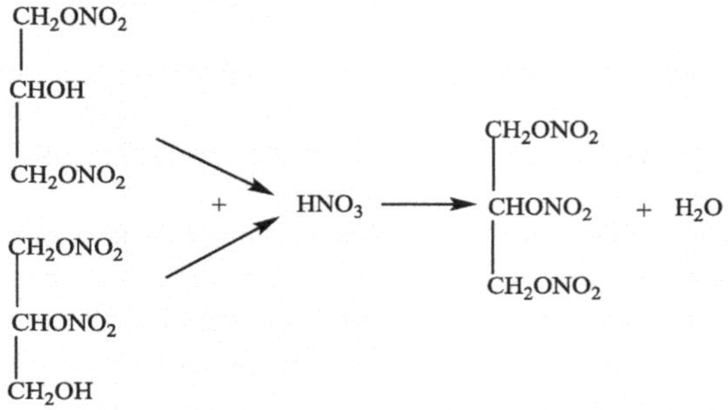

The reaction between the primary hydroxyl and nitric acid is a fast zero-order reaction and that between the secondary hydroxyl with nitric acid is a slower first-order reaction. The industrial production of NG using mixed nitro-sulfuric acid as the nitrating agent is fast with higher conversion rates, less side reactions, and less nitric acid required due to its strong reduction ability.

The production process of NG mainly includes nitration, separation, washing, and stability treatment. In addition, the raw material preparation, product treatment, and spent acid treatment are also imported procedures. Figure 6.8 shows the basic process of NG production.

The conversion of glycerol is always <100% because nitration is a reversible reaction. Usually, there is 1–2% unreacted glycerol remained in the spent acid. If the theoretical yield of NG from glycerol is 246.7%, the actual yield is then only up to 235%. After a thorough investigation of the NG production process, the authors found that the process in the NG yield of 242.5% in lab-scale could give an NG yield of 240.98% at 100 kg production level (injection nitration process).

In the following section, the control conditions and operation of each part of the process described in Fig. 6.8 will be discussed in detail.

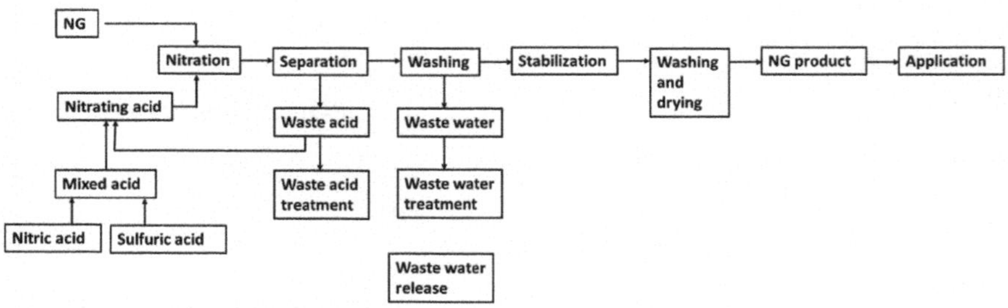

Fig. 6.8 Production process of NG

6.3.2 Preparation of Raw Materials

The raw material preparation includes the preparations of glycerol, mixed acid, nitrating acid, compressed air, refrigerant, hot water for washing, alkali lye, etc.

(1) Glycerol

Glycerol is a raw material of NG, which can significantly affect the production safety and quality of NG. The quality of glycerol must be analyzed before the production process. Table 6.23 lists the specifications of glycerol required for nitration.

(2) Nitrating acid

The quality and composition of nitrating acid can not only influence the yield and stability of the produced NG but also affect its production safety. Therefore, the quality of nitrating acid should be strictly controlled. Table 6.24 shows the specifications of nitrating acid required for NG production.

Highly amounts of N_2O_4 can cause the oxidation of glycerol, producing brown smoke that can be mistakenly considered as the sign of NG decomposition. Therefore, its content cannot be high. The specifications of sulfuric acid and fuming sulfuric acid required for NG production are shown in Tables 6.25 and 6.26, respectively.

Different nitration methods use the nitrating acid with different compositions. Most air-agitated nitration methods use anhydrous mixed acid, and injection nitration methods use the mixture of mixed acid and spent acid. In general, the mixed acid is composed of <1.0% H_2O, <0.3% N_2O_4, 49.00–51.00% HNO_3, and 50.00–52.00% H_2SO_4, and varies with the equipment and production process. The sulfuric acid content in the mix acid should be slightly higher, usually 1% higher

Table 6.23 Quality of glycerol required for NG production

Property	Specification
Color	$\ngtr$0.8 (mL of iron (III) chloride standard solution caused color in 50 ml distilled water)
Transparency	Transparent
Smell	No odor as boiled in water for 10 min
Specific gravity	25/25°C $\nless$1.2582 (equivalent to 98.5%)
Acidity or alkalinity	$\ngtr$1.2 mL 0.1 N HCl or 0.1 N NaOH/50 g NG
Saponification value	$\ngtr$0.5 mg NaOH/1 g NG
Nonvolatile residue	$\ngtr$0.1% (organic and inorganic)
Ash	$\ngtr$0.02%
Propanol and reducible compounds	$\ngtr$50 ppm, glucose caused "silver gel" turbidity
Ammonium and protein	None
Chlorides (as Cl)	$\ngtr$0.0005%

Table 6.24 Specifications of nitrating acid required for NG production

Indicator	Specification	
	1st grade	2nd grade
HNO_3 (%) ≥	98	97
H_2SO_4 (%) ≤	0.08	0.12
N_2O_4 (%) ≤	0.3	0.04
Residue on ignition (%) ≤	0.03	0.05

Table 6.25 Specifications of industrial grade sulfuric acid required for NG production

Indicator	Specification
H_2SO_4 (%)	≥ 98
Residue on ignition (%)	≤ 0.1

Table 6.26 Specifications of fuming sulfuric acid required for NG production

Indicator (%)	Specification
Free SO_3	≥ 20
Residue on ignition	≤ 0.1
Iron	≤ 0.03

than the nitric acid content. The appearance of the mixed acid should be colorless or slightly yellow and transparent with no obvious solid residues. The acids should be mixed and precipitated for 3 days before use. The solid residue in the mixed acid should be <0.1%.

Nitrating acid is usually prepared with fresh acids and spent acid and the spent acid produced from nitration should meet the specifications listed in Table 6.27. Table 6.28 lists composition of the mixed nitrating acids prepared with fresh acid and spent acid.

(3) The design and calculation of mixed acid and nitrating acid

Table 6.29 lists the symbols used to calculate the composition of nitrating acid

1. Calculation of mixed acid

Based on the material balance calculation, it can be obtained:

Table 6.27 Specifications of spent acid produced from nitration

Indicator	Specification
Total nitrogen content (as HNO_3, %)	>13
H_2SO_4	>68
HNO_3	>11
H_2O	<18
N_2O_4	<0.6
NG	<2.5

Table 6.28 Specifications of the mixed nitrating acid prepared with spent acid

Indicator	Specification
Total nitrogen content (as HNO_3, %)	29.5 ± 0.5
H_2SO_4	62–64 ± 1
HNO_3	23–28 ± 1
H_2O	8–13 ± 2
N_2O_4 (%)	<0.5–0.6
NG (%)	<1.6–2.0

Table 6.29 Symbols used to calculate the components for the preparation of nitrating mixed acid

Name	Amount	Component				
		HNO_3	H_2SO_4	H_2O	NG	Total
Mixed acid	*M*	*a*	*B*	*c*		100
Spent acid	*W*	*e*	*F*	*g*	*Ng*	100
Nitrating acid	*Q*	*p*	*Q*	*s*	*NG*	
Nitric acid	*X*	*n*				
Sulfuric acid or fuming sulfuric acid	*Y*		*L*			
water	*Z*					

$$\mathrm{M} \cdot \mathrm{a}\% = \mathrm{X} \cdot \mathrm{n}\% \quad \mathrm{X} = \frac{\mathrm{M} \cdot \mathrm{a}}{\mathrm{n}} \tag{6.2}$$

$$\mathrm{M} \cdot \mathrm{b}\% = \mathrm{Y} \cdot \mathrm{L}\% \quad \mathrm{Y} = \frac{\mathrm{M} \cdot \mathrm{b}}{\mathrm{L}} \tag{6.3}$$

If fuming sulfuric acid is used, then

$$\mathrm{L} = 100 + \frac{18}{80}\mathrm{K} = 100 + 0.225\ \mathrm{K}$$

where 18 is the molecular weight of water, 80 is the molecular weight of SO_3, and *K* is the percentage of free SO_3.

$$\mathrm{Z} = \mathrm{M} - (\mathrm{X} + \mathrm{Y}) \tag{6.4}$$

2. Calculation of nitrating acid

The composition of nitrating acid can be calculated based on the material balance of nitric acid, or sulfuric acid, or water as the cardinal number. As nitric acid is used as the cardinal number, the equation can be written as

$$Q \cdot P = M \cdot a + W \cdot e \tag{6.5}$$

$$Q = M + W \tag{6.6}$$

Equation 6.6 is substituted in Eq. 6.5 to obtain

$$(M + W) \cdot P = M \cdot a + W \cdot a + W \cdot e$$

$$M = \frac{W(e - p)}{p - a} \tag{6.7}$$

$$W = \frac{M(p - a)}{e - p} \tag{6.8}$$

Similarly, if $M = Q{-}W$ or $W = Q{-}M$ is substituted in Eq. 6.5, the equation can be rewritten as

$$M = \frac{Q(p-e)}{a-e} \quad Q = \frac{M(a-e)}{p-e}$$
$$W = \frac{Q(p-a)}{e-a} \quad Q = \frac{W(e-a)}{p-a}$$

The composition of the mixed acid for nitration can be then obtained from the above calculation formula. Here, we take the known amount of spent acid as an example to prepare the mixed acid. If the amount of spent acid is 420 kg, how much new mixed acid is needed? Table 6.30 lists the compositions of the fresh mixed acid, spent acid, and mixed acid for nitration.

The calculation using nitric acid as the cardinal number can be expressed as

$$M = \frac{W(e - p)}{p - a}$$

The data in Table 6.30 are substituted in the equation to obtain

$$M = \frac{420(11.3 - 26.1)}{26.1 - 61.3} = 176.3\ \text{kg}$$

Table 6.30 Composition of known mixed acid, spent acid, and nitrating acid

Acid	HNO_3 (%)	H_2SO_4 (%)	H_2O (%)
Fresh mixed acid	61.3	43.4	−4.7
Spent acid of nitration	11.3	70.8	15.3
Mixed acid for nitration	36.1	62.7	9.4

The nitrating acid in NG production plants is usually prepared using the following equation:

$$\text{Mixed acid (L)} = \frac{\text{Spent acid (L)} \times (\text{Total nitrogen content of nitrating acid} - \text{Total nitrogen content of spent acid})}{\text{Nitric acid content in mixed acid} - \text{Total nitrogen content in nitrating acid}}$$

3. Preparation of mixed acid and nitrating acid

The preparation of mixed acid is an exothermic process, which, therefore, should be conducted with sufficient cooling facilities. To avoid large amounts of evaporation and decomposition of nitric acid, the operation temperature is usually maintained at 60 °C. The spent acid must be frequently stirred to prevent the precipitation of residual NG that could cause accidents. The spent acid should be completely disposed, and the pipes and equipment are strictly cleaned during the break.

The composition of mixed acid should be strictly controlled and it should be precipitated and filtered to remove any mechanical impurity, which is significantly important for the injection and separation. The prepared mixed acid should be precipitated for 24–72 h and cooled to 0–5 °C (0 ± 3 °C is required in some cases) for the NG production process.

4. Compressed air

Compressed air is required for the air-agitated nitration. It is also used to deliver the produced NG, spent acid, and the reagents for stability treatment and is required for pneumatic automatic instruments. It also plays a very important role in the safe production of NG. The injection nitration and centrifuge separation of a 500–600 g/h NG production line can consume 60 m^3/h air. The discontinuous production line needs even more air. Therefore, the pressure facilities are significant important for the NG production.

The compressed air for the NG production should be dry and contain no mechanical impurities or trace grease. Its total pressure is usually controlled at 0.5–0.6 MPa.

5. Coolant

The nitration of glycerol is an exothermic reaction, which therefore needs to be cooled to ensure production safety and product quality. The NG production can be cooled with coolants, such as freon-12, directly, or using cooling media, such as brine. The former can reduce the heat loss of 10–20%, while saving the workspace and avoiding cooling salt water system. However, the cooling process needs to be strictly controlled to prevent the solidification of NG that is very dangerous. The later uses a cooling medium (brine) that is pre-cooled to a certain temperature to control the temperature of NG production process. NaCl, $CaCl_2$, $NaNO_3$, Ca $(NO_3)_2$, and other cheap salts can be used as the cooling medium.

The temperature of the cooling brine can be adjusted based on the production process, typically −15 to −20 °C for the air-agitated production equipment and −5 to 5 °C for the injection nitration.

Table 6.31 Specifications to be met by Na_2CO_3 for stability treatment of NG

Specification	Special industry	Regular industry	
		1st grade	2nd grade
Appearance	White crystal powder	White crystal powder	White crystal powder
Total alkali (as Na_2CO_3 (%)	≥99.0	≥98.5	≥98.0
NaCl (%)	0.5–1.0	1.0	1.2
Fe as Fe_2O_3 (%)	≤0.003–0.010	≤0.01	≤0.02
Water-insoluble matter (%)	≤0.05–0.15	≤0.15	≤0.20
Na_2SO_4 (%)	≤0.05	≤0.08	≤0.10
Loss on ignition (%)	≤0.5	≤0.5	≤0.7

6. Lye for stability treatment

The produced NG should be stabilized by washing with the Na_2CO_3 aqueous solution of a certain concentration at certain temperature. Table 6.31 lists the specifications to be met by Na_2CO_3 for the stability treatment of NG.

The specifications of lye for the stability treatment of NG vary with the washing equipment and production process conditions. The lye concentration is generally controlled at 1–2 wt%, no more than 2.5 wt%. The key step of lye preparation is to make appropriate concentrations of Na_2CO_3. The temperature of lye is usually controlled at 50–60 °C. Na_2CO_3 should be stored under appropriate conditions to prevent the formation of $NaHCO_3$ due to the absorption of water and carbon dioxide in air, which can affect the safety treatment.

7. Preparation of hot water

According to the production process of NG, water of different temperatures is required for the washing, transportation, and insulation of NG.

8. Other auxiliary materials

In addition to the gravity separation of NG from the nitration spent acid, the separation agent, paraffin oil, is also needed. The quality of the paraffin oil shall comply with specifications listed in Table 6.32.

Table 6.32 Specifications to be met by paraffin oil the separation of NG

Specific gravity (20/4 °C)	0.835–0.855
Mechanical impurity	N/A
Water	N/A

The paraffin oil is usually mixed with the mixed acid at a volume ratio of 1:1, shaken for 3–5 min and kept for 5–10 min. The procedure is repeated 2 times. The color and temperature of the paraffin oil should be maintained constant. The paraffin oil can be substituted with medical vaseline oil if unavailable.

6.3.3 The Production Process of NG

Various auxiliary processes for the production of NG have described above. The production process of NG has been well established. Based on the production and production scale, the process can be divided into four kinds as discussed below [13].

(1) The control process setup and conditions for the NG production by discontinuous nitration

The control process setup for NG production by discontinuous nitration is shown in Fig. 6.9.

1. Nitration and separation

Composition of mixed acid:

HNO_3 48–52%.
H_2SO_4 49–52%.
H_2O 0.2–0.5%.
SO_3 $\ngtr$2.0%.
Ash <0.5%

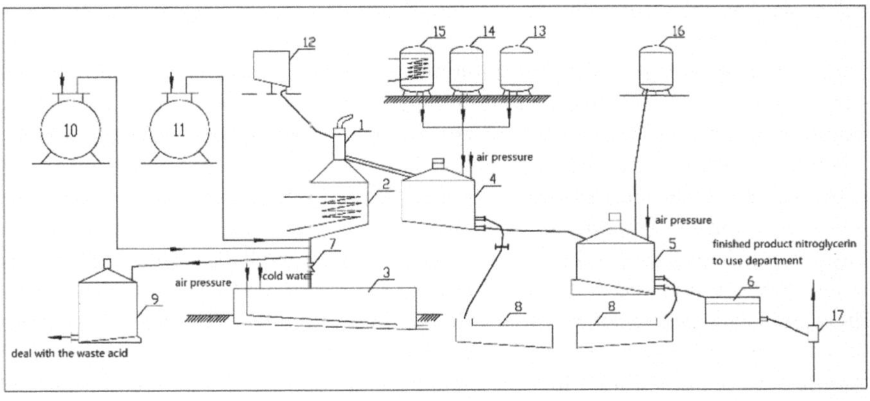

Fig. 6.9 Flowchart of the NG production process by discontinuous nitration. 1—Glycerin sprayer; 2—Nitrator; 3—Water reservoir; 4—Prewash tank; 5—Wash tank; 6—Filter; 7—Switch; 8—Waste water receiver; 9—Spent acid separator; 10—Mixed acid controller; 11—Exchanged acid; 12—Glycerin controller; 13—Cool water tank for prewash; 14—Lye tank; 15—Warm water; 16—Hot water tank; 17—NG conveyer

The nitration coefficient is in the range between 4.9 and 5.3.
Temperatures for nitration:

Onset temperature	<5°C.
Reaction temperature	15–18°C.
Final temperature	<22°C.
Temperature of glycerol	30–35°C.
Temperature of coolant	−5 to −15°C.
Feeding speed	10–13 kg/min.
Air agitation speed	2–3 min/ET

Composition of spent acid:

HNO_3	$\nless$8%.
H_2SO_4	$\nless$67%.
H_2O	$\ngtr$18%.
N_2O_4	$\ngtr$0.3%.
NG	$\ngtr$2.2%.
Specific gravity	1.70–1.71

2. Washing process

Cold water washing is usually conducted at 20–25 °C for 6 min, followed by warm water washing at 30–35 °C for 6 min. The temperature for stability washing is usually set to 40–45 °C and washing time is 6 min.

(2) Process control conditions for the NG production by injection nitration

For the injection nitration, the injection liquid is sucked and ejected by the conversion between static pressure and kinetic energy driven by the nitrating mixed acid flow. The static pressure is converted into dynamic energy as the liquid is ejected at a high speed from a nozzle, resulting in a low pressure or a certain degree of vacuum, that suck glycerol. The nitrating acid and glycerin are thoroughly mixed as they pass through the throat, and rapidly react, which can be finished in 0.5–1 s. The nitrating mixture stays in the injector for ~0.5 s to complete the reaction.

Nitration coefficient has a linear relationship with temperature, which can be used to control nitration temperature, and thus simplify the nitration process. The homogenous injection mixing and good dispersion of glycerol allow the rapid reaction of nitration, which shortens the reaction stay, eliminates the local overheat, improves the NG productivity, and thus is conducive to the reaction. However, the short stay time significantly reduces the system volume. For example, the system volume of an 800 kg/h nitration system is only 15 kg. The injection nitration uses the negative pressure formed by the high flow of nitrating mixed acid to suck glycerin into the ejector, which reduces the risk caused by the unbalanced nitration coefficient. Figure 6.10 shows the flowchart of the continuous injection nitration process.

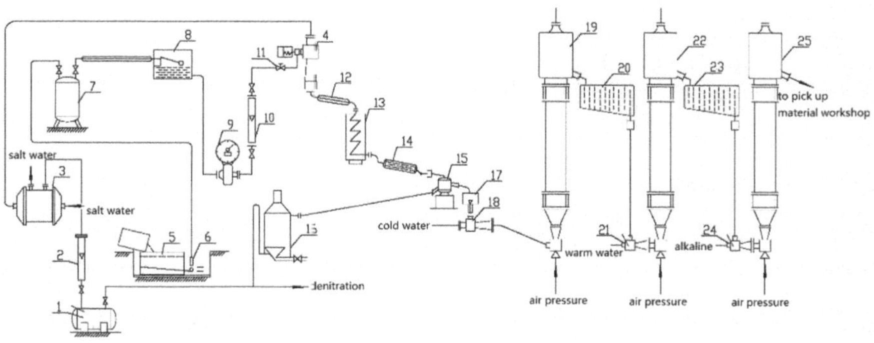

Fig. 6.10 Flowchart of the NG production process by continuous injection nitration. 1—Mixed acid tank; 2—Flow meter; 3—Chiller; 4—Nitration ejector; 5—Glycerin; 6—Conveying pump; 7—Glycerin feed tank; 8—High glycerin tank; 9—Metering pump; 10—Flow meter; 11—Regulatory valve; 12—Cooling jacket; 13 and 14—Chiller; 15 and 16—Centrifuge separator and post-separator; 17—Funnel; 18—Cold water controller; 19, 22 and 25—Scrubber; 20 and 23—Oil skimmer; 21 and 24—Stability treatment

1. Composition of mixed acid

HNO_3	48–50%.
H_2SO_4	50–52%.
H_2O	<0.2%.
N_2O_4	<0.6%.
SO_3	$\ngtr$2.0%.
Ash	<0.1%

2. Control process conditions

Nitration coefficient	12–15.
Nitration temperature	45–48 °C.
Vacuum	280–350 mmHg (during nitration).
Temperature of mixed acid	0 ± 1 °C.
Mixed acid pressure at ejector inlet	2.6–3.2 kg/cm^2.
Total nitrogen content of mixed acid	29.5 ± 0.5%.
Nitration velocity	3–4 kg/min.
Total nitrogen content of spent acid	14.0 ± 0.3%

3. Separation

Separation temperature	16 ± 1 °C.
Centrifuge speed	$\nless$2900 r/min

4. Washing condition

Cold water wash	oil to water ratio, 1:1.2–1.5 (Temperature $\ngtr$30 °C).
Warm water wash	oil to water ratio 1:0.7–1.2 (Temperature: 40 ± 5 °C).
Stabilization treatment	oil to water ratio 1:0.7–1.2 (Temperature 40–45 °C, l ye concentration 80 ± 0.1%).
Mixing pressure in scrubber	2 ± 0.2 kg/cm^2

(3) NG production process by container air-agitated continuous nitration

Figure 6.11 shows the process control conditions of the container air-agitated continuous nitration for NG production.

(4) NG production process by diluted continuous nitration

Figure 6.12 shows the diluted continuous nitration process of NG production.

The NG productivities of the production processed by discontinuous and injection nitration are different. The former can produce 300–400 kg NG per hour, while the later produces up to over 600 kg NG per hour.

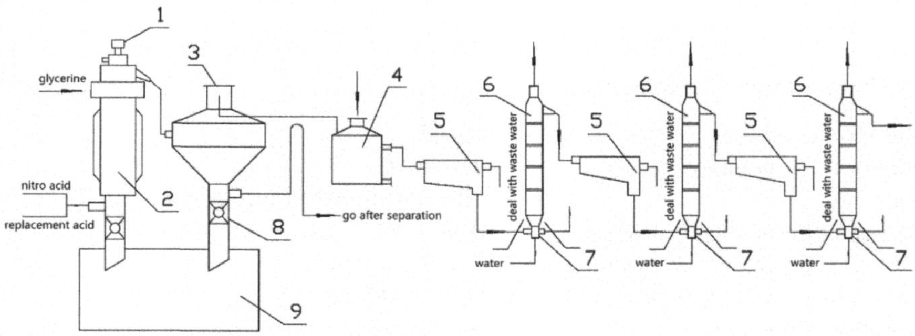

Fig. 6.11 Flowchart of NG production process by container air-agitated continuous nitration. 1—Glycerin sprayer; 2—Nitrator; 3—Hydrocyclone; 4—Prewasher; 5—Separator; 6—Scrubber; 7—Washer; 8—Safety valve; 9—Water reservoir

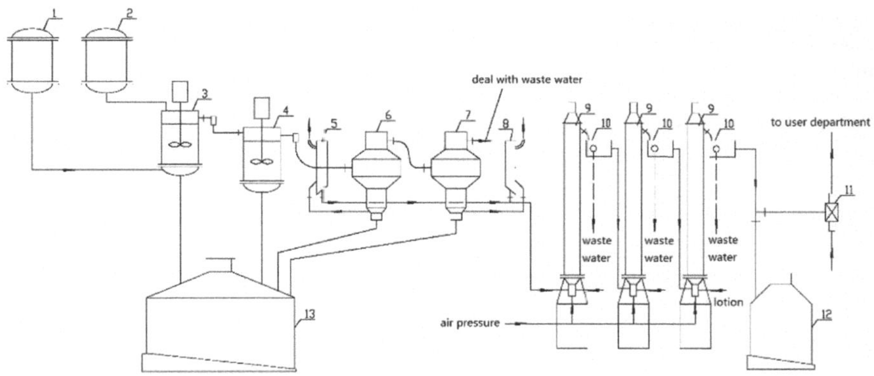

Fig. 6.12 Flowchart of NG produc‐ion process by diluted continuous nitration. 1—Mixed acid; 2—Glycerin; 3—Nitrator; 4—Dilutor; 5 and 8—Regulator; 6—Separator; 7—Clarifier; 9—Scrubber; 10—Oil skimmer; 11—Ejector; 12—Scrubber; 13—Water reservoir

6.3.4 Separation of NG

The separation of NG during its production includes the separation of NG from nitration spent acid and the separation of washing waste water from NG.

(1) Separation of NG from spend acid

Ng can be separated from spent acid by the gravity and centrifugation.

1. Gravity separation

The traditional intermittent gravity separation has been outdated. However, hydrocyclone is still used in the container air-agitated continuous NG production process worldwide to separate NG from the spent acid. Figure 6.13 shows the schematic diagram of the key part of hydrocyclone used in the gravity separation of NG product.

The NG emulsion flow enters in the hydrocyclone from inlet a and is centrifuged to separate. NG and the spent acid stay in the hydrocyclone for 10–15 min and 40–70 min, respectively, and are then brought out, respectively, from outlet b and c when the liquid level is adjusted.

The other type of gravity separator is equipped with an inclined cooling tube array or a corrugated board, as shown in Fig. 6.14. The tube array can shorten the coalescence and ascending of NG drops. The small NG droplets aggregate to form large drops that rise along the tube (or plate) wall. The low cooling temperature can also reduce the solubility of NG in the nitration spent acid. The compressed air in the compressed air mixing tube slightly stirs the mixture on the bottom, which promotes the coalescence and ascending of NG, prevents local overheating, and thus is conducive to the separation.

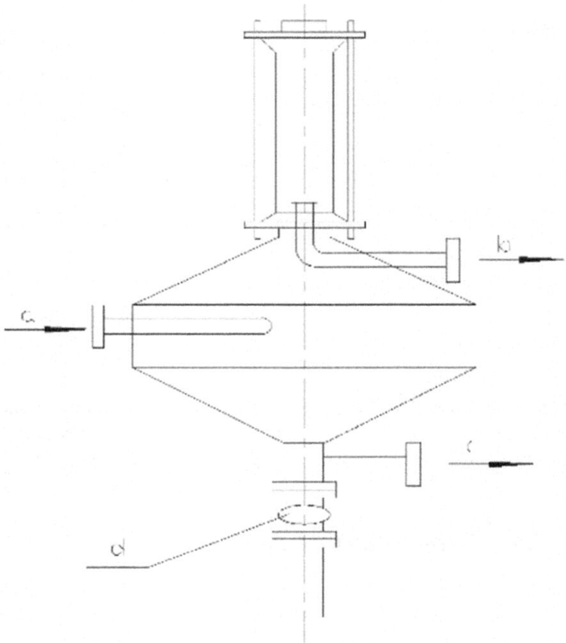

Fig. 6.13 Schematic diagram of the hydrocyclone. **a** Inlet of nitration mixture; **b** inlet of NG; **c** outlet of spent acid; **d** safety valve

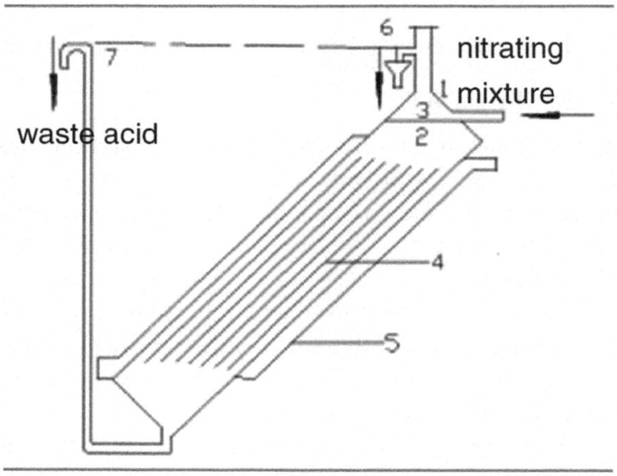

Fig. 6.14 Schematic diagram of array tube separator. 1—Vessek (Inlet of nitration mixture); 2—Spent acid layer; 3—NG layer; 4—Baffle (or corrugated plate); 5—Cooling jacket; 6—Outlet of NG; 7—Siphon (Outlet of spent acid)

2. Centrifugation

The setup and principle of centrifugal separation depend the centrifuge structure. The separation factor of a centrifuge is mainly decided by the diameter of its rotary drum and speed, which is the key to achieve the centrifugal force. In addition, the position of the neutral aperture can also affect the separation of NG. The light liquid component withstands lower centrifugal force and the heavy liquid component withstands higher centrifugal force as the neutral layer is near the rotary drum diameter center. However, the light liquid component is affected more significantly as the neutral layer is away from the drum diameter center. It has been found that the centrifuge with a 52.5 mm radius heavy liquid outlet and 85 mm radius neutral hole results in 7.286% spent acid in the light liquid (NG) and 0.96% NG in the heavy liquid (spent acid). However, as the radius of neutral aperture increased to 105 mm and that of heavy liquid outlet remained unchanged, the light liquid can take 9.6% spent acid and the heavy liquid contains 1.50% NG. Therefore, it can be concluded that the NG content in spent acid can be reduced using the neutral layer with smaller radius to ensure the yield of NG and production safety.

The position of neutral aperture in an equipment is usually fixed. The position of neutral layer, also called neutral plane, can be adjusted by a gravity plate to the position of neutral aperture. Therefore, gravity plate is a key component affecting the separation of NG. It can control the radius of the heavy liquid outflow and balance the position of the neutral layer and the neutral aperture. The diameter of gravity plate is crucial for achieving desired separation. The separation of the light and heavy liquids with smaller gravity difference needs a larger gravity plate. In general, gravity plates with diameters of φ90–105 mm are used to separate NG from nitration spent acid. The gravity plate is chosen based on the separation effects, with <0.1%, usually 0.07%, free NG in the spent acid. The optimum size of gravity plate can be calculated using the following formula:

$$R = \sqrt{R_2^2 - \frac{(R_2^2 - R_1^2)\gamma_1}{\gamma_2}} \tag{6.9}$$

where R is the radius of gravity plate (mm), R_1 is the radius of light liquid outlet (mm), R_2 is the radius of the neutral aperture, γ_1 is the specific gravity of light liquid, and γ_2 is the specific gravity of heavy liquid.

The separation efficiency and yield are directly related to the distance between the discs and the disc number. Smaller distance results in shorter retention time, causing incomplete separation. Extremely large distance causes a longer liquid transportation route, and thus longer separation time, and low yields. Figure 6.15 shows the schematic diagram of the centrifugal separator.

At certain flow rate, increasing the number of discs can increase the separation surface, and thus can completely separate NG from the spent acid. In general, the centrifugal separators of NG and nitration spent acid contain 70–80 discs with both thickness and distance of 0.5 mm. Table 6.33 lists the main parameters of industrial centrifugal separators of NG production process.

The post-separation can be conducted continuously or intermittently, which has been gradually outdated due to the long separation time up to 48 h, bulk equipment, and large space occupancy. The modern NG production processes usually adopt continuous hydrocyclone to separate NG from spent acid. The spent acid is usually

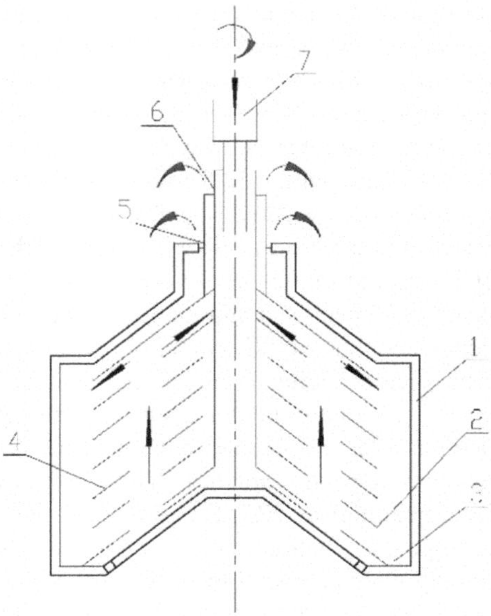

Fig. 6.15 Structure of centrifugal separator. 1—Rotary drum; 2—Disc; 3—Precipitated impurity; 4—Neutral aperture; 5—Heavy liquid outlet; 6—Light liquid outlet; 7—NG emulsion

Table 6.33 Main parameters of the centrifugal separators of NG production process

Parameter	Model of centrifugal separator	
	NGB 207-10B	D-NGA-368
Rotation speed (r/min)	3200	2910
Radius of rotary drum (mm)	295	368
Centrifugal force (MPa)	171	173
Diameter of neutral layer (mm)	192	210
Diameter of light liquid outlet (mm)	76	90
Diameter of heavy liquid outlet (mm)	95	105
Thickness of separation disc (mm)	0.35	0.8
Number of separation disc	77	84–88
Diameter of emulsion inlet (mm)	38	27
Distance between separation discs (mm)	0.5	0.65
NG separation capacity per hour (kg)	600	400

kept in the separator for 3 h to avoid the precipitation of NG. The separated spent acid is added with 2% water to increase its NG solubility.

(2) Separation of NG and waste water

The specific gravities of NG and water are significantly different, and thus they can be easily separated. The global NG production processes adopt hydrocyclone separators to separate NG from waste water.

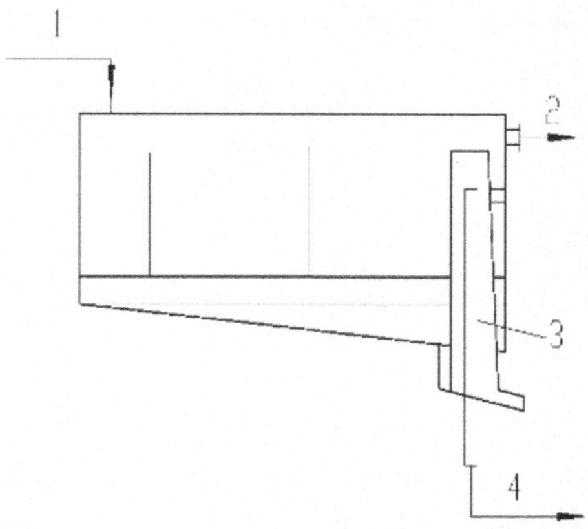

Fig. 6.16 Schematic diagram of a tunnel oil–water separator. 1—Inlet of oil/water; 2—Outlet of waste water; 3—Oil seal tube; 4—Outlet of NG

1. Tunnel-type oil–water separator

Figure 6.16 shows the schematic diagram of a tunnel oil–water separator. The height of the oil seal tube significantly affects the separation of NG and waste water. Appropriate oil seal tube should be chosen. High oil seal tube retains too much oil and low oil seal tube cannot retain enough oil to seal, which thus affects the separation efficiency.

2. Conical oil–water separator

Figure 6.17 is the schematic diagram of a typical conical oil–water separator.

The separation in conical oil–water separator requires longer time and large amount of oil stock with lower separation efficiency than that in tuner separator. However, it has been used in NG production due to its simple structure and easy manufacture.

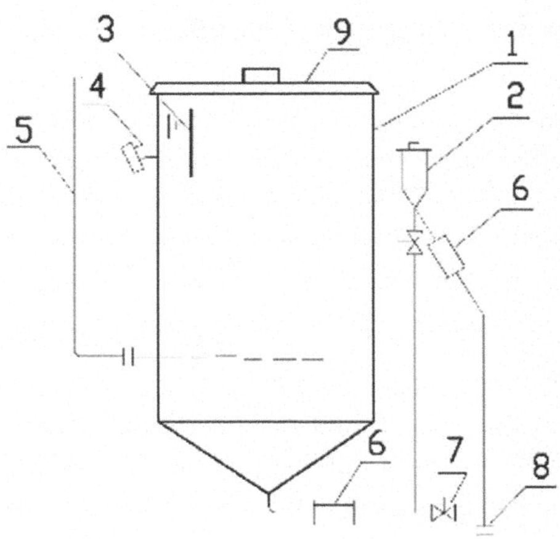

Fig. 6.17 Schematic diagram of a conical oil–water separator. 1—Separator body; 2—Liquid level regulator; 3—Baffle; 4—Outlet of waste water; 5—Feed line of oil-containing waste water; 6—Sighting tube; 7—Vent valve; 8—Outlet of NG; 9—Separator cover

Table 6.34 Comparison between co-current washing and countercurrent washing

washing	Per 100 kg NG				
	Water consumption (kg)	Temperature (°C)	Alkaline consumption (kg)	Amount of NG dissolved in water (g)	Yield of NG (%)
Co-current	260.1	30	2.6	470	233
Count-current	60	25–30	1.0	100	236

3. NG washing

The NG separated from spent acid contains ~8% nitric acid, 0.1–0.3% sulfuric acid, small amounts of dinitroglycerin, and other impurities. It can be refined by a three-step procedure of cold water prewashing, warm water washing, and alkaline washing or a two-step procedure of water washing and alkaline washing. There are can co-current washing and countercurrent washing processes. For the former, new washing solution is added and the waste water is discharged and used for the following prewashing. New washing solution is added if the waste water is not enough for the prewashing. The countercurrent washing can reduce the amount of washing solution and the dissolution loss of NG. Table 6.34 shows the comparison between co-current washing and countercurrent washing.

NG is usually washed in tower scrubbers. Figure 6.18 shows the primary structure of a typical tower scrubber.

The main parameters of an NG scrubber include tower diameter, effective height, plate number, plate diameter, hole number, etc. The currently available NG tower scrubbers are plastic towers with diameters of 150–200 mm, a height of 1760 mm, producing capacities up to 600 kg NG per hour, glass towers with a diameter of 250 mm, height of 1670 mm and hourly processing capacity of 400 kg NG, and stainless steel towers with a height of 2800 mm, and processing capacity of 600 kg NG per hour.

6.3.5 Stability Treatment of NG

In the stability treatment, the equilibrium between NG and water is significantly affected by the dinitroglycerin content and the concentration of acid water. The acidic NG separated from waste water usually contains dinitroglycerin that is more soluble in water, acid, and alkali aqueous solutions than NG. This is the major issue of stability treatment.

(1) Solubility issue in the stability treatment of NG

To better understand the solubility of NG and dinitroglycerin in neutral, acidic, and alkaline water, and their partition in the oil phase and the aqueous phase during washing, the solubility and partition of NG and dinitroglycerin in water, nitric acid aqueous solution and sodium carbonate aqueous solution are discussed as follows.

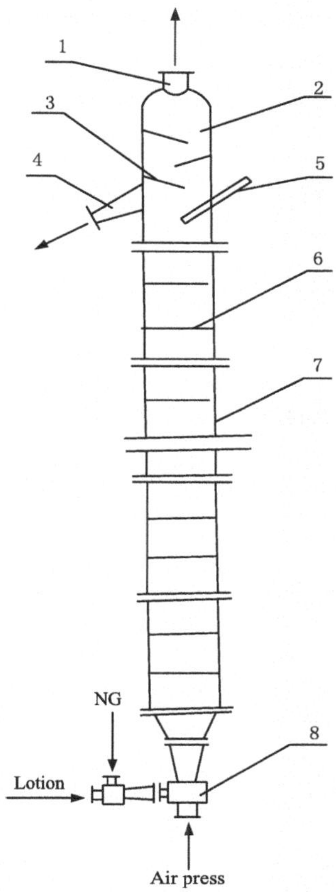

Fig. 6.18 Schematic diagram of a typical tower scrubber. 1—Tower top; 2—Thermometer jacket; 3—Tower plate; 4—Tower body; 5—Washing jet; 6—Vent; 7—baffle; 8—Discharge outlet

1. Solubility of NG in neutral water

Table 6.35 lists the solubility of NG and dinitroglycerin in 100 g neutral water at different temperatures. As can be seen, the solubility of dinitroglycerin in neutral water is almost 60 times higher than that of NG. Therefore, dinitroglycerin can be more easily removed with water.

2. Solubility of NG in dilute nitric acid solution

Table 6.36 shows the solubility of NG in dilute nitric acid solutions. Its solubility increases rapidly with the increase of nitric acid concentration in water. Therefore, the nitric acid content in the washing solutions should be strictly controlled to reduce NG loss. In addition, no hydrolysis or denitration occurs in the nitric acid content range of 0–24%. However, the solubility of NG is significantly increased in the 29% nitric acid aqueous solution after 16 h mixing, possibly due to the hydrolysis of NG by nitric acid to dinitroglycerin. It has been found that 8% NG is hydrolyzed in 48 h, e.g., 1% per hour. These data indicate that the hydrolysis of NG

Table 6.35 Solubility of NG and dinitroglycerin in water

Temperature (°C)	Solubility of nitrates per 100 g water (g)	
	NG	Dinitroglycerin
10	0.14	8.0
20	0.15	8.5
30	0.16	9.0
40	0.18	9.5
50	0.21	10.5
60	0.25	11.8

Table 6.36 Solubility of NG in diluted nitric acid

Final nitric acid concentration (%)	NG in aqueous phase (%)			
	1 h mixing	3 h mixing	8 h mixing	16 h mixing
0	0.15	–	–	–
4.97	0.21	0.24	0.24	0.24
9.85	0.29	0.32	0.32	0.32
14.65	0.45	0.45	0.45	0.45
19.53	0.60	0.60	0.60	0.60
24.36	0.72	0.77	0.77	0.77
29.11	0.93	0.99	0.99	1.01

in 29% nitric acid is not significant at the beginning and does not affect the solubility results. However, the hydrolysis becomes significantly vigorous after a few days. Figure 6.19 shows the relationship between NG solubility and nitric acid concentration.

Table 6.37 lists the solubility of dinitroglycerin in diluted nitric acid. The solubility of dinitroglycerin in dilute nitric acid solution is much higher than in neutral water. Therefore, the dinitroglycerin impurity is removed during the first washing step. In addition, the dinitroglycerin content in 30% nitric acid is significantly higher, indicating that nitroglycerin is hydrolyzed and denitrated. The hydrolysis reaction may occur simultaneously in the aqueous and oil phases. No significant hydrolysis occurs at nitric acid concentrations lower than 20%. Figure 6.20 shows the solubility of nitrates in the aqueous phase as a function of nitric acid concentration.

As can be seen from Fig. 6.20, the solubility of both NG and dinitroglycerin increase with the increase of nitric acid content and the solubility of dinitroglycerin increases more rapidly than that of NG.

Fig. 6.19 Solubility of NG in nitric acid solutions

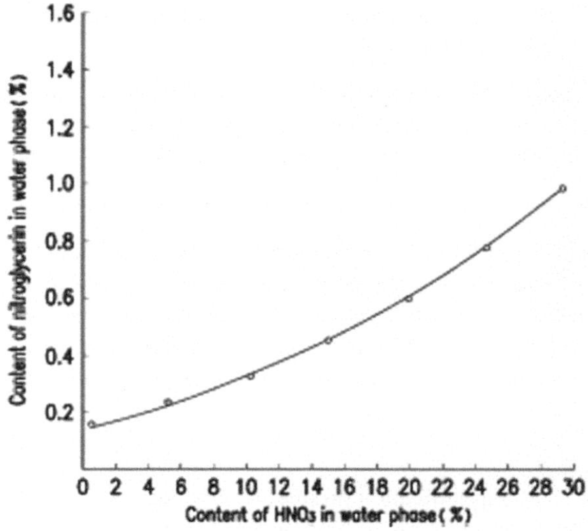

Table 6.37 Solubility of dinitroglycerin in diluted nitric acid (20–22 °C)

Nitric acid in water before stirring (%)	Dinitroglycerin content in aqueous phase (%)		
	1 h stirring	2 h stirring	3 h stirring
0.00	7.6	7.9	7.9
5.00	10.0	10.4	10.1
9.89	12.8	12.8	13.1
19.82	18.8	20.4	20.6
29.38	23.8	29.6	31.8

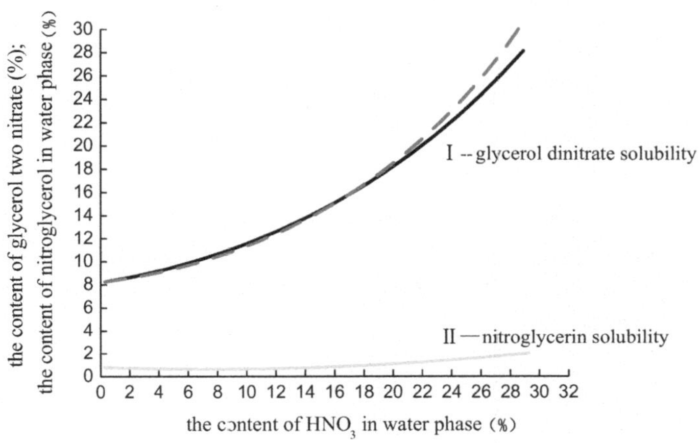

Fig. 6.20 Solubility of nitrates in diluted nitric acid

3. Effect of trace amount of sulfuric acid on the solubility of dinitroglycerin in diluted nitric acid

Acidic NG also contains a small amount of sulfuric acid that can enter the aqueous solution during washing. The experimental results shown in Table 6.38 suggest that sulfuric acid has no significant effect on the solubility of dinitroglycerin.

4. Solubility of NG in sodium carbonate solution

The solubility of NG and dinitroglycerin in sodium carbonate solutions are shown in Tables 6.39 and 6.40 and Fig. 6.21. As can be seen, the solubility of NG decreases with the increase of sodium carbonate concentration.

It is worth noting that NG undergoes vigorous saponification and decomposition as treated with strong alkali or concentrated alkali as shown below:

$$C_3H_5(ONO_2)_3 + 5KOH \rightarrow KNO_3 + 2KNO_2 + CH_3COOK + HCOOK + 3H_2O$$

The saponification product is very complex. In addition to the products shown in the above equation, other compounds, such as oxalic acid, ammonia, and so on, can also be formed. The NG and dinitroglycerine consumed by the saponification in the aqueous phase can be continuously compensated by those in the oil phase. Similar reaction can occur in concentrated sodium carbonate solutions. For the NG production, 1% sodium carbonate solution is usually used as the washing solution, which has no significant effect on NG.

Table 6.38 Solubility of dinitroglycerin in diluted nitric acid solutions containing sulfuric acid

Nitric acid (%)	Sulfuric acid (%)	Dinitroglycerin (%)	Temperature (°C)
8.60	2.82	12.4	20
8.64	0.62	12.7	
8.77	0.00	12.8	
8.64	0.62	12.5	10
8.87	0.00	12.7	

Table 6.39 Solubility of NG in sodium carbonate solutions at 20–22 °C

Na_2CO_3 (%)	NG in aqueous phase (%)		NG content after 3 h shaking and ethyl ether extraction (%)
	After 1 h shaking	After 3 h shaking	
1.39	0.10	0.12	0.10
4.95	0.05	0.06	0.03
10.31	0.02	0.05	0.01

Table 6.40 Solubility of dinitroglycerin in Na_2CO_3 solution at 20–22 °C

Na_2CO_3% before shaking	Na_2CO_3% after 1 h shaking	Dinitroglycerin content in aqueous determined by dichromate method (%)		
		1 h shaking	1 h shaking after 24 h	2 times of 1 h shaking
0.00	0.00	7.6	7.9	7.9
1.30	1.30	6.6	6.7	7.0
12.40	12.20	1.82	2.20	3.21

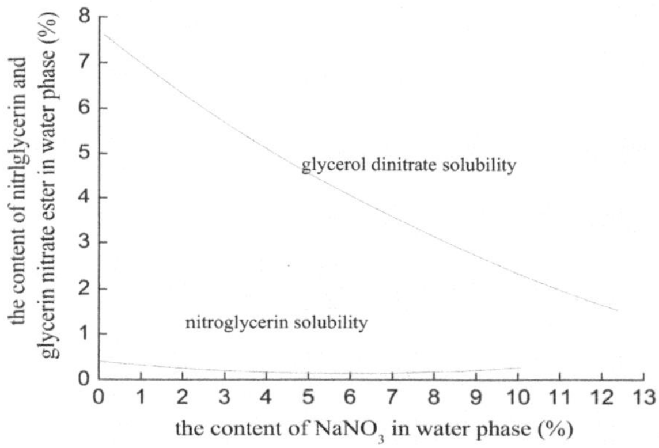

Fig. 6.21 Solubility of NG and dinitroglycerin in sodium carbonate solutions

(2) Partitions of nitric acid and dinitroglycerin between oil phase and aqueous phase

The partition coefficient of a molecule between two immiscible two phases is constant at low concentrations in lieu of no complexing.

1. The partition of nitric acid between the oil phase and water phase

Table 6.41 and Fig. 6.22 show the partition of nitric acid between NG and water.

As can be seen from Table 6.41, the partition coefficient of nitric acid in NG and water is constant. The ionization of nitric acid in water can be expressed as

Table 6.41 Partition of nitric acid between NG phase and water phase

HNO_3 in water (%) C_W	HNO_3 in NG (%) C_O	Partition coefficient C_O/C_W	Partition coefficient $\sqrt{C_O}/C_W$
4.98	0.005	0.001	0.001
9.77	0.012	0.001	0.011
19.53	0.039	0.002	0.010
29.00	0.170	0.006	0.014

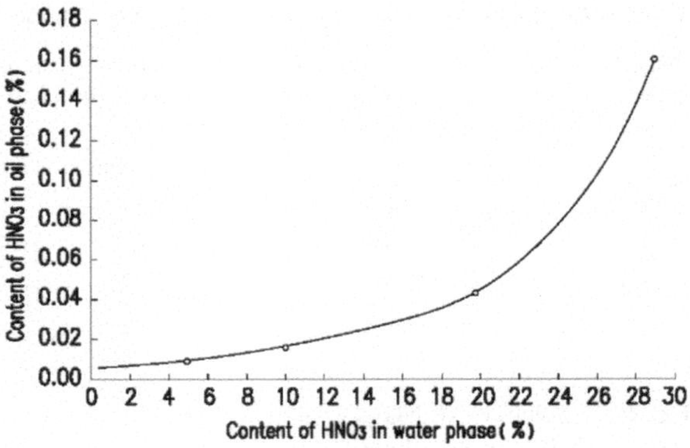

Fig. 6.22 Partition of nitric acid between nitrate containing 4.5% dinitroglycerin and an aqueous phase

$$HNO_3 \underset{k2}{\overset{k1}{\rightleftharpoons}} H^+ + NO_3^-$$

(The oil phase) (The aqueous phase)

Concentration : C_0 C_W C_W

As the partition equilibrium reached

$$C_0/C_W^2 = \text{constant}$$

$$\sqrt{C_0}/C_W = \text{Kpartition}$$

The partition coefficient C_0/C_W is not a constant because of the ionization of nitric acid in water. Instead, $\sqrt{C_0}/C_W$ can be used as the partition constant.

As shown in Fig. 6.22 for the partition of nitric acid between NG phase and an aqueous phase, the solubility of nitric acid in the aqueous phase is higher than in NG phase as its content in the aqueous phase is <20%, and vice versa, the nitric acid content in the organic phase is rapidly increased as the nitric acid content in the aqueous phase is greater than 20%.

2. Partition of dinitroglycerin between NG phase and aqueous phase.

Table 6.42 lists the partitions of dinitroglycerin between NG phase and aqueous phase.

As shown in Table 6.41, the partition coefficient is not a constant due to the molecular association of dinitroglycerin as shown below:

Table 6.42 Partition of dinitroglycerin between NG phase and aqueous phase at 20–22 °C

Weight of NG mixed with 5 ml water (g)	Dinitroglycerin in sample (%)	Dinitroglycerin in NG phase (%) (calculated value), C′	Dinitroglycerin in aqueous phase (calculated value) (%)C	Dinitroglycerin in aqueous phase (%)	NG in aqueous phase (%)	NG in aqueous phase (calculated value) (%)	Partition coefficient of dinitroglycerin between oil and aqueous phases (C′/C)
10.16	1.09	0.96	0.36	0.27	0.11	0.45	3.56
11.72	4.65	4.23	1.06	1.01	0.06	1.32	4.19
11.38	9.08	8.36	1.76	1.74	0.02	2.19	4.80
11.77	14.56	13.63	2.47	2.47	0.00	3.07	5.50
11.26	35.55	34.31	4.09	4.09	0.00	5.10	8.4
12.55	45.72	44.66	4.57	4.57	0.00	5.70	9.8

$$2\text{Din} \rightleftharpoons \text{Din}_2$$

where Din is dinitroglycerin.

Osmotic pressure can affect the molecular association. The tendency of the molecular association increases with the increase of dinitroglycerin concentration.

The dinitroglycerin content in the aqueous phase is extremely low and the molecular association can be ignored. The high concentration of dinitroglycerin in NG is subject to the molecular association reaction as shown below:

$$\begin{array}{lcc} & 2\text{Din} & \rightleftharpoons \text{Din}_2 \\ & \text{(The aqueous phase)} & \text{(The oil phase)} \\ \text{Concentraion} & \text{C} & \text{C}' \end{array}$$

Therefore,

$$\text{C}'^{1/2}/\text{C} = \text{constant}$$

or

$$\sqrt{C_0}/C_W = K_{\text{partition}}$$

Table 6.43 Partition coefficient ($C/\sqrt{C'}$) of dinitroglycerin between NG oil phase and neutral, acidic, and basic aqueous phases

Dinitroglycerin in aqueous phase (%) (C)			Dinitroglycerin in oil phase (%) (C′)			Partition coefficient ($C/\sqrt{C'}$)		
Neutral	Acidic	Basic	Neutral	Acidic	Basic	Neutral	Acidic	Basic
0.27	–	–	0.96	–	–	0.37	–	–
–	–	0.45	–	–	2.00	–	–	0.32
–	0.60	–	–	1.60	–	–	0.47	–
–	0.95	–	–	2.59	–	–	0.59	–
1.01	–	–	4.23	–	–	0.51	–	–
–	–	1.40	–	–	8.15	–	–	0.49
–	1.70	–	–	4.96	–	–	0.77	–
1.74	–	–	8.36	–	–	0.61	–	–
2.47	–	–	13.63	–	–	0.67	–	–
–	2.81	–	–	9.30	–	–	0.93	–
4.09	–	–	34.31	–	–	0.69	–	–
–	4.50	–	–	16.24	–	–	1.12	–
4.57	–	–	44.66	–	–	0.68	–	–

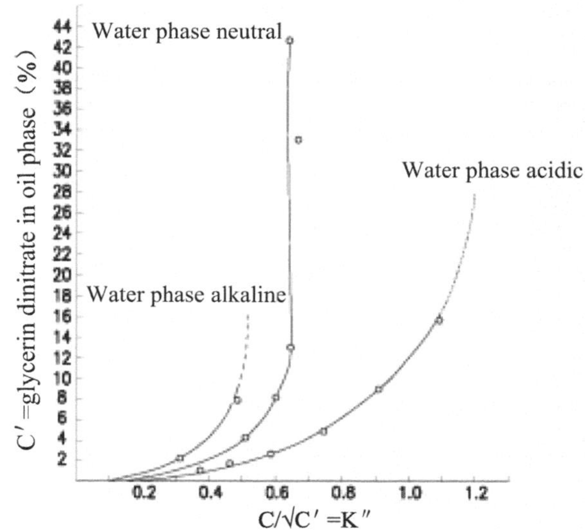

Fig. 6.23 Partition of dinitroglycerin between NG oil phase and neutral, acidic, and basic aqueous phases

where C is the concentration of dinitroglycerin in aqueous phase and C' is the total concentration of free dinitroglycerin and associated dinitroglycerin (It is independent on the acidity and alkalinity of the aqueous phase).

The partition parameters of dinitroglycerin between NG phase and neutral, acidic, or basic aqueous phase are shown in Table 6.43 and Fig. 6.23.

As can be seen from Fig. 6.23, the partition coefficient K″ in neutral aqueous phase becomes a constant as the dinitroglycerin content in NG oil phase increased to 13% (the middle vertical line in Fig. 6.23). The partition coefficient K″ in basic aqueous phase can more rapidly reach the constant (left line on Fig. 6.23) due to the neutralization of the acid. In contrast, the high concentration of H^+ in acidic aqueous phase slows molecular association of dinitroglycerin. The partition coefficient becomes a constant (right curve in Fig. 6.23) much lower than in neutral and

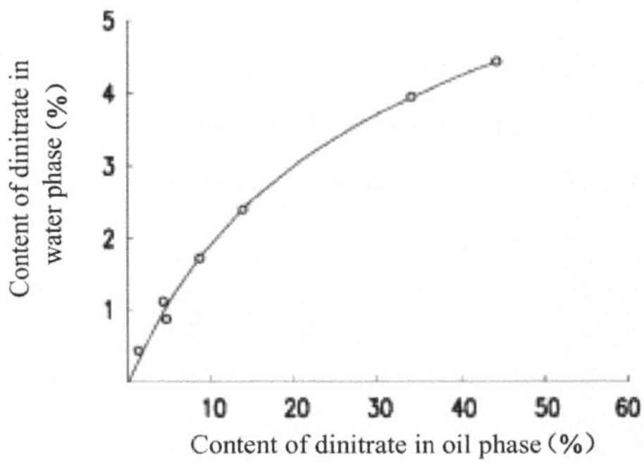

Fig. 6.24 Partition of dinitroglycerin between NG oil phase and a neutral aqueous phase

Table 6.44 Partition of dinitroglycerin between NG and 10% aqueous nitric acid at 20–22°C

Dinitroglycerin content in NG sample (%)	Dinitroglycerin content in NG phase (%) (C′)	Dinitroglycerin content in aqueous nitric acid phase (%) (C)	Partition coefficient (C′/C)
2.79	1.60	0.60	2.67
4.45	2.59	0.95	2.73
8.23	4.96	1.70	2.91
14.51	9.30	2.81	3.31
24.07	16.24	4.50	3.61

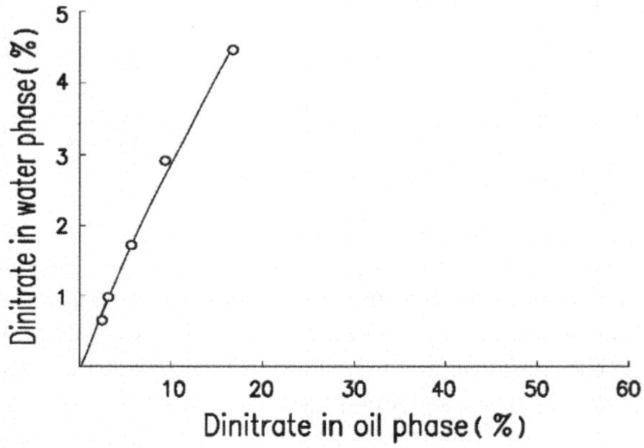

Fig. 6.25 Partition of dinitroglycerin between oil phase and 10% aqueous nitric acid phase

basic phases. Figure 6.24 shows the partition of dinitroglycerin between oil phase and a neutral phase.

The partition of dinitroglycerin between NG oil phase and 10% aqueous nitric acid phase is shown in Table 6.44 and Fig. 6.25.

Similar to that in acidic phase, the NG content in the neutral aqueous phase decreases with the increase of dinitroglycerin content. For example, the maximum NG content in 10% nitric acid aqueous phase is only 0.32%. The NG content decreases with the increase of dinitroglycerin content, until almost zero. The total glycerin nitrate content in nitric acid aqueous phase is a function of nitric acid content. The dinitroglycerin content in the oil phase decreases with the increase of nitric acid content in the aqueous phase.

3. Partition of dinitroglycerin between NG oil phase and aqueous alkaline phase

The NG solubility in 1.39% aqueous Na_2CO_3 solution is very low, only 0.1 g/ 100 g. The experimental results of partition of dinitroglycerin between NG and aqueous alkaline phase are shown in Table 6.45 and Fig. 6.26.

Based on the solubility of NG and dinitroglycerin in neutral, acidic, and basic aqueous solutions and the partitions of nitric acid and dinitroglycerin between NG

Table 6.45 Partition of dinitroglycerin between oil phase and 1.39% aqueous Na_2CO_3 phase

Dinitroglycerin content in NG sample (%)	Dinitroglycerin content in NG phase (%) (C′)	Dinitroglycerin content in aqueous Na_2CO_3 phase (%) (C)	Partition coefficient of dinitroglycerin between NG and 1.39% aqueous Na_2CO_3 solution (C′/C)
0.00	0.00	0.00	–
2.37	2.00	0.45	4.45
9.53	8.15	1.40	5.8

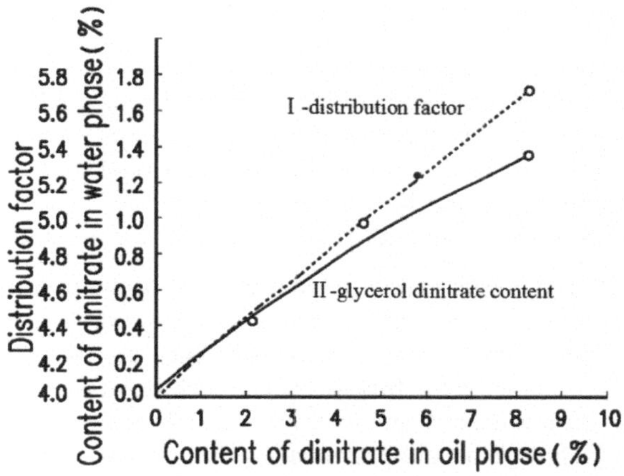

Fig. 6.26 Partition of dinitroglycerin between NG and 1.39% Na_2CO_3 aqueous solution

and the aqueous solution, the parameters for the NG stability treatment including the partitions between NG phase and aqueous solutions, amounts of washing solutions, washing times, etc. can be determined, which provides a reference for the NG production process.

6.3.6 Post-treatment of Spent Acid

Even the spent acid and NG are well separated, there is still trace amount, ~0.1% free NG, in addition to the dissolved NG, in the spent acid. Therefore, small amounts of NG can be precipitated during spent acid storage (1–2 days). The precipitated NG is very dangerous for the transport, storage, and operation of spent acid.

6.4 Issues in NG Production

6.4.1 *Factors of NG Stability Treatment*

According to the relationship between thermodynamics and heat transfer, the extraction rate of NG is proportional to the concentration difference of NG in the spent acid, and inversely proportional to the resistance. Therefore, under the normal separation conditions, the extraction rate can be expressed as

$$\frac{dG}{dt} = KF(C^* - C)$$

where G is the amount of extractable compound (acid) (kg), t is extraction time (h), C is the concentration of the extractable compound in the aqueous extraction phase (kg/kg), C^* is the concentration of the extractable compound in the aqueous extraction phase at the equilibrium state (kg/kg), F is the contact area between two phases (m^2), and K is the mass transfer coefficient (kg/h m^2). Therefore, the following factors should be considered for the separation of NG.

(1) The concentration difference is the major factor promoting the extraction efficiency. Increasing the amount of washing water or countercurrent washing can increase the concentration difference. It is worth noting that the washing water volume for the first wash is critical. However, extremely large amounts of water can accelerate the hydrolysis and increase of NG loss.
Hydrolysis and esterification reaction can still occur in the NG separated from the spent acid. The addition of water can increase water content in NG, break the equilibrium, and accelerate the hydrolysis. Therefore, the amount of water for the first washing must result in the nitric acid content in the aqueous phase less than 29% to prevent serious hydrolysis of NG. In addition, the speed of water addition has also a certain impact on the hydrolysis. Slowly adding water can increase the water content in NG, which promotes the hydrolysis. The rapid addition of water can extract nitric acid quickly with less influence on the oil phase.
In addition to the hydrolysis of NG during the stability treatment, the heat of dilution should be considered to prevent the rapid temperature rise. Therefore, the water volume for the first washing should be large enough. However, high amounts of water can dissolve more NG, reducing the productivity.

(2) The contact surface area between the phases is proportional to the extraction rate. Increasing the contact surface area can promote the extraction. The drop size of extraction solution can significantly affect the mass transfer. Extremely large drops cause slow internal diffusion. The contact surface area can be increased by increasing the dispersion of NG in the extraction solution.
(3) Increasing the velocity of the inter-phase motion can reduce the thickness of fluid boundary layer on the phase boundary, and thus reduces the diffusion resistance. Agitation can increase the amount of liquid passing through the

porous plate, promote the surface refreshing, and thus decrease the thickness of the boundary layer.

(4) Extending the extraction time can increase the extraction efficiency. However, extremely long extraction time can reduce the NG productivity.

In addition, appropriately increasing separation temperature can promote the liquid dispersion and increase the solubility of impurities in water, and thus is conducive to extraction. However, high temperature can also promote the thermal decomposition of NG and increase the NG solubility in water, causing issues in production safety and productivity.

The residual acid in NG cannot be completely removed by water washing. To remove acid residues and ensure production safety, NG must be washed with diluted Na_2CO_3 solution to achieve neutral or alkaline NG for safe storage and operation.

6.4.2 Factors Affecting Nitration Process

(1) Factors related to nitrating acid

The composition and amount of nitrating acid have been discussed above. The composition of final spent acid decides the nitration reaction equilibrium, and thus significantly affects the reaction rate, nitration degree, product yield, stability treatment, and product quality. Therefore, the composition of nitrating acid is mainly determined by the spent acid composition.

1. Effects of nitrating acid composition on NO^{2+}

The NO^{2+} concentration increases with the increase of nitric acid content and its conversion rate increases with the increase of sulfuric acid content in the nitrating acid, which is conducive to the nitration. The conversion rate of NO^{2+} decreases with the increase of water content in the nitrating acid, which reduces the nitration rate, and thus is unconducive to the reaction.

2. Effects of reaction rate

The nitration reaction is very fast when the nitric acid content in the spent acid is more than 8% and [H2O]: [H2SO4] $\leq$ 1. The dinitroglycerin content in NG increases with the increase of water content in the spent acid. The minimum dinitroglycerin content can be achieved 13.5–14% with water (equivalent to [H2O]: [H2SO4] = 1) and 11.5–13.3% nitric acid in the spent acid.

3. Effects of nitric acid in NG

The least loss of NG including dinitroglycerin can be achieved with the nitric acid content of 11.5–13.5%, and water content of 13.5–14% in the spent acid. The nitric acid content in the oil phase increases with the nitric acid content in the spent acid. With the same amount of nitric acid, a water content of 13.5–14% in spent acid gives the highest nitric acid content in oil phase.

4. Effects of water content in oil phase

The water content in oil phase is constant if $[H_2O]:[H_2SO_4] \leq 1$, and increases with the increase of water content in the spent water if $[H_2O]:[H_2SO_4] > 1$.

5. Effects on the stability of acidic NG

The decomposition of NG becomes faster with the increase of water content in the spent acid with $[H_2O]: [H_2SO_4] > 1$. The explosion temperature decreases with the increase of water content in spent acid as shown in Table 6.46. Low ratios of $[H_2O]:[H_2SO_4]$ are favorable for the production safety. The optimum ratio is 1:1 and those higher than higher than 1.5 may affect production safety.

(2) Controlling measures

To obtain high-quality products, the following parameters should be taken into consideration for the NG production process.

1. Molar ratio of $[H_2O]:[H_2SO_4]$ in the spent acid

The concentration of sulfuric acid in the spent acid is often expressed as "dehydration value" that is the molar ratio of sulfuric acid to water in spent acid. In general, the ratio of $[H_2O]: [H_2SO_4]$ should be 1.2–1.36.

2. Excess nitric acid

High amounts of nitric acid, e.g., high amounts of NO^{2+}, can promote reaction rate, decrease side reactions, increase NG yield, and ensure production safety. However, increasing nitric acid amounts also increases the production cost. Therefore, the nitric acid amount in nitrating acid is usually controlled at ~20% more than required, which results in 10–13%, at least more than 8% nitric acid in the spent acid,

3. Nitric oxide content

Nitric oxide is one of the main factors that cause the decomposition of NG. The brown smoke of high concentrations of nitric oxide can affect the correct judgment on the decomposition of NG in production process. To ensure safe production, the less nitric oxide in the system, the better. In general, the N_2O_4 content should be

Table 6.46 Relationship between explosion temperature and water content in spent acid

$[H_2O]$: $[H_2SO_4]$	Explosion temperature of NG (°C)	$[H_2O]$: $[H_2SO_4]$	Explosion temperature of acidic NG (°C)
1.02	155	1.85	107
1.24	121	2.11	95
1.49	110	2.67	75

less than 0.6%, or the nitrite content in spent acid should be less than 0.2% for a practical NG production process.

4. Amount of nitrating acid

Lower nitration coefficient and less NG loss can reduce the cost of spent acid treatment. The nitration coefficient of the air-agitated nitration method should be controlled to be between 4.8 and 5.2 and that of injection nitration is larger, usually between 10 and 15, but with similar amount of fresh mixed acid.

The composition of spent acid for preparing the nitrating acid is as follows:

HNO_3	10%,
H_2O	17%,
H_2SO_4	73%;
$[H_2O]:[H_2SO_4]$ =	1.25

Based on the nitration equation

$$C_3H_5(OH)_3 + 3HNO_3 \rightleftharpoons C_3H_6(ONO_2)_3 + 3H_2O$$

92	3 × 63	227	3 × 18
1.00	2.05	2.47	0.587

Then the amount of spent acid produced from the production of 1000 g NG is

$$0.587/0.17 = 3.5\ \text{kg}$$

$$\text{where}\quad H_2SO = (0.587/18) \times (98/1.25) = 2.55\ \text{kg}$$

$$HNO_3 = 3.5 \times 0.1 = 0.35\ \text{kg}$$

Therefore, the amount of required nitrating acid is

$$G_M = (2.05 + 0.35) + 2.55 = 4.95\ \text{kg}$$

Thus, the nitration coefficient is 4.95.

The composition of the mixed acid is

$$HNO_3\% = \frac{2.40}{4.95} \times 100\% = 48\%$$

$$H_2SO_4\% = \frac{2.55}{4.95} \times 100\% = 52\%$$

Table 6.47 lists the compositions of the mixed acid and spent acid for different production processes of NG.

Reaction contact surface

The nitration of glycerol with mixed acid is a heterogeneous reaction. The reactants must contact each other well to achieve desired reaction performance. Increasing

Table 6.47 Composition of mixed acid and spent acid

Nitration		Composition of mixing acid (%)					Composition of spent acid (%)					$\frac{[H_2O]}{[H_2SO_4]}$
		HNO_3	H_2SO_4	H_2O	N_2O_4	NG	HNO_3	H_2SO_4	H_2O	N_2O_4	NG	
Injection	A plant in Sweden	27.4	61.9	9.0		1.7	11.0	70.8	15.5	2.7		1.2
	A plant in China	Total nitrogen content 28.1 29.5 ± 0.5	62 ± 1	8 ± 2	<0.6	<1.6	13.16	71.32	12.73	0.77	3.62	1.0
Discontinuous		49 ± 1	51 ± 1	0.5	0.6		9.2	71.5	16.44	2.86		1.25
Schmid process		50	50				12	72	14	2		1.1
Biazzi process		50	50				10	73	15	2		1.1

the contact surface area between the two phases can not only improve the reaction rate but also promote the dehydration efficiency of sulfuric acid, and thus reduce the local overheating caused by the heterogeneous reaction.

Spraying glycerol into the mixed acid can increase the dispersion of glycerin with average droplet diameter as low as $\sim 0.65 \times 10^{-2}$cm, which improves the contact surface area. In addition, using the ejector or T-tube reactor can spray a high-speed flow of mixed acid to form a negative pressure zone on the nozzle that can suck the glycerol for the reaction. Glycerol can be dispersed into 0.37–2.8×10^{-2} cm droplets by the high-speed turbulence of the reaction mixture. The contact surface area of the emulsion sprayed with an injector can be increased to 212–640 cm^2/cm^3 and the dispersion of glycerin several times of that achieved by mechanical mixing.

The injection nitration method uses the mixed acid to continuously suck and discharge glycerol, which puts the mixture flows under the turbulence continuously. Therefore, the droplet surface is constantly refreshed. For the container air-agitated nitration method, the contact surface is refreshed by the strong mechanical agitation.

6.4.3 Factors Affecting the Separation of NG and Mixed Acid

The importance and challenges of the separation between NG and mixed acid have been described above. The separation of a viscous substance mucus is affected by its properties. In addition to the separation equipment, the raw material for the NG production can also affect its separation. Numerous industrial production experiences suggest that the qualities of raw materials, glycerol and nitrating acid, have a great impact on the separation rate. For example, high fatty acid contents in glycerol and large amounts of lead and iron salt suspensions can slow the separation. The rapidest separation has been achieved with the spent acid composition of 17% HNO_3, 70% H_2SO_4, and 13% H_2O.

The addition of a small amount of separating agent, such as paraffin oil, vaseline, and so on, to the emulsion can form a thin interfacial film on the NG droplets, change droplet interface tension and self-adhesive properties, and thus accelerate the aggregation of NG droplets and promote the separation. The nitration raw materials containing sodium fluoride and silicon can also result in significantly improved separation of NG.

$$4NaF + SiO_2 + 2H_2SO_4 \rightarrow 2Na_2SO_4 + 2H_2O + SiF_4 \uparrow$$

When there is sodium fluorosilicate in the reactant mixture, the following reaction occurs:

$$Na_2SiF_6 + H_2SO_4 \rightarrow NaSO_4 + 2HF\uparrow + SiF_4\uparrow$$

The mixture of sodium fluoride and sodium silicate can eliminate the charges of NG droplets and thus promote the aggregation of NG droplets. HF and SiF_4 bubbles formed in the emulsion rise up and thus promote the coalescence and floating of NG.

Talc of 0.02–0.05% can also be used as a separation agent to accelerate the separation. The addition of 1% urea, acetamide, or dicyandiamide to the spent acid can release gas bubbles, and thus promote the separation of NG. Most of the NG production lines use paraffin oil, or vaseline as paraffin oil unavailable, as the separating agent.

Compared with the container air-agitated nitration method, the injection nitration method can achieve 2–3 times larger nitration coefficient and result in less NG in the spent acid with higher degree of dispersion. Therefore, the mixed acid obtained by an injection nitration process cannot be separated by gravity separation, but separated by centrifugation. The spent acid can be completely separated from NG in 30 s without any separation agent needed, which reduces the contact time of NG with acid and the residence time of acidic NG, and thus ensures safety, quality, high yield, and low consumption.

6.4.4 Equilibrium in Glycerin Nitration

Glycerol is insoluble in the mixed acid, and the solubility of NG in acid is also very low. Therefore, the nitration of glycerol in mixed acid is a heterogeneous reaction under intense turbulence of two phases. Acid is a dispersion medium and continuous phase, and NG is a discontinuous dispersion phase, resulting in two equilibriums in the two-phase system: chemical reaction equilibrium and phase equilibrium. The chemical reaction equilibrium is related to the composition of spent acid. Under normal reaction conditions, in addition to the equilibrium between nitration and hydrolysis, there is the equilibrium between the esterification and hydrolysis of glycerol and sulfuric acid, and the reaction equilibrium between nitric acid.

The main components of nitration products in oil phase are NG, dinitroglycerin, nitric acid, and a small amount of water. In general, there is 0.1–0.3% sulfuric acid in the oil phase, which is negligible. The reaction equilibrium in the oil phase includes the equilibrium between the esterification and hydrolysis of glycerol and nitric acid.

$$C_3H_5(ONO_2)_3 + H_2O \rightleftharpoons C_3H_5(ONO_2)_2(OH) + HNO_3$$

One hundred grams of acidic NG containing 10% nitric acid in a glass bottle with stopper was added with water under constant shaking in 5–10 min. The solution was continuously sampled and analyzed for the formation of

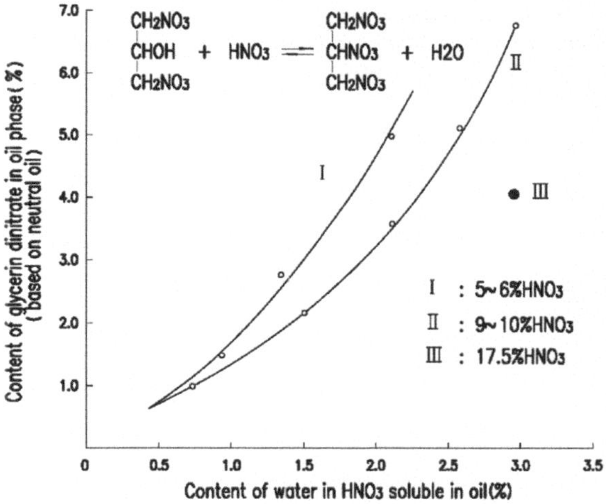

Fig. 6.27 Reaction equilibrium in oil phase

dinitroglycerin in the water. It was found that a large amount of dinitroglycerin was produced and there was 15% dinitroglycerin as 3.2 g water added and 50% dinitroglycerin as 7.4 g water added in the mixture. In contrast, there was only 2.0% dinitroglycerin in the acidic NG before washing. However, if all water was added rapidly in the test, two phases would be formed and the aqueous phase dissolved most nitric acid. No significant change was found on the hydrolysis rate of NG in the aqueous phase containing 30% nitric acid, e.g., 25 g water added in the test, indicating that the hydrolysis equilibrium in oil phase was mainly related to the water content in the acid NG. As shown in Fig. 6.27, the hydrolysis is accelerated with the increase of water content in the oil phase and decreased with the increase of dinitroglycerin and nitric acid contents. The dinitroglycerin content in oil phase decreases with the increase of nitric acid content at the same water content.

The water content in the spent acid can not only affect the reaction equilibrium in the spent acid phase but also affect the water content of nitric acid in the oil phase, and thus influences the phase equilibrium in the oil phase. The relationship between the amount of nitric acid in oil phase and the water content in spent acid is shown in Fig. 6.28 where the numbers on the curve are the percentage of nitric acid in the spent acid.

The numbers on the curve are the percentages of HNO_3 in the spent acid.

The phase equilibrium is gradually formed during the nitration. The equilibrium conditions change constantly and eventually, the whole system evolved to the final equilibrium state and the nitration is completed. The final state is decisive for the yield of NG and the damage to raw materials. The final equilibrium is only affected by the final equilibrium conditions, such as the composition of spent acid at equilibrium, temperature at equilibrium, and so on, regardless the way to establish such an equilibrium state. The nitration process can be divided into the following two stages. The first stage is the initial stage, which can be expressed as

Fig. 6.28 Relationship between the water content and the ratio of $[H_2O]:[H_2SO_4]$ in spent acid

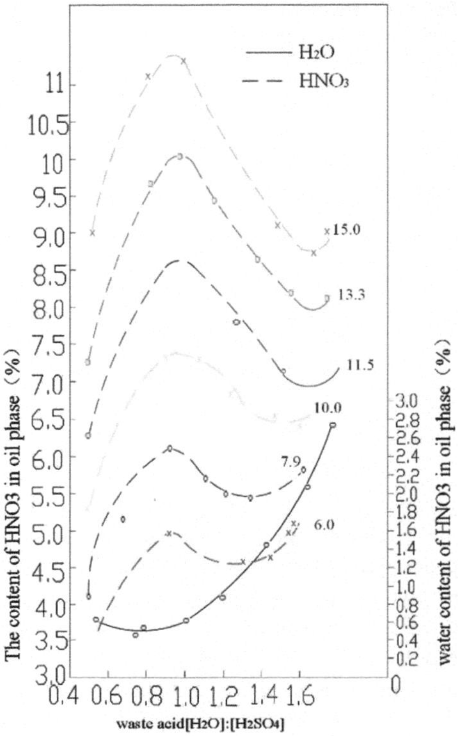

$C_3H_5(OH)_3$ + nitrating acid (nitric acid, sulfuric acid and water)

$C_3H_5(ONO_2)_3$ + spent acid (nitric acid, sulfuric acid and water)

Initial stage

During this stage, glycerol is nitrated by excess nitric acid in the nitrating acid to form NG, a small amount of dinitroglycerin and spent acid containing no organic compounds. The two phases continue to interact with each other until reaching the equilibrium. The equilibrium state of the second stage can be expressed as

Initial stage → Final stage

$C_3H_5(ONO_2)_3$ + *Spent acid*
- *Oillphase*
  - *Nitroglycerin*
  - *Dinitroglycerin*
  - *Nitric acid*
  - *Water*
- (*small amounts of sulfuric acid*)
- *Spent acid*
  - *Nitric acid*
  - *Sulfuric acid*
  - *Water*
  - *Nitroglycerin*
- (*including dinitroglycerin*)

The oil phase equilibrium at this stage results from the mass composition of spent acid of the initial stage. The oil (NG and dinitroglycerin) distributed in the spent acid phase, assuming it is in the physically dissolved state, is at trace level and cannot be accurately quantified. Therefore, in the actual or theoretical treatment of a

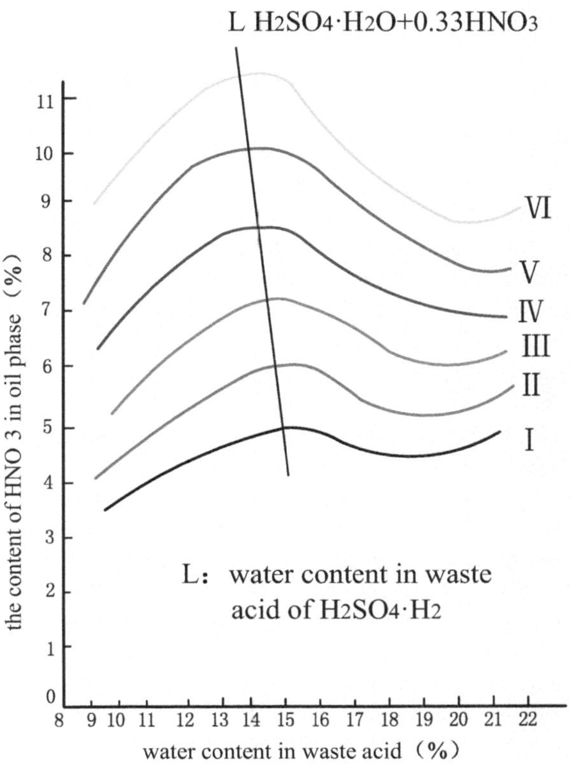

Fig. 6.29 Relationship between the nitric acid content in oil phase and the water content in spent acid. I—6.0%HNO_3; II—8.0% HNO_3; III—10.0%HNO_3; IV—11.5%HNO_3; V—13.5% HNO_3; and VI—15.0%HNO_3

Fig. 6.30 Relationship between the dinitroglycerin content in oil phase and the water content in spent acid

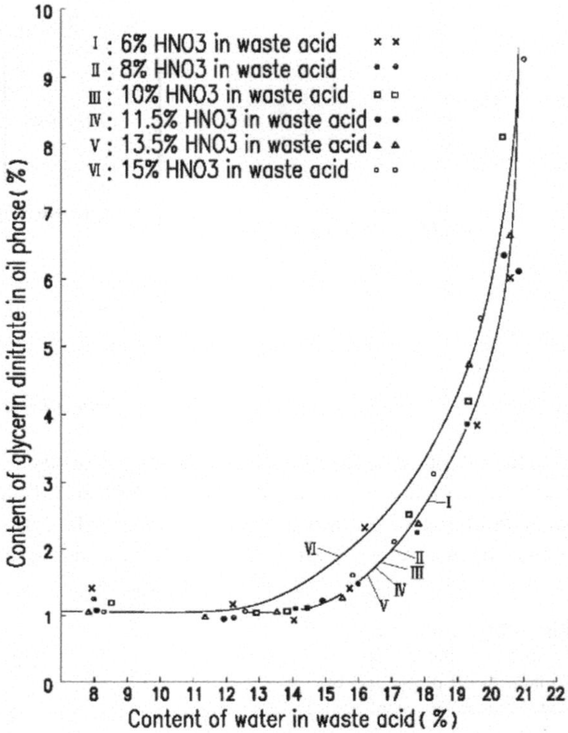

two-phase system of spent acid and NG, it is assumed that the reaction equilibrium of the oil phase is independent of the physical state of the two-phase system.

Based on the oil phase equilibrium, it can be concluded that the ratio of water to nitric acid determines the dinitroglycerin and NG contents in oil phase. Figure 6.29 shows the nitric acid content in oil phase as the function of the water content in spent acid. Each curve in Fig. 6.29 has the maximum value at $[H_2O]:[H_2SO_4] = 1$, e.g., H_2SO_4 and H_2O in the spent acid phase are in the "hydrate" form $H_2SO_4 \cdot H_2O$. These "hydrate" points representing the maximum can be then linked using a straight line L. The minimum value is roughly at the molecular ratio between H_2SO_4 and H_2O of 2:3 in spent acid, that is, ~21% water.

The partition of water between the two phases has been shown in Fig. 2.12. The water content in oil phase refers to the water content of nitric acid in oil phase. The water content in oil phase remains constant as long as the water in spent acid phase is insufficient to convert all sulfuric acid to the "hydrate" form H_2SO_4 H_2O. When the water content in spent acid exceeds the "hydrate" H_2SO_4 H_2O limit, the nitric acid in the oil phase rapidly becomes an aqueous acid. The water content in the oil phase immediately exceeds this limit and increases with the increase of water content in spent acid phase (Fig. 2.12). The sulfuric acid is completed converted into hydrate H_2SO_4 H_2O at the turning point on the curve of water content.

The nitration degree can be determined from Figs. 6.28 and 6.29. The relationship between the degree of nitration and the water content in the spent acid is shown in Fig. 6.30.

The dinitroglycerin content in oil phase is related to the water content in spent acid, but is not affected by the nitric acid content in the spent acid. The distributions of nitric acid and water in the spent acid containing 6–13.5% nitric acid restrict each other. The experiential water content limit in spent acid is 14%. The water contents higher than this limit can cause significant increase in dinitroglycerin content and those lower than the limit result in acceptable dinitroglycerin contents because of the hydrolysis caused by the increased water content in oil phase with the increase of water content in spent acid.

The hydrate actually divides the system into two parts. The water content in the spent acid phase is insufficient to convert all sulfuric acids to the hydrate at the phase equilibrium of oil phase and spent acid, resulting in the minimum water content in the oil phase and the maximum degree of nitration. As the water content in spent acid exceeds the limit to form the hydrate, the binding capacity of sulfuric acid with water is significantly reduced. At this point, increasing water content can increase the water content in oil phase due to the equilibrium, which shifts the dominant nitration equilibrium, reduces the nitration degree, and increases the formation of dinitroglycerin. To achieve the optimal nitration process, the nitration

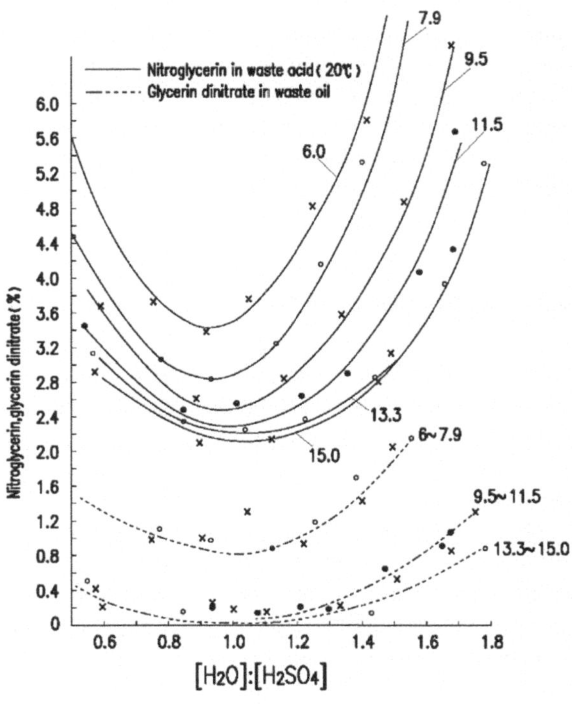

Fig. 6.31 Relationship between nitrate content and water content in the spent acid

mixture must be well mixed, and thus the free sulfuric acid in spent acid can well combine with water.

In conclusion, the dinitroglycerin content reaches the minimum at $[H_2SO_4]$:$[H_2O] = 1$ in spent acid. The hydrolysis is accelerated, and the dinitroglycerin content is increased at $[H_2SO_4]$:$[H_2O] > 1$ in spent acid. The possibility to produce the mixture of nitrate and sulfate is increased at $[H_2SO_4]$:$[H_2O] < 1$ in spent acid.

Increasing nitric acid content promotes the reaction equilibrium to form NG.

The amount of nitrates in spent acid increases with the increase of temperature. For example, the solubility of NG in spent acid at ~10 °C is 25% less than that at 20 °C.

In the whole reaction system, the two equilibriums of oil phase and spent acid phase are affected by the composition of spent acid. The nitric acid content and water content in spent acid play an important role in the reaction equilibrium and phase partition equilibrium. As shown in Fig. 6.31, the solubility of NG in spent acid decreases with the increase of the nitric acid content in spent acid, and reaches the lowest at the nitric acid content of 13.3–15.0%. In contrast, the solubility of NG increases with the increase of the water content in spent acid. At the same nitric acid content, the 1:1 molar ratio of water to sulfuric acid results in the lowest NG solubility.

The yellow dotted curve in Fig. 6.31 shows the relationship between the composition of spent acid and dinitroglycerin content. The lowest solubility of dinitroglycerin is at $[H_2SO_4]$:$[H_2O] = 1$. Therefore, the ratio of water to sulfuric acid in spent acid plays a decisive role in the reaction and phase partition equilibriums.

Appropriately increasing nitric acid content or reducing water content, while considering the production economy, can reduce the loss of NG on dissolution and improve safety. For example, the spent acid with $[H_2SO_4]$:$[H_2O] = 1$ and 15% HNO_3 can give a low dinitroglycerin content and low loss pf MG on dissolution, which is favorable to the production safety. However, the nitric acid content in NG is relatively high under such conditions, which increases the loss of NG on dissolution during washing, and thus leads to high production cost.

6.5 Application of NG

The application of NG in mining was first proposed by the Swedish engineer Alfred Nobel. In 1862, Nobel built the first "explosive oil" factory. Later, many countries, including Russia in 1854, used NG as a high-quality explosive to fill the projectile. Nobel also used the "explosive oil" to substitute black powder (smoky explosive) for mining.

High explosives can be prepared by separating NG from the water–alcohol mixture. A large number of experiments suggest that NG can be absorbed by diatomaceous earth to form diatomaceous earth explosives. In 1867, Nobel successfully prepared a diatomaceous earth explosive containing 75 parts of NG and 25 parts of diatomaceous earth. Mixing NG with solid magnesium carbonate adsorbent can reduce the shock sensitivity of NG. NG has been used in the gold mines of Upper-Uspleaskii in Trans-Baykal of Siberia since 1867.

In 1875, Nobel invented gelignite by dissolving nitrocellulose in nitroglycerin [7]. The gelignite can be mixed with oxygen carriers, such as ammonium nitrate and sodium nitrate, to form "straight" dynalite that is still widely used nowadays. In 1888, Nobel invented "nitroglycerin" smokeless powder (cordite). Since then, NG has been widely used as a component of ammonium nitrate explosives for mining.

The wide application of NG in medicine, clinic, aviation, aerospace, and military has been reviewed in many literature [14–19] and will not be discussed here.

6.6 Toxicity of NG

NG is a toxic chemical causing chronic poisoning to the blood system, liver, and reproductive organs with nonspecific reactions in weight, appetite, and behaviors. In addition to the subchronic poisoning to liver, NG can also dilate blood vessels and reduce blood pressure. NG can cause poisoning through percutaneous or mucosal penetration after inhalation. Different individuals have different sensitivities to NG rent [20].

The main symptoms of NG poisoning are severe headaches with a feeling of grinding. The alcoholics can be accustomed to NG quickly in a couple of days after contact with NG. However, such adaptability will disappear as soon as the contact with NG interrupted. It takes time to resume such adaptability as the body contacts NG again. Work in double-based propellant production lines used to prepared and sieve dynamite (NG) with bare hands without any obvious injury observed. NG does not produce chronic poisoning. Thus, the chemical should not be considered an industrial toxicant [21]. Some people may be allergic to NG and cannot contact NG. The occasional poisoning symptoms include headache, vomiting, cyanosis, visual disorders, limb itching, swelling, etc. [22]. So far, no fatal poisoning of NG has been reported. The headache caused by NG poisoning detoxified by the intramuscular injection of caffeine and sodium benzoate or the oral admission of phenylaminopropane sulfate.

References

1. Marsh N (2000) A short history of nitroglycerine and nitric oxide in pharmacology and physiology. Clin Exp Pharmacol Physiol 27(4):313–319
2. Günther A, Tollens B, Chalmot G et al (1891) Die bestimmungen der enta-glykosen. Z Anal Chem Freseniu 30(1):519–520
3. Li Z, van Beilen JB, Duetz WA et al (2002) Oxidative biotransformations using oxygenases. Curr Opin Chem Biol 6(2):136–144
4. Zhang T, Zhao C, Liu X (1978) Procedure control of nitroglycerin production. Fire Explos 01 (1):5–12

5. Li LX (1984) Safety production technology of nitroglycerin. Chin J Explos Propellants 3:27–34
6. Ban DM, Liu JP (2011) Catalytic synthesis of nitroglycerin by solid acid. In: Hazardous materials and security emergency technology, pp 75–77
7. Department of defense test method standard (2001) Safety and performance tests for the qualification of explosives MIL-STD-1751A
8. Fedoroff T, Sheffield OE (1972) Encyclopedia of explosives and related items, vol 8, pp 86–121
9. Urbanski T (1976) Chemistry and technology of explosives (II). National Defense Industry Press, Beijing
10. Liu JP (2015) Liquid explosives. Berlin, Germany
11. Ban DM, Lu XJ, Liu JP (2012) Research on catalytic synthesis of nitroglycerin. Theory and practice of industrial explosive materials. Ordnance Industry Press, Beijing, pp 512–514
12. Yang RH (2010) Preparation of typical nitrate esters. Beijing Institute of Technology
13. Wu YC (1980) Nitroglycerine chemical technology. National Defense Industry Press, Beijing
14. Maruhashi T, Nakashima A, Matsumoto T et al (2014) Relationship between nitroglycerine-induced vasodilation and clinical severity of peripheral artery disease. Atherosclerosis 235:65–70
15. Reinmutha N, Meyerb A, Hartwigsenc D et al (2014) Randomized, double-blind phase II study to compare nitroglycerin plus oral vinorelbine plus cisplatin with oral vinorelbine plus cisplatin alone in patients with stage IIIB/IV non-small cell lung cancer (NSCLC). Lung Cancer 83:363–368
16. Lindner V (2000) Propellants. In: Kirk-othmer encyclopedia of chemical technology, vol 22
17. Reese DA, Groven LJ, Son SF (2014) Formulation and characterization of a new nitroglycerin-free double base propellant. Propellants Explos Pyrotech 39:205–210
18. Drake GW, Bolden S, Dailey J (2012) New takes on nitrate ester chemistry: salts with oxygen-rich ammonium cations. Propellants Explos Pyrotech 37:40–51
19. Urbandki T (1984) Chemistry and technology of explosives. Peramon Press, New York
20. Feng YZ, Zhang HM, Lu MH (1991) Study on toxicity of nitroglycerine. Ind Health Occup Dis 17(4):207–210
21. Carmichael P, Lieben J (1963) Sudden death in explosives workers. Arch Environ Health 7 (4):424–439
22. Ellis HV, Ellis HV, Hong CB (1984) Subacute and chronic toxicity studies of trinitroglycerin in dogs, rats, and mice. Appl Toxicol 4(2):248

Chapter 7
Pentaerythritol Tetranitrate

Pentaerythritol tetranitrate, also known as nitrated pentaerythritol, is abbreviated as PETN. PETN was first produced in 1891 by Tollens [1]. PETN is the most stable nitrate and more powerful than RDX. It is an explosive with a wide range of applications. After the World War I, the synthesis, production, and application range of pentaerythritol tetranitrate were extensively studied in many countries, and with successively widespread applications in the military. From the mid-1930s to 1940s, its annual production in the United States was more than 6000 tons, and the monthly production capacity in Germany was 1600 tons. From 1940s to 1950s, an annual production in the former Soviet Union had been up to 1 million tons. Because the raw materials to synthesize pentaerythritol, including formaldehyde and acetaldehyde, can be obtained from by-products of oil cleavage or carbohydrates, the synthesis of pentaerythritol tetranitrate has extensive material source and is easy to implement. Because the mechanical sensitivity, physical stability, chemical stability and service performance, and other properties of pentaerythritol tetranitrate are inferior to that of ammonium nitrate explosives, it is possible that pentaerythritol tetranitrate will be replaced by the second-generation explosives (RDX and HMX) or third-generation new ones (CL20, TKX-50).

7.1 Physicochemical Properties of Pentaerythritol Tetranitrate

PETN is a white crystalline with two crystal forms and belongs to the tetragonal crystal system. Its crystal shape varies with different solvents during the crystallization, including needle, bevel square, stable α-cubic, and unstable β-cubic [2–4].

J. Liu, *Nitrate Esters Chemistry and Technology*,
https://doi.org/10.1007/978-981-13-6647-5_7

Chemical formula: $C_5H_8N_4O_{12}$;
Structure:

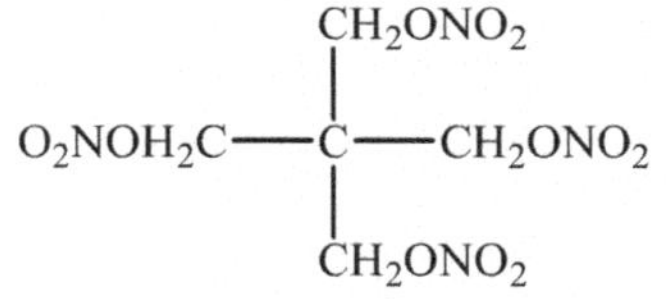

Relatively molecular weight: 316.12;
Oxygen balance: −10.2% (according to the produced CO_2);
Oxygen content (%): 17.72;
Nitrogen content (%): 18.21;
Density (g/cm^3): 1.77;
Melting point (°C): 141.3; and
Moisture absorption: PETN cannot adsorb water.
Solubility: PETN is almost insoluble in water, but soluble in acetone, methyl acetate, and dimethyl. In alcohol, ether, and benzene carboxylic acid, it is insoluble.

7.1.1 Physical Properties of Pentaerythritol Tetranitrate

Pentaerythritol tetranitrate is a crystalline substance that is insoluble in water but with different degrees of dissolution in acid and polar reagents. PETN solubility in solvents are shown in Table 7.1.

As listed in Table 7.1, the solubility of pentaerythritol tetranitrate in acetone is the largest. With the change of water content in acetone, the solubility of PETN is different, which are shown in Fig. 7.1.

PETN has relatively good infiltration in liquid or melted aromatic nitro compounds, and eutectic is formed after the infiltration. The solubility of PETN in TNT and the melting point of its formed eutectic in aromatic nitro compounds are listed in Tables 7.2 and 7.3.

PETN and nitroglycerin can be miscible with each other to form NG eutectic, whose composition and freezing point are shown in Table 7.4.

In addition to the mentioned above, PETN can also form a ternary eutectic with some nitro compounds, whose composition and freezing point are shown in Table 7.5.

Pentaerythritol tetranitrate has four -CH_2ONO_2 groups, which are uniformly located around the central carbon atom to form a symmetrical structure, so PETN is quite stable and is more stable than other polyol nitrates. At 100 °C, the military PETN lost weight only 0.1% in 40 hours, the amount of gas released by the vacuum stability test under the same conditions was 0.5 ml/g.Unlike other liquid nitrates, it does not form colloidal solutions with nitrocellulose.

PETN is soluble in nitric acid and its solubility increases with increasing temperature. The solubility of PETN in nitric acid is shown in Fig. 7.2.

PETN's heat capacity is 1.67 J/(g °C), its burning heat is 8259.2 J/g, its melting enthalpy is 96.2–100.4 kJ/mol, and its enthalpy of sublimation is 151.9 kJ/mol. Its dipole moment in a dilute solution of benzene is 2.0 Debye, while it is 2.5 Debye in

Table 7.1 Pentaerythritol tetranitrate solubility in organic solvents

Solvent	Content									
	The PETN dissolution capability of solvent (100 mL) at different temperatures									
	0	20	25	30	40	47	50	60	80	100
Acetone	14.37	24.95	–	–	30.56	–	–	42.68	–	–
Benzene	0.15	0.45	–	–	1.16	–	–	3.35	–	–
β-ethoyxlethyl acetate	–	1.5	–	4.1	7.6	–	11.2	14.2	–	–
Tetrachloromethane	–	0.096	–	0.108	0.118	–	0.121	–	–	–
Chloroform	–	0.09	–	–	–	–	–	–	–	–
Ethylene epoxide	–	0.35	–	2.8	6.1	–	9.2	12.2	–	–
Ethyl ether	0.20	0.34	–	–	–	0.45	–	–	–	–
Ethanol	0.07	0.195	–	–	0.415	–	–	11.6 (61 °C)	–	–
Dichloroethane	–	0.9 (10 °C)	–	1.25	5.02 (43 °C)	–	8.28	13.8 (65 °C)	–	–
Methanol	–	0.46	–	–	1.15	–	–	2.60	–	–
Ethyl formate	–	13.0	–	17.0	22.0	–	31.0	–	–	–
Ethyl acetate	–	6.322 (19 °C)	–	–	–	–	17.863	–	–	–
2-Methyl-1-propanol	–	0.27	–	0.31	0.39	–	0.52	–	–	–
Nitromethane	3.36	–	8.89	–	–	–	–	–	8.89	–
Tetrachloroethane	–	0.18	–	0.27	0.40	–	0.58	–	–	–
Toluene	0.15	0.43	–	–	0.62	–	–	2.49	–	15.92
Dimethyl Formamide	–	–	–	–	40	–	–	50	–	70
Pyridine	–	5.436 (19 °C)	–	–	–	–	8.567	–	–	–

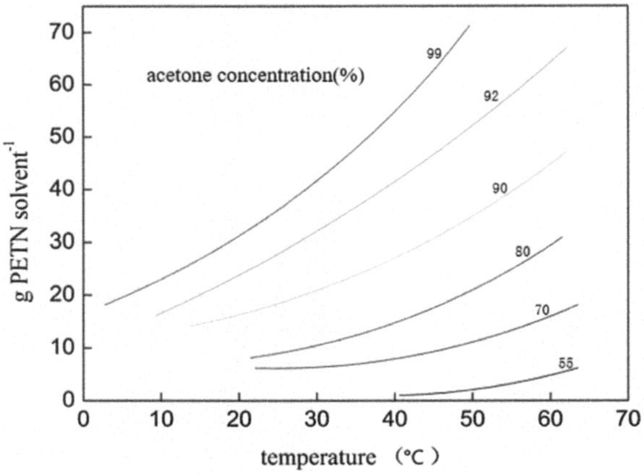

Fig. 7.1 The solubility of PETN in acetone with different concentrations

Table 7.2 The solubility of PETN in TNT at different temperatures (g • 100 g TNT-1)

Temperature (°C)	80	85	90	95	100	105	110	115	120	125
Solubility	19.3	25.0	32.1	39.5	48.6	58.2	70.0	87.8	115.0	161.0

Table 7.3 The eutectic formed from PETN and other aromatic substrates

Content of aromatic nitro compound (%)	Content of PETN (%)	Melting point (°C)
m-Dinitrobenzene 80	20	82.4
2,4-Dinitrotoluene 90	10	67.3
α-Trinitrotoluene 87	13	76.1
Tetralite 70	30	111.3
Nitromannite 80	20	101.3
Diethyl diphenyl urea 88	12	68.0

Table 7.4 Composition of eutectic from PETN and NG

Item (%)	Content (%)	Melting point (°C)
NG content in the eutectic	98.5	98.9
PETN content in the eutectic	1.5	1.1

1,4-dioxocyclohexane. Ultraviolet absorption spectrum characteristic of PETN crystal is that it can entirely absorb ultraviolet light with wavelengths of below 280 nm. The absorption is shown in Fig. 7.3.

After the stability treatment and refining of PETN, it is stable at room temperature. In order to assess the thermal stability of PETN, the following experiment can be done to verify its stability. 0.5 g of PETN is placed in a tiny glass container under the helium protection. This container was heated up under vacuum to observe

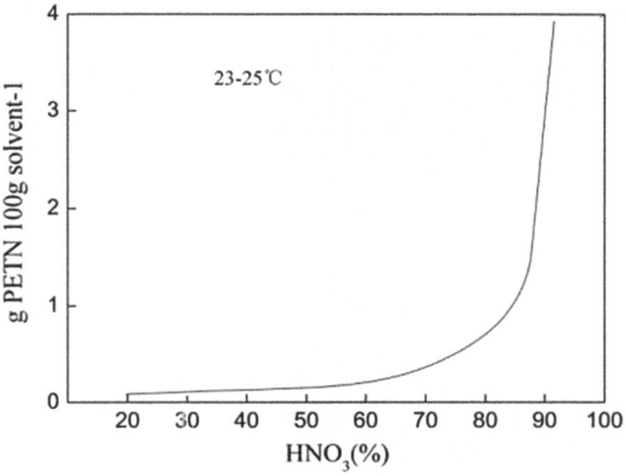

Fig. 7.2 Solubility of PETN in HNO_3 with different concentrations

Table 7.5 Freezing points of ternary eutectic of PETN with some nitro compounds

Composition of ternary eutectic	Freezing points (°C)
PETN (16%), trinitro-m-xylene (8.5%), m-dinitrobenzene (75.5%)	79.5
PETN (12.5%), 1,8-dinitronaphthalene (16.5%), m-dinitrobenzene (71.0%)	69

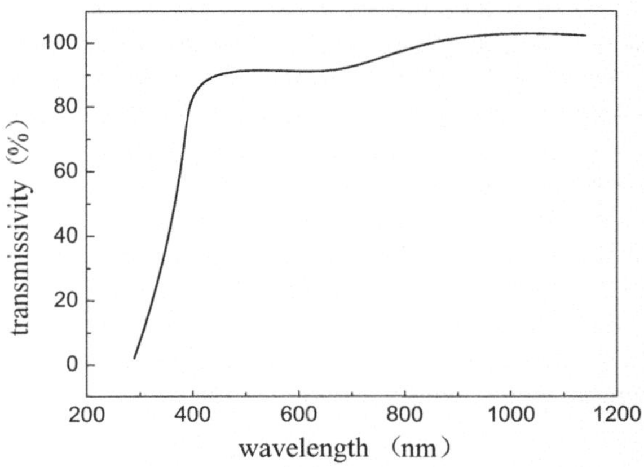

Fig. 7.3 Ultraviolet absorption spectra of PETN single crystal

the decomposition of PETN under at temperatures. The relationship between PETN decomposing fraction and time at different temperatures is shown in Fig. 7.4.

Figure 7.4 shows that PETN has decomposed slightly at 60 °C, but with very long induction period. With the increase of temperature, its decomposition rate is increased and the decomposition rate is accelerated with the increase of liquefaction

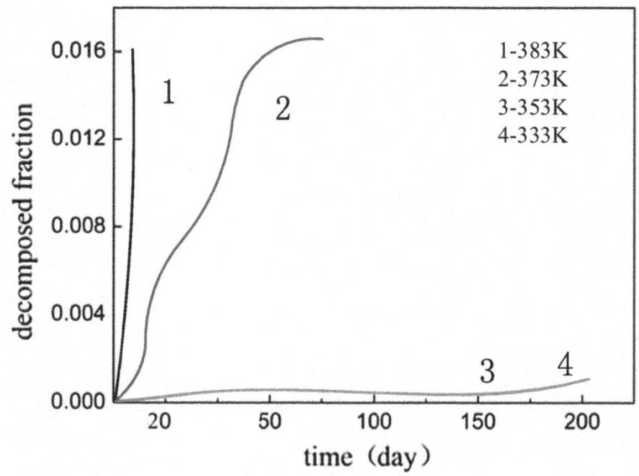

Fig. 7.4 The relationship between PETN decomposing fraction and time [3]

Table 7.6 ΔE and A of the thermal decomposition of PETN

Temperature (°C)	State	ΔE (KJ/mol)	A (s^{-1})
108–120	Solid	212.97	$10^{20.6}$
137–157	Liquid (melting point)	218.82	$10^{23.1}$

degree, thus to significantly increase the decomposition acceleration. The effect of hydrolysis is also observed, in which the generated water is increased with the increase of the acid product from the decomposition.

PETN is decomposed to release nitric oxide at 140–145 °C, to release yellow smoke when it is heated to 175 °C, to decompose violently at 190 °C, and to explode violently at 202–205 °C. In the thermal decomposition rate constant equation, the relationship between A or ΔE and temperature is shown in Table 7.6 [5, 6].

Acetone, acid, and alkali all can accelerate the hydrolysis of PETN, and Fig. 7.5 shows the 2-h hydrolysis results of 10 g of PETN in 500 mL of medium.

As shown in Fig. 7.5, after PETN is heated in water at 100 °C for 2 h, it has almost no weight loss. But, when the temperature rises to 125 °C under pressure, PETN hydrolyzes very rapidly. When acetone is added, 0.1% of nitric acid or sodium hydroxide can increase the hydrolysis rate of PETN, in which sodium hydroxide affects PETN hydrolysis most [6]. Hydrolysis product in alkaline medium is mainly pentaerythritol dinitrate.

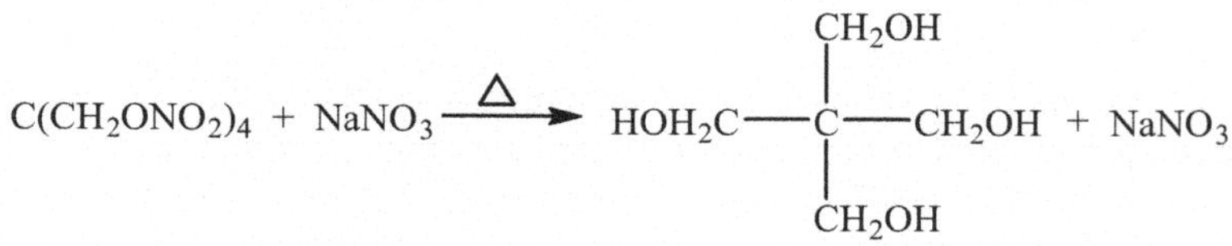

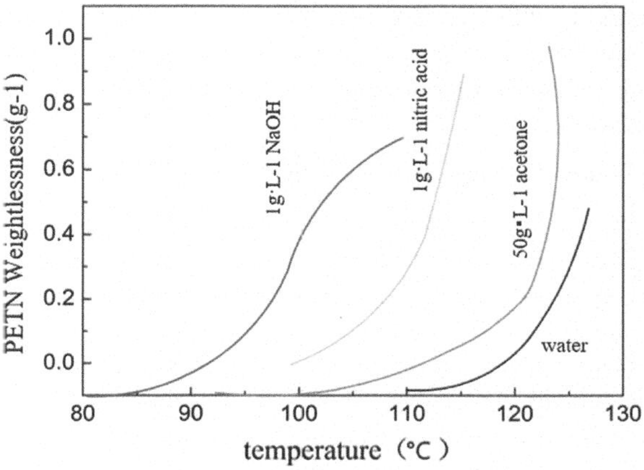

Fig. 7.5 The relationship of PETN weight loss with temperature in different media [3]

Table 7.7 Hydrolysis products of PETN in the presence of hydrochloric acid (%)

Chemical	After 4 h	After 5 h	After 6 h
Pentaerythritol	0	7.7 ± 1.1	31.9 ± 0.7
Pentaerythritol nitrate	0	13.4 ± 1.7	25.6 ± 1.9
Pentaerythritol dinitrate	23.0 ± 1.7	34.7 ± 0.8	27.7 ± 2.1
Pentaerythritol trinitrate	38.9 ± 0.9	27.8 ± 2.1	11.1 ± 1.4
PETN	38.9 ± 0.9	14.8 ± 1.9	5.6 ± 0.1

The hydrolysis results of PETN in dioxocyclohexane solutions (3:1) with 1 N hydrochloric acid as catalyst. And the hydrolysis products formed are listed in Table 7.7.

The refined PETN crystals from aqueous acetone have remnant traces of water, which should be removed through drying because the PETN crystal with water has relatively low stability and is not easy to store. The crystal is crushed and dried to remove moisture and improve the stability of PETN.

7.1.2 Chemical Properties of Pentaerythritol Tetranitrate

Pentaerythritol tetranitrate has symmetrical structure and cannot react to many chemicals. Unlike most nitrates, pentaerythritol tetranitrate neither reduces Fehling's reagent, also form adducts with aromatic nitro compounds. When pentaerythritol tetranitrate is long-term treated in water at 100 °C, there is hydrolysis. When the water temperature rises to 125 °C and pressurized, the hydrolysis is very rapidly. Under this condition, if there is 0.1% nitric acid, the hydrolysis is significantly accelerated. The hydrolysis rate of pentaerythritol tetranitrate with dilute NaOH is faster than that with acid. No matter what the hydrolysis conditions of pentaerythritol tetranitrate, the final product is pentaerythritol dinitrate in all cases.

(1) Oxidization of pentaerythritol tetranitrate

The oxygen balance of pentaerythritol tetranitrate is −10.2%, and the oxygen content is 60.74%/mol; thus, pentaerythritol tetranitrate is nitrate with a negative oxygen balance. When it meets with strong oxidizing agents, such as hypochlorite, reaction can lead to the nitrogen–oxygen bond cleavage quickly on $-ONO_2$ group, thus to cause the explosion and violent reaction. However, perchlorate (AP) can chemically react with PETN. After ammonium perchlorate and PETN are mixed evenly according to a certain ratio, the total energy of PETN can be significantly improved. Due to the addition of ammonium perchlorate, PETN sensitivity has been increased as well.

When pentaerythritol tetranitrate reacts with hydrogen peroxide, the reaction is more violent than that with sodium hypochlorite, and its decomposition releases nitrogen oxide. When a few drops of hydrogen peroxide are added into pentaerythritol tetranitrate, gas bubbles are immediately popped up with the smell of nitrogen oxides. With large amounts of hydrogen peroxide, an explosion can be caused.

$$C(CH_2ONO_2)_4 + 6H_2O_2 \rightarrow 5CO_2 + 4NO + 10H_2O$$

When PETN reacts with excess hydrogen peroxide, the final hydroxide from decomposition is mainly tetravalent nitrogen.

$$C(CH_2ONO_2)_4 + 10H_2O_2 \rightarrow 5CO_2 + 4NO_2 + 14H_2O$$

(2) Reaction of pentaerythritol tetranitrate with reducing agent

Pentaerythritol tetranitrate usually do not react with hydrogen, but its $-ONO_2$ can be converted into $-NH_2$ and H_2O when it is heated under pressure.

$$C(CH_2ONO_2)_4 \xrightarrow[\Delta]{\text{catalyst}, H_2} CH_2O + CH_3CHO + 2(NH_4)_2CO_3 + H_2O$$

PETN is normally relatively stable with metal hydride, but it can be reduced and decomposed with released ammonia in the decomposition products when it is heated [7].

Sodium sulfide can slowly decompose PETN, but this decomposing is far less easy than other nitrates. In laboratories, this method can be sued to destruct a small amount of PETN.

In boiling ferrous chloride solution, PETN decomposes very quickly to release nitrogen oxide. So, this method can be used to determine nitrogen content in nitrate explosives.

(3) Chemical stability of pentaerythritol tetranitrate

The high chemical stability of PETN is better than that of all other nitrates, and its Abel's test can be up to a few hours.

When PETN and TNT or other nitro compounds are mixed, it can also be accelerated to decompose at low temperature. At 110 °C, PETN and TNT mixture begin autocatalytic decomposition after a certain time of induction period, which is easier than pure PETN solid. Among mixture series and PETN with TNT, the mixture with 20% of TNT is the easiest one to decompose.

The stability of PETN is the key to directly relate to whether we can expand its applications. Therefore, how to improve the stability of PETN is important. American army had used polyvinylpyrrolidone with molecular weight of about 90,000 or phenolic resin with polymerization degree of 100 to cover PETN as mainly loaded explosive during a variety of fighting, leading to significant improvement of thermal stability of PETN explosives. At 120 °C, during the stability test under vacuum, the released gas volume within 20 h is about 1/10 of the untreated PETN under the same conditions. After nitrobenzene, dinitro, trinitrobenzene, and other nitroaromatics are added into PETN, its stability is decreased and the PETN mixtures can be heated at 120–135 °C. PETN stability modification results with some materials are listed in Table 7.8.

(4) The toxicity of PETN

PETN has a lower toxicity than nitroglycerine [8] with a very low vapor pressure and very small solubility, and thus no one can be poisoned through inhalation of vapor. Even a small amount of PETN dust particles is inhaled, and it will not cause harmful effects. PETN is a long-term coronary drug to use to treat angina in medicine.

Table 7.8 The stability of PETN modified with some mixtures

Additive	pH of extracted water sample at 120 °C			
	1 h	2 h	3 h	4 h
Pure PETN	7.16	7.15	6.33	5.86
p-Nitrotoluene (9.1%)	5.34	3.50	3.19	1.92
2,4-Dinitrotoluene (9.1%)	6.30	4.20	3.13	2.92
d_4^{20}-Trinitrotoluene (9.1%)	6.54	5.70	3.32	3.03
Phenolic resin (2.0%)	7.12	7.10	7.08	7.07

7.2 The Explosion Properties of Pentaerythritol Tetranitrate

The oxygen balance of PETN is −10.1% with a nitrogen of 17.72%, and its explosion decomposition products are different with different detonation methods. Its explosion decomposition reaction is done according to the following formula:

$$C_5H_8N_4O_{12} \rightarrow CO_2 + CO + 3C + NO_2 + 3NO + 4H_2O$$

Different detonation methods have different PETN explosion reactions in different ways, and the percentage composition of collected gas is listed in Table 7.9.

PETN is hard to ignite. When it is mounted in the pipe with a diameter of 18 mm, it can be ignited only when it is warmed up to 95 °C and its burning is very low. Few of PETN or very thin layer of PETN can be ignited in air and burnt smoothly. When piles of PETN are ignited, especially when it exceeds 1000 g, PETN is burnt on its surface first, and this surface burning turns into detonation quickly. When PETN is ignited or overheated in the closed container, even a little of PETN turns into detonation [9, 10]. The detonation point of PETN is 202–205 °C.

PETN has a high impact sensitivity, and its limit fall of 2 kg drop weight is 17 cm. With 10 kg drop weight, its 25 cm high explosion ratio is 100%. The sensitivity of PETN is higher than that of RDX and HMX. Therefore, when PETN is charged as the main ammunition, it is needed to passivate before using. When passivating reagents are added into the PETN, it is inevitable to decrease the explosion property. Therefore, the main developing field of PETN is to find appropriate passivating reagents of PETN [11].

The friction sensitivity of PETN is relatively high, and its detonation sensitivity is very high. Its limit dose of 0.01 g of lead azide and that of tetryl is 0.025 g under the same conditions. The sand-loaded lead expansion value of PETN is 500 cm^3, and its water-filled lead expansion value is 560 cm^3, indicating that PETN is a good booster primer and can also be used as secondary charged drug and booster primer of detonator. Different from many other explosives, PETN does not lose its explosive power and its explosion rate is also increased after its intercrystal air is removed and PETN is used as nonexplosive liquid replacement. PETN explosive characteristics under different explosion conditions are shown in Table 7.10.

Table 7.9 Gas products of PETN explosive decomposition of (%)

Detonation method	NO_2	NO	N_2O	N_2	CO_2	CO	H_2	O_2
Detonator	–	5.3	–	22.8	37.0	26.7	6.8	1.4
Impact	–	24.3	5.3	9.4	19.1	35.4	6.5	–
Thermal decomposition at 210 °C	12.0	47.6	9.5	1.6	6.3	21.0	2.0	–

Table 7.10 Conditions of PETN explosion rate test under

Explosion rate when the loading density is 1.00 g/cm^{-3} (m/s)		Explosion rate when the loading density is 1.40 g/cm^{-3} (m/s)	
Dry PETN	PETN with 10% water	Dry PETN	PETN with 20 gb glycerin
7295	7445	7125	7825

PETN with 40% of water can still be detonated with 8# detonator, but TNT with 15% of water cannot be detonated, which is very different from that the slurry of nitroglycerine and water that is hard to detonate.

When PETN is heated to above its melting point (205–225 °C), it explodes. At 161 °C or above, PETN is slightly decomposed and the activation energy E = 196.65 J/mol and $\log_{10}B$ = 19.8 at the first stage of decomposition. Typically, the normal decomposition of PETN is

$$C_{15.8}H_{25.3}O_{38.0}N_{12.7} \rightarrow 12.13CO + 3.38CO_2 + 10.35H_2O + 0.1CH_4 + 1.71H_2 + 0.1C_2H_2 + 0.2NH_3 + 0.11HCN + 6.23N_2$$

PETN is sensitive to shock. With 0.02 MPa kinetic energy impact on the surface of PETN, its risk of explosion is up to 50%. The friction sensitivity of PETN is not high although its explosive detonation sensitivity is extremely high. When the wax in PETN is 10%, it is still very easy to detonate. The electric spark sensitivity of PETN is very high. When capacitor discharges between 2000 and 12,000 V, PETN can be detonated.

Ultraviolet rays can detonate PETN. When PETN is strongly radiated for 1.2 ms with ultraviolet (900 J of energy), it rapidly decomposes with the same explosion decomposition under the heat. If the ultraviolet light with 480 J of energy, it can explode within 20 ms. γ rays can also make PETN to decompose.

The flame insensitivity of PETN is low, and it is difficult to ignite with a flame, which is significantly different from the straight-chain nitrate. When seamless steel tubes (Φ18 mm) are loaded with PETN, it cannot be ignited. The only way to ignite PETN is to heat the tube to above 95 °C, which can cause the fire while its burning rate is only 0.047 cm/s. When the pressure is increased, burning with a constant speed can be reached. The relationship between burning rate and pressure is $U = 0.02 + 0.018P$.

In addition to the abovementioned sensitivity of PETN, its detonation sensitivity is also associated with the crystal results and status [12]. Explosion rate of PETN under different loading densities is shown in Table 7.11.

Table 7.11 PETN explosion rate with different loading densities

Loading density of PETN (g/cm^{-3})	Explosion rate (m/s)
0.85	5330
1.50	7600
1.70	8300
1.73	8500

Table 7.12 Other characteristic data of PETN

Item	Parameter
Explosion heat (J/g)	6401.52
Specific volume (ml/g)	768
Explosion temperature (°C)	4230
Brisance (mm)	The compression value of lead column with 25 g of dosage is 14–16

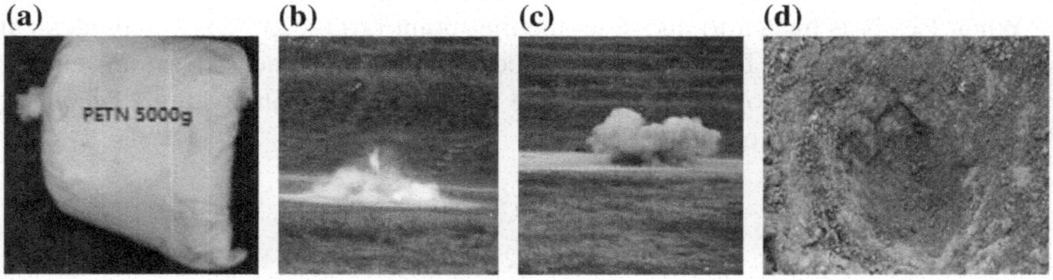

Fig. 7.6 Scene of the explosion of 5000 g PETN

Explosion rate of PETN is high and its critical diameter is small (When the density is 1 g/cm^3, the critical diameter is 3 mm). Thus, explosion rate of detonating wire made of PETN is very small. Even it is bended, the change of explosion rate is within 0.8–1.0%, which is the reason it is widely used in detonating wire. Other characteristics of PETN are shown in Table 7.12.

When PETN is loosely loaded, its explosive power is less than that with loaded PETN with a certain charging pressure. Figure 7.6 shows the PETN explosion scene of 5000 g with a certain bulk density.

In Fig. 7.6a is 5000 g of PETN with random package, b is the light at the moment of an explosion, c is the instant smoke from the explosion, and d is the surface then after the explosion of PETN (5000 g).

The explosion performance of pressed PETN with a certain density is totally different from that of loose PETN. Figure 7.7 is the explosion effect of 1000 g PETN with a density of 1.75 g/cm^3 on the same ground of Fig. 7.6.

In Fig. 7.7a is 1000 g of PETN with a density of 1.75 g/cm^3, b is the instant smoke from the explosion, and c is the dent on the ground from the explosion.

As shown in Figs. 7.6 and 7.7, PETN has different explosion performances with different densities. After the explosion of 5000 g of PETN with regular packing density (ρ = 0.42 g/cm^3), the volume of explosion dent is 0.67 m^3 and the bottom of the dent is pressed to very tight, while the dent of only 1000 g of the latter PETN with a density of 1.75 g/cm^3 is 0.73 m^3. And the dent depth is 276 mm after the removal of loose soil, but the former dent depth is only 167 mm. Through this simple comparison test, it is shown that PETN explosive power is directly related to its loading density.

Fig. 7.7 Scene of the explosion of 1000 g PETN with density of 1.75 g/cm^3

Table 7.13 Sensitivity and power of passivated PETN

Passivating reagents	The dropping limit of 2 kg drop hammer (cm)	Power (mL)	Passivating reagents	The dropping limit of 2 kg drop hammer (cm)	Power (mL)
Unpassivated PETN	30	500	Dinitrotoluene (5%)	–	430
Wax (5%)	35	415	Dinitrotoluene (10%)	57	425
Wax (10%)	60	340	Dinitrotoluene (15%)	–	355

Passivating reagents of PETN include both nonexplosive substances and substances with low explosive power. Explosive properties of passivated PETN are still better than that of TNT and tetryl. Sensitivity and power of passivated PETN are shown in Table 7.13.

7.3 Production Technology of Pentaerythritol Tetranitrate

Whether industrially or in laboratory, pentaerythritol tetranitrate is usually synthesized through the direct nitration of pentaerythritol with nitric acid or the mixed acid from nitric acid and sulfuric acid, and the basic reaction is

$$C(CH_2OH)_4 + 4HNO_3 \rightarrow C(CH_2ONO_2)_4 + 4H_2O$$

According to the difference in the operation and nitrating reagents, there are four methods to prepare pentaerythritol tetranitrate from pentaerythritol, and the four methods are the nitric–sulfuric mixed acid method, nitric acid–sulfuric acid method, sulfuric acid–nitric acid method, and nitric acid method. However, in most industrial manufacturers, PETN is still prepared through the direct nitration of pentaerythritol with concentrated nitric acid, and nitrated product of pentaerythritol and its purity depend on the purity of pentaerythritol. When concentrated nitric acid is used to nitrate pentaerythritol to prepare pentaerythritol tetranitrate with the pentaerythritol purity of 98%, product is relatively easy to precipitate out once the concentration of wasted nitric acid is diluted to around 30% after the nitration. If the purity of pentaerythritol is 95%, the wasted acid is diluted to that with 50–60% of nitric acid concentration after the nitration. If pentaerythritol purity is below 95%, the product is directly filtrated after the nitration. When the concentration of waste acid is higher, nitric acid concentration is higher, and its solubility of impurities and PETN is larger. With a lower concentration of nitric acid after the dilution, the dissolved substances are less, and the yield of crude products is larger, and the wasted acid is relatively easier to deal with. According to the needs of environmental protection and green manufacturing, this section is focused on the preparation process of pentaerythritol tetranitrate with nitric acid method [2, 3, 6, 7, 13–16].

7.3.1 Nitric Acid Methods to Prepare Pentaerythritol Tetranitrate

Environmentally friendly and green manufacturing process to prepare pentaerythritol tetranitrate with nitric acid methods, and the process is shown in Fig. 7.8.

The process of nitric acid method to produce PETN has intermittent and continuous processes. Now, continuous process is used to produce PETN in every country.

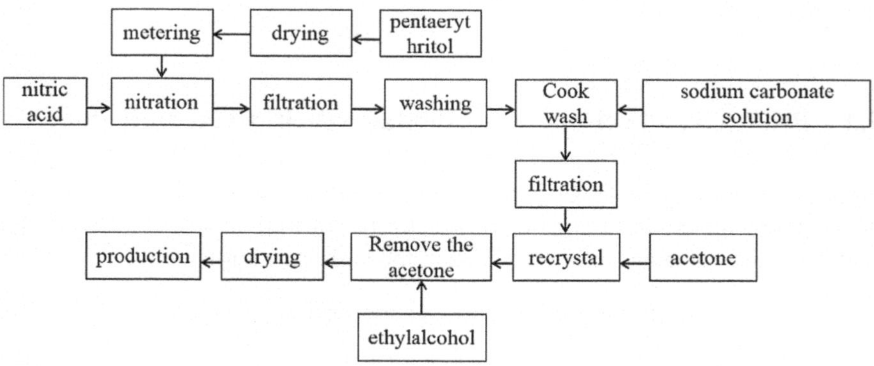

Fig. 7.8 Preparation process of pentaerythritol tetranitrate with nitric acid method

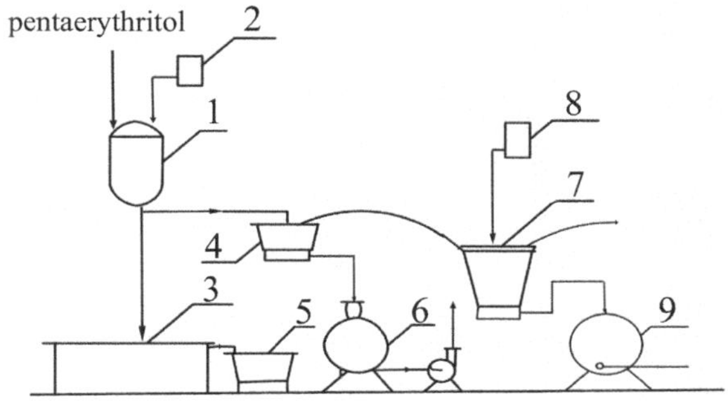

Fig. 7.9 The process of intermittent nitric acid method to produce PETN [6]. 1. Nitration reactor, 2. nitric acid measurement tank, 3. safety pool, 4 and 5. vacuum filters, 6. vacuum collector of wasted acid, 7. washing tank, 8. sink, 9. vacuum collector of washing water

(1) Manufacturing process of intermittent method to produce PETN

Intermittent manufacturing process to directly prepare PETN with nitric acid is a general method with low production instrument investment, easy operation, simple technology, and relatively safe production process. During the production, water in pentaerythritol is removed first, then pentaerythritol is shattered into 200 mesh, and the material is loaded into the reactor. After the nitration, stability treatment is done following the vacuum filtration, and the raw product is recrystallized in acetone to obtain PETN. The process of intermittent nitric acid method to produce PETN is described in Fig. 7.9.

300 kilograms of nitric acid (93–95%) was added into the nitrating reactor. With constant stirring (120 RPM) and cooling, pentaerythritol (60 kg) is gradually added which is dried, crushed, and sieved in advance, in which the nitrator's temperature is kept below 20 °C. The temperature is kept for another 30 min after the chemicals feeding, PETN is crystallized out from wasted acid with a density of 1.43 g/cm^3. The nitrating reagent is added into the vacuum filter 4. The filtering period is 10–15 min to make the acid content in the filtered cake not more than 25%. The wasted acid concentration is 78% and added into the vacuum collector 6.

Acidic PETN is sent to antiacid steel washing tank 7, and the washing tank is equipped with blender and porous ceramic fake bottom (as the filtering board). The tank is filled with water with 6.5 times of PETN volume, PETN is added under the stirring. The stirring is continued for another 15 min after the addition of PETN is finished. Then, acidic water is filtered out with pumping. Water is used to wash again for three times, in which water is 1.5 times of PETN each time, to make acidity of rough products to fall down to around 1%. The obtained PETN is sent to boiling–washing tank with 1% sodium carbonate solution that is 8–10 times of PETN, which is neutralized for 1 h at 85–90 °C. The solution is always kept as alkaline to remove the acid among PETN crystals. The final acidity of PETN crystal is only 0.3–0.5. After the removal of water with vacuum filtration, recrystallization in acetone is done.

During the refining, 2.2 kg acetone per kg PETN is used to dissolve PETN at 58 °C. Carbonate ammonium that is 1 time of calculated volume is added to neutralize the residual acid. After 1 h of stabilization processing, the solution goes

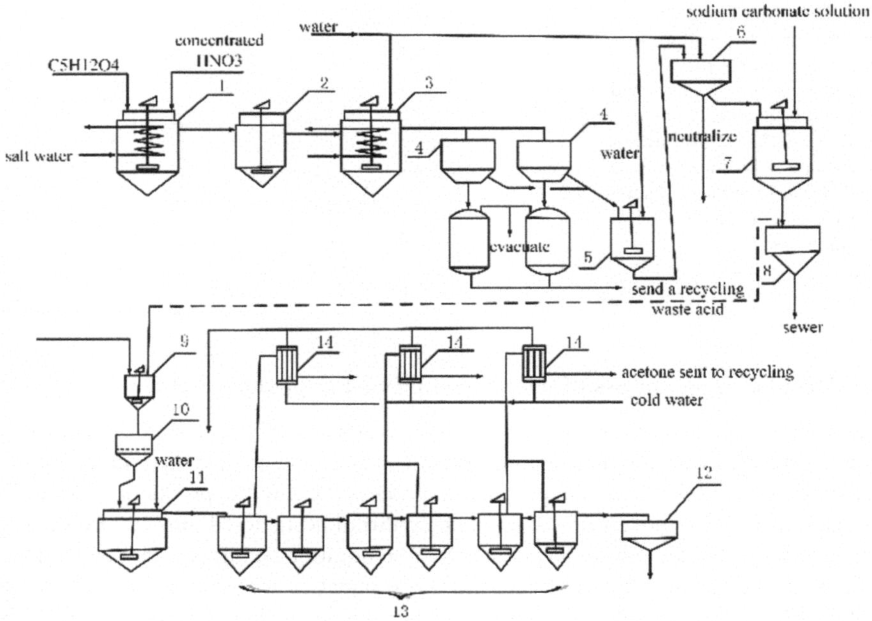

Fig. 7.10 Process flow of continuous production of PETN with nitric acid method [3]. 1. nitrator, 2, 3. diluting tank, 4. filter, 5. washing tank, 6, 8, 10 and 12. filter, 7. filtering tank, 9. dissolution tank, 11. diluting tank, 13. distillation equipment, 14. steam condenser

through the filter that is preheat to 50–55 °C and enter into the crystallization reactor that is preheat to 50 °C to cool down and crystallize slowly, in which the initial cooling rate is 1 °C every 5 min followed by 2–2.5 °C/min. The whole crystallizing process needs 2 h. In this process, the hot acetone solution cannot be added into cold container or cooled down too fast; otherwise, PETN will be crystallized on the wall of crystallizing reactor and pipe to form a hard shell that affects the crystallization.

The purified PETN is separated from solvent acetone in the vacuum filter. After the separation, there is still 15–20% of acetone in PETN, which is then washed with a small amount of ethanol to remove the residual acetone. The PETN containing 15–20% ethanol is dried for 12 h at 40 °C in the drying box. After the discharge of product, PETN is sieved and loaded in cloth bags.

(2) Processes of continuous nitration, dilution crystallization method are shown in Fig. 7.10

This continuous nitration process is started with an intermittent nitration. The concentrated nitric acid (98–99%) is added into the nitrator 1, and pentaerythritol with 1/5 of nitric acid is added through a screwing conveyor within about 40 min. The temperature is maintained at 15–20 °C through cooling.

Then, the continuous nitration is begun. Concentrated nitric acid and pentaerythritol are added continuously according to their ratio, the nitration solution spills into the nitrator 2 and is cooled down to 10 °C, and the nitration solution with the suspension of this PETN crystal with 80% nitric acid flows into dilution tank 3. Water is added to dilute nitric acid concentration to 30% and the temperature is maintained at 15–20 °C. The solution flows into the vacuum filter 4, and these two filters are used alternatively. A filtered-out acid containing PETN is washed into the washer unit.

After the PETN was washed in the washing tank 5, it is sent to the filter 6. After the removal of the acidic washing water, PETN is sent into neutralizing tank 7, in which sodium carbonate at 60 °C is used to wash for 1.5 h. After the filtering through vacuum filters 8, filtrate should be alkaline.

After PETN washed with base, it is refined with continuous crystallization method. In dissolving tank 9, PETN with 10% water is dissolved in acetone at about 45 °C, and a small amount of sodium carbonate (500 L of acetone and 1.5 kg of sodium carbonate per 100 g pentaerythritol) is added. After the temperature is increased to 50 °C, the solution flows through the filter 10 and enters into diluting tank 11. At 60 °C, a certain amount of water is added to dilute the acetone concentration to 50–55%. This acetone with suspension containing PETN grains flows into distiller groups 13 with blenders and heaters, in which acetone is distilled out and the steam is condensed in the steam condenser 14. The temperature of the first distiller is kept at 85 °C, and its concentration of acetone is maintained at 12–15%. The temperature of the second distiller is kept at 90 °C, and its concentration of acetone is maintained at 3–5%. The temperature of distillers 3, 4, and 5 is kept at 100 °C, and only traces of acetone are left. When acetone is distilled out, PETN crystallizes and becomes its final shape, PETN crystals from this method are in large size and uniform, and it is sent to dry after water which is separated through the filter 12.

7.3.2 *Purification of Pentaerythritol Tetranitrate*

Acidic PETN is sent to antiacid steel washing tank, and the washing tank is equipped with blender and porous ceramic fake bottom (as the filtering board). The tank is filled with water with 6.5 times of PETN volume, and PETN is added under the stirring. The stirring is continued for another 15 min after the addition of PETN is finished. The obtained PETN is sent to boiling–washing tank with 1% sodium carbonate solution that is 8–10 times of PETN, which is neutralized for 1 h at 85–90 °C. The solution is always kept as alkaline to remove the acid among PETN crystals. The final acidity of PETN crystal is only 0.3–0.5. After the removal of water with vacuum filtration, recrystallization in acetone is done.

During the refining, 2.2 kg acetone per kg PETN is used to dissolve PETN at 58 °C. Carbonate ammonium that is 1 time of calculated volume is added to neutralize the residual acid. After 1 h of stabilization processing, the solution goes through the filter that is preheated to 50–55 °C and enters into the crystallization reactor that is preheated to 50 °C to cool down and crystallize slowly, in which the initial cooling rate is 1 °C every 5 min followed by 2–2.5 °C/min. The whole crystallizing process needs 2 h. In this process, the hot acetone solution cannot be added into cold container or cooled down too fast; otherwise, PETN will be crystallized on the wall of crystallizing reactor and pipe to form a hard shell that affects the crystallization.

The purified PETN is separated from solvent acetone in the vacuum filter. After the separation, there is still 15–20% of acetone in PETN, which is then washed with a small amount of ethanol to remove the residual acetone. The PETN containing 15–20% ethanol is dried for 12 h at 40 °C in the drying box. After the discharge of product, PETN is sieved and loaded in cloth bags.

After PETN washed with base, it is refined with continuous crystallization method. In the dissolving tank 9, PETN with 10% water is dissolved in acetone at about 45 °C, and a small amount of sodium carbonate (500 L of acetone and 1.5 kg of sodium carbonate per 100 g pentaerythritol) is added. After the temperature is increased to 50 °C, the solution flows through the filter 10 and enters into diluting tank 11. At 60 °C, a certain amount of water is added to dilute the acetone concentration to 50–55%. This acetone with suspension containing PETN grains flows into distiller groups 13 with blenders and heaters, in which acetone is distilled out and the steam is condensed in the steam condenser 14. The temperature of the first distiller is kept at 85 °C, and its concentration of acetone is maintained at 12–15%. The temperature of the second distiller is kept at 90 °C, and its concentration of acetone is maintained at 3–5%. The temperature of distillers 3, 4, and 5 is kept at 100 °C, and only traces of acetone are left. When acetone is distilled out, PETN crystallizes and becomes its final shape, and PETN crystals from this method are in large size and uniform. After PETN is separated with water in filter 1, it is sent to dry.

Now, the simple cooling crystallization method is used generally, whose mother liquid can be reused, but contents of water and PETN in the mother liquor change in a wide range to make the process difficult to control. Thus, diluting crystallization method is usually used. In the dilution crystallization method, solution is diluted to that with acetone content of 55%, cooled to around 30 °C, and filtered. After the dilution, acetone is removed before the filtration.

Product quality requirements are different according to the different applications. PETN made in the USA for various applications is divided into four categories. The first is used as a detonating wire, the second is used as the booster in the ammunition of small arms, the third is used in PETN/TNT, and the fourth is used to make a detonator. The specification of PETN made in the USA is shown in Tables 7.14 and 7.15.

Table 7.14 Indicators of PETN as explosive

Item	Property indicator
Color	White crystal or powder
m.p. (°C)	$\geq$ 140.0
Nitrogen content (%)	$\geq$ 17.50
Compound insoluble in acetone (%)	$\leq$ 0.1
Insoluble particle (sieve number)	No particle left on 40# sieve
Acidity basicity (%)	$\leq$ 0.003
Vacuum stability (120 °C)	$\leq$ 5 mL (20 h)
Particle size	Determined according to customer's requirements

Table 7.15 Particle size of PETN made in USA

USA standard sieve number	Percentage of going through			
	The first group	The second group	The third group	The fourth group
30	–	–	95_{min}	
80	100	–	–	–
100	–	–	–	20_{max}
100	85_{min}	95_{min}	–	5_{min}
140	85_{min}	–	–	–
200	30_{max}	80_{max}	30_{max}	–
200	–	65_{min}	–	–

7.3.3 Passivation of Pentaerythritol Tetranitrate

Passivation of PETN is to cover the refined PETN with a layer of passivating reagent, such as wax oil yellow soluble, wax oil red soluble, to reduce the sensitivity of PETN. In the method, water is first added to round passivating machine equipped with jacket and stirring equipment. After stirring is started, refined PETN is added and heated up to 95 °C above, followed by the addition of 0.2% of SOAP flakes. Then, melted passivating reagents that are heated to 90 °C above are added, and kept for 20–60 min at 95–99 °C before the mixture is cooled down to 80 °C. Then, water is added to the machine to cool down at 45 °C or lower, and filter is added to remove passivated wastewater. After the wastewater removal and filtration, the as-obtained crude PRTN is sent to dry, and the drying temperature is the same as nonpassivated purified PETN, which is 55–70 °C.

7.3.4 *Influence Factors in Nitric Acid Method to Prepare Pentaerythritol Tetranitrate*

(1) Effects of nitric acid concentration and the loading ratio

Production practice of PETN has proved that the nitrating rate of pentaerythritol with nitric acid is very fast. When the nitric acid concentration exceeds 80%, esterification reaction can be carried out completely and the product is PETN. When the nitric acid concentration is 60–80%, violent oxidation reaction will occur. When nitric acid concentration is lower than 60%, PETN cannot be obtained. Therefore, concentrated nitric acid is used in esterification, and the nitric acid concentration should be not less than 80% after the nitration is completed.

When the melting point of pentaerythritol is 249.2 °C, the concentration of nitric acid is 96.71% (N_2O_3 content is 0.87%), and nitration temperature is 19–20 °C, effects of different feeding ratios on the melting point and yield of product are listed in Table 7.16.

As listed in Table 7.16, nitric acid feeding ratio is more than 5; reaction results, products quality, and the yield are relatively good. When the feeding ratio falls down to 4.2:1, nitration reaction is still stable, product yield and its melting point begin to decrease significantly, the nitration smoke is increased, and its oxidation is enhanced. When the feeding ratio continues to decrease, the abovementioned phenomenon is further increased, the yield falls to 90%, the product is light green, white viscous material is found in the waste acid, the product is very unstable and can quickly decompose with large amount of smoke, and its color turns from yellow

Table 7.16 Relationship between material feeding ratio and the yield and quality of PETN in pentaerythritol nitration

Feeding ratio of nitric acid over pentaerythritol	Nitric acid (%)	Product yield (%)	m.p. (°C)	Waste acid component (HNO_3 %)
4.97	159	98.3	136.8–140.7	78.77
5.0	159	97.5	137.1–140.2	80.78
4.52	136	97.8	137.2–140.2	79.43
4.49	134	97.4	138.5–140.2	78.08
4.19	119	97.4	137.1–139.9	76.87
4.18	119	97.7	138.1–140.1	75.06
4.21	119	98.0	137.3–140.8	76.64
3.93	106	93.6	135.5–139.8	71.63
3.93	106	94.8	136.5–139.9	–
3.93	106	93.9	135.9–139.6	–
3.73	95	89.3	135.5–139.9	66.60
3.75	95	89.5	136.1–139.4	67.82
3.71	95	89.2	136.3–139.9	65.34

Table 7.17 NO_2 content in nitrating reagent on the yield of PETN

NO_2 content in nitric acid (%)	PETN yield (%)	NO_2 content in waste acid (%)
0.2	97.2	0.7
1	97.3	2
2	94.2	4
3	88.2	8.8
4	88.2	12.1

to green. Therefore, when feeding ratio is not less than 4.2, in order to ensure smooth production, pentaerythritol:concentrated nitric acid of 1:4.5–6 generally is used. The usually used ratio is 1:5, with about 140–220% of excessive nitric acid and above 78% of nitric acid concentration in waste acid (usually 78–86%).

The concentration of nitric acid used is 98.7%, chlorine dioxide content is 0.2%, there are 50 g of pentaerythritol and 250 g of nitric acid, and feeding temperature is 18–20 °C. Feeding time is 30 min, and the stirring is continued for 30 min at 20 °C after the feeding. When NO_2 content in nitric acid is less than 1%, there is no effect on the yield and stability of product. When NO_2 content in nitric acid is greater than 1%, there is an impact on the yield but without effect on its stability. Table 7.17 lists the effect of nitrogen oxides on the nitration results.

(2) Effect of nitration temperature

When pentaerythritol is nitrated with 97.7% of nitric acid, pentaerythritol melting point at 225–245 °C, feed ratio of 1:5, and results with various nitration temperatures are shown in Table 7.18.

As listed in Table 7.18, at −1 to +1 °C, the yield is the highest and product quality is also relatively good, but the temperature control is strict. When pentaerythritol quality is relatively high, the nitration temperature can be increased accordingly. Experiments show that, when pentaerythritol with melting point of 250 °C is used, nitric acid concentration is 97%, feeding ratio is controlled at 1:6, and nitration temperature is controlled at 25–41 °C, the yield of PETN can still reach to 96–97%, and the melting point of raw PETN is above 138.5 °C. Under the same conditions, when reaction temperature is controlled at 19–20 °C, the yield is 97% and the melting point of raw product is over 137 °C. When the nitration temperature of pentaerythritol is over 20 °C, although its effect on the yield of PETN and the quality is very limited, more smoke will produce at a higher nitration reaction temperature during the nitration. At 20 °C, a certain amount of smoke is

Table 7.18 Effect of temperature in pentaerythritol nitration on the yield of PETN

Temperature (°C)	Yield (%)	PETN m.p. (°C)
−10 to −7	71.6	119–137.2
−1 to +1	98.3	126.5–138.5
12–15	96.1	126–138.5
28–31	95.7	126–138.5

released only after the feeding. In order to control nitration reaction to release smoke to air, nitration temperature of PETN production is controlled within 10–22 °C in most countries.

(3) Effect of pentaerythrit-2-ol content

Pentaerythrit-2-ol is the pentaerythritol impurity that generates hexanitropentaerythrit-2-ol with nitric acid, which affects the thermal stability and explosive performance of PETN. Thus, its content should be controlled strictly, which requires the control on the melting point of pentaerythritol. Production practice proves that, when 5 g of mixture from pentaerythritol with purity of above 99.9% and pentaerythrit-2-ol is nitrated with 97.7% of nitric acid at −1 to 2 °C and nitric acid feeding ratio of 1:9.8, more white stick rubber-like material in the product is generated from mixture with more pentaerythrit-2-ol after the water dilution when the reaction is completed. The white stick resin-like material can be dissolved in concentrated nitric acid. The experiment results are listed in Table 7.19.

With the increase of pentaerythrit-2-ol, the product yield is decreased and the melting point declines. However, some data in literature shows that a small amount of pentaerythrit-2-ol in the raw material (about 2–3%) can improve the crystal size and fluidity of PETN without large effect on the explosion performance. Therefore, less than 5% of pentaerythrit-2-ol hexanitrate in PETN explosive used in industry are allowed generally. The melting point of pentaerythritol raw material is over 240 °C (above 245 °C in most cases). Under the conditions described above, product with melting point at 137–140 °C can be obtained with the crude PETN yield of over 97%. This crude product can meet specifications after the refining.

(4) Acid remover and crystallizing method in the refining

In order to improve product quality and stability, crude PETN is recrystallized in solvent after washing. The used solvent is usually acetone. The PETN is dissolved in hot acetone, then antacids is added to neutralize the acid remained, and the acid removing reagent can be sodium carbonate, ammonium carbonate, sodium bicarbonate, or ammonium bicarbonate.

However, under heating, ammonium carbonate or bicarbonate can be decomposed into ammonia and carbon dioxide, which go in the reflux condenser along with acetone vapor and then cooled down to generate ammonium carbonate to block the pipe, and then to affect the performance of ventilation equipment.

Table 7.19 Effect of pentaerythrit-2-ol content on the yield and mass of raw PETN

Pentaerythrit-2-ol content (%)	Mass of product (g)	m.p. (°C)
0	11.5	129–139
10	11.4	123–138
20	11.1	104–135
30	10.8	95–133

In addition, it can also cause the condensation of acetone as follows:

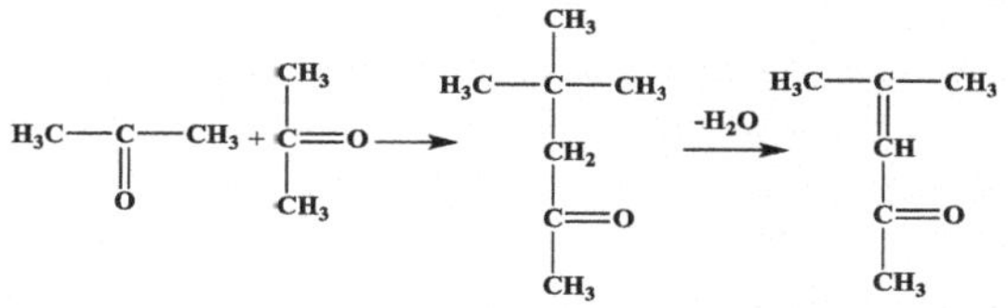

The produced methyl isobutenyl ketone is very active to react with the generated ammonia from decomposition, then to generate a yellow 4-amino-methyl-pentaketone-2.

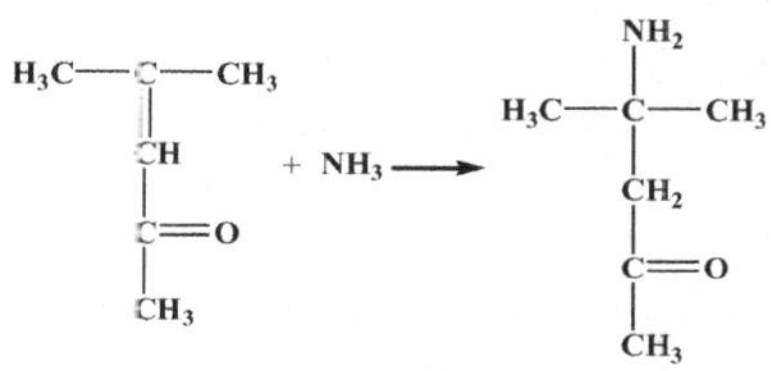

This yellow residue in PETN affects the quality and appearance of PETN. Thus, sodium carbonate or bicarbonate is usually used as an acid remover. In addition, aniline, sodium hydroxide, and ammonium hydroxide are also used as acid removers.

7.3.5 Waste Acid Treatment in the Production of PETN

Composition of waste acid generated from the nitration of pentaerythritol with concentrated nitric acid to produce PETN varies with different technology methods. The components in the isolated waste acid directly after the separation of PETN were 80–82% of HNO_3, 0.5–0.7% of N_2O_3, and 1.0–1.5% of nitrates. After PETN was diluted and crystallized out from nitration solution, in the waste acid nitric acid concentration was 45–60% with less than 0.55% of nitrate and a small quantity of nitric oxide. Its treatment methods were also not the same.

If the concentration of waste acid was high, the concentration of PETN, incomplete nitrates, and condensed pentaerythritol nitrate included was high, and the direct enrichment was not safe. Thus, it was needed to stabilize first, and then enrich. Stabilization treatment was to precipitate waste acid in the aluminum tank for more than 24 h in order to chromium nitrate content was decreased to below 0.75%. Then the waste acid flowed over five continuous towers, whose jackets were flushed with steam to heat these tower to 90–100 °C. The suitable amount of water was added into the third tower to let waste acid decompose large amounts of nitrogen oxides. The strong oxidation causes the complete decomposition of a small amount of nitrate that was not completely decomposed, in the waste acid in the first or second towers. After this treatment, the waste acid concentration was 65–70% almost without nitrate, and it could be sent to concentrate.

Because the diluted acid concentration after the nitration is low and contents of PETN and impurities are low, the treatment of waste acid is relatively simple and safe, in which the waste acid can be concentrated safely after it is only needed to store at room temperature for 3–5 days or is kept at relatively high temperature for a few minutes. The generally used method is to precipitate the waste acid first for a certain period of time. Then, the waste acid was sent into ferrosilicon casing thermal decomposition device to heat to around 90–100 °C, in which its flow is controlled to make waste acid to stay in thermal decomposition device for about 20–30 min before its entering into ferrosilicon casing cooling device. After it is cooled down to 25–30 °C, it flows in the storage tank. Because the ester content in waste acid after this treatment is less than 0.1%, the waste acid can be pumped into nitric acid concentrating workshop to concentrate separately or with diluted nitric acid and the RDX waste acid.

During the thermal stability treatment, waste acid produces nitric acid vapor and nitrogen oxide gases, which are separated through vapor–liquid separator before entering to the sieve plate absorption tower with the nitric acid smoke from nitrating machine, maturing machine, diluting machine, and filter. After the water absorption, diluted nitric acid with more than 40% concentration is produced and then concentrated.

7.3.6 Solvent Recovery

Recycling method is determined according to the refining process of PETN. If it is mother acetone liquor from direct cooling crystallization method, it can be reused after it is mixed with a proper proportion of refresh acetone. According to the reported results, after it can be used for three times, the remained is distilled and recovered. While the mother liquor from direct dilution crystallization method is mixed and reused with the washing water from the purification, the acid is distilled until its concentration is 94–98% to reuse in the PETN purification.

The distillation method in detail is the waste acetone mother liquor, and washing water from purification are mixed together after a certain period of precipitation to obtain a mixture with about 20–40% of acetone, which is preheated to 60–85 °C in the slot preheating tank before entering into the distillation tower whose temperature at the bottom is kept at 95 °C above. The steam enters into the distillation column, whose temperature is kept at 50–57 °C on the top, and the acetone gas distilled out from the tower is condensed into liquid, whose density is determined in order to control its concentration above 94%. If temperature on the top of tower is over the regulated value, acetone density is not qualified, and all acetone should be refluxed until it is qualified to discharge. According to the quality control, the refluxing ratio is controlled accordingly in order to maintain the normal operation.

Acetone from PETN refining processes can be recovered through adsorption, in which the gas goes through the adsorption tower at room temperature, and then is adsorbed with activated carbon at 45 °C, and directly heated to 130 °C with steam after the adsorption is saturated. Then, the obtained desorption liquid is distilled with mother acetone liquid together to concentrate. After desorption, the adsorption tank is dried at about 120 °C with hot air first, then cooled down to 35 °C, and reused in the following.

7.4 Other Preparation Technologies of Pentaerythritol Tetranitrate [17–19]

7.4.1 Nitric–Sulfuric Mixed Acid Process to Produce PETN

Nitric–sulfuric mixed acid process to nitrate alcohol to produce the corresponding nitrate is a common reaction as shown in Fig. 7.11.

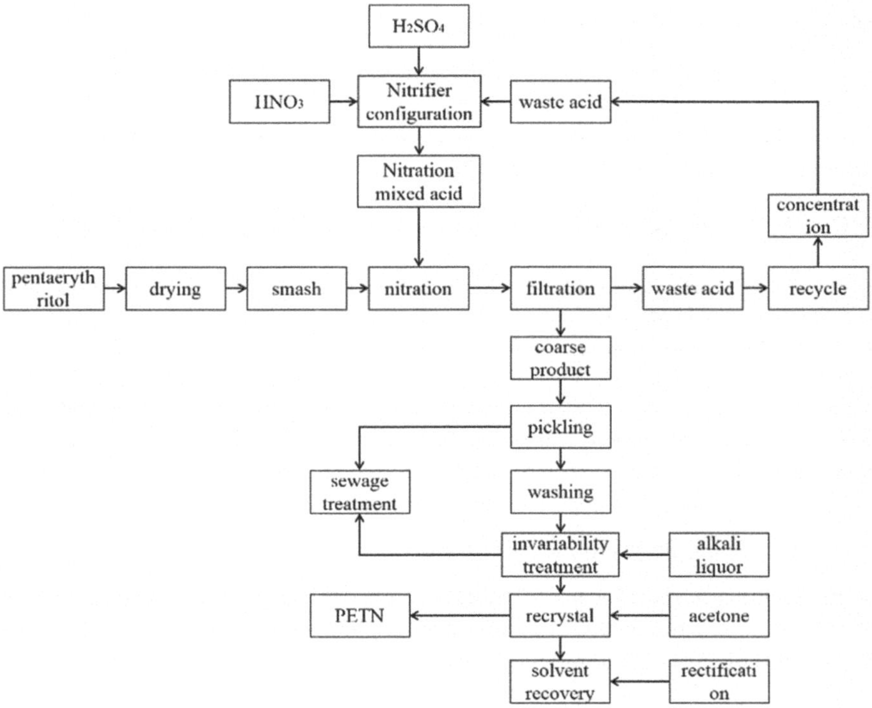

Fig. 7.11 Process of nitric–sulfuric mixed acid method to prepare pentaerythritol tetranitrate

With pentaerythritol as raw materials, it is nitrated with nitric–sulfuric mixed acid and after it is dried and crushed. After the filtration, water washing, and stability treatment, the crude product is obtained. After the recrystallization and purification, the white crystal is pentaerythritol tetranitrate with a final yield of 80%. Preparation of pentaerythritol tetranitrate prepared with this method is not easy to passivate and its stability is poor.

7.4.2 *Other Preparation Technologies to Prepare Pentaerythritol Tetranitrate*

(1) Sulfuric acid precipitation method

Concentrated nitric acid with 6 times of pentaerythritol quality at 55 °C is used to nitrate pentaerythritol, and then concentrated sulfuric acid with 6–7 times of pentaerythritol quality is added to precipitate PETN completely.

(2) Sulfonation nitration method

Pentaerythritol is dissolved into the concentrated sulfuric acid first to sulfonate, where pentaerythritol bisulfonate is formed.

$$C(CH_2OH)_4 + 2H_2SO_4 \rightarrow C(CH_2OSO_3H)_2(CH_2OH)_2 + 2H_2O$$

After the sulfonation, a certain amount of concentrated nitric acid is added to nitrate pentaerythritol to produce PETN.

$$C(CH_2OSO_3H)_2(CH_2OH)_2 + 4HNO_3 \rightarrow C(CH_2ONO_2)_4 + 2H_2SO_4 + 2H_2O$$

Due to the presence of sulfuric acid, the above reaction is conducted in two steps and the reaction heat is released in two times; thus, the nitration can occur at a relatively high temperature (50–60 °C) because the cooling and refrigerating equipment is not required. This reaction is suitable for pentaerythritol with a relatively low quality (such as that with a melting point of 190–195 °C). Particle size of the obtained crude products from this reaction is too thin to filter. And impurities in the crude product are the mixture of nitrate sulfate with poor stability. Therefore, impurities are removed completely during the safety treatment.

(3) nitration with nitric/sulfuric mixed acid

At 30 °C, the mixed acid is used to nitrate pentaerythritol having a melting point of 257–258 °C, and the amount of nitric acid in the mixed acids is 6 times of that pentaerythritol. After the nitration, the waste acid is diluted to that with total acidity

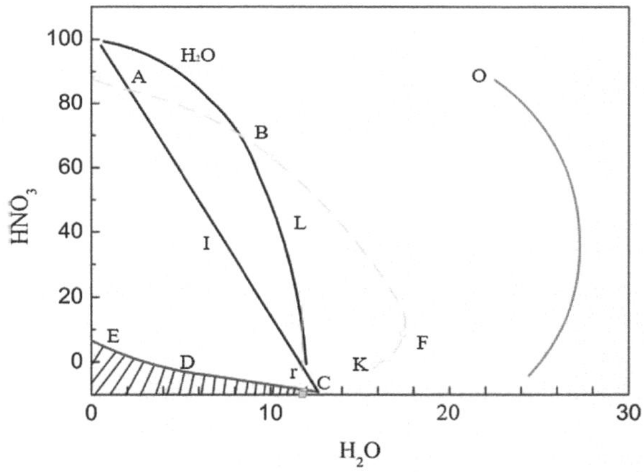

Fig. 7.12 Region with the highest yield of PETN in the mixed acid nitration

(in terms of sulfuric acid) of 20–30%, and the yield of PETN is up to 97.5%. The mixed acid component to get the highest yield is shown in Fig. 7.12.

As shown in Fig. 7.12, mixed acid between BF and OK can generate the highest yield (BF represents the mixed acid component, and OK represents the waste acid component). Curve ALC is the spectrum-measurable limit line of nitryl ion, and shadow under the EDC is the region where all nitric acid is ionized into nitryl ions. In order to get the highest yield, the mixed acid should have a certain amount of water, and this water content is related to the content of nitric acid in the mixed acids. If nitric acid in mixed acids is 10%, the optimum water content is 13–21%. In addition, an appropriate ratio between nitric acid and sulfuric acid should be kept. Otherwise, an ester mixture or sulfate will be produced.

Because six equivalents of nitric acid are needed to nitrate every pentaerythritol and mixed acid consumption is very large, therefore, this method has no practical value.

7.5 Other Methods to Prepare Pentaerythritol Nitrate [17–19]

7.5.1 *Pentaerythritol Hexanitrate*

Pentaerythrit-2-ol hexanitrate is an extremely strong explosive, and many of its chemical properties are similar to those of PETN. And its code is DPEHN.

Chemical formula: $C_{10}H_{16}N_6O_{19}$;
Structural formula:

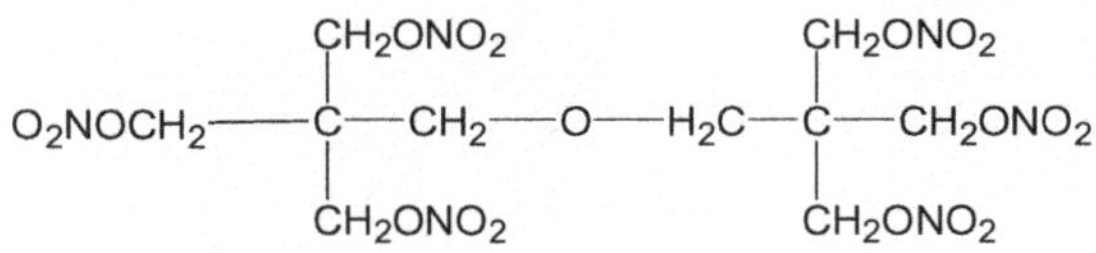

Relative molecular weight: 523.8;
Oxygen balance: −27.49% (calculated according to CO_2);
Nitrogen content: 16.04%; and
Density: 1.63 g/cm^3.
Crystalline and melting point: The pentaerythrit-2-ol hexanitrate is white crystal with two crystal forms. The melting point of its stable form is 75 °C, which belongs to long cubic system with a sheet shape. Its unstable form has a melting point of 72.5 °C.
Solubility: The solubility of pentaerythrit-2-ol hexanitrate in acetone is larger than that of PETN, which can be used to separate these two nitrates. Its solubility in concentrated nitric acid is also larger than that of PETN (Table 7.20), which can be used to control the purity of crude PETN during the manufacture of PETN through controlling the concentration of waste acid.

The explosion performance of pentaerythrit-2-ol hexanitrate is worse than that of PETN, and its stability is also worse than that of PETN, but its mechanical sensitivity is higher than that of PETN. According to explosion properties and stability of PETN, it is harmful impurities of PETN. However, when acetone is used as a solvent to purify PETN, a small amount of pentaerythrit-2-ol hexanitrate can improve the shape and fluidity of PETN crystal, but it shall not exceed 5% in PETN.

Because industrial pentaerythritol contains pentaerythrit-2-ol, pentaerythrit-2-ol hexanitrate is the by-product during the manufacture of PETN. After the nitration of pentaerythritol, the content of pentaerythrit-2-ol hexanitrate in the crude products filtered out depends on the concentration of waste acid. The contents of pentaerythrit-2-ol hexanitrate in PETN with different treatment methods are listed in Table 7.21.

Crude PETN is added into an equal amount of acetone to dissolve. After a certain time, the undissolved PETN was filtered out. In the solution, pentaerythrit-2-ol hexanitrate content is more than that of PETN. Water is slowly added to this solution to separate the solution into two layers, in which the upper is a diluted acetone and the lower is pentaerythrit-2-ol hexanitrate solution with water

Table 7.20 The solubility of DPEHN in nitric acid (25 °C, g/100 g solvent)

Concentration of nitric acid (%)	PETN	DPEHN
97.7	2.17	–
91.8	1.28	–
81.7	0.81	4.70
72.8	0.55	1.60
60.5	0.14	0.72

Table 7.21 DPEHN contents in PETN with different treatment methods

Purity of pentaerythritol (%)	Treatment method	Yield (%)	m.p. (°C)	DPEHN content (%)	The final PETN content (%)
86.3	Nitrating solution is diluted	93.9	117	15.9	84.1
86.3	Nitrating solution is not diluted, and procuct is washed with water	90.5	126–140	10.6	89.4
86.3	Nitrating solution is not diluted, and procuct is washed with 97% nitric acid	80.2	131–138.5	1.0	99

and acetone. After the separation of the two layers, the upper layer is treated with alcohol and the pentaerythrit-2-ol hexanitrate is precipitated out.

Another method is the direct nitration of pentaerythritol with concentrated nitric acid.

Because of the high mechanical sensitivity and relatively strong explosive power of pentaerythrit-2-ol hexanitrate, it can be used as a component of primer or igniter.

7.5.2 Pentaerythrit-3-ol Octanitrate

Pentaerythrit-3-ol octanitrate is a high nitrogen-containing derivative of pentaerythritol and a white crystal with four crystal forms. Its code is TPEON.

(1) Properties of pentaerythrit-3-ol octanitrate

Formula: $C_{15}H_{24}N_8O_{26}$;
Structural formula:

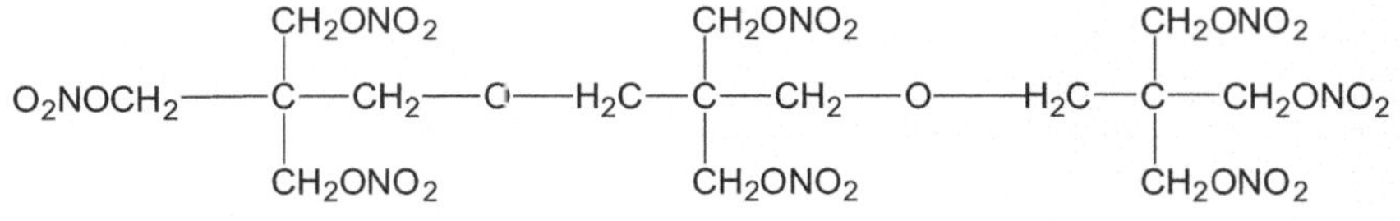

Relatively molecular weight: 732;
Oxygen balance: −34.97 (calculated according to CO_2); and
Nitrogen content: 15.31%.
Melting point: Melting points of TPEON with different crystalline morphologies are shown in Table 7.22.
Solubility: Pentaerythrit-3-ol octanitrate cannot be dissolved in water, but is soluble in ethanol and chloroform, and easily soluble in hot benzene and acetone.

Crystal I is stable with a crystalline density of 1.58 g/cm^3.

Table 7.22 Melting points of TPEON with different crystal types (°C)

I	II	III	IV
83	76–77	68–70	64–65
83.3	74.6	72.1	69.0

Table 7.23 Melting point of DPEHN-TPEON

TPEON (%)	100	95	90	50	20	10	0
DPEHN (%)	0	5	20	50	80	90	100
m.p. (°C)	83	68	73	60 eutectic mixture	63	69	73

The melting point of the mixture of pentaerythrit-2-ol hexanitrate with pentaerythrit-3-ol octanitrate is listed in Table 7.23.

The explosion properties of pentaerythrit-3-ol octanitrate are listed in Table 7.24.

(2) Preparation of pentaerythrit-3-ol octanitrate

With stirring and cooling, 220 g of pentaerythrit-3-ol (its purity is over 99%) is added to 1760 g of 99% nitric acid, in which the amount of nitric acid is 6.3 times of the theoretical amount. The material feeding temperature is −25 to 0 °C, and the material is continuously stirred to react for 1 h at 0–5 °C. The reaction mixture is poured into crushed ice with the volume of no less than 8 times that of the reaction mixture. After left overnight, the product in the water is crushed, then filtered, washed with water, and then stirred with 1% ammonium carbonate with a solution having the amount of 5 times. After the filtration, it is washed with water again until

Table 7.24 Explosion properties of PETN, DPEHN, and TPEON

Property	PETN	DPEHN	TPEON
Oxygen balance (%)	−10	−25	−35
Impact sensitivity, falling height of 2 kg drop hammer (the lower limit) (cm)	17	14	–
Picatinny Arsenal, falling height of 2 kg drop hammer (the lower limit) (cm)	15	10	23
Explosion point (5 s of delay) (°C)	225	255	225
Vacuum stability test (ml g^{-1} 40 h^{-1}) 100 °C 120 °C	0.5 11	3.7 11	2.45 specifically purified product 1.94
Explosion heat (kcal kg^{-1})	1385	1092	1085
Power (lead expansion, TNT is 100)	173	17328	–
Explosion rate (m s^{-1})	(ρ = 1.70) 8300	(ρ = 1.59) 7410	(ρ = 1.56) 7650

the solution is neutral (determined with litmus). The product is dried in air to obtain 422.3 g of pentaerythrit-3-ol octanitrate with a yield of 97.5% and melting point of 73–74 °C, which can be recrystallized with chloroform.

Because the low melting point of pentaerythrit-3-ol octanitrate is low, it can be used as the coating reagent of ammonium nitrate and PETN, or plasticizer of nitrocellulose.

7.5.3 Nitrates of Pentaerythritol Derivatives

(1) Pentaerythritol trinitrate

Its code is Petrin.
Molecular formula: $C_5H_7N_3O_{10}$;
Relatively molecular weight: 271;
Structural formula:

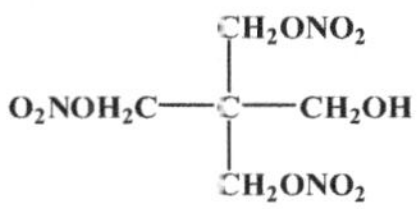

Oxygen balance: −27% (calculated according to CO_2 generated);
Nitrogen content: 15.5%;
Melting point: 26–28 °C (white solid with low melting);
Density: 1.54 g/cm;
Impact sensitivity: lower than that of PETN;
100 °C vacuum stability test: the amount of gas released is 2.54–5.69 ml/g 40 h;
Explosion heat: 5037.54 kJ/g; and
Usage: It can be used as explosive, propellant or low melting point components in gunpowder.

The earliest method to prepare pentaerythritol trinitrate is to add pentaerythritol to 80% nitric acid, followed by dropwise addition of 80% sulfuric acid, in which the temperature is kept at 0–5 °C. After the temperature is kept for 2.5 h, the reaction mixture is poured into cold water, and the product is filtered out that contains PETN. Filter cake is dissolved in acetone and neutralized with added ammonium carbonate, and then water is added to precipitate PETN. Pentaerythritol trinitrate is remained in the filtrate. After the removal of solvent under vacuum, the average yield of trinitrate is 40%, and this is called Method A.

In another method, the reaction mixture is diluted with water after the reaction followed by dichloromethane extraction. The organic phase is separated, and the waste acid is neutralized with sodium bicarbonate solution. After the removal of solvent under vacuum, pentaerythritol trinitrate is directly obtained with a yield of about 50%. And the quality of product is much better than the Method A. This method is known as Method B.

Now, Method C is generally used to manufacture such products. Method C is similar to the above method except that methylene chloride (equal to the sum of reactants) is added before sulfuric acid. After the reaction, solvent is separated, and then dichloromethane is used to extract for one or several times. The extract liquids are merged together to neutralize and then concentrate. Because of the large amount of solvent, PETN in the concentrated solution should be crystallized and filtered out, and finally the solvent is vapored. The yield of pentaerythritol trinitrate is 65–67%.

(2) Trinitroxypentaerythritol acrylate (Petrin Acrylate)

Molecular formula: $C_8H_{11}N_3O_{11}$;
Relatively molecular weight: 324.9;
Structural formula:

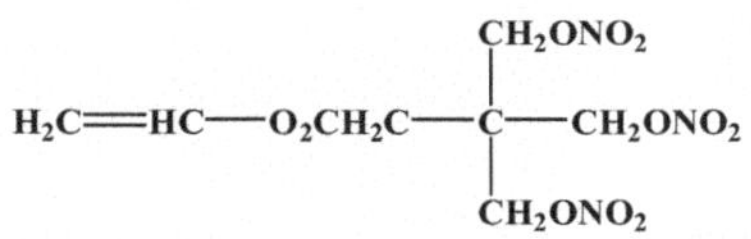

Oxygen balance: −51.7% (calculated according to CO_2 generated);
Nitrogen content: 12.9%;
Density of the white solid: 1.562 (d_4^{20});
Melting point: 78–79 °C;
Formation enthalpy: −52231.75 J/g;
Explosion heat: 2971.5 J/g;
Explosion point: 168 °C with the heating rate of 5 °C/min;
Impact sensitivity: relatively low, its 5000 g hammer falling height is 43 cm;
Thermal stability: 35 min (1.3 g sample) in Abel test at 82 °C with good thermal stability;
Specific impulse: the specific impulse of a single base propellant from it is 214 s;
Burning rate: 5 mm/s. With the addition of oxidants, the specific impulse and the burning rate can be increased; and
Solubility: easily soluble in acetone and ethyl acetate, soluble in ethanol, diethyl ether and benzene, and insoluble in water.

As listed in Table 7.25, when trinitroxypentaerythritol acrylate is used in propellants and the propellant density is 1.3 g/cm^3, its specific impulse is 238 s.

Table 7.25 Trinitroxypentaerythritol acrylate

Component	Component content (%)
Petrin Acrylate (purity > 97%) (adhesive reagent)	34.3
Triethylene glycol dinitrate (plasticizer)	11.8
Ethylene glycol diacrylate (crosslinker)	3.9
Ammonium perchlorate (oxidant)	51.0
Hydroquinone (inhibitor)	0.014

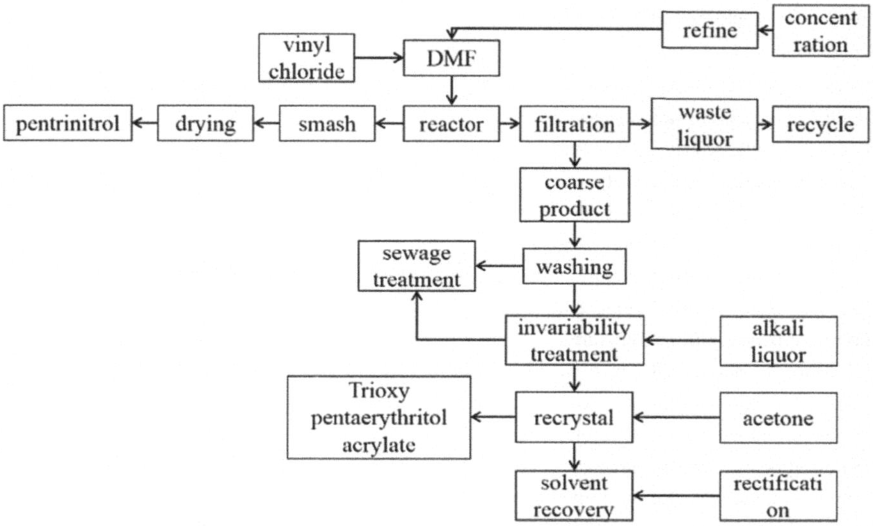

Fig. 7.13 The preparation process of trinitroxypentaerythritol acrylate

Trinitroxypentaerythritol acrylate can be polymerized into the polymer with a specific density of d_4^{20} 1.492 g/cm^3, which is insoluble in all solvents, but can be swollen in acetone and ethyl acetate.

The main reaction to prepare trinitroxypentaerythritol acrylate is

$$HOH_2C-C(CH_2ONO_2)_3+H_2C=CHCOCl+C_6H_5N(CH_3)_2\rightarrow$$

$$H_2C=CHCOOCH_2\cdot C(CH_2ONO_2)_3+C_6H_5N(CH_3)_2\cdot HCl$$

In this reaction, there is white smoke released when vinyl chloride is used. The process to prepare pentaerythritol acrylate from the reaction between pentaerythritol trinitrate and vinyl chloride is shown in Fig. 7.13.

(3) Nitrates of other pentaerythritol derivatives

Other mixed nitrates of pentaerythritol are listed in Table 7.26, and most of them are explosive.

Table 7.26 Pentaerythritol derivatives

Compound	m.p. (°C)
Trinitroxypentaerythritol formate	62–63
Trinitroxypentaerythritol acetate	87–88
Trinitroxypentaerythritol propionate	39–40
Trinitroxypentaerythritol pentanoate	–
Trinitroxypentaerythritol caproate	–
Trinitroxypentaerythritol chloroacetate	52–63
Trinitroxypentaerythritol oxalate	97–100
Trinitroxypentaerythritol pentanedioate	87–88
Trinitroxypentaerythritol butyrate	88–90
Trinitroxypentaerythritol dimethylpentanedioate	71.5–72.0
Trinitroxypentaerythritol oxalate	82–83
Trinitroxypentaerythritol benzoate	89–90
Trinitroxypentaerythritol phthalate	125.0–125.5
Trinitroxypentaerythritol o-nitrobenzoate	94–95
Trinitroxypentaerythritol m-nitrobenzoate	102–103
Trinitroxypentaerythritol p-nitrobenzoate	106–108
Trinitroxypentaerythritol 3,5-dinitrobenzoate	109–110
Trinitroxypentaerythritol 4-chloro-3-nitrobenzoate	112–113
Trinitroxypentaerythritol benzenesulfonate	116–117
Trinitroxypentaerythritol p-methylbenzenesulfonate	97–100
Trinitroxypentaerythritol hydrochloric acid ester	65–66
Trinitroxypentaerythritol hydrobromic acid ester	89–90
Trinitroxypentaerythritol hydrogen iodide ester	108.5–110.5
Dinitroxypentaerythritol hydrochloric acid ester	–
Dinitroxypentaerythritol monohydrobromic acid ester	–
Dinitroxypentaerythritol bihydrobromic acid ester	75
Nitroxypentaerythritol trihydrobromic acid ester	58–60

References

1. Tollens B, Wigang P (1891) Annalen 265:318
2. Liu JP (1987) Basic chemical of propellants and explosives. GuiZhou Press, Guiyang
3. Sun RK, Ren TS et al (1981) Chemistry and technology of high explosive. National Defense Industry Press
4. Faubion BD (1971) Analyt Chem 43:241
5. Gong XD, Xiao HM (1997) The molecular structure and thermolysis mechanism of pentaerythritol tetranitrate. Chin J Org Chem 17:518–519
6. Ou YX (2014) Explosives Beijing Institute of Technology Press
7. Urbanski T (1976) Chemistry and technology of explosives (II). National Defense Industry Press, Beijing

8. Feng YZ et al (1991) Study on toxicity of nitroglycerine. Ind Health Occup Dis 17(4):207–210
9. Li D, Cheng XL, Yang XD (2005) Numerical simulation of detonation parameters for PETN Explosive. J Sichuan Norm Univ (Natural Science) 28(4):448–451
10. Yang L et al (2013) Theoretical study on detonation performances and pyrolysis of pentaerythritol tetranitrate and its analogue compounds. Chin J Energ Mater 21(5):570–577
11. Geng X (2012) Influence of particle size on mechanical sensitivity of PETN. Chin J Explos Propellants 35(4):30–32
12. Liu D et al (2013) Sensititvity criterion mechanical properties predieton of PETN crystals at different temperatures by molecular dynamics simulation. Chin J Energ Mater 21(5):563–569
13. Mitrofanov A et al (2016) Sensitization of PETN to laser radiation by opaque film coating. Combust Flame 172:215–221
14. Murray JS, Lane P, Nieder A et al (2010) Enhanced detonation sensitivities of silicon analogs of PETN: reaction force analysis and the role of r–hole interactions. Theor Chem Acc 127:345–354
15. Zhai M, McKenna GB (2016) Mechanical properties of pentaerythritol tetranitrate(PETN) single crystals from *nano-in*dentation: depth dependent response at the nano meter scale. Cryst Res Technol 51(7):414–427
16. Urbandki T (1984) Chemistry and technology of explosives. Peramon Press, New York
17. Bhattacharia SK et al (2012) Sublimation properties of pentaerythritol tetranitrate single crystals doped with its homologs. Propellants Explos Pyrotech 37:563–568
18. Bhattacharia SK et al (2014) Effect of homolog doping on surface morphology and mass-loss rates from PETN crystals: studies using atomic force microscope and thermogravimetric analysis. Propellants Explos Pyrotech 39:24–29
19. Yan Q et al (2013) The effect of molecular structure on thermal stability, decomposition kinetics and reaction models of nitric esters. Thermochim Acta 566:137–148

Chapter 8
Multivariate Nitrates

Multivariate nitrates mainly include the compounds containing three or more nitrates except for nitroglycerin and pentaerythritol tetranitrate [1]. These compounds have special explosive properties, and thus will be discussed here in detail.

8.1 Nitroisobutylglycerol Trinitrate

Nitroisobutylglycerol trinitrate (Nb) is also called trimethylol nitromethane trinitrate or nitroisobutametriol trinitrate.

8.1.1 Properties of Nitroisobutylglycerol Trinitrate

Nitroisobutylglycerol trinitrate is a pale-yellow viscous oily liquid. The industrial products are dark yellow or with a deeper color because of impurities in feedstock. Some products have abnormal brown or light green color because of different production processes.

Formula: $C_4H_6N_4O_{11}$
Chemical structure:

$$\begin{array}{c} CH_2ONO_2 \\ | \\ O_2N-C-CH_2ONO_2 \\ | \\ CH_2ONO_2 \end{array}$$

Molecular weight: 286.11
Oxygen balance: 0
Nitrogen content: 19.58%
Density: 1.6171 (20 °C)

J. Liu, *Nitrate Esters Chemistry and Technology*,
https://doi.org/10.1007/978-981-13-6647-5_8

Viscosity: 308.55 cP (20 °C)
Freezing point: −39 °C
Refractive index: $n_d^{25} = 1.4896$
Heat of formation: 702.08 J/g (constant volume)/200.83 kJ/mol (constant volume) or 226.35 kJ/mol (constant pressure)
Heat of combustion: 2217.52 kJ/mol (constant volume) or 7765.5 J/g (constant volume).

Vapor pressure and volatility: Less than that of nitroglycerin. Nitroisobutylglycerol trinitrate is slightly volatile without emitting any odor at room temperature. Trace amounts of volatile, tar smell, and spicy taste occur at 30 ° C. Its volatility increases with temperature. Nitroisobutylglycerol trinitrate becomes highly volatile at 50 °C. Its volatility at 25 °C is 0.127 × 10 mg/cm 24 h. Figure 8.1 shows the relationship between the volatility of nitroisobutylglyceril trinitrate and temperature. The vapor pressures of nitroisobutylglyceryl trinitrate at different temperatures are listed in Table 8.1.

Surface tension: Table 8.2 lists the surface tensions of nitroisobutylglycerol trinitrate at different temperatures. As shown in Fig. 8.2, its surface tension linearly decreases with the increase of temperature in the range of 20–60 °C.

Resistance to freezing: The freezing point of nitroisobutylglycerol trinitrate is −35 °C. It is highly viscous at 0 °C, and like the melted glass, can be drawn into a wire. At the freezing point, it becomes a hard wax like solid with similar mechanical properties to that of the liquid phase [1].

Solubility: The solubility of nitroisobutylglycerol trinitrate in water is 0.081% at 20 °C, lower than that of nitroglycerin. It is miscible with many organic solvents including ether, acetone, benzene, ethyl acetate, tetrachloromethane, dichloromethane, nitrobenzene nitromethane, etc. Nitroisobutylglycerol trinitrate has the ability to dissolve aromatic nitro compounds, which can be used for the production of double-base or multi-base propellants. It can dissolve low nitrogenous celluloses

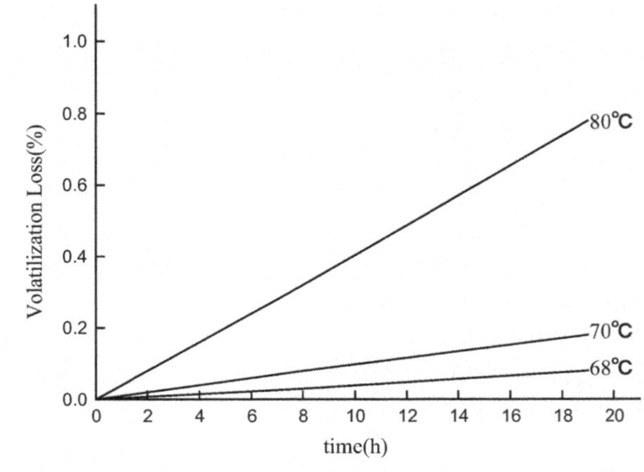

Fig. 8.1 Volatilization losses of nitroisobutylglycerol trinitrate with time at different temperatures [1]

Table 8.1 Vapor pressures of nitroisobutylglycerol trinitrate at different temperatures

Temperature (°C)	40	50	60	70	80
Vapor pressure (mmHg)	2.74×10^{-4}	5.00×10^{-4}	14.8×10^{-4}	33.6×10^{-4}	79.3×10^{-4}

Table 8.2 Surface tensions of nitroisobutylglycerol trinitrate at different temperatures

Temperature (°C)	20	30	40	50	60
Surface tension (dyne/cm)	56.62	55.87	55.13	54.38	53.64

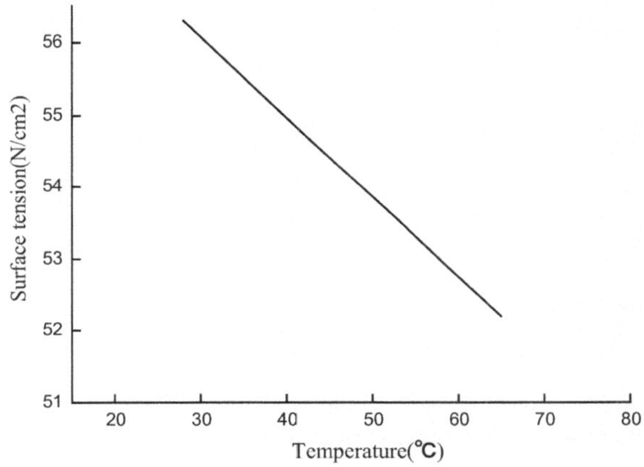

Fig. 8.2 Relationship between the surface tension of nitroisobutylglycerol trinitrate and temperature

(nitrogen content of 11.8–12.2%), which meets the requirements for the production of double-base propellants.

Reactions with acids and bases: Nitroisobutylglycerol trinitrate is soluble in concentrated sulfuric acid where it is dissociated to produce nitric acid. It is sensitive to bases and can be saponified by alkalis, especially alkali metals and alkali-earth metals. For example, nitroisobutylglycerol trinitrate can be broken by NaOH. It is difficult to be hydrolyzed in water or diluted acids, but can be easily hydrolyzed to a yellow solution in diluted bases with the hydrolysis constant of 1.81×10^{-3}.

Thermal decomposition properties: Nitroisobutylglycerol trinitrate is less thermally stable than nitroglycerin. It is slowly decomposed during the storage. Its decomposition can be accelerated by acids, bases, impurities or heating to release nitrogen oxide, leading to self-catalyzed reaction. The low thermal stability of nitroisobutylglycerol trinitrate is attributed to the asymmetric nitro group 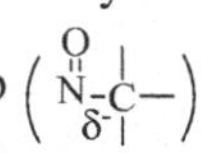 that affects the thermal stability of the ester groups 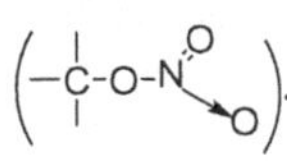. As shown in

Figure 8.3, the thermal decomposition of nitroisobutyl glycerol trinitrate is more obvious than that of nitroglycerin (NG) under the same conditions. The decomposed amount of nitroisobutylglycerol trinitrate at 5–6 h is 4 times of that of nitroglycerin.

8.1.2 Explosive Properties of Nitroisobutylglycerol Trinitrate

Nitroisobutylglycerol trinitrate can completely explode to gases by the reaction shown below:

$$C(CH_2ONO_2)_3NO_2 \rightarrow 4CO_2 + 3H_2O + 2N_2$$

Therefore, the oxygen balance of the explosive reaction is zero.

Heat of explosion: 6924 J/g (as water is gaseous) or 7389 J/g (as water is liquid);

Explosion temperature: 4870 °C (Some literatures report 4627 °C and 4800 K);

Specific volume: 705 ml/g;

Detonation velocity: 7860 m/s;

Brisance compression height: Table 8.3 lists the brisance compression heights of nitroisobutylglycerol trinitrate.

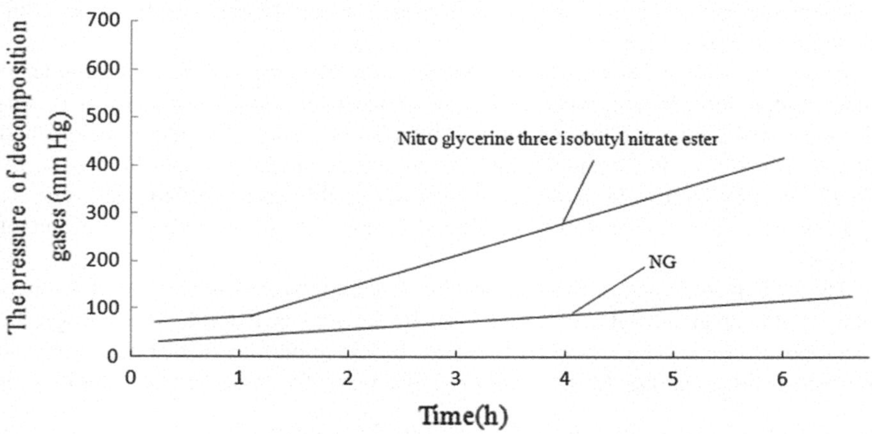

Fig. 8.3 Thermal-decomposition curves of nitroisobutylglycerol trinitrate and nitroglycerin (NG) [1]

Table 8.3 Brisance compression heights of nitroisobutylglycerol trinitrate

Compression height of copper cylinder (mm)	Relative pressure (kg/cm^2)
12.07	2493
11.97	2552
12.64	2156
12.55	2210

Testing conditions: (1) measurement by Caste brisance meter; (2) Material of the sample container: galvanized steel, wall thickness: 1 mm, loading amount: 40 g, initiation: #8 Tetryl detonator; (3) Copper cylinder: 10 × 15 mm, without pre-compaction.

Power: Calculated to be 184 by the Custer equation with TNT as 100.
Heat coefficient h_i: 77.09 J/g.
Explosion coefficient: 54.39.
Specific volume coefficient (Γ): 7.049 ml/g.
Sensitivity: lower explosion limit, 15 cm (2 kg hammer); 100% explosion, 25 cm; friction pendulum test, 6% (that of nitroglycerin is 16%); 5 s delay of the explosion point, 255 °C.

8.1.3 Preparation of Nitroisobutylglycerol Trinitrate

The production process of nitroisobutylglycerol trinitrate is different from those of most nitrates due to the high viscosity of isobutylglycerol [2–5]. Nitroisoglycerol is first converted into nitroisoglycerol sulfate that is then nitrated with mixed acid to afford nitroisobutylglycerol trinitrate as shown below.

$$O_2N\text{-}\underset{CH_2OH}{\overset{CH_2OH}{C}}\text{-}CH_2OH + 3H_2SO_4 \rightleftharpoons O_2N\text{-}\underset{CH_2OHSO_3}{\overset{CH_2OHSO_3}{C}}\text{-}CH_2OHSO_3 + 3H_2O$$

$$O_2N\text{-}\underset{CH_2OHSO_3}{\overset{CH_2OHSO_3}{C}}\text{-}CH_2OHSO_3 + 3HNO_3 \xrightleftharpoons{H_2SO_4} O_2N\text{-}\underset{CH_2ONO_2}{\overset{CH_2ONO_2}{C}}\text{-}CH_2ONO_2 + 3H_2SO_3 + 3H_2O$$

Both of the esterification reactions are reversible, and thus many factors, such as concentration and amounts of nitric acid and sulfuric acid, reaction temperature and so on, can affect the purity and yield of the final product.

Table 8.4 Specifications to be met by nitroisobutylglycerol [1]

Specification	Value
Appearance	White or light yellow crystal, no visible impurities
Purity (%)	>96
Melting point (°C)	>130
Moisture (%)	<1.0
Ash (%)	<0.5

Table 8.5 Specifications to be met by Na_2SO_3

Specification	Value	
	First class	Second class
Na_2SO_3 (%)	≥ 96	≥ 93
Iron (%)	≤ 0.02	≤ 0.02
Water insoluble (%)	≤ 0.03	≤ 0.03
Free alkali (%, Na_2SO_3-based)	≤ 0.6	≤ 1.0

(1) Production of nitroisobutylglycerol trinitrate

The qualities of raw materials can significantly affect the quality of nitroisobutylglycerol trinitrate. Tables 8.4 and 8.5 list the specifications to be met by nitroisobutylglycerol and Na_2SO_3, respectively, for the production of nitroisobutylglycerol trinitrate.

The preparation process of nitro isobutylglycerol trinitrate is shown in Fig. 8.4.

For the production process shown above, the sulfuric acid concentration and amount should not only ensure the raw material is converted into sulfuric acid ester,

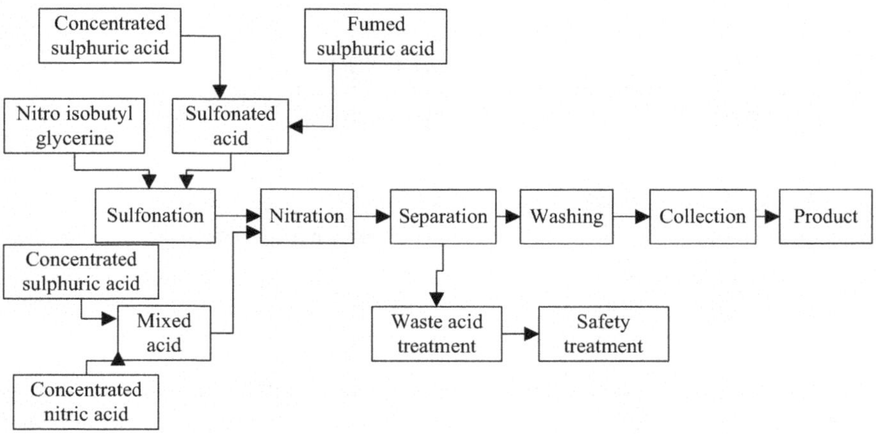

Fig. 8.4 Preparation process of nitroisobutylglycerol trinitrate

but also not excessive to produce mixed sulfate and nitrate mixed esters during the esterification with mixed acid. The nitric acid concentration should be high enough to produce desired amounts of NO_2 that can completely convert sulfate to nitrate with minimized side reactions, such as oxidation, resinification, and so on. Reaction temperature can affect the quality and yield of product, production safety, etc. Low reaction temperatures reduce the reaction rate of sulfonation of nitroisobutylglycerol, while high temperatures can cause dehydration condensation. The sulfonation of nitroisobutylglycerol is usually conducted at 45–50 °C. For the container air-agitated nitration process, the optimum nitration temperature is 44 ± 1 °C and the reaction time is 10–15 min.

Nitroisobutylglycerol trinitrate is very sensitive to alkalis, which makes its stability treatment very difficult [4]. Nitroisobutylglycerol trinitrate is usually washed with mixtures of Na_2CO_3 and Na_2SO_3 for the stability treatment, which is very effective and can reduce the saponification reaction. It has been reported that saponification and semisolid impurities in the product are closely related to final pH of the washing solution. The washing solution of pH < 7pH indicates no saponification occurred and the washing solution exhibits a milky white or pale-yellow color. The washing solution of pH > 9 suggests saponification occurred and the washing solution shows turbid and dark yellow color with flocs floating on top. The key to triggering the saponification is the concentration of [OH]. Both Na_2CO_3 and $N_{a2}SO_3$ are strong alkali and weak acid salts, and thus can be used to well control and buffer the [OH] concentration. Nitroisobutylglycerol trinitrate is hydrolyzed in water as shown below.

$$Na_2CO_3 + 2H_2O = 2Na^+ + 2OH^- + H_2CO_3$$
$$Na_2SO_3 + 2H_2O = 2Na^+ + 2OH^- + H_2SO_3$$

Because $Na_2CO_3 > Na_2SO_3$, H_2S0_3 produced from the hydrolysis of Na_2SO_3 can further dissociate as shown below.

$$H_2SO_3 = 2H^+ + SO_3^{2-}$$

The dissociation constant of H_2SO_3 is greater than that of H_2CO_3, and thus its ionization is much fast than that of H_2CO_3. The repeated production of H^+ consumes OH^- in the solution to form water, which reduces the alkalinity of the solution and the production of [OH], and thus minimizes saponification.

(2) Production process and control conditions of nitroisobutylglycerol trinitrate

① Sulfonation: H_2SO_4 concentration, 100 ± 0.5%; $NO_2C(CH_2OH)_3$: H_2SO_4 = 1:2.75; Sulfonation temperature, 45 ± 5–55 ± 5 °C.
② Composition of mixed acid: HNO_3 80 ± 1.5%, H_2SO_4 20 ± 1.5%, H_2O 2%, and N_2O < 0.4%.
③ Ratio of sulfuric ester to mixed acid: 3.75: 2.75 (mass ratio).

Table 8.6 Specifications to be met by nitroisobutylglycerol trinitrate

Parameter	Value
Appearance	Colorless or light yellow transparent oily liquid
Abel test at 72 °C	>25 min
Alkalinity (Na_2CO_3-based)	<0.01%
Acidity	None
Moisture	N.A.

④ Nitration temperature: 40 ± 1–46 ± 1 °C; stop temperature 45–49 °C; discharge temperature 51 °C.
⑤ Separation temperature: 35 ± 6 °C; stay time for gravity separation: 35 ± 5 min.
⑥ Washing and stability treatment:

Prewash: Oil: Water = 1:8 (v/v, 20 ± 5 °C); stirring pressure, 0.1–0.2 kg/cm.
Alkali wash: 2% Na_2CO_3 and 6% Na_2SO_3; oil: mixed alkali = 1:2.8–3.8; first scrubber pH 9–10; second scrubber pH = 11–12.

⑦ Vacuum water removal: Water in the nitroisobutylglycerol trinitrate product should be removed before storage. Vacuum water removal is usually conducted at 40 ± 1 °C under vacuum of 0.08 MPa until the moisture content becomes less than 0.2%.

Based on the reaction formula, the theoretical yield is calculated to be 189%. The actual yield is 163.5%, 86.5% of the theoretical yield. Table 8.6 lists the specifications to be met by nitroisobutylglycerol trinitrate.

8.1.4 Application of Nitroisobutylglycerol Trinitrate

Nitroisobutylglycerol trinitrate has been applied to gelignite, which provides better properties than the gelignite made with nitroglycerin. It is suitable for the low-temperature applications, especially the production of liquid explosives and powder explosives due to its high freezing resistance. Its properties, except for the shock sensitive, are better than those of other ammonium nitrate explosives. Nitroisobutylglycerol trinitrate has broad application spectrum in high plastic explosives, plastic explosives, gelignites including mixed explosives, powder explosives, gun propellants, and rocket propellant [4, 5].

Compared with nitroglycerin, nitroisobutylglycerol trinitrate has a lower freezing point, which can improve the low-temperature properties and resistance to freezing of propellants and explosives [6]. Its low sensitivity is conducive to the production and application safety [7]. In addition, the low volatility of nitroisobutylglycerol trinitrate makes it less irritating. However, nitroisobutylglycerol trinitrate is less stable than nitroglycerin, and usually fails the Abel's test. It has 1.5% smaller specific volume, 190 °C higher detonation temperature, and 8–9 times higher viscosity than nitroglycerin, and thus is unsuitable for the preparation of propellants.

8.1.5 Toxicity

Nitroisobutylglycerol trinitrate is a toxic substance, but less toxic than nitroglycerin. Its vapor can cause irradiation to the human body. The contact with liquid nitroisobutylglycerol trinitrate can cause swollen erythema and headache. However, the symptoms are lighter than those caused by nitroglycerin. The prevention and treatment of its poisoning are same as those of nitroglycerin.

8.2 1,2,4-Butanetriol Trinitrate

1,2,4-butanetriol trinitrate (BTTN) is also called butanetriol trinitrate, known as nitrutone trioxide. It has excellent plasticity and has been widely used to prepare double-base propellants or, mixed with nitroglycerin, to prepare high-energy propellants. It can also be directly used as a liquid propellant or liquid explosive [1, 8, 9].

8.2.1 Properties of 1,2,4-Butanetriol Trinitrate

1,2,4-butanetriol trinitrate is a yellow clear liquid.

Formula: $C_4H_7N_3O_9$
Chemical structure:

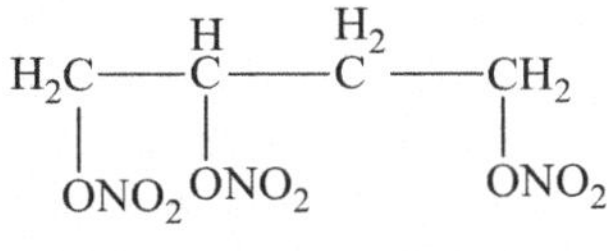

Molecular weight: 241.11
Oxygen balance: −13.3% (Calculated based on the amount of produced CO_2)
Nitrogen content: 17.43%
Specific gravity: 1.52
Viscosity: 2.2 times of that of nitroglycerin
Freezing point: −2.7 °C

Vapor pressure and volatility: ~1/2 of those of nitroglycerin
Hygroscopicity: 0.04% at the humidity of 65%
Solubility in water: 0.08 g/100 g at 20 °C and 0.15 g/100 g at 60 °C

Solubility in solvents: miscible with ether, alcohol, ketone and other organic solvents; similar ability to dissolve nitrocellulose to that of nitroglycerin.

Butanetriol trinitrate has two isomers, 1,2,4-butanetriol trinitrate and 1,2,3-butanetriol trinitrate.

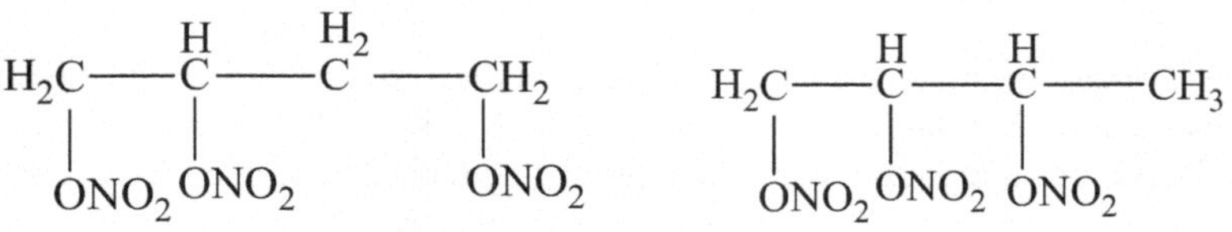

The detonation heat of 1,2,4-butanetriol trinitrate is 6025 J/g (constant volume, water is gaseous) and 5941 J/g (constant volume, water is liquid). Its specific volume is 840 ml/g (water is liquid) and its denotation of 5 s relay is 230 °C. The impact sensitivity of 1,2,4-butanetriol trinitrate is 58 cm (2 kg hammer). Butanetriol trinitrate is a chemically stable.

8.2.2 *Preparation of 1,2,4-Butanetriol Trinitrate*

1,2,4-butanetriol trinitrate can be prepared via the nitration of 1,2,4-butanetriol that can be synthesized from acetylene or obtained via the hydrogenation, polymerization and hydrolysis of ethylene as shown below [4, 8].

$$CH \equiv CH + 2H_2O \xrightarrow{Cu_2C_2} HOH_2C{-}C \equiv C{-}CH_2OH$$

$$H_3COH_2C{-}C \equiv C{-}CH_2OH \xrightarrow[H_2SO_4 + HgSO_4]{+H_2O} HOCH_2{-}C{-}CH_2OH$$

$$HOCH_2{-}C{-}CH_2OH \xrightarrow{H_2} HOCH_2{-}CH_2O{-}CH_2{-}CH_2OH$$

As can be seen from its molecular structure, 1,2,4-butanetriol contains two α-hydroxy and one β-hydroxy. The primary hydroxyl is a strong electron-donating group and thus high negativity, which is favorable to nitration. The secondary hydroxyl (γ-hydroxyl) has a larger steric hindrance, and thus its nitration is slow. The nitration of 1,2,4-butanetriol is

$$HOCH_2{-}CH_2O{-}CH_2{-}CH_2OH + 3HNO_3 \rightarrow O_2NO{-}H_2C{-}CH{-}CH_2{-}CH_2ONO_2 + 3H_2O$$

The nitration, no matter fast or slow, undergoes the following routes.

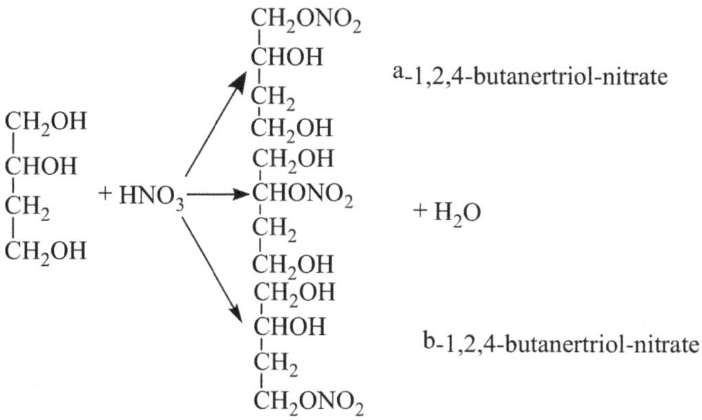

The nitration of different hydroxyl groups produces isoforms that are further nitrated to dinitrates.

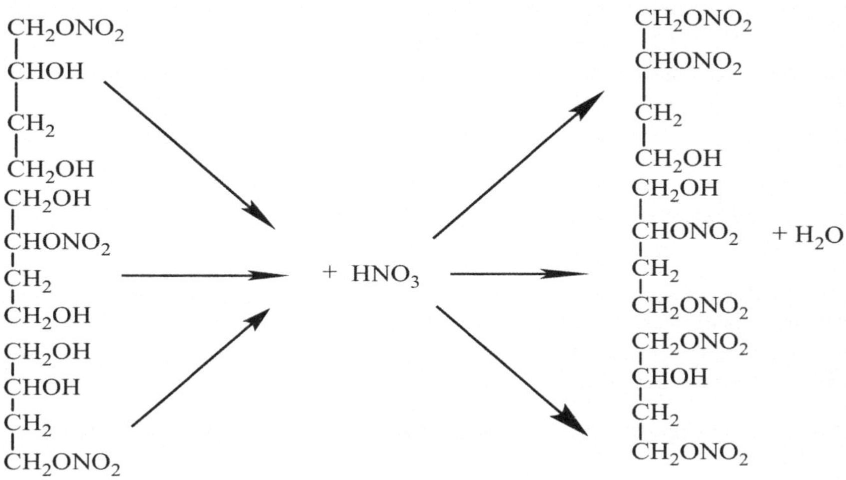

The dinitrates react with nitric acid to afford the final product, 1,2,4-butanetriol trinitrate.

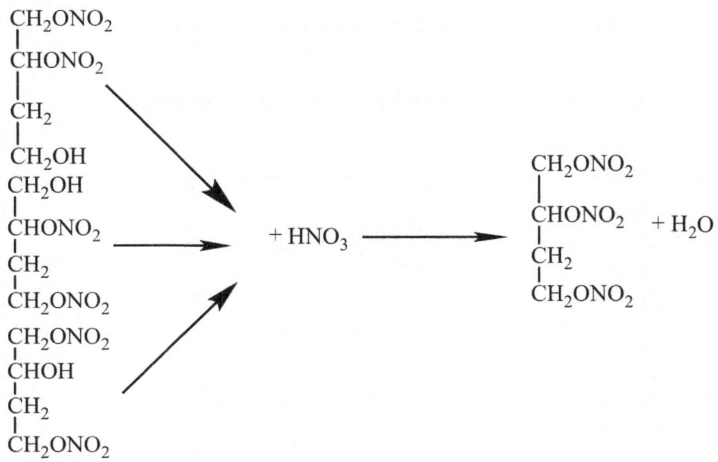

The reaction between primary hydroxyl and nitric acid is a fast zero-order reaction and that between secondary hydroxyl and nitric acid is a slower first-order reaction. Since the nitration is a reversible process, 1,2,4-butanetriol cannot be 100% converted to 1,2,4-butanetriol trinitrate. There is about 0.5–1.5% of unreacted 1,2,4-butanetriol remained in the spent acid. The theoretical yield is then calculated to be 227.36%. However, in the practical production, for example, the China Aerospace factory production line imported from Germany as an example can achieve 216.5% yield. The author optimized synthesis method and control conditions of the production process and was able to achieve 225.5% yield in lab scale and 224.86% yield at 100 kg production level using the injection nitration process.

8.2.3 Production Process of 1,2,4-Butanetriol Trinitrate

The production process of 1,2,4-butanetriol trinitrate mainly includes nitration, separation, washing, and stability treatment. The raw material preparation, product treatment, and spent acid treatment are also imported procedures. Figure 8.5 shows the basic production process of 1,2,4-butanetriol trinitrate.

(1) Nitration mechanism of 1,2,4-butanetriol using nitric acid as the nitrating reagent

Large amounts of experimental results show that the ONO_2 group of nitric acid reacts with the hydroxyl group of 1,2,4-butanetriol to yield water and nitrate. The reaction is an electrophilic substitution reaction where the positively charged proton of nitric acid acts as the attacking reagent (i.e., the nitrating group in nitric acid). Many researchers have found, based on the characterizations with conductivity,

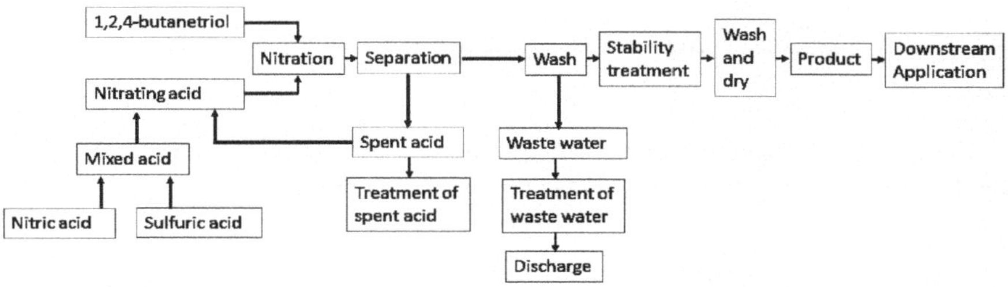

Fig. 8.5 Flow chart of the production process of 1,2,4-butanetriol trinitrate

UV-Vis spectroscopy, and Raman spectroscopy, that nitric acid dissociates and forms different protons at different concentrations. Anhydrous nitric acid mainly exists as nitrate at −40 °C, e.g., HO—N(=O)→O and a small amount is ionized to nitryl cation NO_2^+ and nitrate ion NO_3^- as shown below.

$$2HNO_3 \rightleftharpoons H_2NO_3^+ + NO_3^-$$
$$H_2NO_3^+ \rightleftharpoons NO_2^+ + H_2O$$

The overall dissociation is

$$2HNO_3 \rightleftharpoons NO_2^+ + NO_3^- + H_2O$$
$$96.6\% \quad 1.2\% \quad 1.7\% \quad 0.5\%$$

The electron diffraction of nitric acid vapor and Raman spectrum of anhydrous nitric acid suggest that the conjugate effect of the three-atom unit —N(O)(O) in nitric acid leads to uniformly distributed electron cloud. Therefore, this atomic group is very stable and the N= O bond is difficult to be broken. The electron cloud can transfer between the OH radical and NO_2 radical in the nitric acid molecule $\overset{\delta+}{\text{HO-N}}$(O)(O). Therefore, −OH cannot easily dissociate to a negative form. In contrast, H can easily dissociate to form H^+. Therefore, $-NO_2$ exhibits a high polarity and the oxygen can easily carry negative electron.

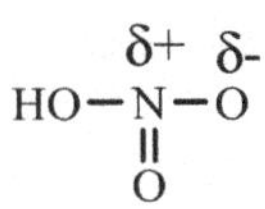

In the anhydrous nitric acid, the hydrogen of one nitric acid molecule dissociates as H^+ and the $-NO_2$ in another molecule accepts the H^+ in the first step as shown below.

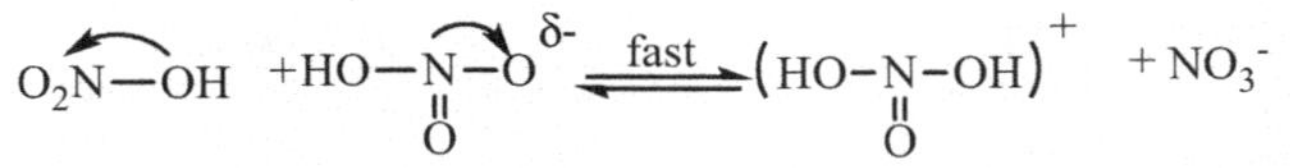

The nitrogen connecting two −OH groups is an unstable structure, and is then stabilized by forming water and nitronium as shown below.

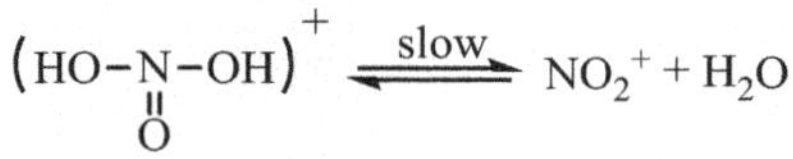

As diluted, nitric acid dissociates as

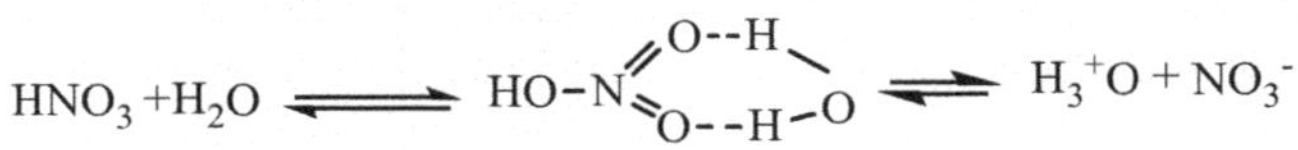

The nitryl cation completely disappears at the nitric acid concentration of 94–95% and the concentration of $H_2NO_3^+$ becomes very low as the nitric acid concentration decreased to 85%. Almost 95% of nitric acid is in the form of NO_3^- at the concentration of 48%. Therefore, it can be concluded that the presence of positively charged protons in nitric acid is strongly related to the concentration of nitric acid.

Nitric acid can form different ions at different concentrations. Among all of positively charged nitrate protons, nitryl cation has the strongest electrophilicity, followed by the hydrogen-containing ion ($H_2NO_3^+$). The ester forms of nitric acid can be polarized as

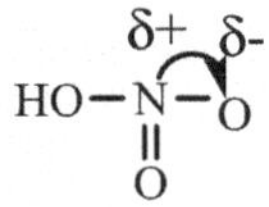

They are thus electrophilic and can act as the attacking reagent of the nitration reactions. The reaction mechanism of the nitration can be expressed as

$$ROH + NO_2{}^+ \rightleftharpoons RONO_2 + H^+$$

$$ROH + H_2NO_3 \rightleftharpoons RONO_2 + H^+ + H_2O$$

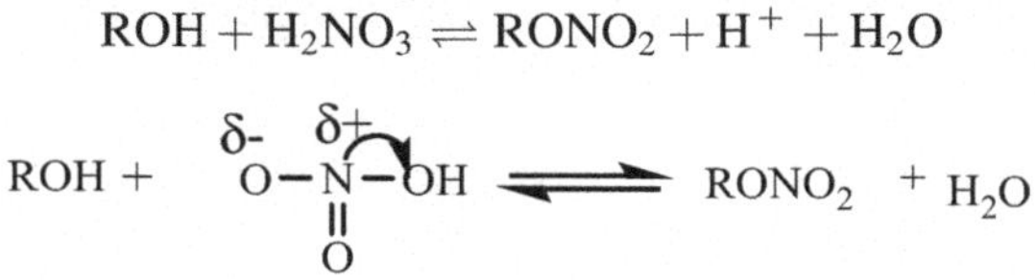

When nitric acid is used as the nitrating reagent alone, the large amounts of NO_2^+ and $H_2NO_3^+$ in the concentrated nitric acid have strong ability to attack the electrophile for the nitration, which promotes the nitration. However, nitration continuously produces water that dilutes the nitric acid, and gradually reduces the concentrations of NO_2^+ and $H_2NO_3^+$. Eventually, the ions disappear. Therefore, the nitrating capacity of nitric acid gradually weakens and eventually disappears. Experiments suggest that the nitrating capacity of nitric acid is completely lost as the nitric acid concentration reduced to 77% where no nitration, but oxidation of glycerol occurs. Therefore, large amounts of concentrated nitric acid are needed to maintain desired nitric acid concentration for a smooth reaction, which not only increases the amount of nitric acid, but also causes high dissolution losses of nitrate in the spent acid and lowers the yield of nitrate. Therefore, the industrial nitration usually uses the mixture of nitric acid and sulfuric acid. The mixed acid has stronger nitrating capacity to accelerate the reaction, promote conversion, reduce side reactions, and decrease the amount of nitric acid needed. Therefore, mixed acid is a better nitrating reagent of 1,2,4-butanetriol.

(2) Nitration mechanism of 1,2,4-butanetriol with mixed acid as the nitrating reagent

Concentrated sulfuric acid (100%) mainly exists in the molecule form, and only ~0.094% of it can dissociate as shown below.

$$2H_2SO_4 \rightleftharpoons H_3{}^+SO_4 + HSO_4 \Leftrightarrow SO_3 + H_3O^+ + HSO_4{}^-$$

The following reactions can occur in the presence of water.

$$H2\ SO4+H2O \rightleftharpoons H_2SO4 \cdot H_2O \rightleftharpoons H_3{}^+O+HSO_4{}^-$$

$$H_2SO4 \cdot H_2O \xrightarrow{H_2O} H_2SO4 \cdot H_2O$$

Nitric acid exists as nitryl cation in the mixture of nitric acid and sulfuric acid. Nitric acid and sulfuric acid forms nitro-sulfuric acid, also known as mixed acid anhydride of nitric acid and sulfuric acid or nitrosylsulfuric acid.

$$H_2SO_4 + HNO_3 \rightarrow HO{-}SO_2{-}ONO_2 + H_2O$$

Nitro-sulfuric acid can dissociate to form nitryl cation (NO_2^+) and hydrogen sulfate anion (HSO_4^-). Sulfuric acid reacts with water to generate hydrogen ion (H_3O^+) and hydrogen sulfate ion. Therefore, when nitric acid and sulfuric acid coexist,

$$HNO_3 + 2H_2SO_4 \rightleftharpoons NO_2 + + H_3O^+ + 2HSO_4{}^-$$

Table 8.7 Relationship between the composition of nitro-sulfur mixed acid and conversion rate to NO_2^+

HNO_3 content (%)	5	10	15	20	40	60	80	90	100
Conversion rate of HNO_3 to NO_2^+ (%)	100	100	80	62.5	28.8	16.7	9.8	5.9	1

The following reaction will occur as the nitrosylsulfuric acid is diluted with water.

$$H_2SO_4 + H_2O \rightleftharpoons H_3O^+ + HSO_4^-$$

The equilibrium reaction to generate nitryl cation then moves to left, and the concentration of NO_2^+ is reduced. Increasing sulfuric acid content or reducing water content can move the reaction to right and increase NO_2^+ content. Most nitric acid or all nitric acid is converted to NO_2^+ with excessive amounts of sulfuric acid or at low water content. For example, 0.2 mol nitric acid in 98–100% sulfuric acid can be completely converted to NO_2^+. In contrast, only 14.7% nitric acid is converted to NO_2^+ in 87% sulfuric acid under same conditions.

For anhydrous mixed acid of nitric and sulfuric acids, the conversion rate of nitric acid to NO_2^+ gradually decreases with the increase of nitric acid content. According to the intensity of the Raman spectral line at 1400 cm^{-1}, the conversion rate of nitric acid to NO_2^+ in the nitro-sulfuric acid has been deduced as shown in Table 8.7. As can be seen from Table 8.7, most of nitric acid is converted to NO_2^+ when there is enough concentrated sulfuric acid. The conversion rate of pure nitric acid to NO_2^+ is very low. It is then can be concluded that sulfuric acid acts as both a dehydrating reagent and catalyst to improve the nitrating capacity of nitric acid.

In the mixed acid, there are two possible formation routes of NO_2^+ via the addition of H^+ from sulfuric acid to HNO_3.

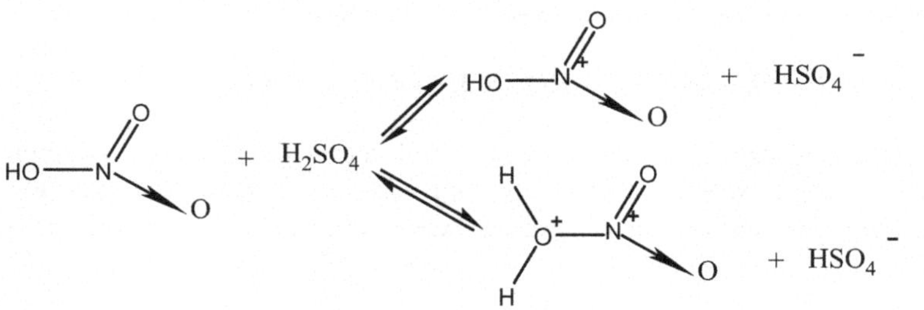

(I)

(II)

Route (I) results in two hydroxyl groups on one atom, which is unstable and subjects to further reaction.

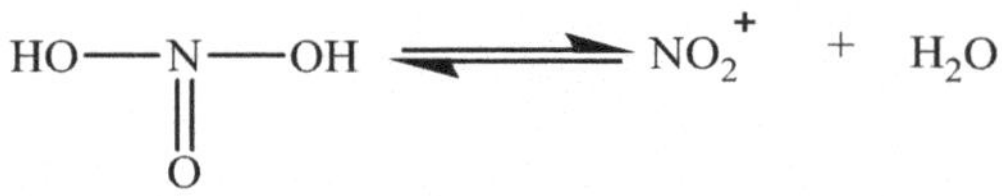

The produced water further reacts with sulfuric acid to form H_3O^+ and HSO_4^-.

$$H^{\delta+}HSO_4 + H_2O \rightleftharpoons H_3O^+ + HSO_4^-$$

The product of route (II) further reacts with sulfuric acid as shown below:

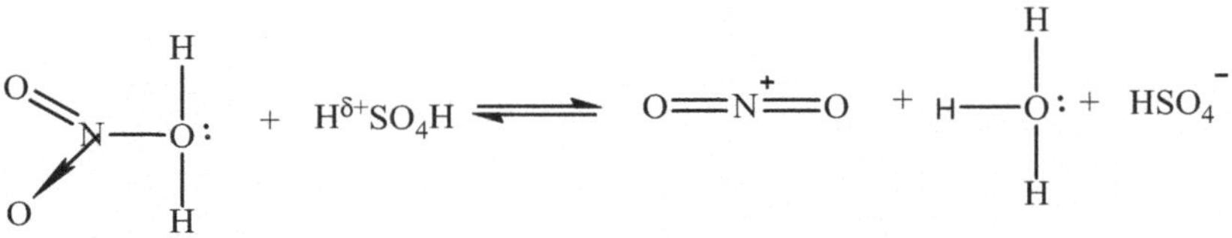

Therefore, both addition reaction (I) and (II) confirm that the presence of sulfuric acid is conducive to the formation of NO_2^+. As can be seen, nitric acid can be dissociated to NO_2^+, either used alone or used with sulfuric acid, as long as the water content is low enough, but the later results in more NO_2^+ [10].

In the ternary mixture of nitric acid, sulfuric acid, and water, NO_2^+ content decreases with the increase of water content, leading to reduced nitrating capacity of the mixed acid. However, sulfuric acid can combine with water to maintain an anhydrous environment for nitric acid. Therefore, even if there is a certain amount of water, NO_2^+ can still be observed and the mixed acid still has sufficient nitrating capacity. NO_2^+ disappears as the water content exceeds 11% in the ternary mixed acid containing 70% nitric acid or 13.5% in the ternary mixed acid containing 30% nitric acid. The ability of this ternary mixture to produce NO_2^+ is equivalent to that of 94–95% of nitric acid alone. Both of them have similar nitrating capacities for 1,2,4-butanetriol because sulfuric acid can remove the excess water.

(3) Production process of 1,2,4-butanetriol trinitrate

Figure 8.6 shows a typical production process of 1,2,4-butanetriol trinitrate.

The production process of 1,2,4-butanetriol trinitrate by the injection nitration method consists of four parts.

1. The composition of mixed acid shown in Table 8.8.
2. The process control conditions are shown in Table 8.9.
3. Separation conditions
   Separation temperature: 16 ± 1 °C
   Rotation speed of centrifuge: 2900 r/min.
4. Washing conditions
   Cold water wash: oil: water 1:1.2–1.5 (Temperature, ~30 °C)
   Warm water wash: oil: water 1:0.7–1.2 (Temperature, 40 ± 5 °C)

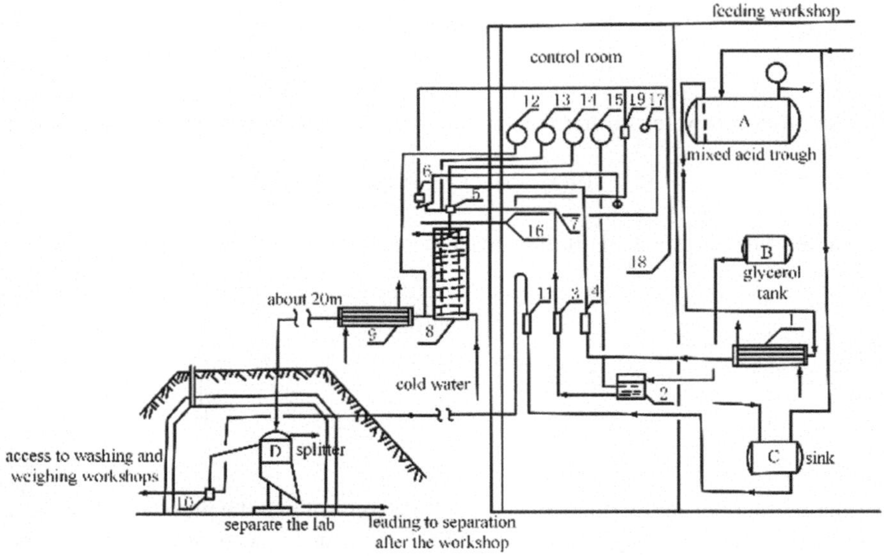

Fig. 8.6 Production and separation processes of 1,2,4-butanetriol trinitrate. 1,9-Chiller; 2-Low temperature tank of glycerin; 3,4-Rotameter; 5-Injection nitrator; 6-Electromagnetic switch; 7-Needle valve; 8-Hydrocooler; 10-Nitrate-H_2O emulsion sprayer; 11-Flow meter; 12,14,15,16-Thermometer; 13-Pressure gauge; 17-Alert; 18-Power switch; 19-Relay

Table 8.8 Composition of mixed acid for injection nitration

Raw material	Content (%)
HNO_3	48–50
H_2SO_4	50–52
H_2O	<0.2
N_2O_4	<0.6
SO_3	≯2.0
Ash	<0.1
Other impurities	<0.1

Table 8.9 Control conditions of nitration process

Parameter	Specification
Nitration coefficient (n)	12–15
Nitration temperature (°C)	45–48
Vacuum (mmHg, nitration)	280–350
Temperature of mixed acid (°C)	0 ± 1
Mixed acid pressure at injection inlet (kg/cm^2)	2.6–3.2
Nitrogen content of mixed acid (%)	29.5 ± 0.5
Nitration velocity (kg/min)	3–4
Nitrogen content of spent acid (%)	14.0 ± 0.3

Stability treatment: oil: water 1:0.7–1.2 (Temperature 40–45 °C, alkaline concentration 0.80 ± 0.1%)
Stirring pressure in scrubber: 2 ± 0.2 kg/cm^2.

8.2.4 *Stability Treatment of 1,2,4-Butyltriol Trinitrate*

The separated 1,2,4-butanetriol trinitrate contains about 7–11% nitration spent acid composed of 70–80% nitric acid, 0.4–0.7% of sulfuric acid and a small amount of water, in addition to the incomplete nitration product of 1,2,4-butanetriol dinitrate. These impurities will affect the stability of 1,2,4-butanetriol trinitrate, posing a risk to the production and use, and thus must be removed. For the 1,2,4-butanetriol trinitrate stability treatment, the acid must be quickly removed first, while minimizing the product loss. Therefore, the stability treatment of 1,2,4-butanetriol trinitrate inevitably involves the treatment efficiency and dissolution loss.

The stability treatment of 1,2,4-butanetriol trinitrate is started with water wash to remove acid, followed by Na_2CO_3 wash to neutralize the residual acids. The ester phase and the aqueous phase contacts each other during washing, resulting in a new equilibrium of the acid concentration between the acidic 1,2,4-butanetriol trinitrate phase and aqueous phase. The acid rapidly diffuses from the 1,2,4-butanetriol trinitrate phase to the aqueous phase due to the concentration difference until the equilibrium is reached.

(1) Solubility of 1,2,4-butanetriol trinitrate and 1,2,4-butanetriol dinitrate in neutral, acidic and alkaline aqueous phase

1. Solubility of, 2,4-butanetriol trinitrate in neutral aqueous phase

The two isomers of dinitrates have same solubility in water. Table 8.10 list the solubility of 1,2,4-butanetriol trinitrate and two isomers of dinitrate in water at different temperatures. As can be seen, the solubility of 1,2,4-butanetriol dinitrate in water is almost 150 times at 0 °C and 100 times at 60 °C higher than that of 1,2,4-butanetriol trinitrate. Therefore, the dinitrate can be easily removed by water wash.

2. Solubility of 1,2,4-butanetriol trinitrate in diluted nitric acid

Table 8.11 shows the solubility of 1,2,4-butanetriol trinitrate in dilute nitric acid. As can be seen, the solubility of 1,2,4-butanetriol trinitrate increases with the increase of nitric acid content in the aqueous phase. Therefore, high acid concentration should be avoided for the actual production process to reduce nitrate loss. No hydrolysis or denitration occurs in the nitric acid concentration range of 0–24% in 960 min. However, the solubility is dramatically increased as the nitric acid content increased to 35% after 960 min wash, which is due to the formation 1,2,4-butanetriol dinitrate from the hydrolysis reaction. About ~18% 1,2,4-butanetriol trinitrate in the washing solution is hydrolyzed and denitrated in 2880 min, corresponding to the

Table 8.10 Solubility of 1,2,4-butanetriol trinitrate and two isomers of 1,2,4-butanetriol dinitrate in water

Temperature (°C)	Solubility (g) in 100 g water	
	1,2,4-butanetriol trinitrate	1,2,4-butanetriol dinitrate
0	0.04	5.90
10	0.05	7.65
20	0.08	8.50
30	0.09	8.90
40	0.11	9.50
50	0.13	10.60
60	0.15	15.11

Table 8.11 Solubility of 1,2,4-butanetriol trinitrate in dilute nitric acid

Nitric acid (%)	1,2,4-butanetriol trinitrate content in aqueous phase (%)			
	60 min wash	180 min wash	480 min wash	960 min wash
0	0.05	0.06	0.08	0.08
4.99	0.09	0.11	0.12	0.12
9.98	0.16	0.18	0.19	0.19
14.99	0.27	0.28	0.29	0.29
19.93	0.37	0.39	0.39	0.39
24.66	0.59	0.60	0.60	0.60
29.85	0.87	0.90	0.91	0.91
34.76	1.02	1.06	1.07	1.08

hydrolysis rate of 0.38% per hour. These data indicate that the hydrolysis of 1,2,4-butanetriol trinitrate in 35% nitric acid is mile at the beginning, and thus does not affect the solubility results. However, the hydrolysis can become very vigorous in a few days.

Figure 8.7 shows the relationship between the solubility of 1,2,4-butanetriol trinitrate and the concentration of nitric acid. It is clear that the solubility of 1,2,4-butanetriol trinitrate increases with the increase of nitric acid concentration.

The solubility of 1,2,4-butanetriol dinitrate in dilute nitric acid solution is much higher than in neutral water (Table 8.12). Therefore, in the actual production, the first wash can remove most of the impurities. In addition, the amount of 1,2,4-butanetriol dinitrate is dramatically increased at the nitric acid concentration of 35%, indicating the occurrence of the hydrolysis and denitration of 1,2,4-butanetriol trinitrate. The hydrolysis reaction may occur simultaneously in the aqueous and ester phases. No significant hydrolysis of 1,2,4-butanetriol trinitrate occurs at nitric acid concentrations lower than 20%.

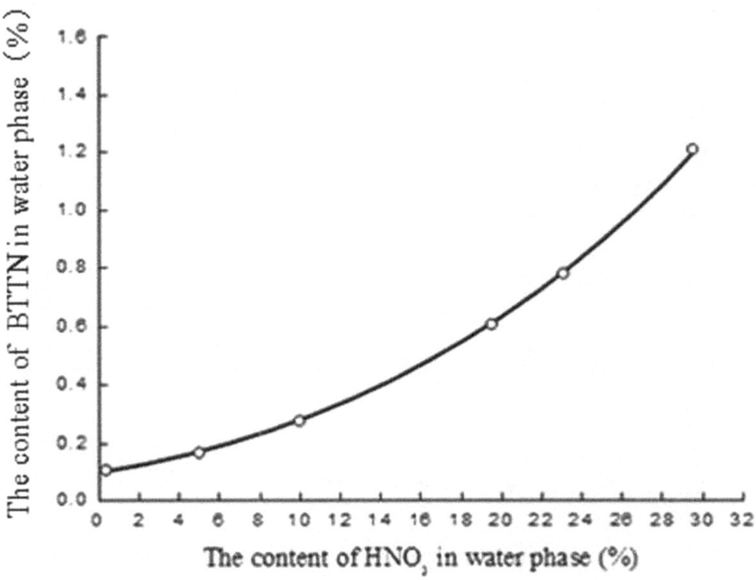

Fig. 8.7 Solubility of 1,2,4-butanetriol trinitrate in aqueous nitric acid solutions

Table 8.12 Solubility of 1,2,4-butanetriol dinitrate in diluted nitric acid (20–22 °C)

Nitric acid in aqueous phase (%)	1,2,4-butanetriol dinitrate in aqueous phase (%)		
	60 min wash	120 min wash	180 min wash
0.00	6.0	6.1	6.1
5.00	10.2	11.1	11.3
9.96	12.1	13.2	13.4
19.92	17.9	20.6	20.9
29.88	26.6	30.3	30.5
35.02	28.3	35.0	35.2

Figure 8.8 shows the solubility of the dinitrate and trinitrate in the aqueous phase as a function of the nitric acid content. As can be seen, the solubility of both nitrates increases with the increase of nitric acid content, and the solubility of 1,2,4-butanetirol dinitrate increases more rapidly than that of 1,2,4-butanetirol trinitrate.

3. Effects of trace sulfuric acid in dilute nitric acid

Acidic 1,2,4-butanetriol trinitrate product also contains trace level sulfuric acid, which can be transferred to the aqueous phase during washing. Table 8.13 lists the experimental data of the effects of the sulfuric acid on the solubility of

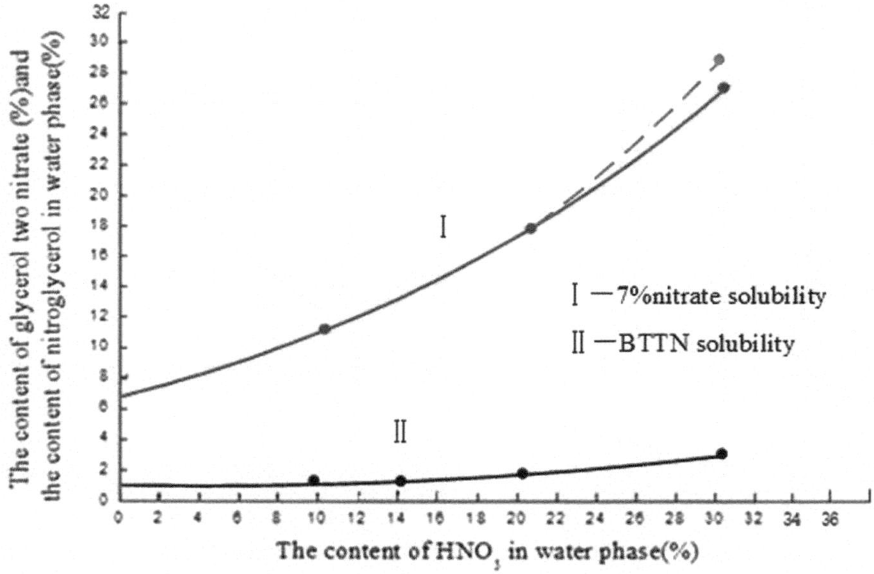

Fig. 8.8 Solubility of 1,2,4-butanetirol trinitrate and dinitrate in nitric acid aqueous solutions

Table 8.13 Effects of sulfuric acid on the solubility of 1,2,4-butanetriol dinitrate

Nitric acid (%)	Sulfuric acid (%)	1,2,4-butanetriol dinitrate (%)	Temperature (°C)
3.50	1.50	7.21	20
8.50	2.92	11.4	
8.72	0.68	12.7	
8.90	0.00	11.8	
8.72	0.68	12.5	10
8.90	0.00	11.7	
10.00	0.00	10.5	
10.00	2.00	11.9	

1,2,4-butanetriol dinitrate. As can be seen, the trace amounts of sulfuric acid have no significant effect on the solubility of 1,2,4-butanetriol dinitrate.

4. Solubility in sodium carbonate solutions

Tables 8.14, 8.15 and Fig. 8.9 show the solubility of 1,2,4-butanetriol trinitrate and 1,2,4-butanetriol dinitrate in aqueous sodium carbonate solutions at different concentrations.

As can be seen, the solubility of 1,2,4-butanetriol trinitrate decreases with the increase of sodium carbonate content in the aqueous phase. It is worth noting that

Table 8.14 Solubility of 1,2,4-butanetriol trinitrate in aqueous sodium carbonate solutions at 20–22 °C

Na_2CO_3 content (%)	1,2,4-butanetriol trinitrate in aqueous phase (%)		1,2,4-butanetriol trinitrate content in ethyl ether after 3 h wash
	60 min wash	180 min wash	
0.50	0.055	0.060	0.090
1.00	0.050	0.050	0.080
1.50	0.045	0.050	0.070
4.96	0.035	0.035	0.040
10.02	0.020	0.025	0.020

Table 8.15 Solubility of 1,2,4-butanetriol dinitrate in sodium carbonate solutions at 20–22 °C

Na_2CO_3 content (%)	Na_2CO_3 after 60 min wash (%)	Content of 1,2,4-butanetriol dinitrate in aqueous phase measured by dichromate method (%)		
		60 min wash	Keep still for 1440 min, followed by 60 min wash	120 min wash
0.00	0.00	6.1	7.2	7.2
1.00	0.88	5.8	7.0	6.0
1.30	1.19	5.3	5.9	6.3
12.00	11.73	2.02	2.43	2.71
15.00	14.28	1.37	2.06	2.08

strong saponification of 1,2,4-butanetriol trinitrate can occur in strong alkalis or concentrated alkalis.

$$C_4H_7(ONO_2)_3 + 5KOH \rightarrow KNO_3 + 2KNO_2 + 2CH_3COOK + 3H_2O$$

The saponification products are very complex. In addition to the abovementioned products, oxalic acid, ammonia, and other compounds can also be produced. The 1,2,4-butanetriol trinitrate and 1,2,4-butanetriol dinitrate consumed by saponification in aqueous phases can be compensated by the ester phase. Similar reactions can occur in concentrated sodium carbonate solutions. In industrial production, the stability treatment of 1,2,4-butanetriol trinitrate is usually conducted with 1% sodium carbonate aqueous solution.

(2) Partitions of nitric acid and 1,2,4-butanetriol dinitrate between the ester and aqueous phases

According to the Nernst's distribution law, at a constant concentration, a molecule distributes itself between two immiscible solvents follows the distribution law.

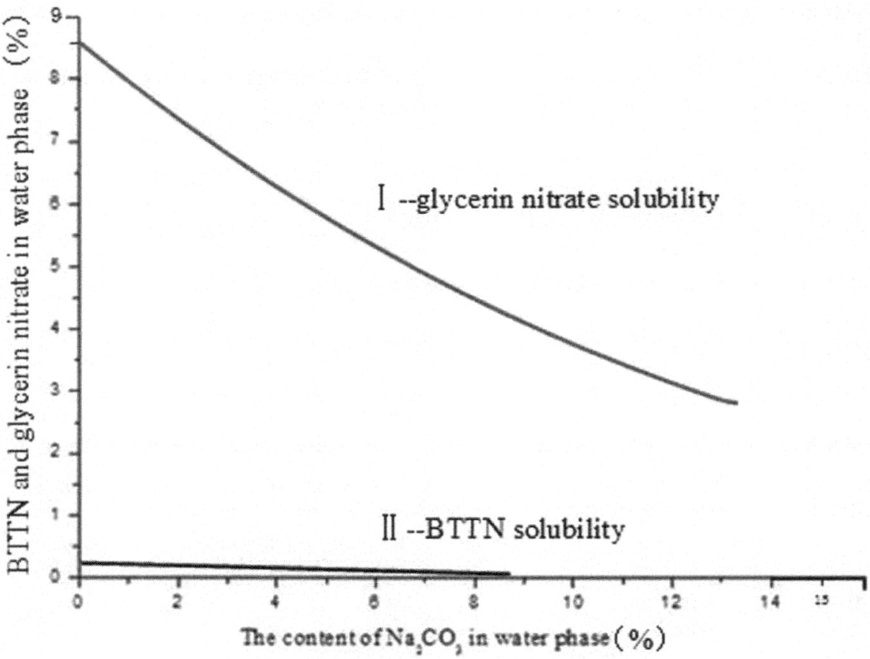

Fig. 8.9 Solubility of 1,2,4-butanetriol trinitrate and 1,2,4-butanetriol dinitrate in sodium carbonate solutions

1. Partition of nitric acid between ester and aqueous phases

The partition parameters of nitric acid between 1,2,4-butanetriol trinitrate phase and aqueous phase is shown in Table 8.16 and Fig. 8.10.

As shown in Table 8.16, the partition coefficient, C_0/C_W, is a variable, instead of a constant, due to the dissociation of nitric acid in the aqueous phase.

$$HNO_3 \underset{k_2}{\overset{k_1}{\rightleftharpoons}} H^+ + NO_3{}^-$$

(The oil phase) (The aqueous phase)

Concentration : C_0 C_W C_W

As the partition equilibrium reached, C_0/C_W^2 is a constant or $\sqrt{C_0}/C_w = \text{Kallocation}$.

Because of the dissociation of nitric acid in the aqueous phase, the partition coefficient $\sqrt{C_0}/C_W$, instead of C_0/C_W, is a constant, consistent with the data listed in Table 8.16.

As shown in Fig. 8.10 for the partition of nitric acid between the ester phase and aqueous phase, the nitric acid content in aqueous phase is much higher than that in

Table 8.16 Partition of nitric acid between 1,2,4-butanetriol trinitrate phase and aqueous phase

HNO_3Content (%), C_W	HNO_3 content in ester phase (%), C_O	Partition coefficient, C_O/C_W	$\sqrt{C_O}/C_W$ partition coefficient
5.00	0.005	0.001	0.001
9.97	0.012	0.001	0.011
20.00	0.039	0.002	0.010
30.00	0.170	0.006	0.014
35.00	0.210	0.008	0.016

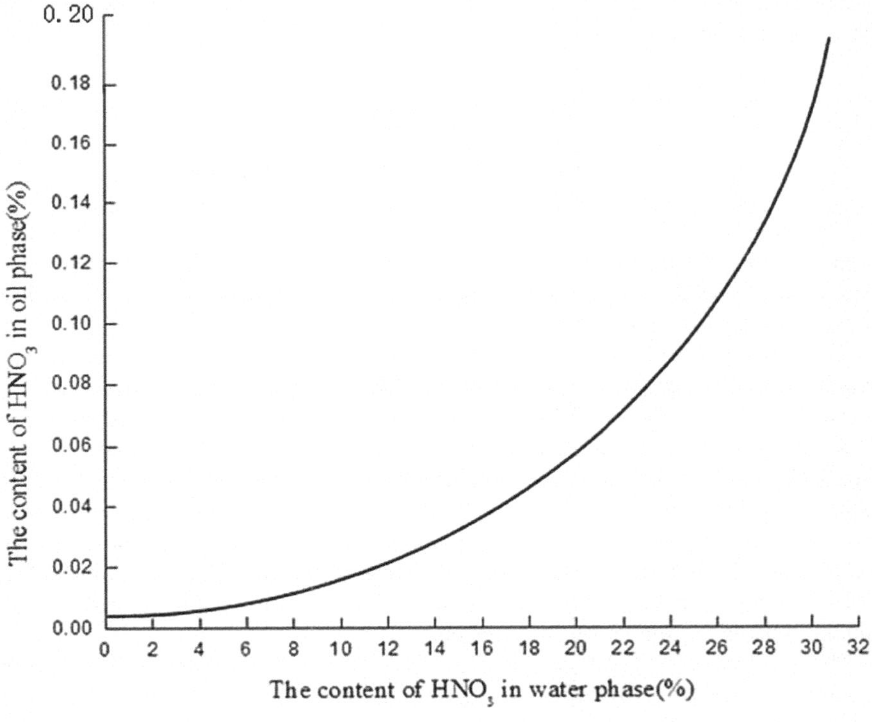

Fig. 8.10 Partition of nitric acid between the 1,2,4-butanetriol trinitrate phase containing 5% 1,2,4-butanetriol dinitrate and aqueous phase

ester phase at the nitric acid content in aqueous phase less than 20%, and the nitric acid content in ester phase is rapidly increased as the nitric acid content in aqueous phase increased to over 20%.

2. Partition of 1,2,4-butanetriol dinitrate between 1,2,4-butanetriol trinitrate phase and aqueous phase

Table 8.17 lists the partition parameters of 1,2,4-butanetriol dinitrate between the ester phase and aqueous phase.

$$2\text{Din} \rightleftharpoons \text{Din}_2$$

As can be seen, the partition coefficient, C'/C, is not a constant due to the association of 1,2,4-butanetriol dinitrate. Where Din is 1,2,4-butanetriol dinitrate.

The association is affected by the osmotic pressure. Therefore, the osmotic pressure is decided by the concentration of 1,2,4-butanetriol dinitrate and the tendency to association is higher at higher concentrations. The aqueous phase contains a low amount of 1,2,4-butanetriol dinitrate, and thus the association can be ignored. The high amount of 1,2,4-butanetriol dinitrate in the ester phase associates to reach the partition equilibrium as shown below:

$$\begin{array}{lcc} \text{Concentration} & C & C' \\ & 2\text{Din} \rightleftharpoons & \text{Din} \\ & \text{(The aqueous phase)} & \text{(The oil phase)} \end{array}$$

Therefore, $C'^{1/2}/C$ is a constant or $\sqrt{C_0}/C_W = K_{\text{allocation}}$, where C is the concentration of 1,2,4-butanetriol dinitrate in aqueous phase, and C' is the total concentration of free Din and the associated Din_2, which is independent on the acidity and alkalinity of the aqueous phase.

The partition coefficients of 1,2,4-butanetriol dinitrate in the neutral, acidic and alkaline aqueous phases are summarized in Table 8.18 and Fig. 8.11.

The partition coefficient (K") in neutral aqueous phase becomes constant as the 1,2,4-butanetriolate dinitrate content in ester phase reaches 13% (middle curve in Fig. 8.11). K" becomes constant more quickly in alkaline aqueous phase than in neutral aqueous phase as shown by the left curve in Fig. 8.11 because the sodium carbonate can neutralize the acid. In acidic aqueous phase, hydrogen ions, such as nitric acid, inhibit the association of 1,2,4-butanetriol dinitrate molecules. Therefore, K" tends to become constant in acidic phase slower than in neutral and alkaline phases, as shown by the right curve in Fig. 8.11.

The partition of 1,2,4-butanetriol dinitrate between ester phase and neutral aqueous phase and that between ester phase and 10% nitric acid aqueous solution are shown in Figs. 8.12, 8.13 and Table 8.19.

The 1,2,4-butanetriol trinitrate contents in both neutral and acidic aqueous phases decrease with the increase of 1,2,4-butanetriol dinitrate content. For example, the maximum 1,2,4-butanetriol trinitrate content in 10% nitric acid is only 0.32–0.35%, which decreases with the increase of 1,2,4-butanetriol dinitrate content until approaches zero. The total amount of nitrates in acidic aqueous phase is a function of the nitric acid content in aqueous phase and 1,2,4-butanetriolate dinitrate content in ester phase decreases with the increase of nitric acid content in aqueous phase.

Table 8.17 Partition parameters of 1,2,4-butanetriolate dinitrate between 1,2,4-butanetriol trinitrate phase and aqueous phase at 20–22 °C

1,2,4-butanetriol trinitrate washed with 5 mL water (g)	Dinitrate in sample (%)	Dinitrate in ester phase (%, calculated value) C'	Dinitrate in aqueous phase (%, calculated value) C	Dinitrate in aqueous phase (%)	1,2,4-butanetriol trinitrate in aqueous phase (%)	1,2,4-butanetriol trinitrate in aqueous phase (%, calculated value)	Partition coefficient (C'/C)
10.00	1.09	1.32	1.29	1.36	0.04	0.09	2.76
10.50	4.65	1.57	1.39	1.41	0.03	0.12	4.19
11.00	9.08	2.36	1.76	1.51	0.02	2.14	4.70
11.50	14.56	4.57	1.93	1.87	0.01	0.16	5.45
12.00	35.55	6.99	2.76	2.16	0.01	0.16	8.02
13.00	45.72	12.26	4.57	4.27	0.01	0.16	9.30

Table 8.18 Partition coefficients ($C/\sqrt{C'}$) of 1,2,4-butanetriol dinitrate in neutral, acidic and basic aqueous phases

1,2,4-butanetriol dinitrate content in aqueous phase (%, C)			1,2,4-butanetriol dinitrate content in ester phase (%, C')			Partition coefficient ($C/\sqrt{C'}$)		
Neutral	Acidic	Alkaline	Neutral	Acidic	Alkaline	Neutral	Acidic	Alkaline
0.27	–	–	0.96	–	–	0.37	–	–
–	–	0.45	–	–	2.00	–	–	0.32
–	0.60	–	–	1.60	–	–	0.47	–
–	0.95	–	–	2.59	–	–	0.59	–
1.01	–	–	4.23	–	–	0.51	–	–
–	–	1.40	–	–	8.15	–	–	0.49
–	1.70	–	–	4.96	–	–	0.77	–
1.74	–	–	8.36	–	–	0.61	–	–
2.47	–	–	13.63	–	–	0.67	–	–
–	2.81	–	–	9.30	–	–	0.93	–
4.09	–	–	34.31	–	–	0.69	–	–
–	4.50	–	–	16.24	–	–	1.12	–
4.57	–	–	44.66	–	–	0.68	–	–

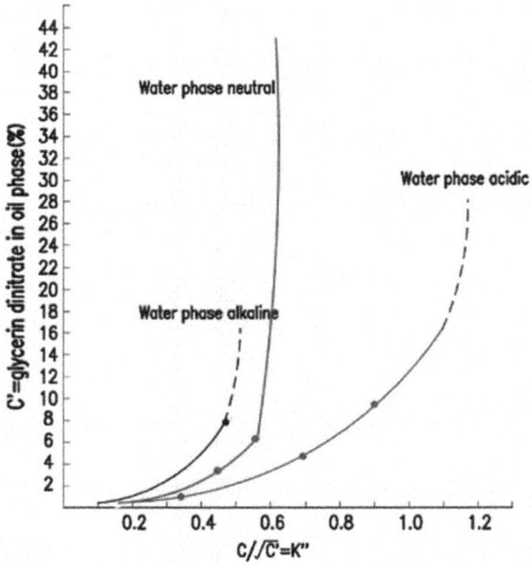

Fig. 8.11 Partition coefficients of 1,2,4-butanetriol dinitrate in acidic, neutral and alkaline aqueous phases

3. Partition of 1,2,4-butanetriol dinitrate between 1,2,4-butanetriol trinitrate and alkaline aqueous phase.

The solubility of 1,2,4-butanetriol trinitrate in 100 g of 1.0% sodium carbonate solution is 0.05 g and that in 100 g 10.0% sodium carbonate solution is only

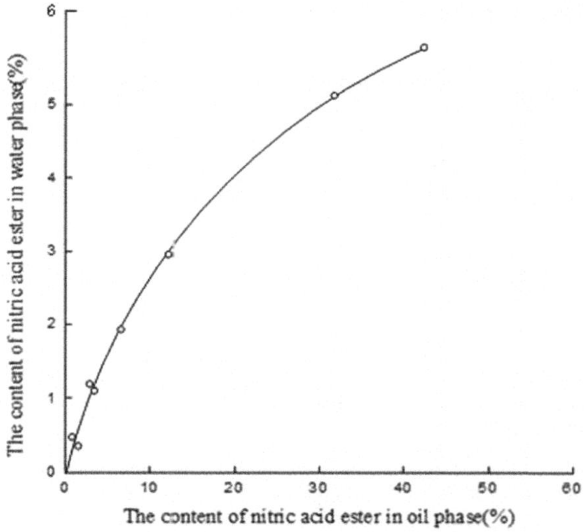

Fig. 8.12 Partition of 1,2,4-butanetriol dinitrate between ester phase and neutral aqueous phase

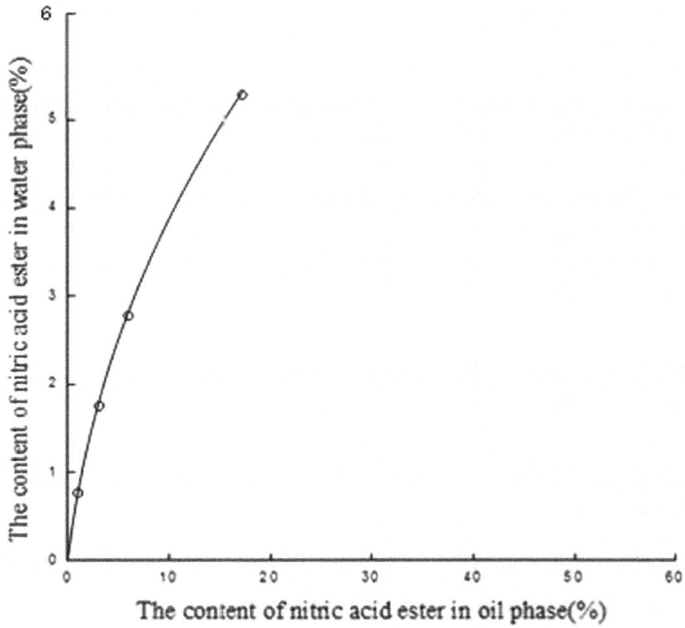

Fig. 8.13 Partition of 1,2,4-butanetriol dinitrate between ester phase and 10% aqueous nitric acid phase

Table 8.19 Partition of 1,2,4-butanetriol dinitrate between ester phase and 10% aqueous nitric acid (20–22 °C)

1,2,4-butanetriol dinitrate in ester sample before washing (%)	1,2,4-butanetriol dinitrate in ester phase after washing (%, C')	1,2,4-butanetriol dinitrate in nitric acid aqueous phase (%, C)	Partition coefficient of (C'/C)
2.79	1.60	0.60	2.67
4.45	2.59	0.95	2.73
8.23	4.96	1.70	2.91
14.51	9.30	2.81	3.31
24.07	16.24	4.50	3.61

Table 8.20 Partition parameters of 1,2,4-butanetriol dinitrate between ester phase and 1.5% Na_2CO_3 aqueous phase at 20–22 °C

1,2,4-butanetriol dinitrate in ester sample before washing (%)	1,2,4-butanetriol dinitrate in ester phase after washing (%, C')	1,2,4-butanetriol dinitrate in alkaline aqueous phase (%, C)	Partition coefficient (C'/C)
0.00	0.00	0.00	–
2.37	2.00	0.45	4.45
9.53	8.15	1.40	5.8

0.02 g. Table 8.20 and Fig. 8.14 are the experimental results of the partition of 1,2,4-butanetriol dinitrate between the alkaline aqueous phase and ester phase.

The water amount, washing time, and washing efficiency depend on the solubility of 1,2,4-butanetriol trinitrate and 1,2,4-butanetriol dinitrate in aqueous solutions (neutral, acidic and alkaline), as well as the partitions of nitric acid and 1,2,4-butanetriol dinitrate between the ester phase and aqueous phase, partition between the ester phase and the aqueous phase during stability treatment.

8.2.5 *Main Factors Affecting the Production Process of 1,2,4-Butanetriol Trinitrate*

The production process of 1,2,4-butanetriol trinitrate is mainly affected by the nitration process and stability treatment.

(1) Factors affecting nitration and solutions

1. Factors affecting nitration

The factors affecting the product quality, recovery rate, spent acid utilization, production safety, etc. of nitration are summarized as follows.

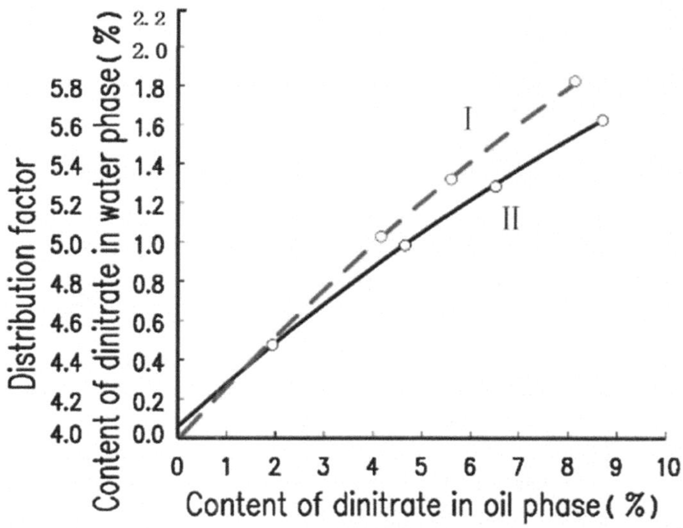

Fig. 8.14 Partition of 1,2,4-butanetriol dinitrate and trinitrate in 1.5% Na_2CO_3 aqueous solution. I-Partition coefficient; II-1,2,4-butanetriol dinitrate content (%)

① Composition and amount of nitrating acid

The composition of spent acid at the end of reaction decides the final equilibrium of the reaction, which can significantly affect the reaction rate, nitration degree, yield, stability treatment, and product quality. Therefore, the composition of nitrating acid is mainly determined based on the composition of spent acid.

② Effects of nitrating acid composition on NO_2^+

The concentration of NO_2^+ increases with the increase of nitric acid content in the nitrating acid. The conversion to NO_2^+ increases with the increase of sulfuric acid content, which is conducive to nitration. Increasing water content causes negative effects.

③ Effects on reaction rate

High reaction rates can be achieved at the nitric acid contents in the waste acid over 8% and $[H_2O]:[H_2SO_4] \leq 1$.

④ Effects on nitration degree

The 1,2,4-butanetriol dinitrate content increases with the increase of water content in spent acid, but decrease with the increase of nitric acid content. The lowest 1,2,4-butanetriol dinitrate content can be obtained at the water content of 13.5–14% (equivalent to $[H_2O]:[H_2SO_4] = 1$) and nitric acid content of 11.5–13.3%.

⑤ Effects on dissolution loss

The least dissolution loss of 1,2,4-butanediol dinitrate has been found at the nitric acid content of 11.5–13.5% and water content of 12–15%.

⑥ Effects on the nitric acid content in 1,2,4-butanetriol trinitrate

The nitric acid content in ester phase increases with the increase of nitric acid content in the spent acid. At the same nitric acid content, 13.5–14% water results in the highest nitric acid content in ester phase.

⑦ Effect on the water content in ester phase

The water content in ester phase is a constant as $H_2O]:[H_2SO_4] \leq 1$ and increases with the increase of water content in the spent acid.

⑧ Effects of acid content in 1,2,4-butanetriol trinitrate on its stability

At $[H_2O]:[H_2SO_4] > 1$, the decomposition rate of 1,2,4-butanetriol trinitrate increases and its flashpoint decreases with the increase of water content in the spent acid, as shown in Table 8.21. Therefore, low $[H_2O]:[H_2SO_4]$ ratios are favorable to production safety. In practical production, the ratio is usually controlled at about 1:1. $[H_2O]:[H_2SO_4] > 1.5$ may cause production safety issues.

2. Control measures

① Appropriate $[H_2O]:[H_2SO_4]$ molar ratio in spent acid

The concentration of sulfuric acid in spent acid is often expressed as "dehydration value" that is the molar ratio of sulfuric acid to water. Generally, the value of [H2O]:[H2SO4] is controlled at 1.2–1.36.

② Sufficient nitric acid

High amounts of nitric result more NO_2^+, which can accelerate the reaction with less side reactions, high yields and safe production. However, large amounts of nitric acid would increase production cost. Therefore, the amount of nitric acid is usually 20% more than needed for the nitration, and results in ~10–13%, no less than 8%, in the spent acid.

Table 8.21 Relationship between the flashpoint of acidic 1,2,4-butanetriol trinitrate and the water content in spent acid

$[H_2O]$: $H_2SO_4]$ in spent acid	Flashpoint of acid 1,2,4-butanetriol trinitrate (°C)	$[H_2O]:[H_2SO_4]$ ratio spent acid	Flashpoint of acid 1,2,4-butanetriol trinitrate (°C)
1.02	183	1.85	121
1.24	162	2.11	109
1.49	132	2.67	97

③ Nitrogen oxide content

Nitrogen oxides are one of the main factors causing the decomposition of 1,2,4-butanetriol trinitrate. High amounts of nitrogen oxides can form serious brown smoke that interferes the judgement on the decomposition of nitrates during production process. To ensure safe production, the less nitrogen oxides, the better. In practical production, the N_2O_4 content should be not more than 0.6%, or the nitric acid content in spent acid should be ~0.2%.

④ Amount of nitrating acid

Low nitration coefficient can decrease the dissolution loss of 1,2,4-butanetriol trinitrate and reduce the cost of spent acid treatment. The nitration coefficient of the container nitration method is controlled between 4.8 and 5.2, and that of the injection method can be higher, but with the similar amount of fresh mixed acid required.

The composition to be met by spent acid for the preparation of nitrating acid is

$$\begin{gathered} HNO_3, 10\%; \\ H_2O, 17\%; \\ H_2SO_4, 73\%; \\ [H_2O] : [H_2SO_4] = 1.25 \end{gathered}$$

Based on the nitration reaction of 1,2,4-butanetriol,

$$\begin{array}{llll} C_4H_7(OH)_3 + 3HNO_3 & \rightleftharpoons C_4H_7(ONO_2)_3 & + 3H_2O & \\ 106 & 3\times 63 & 241 & 3\times 18 \\ 1.00 & 1.78 & 2.27 & 0.50 \end{array}$$

The amount of spent acid produced from the nitration of 1 kg 1,2,4-butanetriol is

$$0.587/0.17 = 3.5\,\text{kg}$$

where $H_2SO_4 = (0.587/18) \times (98/1.25) = 2.55$ kg, and $HNO_3 = 3.5 \times 0.1 = 0.35$ kg.

Therefore, the amount of required nitrating acid is

$$G_M = (2.05 + 0.35) + 2.55 = 4.95\,\text{kg}$$

The nitration coefficient is calculated to be 4.95.

The composition of the mixed acid is

$$HNO_3\% = \frac{2.40}{4.95} \times 100\% = 48\%$$

$$H_2SO_4\% = \frac{2.55}{4.95} \times 100\% = 52\%$$

Table 8.22 lists the compositions of nitrating acid and spent acid used for the production of 1,2,4-butanetriol trinitrate.

3. Increasing the contact surface area for reaction

The nitration of 1,2,4-butanetriol with mixed acid is a heterogeneous reaction. The reactants must contact each other well to achieve desired reaction performance. Increasing the contact surface area between the phases can not only increase reaction rate, but also improve the dehydration efficiency of sulfuric acid, and thus reduce the local overheating caused by the heterogeneous reaction.

Spraying 1,2,4-butanetriol into mixed acid can increase their contact surface area by forming droplets with the average diameter of as low as $\sim 0.65 \times 10^{-2}$ cm. In addition, an ejector or T-tube reactor can eject a high-speed flow of mixed acid to form a negative pressure zone in the nozzle, which sucks the 1,2,4-butanetriol for the reaction. 1,2,4-butanetriol can be then dispersed and form $0.37–2.8 \times 10^{-2}$ cm droplets by the high-speed turbulence of reaction mixture. It has been reported that the contact surface area of the emulsion sprayed with an ejector can be increased to 212–640 cm^2/cm^3, several times of that achieved by mechanical mixing. The injection nitration method uses mixed acid to continuously suck and discharge 1,2,4-butanetriol, which continuously exposes the mixture flow under the turbulence. Therefore, the droplet surface is constantly refreshed. For the container air-agitated nitration method, the contact surface is refreshed by strong mechanical agitations.

(2) Factors affecting stability treatment

According to the relationship between thermodynamics and heat transfer, the extraction rate is proportional to the concentration difference, and inversely proportional to the resistance. Therefore, under normal conditions, the extraction rate can be expressed as

$$\frac{dG}{dt} = KF(C^* - C)$$

where G is the amount of extractable compound (acid) (kg), t is extraction time (h), C is the concentration of the extractable compound in the extraction phase (aqueous) (kg/kg), C^* is the equilibrium concentration of the extractable compound (acid) in the aqueous extraction phase (kg/kg), F is the contact area between two phases (m^2), and K is the mass transfer coefficient (kg/hm^2). Therefore, the key to

Table 8.22 Compositions of mixed acid and spent acid usually used in industrial production

Nitration method		Mixed acid (%)					Spent acid (%)					$\frac{[H_2O]}{[H_2SO_4]}$
		HNO_3	H_2SO_4	H_2O	N_2O_4	NG	HNO_3	H_2SO_4	H_2O	N_2O_4	NG	
Injection	Plant in Sweden	28.4	63.9	6.0	–	1.7	12.0	72.6	12.6	2.8	–	1.3
	Plant in China	28.1 total nitrogen content 29.5 ± 0.5	62 ± 1	8 ± 2	<0.6	<1.6	13.16	71.32	12.73	0.77	3.62	1.0
Discontinuous		49 ± 1	51 ± 1	0.5	0.6	–	9.2	71.5	16.44	2.86	–	1.25
Schmid process		50	50	–	–	–	12	72	14	2	–	1.1
Biazzi process		50	50	–	–	–	10	73	15	2	–	1.1

produce high-quality products is appropriate water washing and alkaline washing for the stability treatment. The following factors should also be considered.

1. The concentration difference is the key factor to promote the extraction efficiency. Increasing the amount of washing water or countercurrent washing can increase the concentration difference. It is worth noting that the volume of washing water for the first wash is critical. However, extremely large amounts of water can accelerate the hydrolysis, leading to 1,2,4-butanetriol trinitrate loss.

Hydrolysis and esterification reactions can still occur in the 1,2,4-butanetriol trinitrate separated from spent acid. The addition of water increases water content in the nitrate, break the equilibrium, and accelerate the hydrolysis. Therefore, the water amount for the first washing must result in nitric acid contents that cause no hydrolysis of 1,2,4-butanetriol trinitrate, usually less than 25% to prevent serious hydrolysis. In addition, the speed of water addition has a certain impact on the hydrolysis. Slowly adding water can increase the water content in the ester phase, and thus accelerates the hydrolysis. Rapidly adding water can extract nitric acid quickly with less influence on the ester phase.

In addition to the hydrolysis, the heat of dilution should also be considered to prevent the rapid temperature rise. Therefore, the water volume for the first washing should be large enough. However, high amounts of water can dissolve more 1,2,4-butanetriol trinitrate, reducing the productivity.

2. The interfacial surface area between the aqueous phase and ester phase is proportional to the extraction rate. Increasing contact surface area can promote the extraction. The droplet size of extraction solution significantly affects the mass transfer. Extremely large droplets cause slow internal diffusion. The interfacial surface area can be increased by the dispersing 1,2,4-butanetriol trinitrate in the extraction solution.

3. Increasing the interphase motion velocity can reduce the thickness of interfacial layer on the phase interface, and thus reduce the diffusion resistance. Agitation can increase the amount of liquid passing through the porous plate, promote the surface refreshing, and thus decrease the thickness of the interfacial layer.

4. Prolonging the extraction time can increase the extraction efficiency. However, extremely long extraction time can reduce the productivity. Appropriately increasing extraction temperature can increase the solubility of impurities in water, and thus is conducive to extraction. However, high temperature can also promote the thermodecomposition of 1,2,4-butanetriol trinitrate and increase its solubility in water, causing issues in production safety and productivity.

The residual acid in 1,2,4-butanetriol trinitrate cannot be completely removed by water wash. To remove acid residues and ensure production safety, 1,2,4-butanetriol trinitrate must be washed with diluted Na_2CO_3 solution to achieve weak alkaline products for safe storage and operation.

8.2.6 Application of 1,2,4-Butanetriol Trinitrate [11, 12]

The modified double-base propellant with improved mechanical properties by diisocyanate crosslinking contains about 28.04% of 1,2,4-butanetriol trinitrate. The co-polymer propellant of ethylene oxide decaborane methacrylate and acrylic acid contains 16.3% of triethyleneglycol dinitrate and 10.8% of 1,2,4-butanetriol trinitrate. Because of their similar properties, 1,2,4-butanetriol trinitrate can be used to substitute nitroglycerin in explosives. The mixture of nitroglycerin and 1,2,4-butanetriol trinitrate is also used as the binder of well-known high-energy solid propellant NEPE to improve sensitivity, energy and mechanical properties of the propellant.

8.3 Pentaerythritol Trinitrate [1, 4, 13, 14]

Pentaerythritol trinitrate with the code of ITMETN, also called 1, 1, 1-trimethylolethane trinitrate, trimethylol methane trinitrate, dimethylol-2-propanol-1-trinitrate, methyl isobutanetriol trinitrate, methyl isobutyl glycerin trinitrate, or methyl tertbutanetriol trinitrate, is a very important nitrate ester.

8.3.1 Chemical Physical Properties of Pentaerythritol Trinitrate

Pentaerythritol trinitrate is a pale-yellow sticky liquid with an odor of burned alcohols.

Formula: $C_5H_9N_3O_9$
Chemical structure:

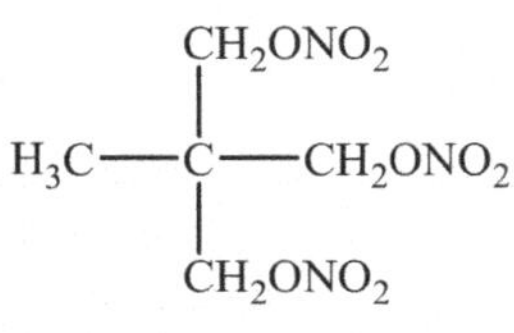

Molecular weight: 255.14
Oxygen balance: −34.49%
Nitrogen content: 16.47%
Specific gravity: 1.4685(20 °C)
Freezing point: −60 °C, lower than that of nitroglycerin
Refractive index: 1.4760 (17.5 °C)
Viscosity: 71 times of that of nitroglycerin

Volatility: 1.21 mg/dmh at room temperature, lower than that of nitroglycerin; 1/3 of that of nitroglycerin at 60 °C
Hygroscopicity: 0.07% at the relative humidity of 65% and 0.14% at the relative humidity of 90%.

Solubility in water: 0.516 g/L at 19 °C and 0.685 g/L at 36 °C; soluble in organic solvents, such as ether, acetone and vinyl chloride and concentrated nitric acid and sulfuric acid. It can dissolve nitrocellulose, but with weaker plasticity than nitroglycerin. Its solubility in 65% acetic acid is 11.4% and that in 96% ethanol is 4%.

8.3.2 Explosion Property

The explosion heat (constant volume) of pentaerythritol trinitrate is 5271.8 J/g with liquid water. The specific volume is 967 ml/g and explosion temperature is 3500 K. The lead block value is 480 mL. The heat coefficient is 2824.2 J/g and the detonation coefficient is 49.91 J/g. The coefficient of specific coefficient is 9.661 L/kg. The detonation velocity of pentaerythritol trinitrate in an open vessel with a diameter of 30 mm is listed in Table 8.23.

The detonation point of pentaerythritol trinitrate is 20 °C. Its shock sensitivity is much weaker than nitroglycerine, which characteristic height is 4 cm for 2 kg falling weight, but the height is 20–40 cm, or up to 47 cm in some references, for pentaerythritol trinitrate. An ester mixture of pentaerythritol trinitrate, triglycol nitrate, triaceticanhydride, and stabilizer can stand in a jolting truck without explosion. It is a good property for production and application.

The thermal stability of pentaerythritol trinitrate is better than nitroglycerine. The duration time in Abel test at 72 °C is 40–45 min.

8.3.3 Preparation of Pentaerythritol Trinitrate

Pentaerythritol trinitrate is prepared by a condensation reaction between formaldehyde and propionic aldehyde to pentaerythritol, which is further nitrated by a mixed acid of nitric acid and sulphuric acid. The preparation condition is nitration coefficient of 5–6, nitration temperature of 20–25 °C, nitration time of 25–30 min. The nitration yield at 10 °C with a mixed acid of 45% nitric acid and 55% sulphuric acid is 197–198%, about 93% of the theoretical yield. If 20% fuming

Table 8.23 Detonation velocity of pentaerythritol trinitrate with different densities

Loading density (g/cm^3)	1.39	1.48	1.50
Detonation velocity (m/s)	6750	7040	7060

Table 8.24 Parameters of stability treatment of pentaerythritol trinitrate in water-washing process

Washing solution	Temperature (°C)	Agitation time (min)	Washing times
Water	15	5	1
Water	50	5	1
5% Na_2CO_3	70	15	2
Water	50	5	1
Water	30	5	1
Water	15	5	1

sulphuric acid is used, the yield can be increased to 97–98% of the theoretical yield. The parameters of stability treatment with water-washing coefficient of 5 are listed in Table 8.24.

In summary, the product can be obtained by sulfonation first and nitration later by a mixed acid with a yield of no less than 90% theoretical yield.

8.3.4 Application of Pentaerythritol Trinitrate

Pentaerythritol trinitrate is used in explosives and propellants, partly or all replacing nitroglycerine. Americans believe that there is 2.0% pentaerythritol trinitrate in modified double-base propellants with cellulose nitrate and an adhesive of polyglycol.

8.3.5 Toxicity

The toxicity of pentaerythritol trinitrate to human body is similar with nitroglycerine, but slightly lighter. Its treatment is also similar with nitroglycerine.

8.4 Diglycerol Tetranitrate

8.4.1 Properties of Diglycerol Tetranitrate [1, 4]

Diglycerol tetranitrate, also called nitrodiglycerin, can be prepared by the nitration of diglycerol.

Formula: $C_6H_{10}N_4O_{13}$
Chemical structure:

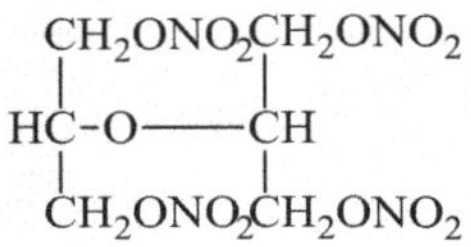

Molecular weight: 346.14
Oxygen balance: 18.49% (Calculated based on the amount of produced CO_2)
Nitrogen content: 16.18%

Compared with nitroglycerin, diglycerol tetranitrate is a more viscous liquid. It is insoluble in water and non-hygroscopic, but soluble in most organic solvents. It can partially dissolve collodium. The chemical stability of diglycerol tetranitrate is comparable to that of nitroglycerin, but with a lower freezing point of −35 °C.

Diglycerol tetranitrate is one of most powerful known explosives. The combustion heat of the propellant containing 75% nitroglycerin and 25% diglycerol tetranitrate is similar to that of pure nitroglycerin. It is insensitive to impacts with the drop height of 8–10 cm (2 kg hammer). The explosive decomposition of diglycerol tetranitrate is:

$$C_6H_{10}N_4O_{13} \rightarrow 2CO_2 + 4CO + 5H_2O + 2N_2 + 5564.7\,J/g$$

8.4.2 Preparation of Diglycerol Tetranitrate [1, 15]

Diglycerol can be prepared by heating glycerol in the presence of a small amount of sulfuric acid at 130–160 °C. Alternatively, heating glycerol with a small amount of sodium carbonate or sodium sulfite (~0.5%) at 290 °C can also afford the diacyl ether, 2,2',3,3-tetrahydroxy ether or symmetrical bis(2,3-dihydroxypropyl) ether, along with water as shown below.

$$C_3H_5(OH)_3 \rightarrow CH_2OH{-}CHOH{-}CH_2{-}O{-}CH_2{-}CHOH{-}CH_2OH + H_2O$$

The reaction also produces 4–6% triglycerin and polyglycerol, leaving 30–40% of the unpolymerized glycerol. The main component of the product mixture is diglycerin with the density of 1.33 g/cm^3, viscosity over 11 times of that of glycerol and boiling point of 247–249 °C (8 mmHg).

The resultant diglycerin is nitrated with nitric acid/sulfuric acid to form diglycerin tetranitrate. The nitration process of diglycerin is usually same as that of glycerol with some modifications due to their different properties. The main reaction is:

$$CH_2OH{-}CHOH{-}CH_2{-}O{-}CH_2{-}CHOH{-}CH_2OH \rightarrow$$
$$CH_2ONO_2{-}CHO\,NO_2{-}CH_2{-}O{-}CH_2{-}CHO\,NO_2{-}CH_2O\,NO_2 + H_2O$$

The nitration of diglycerol needs slightly less mixed acid than that of glycerol. Diglycerin is usually preheated at 50–60 °C due to its high viscosity and then added into the nitrator by the upper addition method. The nitrating mixture is bubbled with inert gas to ensure a homogenous reaction.

The separation of diglycerin dinitrate is much slower than that of nitroglycerin, mainly due to its high viscosity. The more impure the diglycerin, the easier it is to form an emulsion. Therefore, the separated diglycerin nitrate is then washed with a NaCl solution to prevent the emulsification. The moisture produced by the washing is more difficult to remove than that in nitroglycerin.

One hundred kilograms of diglycerin can produce 181.7 kg of diglycerin nitrate in the yield of 87.6%.

Diglycerin nitrate is mainly used in the dynamites of low coagulation point and the production of other liquid mixed explosives. Its performance is better than chloroglycerin dinitrate [15] because its explosion does not produce hydrochloride, and it is insoluble in water. The mixture of diglycerin and glycerol can be easily prepared, which provides raw materials of low costs. However, diglycerin nitrate is highly viscous, which makes the residual acid difficult to be removed. In addition, the pure water washing of its nitration product usually results in emulsification.

Diglycerin nitrate can be mixed with ethyleneglycol dinitrate, other nitrates, and combustibles to form new blasting explosives for civil engineering and mining.

8.5 Other Nitrate Esters

8.5.1 D-Sorbitol Hexanitrate [1, 3]

Sorbitol hexanitrate or nitrosorbitol is an isomer of mannitol nitrate. It is a crystal under normal conditions and melted at 55 °C. Sorbitol hexanitrate has been used as an additive for nitroglycerin to lower the freezing point of dynamite in the United States.

Sorbitol reacts with five equivalents of concentrated nitric acid (d = 1.52/0 °C) and the product is precipitated with 10 equivalents of 20% fuming sulfuric acid (+5 °C) to afford hexanitrate with a low amount of pentanitrate contamination. An oily mixture of sorbitol hexanitrate and pentanitrate can be formed as the nitration and precipitation temperature set to 0–10 °C. Sorbitol pentanitrate is generally in an oily form. The explosive properties of sorbitol nitrate esters are similar to those of mannitol nitrate esters.

Sorbitol is usually produced by the catalytic reduction of D-glucose in industrial scale. Therefore, the nitration products may have great practical values.

The 1, 4, 3, 6-dinitrates of D-mannitol, D-glucitol and L-iditol have been prepared for the production of the vasodilator, 1,4,3,6-dihexnol 2,5-dinitrate. Table 8.25 lists the melting points of several dinitrates.

Table 8.25 Melting points of dinitrate esters

Nitrate ester	Melting point (°C)
Isomannide dinitrate	64.5–65.5
Isosorbide dinitrate	50.5–51.5
Isoidide dinitrate	68–69

8.5.2 Polyvinyl Nitrate [1–3, 16–18]

Polyvinyl nitrate is a synthetic polymer, a substitute of nitrocellulose. It was first prepared in 1929. Polyvinyl alcohol was dissolved in concentrated sulfuric acid at ~0 °C, which was slowly added to a mixed nitrating acid and gradually heated to 40–50 °C to afford polyvinyl nitrate. The plastic precipitate became brittle and was easily pulverized after filtration and washing. The nitrogen content of the product was ~10%. Polyvinyl alcohol reacts with nitric acid at 10–20 °C to afford nitrate esters. Polyvinyl alcohol is easily oxidized and measures must be taken to prevent such oxidation.

Neither nitric/sulfuric mixed acid nor nitric/acetic mixed acid can esterify all hydroxyl groups due to the simultaneous hydrolysis. High purity nitrate esters can be obtained with a mixture of nitric and acetic anhydride or a nitric acid solution in carbon tetrachloride and is then softened at 40–50 °C. The lead block expansion value of polyvinyl nitrate with the nitrogen content of 13–15% can reach up to 153–342 cm. The nitrogen content of polyvinyl nitrate is usually 13.5–14.5% with theoretical maximum value of 15.75%. The detonation velocity of polyvinyl nitrate is similar to that of nitrocellulose with same nitrogen contents. Table 8.26 lists the detonation velocities of polyvinyl nitrate with 13.4% nitrogen content in cardboard drums of a diameter of 30 mm.

Polyvinyl nitrate is an unmeltable flammable material. It is rarely used in industry because of its poor chemical stability. Unlike nitrocellulose that can form a membrane with high mechanical strength after the solvent of its solution evaporated, polyvinyl nitrate cannot be dissolved to from a solution, possibly due to its low molecular chain orientation.

Table 8.26 Relationship between the loading density and detonation velocity of polyvinyl nitrate

Loading density (g/cm^3)	Detonation velocity (m/s)
0.3	2030
0.6	3450–3520
1.0	4920–5020
1.4	6090
1.5	6560

8.5.3 Anhydroenneaheptitol Pentanitrate [3, 19]

Anhydroenneaheptitol pentanitrate, also called as 3,3,5,5-tetramethylol-4-hydroxypyran pentanitrate, is a derivative of anhydroenneaheptitol produced by the its esterification with nitric acid. It is a pale-yellow solid powder and highly explosive.

(1) Properties of anhydroenneaheptitol pentanitrate

Formula: $C_9H_{13}N_5O_{16}$
Chemical structure:

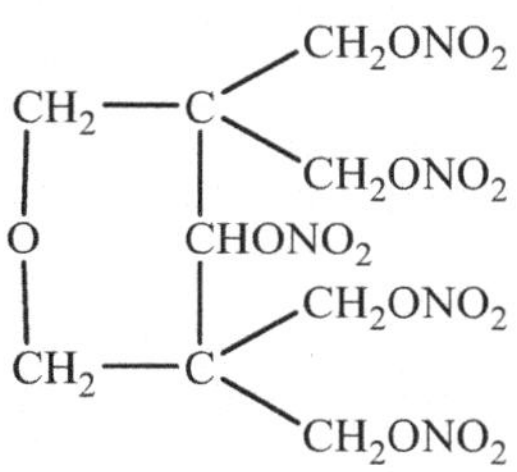

Molecular weight: 447.24
Oxygen balance: −30.4% (Calculated based on the amount of produced CO_2)
Nitrogen content: 15.66%
Density: 1.63 g/cm^3
Melting point: 159 °C

The explosion of anhydroenneaheptitol pentanitrate is more powerful than pentaerythritol tetranitrate. It is much less sensitive than pentaerythritol tetranitrate and similar to TNT.

(2) Preparation of anhydroenneaheptitol pentanitrate

Anhydroenneaheptitol pentanitrate can be prepared by the following method. Acetone reacts with formaldehyde in the presence of $Ca(OH)_2$ at first to produce vinyl methyl ketone that is polymerized to afford anhydroenneaheptitol with the pyran ring structure in the yield of up to 70–73%. Anhydroenneaheptitol s then nitrated with the mixed acid at 5–15 °C to afford dehydrated 9-heptanol pentanitrate in the yield of 91.3%. The main reactions for the preparation of anhydroenneaheptitol pentanitrate are:

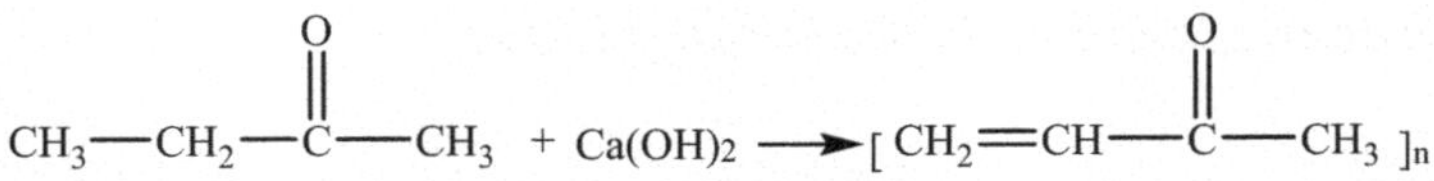

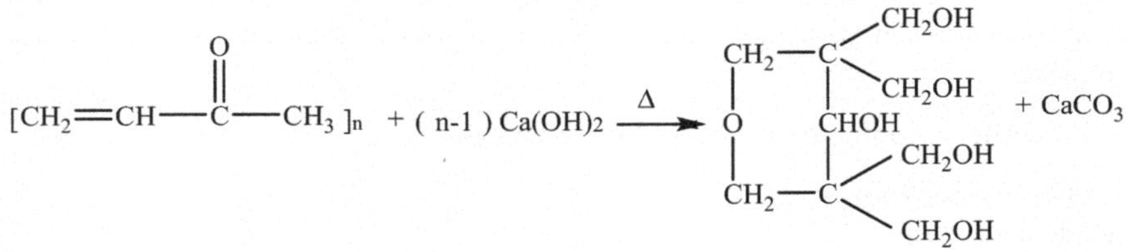

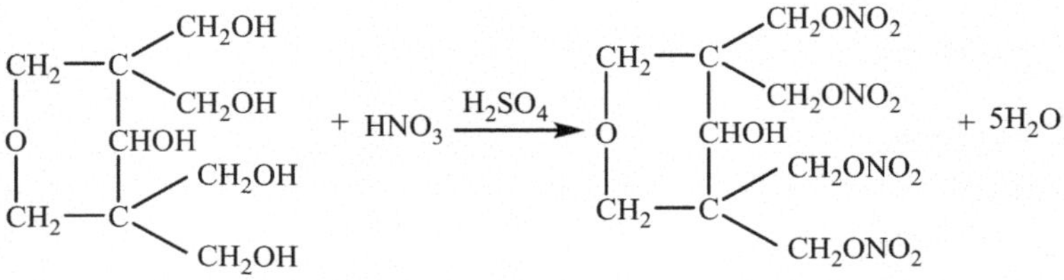

The nitration product is then separated from the spent acid, stabilized by water washing and alkali washing etc., and vacuum-dried to afford the target product.

8.5.4 Inositol Hexanitrate [3, 20, 21]

Inositol hexanitrate is a white solid product of the nitration of cyclohexanol under the protection of a protection reagent.

(1) Properties of inositol hexanitrate

Formula: $C_6H_6O_{18}N_6$
Chemical structure:

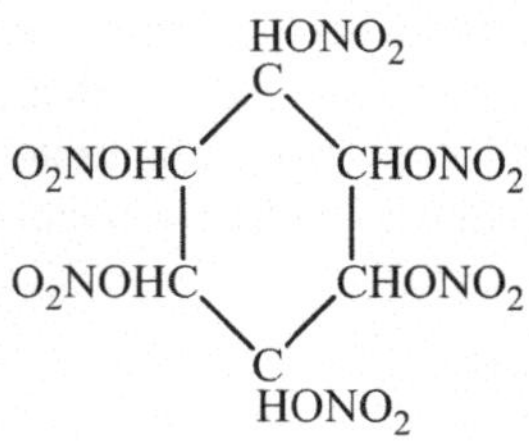

Molecular weight: 450.15
Oxygen balance: + 30.4% (Calculated based on the amount of produced CO_2)
Nitrogen content: 18.67%
Melting point: 132.5 °C
Density: 1.66 g/cm^3

In addition to the above structure, there are isomers of "alcohol nitrate" of inositol hexanitrate. Inositol hexanitrate can substitute dynamite in detonators and bursting tubes. Its explosive power, mechanical sensitivity and safe properties are similar to those of mannitol nitrate.

(2) Preparation of inositol hexanitrate

Inositol must be protected before nitration because the ring is easily oxidized. Some of the nitrate groups can be oxidized by strong oxidizing acid, leading to decomposition. The protection reagent should be able to absorb, but not to react with nitrogen oxides. The major reaction for the preparation of inositol hexanitrate is:

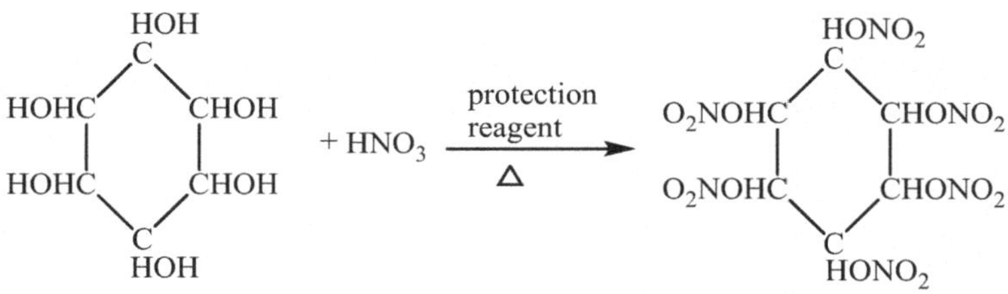

where ammonium nitrate is the protection reagent and it can be substituted with dicyandiamide and urea etc. The reaction is initiated at 50–60 °C and the reaction temperature can increase to 90 °C at the late nitration process. The reaction can be finished in 20 min. The product is distilled with a water vapor to give a crude product. The crude product is then subject to the stability treatment including water washing to neutral, rinsing with 1.0–1.5% Na_2CO_3, and washing with water to neutral to achieve a product of gray solid powder. Figure 8.15 shows the production process of inositol hexanitrate from inositol.

8.5.5 Tetramethylolcyclohexanol Pentanitrate [22]

Tetramethylolcyclohexanol pentanitrate, also known as nitrohexanol, is a pale-yellow solid product of the esterification of tetramethylolcyclohexanol with mixed nitrating acid. Its explosive property is stronger than that of hexogen.

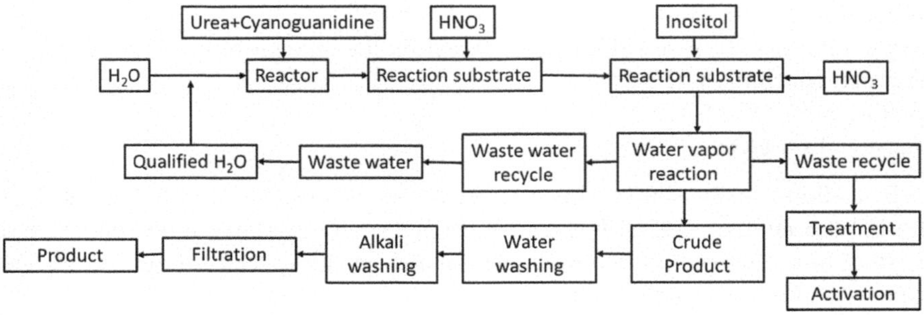

Fig. 8.15 Production process of inositol hexanitrate

(1) Properties of tetramethylolcyclohexanol pentanitrate

Formula: $C_{10}H_{17}N_5O_{15}$
Chemical structure:

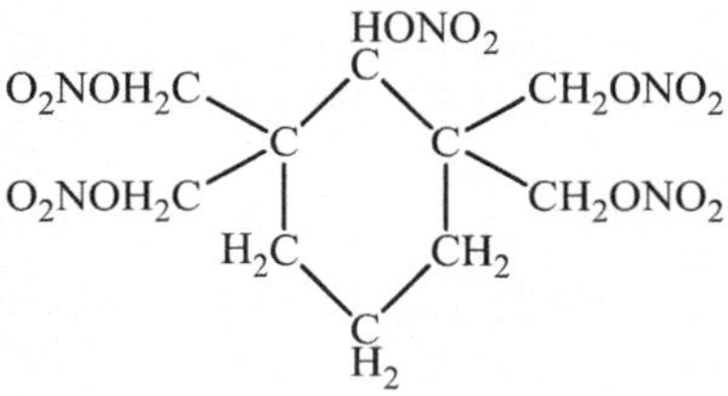

Molecular weight: 447.27
Oxygen balance: −48.29% (calculated based on the amount of produced CO_2)
Nitrogen content: 15.66%
Density: 1.44 g/cm^3
Heat of explosion: 3765.6 J/g, f = 11700 m
Melting point: 122.5 °C

(2) Preparation of tetramethylolcyclohexanol pentanitrate

The addition reaction of cyclohexanone with formaldehyde gives an intermediate, tetramethylolcyclohexanone (I) that is then reduced to produce tetramethylolcyclohexanol (II). Compound II is nitrated to afford tetramethylolcyclohexanol pentanitrate. The major reactions are:

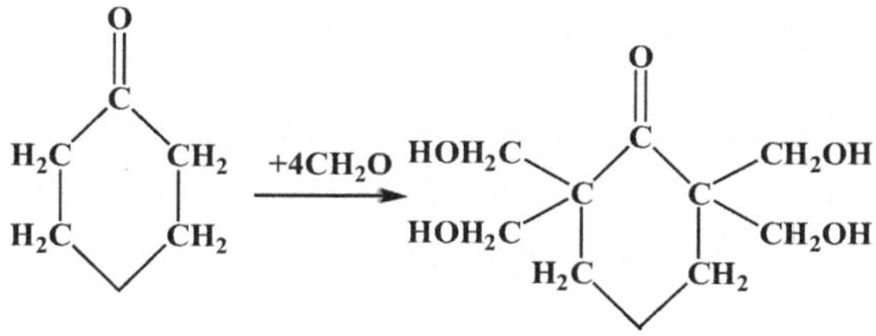

(I)

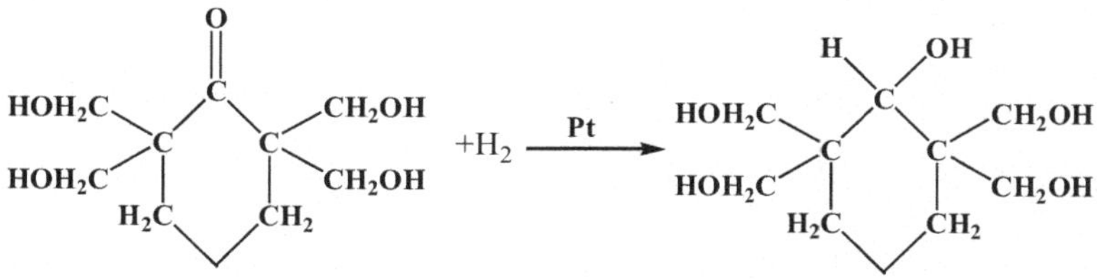

(II)

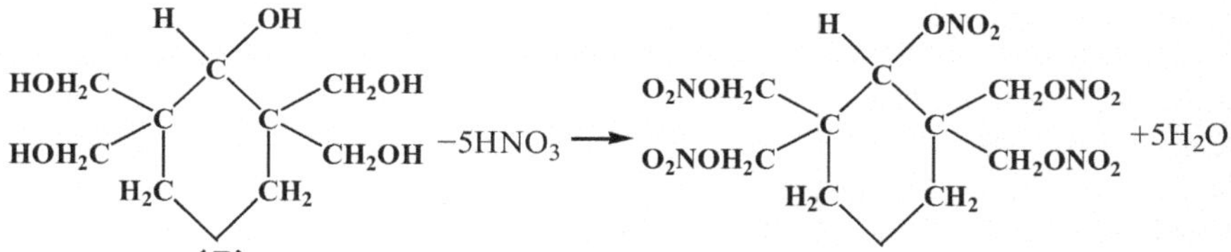

(II)

Compound (I) can also be nitrated to form cyclohexanone tetranitrate.

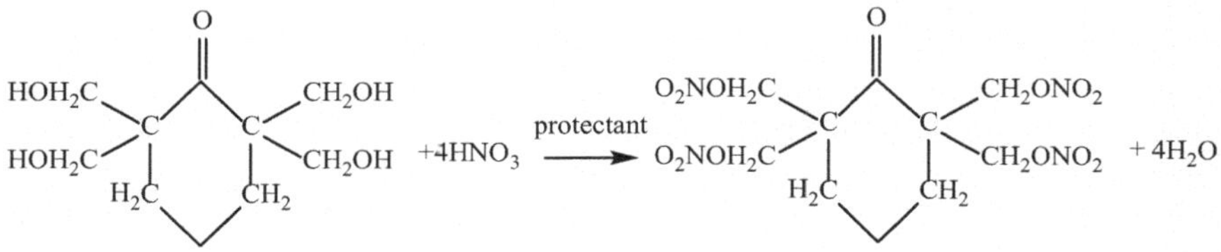

The melting point of cyclohexanone tetranitrate is 66 °C. Its explosion heat is 3451.8 J/g, f = 12110 m, and its detonation velocity is 7740 m/s at the density of 1.51 g/cm^3.

8.5.6 Tetramethylolcyclopentanol Pentanitrate

Tetramethylolcyclopentanol pentanitrate, also known as nitrocyclopentanone [23] and cyclopentanone pentanitrate, is the strongest explosive, after hexogen.

(1) Properties of tetramethylolcyclopentanol pentanitrate

Formula: $C9H_{13}N_5O_{15}$
Chemical structure:

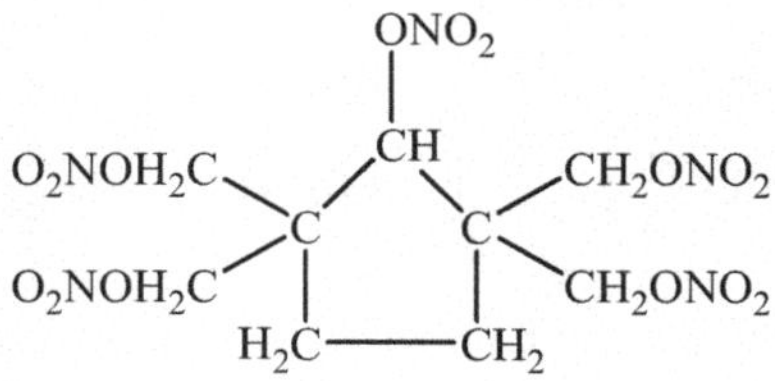

Molecular weight: 431.28
Oxygen balance: −35.24% (Calculated based on the amount of produced CO_2)
Nitrogen content: 16.24%
Density: 1.61 g/cm^3
Heat of explosion: 4330.44 J/g
Detonation velocity: 7360 m/s, f = 12550 m
Melting point: 79 °C
Solubility: insoluble in water and soluble in most organic solvents

(2) Preparation of tetramethylolcyclopentanol pentanitrate

Tetramethylolcyclopentanol pentanitrate can be prepared with cyclopentanone as the raw material. The addition reaction of cyclopentanone with formaldehyde affords tetramethylolcyclopentanone that is then added with hydrogen to yield tetramethylolcyclopentanol. The nitration of tetramethylolcyclopentanol with mixed acid or nitric acid under the protection of nitrogen-containing protection reagent affords tetramethylolcyclopentanol pentanitrate. The crude product is purified and stabilized to achieve a stable product. The yield of nitration with mixed acid is 69.2%, and that with nitric acid is 77.53%. The major reactions for the synthesis of tetramethylolcyclopentanol pentanitrate are:

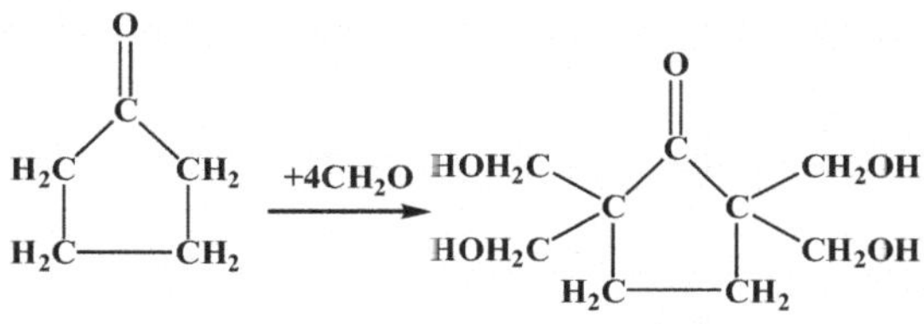

(III)

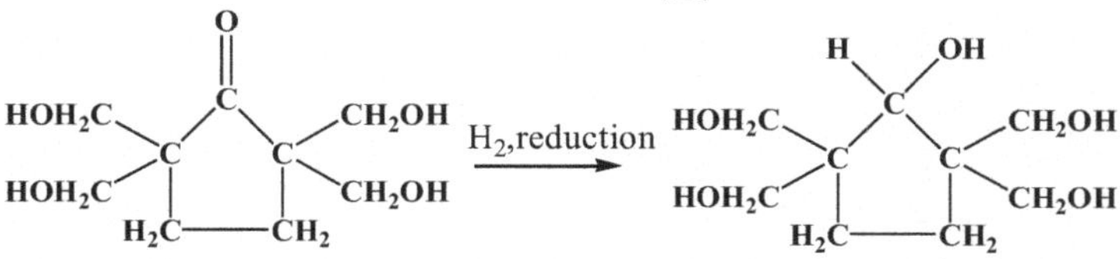

(IV)

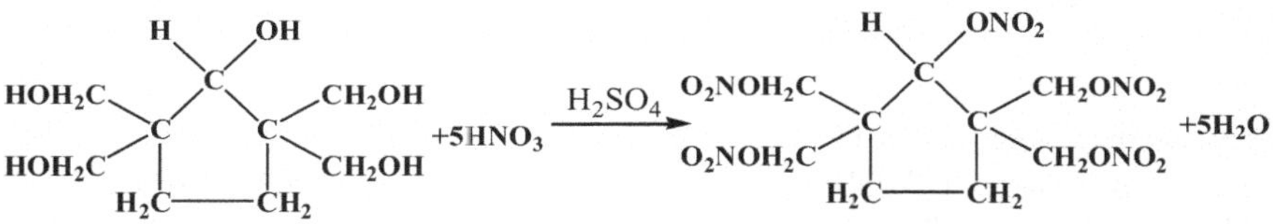

The intermediate, tetramethylolcyclopentanone (III), can also be nitrated to produce tetramethylolcyclopentanone tetranitrate.

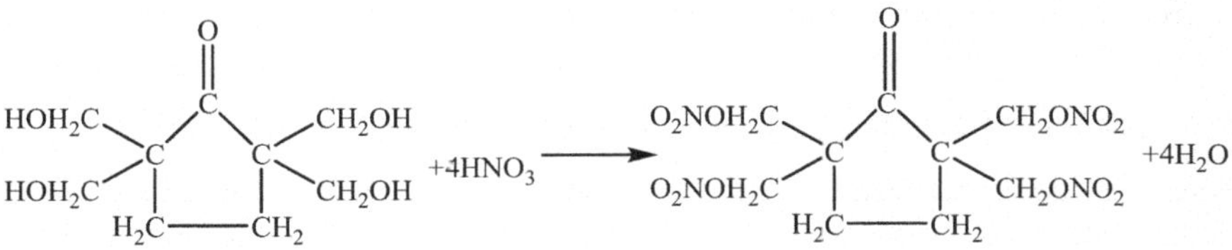

The melting point of tetramethylolcyclopentanone tetranitrate is 74 °C. Its explosion heat is 3430.88 J/g, f = 10680 m and its detonation velocity is 7940 m/s at the density of 1.59 g/cm^3.

8.5.7 1,4-Dinitrocyclohexane-2,3,5,6-Tetranitrate

1,4-Dinitrocyclohexane-2,3,5,6-tetranitrate is a white solid with the density of 1.97 g/cm^3. It is one of the nitrate esters with the highest energy density so far [24]. Its incredible detonation velocity and explosion pressure provide it great practical application potentials. Its synthesis involves three compounds with high explosive energies.

(1) Properties of 1,4-dinitrocyclohexane-2,3,5,6-tetranitrate

Formula: $C_6H_6N_6O_{16}$
Chemical structure:

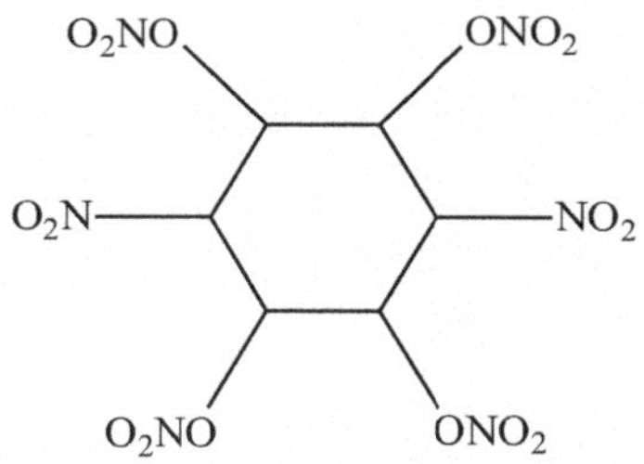

Molecular weight: 418.17
Oxygen balance: + 3.83% (Calculated as the amount of produced CO_2)
Nitrogen content: 20.10%
Density: 1.97 g/cm^3
Heat of explosion: 4330.44 J/g
Detonation velocity: 9616 m/s
Melting point: 124.5 °C
Decomposition temperature: 192.1 °C
Solubility: slightly soluble in water and soluble in most organic solvents.

(2) Preparation of 1,4-dinitrocyclohexane-2,3,5,6-tetranitrate

There are three methods for the synthesis of 1,4-dinitrocyclohexane-2,3,5,6-tetranitrate. The first one uses glyoxal as the raw material to produce intermediate (I) in the presence of nitromethane that is then nitrated to afford the target product. The second method uses 1, 4-dinitro-tetrachlorocyclohexane as the raw material that is decomposed to form intermediate (I) for the nitration. The third method uses 1, 4-dinitro-tetrahydroxycyclohexane as the raw material to afford the intermediate (II) for the nitration. Here, we will focus on the production process using glyoxal as the starting material. The major reactions and production processes for the preparation of 1,4-dinitrocyclohexane-2,3,5,6-tetranitrate from glyoxal are:

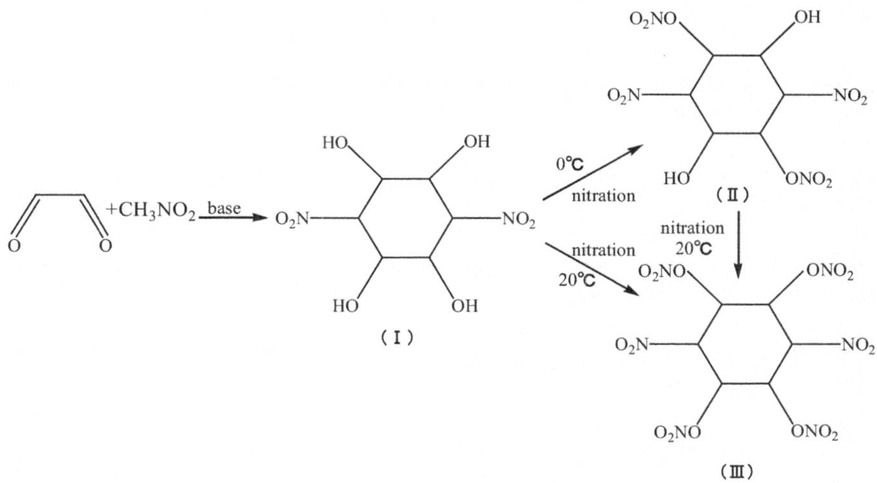

Production processes and operations of each reaction:

1. Preparation of intermediate (I)

Under stirring, 3.6 mL of acetic anhydride is added dropwise to 1.2 mL of concentrated nitric acid (28.6 mmol), followed by the portionwise addition of compound (I) (0.476 g, 2 mmol) at 0 °C. The reaction mixture is slowly warmed to 20 °C, allowed to react for 1.5 h, then poured into 20 g of ice water, filtered, washed with water and dried to produce 0.59 g intermediate (I), 1,4-dinitro-tetramethylolcyclohexane, as a white solid in the yield of 70.57%.

2. Preparation of target products (II) and (III)

Under stirring, 3.6 ml of acetic anhydride is added dropwise to 1.2 mL of concentrated nitric acid (28.6 mmol), followed by the addition of compound (I) (0.476 g, 2 mmol) as the temperature controlled at 0 °C. The reaction is allowed to proceed for 1.5 h. The reaction mixture is then poured into 100 g of ice water, filtered, washed with water and dried to afford 0.59 g target product (II), 3,6-dihydroxy-2,5-dinitrate-1, 4-dinitrocyclohexane, as a white solid in the yield of 89.80%.

Five grams of compound (II) are fed into the nitric acid/acetic anhydride system, heated to 20 °C, and stirred for 20-min. The reaction mixture is then poured into ice water, washed repeatedly, filtrated, and washed with water to afford 3.77 g of compound (III), 1,4-dinitrocyclohexane-2,3,5,6-tetranitrate, as a white solid in the yield of 75.31%.

Compound (III) can also be synthesized by a stepwise one-pot reaction. The reaction is conducted at 0 °C for 1.5 h. The reaction temperature is then increased to 20 °C as nitrating acid further added for the nitration and the reaction is allowed for 15 min. The product is washed, filtrated, washed with water, and dried to afford compound (III) in a similar yield.

The characterization data of the products are show below. In the IR spectra, the C-H stretching vibration of cyclohexane results in absorption peaks at 3107 cm^{-1} (compound II) and 2996 cm^{-1} (compound III). The introduction of nitrates is confirmed by the characteristic peaks of nitrate at 1671 cm^{-1} and 1296 cm^{-1} for compound II, and 1703 cm^{-1}, 1676 cm^{-1} and 1295 cm^{-1} for compound (III). The absorption peak of compound II at 3556 cm^{-1} is attributed to its two hydroxyl groups. The characteristic peaks of hydroxyl groups disappeared in compound (III) and the characteristic peaks of nitrate group appeared at 3300–3600 cm^{-1}, indicating that all of its four hydroxyl groups were nitrated to nitrate groups. Compound (II) exhibits four groups H NMR signals with equal peak areas, including hydroxy H (6.88), C3 and C6 H (5.94), C2 and C5 H (5.60), C1 and C4 and C5 H (77.10), and C1 and C4 H (67.46). Compound III exhibits three sets of C HMR signals including C3 and C6 (81.85), C2 and C5 (77.1 0), and C1 and C4 (67.46). The chemical shifts of H and C shift to the lower field because the significance of the electron attracting effect is in the same order as that of by the electronegativity, e.g. nitro > nitrate ester > hydroxyl. The elemental analysis suggests that the experimental compositions of compounds II and (III) are consistent with the theoretical values. In all, the structures of compounds II and (III) are confirmed by IR, NMR and elemental analyses.

(3) Thermal properties of 1,4-dinitrocyclohexane-2,3,5,6-tetranitrate

Figure 8.16 shows the DSC curves of compounds (II) and (III). Both of them are relatively stable at temperatures below the decomposition temperatures. Compounds II and III exhibit exothermic peaks at 241.1 °C and 192.0 °C, respectively. The exothermic peaks are sharp with narrow temperature spans, indicating that the exothermic decomposition occurred. Figure 8.17 is the TG-DTG curve of compound (II). As can be seen, compound (II) is stable at temperatures

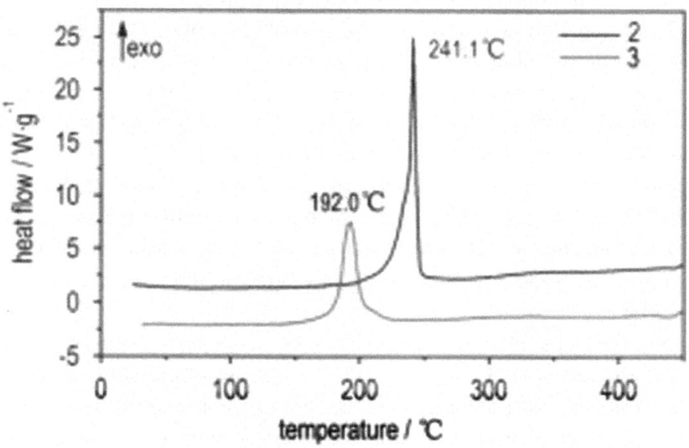

Fig. 8.16 DSC curves of compounds II and III

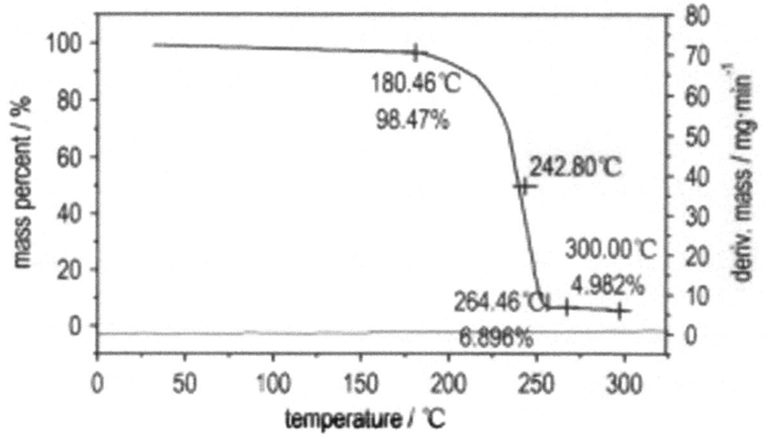

Fig. 8.17 TG-DGT curve of compound II

below 180.46 °C. The maximum weight loss occurred at 242.80 °C. As heated to 264.46 °C, 93.10% of compound II is decomposed and it is completely decomposed at 300.00 °C with 4.982% of residue left. As shown in Fig. 8.18 for the TG-DGT curve of compound (III), it is stable at temperatures below 124.46 °C and the maximum weight loss peak occurs at 199.98 °C. As temperature increased to 242.90 °C, 94.37% compound (III) is decomposed and it is completely decomposed at 292.60 °C, leaving 2.628% residue. A comparison between DSC and TG-DTG curves suggests the two compounds did not undergo any melting process, and were directly decomposed. The sharp exothermic peaks and narrow temperature spans

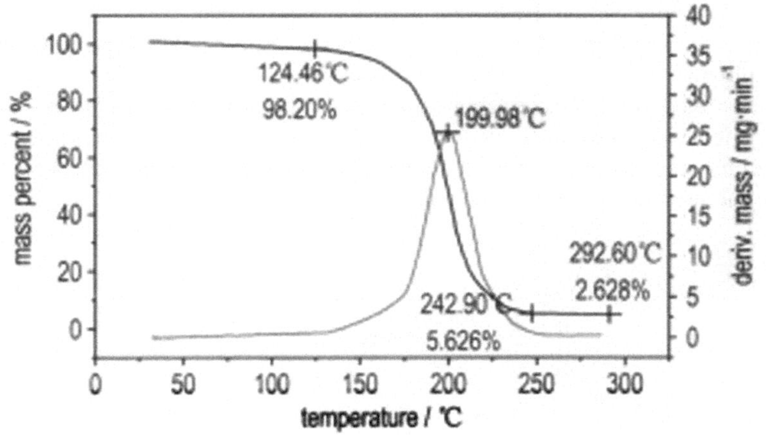

Fig. 8.18 TG-DGT curve of compound III

indicate that they are thermally stable. The thermal stability of compound (II) is better than that of compound (III) due to its intramolecular hydrogen bonds and less nitrate groups.

8.5.8 2,3-Dinitro-isobutyl Ester [25–27]

David E [26] of the Los Alamos National Laboratory of United States synthesized a new energetic nitrate, $C_6H8O_{16}N_6$, with a low melting point.

(1) Properties of 2,3-dinitro-isobutyl ester

Formula: $C_6H_8O_{16}N_6$
Chemical structure:
Molecular weight: 419.98
Oxygen balance: 0% (Calculated based on the amount of produced CO_2)
Nitrogen content: 20.00%
Density: 1.973 g/cm^3
Heat of explosion: 7143 J/g
Detonation velocity: 9165 m/s
Melting point: 85–86 °C
Decomposition temperature: 141 °C

(2) Forms and preparation of 2,3-dinitro-isobutyl ester

David E [26] report Synthesis of an Energetic Nitrate Ester in 2008. A more practical material would be a melt-castable explosive, which would allow the simplification of manufacturing processes. Herein we describe the synthesis of a new energetic nitrate ester 1 that is a solid at ambient temperature, has a melting point range of 85–86 °C, and has the highest density of any known nitrate ester composed of only carbon, hydrogen, nitrogen, and oxygen.

As part of our efforts to develop new energetic materials, we became interested in employing the Kaplan–Shechter reaction to construct novel energetic gem-substituted nitro-N-heterocyclic moieties. The Kaplan–Shechter reaction is one of the most convenient methods for the synthesis of molecules containing gem-dinitro substituents, and involves the treatment of a nitronate salt with nitrite ions in the presence of an oxidant, such as silver nitrate.

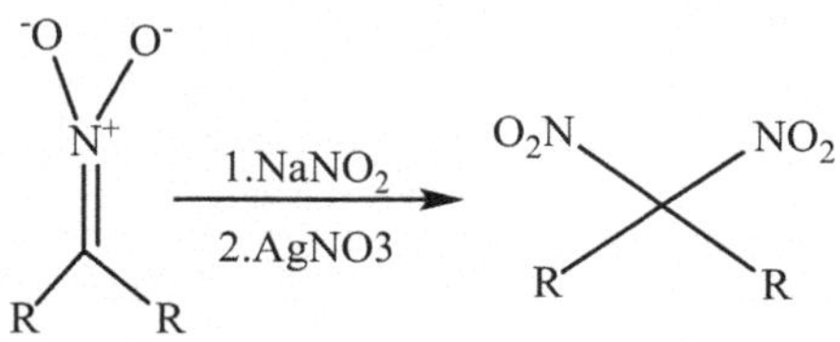

David E original goal was to use nitro-compound 2 as a substrate for oxidative coupling reactions with heterocyclic nucleophiles, such as 5-aminotetrazole.

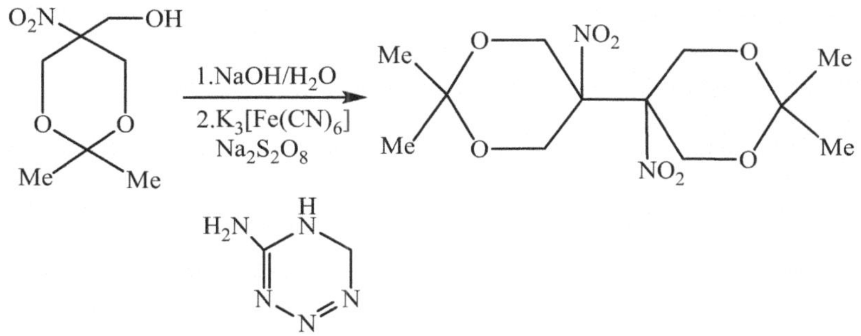

We investigated the modified Kaplan–Shechter [27, 28] reaction as a method to reach this target. The reaction conditions involve deformylation of 2 by treatment with hydroxide in the presence of potassium ferricyanide and sodium persulfate, followed by addition of 5-aminotetrazole. The overall result was the production of a white precipitate in low yield. ^{1}H NMR spectroscopy showed that no coupling had occurred between 2 and 5-aminotetrazole. Instead, homocoupling of the nitronate salt of 2 appeared to have taken place, albeit inefficiently (12% yield), to provide 3. The homocoupling of nitronate salts has previously been reported [29], although the use of 3 as a homocoupling substrate under the modified Kaplan–Shechter conditions has not been reported.

Alternative methods for the synthesis of 3 have been reported and include homocoupling of the 5-bromo-5-nitro derivative of 2 in the presence of tert-butyl thiol [30], and the use of an $S_{RN}1$ substitution reaction to provide the product in 30–50% yield [31–33]. Other methods have also been reported [34, 35].

During the examination of the homocoupled product, we reasoned that it would be possible to hydrolyze the dimethyl ketal protecting group to access the tetraol product 4, which has not been reported to date. The novel nitrate ester molecule 1 can then be accessed by exhaustive further nitration of the tetraol 4.

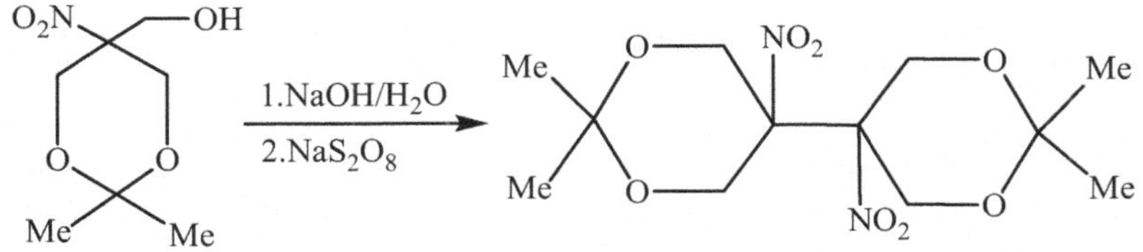

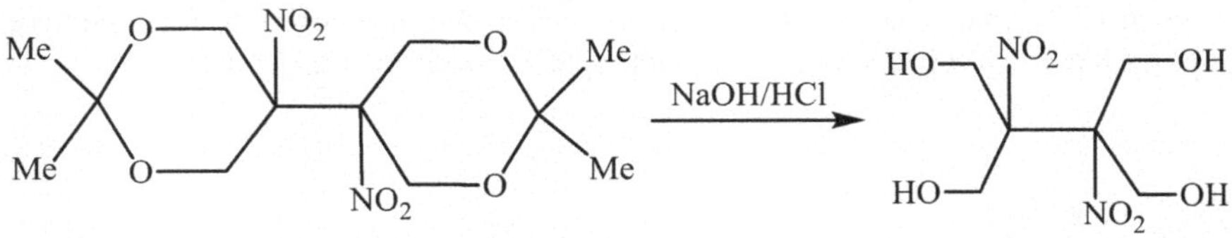

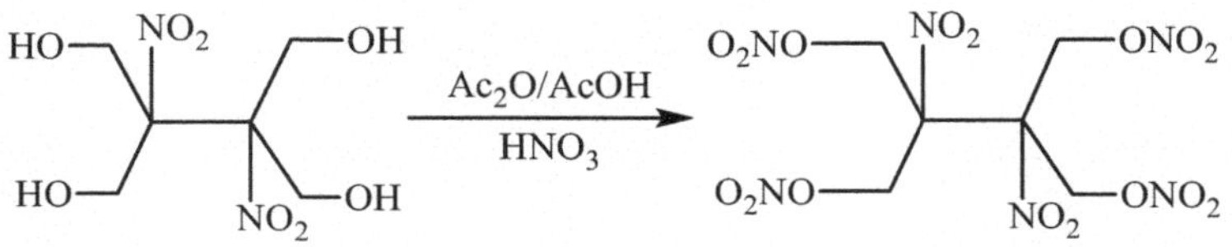

However, for the method to be feasible, a dramatic improvement in the yield of the oxidative coupling step was needed.

Starting from the commercially available dioxane 2, we studied the effects of concentration, oxidizer, and nitronate formation on the yield of 3. The amount of the potassium ferricyanide catalyst was also varied. Interestingly, we found that increasing amounts of potassium ferricyanide led to reduction in the yield of the homocoupled product 3, whereas the yield of 3 increased in the absence of the catalyst. After optimization of temperature, concentration, and number of equivalents of base, we were able to improve the yield of the homocoupled product to 65%.

With an improved method for the preparation of 3 in hand, we turned our attention to the remaining two steps. Ketal deprotection was easily accomplished using methanolic HCl. Removal of the solvent led to 4 in good yield (85%). The nitration of the tetraol was found to proceed smoothly in the presence of acetyl nitrate in acetic acid to provide the corresponding tetranitrate ester 1 in 85% yield.

Compound 1 was characterized spectroscopically and thermally. The material begins to decompose at 141 °C with a decomposition energy release of 1818 J/g. Interestingly, the material has a melting point range of 85–86 °C. This property may allow the material to be used in melt-castable explosives applications. The heat of formation was measured to be 371 kJ/mol by combustion calorimetry using a Parr 6300 bomb calorimeter. The tetranitrate ester 1 can be recrystallized from a variety of solvents, and large hexagonal crystals can be obtained from ethanol. The density is 1.973 g/cm^{3}A digital photograph of the crystal morphology is shown in Fig. 8.19.

The sensitivity of material 1 towards destructive stimuli, such as impact, spark, and, friction was also investigated. It was determined that the sensitivity properties of 1 were very similar to those of pentaerythritol tetranitrate (see Table 8.27).

Explosive performance calculations were performed with the CHEETAH thermochemical code, [36] using the experimental crystal density and measured heat of formation value as the input data. As displayed in Table 8.28, the performance of 1

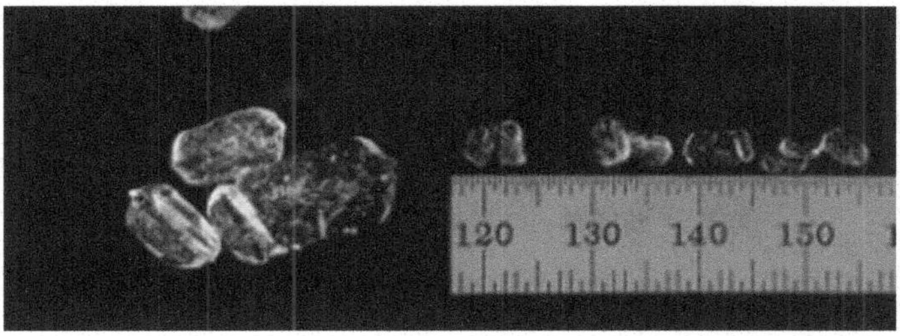

Fig. 8.19 Photographs of crystals of 1 displaying the crystal morphology of crystals grown from ethanol. The ruler scale is in mm

Table 8.27 Sensitivity properties of 1

	Impact[a]	Spark[b]	Friction[c]	DSC[d]
1	2.7 J	0.625 J	74.5 N	140 °C
PETN	2.9 J	0.625 J	56.8 N	160 °C

[a]LANL type 12, 50% drop height, 2.5 kg
[b]ABL spark threshold initiation level (TIL)
[c]50% load Bruceton up/down method
[d]10 8C ramp rate

Table 8.28 Predicted performance properties of 1

	V_{Det} [kms^{-1}]	P_{C1} [GPa]
1[a]	9.1	40
HMX[b]	9.1	39

[a]Cheetah 5.0 calculation
[b]Livermore explosives handbook

is predicted to be equal to that of HMX, a well-characterized high-performance explosive. Unlike HMX however, nitrate ester 1 has a low melting point, which may provide a unique opportunity for melt-castable explosive components. Additionally, 1 offers the possibility of use as a high-energy plasticizer or double-base propellant ingredient. Further experiments are under way to fully characterize the explosive performance and properties of this novel nitrate ester. Additionally, compound 4 can serve as a synthon for new energetic materials and we are currently investigating the synthetic utility of 4.

8.6 Nitro-Aryl Nitrates [1, 25, 37, 38]

A series of explosives have been prepared by the nitration of aromatic alcohols. Their explosive powers can reach or exceed TNT. In the following section, the application potentials of nitro-aryl nitrates will be discussed.

8.6.1 Trinitro-Phenoxyethanol-ω-Nitrate

Trinitro-phenoxyethanol-ω-nitric acid ester is a white crystal explosive with stronger explosive power and slightly lower stability than TNT. The nitrate group is decomposed at high temperature to release acid that can further catalyze the decomposition, leading to decomposition at temperatures above 110 °C.

(1) Properties of trinitro-phenoxyethanol-ω-nitrate

Formula: $C_8H_6N_4O_{10}$
Chemical structure:

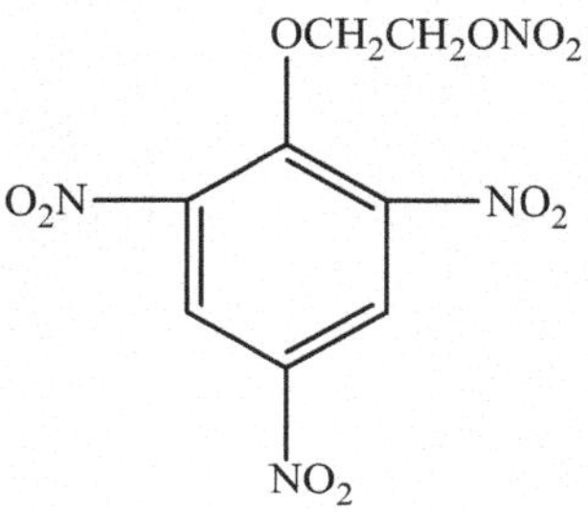

Molecular weight: 302.21
Oxygen balance: −52.94% (Calculated based on the amount of produced CO_2)
Nitrogen content: 18.54%
Melting point: 104.5 °C
Solubility: soluble in acetone and most organic solvents and slightly soluble in water; able to gelatinize collodium.

(2) Preparation of trinitro-phenoxyethanol-ω-nitrate

Trinitro-xyphenoxyethanol-ω-nitrate can be prepared by the nitration of dinitroxyphenoxyethanol that is synthesized by the 2 h reaction between dinitrochlorobenzene and sodium hydrogencarbonate at 180 °C. The nitration is conducted at 15 °C for 40 min. The crude product is washed, purified and stabilized to afford the target product, trinitro-phenoxyethanol-ω-nitrate, in the yield of 81.39%. Alternatively, trinitroxy-phenoxyethanol can be synthesized from trinitrochlorobenzene and is nitrated to afford trinitro-phenoxyethanol-ω-nitrate in a yield of 67.83%. The main reactions for the synthesis of trinitro-phenoxyethanol-ω-nitrate are:

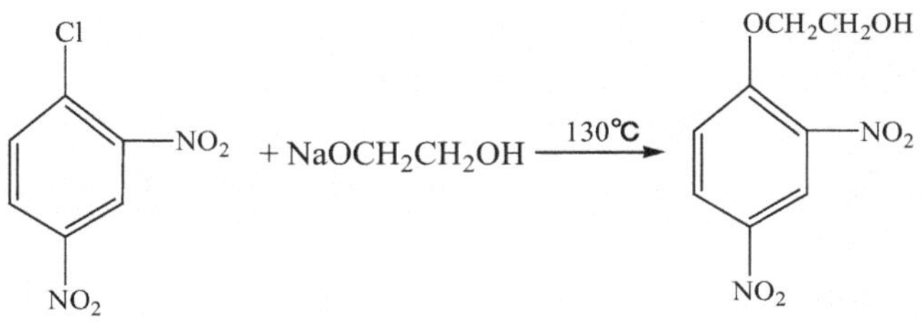

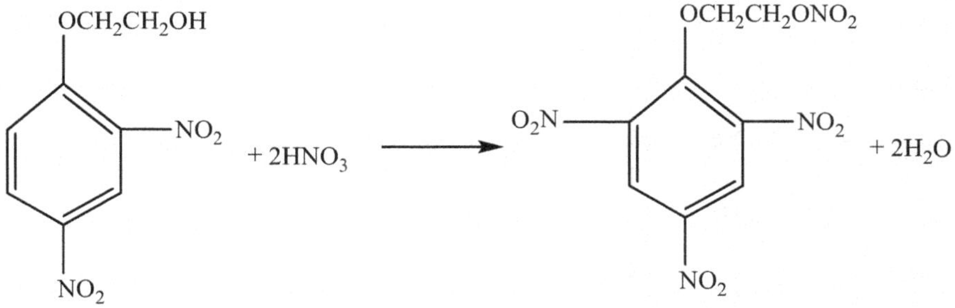

The alternative method for the synthesis of trinitro-phenoxyethanol as the raw material for the preparation of trinitro-phenoxyethanol-ω-nitrate is:

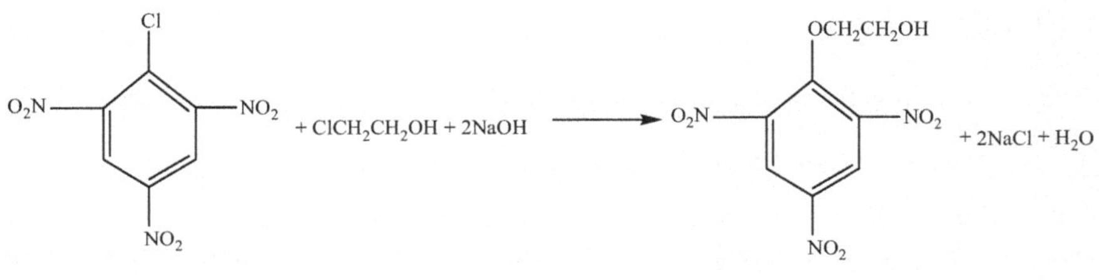

Trinitro-phenoxyethanol-ω-nitrate is more sensitive to impact than picric acid and less sensitive than dynamite. Its explosion power is similar to that of dynamite. The lead block expansion value of trinitro-phenoxyethanol-ω-nitrate is 350–355 cm^3.

8.6.2 *Trinitrophenyl-α-Glycero-β, γ-Dinitrate (α-Glyceryl-Trinitrophenyl Ether β, γ-Dinitrate)*

Trinitrophenyl-α-glycero-β, γ-dinitrate is a yellow crystalline compound with a higher explosive power than that of TNT.

(1) Properties of trinitrophenyl-α-glycero-β, γ-dinitrate

Formula: C9H7N5O13
Chemical structure:

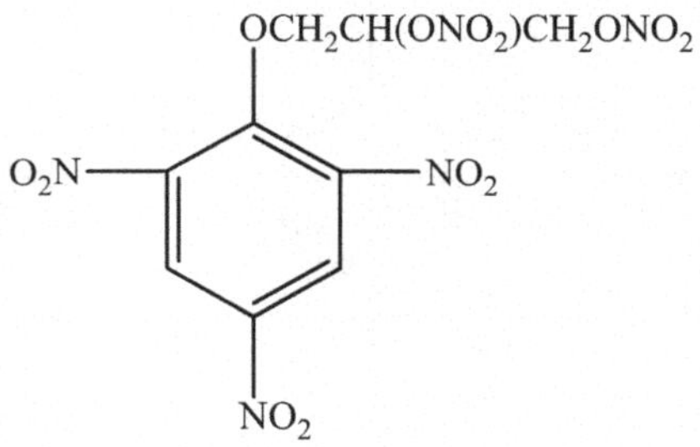

Molecular weight: 393.23
Oxygen balance: −34.58% (Calculated based on the amount of produced CO_2)
Nitrogen content: 17.81%
Density: 1.65 g/cm^3
Melting point: 125–127 °C
Solubility: insoluble in water, and miscible with most organic solvents.

(2) Preparation of trinitrophenyl-α-glycero-β, γ-dinitrate

Dinitrochlorobenzene and glycerol react in the presence of sodium hydroxide to form dinitrophenyl-α-glycerol (II) that is then nitrated to afford trinitrophenyl-α-glycero-β, γ-dinitrate as shown below.

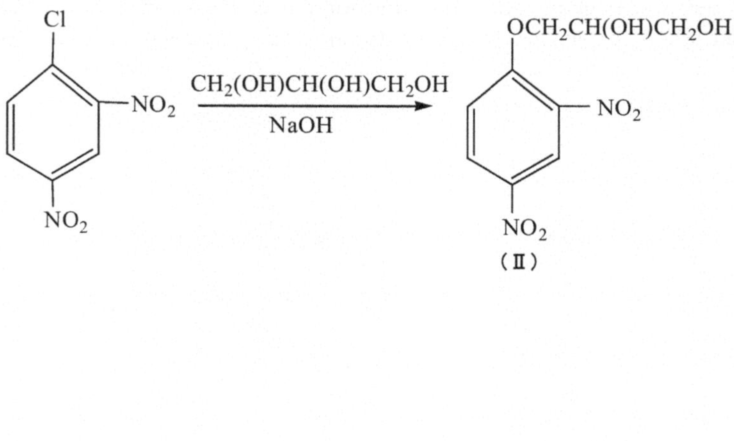

(II)

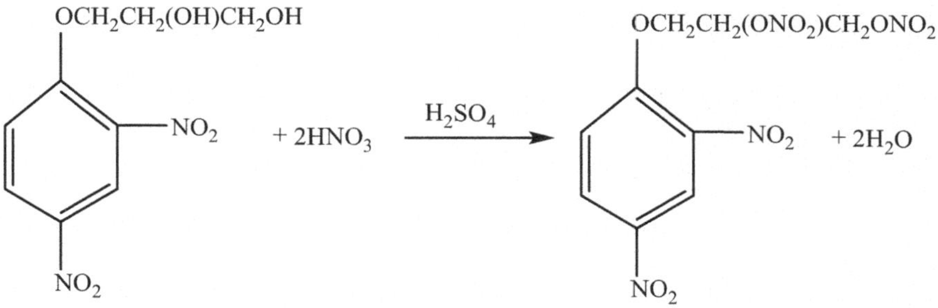

The lead block expansion value of product (I) is 25–30% greater than picric acid. The dinitrate compound (II) can also be formed by the nitration.

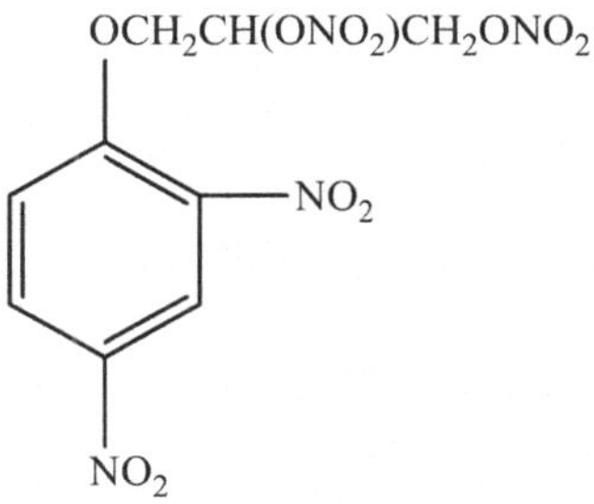

The melting point of the compound (II) is 119–120 °C.

8.6.3 *Trinitrophenylethanol Nitrate*

Trinitrophenylethanol nitrate, also known as trinitrobenzene-β-ethyl nitrate, melts at 83 °C. It can be prepared from trinitrotoluene. Trinitrotoluene reacts with formaldehyde in an alkaline medium to form the corresponding alcohol (melting point, 112 °C) that is then nitrated to form trinitrophenylethanol nitrate.

(1) Properties of trinitrophenylethanol nitrate

Formula: $C_8H_6N_4O_9$
Chemical structure:

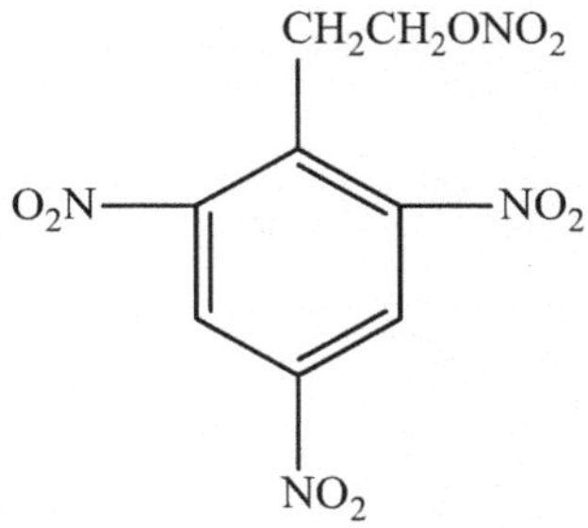

Molecular weight: 302.21
Oxygen balance: −52.99% (Calculated based on the amount of produced CO_2)
Nitrogen content: 18.54%
Density: 1.59 g/cm^3
Melting point: 83 °C
Solubility: miscible with most organic solvents, 22.3% in hot water and 2.65% in cold water.

(2) Preparation of trinitrophenylethanol nitrate

Trinitrophenylethanol nitrate is usually prepared using trinitrotoluene (TNT) as the raw material. Trinitrophenylethanol is first synthesized by the addition reaction between formaldehyde and the methyl group on the benzene ring in an alkaline medium, which is then nitrated with nitric acid to afford the target product. The major chemical reactions are:

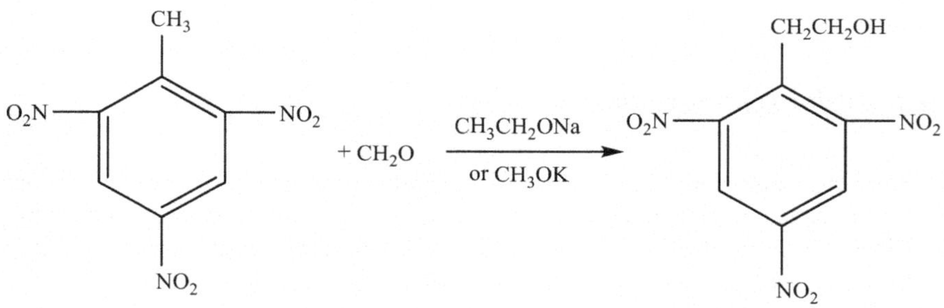

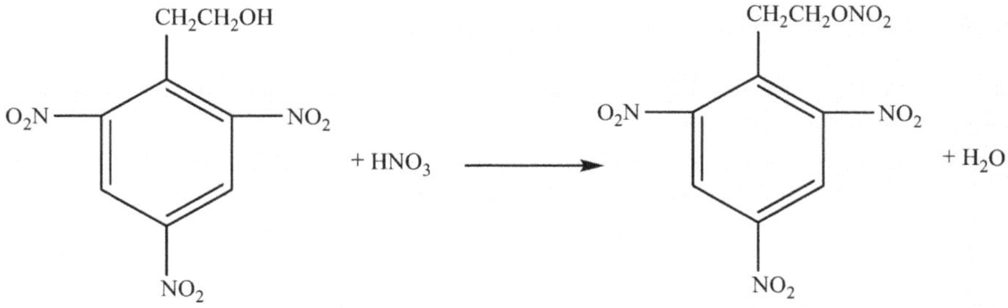

For the reaction, trinitrotoluene (TNT) powder is dissolved in one equivalent of sodium alcohol solution and refluxed at 80 °C under stirring while formaldehyde is slowly added. The reaction mixture is then refluxed at 110 °C for 2 h under stirring. The product is filtrated and washed to afford a pale-yellow solid in 86.1% yield.

8.6.4 Other Nitrate Esters of Trinitrotoluene

(1) 2,4,6-trinitro-3-(β-nitroxyethylamine) toluene

2,4,6-trinitro-3- (β-nitroxyethylamine) toluene is a yellowish crystalline compound prepared from TNT, but with stronger explosive power than TNT.

1. Properties of 2,4,6-trinitro-3-(β-nitroxyethylamine) toluene

Formula: $C_9H_9N_5O_9$
Chemical structure:

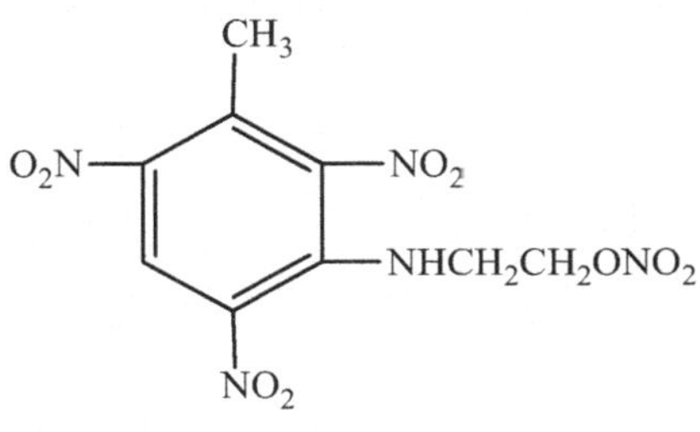

Molecular weight: 331.25
Oxygen balance: −65.20% (Calculated based on the amount of produced CO_2)
Nitrogen content: 21.14%
Density: 1.69 g/cm^3
Melting point: 100–102.5 °C
Solubility: insoluble in water and miscible with most organic solvents.

2. Preparation of 2, 4, 6-trinitro-3- (β-nitroxyethylamine) toluene

The meta-nitro group of TNT is reduced to $-NH_2$ that then reacts with chloroethanol to loss an HCl to afford nitroxylethylamine methylbenzyl ethanol. The ethanol is nitrated with weak nitrating reagent to afford the target compound, 2, 4, 6-trinitro-3-(β-nitroxyethylamine) toluene, as shown below.

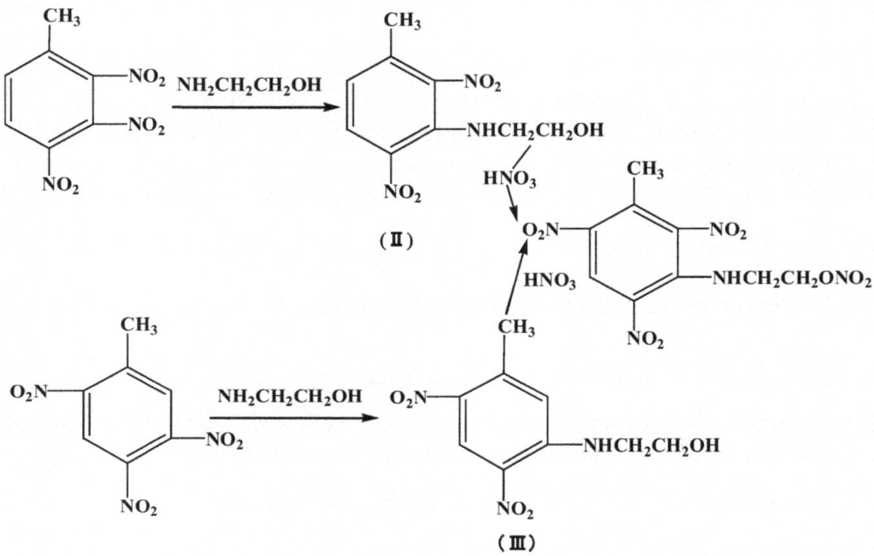

The melting points of compounds (II) and (III) are 85.5–87.5 °C and 115–117.5 °C, respectively. Both of them can be nitrated to form the target compound.

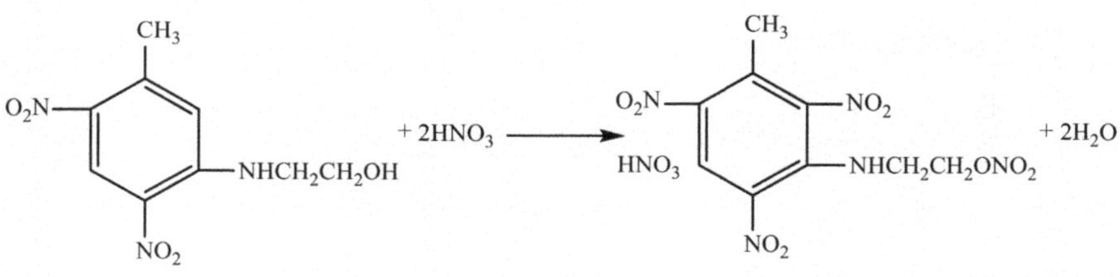

The conversion of the reaction is 61.32%. However, the multiple reactions, by products and isomers make the separation and purification of the target product very difficult.

(2) Hexanitrodiphenylglycerol mononitrate

Formula: $C_{15}H_{10}N_7O_{17}$
Chemical structure:

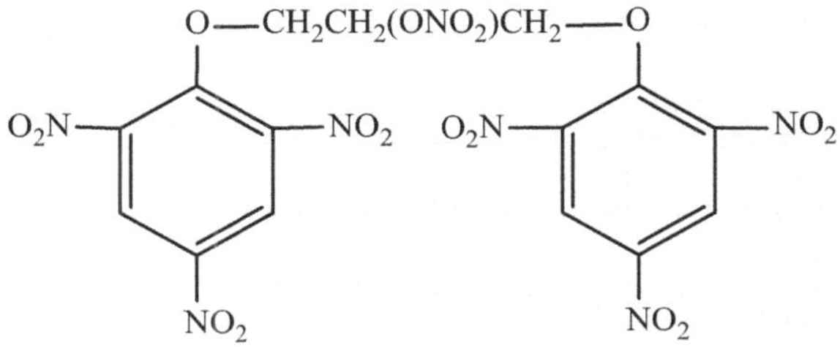

Molecular weight: 560.37
Oxygen balance: −50.14% (Calculated based on the amount of produced CO_2)
Nitrogen content: 17.5%
Melting point: 178–178.5 °C
Solubility: insoluble in water and miscible with most organic solvents.

Hexanitrodiphenylglycerol mononitrate is a straw yellow crystal with the melting point of 178–178.5 °C. It has no practical values due to its low stability.

References

1. Liu JP (2015) Liquid explosives. Springer Press, Heidelberg
2. Liu JP (1987) Basic chemical of propellants and explosives. GuiZhou Press, Guiyang
3. Urbanski T (1965) Chamistry and technology of propellants & explosives (II). Pergamon Press, Oxford
4. Wu CY (1980) The chemical technology of nitroglycerin. National Defense Industry Press, Beijing
5. Li XZ, Li H, Lian P et al (2016) Synthesis crystal structure and thermal properties of two nitrate esters energetic materials of 3,6-dinitrocyclohexane. Chin J Energ Mater 24(7): 664–668
6. Austruy H (1993) Double-base propellants. In: Davenas A (ed) Solid rocket propulsion technology. Pergamon Press, Oxford, pp 404–405
7. Agrawal JP, Singh H (1993) Qualitative assessment of nitroglycerin migration from double-base and composite modified double base rocket propellants: concepts and methods of prevention, propellants. Explos Pyrotech 18:106–110
8. Xianga YQ, Sluggettb GW (2010) Development and validation of a GC method for quantitative determination of enantiomeric purity of a proline derivative. J Pharm Biomed Anal 53:878–883
9. Reppe JW (1949) Acetylene chemistry, P.B. Report 18852s. New York
10. Gouranlou F, Kohsary I (2010) Synthesis and characterization of 1,2,4-butanetrioltrinitrate. Asian J Chem 22(6):4221–4228
11. Luo SG, Chen F, Luo Y et al (1999) Effect of propellant components on degradation of polyether-urethane binder. Chin J Propellant Technol 20(2):1999

12. Krümimel DAG, Christianstadt D (1945) Explosives plants technical report PB 925. US Dept of Commerce, Washington, DC
13. Elrick DE, Gardner WH, Marans NS et al (1954) Preparation of halides of pentaerythritol trinitrate1. J Am Chem Soc 76(5):1374–1375
14. Sun RK, Ren TS (1981) The chemistry and technology of high explosive. National Defense Industry Press, Beijing, pp 700–701
15. Sun YB, Zhou JY (2003) Advance in research of energetic plasticizers. Chem Propellants Polymeric Mater 1(5):20–25
16. Pang WQ, Fan XZ, Zhang LB (2008) Progress in study on the energetic adhesive for solid propellant. Chem Adhes 30(5):46–49
17. Backof E, Cai B (1983) Polyvinyl alcohol nitrate as a component of explosives. Fire Explos 1983(6):49–55
18. Yuhong Zhang, Qichao Zou, Jianjun Yuan et al (2004) Ce ~ (4 +) initiated polyvinyl alcohol grafted methyl methacrylate. Colloids Polym 22(4):14–16
19. Li GW (2000) Study on the explosion properties of nitrate plasticized high energy propellant. J Solid Rocket Technol 3:010
20. Clingher A, Doran CF (2006) Modular invariants for lattice polarized K3 surfaces. Mich Math J 55(2):355–393
21. Jiang H, Ma J, Wu M (2005) Preparation of inositol triphosphate by microwave hydrolysis of phytic acid. Chem Res Appl 17(3):424–426
22. Wen W, Yu Z, Dechang J et al (2002) Study on synthesis of SrBi-2Ta-2O-9 (SBT) powder by sol-gel process using bismuth subnitrate as bismuth source. J Chin Ceram Soc 30(1):117–120
23. Karlsson S, Högberg HE (2002) Catalytic enantioselective 1, 3-dipolar cycloaddition of nitrones to cyclopent-1-enecarbaldehyde. Tetrahedron: Asymmetry 13(9):923–926
24. Li XZ, Li H (2016) Synthesis, crystal structure and thermal properties of two nitrate esters energetic materials of 3,6-dinitrocyclohexane. Chin J Energ Mater 24(7):664–668
25. Guillén F, Martínez AT, Martínez MJ (1992) Substrate specificity and properties of the aryl-alcohol oxidase from the ligninolytic fungus pleurotuseryngii. Eur J of Biochem 209 (2):603–611
26. Chavez DE, Hiskey MA, Naud DL et al (2008) Synthesis of an energetic nitrate ester. Angew Chem Int Ed 47(43):8307–8309
27. Garver L C, Grakauskas V, Baum K (1985) ChemInform abstract: catalyzed oxidative nitration of nitronate salts. Chem InfSdienst 16(41):no-no
28. Kornblum N, Singh HK, Kelly WJ (1983) Substitution reactions which proceed via radical anion intermediates. 26. Oxidative substitution of nitroparaffin salts. Chem InfSdienst 14(28): no-no
29. Kaplan RB, Shechter H (1961) A new general reaction for preparing gem dinitro compounds: oxidative nitration. J Am Chem Soc 83(16):25–28
30. Amrollah-Madjdabadi A, Beugelmans R, Lechevallier A (1987) Sur la réactivité de gem-bromonitroalcanes en présence de thiolates = on the reactivity of gem-bromonitroalcanes in the presence of thiolates. Tetrahedron Lett 28(39):4525–4528
31. Beugelmans R, Amrollah-Madjdabadi A, Frinault T, et al (1995) ChemInform abstract: new and straightforward access to olefins, butenolides and β- nitro γ-butyrolactones from precursors obtained via SRN1 reactions. ChemInform 26(27):no-no
32. De MM, Vanelle P, Bernadini E, et al (1992) Evaluation of the mutagenic and genotoxic activities of 48 nitroimidazoles and related imidazole derivatives by the ames test and the SOS chromotest. Environ Mol Mutagen 19(2):167
33. Crozet MP, Archaimbault G, Vanelle P et al (1985) Reactions S RN 1 en serie heterocyclique: IV: reactivite des sels du dimethyl-2,2 nitro-5 dioxanne-1,3. Tetrahedron Lett 26(42):5133–5134
34. Senkus M (1951) Nitro cyclic acetals. US Patent 2,543,472, 27 Feb 1951
35. Piotrowska H, Poplawska M (1983) ChemInform Abstract: reaction of α-cyanonitro compounds with nitronate salts. ChemInform, 14(16):no-no

36. Bastea S, Fried LE, Glaesemann, Howard KR, Vitello PA, et al (2006) Cheetah 5.0, User's Manual. Lawrence Livermore National Laboratory
37. Jilan Wu, Jinshan Wei (1991) Formation of nitro and nitrate compounds of tributyl phosphate in 30% TBP-N-Dodecane-2NHNO_3 System. Nucl Sci Eng 4:88–93
38. G.Bonner I, Friesel DE (1959) J Chem Soc 3902

Chapter 9
Azido Nitrates

Azido nitrate, with both nitrate and azido ($-N_3$) groups on different positions, is a new energetic material [1], which was mainly used as adhesive and energetic plasticizer currently. Since 1864, scientists have done a large number of studies on structures, properties, syntheses, and applications of azide and azido nitrate [2]. In 1987, scientist in the United States first reported the method to synthesize pentaerythritol diazido dinitrate (PDADN) [3], which caused a strong reaction globally to spark research on azide and became an important direction to enhance the energy of energetic materials. Solvents containing azido groups, intermolecular adhesives, adhesives, plasticizers, antioxidants, and other additives with unique properties play unique roles in the gunpowder, propellant, high-energetic explosives, and other materials not only to increase the energy level of energetic materials but also to give them other excellent properties [4]. Azido nitrates are liquid nitrogen-rich compounds with high nitrogen contents, which have both the characteristics of nitrates and the energy of azide. Typically, these compounds are believed as high-energetic binders or high-energetic plasticizers. They should be the optimal materials as propellants of polymer adhesive explosives in the future with extremely broad application prospects. Apart from the azido nitrate, there are typical azides as: azidoalkanes, n-azidonitro alcohol, azido nitro compounds, azido ethers (ketones), etc.

In the same compound, small molecule alkanes with two or more azido groups or azido nitrate and azido ether (ketone) with multiple azido groups are new energetic plasticizer of high-energetic explosives and propellants [5]. In order to distinguish azides from azido nitrate and use azido nitrate easily, only typical preparation and properties of azido nitrate are described in this chapter.

J. Liu, *Nitrate Esters Chemistry and Technology*,
https://doi.org/10.1007/978-981-13-6647-5_9

9.1 Properties of Azido Nitrate

9.1.1 Physical and Chemical Properties of Azido Nitrate

Azido nitrate compounds are hydrazoic acid (HN_3) derivatives of nitrates with nitroxyl ($-ONO_2$), azido ($-N_3$) and 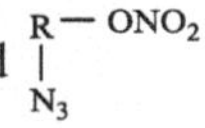 as their molecular structural characteristics. According to Panling's electronegativity concept [6], the electronegativity of azido group is 2.95 and 3.18 [7], which is close to that of chlorine 3.0. With the rapid development in the physical method to determine crystal structures, through experiments, azide can be divided into two groups according to the molecular orbital theory [8]. One is the group of inorganic azides with symmetric azido groups that have ionic bonds to the molecules to some extent; and another is the organic azide group with covalent bonds.

(1) General chemical properties

Azido nitrate is chemically active, which is mainly showing in $-N_3$, and more active than pure nitrate. Under illumination or heating, the azido can be decomposed into triazene that can have many reactions. Azido nitrate is energetic explosive material, and its operation and using must be done carefully. During it preparation, halogenated alkyl, acyl halide, diazonium salts can be used to react with sodium azide on the corresponding nitrate chain, or hydrocarbon-substituted hydrazine and hydrazine react with nitrous acid, to produce azido nitrate. The ring-opening nitration of ring azido hydroxy compound can also be used to prepare azido nitrate.

When the end of azido nitrate is open with a carbon terminal, it is nucleophilic and unstable, and easily emits nitrogen, in which Curtius rearrangement reaction occurs through an intermediate of acyl azide. Some azides or azido nitrates have 1,3-dipolar, thus their cycloaddition reactions can be carried out.

Azido is relatively stable. With acid, it is difficult to decompose but produce hydrazoic acid that reacts with other metal ions to produce more dangerous high-explosive compounds. Azido nitrate is easy to saponificate and decompose, but without danger.

(2) Physical properties

Typically, azido nitrate is waxy solid and insoluble in water. Its melting point rises with the increase of carbon atoms in azido group. Its combustion heat is high and flash point (5s) is increased with the increase of carbon chain. Its density, impact sensitivity, and friction sensitivity increase with the increase of azido group.

9.1.2 Thermal Stability and Explosion Properties of Azido Nitrate

(1) Thermal stability of azido nitrate

The thermal decomposition of azido nitrate is usually divided into covalent and ionic types. In the covalent azides, the N–N bonds in $-MN_3$ do not have the same length because C–N bond is enhanced due to the electronegativity of the $-N_3$. For example, the N–N bond lengths in triazidomelamine are $—N\overset{1.26}{—}N\overset{1.18}{—}N$ with the following decomposition equation:

$$R{-}N_3 \rightleftharpoons RN + N_2$$

In ionic inorganic azides [9], the decomposition is not to break the N–N bond because their N–N bond lengths are equal (regardless of azido polarization), with the following decomposition equation:

$$MN_3 \rightleftharpoons M + N_3$$
$$N_3^- \rightleftharpoons N_3^+ + e$$
$$2N_3^- \rightarrow 3N_2$$

In most cases, the azide is to explosively decompose. But, a very few of them decompose slowly through the thermo-chemistry, photochemistry, or discharging. Through the explosive decomposition, the corresponding element is formed, and the decomposition heat is equivalent to the Standard formation enthalpy of the compound.

Decomposition of some azides can generate nitrogen compounds,

$$3LiN_3 \rightarrow Li_3N + 4N_2 \uparrow$$

Generally, the decomposition temperature of azide is relatively high, particularly that of metal azide. If the decomposition is explosive reaction at 200 °C under low pressure and products are collected on the surface with liquid nitrogen cooling, the reaction is,

$$6HN_3 \rightarrow 7N_2 + H_2 + (NH)_4$$

Decomposition of alkali metal azide is not explosion but slow decomposition only.

$$2NaN_3 \rightarrow 2Na + 3N_2 \uparrow$$

The decomposition of heavy-metal azide is due to excitation of azide ion, causing an electron to jump to the conduction band to produce azide radicals. Dissociation of ground state azide radicals into the ground state of N and N_2 is prohibited according to the election law. However, dissociation of the ground state azide radicals into the excited state of N and N_2 is allowed according to the election law, which needs a 259 kJ/mol of energy, thus it is not important at room temperature. Interaction between two azido radicals is allowed according to the election law, and the process is exothermic. Thus, it can be further considered that this step in the decomposition of solid ion azide is an important step.

Azide can inhibit activities of cytochrome oxidase and a variety of other enzymes, which can lead to abnormal phosphorylation and cell respiration. The main acute toxicity of hydrazoic acid and its sodium salt caused vascular angiotasis to extremely reduce, due to the direct effect on vascular smooth muscle. This effect is similar to and stronger than that of nitrite. The difference is that alkyl azide does not cause high blood red protein in the body. Azides may stimulate breathing and enhance cardiac strength. And large dose of azide can increase blood pressure, leading to convulsion, followed by depression and shock. Within organic azide compounds, ethyl-azide and amyl-azide are valid vessel tension-reducing reagents. However, effects of some aromatic azides are not significant.

(2) Explosion properties of azido nitrate

Hydrazoic acid salts of heavy metals, such as silver azide (AgN_3) and lead azide (PbN_6), are highly explosive. Because lead azide is extremely sensitive to impact, it is used as a detonator. Generally, its alkali salts are not explosive, in which sodium azide is most widely used and decomposes in water to release the hydrolysis product hydrazoic acid. Alkyl azide is stable at room temperature, but it is easy to explode with heating. When temperature is increased, alkyl azide is decomposed to release hydrogen azide (HN_3). Aryl azides are colorful and relatively stable solid, easy to explode under impact, decompose when they are melted to release HN_3. Sodium azide is easy to react with lead or copper to generate very explosive metal azides.

9.2 Properties and Preparation of Diazido Neopentyl Glycol Dinitrate

Diazido neopentyl glycol dinitrate is also known as diazido pentaerythritol glycol dinitrate (PDADN).

9.2.1 Properties of Diazido Neopentyl Glycol Dinitrate

Formula: $C_5H_8N_8O_6$
Chemical structure:

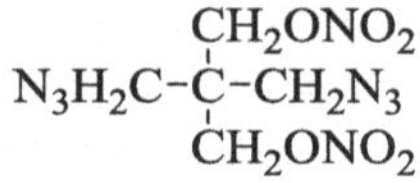

Molecular weight: 276.14
Oxygen balance: −46.35% (Calculated according to formed CO_2)
Nitrogen content: 40.57%
Color appearance: white waxy solid with low melting point
Melting point (°C): 32–34
Density (g/cm^3): 1.507
Solubility: Insoluble in water, soluble in organic solvents such as chloroform;
Mechanical sensitivity: Impact sensitivity (50%) with falling height of 15.4 cm, friction sensitivity (4%) (25 kg-66° angle)
Decomposition temperature (°C): 179, decomposition activation energy E = 123.33 kJ/mol, where a is $10^{11.44}$
Heat of combustion: 14,259 J/g
Heat of formation: 825.42 kJ/mol
Flash point: 225 (5s)
Compatibility: good compatibility with NG, NC, polyether and GAP.

9.2.2 Preparation and Process of PDADN [3]

Since the first reported synthesis of PDADN from the United States, its synthesis technology is based on the raw material pentaerythritol with five steps, including chlorination, cyclization, azidization, nitrolysis, and nitration. While, this study is different from the USA's reported, where pentaerythritol was brominated, azidized and then nitrated to three-step synthesize PDADN, which had few steps and high yield (total yield of 73.02%) to realize the green nitration with limited pollution. Therefore, it is easy to produce in industry.

Pentaerythritol is used as the starting material to PDADN after the three-step reaction of bromination, azidization, and nitration, and its main chemical reactions are

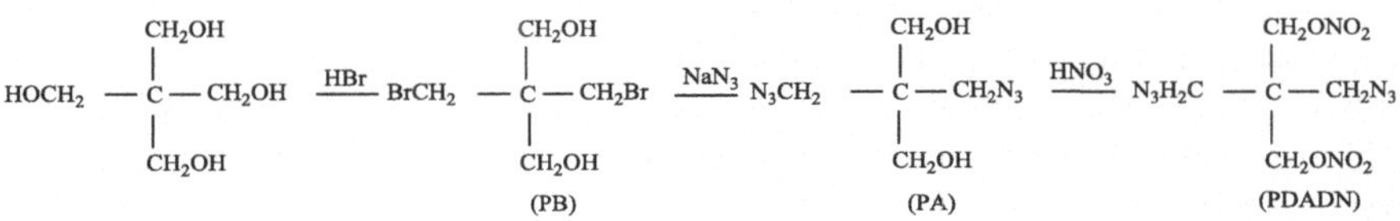

Pentaerythritol (28 g, 0.20 mol), 50% hydrobromide solution (100 g, 0.50 mol), and acetic acid (3 g) were added in the three-neck bottle equipped with mechanical

stirring, thermometer and a refluxing condenser. The solution was warmed up to 115 °C, refluxed to react for 5 h. After the reaction, excess water and hydrogen bromide were removed through vacuum evaporation, cooled to room temperature, neutralized with alkaline to neutral, then treated with mixed toluene and water, filtered and dried to obtain 2,2-bromomethyl-1,3-propanediol (48.94 g, 93.2%) with a melting point of 108–109 °C [4].

Into the 250 mL three-neck flask, 13.1 g (0.05 mol) of PB, 7.25 g (0.11 mol) of NaN_3, and 175 mL of DMF were added and warmed up to 120 °C to react for 3–4 h. After the reaction, DMF was distilled and removed under vacuum, and 100 mL of methylene chloride was added. Through filtering, dichloromethane solution of PA was collected for the next step. The yield was up to 97.5%.

Into a three-neck flask equipped with thermometer and a stirrer with 9.2 g of PA dichloromethane extract, a calculated quality of HNO_3 was dropped at 15–20 °C. After 10 min, ice water was poured into the solution to separate the organic layer. Then, 150 mL of water were used to wash twice, then 80 mL of 1% $NaHCO_3$ solution were used to washed for two times, and finally water was used to wash until it was neutral. Methylene chloride layer containing PDADN was dried with anhydrous $MgSO_4$. After the removal of solvent, 12.11 g of PDADN product were obtained with yield of 89.6% and m.p. of 39–41 °C.

(1) Product characterization

1. IR analysis

There was a $-N_3$ strong absorption peak at 2107 cm^{-1}, and $-ONO_2$ strong absorption peaks at 1641 cm^{-1} and 1275 cm^{-1}.

2. Nuclear magnetic resonance analysis

FX290Q NMR from Japan's JEOL company was used. D3.52 ppm was the chemical shift of CH_2N_3. D 4.43 ppm was the chemical shift of CH_2-ONO_2.

3. Elemental analysis

Italian CARLO ERBA SJRUM EN TA ZION E1106 elemental analyzer was used. Elemental analysis results (%): in C5H8N8O6, calculated values (%) were C 21.74, H 2.90 and N 40.58, and the measured values (%) were C 20.85, H 2.95 and N 40.49. According to IR, ^{1}HNMR and elemental analysis, the as-synthesized product was the designed one with a purity of 98.55% (GC).

(2) Preparation process and controlling conditions of PDADN

In order to study the synthesis of PDADN from pentaerythritol, the chlorination and then bromination replacement of pentaerythritol were done by the authors. Within many reactions, azide radicals was easy to exchange after the bromination, and its yield was high. The preparation process of PDADN from pentaerythritol was shown in Fig. 9.1.

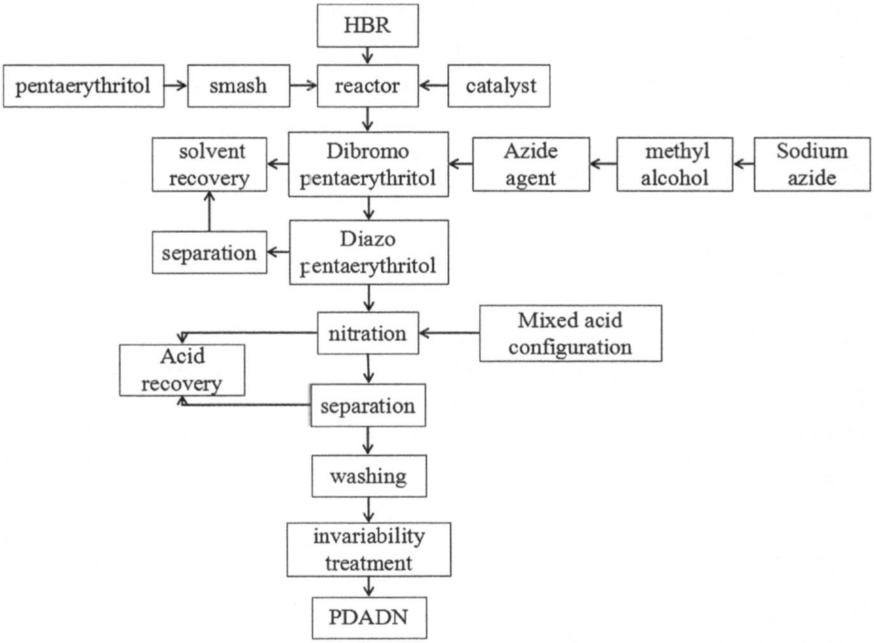

Fig. 9.1 The preparation process of PDADN from pentaerythritol

In this process, the conversion of every reaction step was relatively high. After the recrystallization, the product was white crystal with a melting point of 34.5 °C and a purity of 99.3%.

9.2.3 Problems in the Synthesis of PDADN

(1) Separation of PB

The reaction mixture may contain unreacted pentaerythritol, pentaerythrityl mono-, tri- and tetrabromides. Because PB is insoluble in water and toluene, and pentaerythritol and its monobromide can be dissolved in water, its tri- and tetrabromides are soluble in toluene, PB can be separated easily out with an equal volume of toluene and water mixtures. Then, hot water can be used to purify and recrystallize it.

(2) Effects of nitration coefficient on the yield of PDADN

Due to the significant effects of nitration coefficient on the nitration yield, at the constant temperature, nitration yields with different nitration coefficients are studied and shown in Table 9.1.

Table 9.1 The influence of nitrating coefficient on PDADN yield

Number	Nitration coefficient	Reaction temperature (°C)	Yield (%)
1	2	15–20	50.45
2	4	15–20	70.35
3	6	15–20	80.36
4	8	15–20	82.10
5	10	15–20	82.10
6	12	15–20	82.09

As listed in Table 9.1, with the increase of nitration coefficient, the yield of PDADN was increased continuously. When the nitration coefficient is more than 8, PDADN yield is not increased further. Therefore, it is appropriate to choose 8 as the nitration coefficient of PA, meaning the molar ratio of nitric acid over PA is 8:1.

(3) Effect of reaction temperature on the yield of PDADN

Due to the significant influence of temperature on the nitration reaction, with the fixed nitration coefficient, nitration yields of products at different nitration temperatures are shown in Table 9.2.

As listed in Table 9.2, with the increase of temperature, PDADN yield is also increased. However, when the temperature is more than 20 °C, the yield is not increased again. It is also found that, when temperature exceeds 25 °C, the yield of PDADN declines with some degree. According to the reported, when 1,3-diazido-2,2-propanol is nitrated at a high temperature, azido is gradually converted into the nitrate group with the increase of temperature. Only at an appropriate temperature (−5 to 5 °C), the compound 1,3-diazido-2-2-dinitrate propane could be obtained. Therefore, at a temperature over 25 °C, azido can be converted to nitrate ester group in PA, and the optimum temperature of PA nitration is 15–20 °C.

Table 9.2 The influence of reaction temperature on PDADN yield

Number	Nitration coefficient	Reaction temperature (°C)	Yield (%)
1	8	0–5	50.38
2	8	5–10	72.87
3	8	10–15	80.25
4	8	15–20	82.10
5	8	20–25	82.06
6	8	25–30	81.50
7	8	30–35	80.48

9.3 Properties and Preparation of 1,3-Diazido-2-Nitrate Propane

It is well known that nitroglycerin risk level is high, people just use NG in the past century, and no researchers are dared to reform NG. In this study, our synthesized 1,3-diazido-2-nitrate propane (DANG) is a glycerol-like structure with two ester groups replaced by azido groups, which not only keeps the nitroglycerin structure, but also maintains its energy and reduces its sensitivity.

9.3.1 Properties of 1,3-Diazido-2-Nitrate Propane

Chemical formula: $C_3H_5N_7O_3$
Structure

$$\begin{array}{c} N_3CH_2-CH-CH_2N_3 \\ | \\ ONO_2 \end{array}$$

Relative molecular weight: 187
Oxygen balance: −47.06% (calculated according the generated CO_2)
Nitrogen content: 52.41%
Freezing point: −15 °C
Color appearance: light yellow liquid
Relatively density (20/4): 1.0059
Melting point (°C): 154(774 mmHg)
Refractive index: 1.435820
Solubility: DANG is soluble in water, ethanol, ether and other organic solvents
Thermal decomposition of DANG: decomposed at 188 °C with an endothermic peak at 149.5 °C
Friction sensitivity (%): Its average firing rate is 90
Impact sensitivity (%): Its average firing rate is 50, limit falling height is 31.5 cm (hammer: 1.2 kg), while the explosion probability of nitroglycerin (NG) is 100% with 2 kg drop and 25 cm falling height.
DANG decomposition temperature (°C): At (90 ± 3) °C, the decomposition reaction is up to 12–13 h. The intermediate azide alcohol cannot be decomposed at a relatively high temperature (90 °C) for 10 h, showing that alcohol azide is a thermally stable material.

9.3.2 Preparation of 1,3-Diazido-2-Nitrate Propane

Theoretically, the synthesis route of 1,3-diazido-2-nitrate propane starts from partial azido replacement of nitroglycerin in which three new azido nitrate materials can be derived. In order to avoid the risk of NG as the raw material, three synthesis routes are designed to prepare azido nitroglycerin.

(1) With 1,3-dichloro-n-(2) as the raw material

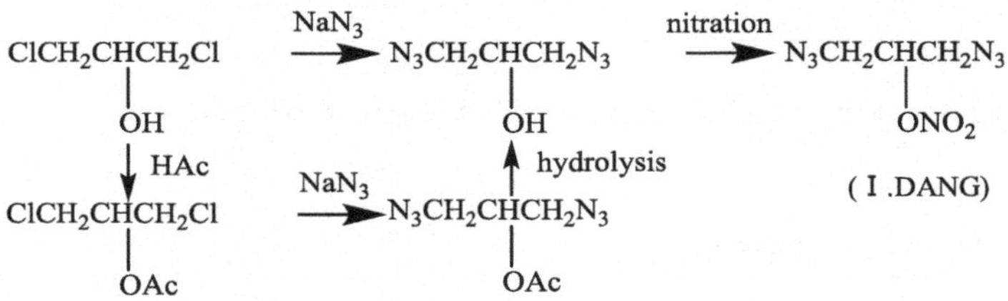

(2) With 2-chloro-1,3-propanediol as raw material

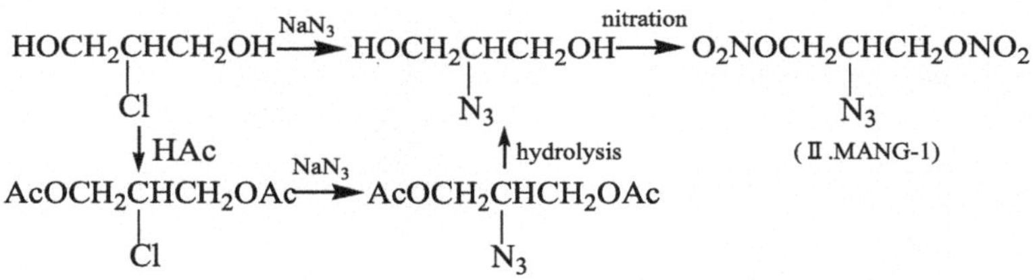

(3) With epoxy chloropropane as the raw material

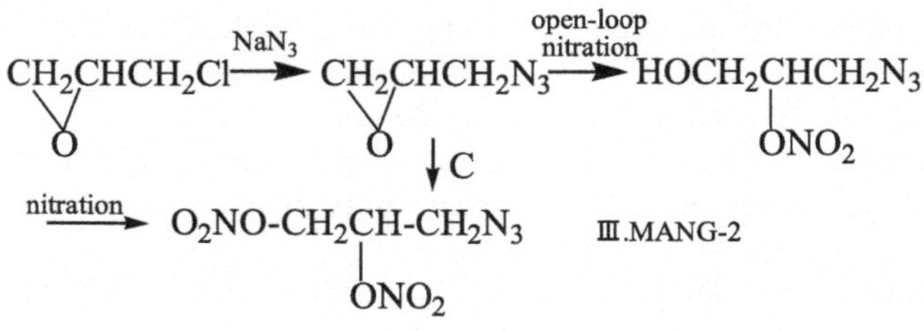

(1) Preparation of 1,3-diazidopropanol

With 1,3-dichloro-n-(2) as the raw material, the main chemical reaction during the synthesis of 1,3-diazidopropanol is,

The main synthesis steps are that: Into a 500 mL three-neck flask with a stirrer and a thermometer, 76 g (0.6 mol) 1,3-dichloro-n-(2) and 150 mL DMF are added. With stirring, 108 g (1.66 mol) NaN_3 is then added and warmed up to (95 ± 2) °C to react for 90 min. The reaction solution is cooled down to room temperature or extracted with CH_2Cl_2 for three times. The combined extracted solution is washed with water for three times, and the organic layer is separated, dried with anhydrous $MgSO_4$, filtered to remove solvent CH_2Cl_2. Then, 66.06 g of 1,3-diazido-n-(2) is obtained with a yield of 86% and a purity of 99.2% (HPLC).

HPLC: 99.2% IR: At 2102 cm^{-1}, there is a strong $-N_3$ absorption peak. At 3422 cm^{-1}, there is a strong –OH absorption peak.

Elemental analysis: The calculated and measured values match. HPLC: 99.2%.

(2) Preparation of 1,3-diazido-2-nitrate propane

Into a 1000 mL three-neck flask (in an ice bath), 200 mL CH_2Cl_2 and 4 g (0.112 mol) 1,3-diazido-(2) is added and cooled to 0 °C. With stirring, 38 mL HNO_3 (about 25 min) at 0–5 °C are dropped into the solution to react for 150 min. After the reaction, H_2O is added and stirred for 15 min to separate the organic layer, which is washed with water for 4 times, dried with anhydrous $MgSO_4$, and filtered to remove solvent CH_2Cl_2. Then, 1,3-diazido-2-nitrate propane (DANG, 17.94 g, 0.094 mol) is obtained with a yield of 85.46%.

Infrared analysis: At 1208 cm^{-1}, there is a strong $-N_3$ absorption peak. At 1649 1275 and 845 cm^{-1}, there are $-ONO_2$ chemical shifts.

Nuclear magnetic resonance spectroscopy: δ (3.6 ppm) (a) is the chemical shifts of $-CH_2N_3$. δ (5.2 ppm) (b) is the chemical shifts of $-CH-ONO_2$.

Elemental analysis (%): The calculated values of $C_3H_5O_3N_7$ are C 19.5, H 2.67, and N 52.4. The measured values are C 22.37, H 3.34, and N 52.1. HPLC: 96%.

The main chemical reaction to synthesize 1,3-diazido-2-nitrate propane is

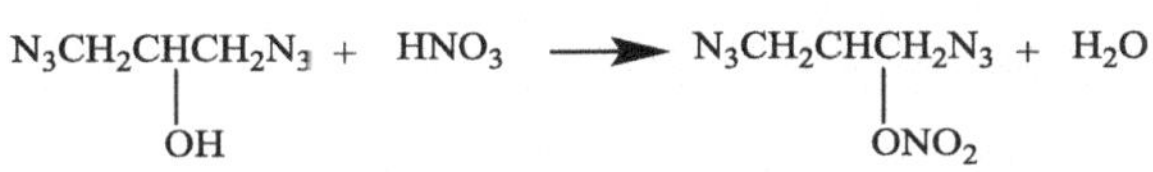

When 1,3-diazidochloropropanol-(2) is nitrated (at 15–75 °C), the formed product is not the objected one, but is nitroglycerin. With the increase of reaction temperature, the obtained NG purity is also increased.

Experiments show that, at a relatively high temperature, azido ($-N_3$) is gradually transformed into nitrate group ($-ONO_2$) with the increase of temperature. Only at an appropriate temperature (–5 to 5 °C), the objected product can be obtained. Experimental results at high temperature show that NG with a content of 93% is able to withstand at 75 °C for 1–3 h without decomposition, thus to provide valuable experiment data to safely handle NG (Table 9.3).

1. Studies show that, with 1,3-dichloropropanol-(2) as the raw material, the Nitration product is 1,3-diazido-2-nitrate propane (DANG) after the azidization and nitration, and its structure is identified to prove the objected product.

Table 9.3 Relationship of reaction temperature and purity of NG

Number	Reaction temperature (°C)	Reaction time (h)	NG purity [%(HPLC)]
1	15		8.25
2	30	1–3	27
3	40		50
4	50		61
5	55		69.4
6	65	1–3	87.3
7	70		90
8	75		93

2. Freezing point of the synthesized DANG is significantly lower than that of NG, which again proved the correctness of the law that, through locally replacing nitrate groups in nitrates with azido groups, new materials can be synthesized with a lower freezing point than that of the original nitrate.
3. The sensitivity of DANG is equivalent or slightly lower than that of NG. After it meets the requirements with other further tests, it can be tested to use.
4. The study on NG azide is the first worldwide. The successful synthesis and preliminary performance tests of DANG show that azidoglycerin is not the traditionally considered and more dangerous materials over nitroglycerin material. After further studies, it is possible to find new material with better general properties.

9.4 Syntheses of 3-Azidomethyl-3-Nitrate Methoxyl Cyclobutane (AMNMO) and Its Polymers

9.4.1 Preparation of AMNMO

(1) Properties of AMNMO

Chemical formula: $C_5H_8N_4O_4$
Structure

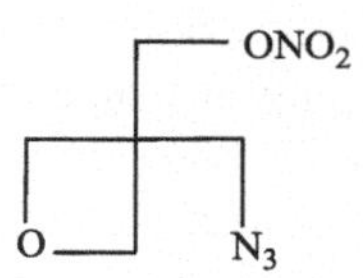

Relatively molecular weight: 188
Oxygen balance: −85.11% (calculated according to the formed CO_2)

Nitrogen content: 29.79%
Freezing point: <−15 °C
Color appearance: Light yellow transparent liquid
Relative density:
Decomposition temperature: 206.4 °C
Refractive index: 1.5029
Solubility: miscible with nitrate with layer separation.

(2) Synthesis of AMNMO

With 2,2-dibromomethyl-1,3-propanediol as the basic raw material, through the three-step reaction, including loop closing, substituting and nitration, AMNMO and its polymers are synthesized. The main chemical reactions in the synthesis process are

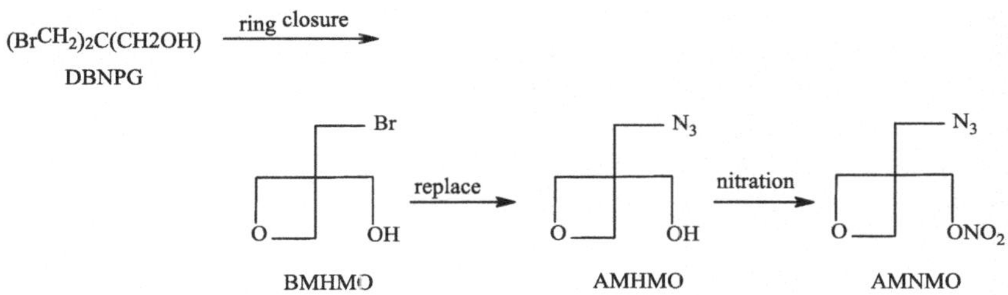

(3) Synthesis of BMHMO

2620 g (10 mol) of DBNPG are added into 7500 mL of anhydrous ethanol under stirring, warmed up to 40 °C, and dissolved into a transparent solution. 400 g of sodium hydroxide (10 mol) are added and the system temperature is increased to around 50 °C automatically. Until the temperature is no longer increased, the system is refluxed in a water bath for 3 h. The solution is frozen in the freezer overnight and filtered out the solid. The filtrate is distilled to remove alcohol under vacuum, and colorless oily liquid containing solids is obtained. 2000 mL of dichloromethane is used to dissolve the as-obtained mixture. After the filtration, 1730 g of raw product is obtained after the evaporation of the solvent, which then is distilled under vacuum at 400–600 Pa to collect fraction within the boiling range of 105–110 °C. And 1300 g of liquid product are obtained with a yield of 72–75%, n_{20}^{D} of 1.5080. Its IR is: $\nu_{max}(cm^{-1})$: 3408(–OH),1041(C–OH),976(cyclo C–O–C),700(C–Br). ^{1}HNMR, $\delta_H(CDCl_3)$: 4.47(s,4H,cyclo H),3.92(d,2H,CH_2OH),3.73(s,2H,CH_2Br).

(4) Synthesis of AMHMO

905 g (5 mol) of BMHMO and 346.5 g (5 mol) of sodium azide are added to the solution containing 1000 mL acetone and 2000 mL distilled water, and then refluxed to react for 12 h. After the reaction is stopped, acetone is removed under vacuum and the remaining is extracted with 1200 mL dichloromethane twice. Organic phases are merged, washed with 500 mL distilled water, dried with

anhydrous magnesium sulfate overnight, distilled to remove solvent to obtain 645 g of yellow transparent liquid with a yield of 90–92% and n_{20}^{D} of 1.5013. Its IR, $\nu_{max}(cm^{-1})$: 3402(–OH), 1046(C–OH), 973(cyclo C–O–C), 2105,1279(–N_3). ^{1}HNMR, $\delta_H(CDCl_3)$: 4.45(s, 4H, cyclo H), 3.80(d, 2H, CH_2OH), 3.68 (s, 2H, CH_2N_3).

(5) Synthesis of AMNMO

75.0 g (1.2 mol) of 100% nitric acid are dropped into 146.6 g (1.3 mol) of acetic anhydride at 0 °C, during which the temperature is controlled to be not higher than 5 °C. After the dropping, the mixture is mixed for 10 min and then cooled to 0 °C. 143.1 g (1.0 mol) of AMNMO methylene chloride solution (400 mL) are added to nitric acid/acetic anhydride solution, during which the temperature is controlled to be not higher than 5 °C. After the dropping, the mixture is stirred for 1.5 h. Sodium carbonate aqueous solution with mass fraction of 10% is used to wash. The oil layer is separated and then washed with distilled water until it is neutral, dried with anhydrous magnesium sulfate overnight. Solvent is distilled out to obtain 120 g of light yellow transparent liquid with a yield of 63–65% and n_{20}^{D} of 1.5029. Its IR, $\nu_{max}(cm^{-1})$: 2105,1279(–N_3), 1637, 865(–ONO_2), 977(cyclo C–O–C). ^{1}HNMR, $\delta H(CDCl_3)$: 4.68(s, 2H, CH_2ONO_2), 4.44(s, 4H, cyclo H), 3.73(s, 2H, CH_2N_3).

9.4.2 *Synthesis of PolyAMNMO*

With (AMNMO) as the raw material, the main chemical reaction to synthesize PolyAMNMO is

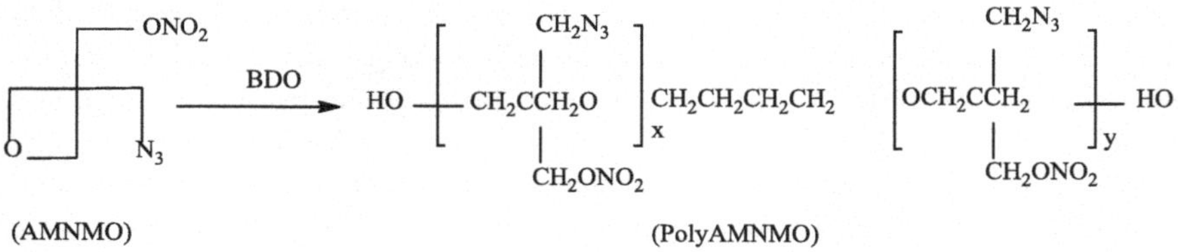

Synthesis steps are: 39 mL of trifluoroboroethyl ether complex are added into the solution with 12 g (0.1333 mol) of 1,4-butanediol and 630 mL of methylene chloride with stirring under nitrogen. After 20 min of stirring at room temperature, 500 g (2.66 mol) of AMNMO are added, in which the temperature should be controlled to be not higher than 25 °C. And the addition, the reaction is continued for 48 h at room temperature, and 100 mL sodium carbonate aqueous solution with 5% of mass fraction is added to terminate the reaction. The organic phase is washed until it is neutral, concentrated to get 480 g of viscous liquid polymers with a yield of 92–94%. Its IR, $\nu_{max}(cm^{-1})$: 3360–3600 (–OH), 2111, 1278 (–N_3), 1637, 870 (–ONO_2), 1106 (C–O–C ether).

9.5 Applications of Azido Nitrate

Energy-containing additives, especially energy-containing plasticizers and binders, are considered as one of the ways to improve the energy level and other technology performance of solid propellants, the propellant explosives and mixed explosives.

9.5.1 Application of Azido Nitrate in Propellant

Energy-containing organic azide compounds have a series of advantages as additives. Researchers at India High Energy Materials Research Laboratory (HEMRL) introduced a variety of new formula of high-energy LOVA chemicals, and effectively improve the energy levels of LOVA with different technical approaches. If energy-containing plasticizers (low-molecular-weight GAP) are used to replace the traditional inert plasticizers, performance of LOVA drugs can be significantly improved, in which explosion heat is increased by 6.6%, explosive power is increased from 1180 to 1225 J/g, mechanical properties are also significantly improved. In the 1970s, the synthesis of organic azide compounds and their applications as energetic materials had been studied widely [10]. Results showed that, in the propellant explosives, propellants, high-energetic explosives and other energetic materials, azide-containing adhesives, plasticizers, antioxidants and other additives are used and not only can improve their energy levels, but also give them many outstanding performances, thus to increase the nitrogen content of the system but not to affect their carbon-to-hydrogen ratios, then to increase their system burning capacity and burning speed of propellants, but not to increase the flame temperature. In recent years, America has introduced compounds with both amino and azido groups. Through the calculation of their theoretical structures and thermodynamic properties, results have proved that these compounds have practical values in the design of propellants. Meanwhile, researchers at Feaunhofer ICT Research Institute in Germany synthesized the azide compounds, namely TAPE-E (triazide neopentanoate ethyl ester) and TAP-Ac (triazido pentaerythritol ester of acetic acid). Because these two energetic plasticizers and azide Binder system have a very good compatibility, it is used as a plasticizer to improve the plastic property and energy of this type of adhesive system. Although explosion heat of azido energetic plasticizer is high, but the detonation velocity is low and the energy is lower than that of nitrate with poor compatibility with NC and NG. They can exclusively be adhesive together with those class azide binders having similar structures. In order to improve the fragility of NEPE propellant at low temperature, BTTN/NG mixed nitrate is used to cooled plasticizer down to −40 °C, which effectively prevents the fragile problem of propellant at low temperature. Besides the used BTTN/NG mixed nitrate, a NG/TMETN(trihydroxymethyl ethylene nitrate)/TMMTN (tri-n-hydroxy-methyl methyl nitrate) mixed homogeneous propellant formulation is used as tank explosive propellant through the compression process. If the class or azido nitrate energetic plasticizers with both high energy and low freezing point can be synthesized directly, it will be potentially important to promote the development of propellants. Azido nitrate compounds combine azide radicals and nitrate radicals together as a whole, not only to deepen and develop the traditional "mixed" concepts, but also to improve and enhance properties of traditional mixed nitrates. It can be used in low characteristic signal propellant, high-energy smokeless propellant, screw extrusion modified double-base propellants, large-caliber anti-tank guns, and other weapons. In nitramine propellants with formula of plasticizer containing azido nitrates, PDADN has high energy and can improve disadvantage of nitramine propellant as the pressure index is greater than 1. The calculated specific impulse of propellant containing 10–15% PDADN is improved by 39.2–49 N s/kg. It can increase the burning rate [11] and has good performance. The trial of the new material PDADN shows that, when the content 10–20%, the powder of explosive is up 1266 kJ/kg with an explosion temperature of 3420–3500 K. The static mechanical properties of propellant column are superior to that of three-base one, and it is smooth and tight. In reported literatures [12], the chemical stability of PDADN and its effects on combustion and thermal decomposition performance of nitramine-modified double-base propellant have been studied. Results shows that the replacement of RDX or NG with PDADN not only increases the energy of formula, but also decrease the amount of water in the burning gas, which is very useful to further reduce the characteristic signal of nitramine-modified double-base propellant. The addition of PDADN is a practical way to increase burning rate and energy of nitramine-modified double-base propellant. Therefore, studies on azido nitrate compounds have a certain meaning to the research and development of weapon equipment and defense technology.

9.5.2 Applications of Azides in Cycloaddition Reaction and Chemical Synthesis

Organic azides can react with amphiphilic molecules, such as olefins, ketones, $X \equiv Y$ triple bonds, $X = Y$ double bond and alkynes to form five-membered heterocyclic rings through cycloaddition reaction. With heating or catalyst, organic azides react with olefins to produce five-membered triazole ring through cycloaddition. With Lewis acid, they react with ketones through cycloaddition, and then react with $X \equiv Y$ triple bond compounds (nitrile) to form a tetrazolium compound and react with heteroatom-containing $X = Y$ double bonds accumulated polyene (such as isocyanate, isothiocyanate, and carbide diimine) through non-catalyst cycloaddition reaction to form oxazole and other nitrogen-containing heterocycles.

The cycloaddition between organic azides and alkyne Huisgen is the reaction between terminal alkynes or internal alkyne through the cycloaddition of 1,3-dipolar to form 1,4- and 1,5-substituted triazole, and the reaction is as follows:

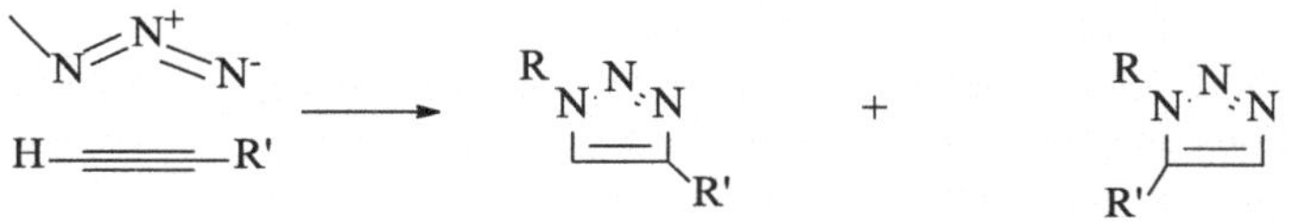

(1,4 substituted triazole) (1,5 substituted triazole)

In the 1960s, based on studies of Huisgen, Huisgen cycloaddition reaction was thoroughly understood and was established as one of the most important new reactions [13]. At the beginning of the twenty-first century, Meldal and Sharpless independently had improved this reaction. With Cu(I) as catalyst, its main product is 1,4-subsituted triazole with enantioselectivity, and this reaction (1) could be completed quantitatively with yield close to 100%, (2) had a very high functional group tolerance (not sensitive to chemical structures of ligand molecules), and did not need protection to other functional groups, (3) could be carried out in all the solvents, (4) could be done at room temperature, and keep the biomolecular activity, (5) not only could be done in homogeneous phase, but also could be done on all interfaces, such a s solid–liquid, liquid–liquid and even solid–solid interfaces. Due to these characteristics, it plays an important role in many fields, such the connection between various ligands and synthesized blocks, surface modification of polymer and nanomaterials, biomolecule coupling and new drug development.

(1) Applications in the synthesis of triazole and amine compounds

Triazole is one of important nitrogen-containing heterocycles with stable chemical properties. Compared with glutamine, triazole ring can be hydrolyzed. It is also different from benzene and other aromatic heterocycles, triazole ring is hardly oxidized or reduced. It is widely used as from fungicides, herbicides, and microbial inhibitor in agriculture, to light stabilizer, fluorescent whitening reagent, and antiseptic flame retardant into industry. In the field of functional molecules, it is used in drug delivery, nano-electronic devices. In the pharmaceutical and biotechnology fields, it is used to prepare oligonucleotides, cell culture, anti-inflammatory and antiviral therapy, can also be used as HIV-1 protease inhibitors and specific ligands in the hairpin loop of messenger RNA (m-RNA). Due to it is metabolic inert and is multiple hydrogen bond receptor, it can be used as the estimated analog of peptide bond. It can be used in the biological effect estimation with other molecules in the diversity-oriented synthesis (DOS) compound library. The cycloaddition of azide with alkyne or activated olefin can produce triazo compounds. The available azide compounds include hydrazoic acid, metal azide, and organic azides. And the reaction with olefin is limited to only the organic azides. In the cycloaddition of azide-alkyne, types of azide and alkyne will affect the distribution of triazole. Aromatic or alkyl azide react with activated alkyne to generate 1,2,3-trizole. With sulfonyl azide, the product is N-sulfonyl trizole that generates trizole, its open-ring isomer and α-azidoimine through rearrangement [14]. With Cu(I) as catalyst, 4-substituted 1-(N-sulfonyl)-1,2,3-triazole is produced selectively. Ilaria [15] also found a special method to synthesize triazole through the cycloaddition of α, β-unsaturated amino ketone and diethyl aluminum azide. Through the reduction of azide compounds, amines can be generated. In the presence of thiols, complex hydrides and Linde amine compounds of borane, the azido groups can be converted to amines. With lithium/2-tert-butylbiphenyls (DTBB), acyl azides can be converted to amines at room temperature. With ferric chloride as catalysts, N-dimethyl hydrazine is a mild reductant to azides. Trimethyl chlorosilane can convert aliphatic, aromatic and benzoyl azide compounds into the amines, but it needs more harsh reaction conditions. If samarium iodide (SmI_2) is catalyst, the conversion conditions are mild. Huang et al. [16] used samarium iodide as catalyst to convert alkyl, aryl, acyl and sulfonyl azide compounds into amines in methanol, and the reaction environment is neutral and yield is high. With Cu (II) as catalyst, zinc borohydride and sodium borohydride can reduce azides to amines. Fe (II) salts and sulfur-substituted aryl Sn (II) complexes are very good reductants too [17].

Curtius rearrangement: The volatile acyl azide from carboxylic acid is heated in an inert solvent to lose nitrogen and rearranged to then form the intermediate isocyanate, which is further hydrolyzed into primary amines. This is the famous Curtius rearrangement.

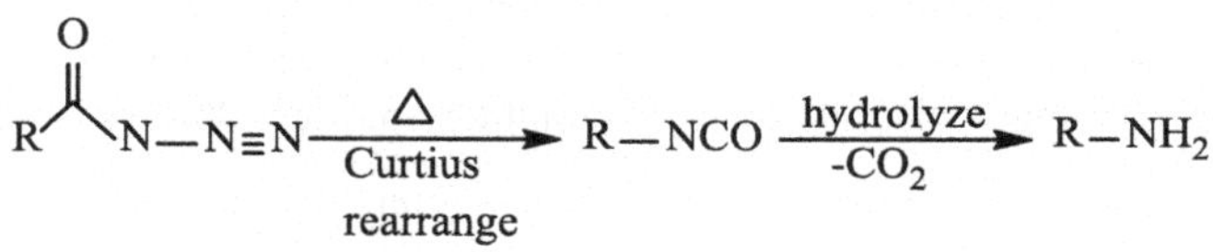

(2) Applications in nitrogen chemistry

Triazene is a reactive intermediate in organic synthesis from the optical or thermal decomposition of organic azides. It is nitrogen analogues of carbine with a strong electrophile, and can be widely used in the cycloaddition, rearrangement, ring expansion, ring shrinking, and insertion reaction.

The intermolecular cycloaddition of azide with many functional groups is the classic method to synthesize nitrogen heterocyclic azole, and the intramolecular cycloaddition product can derive into various organic compounds. Tatjana et al. [18] synthesized hexahydrogen-1,2,3-triazole-1,5-a-pyrazinamide through intramolecular cycloaddition of azido group with olefins, and the released nitrogen formed C–N bond during the cyclization. The product can be used in various condensation synthesis of piperazine derivatives precursors, which can be easy to convert into final or target product with alkylating agents and acylating agent.

Phenyl azide decomposes optically to form a variety of reactive intermediates, and its singlet phenyl nitrogen olefin is the key element. Azidizing reagents and cyclic compounds under the light or heating can have ring expansion or shrinkage, and aryl nitrogen olefin can form n-heterocyclic or nitrile compounds through ring expansion or shrinkage. Phenyl azide in diethylamine can generate 2-diethylamino-3H-acridine through photolysis and ring expansion. In aniline/trimethylamine, derivatives of 2-aniline can be obtained. And 2-amino derivatives can be generated in liquid ammonia [19]. After ring expansion, mesitylphospha-triafulven is changed from ternary ring to four-membered ring (1H-2-iminophosphetes) [20]. Through pyrolysis or photolysis, azide pyrazine compounds change from six-membered to five-membered ring.

Through pyrolysis or photolysis, organic azides can form singlet and triplet triazenes. The former has C–H insertion reaction with saturated hydrocarbons, the later can capture hydrogen atoms in n-alkanes to generate free radicals, and both of them can react with C = C. Through this reaction, functional molecules can be fixed on the surface of substrate to obtain a wide range of functional materials. Available azides include alkyl/aryl, acyl, phosphoryl, trimethylsilyl and sulfonyl azide compounds, and methyl esters of hydrazoic acid. The usable substrates are glass, silicon, metals, nanoparticles, polymer microspheres, inert polymer (film, fiber, nonwoven fabric). Under heating or UV light irradiation, natural polymers, enzymes, cells, fluorescent chromophores, and marking molecules can be fixed, adhered, grafted, or coupled on polymers. Polymer crosslinking and rheological modification, introduction of long chain functionalization of long chain can be realized under the melting conditions.

Yan and Ren [21] used spin coating to realize covalent attachment of ultrathin polypropylene film on a silicon wafer with the help of C–H insertion reaction of polyfluorinated benzene triazene, which is expected to modify the surface of silicon wafer devices for biochemical reactions and medical micro-testing, to improve their bioavailability, adhesion, and robustness. Zhu et al. [22] used azene insertion reaction to attach chitosan molecule onto polylactic acid (PLA) film. After it was combined with heparin, a new biological material had been made, which could hinder the platelet adhesion and enhance the cell adhesion. Hicke et al. [23] used acyl azides to activate copolymer of polyacrylonitrile UF membrane, then fixed starch glucosidase (AG) on the ultrafiltration membrane and synchronically realized the hydrolysis of maltose or starch and product separation. Weng et al. [24] first introduced 3-aminopropylphosphoric acid to obtain amino-containing self-assembled monolayer, and then fixed azide gelatin on the titanium oxide film. The titanium dioxide membrane with fixed gelatin (about 5–20 nm in thickness) has a good biocompatibility, can be used in biomedicine potentially. Mehdinia et al. [25] used azido-N-hexylmercaptan prepared self-assembled monolayer-modified Au electrode first, and then used azido to connect DNA to obtain Au electrode with fixed DNA (DNA/SAM/Au), which can be used in the sensitive sensor to determine paclitaxel in human serum. In the field of functionalization of polymer after the polymerization, Boaen and Hillmyer [26] wrote a detailed review, and Thermo Scientific company also introduced a series of azide crosslinking agent.

(3) Other catalyzed cycloaddition

Many people try to improve and simplify azide cycloaddition reaction. For example, Foking et al. [27] used microwave to activate the reaction of benzyl chloride/NaN_3 with terminal alkynes and synthesized triazole in one step with a yield of 86–93%. Another valuable improvement was carried out by Orguira et al. with elemental copper Cu(0) as catalyst and ammonium salt as cocatalyst. And the yield was more than 95% [28]. Then, Yamamoto et al. used double metal catalysts to produce 1,4,5-substituted triazole from intramolecular alkyne [29]. In 2005, Zhang et al. [30] revealed that 5-methylcyclopentadienyl ruthenium complexes (Cp*Ru) could catalyze cycloaddition reactions, and the product was 1,5-substituted triazole. In addition, tetrahexyl ammonium chloride (THAC) could also catalyze cycloaddition reaction of aryl azides, and the reaction of phenyl azide with methyl methacrylate had a yield of above 90% [31]. With the specific enzyme-catalyzed azide-alkyne cycloaddition reaction in situ, valuable enzyme inhibitors can be synthesized, which is "taking enzyme to catalyze synthesis of its own inhibitors".

Cu-catalyzed 1,3-dipolar cycloaddition reaction of azido alkyne, because of its reliability, specificity and bio-compatibility, has become the flagship of click response. N-heterocycle in the product triazole is rigidly connected, can mimic the atom replacement and electrical properties of peptide bonds. Under the reaction conditions, functional groups of common biological molecules are very stable, azide reacts with alkyne with very high reactivity, but both of them do not react with biomolecular functional groups. This biological orthogonal effect of them makes them especially suitable for the syntheses of natural products and their derivatives, target-inducing synthesis and active protein MS in biological and medicine fields. Because single step can become fast, effective and predictable in multi-step synthesis, with help of combinatorial chemistry, high throughput screening and compound library, the discovery pace of new medicines will be significantly accelerated.

Once Click reaction was established, related reports have emerged explosively. This practical and intelligent reaction is very effective in three types of interrelated areas: (1) polymers and materials science; (2) biological coupling; (3) medicine development. Especially, it is popular in the syntheses of polymers and their modification, self-assembly of molecular single layer on the surface of metals and nanomaterials, syntheses of supramoleculars, molecular-electronic components, sensor probes, nonlinear optical materials, high-performance liquid chromatography (HPLC) column fillers, immobilization of enzymes, adhesion of biological molecules (such as oligo-peptides, polynucleotides, oligosaccharides, proteins, Lipids, cells tissues and organs) and their coupling, labeling and detection, and discovery of new medicines.

Steinke et al. [32] reported first to used diazido nitrogen and double alkyne as building blocks to synthesize polymer with cucurbituril as catalyst.

Sharpless et al. [33] realized the in situ polymerization of diazido nitrogen and alkyne, and the average molecular weight of condensed polymer was less than

12,500. These studies opened a new method in Click reaction to prepare resin-type structure, and different types of diazido nitrogens and alkynes can be polymerized into different crosslinked resins, which then can be used in metal (such as copper block) connection, which formed a new technology to bond metal without epoxy resin. Ankit et al. [34] synthesized acetyl-functionalized polycaprolactone and polylactic acid, which, in the presence of cupric bromide as catalyst, formed miktoarm star polymers through click reaction with azide-functionalized polymers. The click reaction between PVA azide (PVA–N_3) and optically active propargylic can be catalyzed by Cu to form pyrene-functionalized PVA with pyrene fluorescence. After the functionalization of cysteine thiol alkyne on SA, Cu can be used to catalyze the click reaction to effectively couple bovine serum albumin (BSA) on the polystyrene.

Anja et al. [35] used 1-azidoundecane-11-mercaptan mercaptan through method-olefin method to introduce azide groups on the surface of poly divinylbenzene microspheres then realize the hydrophilic modification of microspheres through its click reaction with alkyne-functionalized poly 2-hydroxyethyl methacrylate.

9.5.3 Applications of Azide Compounds in Energetic Materials and Gas Generator

Nitroamine compounds containing azido has been produced in large scales, and there are a lot of work on its properties and applications [36]. It is Reed and Dolah to study these compounds first. Rosher, Morton, and Eimslic synthesized a series of azide nitramine compounds, which were used as propellants because they could significantly increase the burning speed and increase specific impulse, but did not affect pressure index. Then, they found its energy was close to that of HMX, and it had low explosion sensitivity. Thus, it was attempted to apply as low sensitive and smokeless propellant to replace the HMX.

The azido groups have high formation enthalpy (ΔH is 3556 kJ/mol). As propellant component, it can decrease the average molecular weight of combustion products, which can be helpful to improve specific impulse. Meanwhile, the introduction of azido nitrogen increases the nitrogen content of the system, and the relative hydrocarbon content is decreased, which meets the smoke-free or smoke-reduced requirements of propellant in the combustion process. Therefore, studies on azides have been paid more and more attention [37].

Adizo ammonium nitrate and adizo compounds have been produced in large quantities. According to their difference in properties, they can be used as energetic plasticizers and antioxidants [38].

The synthesis of methyl azide ammonium nitrate (1) synthesis is based on the preparation of primary ammonium nitrate through chloromethylation, introduction of halogen atoms and then azidization [39], which can be expressed as follows:

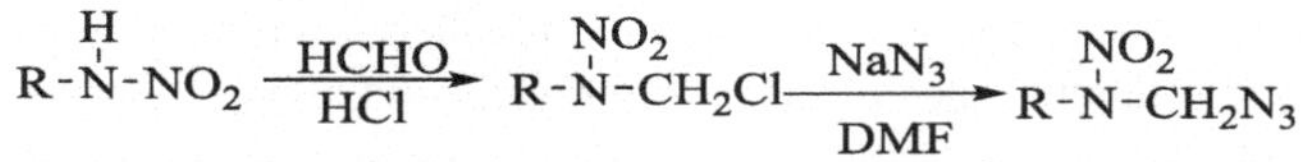

For example, in the synthesis of azido ammonium nitrate 3-nitro-1,5-diazido-3-azidoheterocyclopentane (2), diethanolamine is the raw material. Nitrate is first introduced in the reaction substrate before the azidization, and the reaction is:

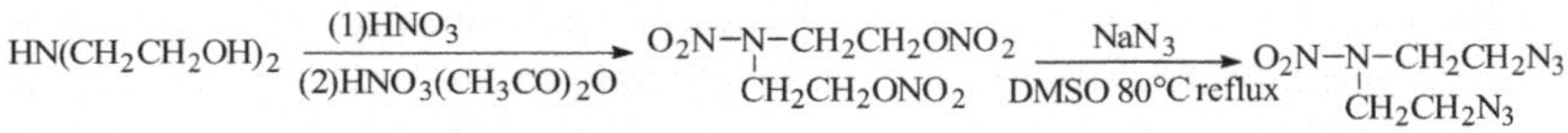

Metal organic azides are new energetic materials. They can be used as igniting agents of solid and liquid propellants. For example, the reaction of diethyl aluminum azide is as follows:

$$NAN_3 + (CH_3CH_2)2AlCl \longrightarrow \text{solvent}(CH_3CH_2)2AlN_3$$

Rapid decomposition can lead to the ignition or detonation of azide explosive. Because azide compounds have characteristics of the fast gas-production, they have been used in automobile airbag.

References

1. Liu JP (2015) Liquid explosives. Springer Press
2. Shao L (1997) Study of one-time detonation FAE. Beijing Institute of University
3. Zhang HD (1998) Preparation of FAE. Beijing Institute of University
4. Tian D, Zhao F, Liu J (2011) Handbook of energetic materials and the related compounds. National Defense Industry Press, Beijing
5. Chen Z, Zheng H, Wang S et al (2007) Status and advance of azide energetic binders. Chem Def Ships 2:1–5
6. Herber RH, Cheng HS (1969) Inorg Chem 8:2145
7. Huheey JE (1966) J Phys Chem 70:2086
8. Treinin A (1971) The chemistry of the azido group, Chaper 1. NewYouk 10(6):6–8
9. Huang PC (1981) The polarization of ions in metal azides and its influence on the properties of azides. J Beijing Univ Technol 2:37–43
10. Edgar R, Wilson X, Frankel B (1985) Synthesis of novel energetic compounds. Azido derivatives of pentaerythritol 50(17):3211–3212
11. Liu Z, Liu Y, Zhao F et al (2003) Thermal decomposition kinetics of PDADN-RDX-CMDB propellant at high pressure. In: Proceedings of the NATAS annual conference on thermal analysis and applications
12. Wang J, Li SF, Zhang XH et al (2001) Effect of azido nitrate on combustion behaviour of nitram in modified double base propellant. Chin J Explos Propellants 24(2):22–25
13. Wang LL, Tang H, Zhang X (2010) Advance on the synthesis and application of organic azides. Application of organic azides in organic synthesis (1). Sci Technol Chem Indus 18 (2):54–58
14. Yoo EJ, Ahlquist M, Kim SH et al (2006) Copper-catalyzed synthesis of N-Sulfonyl-1,2,3-triazoles: controlling selectivity. Angew Chem Inter Ed 46(10):1730–1733

15. Adamo I, Benedetti F, Berti F et al (2003) Unexpected 1,2,3-triazole formation in the reaction of diethylaluminum azide with α-amino-α, β-unsaturated ketones. Tetrahedron Lett 44 (51):9095–9097
16. Huang Y, Zhang Y, Wang Y (1997) Facile reduction of azides to the corresponding amines with metallic samarium and catalytic amount of lodine. Tetrahedron Lett 38(6):1065–1066
17. Roy UK, Roy S (2006) SnCl2 mediated efficient N, N-dialkylation of azides to tertiary-amine via potential stannaimine intermediate. J Organomet Chem 691(8):1525–1530
18. Lukina TV, Sviridov SI, Shorshnev SV et al (2005) Intramolecular [2+ 3]-addition of an azide to a C = C double bond as a novel approach to piperazines. Tetrahedron Lett 46(7):1205–1207
19. Von E, Doering W, Odum RA (1966) Ring enlargement in the photolysis of phenyl azide. Tetrahedron 22(1):81–93
20. Eisfeld W, Slany M, Bergstraβer U et al (1994) Ring enlargement of phosphatriafulvenes with azides to 1H-2-iminophosphetes. Tetrahedron Lett 35(10):1527–1530
21. Yan M, Ren J (2005) Covalent immobilization of propylene thin film. Mater Chem 15 (4):523–527
22. Zhu A, Zhang M, Hua J et al (2002) Covalent immobilization of chitosan/ heparin complex with a photosensitive hetero-bifunctional crosslinking reagent on PLA surface. Biomaterials 23(23):4657–4665
23. Hicke H, Bohme P, Becker M et al (1996) Immobilization of enzymes onto modified polyacrylonitrile membranes: application of the acryazide method. J Appl Polym Sci 60 (8):1147–1161
24. Weng YJ, Ren JR, Huang N et al (2008) Surface engineering of TiO films by photochemical immobilizati on of gelatin. Mater Sci Eng 28(8):1495–1500
25. Mehdinia A, Kazemi HS, Bathaie ZS et al (2008) Electrochemical studies of DNA immobilization onto the azide-terminated monolayers and its interaction with taxol. Anal Biochem. 375(2):331–338
26. Boaen NK, Hillmyer MA (2005) Post-polymerization functionalization of polyolefins. Chem Soc Rev 34(3):267–275
27. Appukkut TP, De Hae NW, Foking VV et al (2004) A microwave assisted click chemistry synthesis of 1,4-disubstituted 1,2,3-triazoles via a copper (I) catalyzed three component reaction. Org Lett 6:4223–4225
28. Orgueira HA, Fokas D, Isome Y et al (2005) Regioselective synthesis of [1,2,3]-triazoles catalyzed by Cu(I) generated in situ from Cu(0) nanosize activated powder and amine hydrochloride salts. Tetrahedron Lett 46(16):2911–2914
29. Kamijos J, Yamamoto Y (2004) Four-component coupling reactions of silylacetylenes, allylcarbonates, and trim-ethylsilylazide catalyzed by a Pd(0)-Cu(I) bimetalliccatalyst. Fully substitued triazole synthesis from seemingly in-ternal alkynes. Tetrahedron Lett 45:689–691
30. Zhang LL, Chen X, Xue P et al (2005) Ruthenium-catalyzed cycloaddition of alkynes and organicazides. J Am Chem Soc 127:15998–15999
31. Molteni G, Ponti A (2006) Regioselectivity of arylazide cycloaddition to methyl propiolate in aqueous media: experimental evidence versus local DFT HSAB principle. General Papers 16:49–56
32. Krasiat C, Steinke JHG (2002) Formation of oligotriazolescatalysed by cucurbituril. Chem Common 7(1):22–23
33. Diaz DD, Punna S, Holzerp P et al (2004) Click chemistry in materials synthesis. 1. Adhesive polymers from copper-catalyzed azide-alkyne cycloaddition. Polym Sci Part A Polym Chem 42:4392–4403
34. Vora A, Singh K, Webste DC (2009) A new approach to 3-miktoarm star polymers using a combination of reversible addition-fragmentation transfer (RAFT) and ring-opening. Polymerization (ROP) via Click chemistry. Polymer 15(13):2768–2774
35. Goldmann AS, Walther A Nebhani Leena et al (2009) Surface modification of poly (divinylbenzene). Microspheres via thiolene chemistry and alkyne-azide click reactions. Macromolecules 42:3707–3714

36. Xiao H, Li Y (1994) Study on inductance and conductivity of alkali metal and heavy metal azide. J Phys Chem 10(3):235–240
37. Zhou ZM, Chen BR (1996) Synthesis and properties of some organic azido compounds. J Energ Mater 4(2):49–55
38. Li Y, Li Y (1998) Azido reaction and its application in explosives synthesis. J North China Inst Technol 19(1):40–44
39. Liu JP, Zhang J, Shao L (2009) Preparation of azide alkane compounds. ZL200610056342. 0. 0. China patent

Chapter 10
Nitrocellulose

Nitrocellulose, also known as cellulose nitrate or nitrocotton, has an English code name of NC. It has a similar characteristic appearance to cotton. Nitrocellulose is insoluble in water but is dissolved readily in a nonpolar organic solvent. Practically, it is an indispensable and excellent chemical product.

Nitrocellulose is a processed product that utilized cellulose as the basic raw material. The basic preparation process is as follows. The cotton or cotton linter is processed into refined cotton (commonly known as refined cellulose) by cooking, bleaching, washing, drying, and other processes. In general, the refined cotton is nitrated with nitric and sulfuric mixed acid. After the nitration reaction, the crude product is treated with acid removal, water wash, and stability treatment in a semi-high-pressure reaction cooker (commonly known as stability treatment), and finally bleaching, dehydration, and other processes. The appearance of product nitrocellulose containing 28–30% wetting agent is similar to that of cotton [1, 2].

In addition to cotton fibers, cellulose fibers, bamboo fibers, bagasse fibers, and cellulose extracted from other stalk plants can also be used to produce refined cellulose and nitrocellulose by using the similar process described above. The author studied the cotton rod, Eulaliopsis binata, bagasse, and other plant extracts and refined them into cellulose to prepare a thin paper, which can be manufactured to smokeless gunpowder nitrocellulose products after a nitration process [3, 4].

The manufacturing methods of nitrocellulose are different for different applications. For the applications as nitro coatings, adhesives, celluloid, and other aspects, the nitrogen content should be controlled at 11.5–12.2%. For the application of smokeless gunpowder, the nitrogen content of NC should be 12.5% or more. Regardless of the nitrogen content of nitrocellulose, their basic characteristics vary slightly.

J. Liu, *Nitrate Esters Chemistry and Technology*,
https://doi.org/10.1007/978-981-13-6647-5_10

10.1 Structure and Properties of Nitrocellulose

In this chapter, whenever the property and preparation of the NC are related to the raw materials, the NC refers to the nitration product of refined cellulose. The nitration of cellulose is not related to the preparation of cellulose raw material, which is a common property of NC. The reaction of cellulose with nitric acid forms NC. The reaction is shown as follows:

$$[C_6H_7O_2(OH)_3]_n + \gamma^* nHNO_3 \Leftrightarrow [C_6H_7O_2(ONO)_{3-\gamma}]_n + \gamma^* nH_2O$$

γ = 1, 2, 3, this reaction is a typical esterification reaction, by which the resulting product is polyol nitrate, so the correct name should be cellulose nitrate. Because of the traditional and historical reasons, usually, it is called NC or nitrocotton.

10.1.1 The Structure of Nitrocellulose

In addition to the use of a special nitrating agent, the use of nitric acid or other general nitrating agents cannot convert all the hydroxyl groups on the cellulose into the nitrates completely. Indeed, different industrial processes result in NC with different degrees of nitration. Since the safety and explosion property of NC solution is of special significance in the gunpowder industry, it is very necessary to study the structure of the NC. The connection between the macromolecules of nitrocellulose and the structure of (–ONO_2) groups are discussed emphatically. The basic structure of nitrocellulose can be described as

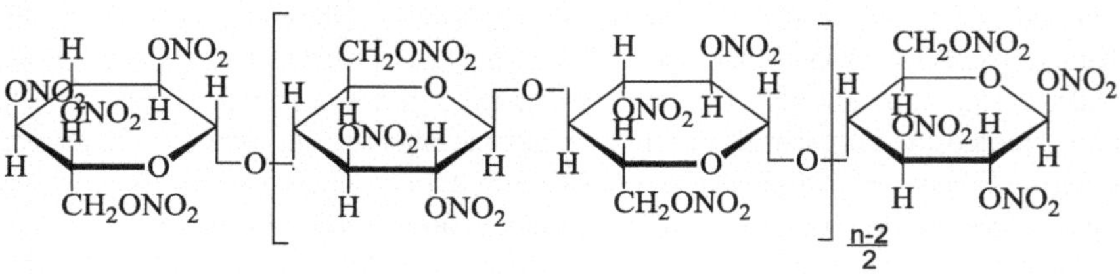

It can be seen clearly from the reaction equation of nitrocellulose that NC is cellulose-derived nitrate. Compared the structure of NC macromolecule with nitrocellulose, their structural similarity is the retention of the basic skeleton of the cellulose macromolecule, whereas their difference is the replacement of hydroxyl group (–OH) with nitrate group (–ONO_2). The number of substituents only reflects the general law from quantitative changes to qualitative changes. Therefore, NC macromolecular structure and cellulose must have a close relationship, but there are essential differences.

Since the (–ONO_2) group replaces the hydroxyl group in the cellulose macromolecule, this reduces the hydrogen bonding. In addition, the steric size of the

(–ONO_2) group is considerably larger than that of the (–OH) group, which makes the nitrocellulose macromolecules to be more flexible than the cellulose macromolecules. Furthermore, it also changes the distance between the cellulose backbones and the interaction between the macromolecules, thus affecting the arranging order of macromolecules. These changes are determined by the number and position of substituents (–ONO_2) [1, 3].

Because of the difference in the degrees of polymerization of the cellulose and the characteristics of the cellulose reaction, the nitrogen content, the degree of polymerization, and the nitrate group distribution in the NC macromolecules prepared under normal nitration conditions are different. For example, for NC having an average degree of cellulose polymerization of 100 and the average degree of substitution of each link of 1.5, there are 1039 different substitution patterns according to the calculation. Therefore, the heterogeneity of NC is much more complex than that of cellulose; only the completely substituted cellulose trinitrate is chemically homogeneous. The average degrees of polymerization of nitrocellulose of No. 1, No. 2, and No. 3 for manufacturing of gunpowder are generally in the ranges of 500–700, 700–1000, and 250–400, respectively [2].

According to X-ray diffraction studies, NC with nitrogen content of 7% or less still exhibits the characteristics of swelling cellulose [5]; NC with nitrogen content of 8.99% shows a mixed characteristic of natural cellulose and trinitrate cellulose; in contrast, when the nitrogen content of NC is 12.73% or more, NC after a stability treatment only exhibits the characteristics of cellulose trinitrate. This result also indicates that when the degree of nitration of nitrocellulose derived from cellulose is less than or equal to approximately 1.00, the molecular group structure of NC does not have a significant difference than that of cellulose; when the nitration degree is in the range of 2.00–2.25, the structure of NC is usually irregular with a possible small amount of unevenly distributed hydrogen bonds, decreased intermacromolecular interaction, and increased the flexibility; when the nitration degree is 2.50 or more, nitrocellulose exhibits a completely repeated and symmetric macromolecular structure with orderly arrangement, reduced chain flexibility, and increased intermolecular interaction. These results also indicate that the arrangement of macromolecules in the aggregation state of NC is still composed of the oriented region and the non-oriented region, and the ratio between them changes with the different nitrogen content and the distribution of nitrate groups. In fact, the relationship between the degree of substitution and structural change should be linear and reflect the trend from quantitative change to qualitative change [6, 7]. The effect of substitution on the molecular symmetry of cellulose is shown in Fig. 10.1.

The structure of nitrate in nitrocellulose can be illustrated by study the structure of methyl nitrate. The structure of the methyl nitrate (CH_3ONO_2) shows that the structure of the –ONO_2 group is on the same plane, and there exists a symmetry plane defined by the C–O–N bond. The bond length and bond angle of –ONO_2 are shown in the following text.

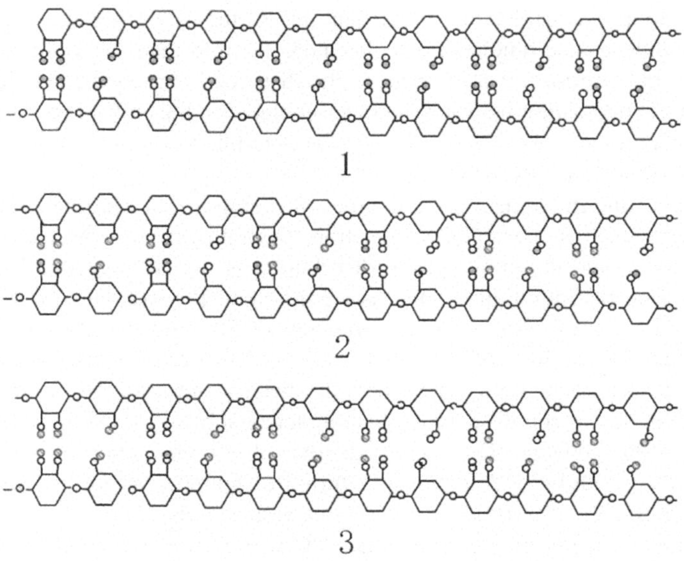

1-non-nitrated cellulose; 2-partially nitrated cellulose derivative;
3-fully nitrated cellulose

Fig. 10.1 Effect of substituents on the molecular symmetry of cellulose [1]

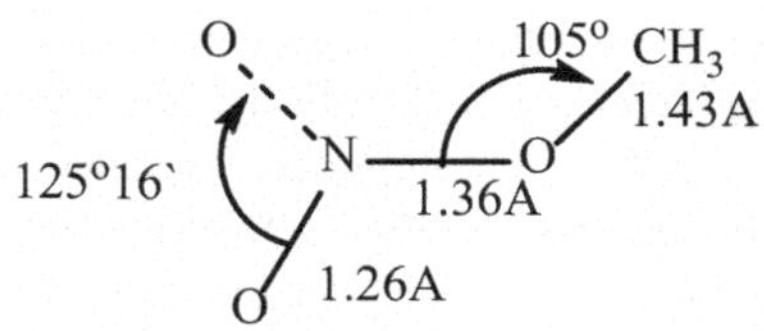

Nitrocellulose has three nitrate groups with different positions. According to the infrared spectrum, their characteristic frequencies are approximately 840, 1280, and 1660 cm^{-1}.

The fiber morphology of nitrocellulose made of refined cellulose is very similar to that of refined cellulose. Similarly, the fiber morphology of nitrocellulose prepared from thin paper made of wood, bamboo, cotton spoon, willow grass, bagasse, and other extracts from plants is still highly similar to that of refined cellulose before nitration [3–6]. After the general industrial nitration and boiling treatment, the nitrocellulose still retains the tubular structure of the fiber and has a large inner surface but swollen slightly with a large fiber diameter and a small inner cavity. From a morphological perspective, the nitrocellulose fiber exhibits hair surface, multilayer, break in horizontal shaft fiber, slightly hard without curling, and a slight change in size. However, if NC is processed via fine-breaking, the fibrous macroscopic structure is completely changed. The change is determined by way of the cutting off, but it can be divided into two categories. One type is the horizontal cutting of NC, which almost retains the basic structure of the original fiber. This process is also called "crushing" or "free cutting off"; another method is "coarse crushing" or crosscutting off by the longitudinal division, crushing, fibrillation and compressing, which significantly changes the original fibrous macroscopic structure. The industrial fine-breaking equipment has two roles, but which one plays the major role is different. In the fine-breaking process, the NC is divided into small structural units corresponding to microfibers and fibrils. After cutting off, except for a multifold increase in specific surface area and decreased in the molecular weight of NC, other properties do not change significantly, including macromolecule structure and molecular cluster structure.

10.1.2 Physical and Chemical Properties of Nitrocellulose

(1) The physical properties of NC

NC is an odorless and tasteless solid polymer. The appearance of pure NC is white with a tubular structure of the fiber. When in contact with acid, NC is not as soft as cotton. The fiber of NC is slightly hard and glossy without curl. When torn, NC fiber gives a noticeable sound.

The density of NC is greater than that of cotton. With increasing nitrogen content, the density of NC is generally in the range of 1.65–1.66.

NC does not melt or evaporate with a specific heat of 1.38–1.68 J/yc. Dried NC is a poor conductor of electricity. The friction of NC is more prone to generate static electricity than cellulose. When the water is contained, the conductivity of the NC is increased because of the conductive property of water. Therefore, fire and explosion caused by the electrostatic should be avoided to ensure the safety of production.

NC is insensitive to heat. It can be stored in the temperature range below 100 °C without decomposition for a long term. It can also be dried at about 110 °C.

(2) The chemical property of NC

Under normal circumstances, NC is a very stable and safe chemical product in terms of storage and transportation, especially when 28–30% of the wetting agent is present in the NC. The reactions of NC with various chemical reagents are as follows:

1. The reaction with acid

Under the catalysis of acid, nitrates will be hydrolyzed to cleave the glycoside bond. Nitrate hydrolysis is commonly called denitration. The reaction of an acid with NC depends on various factors such as the concentration of acid and the reaction temperature. The reaction equation is RONO + HOROH + HNO. If the product

NC contains a small amount of acid, it will cause the local acid concentration is too high, resulting in denitration hydrolysis. The resulting products, in turn, promote the further decomposition of the NC. This catalytic chain process will eventually lead to combustion explosion. For this reason, the finished product NC should not contain acid.

2. The reaction with alkali

The effect of alkali on NC is much more sensitive than acid, especially the effect of strong alkali due to the saponification effect of alkali. The effect of alkali on the NC will be more intense with increasing concentration of alkali, temperature, and reaction time. Since the presence of alkali can also lead to an increase in the color of NC, the finished product NC should not contain alkali either.

3. The reaction with reducing agent

Since the nitrogen atom on the nitrocellulose nitrate has the highest oxidation state, which is same as that of nitric acid, it is unstable and has a strong oxidizing capability. When the nitrocellulose reacts with reducing agent, it can be reduced to cellulose via denitration reaction. However, this reduction reaction is not a reversible reaction of nitration because the reduction is accompanied by a large number of side effects. As a result, the morphology of cellulose is destroyed so that the nitrocellulose cannot be reduced to the original cellulose.

4. The reaction with oxidants

Nitrocellulose is very stable to oxidants, such as potassium permanganate and bleaching powder will not cause significant changes in the properties of NC. Therefore, in the manufacture of high-grade plastics with high whiteness requirements of celluloid cotton and sprayed cotton, a lot of bleaching reagents can be used, such as potassium permanganate, hypochlorous acid, and ammonia solutions.

5. The reaction with air

Since the oxygen in the air does not react with NC at all, NC is safe in the air. Even at the high temperature, the impact of oxygen on NC is also negligible.

(3) Plasticization of nitrocellulose

NC is a rigid molecule that does not exhibit a "viscous flow" temperature (commonly known as a softening point) before the decomposition. As a result, general thermoplastics method that increases the temperature to above "viscous flow" temperature for plasticization cannot be used in the NC molding process, in which only the addition of plasticizer can be employed for molding processing. Therefore, this special performance of NC can be fully utilized to manufacture many types of toys and industrial products.

10.1.3 Explosive Properties of Nitrocellulose

The explosion of NC is a special form of chemical reaction, determined by its chemical structure, which is normally initiated by the physical factors such as light, mechanical, heat, sparks, friction, and shock waves.

This type of reaction of NC is typically divided into combustion and detonation, both of which are determined by external factors. The explosion decomposition of NC varies with changes in nitrogen content. The mode of explosive decomposition of NC with 13.1% nitrogen content is shown as below [8].

$$2\left\{\begin{array}{l} C_{24}H_{29}O_9(ONO_2)_{11} \\ C_{24}H_{30}O_{10}(ONO_2)_{10} \end{array}\right. \rightarrow 36CO_2 + 47CO + 4CH_4 + 39H_2O + 2C_2H_4 + 3HCN + 3.5H_2 + 18.5N_2 + 2NH_4HCO_3$$

The explosion reaction proceeds with an explosion heat of 4288.6 J/g, a specific volume of 765 ml/g, a detonation temperature of 3100 °C, a detonation velocity of 6300 m/s (ρ = 1.3 g/m^3), and a lead block expansion value of 375 cm^3.

Nitrocellulose with the same content of nitrogen, the one containing 16% water exhibits declined explosive properties, such as explosion heat and specific volume [9].

$$2\left\{\begin{array}{l} C_{24}H_{29}O_9(ONO_2)_{11} \\ C_{24}H_{30}O_{10}(ONO_2)_{10} \end{array}\right. + 48H_2O \rightarrow 41CO_2 + 37CO + 5CH_4 + 63H_2O + C_2H_4 + HCN + 7.5H_2 + 15.5N_2 + 10NH_4HCO_3$$

The explosion reaction of NC containing 16% water proceeds with an explosion heat of 3661 J/g, a specific volume of 720 ml/g, a detonation temperature of 2260 °C, a detonation velocity of 6800 m/s (ρ = 1.3 g/m^3), and a lead block expansion value of 280 cm^3.

The explosion heat of NC increases with increasing nitrogen content. In contrast, the specific volume, the heat of combustion, and the heat of formation decrease with the increase of nitrogen content. The explosion heat, combustion heat, and explosive decomposition products of nitrocellulose with different nitrogen contents are shown in Tables 10.1 and 10.2.

With increasing nitrogen content, the combustion heat of NC decreases while the heat of formation increases. Table 10.2 shows the combustion heat and heat of formation of NC with different nitrogen contents.

Both the catalytic activity of the decomposition products on the NC and the accumulation of heat generated by decomposition can accelerate the decomposition of the NC. The mutual promotion effect can reach the ignition point of NC, which may eventually lead to the combustion explosion.

Table 10.1 The decomposition products and their masses of NC with different nitrogen contents

	The nitrogen content of NC (%)				
	12.62	13.0	13.15	13.2	13.45
Explosion heat (J/g) (liquid H_2O)	4071	4288.6	4376.5	4414.1	4585.7
Explosion heat (J/g) (vapor H_2O)	3619.2	3870.2	3958.1	3995.7	4029.2
Specific volume V_o (ml/g) (liquid H_2O)	900	880	874	868	857
Gas composition (liquid H_2O)					
CO_2	21.2	23.8	24.3	24.4	27.0
CO	45.9	43.9	43.7	43.5	41.1
H_2	18.7	17.3	16.7	16.5	16.1
CH_4	0.4	0.4	0.4	0.4	0.4
N_2	13.8	14.6	14.9	15.1	15.6
Detonation temperature (°C)	2840	3025	3095	3130	3245

Table 10.2 The combustion heat and the heat of formation of NC with different nitrogen contents

The nitrogen content of NC (%)	The combustion heat (J/g)	The heat of formation (J/g)
7.66	12849.07	4012
10.14	11367.93	3364.36
11.06	10928.61	3105.79
12.45	10183.86	2699.52
12.88	9999.76	2579.44
13.53	9773.83	2485.30
13.58	9489.32	2379.44
13.92	9367.98	2146.4
14.12	9238.28	2108.74

10.1.4 The Representation Method and Classification of Nitrocellulose Nitrogen Content

(1) The representation method of nitrocellulose nitrogen content

The nitration degree of cellulose can be expressed in terms of the percentage of nitrogen content, nitration degree, and esterification degree.

Percentage of nitrogen content—refers to the mass percentage of nitrogen atoms in the molecule, expressed as: "N%".

The degree of nitration—refers to the volume of milliliters of nitric oxide (NO) gas released after 1 g NC is completely decomposed under the standard conditions (0 °C, 760 mmHg), expressed in NO (ml/g).

The degree of esterification—refers to the number of hydroxyl groups esterified on each glucose residue in the cellulose macromolecule, expressed as "γ". The relationship of the percentage of nitrogen content, nitration degree, and esterification degree is as follows:

$$Hardness\ NO\,(ML/g) = \frac{N}{100 \times 14} \times 22.4 \times 1000 = 16N$$

$$Nitrogen\ content\ N\% = \frac{14\gamma}{162 + 45\gamma} \times 100 = \frac{31.11\gamma}{3.6 + \gamma}$$

$$degree\ of\ esterification\ \gamma = \frac{3.6N}{31.11 - N}$$

where

14 is nitrogen atomic weight.

162 is the molecular weight of a glucose residue.

45 is the increased molecular weight after the addition of one (ONO_2) ester group to the glucose residue.

The nitrogen content of the NC indicated by the above three methods is only a statistical average. That is, it represents only the number of grams of nitrogen contained in a given weight of NC or the number of total hydroxyl groups that are nitrated rather than one or each macromolecule contains the amount of nitrogen and the number of nitrated hydroxyl groups.

The corresponding degrees of esterification and nitration of certain percentages of nitrogen content are shown in Table 10.3.

The degree of nitration is generally used in the industry to represent the nitrogen content of the NC. This is because the nitrogen content of the NC is measured by using five-tube nitro meter [10]. The measured millimeter number of nitrogen oxide gas per gram of NC decomposition is the value of nitration degree. When the error of nitrogen content is only 0.1%, the error of nitration degree can reach 1.6 ml/g. Therefore, the use of nitration degree to represent the nitrogen content of NC is

Table 10.3 Corresponding relation in the degree of esterification, the degree of nitration, and the percentage of nitrogen content

N %	NO (ml/g)	γ
10.37	165.40	1.79
10.75	171.46	1.90
11.13	177.52	2.00
11.46	182.79	2.10
11.97	190.92	2.25
12.44	198.42	2.40
12.90	205.76	2.55
13.28	211.82	2.70
13.61	217.08	2.80
13.75	219.00	2.85
13.88	221.29	2.90
14.14	225.53	3.00

more accurate and convenient. With the improvement of the test technology, the use of DeWald alloy reduction method for determining the amount of nitrogen content has significantly improved the nitrogen content test accuracy of nitrocellulose [11].

(2) Classification and application of nitrocellulose

In the past, the nomenclature and classification of nitrocellulose are simply according to the number of nitro group. This classification cannot meet the needs of industrial production of multispecies nitrocellulose. Nitrocellulose is widely used in industry, such as for the manufacture of gunpowder, explosives, film, paint, and plastic. Generally in the industry, the classification of nitrocellulose is according to the average degree of nitration and their application.

NC with a nitration degree of 210 ml/g or more is denoted by 1#NC, also known as No. 1 strong nitrocotton or simply No. 1 nitrocotton, which is used to formulate mixed NC.

NC with a nitration degree of 190–198 ml/g is denoted by 2#NC, also known as No. 2 strong nitrocotton or No. 2 nitrocotton, which is used to formulate mixed NC.

Mixed NC is also known as blended NC. It is based on the requirements of the application to mix 1#NC and 2#NC in a certain ratio, which is used for the production of single-base propellant.

Colloidal NC with a nitration degree of 200–203.2 ml/g (nitrogen content of 12.6 ± 0.1%) is also known as collodion, Pirro collodion, and Mendeleev cotton, which can be individually plasticized to produce single-base propellant.

NC with a nitration degree of 188–193.5 ml/g is denoted by 3#NC, also known as "H" type NC, No. 3 weak nitrocotton, and abbreviation of No. 3 nitrocotton. It is used for the production of double base propellant.

Colloidal explosive NC has a nitration degree of 191–197 ml/g, also known as colloidal explosive nitrocotton, which is used in the manufacture of colloidal explosives.

The film NC, which has a nitration degree of 182–192 ml/g, is used for the manufacture of transparent films and films in the field of firework, such as sheets, kits, capsules, and caps.

NC with a nitration degree of 185–195 ml/g for spray paint is abbreviated as painted nitrocotton or lacquered nitrocotton, which is used for the manufacture of nitrocellulose varnish, nitro-enamel, nitro-based adhesives, and so on.

NC with a nitration degree of 172.8–179.2 ml/g (nitrogen content of 11.0 ± 0.2%) for plastic application is abbreviated as plastic NC, which is also known as celluloid cotton or Etlo cotton. It is widely used in the manufacture of celluloid and other plastics.

In addition to the abovementioned NC, NC with a nitration degree less than 170 ml/g is collectively referred to friable nitrocotton, which does not have the industrial value.

In other countries, the classification of military NC is not the same. Table 10.4 is the classification used in the United States. The classification used in Russia is similar to that of China. In Germany, four types of NC with nitrogen contents of 13.3, 12.6, 12.2, and 11.0% are used to formulate seven mixed NC, which are used

Table 10.4 The classification of military NC in the US

Name	N %	Application
1#NC	13.35–13.55	Preparation of mixed NC
Colloidal explosive NC	12.5–12.7	Preparation of mixed NC
Mixed NC	13.15–13.20	Preparation of propellant for gun, cannon, and rocket
Colloidal NC	11.8–12.2	Preparation of double base propellant

in the manufacture of dual-based gunpowder. Britain, France, Japan, and other countries also produce 1#NC, 2#NC, and plastic NC for the manufacture of gunpowder.

NC for various applications has different product grains and quality requirements. The quality and grain requirements of NC products for different applications are shown in Table 10.5.

10.2 Refining Cellulose

10.2.1 Classification and Structure of Cellulose

Pure cellulose is composed of three elements, carbon, hydrogen, and oxygen. According to the experimental result, the mass percentages of carbon, hydrogen, and oxygen in cellulose are 44.44%, 6.17%, and 49.39%, respectively. Their corresponding atomic ratios can be obtained by dividing their mass percentages with carbon, hydrogen, and oxygen atomic weights.

$$\text{Carbon number} = \frac{44.44}{12} = 3.70$$

$$\text{hydrogen number} = \frac{6.17}{1} = 6.17$$

$$\text{oxygen number} = \frac{49.39}{16} = 30.09$$

These numbers are divided by 0.617 and rounded to afford the ratio of the three elements as follows: C: H: O = 6: 10: 5. Therefore, the empirical formula of cellulose is $C_6H_{10}O_5$. The empirical formula only represents the species and ratio of elements in one compound, which is different than a molecular formula.

Table 10.5 Quality requirements for various NC items

Criterion		The degree of nitration (ml/g)	Solubility in alcohol ether (%)	Solubility in ethanol (%)	Viscosity (Engler degree)	Beating degree (ml)	Ash content (%)	Alkalinity (based on $CaCO_3$) (%)	Stability			
									Degree of stability (ml/g)	Vieille test (h)	Abel test (min)	Ignition point (°C)
1#NC		More than 210	Less than 15	Less than 4	3–20	Less than 90	Less than 0.5	≤0.25	Less than 3.5	>7		
2#NC		190–198	More than 97	Less than 7	3–20	Less than 80	Less than 0.5	Less than 0.25	Less than 2.5	>7		
Mixed NC		More than 208 204–207.5	22–34 34–45	Less than 5	3–20	Less than 85	Less than 0.5	Less than 0.25	Less than 3.0	>7		
Colloidal NC		200–203.2	More than 95									
3#NC		188–193.5	More than 98	Less than 12	1.9–2.6	Less than 65	Less than 0.5	Less than 0.2	Less than 2.5	>7		
Colloidal explosive NC	Regular	191–197	More than 97	Less than 4	>50	80–105	Less than 0.5	Less than 0.25 (based on $NaCO_3$)			65 °C >10	
	Antifreeze				>160							
Film NC		182–192	More than 99.2		Falling ball viscosity 5–20 s		Less than 0.3			>7		>180 °C
Spray paint NC		185–195	More than 98.5	Less than 9	Falling ball viscosity 0.25–40 s		Less than 0.2	Acidity < 0.1			80 °C >10	>180 °C
Plastic NC	First grade	172.8–179.2	More than 98		2.5–4.5		Less than 0.2	Less than 0.05		>7		>180 °C
	Second grade		More than 96		1.5–6.0		Less than 0.3					

10.2.1.1 Classification of Cellulose

Cellulose is the main component of higher plants, widely distributed in nature. Through the photosynthesis, about billions of tons of natural cellulose are generated each year. According to the growing zone, shape, and content of cellulose, cellulose can be divided into four categories:

(1) Seed fiber

Seed cellulose is clustered in the outer skin or fruit of the seed, such as cotton, kapok, and coconut fiber. Cotton has the highest cellulose content in the plant, up to 94% or more.

(2) Phloem fiber

Phloem fibers are gathered in the phloem of plants, such as hemp, flax, jute, ramie, and other wild hemp. Hemp contains a relatively high content of cellulose, which is generally in a range of 60–85%. However, it is typically not used to extract cellulose because of its low yield and high economic value.

(3) Stem fiber

Cellulose grows in the stem, stalk, branches, and leaves of plants, such as Eulaliopsis binata, bamboo, bagasse, reed, wheat straw, and rice straw. Their cellulose contents are in the range of 30–68%. This kind of cellulose has an abundant and wide range of resources, and cheap, which are good raw materials for the manufacture of paper and extraction of cellulose, of which Eulaliopsis is the most prominent.

(4) Wood fiber

All kinds of wood are composed of cellulose. The cellulose content of wood is typically in the range of 50–65%, which is an important raw material for the preparation of cellulose and papermaking.

Methods for cellulose extraction are different for raw materials with different sources. Moreover, cellulose purity and extraction methods are related to many factors such as plant species and growth conditions. The cellulose content in different plants is shown in Table 10.6.

Cellulose refined from stem fiber and wood fiber is not only heavily used in the household paper and textile industry but also has a very wide range of applications in industrial and agricultural production and defense industry, which is an important source of raw materials for the production of nitrocellulose.

Cotton has the highest cellulose content with the low content of non-cellulose impurities, which is easy to be refined. Although cotton has limited planting region and higher cost, the production of NC and other products with high-quality requirements still use cotton fiber, cotton linter, and cotton hair as the main raw materials for the preparation of cellulose. The preparation of NC also utilizes other raw materials for the extraction of cellulose, including wood fiber, Eulaliopsis binata fiber, bamboo fiber, and hemp fiber.

Table 10.6 The cellulose content in different plants [3]

Plant name	Growing zone of fiber	Cellulose content %
Cotton	Seed	>90
Kapok	Follower and seed	>83
Coconut fiber	Seed	>81
Ramie	Phloem	70–84
Hemp	Phloem	70–80
Flax	Phloem	70–80
Jute	Phloem	60–65
Eulaliopsis	Stem	49–60
Bamboo	Stem	40–55
Reed	Stem	40–50
Bagasse	Stem	35–50
Wheat straw	Stem	35–45
Conifer	Stem, branch	50–60
Broadleaved tree	Stem, branch	45–55

Cotton fiber is already known as cellulose. In the following introduction, it involves the stalk plant cellulose represented by Eulaliopsis binata and the extraction of cellulose from wood [3, 5, 6].

Eulaliopsis binata (Retz) is also called Chinese alpine rush, which is a perennial herb of the family Gramineae. The stems of Eulaliopsis are tightly clustered, upright, hard epidermis, and leaf is turning slender near the roots.

Eulaliopsis fiber has a capillary-like structure. It grows on the inner surface of pectin in the straw, in which one end is open, and the other end gradually shrinks until the fiber chamber close to a duckbill shape. Eulaliopsis fiber is slender, which is typically in the range of 630.5–3705.8 μm with an average length of more than 2 mm. The diameter of the fiber is in the range of 5.3–19.8 μm with an average diameter of 12.7 mm. The thickness of cell wall is in the range of 4.3–7.6 μm with an average of 5.8 μm.

Similar to cotton fiber, Eulaliopsis and other wild plant fiber consist of a primary wall, secondary wall, and hollow fiber chamber. With the growth of plants, the fibers continue to grow, and oxidized fiber and sugar outside the plant are converted into colloidal deposited on the surface of the plant, increasing the thickness of the plant's surface. At this point, lignin and fiber grew. During the growth of lignin, pectin, and fiber, cellulose, a type of sugar material, deposits on the primary wall from outside to inside to increase the thickness of primary wall gradually, forming some concentric circles. These concentric circles are called secondary walls.

Lignin, pectin, and primary wall do not swell in the copper ammonium solution, which is detrimental to the nitration treatment. Therefore, the destruction of pectin, lignin, and the primary wall is the basic task and important process in the refining process of Eulaliopsis cellulose.

Primary wall and secondary wall are collectively referred to the cell wall. The thickness of cell wall is related to the maturity of the fiber. When the fiber inside the plant is mature, the death of protoplasm forms a fiber cavity. As the fiber skin has the property of semipermeable membrane, in general, the solution and reagents are more difficult to penetrate into the fiber cavity. Therefore, it is of great significance that the fiber cavity is a duckbill-shaped opening at the end. The shorter the cellulose, the shorter the cavity, which is easy for the solution and the reagent enter into the cavity to react with the cellulose. The wall of the wild plant fiber is elongated and narrow with a duckbill opening at the end, so the reactivity is lower than that of the cotton cellulose. This is the reason why the properties of wild plant cellulose are not as good as those of cotton cellulose, as shown in Table 10.7.

Since cellulose contains many impurities, it must be refined and dried before it can be used as a raw material for making NC.

10.2.1.2 The Macromolecular Structure of Cellulose

In organic chemistry, polyfunctional compounds containing only carbon, hydrogen, and oxygen are commonly referred to as sugars. The most important of the functional groups are –OH (hydroxyl), aldehyde and keto. Since the ratio of hydrogen to oxygen atoms is 2: 1 in most of the carbohydrates, the composition of sugar molecules is generally expressed in the form of $C_xH_{2y}O_y$.

There are monosaccharides, disaccharides, and polysaccharides in nature. Monosaccharides are small molecules of sugar that can no longer be hydrolyzed to other sugars, such as fructose and glucose. Polysaccharides are macromolecules composed of dehydrated monosaccharides, which are linked by glycosidic bonds. The hydrolysis of each molecule polysaccharide can produce multiple monosaccharides. Usually, the glycosidic bond is called the glycoside bond, which is the bond connecting two carbon atoms of two dehydrated monosaccharides via an oxygen bridge. The general form of the type of bond is –C–O–C–.

Whether seed cellulose or other wild plant cellulose, it is polysaccharide composed of a large number of dehydrated glucose molecules and can produce glucose after hydrolysis.

(1) The chemical structure of cellulose macromolecules

In the 1890s, Russian gunpowder researchers put forwarded that [5] cellulose is a polymer with large molecular weight. Regardless the type of raw materials and the process of making cellulose, their chemical compositions are identical, consisting of carbon, hydrogen, and oxygen; three elements with a mass ratio of C = 44.44%, H = 6.17%, and O = 49.39%. After the division of the mass ratio with the atomic weight for each element, the ratio of the elements in the cellulose is C: H: O = 6: 10: 5. Therefore, the empirical formula of cellulose is $C_6H_{10}O_5$. Since the molecular weight of the cellulose is very large, the molecular formula of the cellulose is generally written as $(C_6H_{10}O_5)_n$, where n represents the degree of polymerization of the cellulose. According to the literature, the polymerization degrees of cotton

Table 10.7 Comparison of chemical composition of some plants

Chemical composition (%)	Cotton	Eulaliopsis	Bamboo	Bagasse	Conifer	Wheat straw	Picea glauca	Birch	Ramie
α-Cellulose	93.87	56.78	44.58	42.21	51.83	40.70	48.38	44.58	74.33
Pectin	0.03	1.97	2.39	2.26	2.18	3.14	1.33	4.16	3.22
Lignin	None	16.35	22.14	19.00	26.46	22.56	27.10	18.90	7.57
Acetaldehyde	None	0.53	0.42	2.23	1.76	2.86	1.30	4.40	0.29
Ash	0.92	4.49	2.20	3.66	2.28	4.59	0.30	0.20	1.86
Protein	0.87	0.33	0.36	2.87	1.39	2.71	3.60	0.13	0.43
Waxy fat	0.68	8.68	7.03	1.70	2.10	1.90	2.12	4.60	1.01
Water-soluble	2.34	0.68	1.02	1.27	0.20	0.93	–	0.72	2.76
Pentosan	1.26	9.87	19.86	24.71	11.80	20.61	15.87	22.31	8.53
Hot water extract	–	9.01	6.78	16.80	6.77	–	–	8.87	–
1% NaOH dissolved substances	–	38.18	31.24	26.26	22.47	–	–	28.56	–

cellulose, coniferous cellulose, and Eulaliopsis are 15,000, more than 20,000 and about 18,000–23,000, respectively.

The chemical structure of cellulosic macromolecules has been well elucidated long time ago, which is believed that the basic unit of ideal cellulosic macromolecules is dehydrated D-glucose linked on 1,4-carbon atoms via beta-glycosides (also known as oxygen bridges) [6]. The 1,4-β-glycosidic bond can be considered as the condensation product of the hydroxyl group of the first carbon atom on the β-D-glucose with the hydroxyl group on the fourth carbon atom of other β-D-glucose.

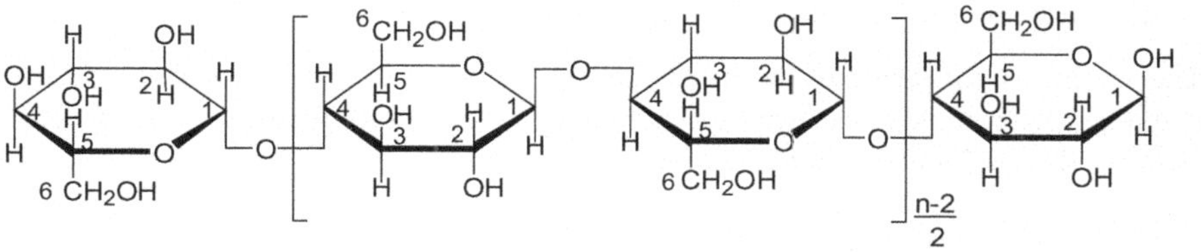

In the structure of cellulose, n represents the degree of polymerization, indicating that the number of identical basic unit in the cellulose macromolecule. In general, the value of n is very large. Therefore, the molecular formula of cellulose can be simplified as $[C_6H_7O_2(OH)_8]_n$. Cellulose macromolecules are also aldehyde acetal linked linear polymer.

This structure has been confirmed by the data accumulated in various research fields, such as chemistry, polarized light, kinetics, and pyrolysis products [3].

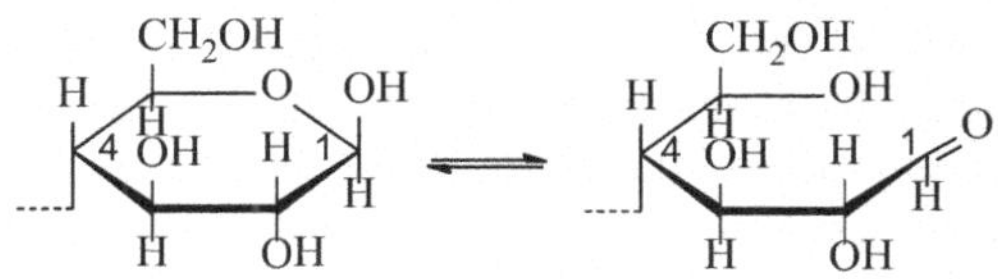

As the cellulose is a natural polymer, different plants and different growth conditions can alter the chemical structure of natural cellulose compared with the standard molecular formula. Some other monosaccharides, such as pentose, pectic acid, and glucose residues, are also present in the cellulosic macromolecule structure. Cellulose macromolecule structures containing other monosaccharides are shown as follows:

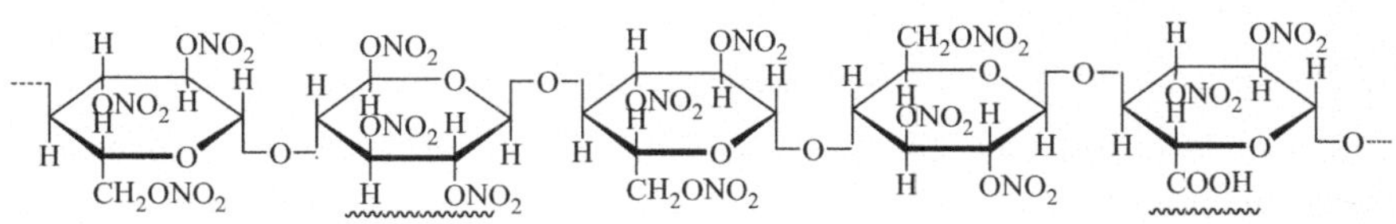

Pentose residues Pectin acid residue

Cellulose macromolecule structure containing open glucose residues

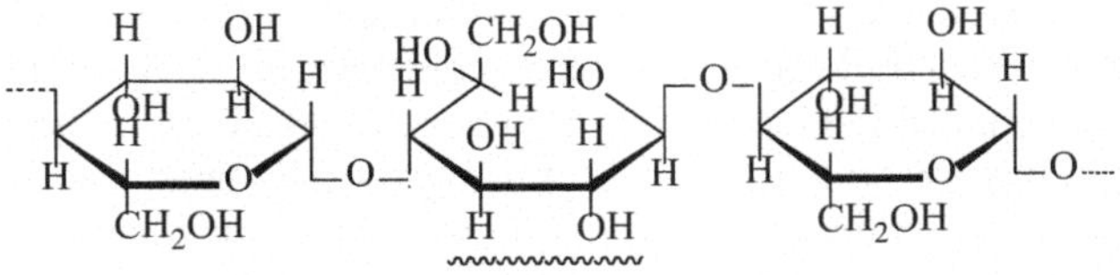

open residue

(2) Conformation of cellulose macromolecule

The spatial conformation of the β-D hexacyclic glucose is a chair structure, and the cellulose macromolecules form a syndiotactic spatial configuration from the glucose residues with chair conformation.

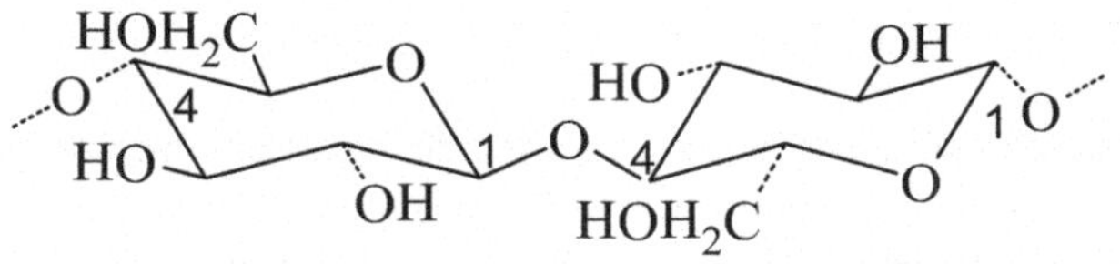

Cellulose macromolecules can be described by asymmetry degree. The so-called asymmetry degree refers to the ratio of the length to the diameter of the macromolecule with a certain shape, which represents the asymmetry degree of the macromolecule. The high molecular weight of the cellulose macromolecule bonds is very rigid due to its chemical structure and spatial configuration making the rotation of the cellulose macromolecule very difficult. The factors that contribute to the difficulty of rotation include the existence of many extremely polar hydroxyl groups in β-D-glucose residues containing, which form internal hydrogen bonds and cause the difficulty for internal rotation. Since the theoretical ratio of length to diameter of cellulose macromolecules are very large and the asymmetry is very high, the cellulose macromolecules have a certain degree of flexibility rather than a completely stiff rod.

The shape of the cellulose macromolecule is not immutable. If the intermolecular and intermolecular interactions change, or the degree of macromolecule polymerization changes, their effect on the flexibility is changed, which leads to the change in the shape of the macromolecule.

(3) The structure of the cellulose macromolecule in the aggregation state

In the cellulosic structure, the macromolecules are bound by two types of bonds, such as the primary and secondary bonds. These two types of bonds have different energies and intensities. In the cellulose, the main valence energies of –C–C– and –C–O–C– are between 334.92 and 376.56 kJ/mol. A small number of primary bonds in macromolecules are mostly ester and glycoside bonds. The ester bond is formed by the reaction between the carboxyl group in the macromolecule and the hydroxyl group of the adjacent macromolecule, whereas the glycosidic bond is formed by the hydroxyl group on the open ring of the basic unit of the macromolecule and the aldehyde group on another macromolecule.

The secondary bond force is the intermolecular interaction, including van der Waals force and hydrogen bonding force. The type and intensity of the secondary valence bond force vary with the distance between molecules.

In the structure of the cellulose, the aggregation and phase of cellulose are determined by the interactions among macromolecules.

Cellulose is a continuous structure formed by the monomer of many macromolecules. In the most densely arranged macromolecules, they are arranged in parallel and oriented orderly, forming the oriented region of the cellulose. The binding force between the macromolecules increases with shortening distance, and the binding force is the maximum in these places where the distance is shortened. In the loss region, the binding force of macromolecules is also weak with a larger gap between macromolecules. Their distribution is not completely parallel, which form the non-oriented region of cellulose. A macromolecule may have a portion in the oriented region of the cellulose, while the other part is in a non-oriented region or re-penetrated into other regions. Because of a large number of alternating orientations, the cellulose has a crystal structure characteristic in the X-ray diffraction pattern, as shown in Fig. 10.2.

In 1858, the micelle theory of cellulose material was proposed [5]. It is believed that the cellulose fibers are bundled in parallel with each other along the cellulose axis, forming the so-called micelle. According to the X-ray measurement, a micelle consists of 100–170 simple cellulose molecular chains. It has a length of at least 600 Å with a width of 50–60 Å (Fig. 10.3).

Fig. 10.2 Unit cell structure of cellulose ~ II [3]

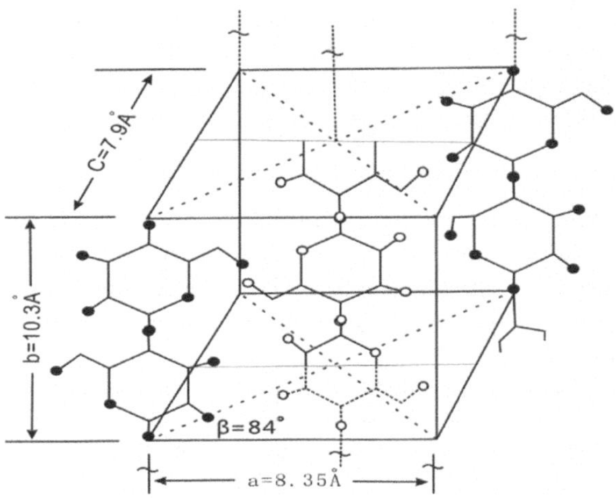

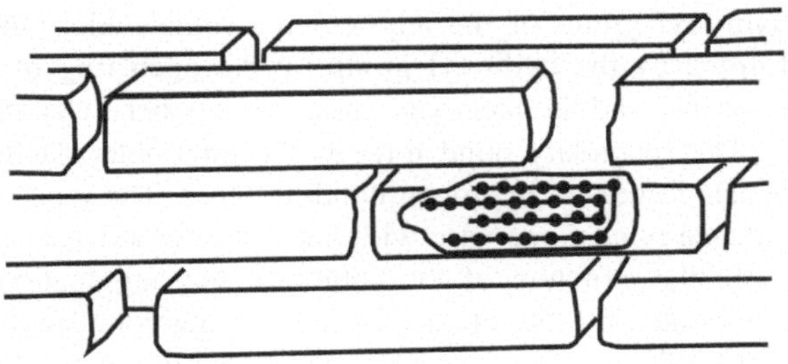

Fig. 10.3 Micelle structure of cellulose [3]

10.2.2 Physical and Chemical Properties of Cellulose

(1) The physical properties of cellulose

Cellulose is a white filamentous solid polymer with the soft handle while non-bleached cellulose is light brown. Because of different methods of refining or measuring methods, the specific gravity, the specific heat, and the heat of combustion of cellulose are in the ranges of 1.40–1.52, 1.3–1.38 J/g °C, and 18200.4–18263.16 J/g, respectively. The ignition point of cellulose is 290 °C. Cellulose is generally insoluble in water and organic solvents, but soluble in copper ammonium solution and some metal complexes.

Cellulose has a birefringence phenomenon. Because of the existence of oriented and non-oriented regions, the cellulose structure exhibits anisotropy.

Dry cellulose is a poor conductor. When the hot air passes through the dry cellulose, there will be static charge accumulation. In serious conditions, the discharge will occur.

Cellulose is a porous tissue material. Both the outer surface and a great inner surface of fiber can be seen under the microscope. The pore volume of cotton fiber accounts for 35–40% of the total volume with an inner surface of 2300–3200 cm^2/g. Since the wild plant fiber is shorter than the cotton fiber, the total inner surface of wild plant fiber is larger than the total inner surface of the cotton fiber.

(2) The chemical properties of cellulose

The chemical properties of cellulose are mainly manifested in the properties of hydroxyl and glycoside bonds. Cellulose macromolecules are polyols that can be esterified, etherified, and oxidized to aldehydes, acids, and ketones by oxidants. The potential aldehyde group at one end of the cellulose can be reduced by the reducing agent.

Cellulose is biodegradable because the glycosidic linkage in the cellulose macromolecule is a weak bond that can be cleaved under the action of chemical or physical factors to shorten the macromolecular chain.

1. The reaction with acid

The interaction of cellulose with acid involves various reactions, including swelling, hydrolysis, and esterification with an acid having an esterification capacity, and oxidation with an acid having the oxidizing capability. Different acids or the same acid under the different conditions, such as temperature, pressure, time, and concentration of the solution, their reaction or interaction with cellulose are different.

The ability of a hydrochloric acid to swell the cellulose is weak, but it can hydrolyze cellulose at any concentration. Hydrochloric acid cannot esterify or oxidize cellulose.

When the concentration of sulfuric acid is 1–58%, it can hydrolyze cellulose; when the concentration is 60–70%, it can dissolve the cellulose and form cellulose sulfate. In the process of sulfate ester formation, the glycosidic bonds of cellulose macromolecule will be significantly destroyed. During the reaction of cellulose with sulfuric acid, carbonization phenomenon would occur if the wetting is poor.

When the concentration of nitric acid is 60–65%, the cellulose can be swollen, and small molecular compounds can be formed; when the concentration of nitric acid is 70–75%, the cellulose can be partially dissolved; when the concentration of nitric acid reaches 78–85%, the cellulose is completely dissolved and forms cellulose nitrate. However, a further increase in the concentration of nitric acid reduces the ability to dissolve cellulose.

Phosphoric acid has a strong swelling effect on cellulose, but its hydrolysis capacity is weak. Under the same conditions, the swelling ability of the acids on cellulose is in the following order: $H_3PO_4 > HNO_3 > H_2SO_4 > HCl$. In contrast, the hydrolysis ability of acid is in a reversing order: $HCl > H_2SO_4 > HNO_3 > H_3PO_4$. The hydrolysis of cellulose under the action of acid increases with increasing temperature and extended reaction time.

2. The reaction with alkali

The glycosidic linkages in the cellulose macromolecules have high stability to the base, which is illustrated by the fact that treatment of cellulose with 2–4% NaOH does not cause extensive damage. However, although cellulose is insensitive to the base, the base still has a degradation effect on cellulose, especially when the temperature and alkalinity are significantly high. In the industry, the definition of α-cellulose and classification of cellulose are according to the properties of cellulose in an aqueous solution of NaOH in a concentration of 17–18% (mass ratio) at 20 °C.

Alkali soaked cellulose can be oxidized under the exposure of air for an extended time. After the degradation of cellulose, the viscosity decreases.

3. The reaction with oxidants

Cellulose can be oxidized by oxidants. The product produced by the oxidation reaction is called oxidized cellulose, which is heterogeneous with cellulose regarding physical properties and chemical properties. For example, NCs made with oxidized cellulose exhibit reduced stability.

4. Thermal interaction

The polymerization degree of cellulose is reduced under the heating at different temperatures and for a long period. Even if it is dried under normal conditions with appropriate temperature and drying time, the cellulose is slightly changed, including changes in surface properties and decreased viscosity. When heated at 140 °C or higher, the degradation of cellulose is accelerated with extended heating time. At the temperature of 162 °C, it is observed that the cellulose begins to be carbonized; at the temperatures above 320 °C, the cellulose was quickly carbonized and burned.

Under the action of heat, the destruction of hydrated cellulose is intensified because water molecules are involved in the chemical reaction of cellulose.

5. Photoreaction

Cellulose exposure to light and air for a long time, the polymerization degree of cellulose will gradually decrease. Meanwhile, the strength of cellulose also decreases significantly. Ultraviolet and visible light can destroy the cellulose. The shorter the wavelength, the greater damaging effect of light.

If the cellulose contains heteroatoms, including oxygen, chlorine, sulfide gas, and nitrogen oxide, the degradation of cellulose by light can be accelerated by metal salts, such as iron salt and manganese salt.

10.2.3 Refining of Cellulose

The refining of cotton fiber is straightforward while the refining of cottonseed hair is relatively complicated. The progress of papermaking technology provides more reference technology for the refining of cellulose, especially the technology to extract cellulose from the wood, ramie, phloem, bamboo, bagasse and other stem plants. When the cellulose is extracted from the wood, the wood is first to cut into pieces and pre-hydrolyzed in a spherical digester, and then mixed with a mixture of sodium sulfide and sodium hydroxide solution to digest and prepare pulp containing 85% α-cellulose. The first stage is the preparation of cardboard, followed by refining and shaping to afford tissue paper (similar to napkins) with high water absorption capability. This stage is similar to the purification of cotton linter. For the extraction of the first stage, please refer to the related literature [4–6, 12, 13]. The following is the description of refining process for cotton linter fiber.

10.2.3.1 Principles of Cellulose Refining

The main purpose of cellulose refining is to remove non-cellulose components while reducing the viscosity. The quality uniformity of crude fiber after refining is to some extent assured. The cellulose refining is achieved through high temperature, high pressure, and alkali treatment, combined with other measures. Customarily, the process for the refining of cotton fiber is called degreasing. Therefore, the refined cotton is known as absorbent cotton.

The base used in the refining cotton is NaOH diluted solution, which has a strong saponification effect on the associated materials of cellulose, such as lignin, pectin, cotton peel, and cottonseed shell. The formed saponified products are not only easy to be scoured off with water but also good emulsifiers for higher alcohol and wax and others in the degreasing process. NaOH will not increase the ash content in the absorbent cotton.

The emulsifier can remove high alcohols and high hydrocarbons because one end of the emulsifier is a nonpolar group (one end of R–) and another end is a polar group (–COONa). The emulsifier interacts with nonpolar high alcohols and high hydrocarbons of cellulose by surrounding them with nonpolar ends to remove them from the cellulose. In the meantime, the polar end of the emulsifier interacts with polar water molecules via the formation of the emulsion, thus removing from the cellulose. To improve the wettability of NaOH on the fiber, enhance the reactivity, and homogeneity of the reaction, a small amount of emulsifier should be added to the lye. The emulsifier removes some of the high alcohols that are not reactive with the base, some of the wax, water-insoluble saponified products. The commonly used emulsifier is rosin soap that prepared from rosin and NaOH.

The methods of removing non-cellulose impurities in the raw materials with different sources depend on their properties. Among them, the rosin soap can remove non-cellulose impurities completely and efficiently.

The polymerization degree of pentosan is much less than that of cellulose. In addition, pentosan is easily hydrolyzed and soluble in diluted lye (4–5%). Pectin is not easy to be hydrolyzed but can be hydrolyzed into the corresponding sodium salt with alkali and pressurized by digestion. Under the high temperature and high pressure, fat and wax can be hydrolyzed to a soluble sodium salt in diluted lye. Inorganic salt ash and mechanical impurities are removed by combing the raw materials, followed by washing with hot water and treatment with diluted acid, and other methods. Tannins and pigments can be removed with lye under high temperature and high pressure. Therefore, relatively pure cellulose can be obtained with diluted alkali treatment under high temperature and high pressure, combined with other measures.

10.2.3.2 Refining Production Process of Cellulose

Because of its high α-cellulose content and less impurity content, the refining process of cotton linter to produce refined cotton is relatively simple and easy to control the quality. The process of refining cellulose from cotton linter is shown in Fig. 10.4.

In the process of Fig. 10.4, digestion is the core of the entire cellulose refining process. In the refining of wood cellulose and stem plant cellulose, in addition to the

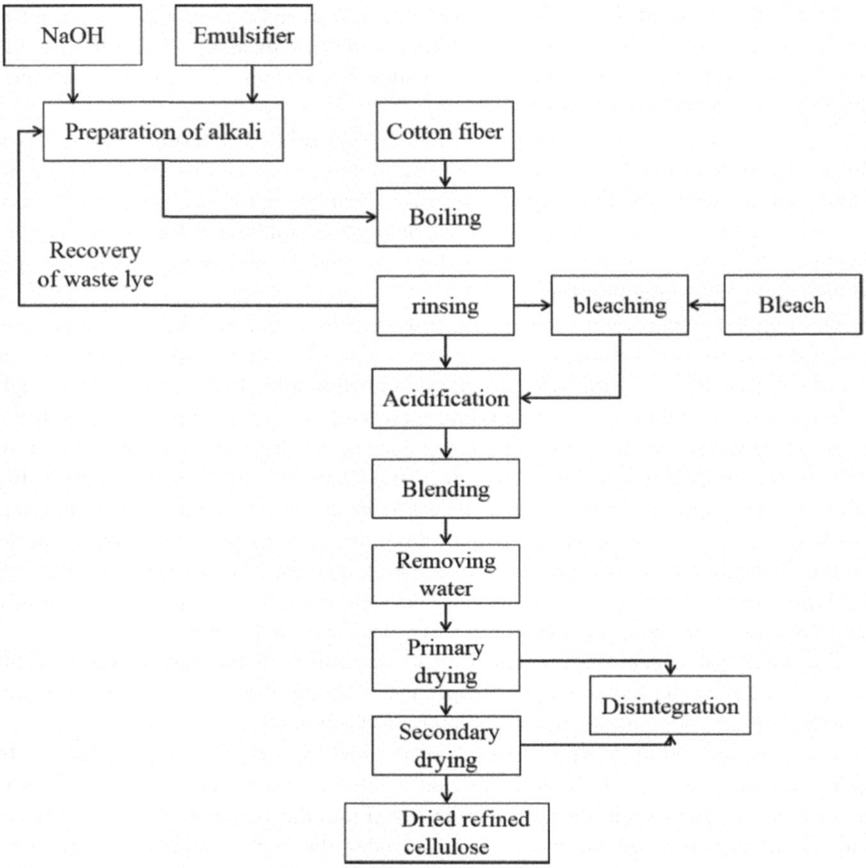

Fig. 10.4 The process of refining cellulose from cotton linter and cottonseed hair

introduction of multistage treatment, the digestion process is still the core technology to determine the quality of cellulose.

After combing by fiber opener, cotton linter is delivered to digester by air flow and digested with 2–4% lye. Upon the completion of digestion, it is rinsed with hot water and cold water. Waste lye is recycled for reuse. The rinsed cotton material is acidified and washed, and then combined with other batches. After a lot of water is removed from the combined batch of refined cellulose, it is treated with spray air drying to control the moisture of refined cellulose below 10%. Subsequently, it is treated with the secondary air drying to reduce the water below 5%. The dried and refined cellulose is directly delivered to the feed port of nitration reactor with air flow.

It is relatively easy to accomplish the refining of cotton cellulose for preparation of nitrocellulose. In the foregoing, it is mentioned that the process of extracting cellulose from wood fiber and plant stem fiber in the pre-refining stage is similar to that of the papermaking pulping process; in the latter stage (i.e., when the content of α-cellulose reaches 95% or more), the refining process is basically identical to that of refining process of cotton cellulose. The author systematically studied the process of extracting cellulose from the stem plants such as Eulaliopsis binata, cotton stalk, bagasse, and bamboo. The process of refining the cellulose from the stem plant is shown in Fig. 10.5.

As the wood, bamboo, sugar cane bagasse, and Eulaliopsis fibers are different than cotton fiber, cellulose refining processes for cottonseed hair, cotton linter, ramie, and others are different. When the cellulose is refined, the cotton linter is pre-hydrolyzed with a sulfite saponification method, and the refined cellulose meets the technical specifications for preparation of nitrocellulose. Experimental results demonstrate that the process of Fig. 10.5 can be used to produce higher quality Eulaliopsis cellulose.

Fibers from stem plants such as Eulaliopsis are difficult to form flocculent. To facilitate the following nitration reaction, these types of fibers are prepared into thin papers. The refining process of Eulaliopsis fiber is not only relatively straightforward but also the high yield of cellulose. Stem fiber has promising application prospects, which can replace cotton fiber for manufacturing NC and propellant. The author invented the process for the production of NC from the stem fibers, which accounts for 20% of the NC production in China. In addition, the stem fiber is also an important fiber raw material for textiles and daily necessities.

(1) Digestion

The digestion of cotton linter with NaOH solution is the core process in the production of refining cellulose. In the digestion process, cotton linter is chemically processed with NaOH solution containing a small amount of emulsifier under high temperature and high pressure to remove non-cellulose impurities, such as fat, wax, pectin, protein, and ash, for the purpose of improving the content of α-cellulose and the degree of moisture absorption and reducing the content of sulfuric acid and ash content. In the meantime, the primary wall of fiber, intermolecular hydrogen bonding in cellulose molecules, and the structures of cellulose macromolecules are partially degraded, which reduces the viscosity of cellulose and improves the nitration reactivity, thus achieving the quality uniformity of refined cotton products.

The digestion processes are divided into batch digestion and continuous digestion. The equipment used in the batch digestion process includes a vertical digester and a horizontal digester. The digestion of cotton linter, cottonseed hair, and ramie typically employs a vertical digester for refining. The extracting cellulose from stem fibers, such as wood, bamboo, bagasse, and Eulaliopsis, is first processed with a horizontal digester to obtain a crude fiber in a cardboard shape and then transferred to a vertical digester for further refining. The heating method for refining cellulose includes direct heating and indirect heating.

The advantages of direct heating include quick heating, short digestion cycle, insignificant heat, and steam loss. However, the condensed steam continuously dilutes the lye when using steam for direct heating, which significantly reduces the

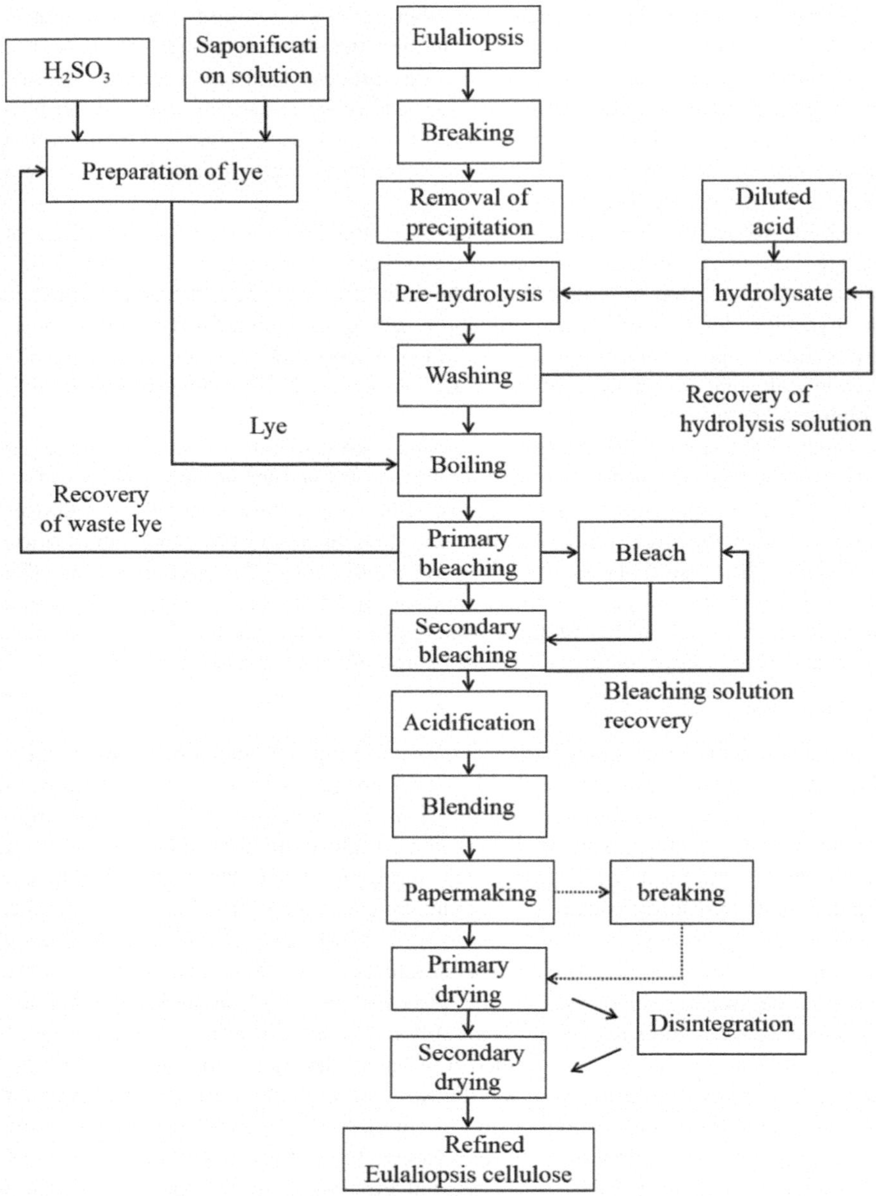

Fig. 10.5 Cellulose refining process for stem plants with Eulaliopsis as raw material [3, 4]

concentration of the digestion lye and negatively affects the product quality. For this reason, the refining of cotton linter normally employs indirect heating, which can maintain a stable lye concentration and a better lye circulation but with high steam

(a) the structure of horizontal digester (b) the structure of vertical digester with indirect heating

Fig. 10.6 Structures of different digesters

consumption. The structure of the horizontal digester and the vertical digester with indirect heating are shown in Fig. 10.6a and b, respectively.

The whole digester equipment mainly consists of the digester, circulation system, and heating system. The digestion process is shown in Fig. 10.7.

As shown in Fig. 10.7, the combed cellulose is delivered into the cyclone separator 1, following by feeding into the digester 3 by the screw conveyor 2. In the meantime, alkali in the alkali tank 5 is fed into the digester 3 through the alkaline solution pump 9. Alkaline solution is fed into the heater 4 through the circulation pump 7, heated and then added to the digester. This heating cycle is continuously circulated. After the digestion is complete, it is discharged into the rinsing machine 6, and the rinsed material is pumped out by pump 8 for further acidification or bleach. The digestion process conditions should be determined by three major factors, including removing impurities, reducing viscosity, and a certain rate of production, which can not only remove impurities in cellulose but also maintain a certain polymerization degree of cellulose. The digestion system consists of the separator, conveyor, digester, heater, rinsing machine, and so on. The circulation system of vertical digester can avoid the excessive loss of cellulose caused by side reactions, thus having a high utilization efficiency of equipment.

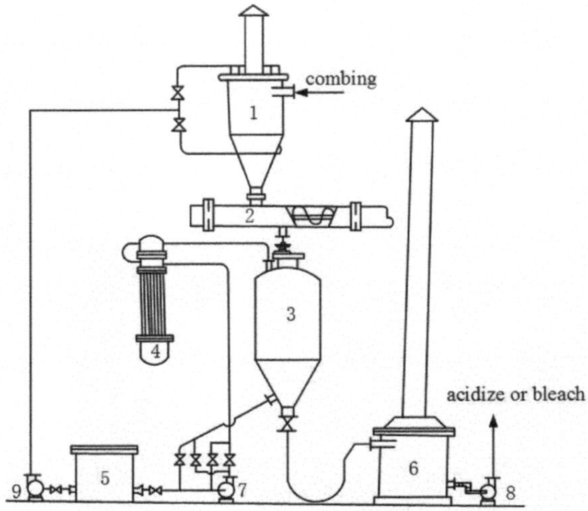

Fig. 10.7 The circulation system of vertical digester for cellulose [2]

1. The concentration of lye for digestion

When the cellulose is refined by using the batch method, the weight ratio of the raw material cellulose (dry amount) to the lye is about 1:7–8, which can ensure that the whole material is immersed in the lye. If the residual pressure is used to discharge the material, the ratio of cellulose/lye is slightly large, about 1:10 or so. The excessive ratio of cellulose to lye will reduce the loading amount of the cotton, reduce the utilization efficiency of the equipment, and increase the alkali consumption. Moreover, it may cause a short circuit of the lye circulation to affect the quality of the product.

The rates of removing impurities and reducing the viscosity would be faster when the concentration of lye is controlled in the range of 2–4% (by weight). Under this condition, the loss of cellulose is usually less than 5%. The relationship between the lye concentration and the viscosity of the cellulose is shown in Table 10.8.

2. The time and temperature of digestion

The time and temperature of cellulose refining are directly related to the concentration of the lye used. With increasing temperature, impurities are easy to remove.

Table 10.8 Relationship between lye concentration and cellulose viscosity

Lye concentration (%)	Digestion time (h)	Digestion temperature (°C)	Viscosity (centipoise)
1.0	5	135	1914
2.0	5	135	821
3.0	5	135	626
4.0	5	135	388

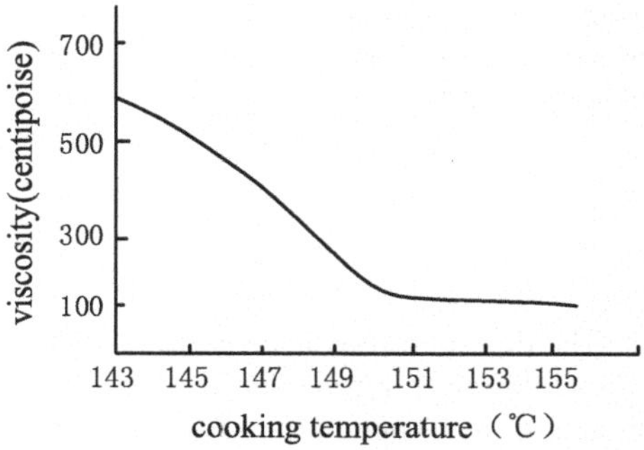

Fig. 10.8 The relationship between the temperature and variation in the viscosity of cellulose

On the other hand, the cellulose macromolecules degrade fast, resulting in decreased viscosity and increased product loss via dissolution. The effect of temperature on the molecular weight of the purified cellulose is shown in Fig. 10.8.

As can be seen from Fig. 10.8, when the temperature of refining increases by 1 °C, the change in viscosity reaches—$\Delta\ \eta$ = 40–45 centipoise, suggesting the change in temperature has a great influence on the molecular weight of the refined cellulose. Therefore, the optimal production temperature is usually 150–156 °C (pressure of 0.4–0.6 Mpa), in which the impurities in the fiber can be removed in a short time. The reduction of polymerization degree of cellulose from 1500 to less than 1000 with 2% NaOH at 150 °C requires 4 h, so the time required to reduce the viscosity is longer. The relationship between digestion time and viscosity is shown in Fig. 10.9.

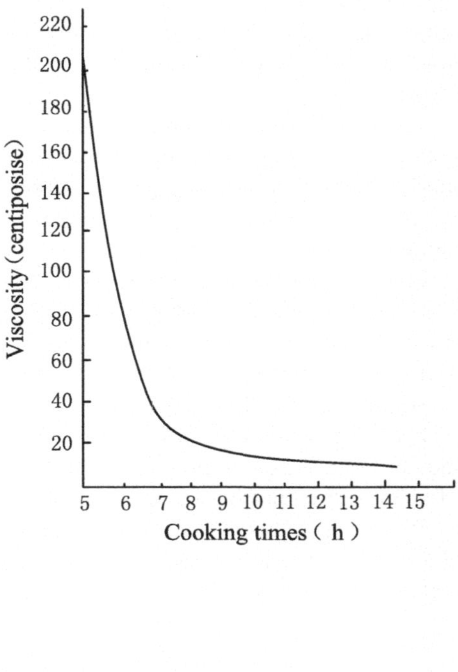

Fig. 10.9 The relationship between digestion time and viscosity of cellulose

Table 10.9 Relationship between cellulose refining time and temperature

NaOH (%)	Temperature (°C)	Saponification time (h)
2.0	130	2–3
2.0	140	1–2
1.5	160	1.5
2.5	120	3

The digestion time is determined by the fiber purity. In general, the refining time of cellulose is controlled at 3–10 h. The relationship between the cellulose refining time and the temperature is shown in Table 10.9.

3. Addition amount of emulsifier

Rosin is typically used as an emulsifier in industrial production. The addition amount of emulsifier is generally controlled in 0.1–0.2% by weight.

(2) Rinsing and acidification

The purpose of rinsing and acidification is to wash away the residual lye in the cellulose and to remove impurities such as ash and sediment. In the meantime, this process can completely remove remaining saponified products and emulsifiers on the surface or capillaries of cellulose.

The industrial rinsing machine consists of two paddle-type agitators, an exhaust pipe, and an oval shape with multiple false ends. Under the false bottom, there is a jet pipe for flushing the mud and sand.

The waste lye of the refined cellulose material from the digester is discharged to the waste alkali collection tank through the rinsing machine on the false bottom. After discharging the lye, the cellulose material is sprayed with a small amount of hot water to replace the adsorbed lye in the cellulose. The collected lye is combined for recovery.

To prevent the oxidation of cellulose, it should be immediately washed with the hot water after digestion. The temperature of hot wash should be controlled in the range of 60–80 °C because pectin and other things can be almost dissolved in hot water. After hot washing, the cellulose is washed with cold water until to neutral and then pumped to the next processor for acidification (or bleaching). During the rinsing, the concentration of cellulose should be controlled in the range of 3–4%.

Acidification is the main process of reducing the ash content of refined cellulose. As in the acidic solution, some of the metal ions adsorbed by the cellulose react with acid to form water-soluble salts, thus reducing the content of ash.

Acidification and bleaching equipment is made of glass fiber reinforced plastic composite groove. The tank is equipped with a stainless steel stirring paddle, and the bottom is false bottom with multiple holes for discharging water. The structures of acidification and bleaching equipment are shown in Fig. 10.10.

When sulfuric acid is used for acidification, the concentration of sulfuric acid should be controlled in the range of 2–4 g/l Acidification with sulfuric acid. The cellulose is washed at 15–25 °C for about 30 min. After acidification, the cellulose

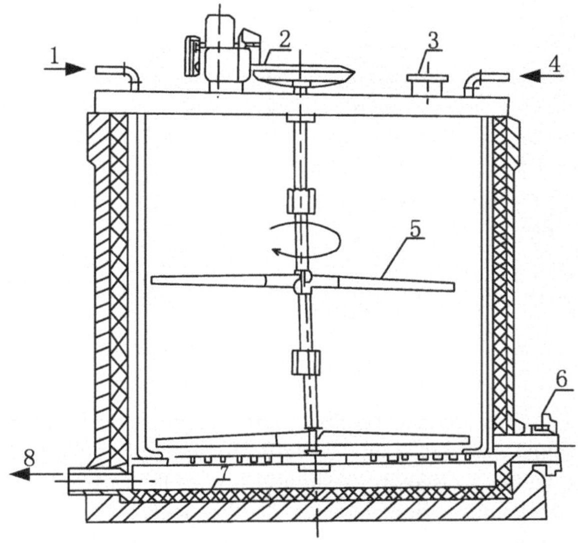

Fig. 10.10 Structures of acidification and bleaching equipment [2]: 1–4 inlet, 2—stirring rotary system, 3—feed port, 5—agitator, 6—outlet, 7—false bottom with multiple holes, and 8—outlet

is washed with hot water with a temperature of 50–60 °C for one time and then washed with cold water until neutral.

(3) Blending

The blending of refined cellulose is mixing several small batches with different quality through the same mechanical stirring to form a large batch with a homogeneous quality. The blending device is a trough made of fiberglass composite or engineering plastic. The blending device for the refined cellulose is shown in Fig. 10.11.

The semicircular bottom of the blending device is made of epoxy resin material. When blending to prepare a large batch, it is necessary to ensure (usually a loading amount of degreasing pot) the homogeneous quality of each small batch. Do not

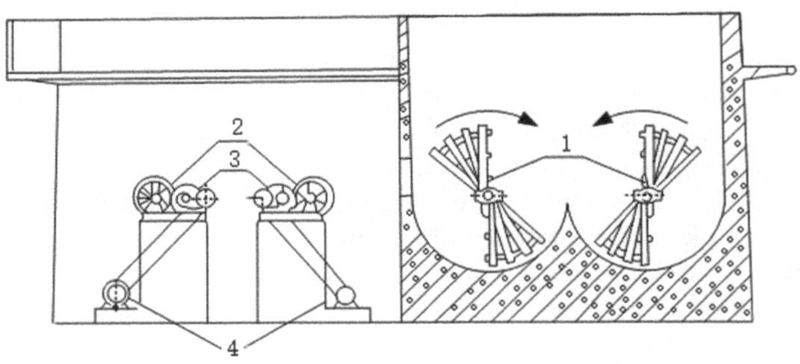

Fig. 10.11 Blending device and working principle [2]: 1—agitator; 2, 3, 4—rotation system

blend small batches with significantly different viscosities, which can minimize heterogeneous quality caused by the production process. The batch combined by blending is called a large batch or total batch.

The viscosity of total batch of refined cellulose is calculated by the weighted average method. For example, four small batches of refined cellulose have masses of 2.6, 3.5, 2.9, and 3.7 tons, respectively, which have viscosities of 21, 23, 26, and 29 centipoises, respectively. The average viscosity of total batch is calculated as follows:

$$\text{The average viscosity of total batch} = \frac{2.9 \times 21 + 3.5 \times 23 + 2.9 \times 26 + 3.7 \times 29}{2.6 + 3.5 + 2.9 + 3.7}$$
$$= 250.02$$

The calculated results of the average viscosity of total batch are very close to the actual analysis result. Usually, for refined cellulose batch with high viscosity, the calculated viscosity is slightly higher than the analytical result.

Blending time depends on the degree of homogeneity. The concentration of cellulose is about 3% (g/100 ml) during the blending operation.

(4) Removal of water and drying

The use of mechanical or other means of pressing or centrifuging to remove most of the moisture in the material is the preparation phase of the material drying. And the heat of hot air or flue gas is used to vaporize the material and evaporate quickly.

To facilitate the storage and transportation of refined cellulose, it is generally required the moisture content of dried cellulose should be below 10%. If the water is too high, the cellulose is easy to grow mold during the storage; if the water content is too low, refined cellulose may adsorb moisture from the air.

Usually, one drying process will reduce the water content of cellulose to below 10%, and the second drying process can reduce the water content from 10 to 5% or less. The second drying process is the first process for the NC production. The refined cellulose after the second drying can be directly used for nitration.

Commonly used methods for water removal include centrifugal, screw pressure, and stick pressure methods, and so on. Air drying is the most commonly used drying method in industrial production. The drying process consisting of a water drive linkage and an up-jet air flow drying is shown in Fig. 10.12.

As shown in Fig. 10.12, the fiber material fed from the mixture is fed into the concentrator 1, in which the concentration is adjusted to 2–3%, and then placed in the receiving hopper 2 of the water drive linkage. Subsequently, the cellulose material flows from the hopper to the rotating screen concentration roller 4. The cellulose is deposited on the surface of the roller, in which water is drained. After the deposited cellulose is flattened by the compression roller 3, the cellulose layer is torn off by tearing roller 5 and sent to the screen conveyor 6. Part of the water is filtered out in the conveyor belt. Next, the fiber material enters the first pair of water displacement roller 8. Under the action of the water displacement roller, most of the

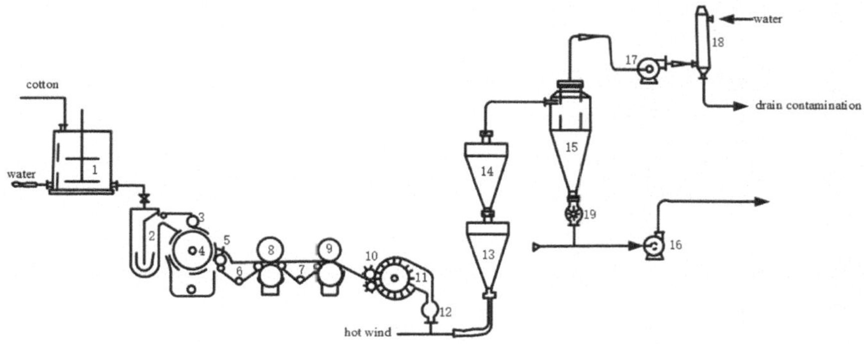

Fig. 10.12 Up-jet air flow drying process [1]

water is removed, and the water content of fiber material is about 65–75%. The fibrous material after first water removal enters the second screen conveyor 7, the water content of the fibrous material is reduced to 60% or less by the second pair of water displacement rollers 9. The cellulose material is fed through the cellulose roller 10 to the disintegrator 11. After the fiber material is loosened, it is fed into the drying pipe through the closing gate 12. The hot air and the cellulosic material are blown into the first drying hopper 13 and the second cup shape hopper 14 by the blower 18. After the fibrous material is dried, it enters the cyclone discharge hopper 15. Finally, it enters the cellulose transportation pipe through the closed gate 16, followed by delivery by blower 16 for secondary drying or packaging.

The drying process generates water vapor and waste gas, which is flowed to the scrubber 19 by the blower 18 via the upper part of the discharge hopper. The exhaust gas from the drying process contains cellulose dust. In the scrubber 19, water is sprayed from the top of the scrubber to remove the cellulose dust via the lower part of the scrubber, and the exhaust gas is released from the top of the scrubber into the atmosphere.

There are two types of dry media, natural gas, and hot air. The ancient production process utilizes flue gas medium for drying. The temperature of hot air is up to 250–350 °C. Generally, the one-time drying process employs hot air with temperature as high as possible because of high moisture content. Hot air drying is often used for the secondary drying, in which air temperature does not exceed 160 °C.

The drying of refined cellulose is the use of the mixture of cold air with high-temperature gas resulted from the liquefied gas combustion in the combustion furnace. Hot air is prepared by heating the filtered air with fin tube heater that is heated by steam flow. The early flue gas is generated by the combustion of anthracite or coke in the industrial furnace, which is used as a drying medium after removal of dust and mixed with cold air. The secondary drying process is generally carried out by airflow, and the air drying process is shown in Fig. 10.13.

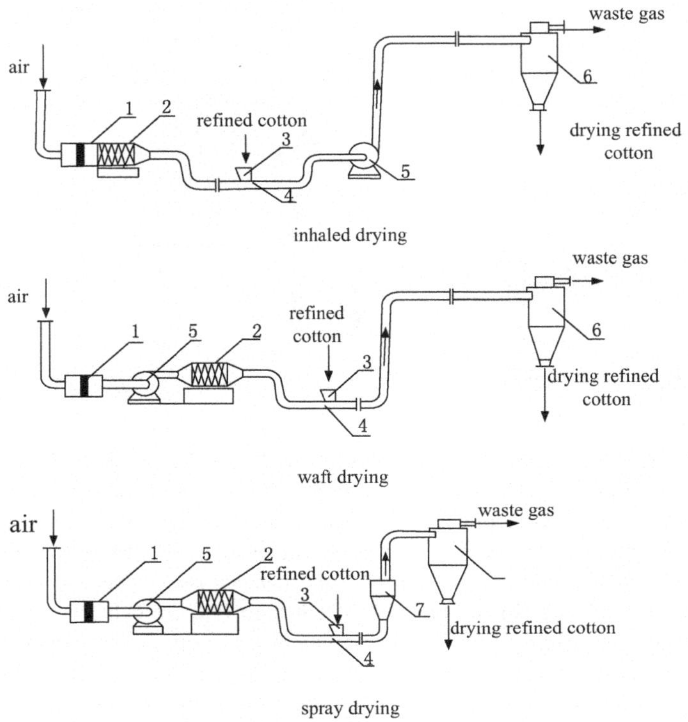

Fig. 10.13 Secondary drying process of refined cellulose [1, 2]: 1—air filter, 2—air heater, 3—loading hopper, 4—drying pipe, 5—blower, 6—cyclone separator, and 7—drying hopper

The advantage of suction type is a small loss of material. The pressure of suction is generally in the range of 300–350 mm water. On the other hand, the disadvantage is the long-term operation of the blower at high temperatures, affecting the service life of blower. In blow-in type, the blower is operated at a low temperature and under the pressure of 500 mm water, but the material loss is significant. In addition, one pipe can only have one feed port. The advantages of up-jet type are extended retention time of fiber material and reduction of drying pipe from the original 150 m to about 20 m because of the addition of a cup-shaped hopper.

The drying medium of secondary drying is hot air or natural gas, and the temperature is 110–150 °C. To prevent the pipeline blockage, the interaction among the feeding amount, hot air temperature, wind pressure, and strengthen procedure should be carefully controlled in the drying process. The standard start-up and shutdown procedures should be abided strictly when start-up and shutdown the equipment. If the pipeline has not been used for a long time, it should be carefully checked before start-up and run a trial by adding a certain amount of refined cellulose to clean up and test the pipeline. After that, the pipeline can be formally

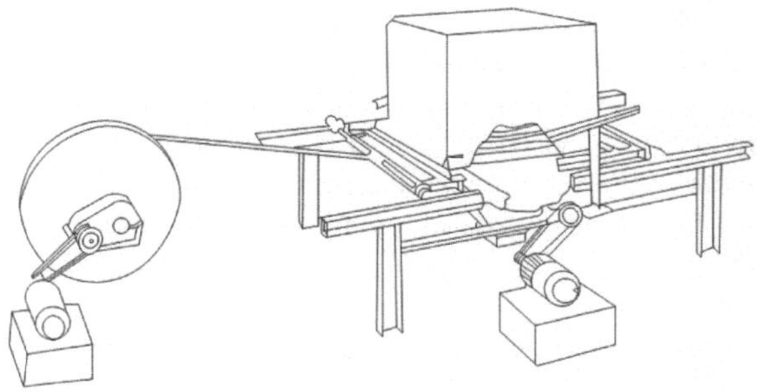

Fig. 10.14 Structure of reciprocating disintegrator

put into production. Electrostatic eliminator should be used to remove static electricity when using drying pipes.

As the refined cotton and NC production process is very long and complex, refined cotton can also be temporarily packaged for storage if the abnormal situation occurs. Nitration using packaged refined cotton, it should be disintegrated by disintegrator before the secondary drying and then sent to the secondary drying pipe.

(5) Disintegration

Once the dried cellulose is stored, it is necessary to compact and pack the fluffy refined cellulose. If the packed refined cellulose is used for nitration, disintegration is required before use to ensure the refined cellulose is fluffy, which is conducive to secondary air drying and nitration. Continuous nitration is normally combined with the continuous disintegrators, which are classified into horizontal and reciprocating disintegrators; horizontal disintegrator has a low disintegration efficiency and affords a poor quality uniformity. Before the end of disintegrating each pack of cellulose, the whole piece of fiber will be transferred into the disintegrator [2]. The structure of the reciprocating disintegrator is shown in Fig. 10.14.

Reciprocating disintegrator not only can move forward and back forward but also can shift in the process of reciprocating to ensure the uniformity and consistency of feeding.

10.2.3.3 Bleaching Process of Refined Cellulose

For the production of NC with low viscosity, in addition to continuous digestion that is used to reduce the viscosity of NC, the bleaching process can be used to reduce further and control the viscosity of NC. For a special industrial application, to achieve a requirement of whiteness, cellulose must be bleached in the refining. For the production of film, paint, and plastic NC, in addition to the use of refined cellulose, the bleaching process is required; for the production of NC with other applications, the cellulose after digestion can be directly acidified and does not need bleaching. In the production of NC that requires refined cellulose with low viscosity by using continuous digestion process, if only using digestion to reduce the viscosity of the refined cellulose, it will affect the α-cellulose content of refined cellulose, alkali consumption of digestion, production cycle, and so on. Therefore, the reduction of viscosity can also be achieved by the bleaching process, which also reduces the ash content and improves moisture absorption capability.

Whether cellulose raw materials come from cotton linter, wood or stalk plants, the mostly used bleaching reagent for cellulose in the industry is sodium hypochlorite.

Sodium hypochlorite is a salt of hypochlorous acid, which can dissolve in water to form hypochlorous acid and NaOH

$$NaClO + H_2O \rightarrow HCl + NaOH$$

When the solution is neutral, hypochlorous acid decomposes to release the primary oxygen and hydrochloric acid

$$HClO \rightarrow HCl + [O]$$

The resulting hydrochloric acid causes the sodium hypochlorite to decompose and produce hypochlorous acid

$$HCl + NaClO \rightarrow HClO + NaCl$$

Sodium hypochlorite solution contains a high content of primary oxygen, which can completely oxidize the impurities on the surface of cellulose surface. When the solution is acidic, sodium hydroxide from the hydrolysis of sodium hypochlorite solution is neutralized by the acid. On the other hand, hypochlorous acid oxidizes sodium hypochlorite to sodium chlorate with the formation of hydrochloric acid.

$$2HCl + NaCl \rightarrow NaClO_3 + 2HCl$$

When hydrochloric acid in the solution increases to a certain level, the excessive hydrochloric acid reacts with sodium hypochlorite or hypochlorous acid to produce free chlorine.

$$NaClO + 2HCl \rightarrow NaCl + Cl_2 + H_2O$$

With increasing acidity in the solution, the concentration of free chlorine in the solution increases, which diminishes the concentration of hypochlorite and weakens the bleaching effect. When the solution is alkaline, the hydrolysis of sodium hypochlorite is affected, thus reducing the concentration of hypochlorite in the solution; the stronger the alkalinity, the less the hypochlorite and the weaker the bleaching effect.

In general, the pH value of the bleaching solution should be strictly controlled in the range of 9–10 to control the bleaching speed, to reduce the formation of chloride, to ensure the stability of bleached cellulose viscosity, to prevent excessive damage to α-cellulose, and so on. Since the pH value of the bleaching solution includes the residual alkali and alkalinity of water in the cellulosic material, it is an overall pH value reflecting several factors. Excessive high pH value is inappropriate because it not only extends the bleaching time but also deteriorates the whiteness of refined cellulose.

The addition of liquid chlorine to a certain concentration of sodium hydroxide solution generates a sodium hypochlorite solution.

$$NaOH + Cl_2 \rightarrow NaClO + NaCl + H_2O$$

The effective chlorine content of 0.5–1.0 g/L in bleach is appropriate. Bleaching time, temperature, and the effective chlorine concentration of the bleach solution are the main process parameters, which are determined by the viscosity of the post-digestion cellulose material and the viscosity required for the finished product. The process parameters are adjusted with the changes of equipment. For example, horizontal bleacher has high efficiency of mixing, rapid and uniform reaction, and mild bleaching conditions can choose to ease some. In contrast, vertical bleacher has a poor mixing efficiency, poor uniformity, more bleach, and more reaction. The bleaching temperature depends on the product quality. The bleaching temperature is no more than 30 °C, and the bleaching time is no more than 4 h. When the temperature rises (above 20 °C), the viscosity of bleached cellulose decreases significantly, and the effective chlorine consumption increases with the increase of temperature. The fiber concentration should be controlled in the range of 3.5–4% during bleaching.

Table 10.10 The quality of the cellulose after refining stem plants

Composition of cellulose	Sample number				
	1	2	3	4	5
α-Cellulose	97.33	95.60	95.25	95.06	96.23
Pentose	1.63	3.88	3.69	3.715	2.273
Viscosity	18.7	24.2	25.8	20.0	26.5
Dissolved matters at −5 °C in 12% NaOH	88.50	87.12	85.10	87.04	84.7
Compounds are not nitrated in 5 min	5.3	5.4	6.50	5.70	6.16
Lignin	0.30	0.32	0.34	0.36	0.41
Resin	0.19	0.21	0.23	0.21	0.26
Moisture absorption (g)	136	119	122	130	128
Amount (g/cm^2)	36.5	38.0	39.0	39.3	40.1
Tightness (g/cm^3)	0.58	0.49	0.51	0.54	0.53
Thickness (mm)	0.125	0.120	0.130	0.134	0.142
Ash content	0.340	0.30	0.27	0.365	0.417
Insoluble matter in sulfuric acid	0.260	0.30	0.29	0.27	0.29
Pectin	0.21	0.24	0.22	0.29	0.410

The quality and source of the cellulose considerably affect the bleaching quality. In addition to the commonly used alkaline bleaching, acidic bleaching is also frequently used (acidity of 0.05 g/l or less based on sulfuric acid). Acid bleaching is characterized by rapid response, short time, and optimal whiteness. However, to prevent severe damage to cellulose by acidic bleaching, cellulose must be subjected to acid pretreatment. The quality of the cellulose after refining stem plants is shown in Table 10.10.

10.3 Nitration of Refined Cellulose

The second stage of nitrocellulose production is nitration. This stage includes three basic processes, namely, nitration, removal of acid, and washing. Nitration stage focuses on the study of nitration reaction process and the effect of refined cellulose structure on the nitration reaction.

The mixed acid nitration is the main mechanism to explain the nitration of cellulose, so it is necessary to study the theory of acid–base proton and the kinetics of nitration.

10.3.1 Nitration of Refined Cellulose

(1) The main chemical reaction of cellulose nitration,

The main chemical reaction of cellulose nitration is nitration of hydroxyl groups (–OH) in the cellulose macromolecules to convert into nitrates under the action of nitric acid, which is similar to the nitration of hydroxyl groups in the low molecules. The reaction process can be expressed as

$$ROH + HONO_2 \Leftrightarrow RONO_2 + H_2O + Q$$

or

$$[C_6H_7O_2(OH)_3]_n + \gamma^* nHNO_3 \Leftrightarrow [C_6H_7O_2(ONO)_{3-\gamma}]_n + \gamma^* nH_2O + Q$$

where Q is reaction heat, and γ is esterification degree.

Nitration/denitration

It can be seen from the nitration reaction of cellulose that it is a reversible reaction. Increasing the concentration of nitric acid, removing the product of water, or removing the heat generated in the reaction is conducive to the formation of nitrate; otherwise, it is against the formation of nitrate. The reverse reaction is a denitration reaction. The possible number for the introduction of nitrate groups for each

Table 10.11 Mixed acid nitration of cellulose and denitration profile of nitrocellulose with 13.33% nitrogen content in the presence of mixed acid

Treatment time (h)	Nitrogen content of nitrated product (%)	Nitrogen content of denitration product (%)
0	0.0	13.33
1	11.81	13.23
4	12.14	13.09
8	–	13.06
24	12.32	12.99
96	12.41	12.94
168	12.62	12.87
336	12.70	–
480	–	12.87
1440	12.64	12.66

glucose residue is 0, 1, 2, and 3, respectively. The limit is 3, where the NC reaches a theoretical nitrogen content of 14.14%. If the two terminals of the cellulose macromolecules having four hydroxyl groups are taken into account, which has maximum substitution degree of 4, the theoretical nitrogen content is slightly higher than 14.14%. The average gamma value for NC is from 0 to 3, including the fractional value. The nitrogen content is in the range of 10–13.5%. The reaction heat varies slightly, and the nitration of each hydroxyl group with 1 mol nitric acid releases heat approximately 14895.28 J.

The nitration rate of cellulose is dependent on the ratio of the positive reaction rate to the reverse reaction rate. However, since the structural characteristics of refined cellulose have a great impact on the reaction rate, the dynamic process of this reaction cannot be understood by the general concept of a small molecule. In the nitration process of cellulose, there is almost no denitration reaction. The denitration profile of nitrocellulose in nitrated acid is shown in Table 10.11.

The experiments in Table 10.11 were performed at lower temperatures with an excessive amount of mixed acid. The denitration process requires absorption of heat, in which H^+ is the catalyst for the process of denitration. If the mixed acid composition changes greatly and the temperature is high, the rate of denitration reaction will be significantly accelerated. The denitration of NC cannot obtain refined cellulose that is same as the one before the nitration reaction primarily because both nitration and denitration processes have side reactions, which significantly destroys the cellulose structure. Therefore, the reverse reaction of nitration cellulose is different from the reverse reaction of the small molecule. In other words, the reverse reaction of nitration cellulose is a relatively reversible process or an incomplete reversible reaction.

The rapid addition of a large amount of cold water to the nitration reaction system of cellulose tends to terminate the reaction, in which the denitration rate is extremely slow. The method of slowing down the denitration rate include: (1) addition of a large amount of water reduces the concentration of acid to very low, which considerably weakens the catalytic activity of H^+; (2) a large number of cold water reduces the temperature, which is not conducive to the occurrence of denitration; and (3) extreme dilution of nitric acid switches the swelling of NC to a state of sudden contraction, which makes the diffusion of adsorbed acid in NC difficult, thus diminishing the corresponding denitration rate. However, it can be speculated that there is a certain denitration on the outer layer of the cellulose or around the capillary in the NC, which leads to heterogeneous of NC quality. Rapid addition of a large amount of water to the acid containing NC can overcome this issue. For example, the addition of a large amount of water after removal acid in the industrial production of NC and hydraulic transmission of NC are typical examples in the industrial production. If a suitable amount of water is added to the system, the reaction proceeds in the direction of denitration. Therefore, during the recovery of acid by replacement method and recovery of acid by centrifugal operation in production, the time, the acid temperature, the amount, and the concentration should be controlled to reduce the denitration.

Since the nitration is conducted in acidic media containing nitric acid, some chemical changes in cellulose are unavoidable. These changes include cleavage of glycosidic bond caused by the shortening of macromolecular chain and the oxidation of cellulose macromolecules by nitric acid. In particular, some of the hydroxyl groups are oxidized to aldehydes, ketones, and acids. The products of these side reactions can be nitrated to a low molecular weight product containing multiple functional groups.

Sulfation also takes place to form cellulose sulfate in addition to oxidation and hydrolysis side reactions if using mixed acid composed of nitric acid, sulfuric acid, and water as nitration reagent. Oxidative and hydrolytic products can also be sulfated. The sulfation reaction is a reversible reaction.

$$R(OH) + H_2SO_4 \Leftrightarrow R(OSO_3H) + H_2O$$

Sulfate is less stable than nitrate. Sulfate releases sulfuric acid after saponification. The residual side products remaining in the purified cellulose may be oxidized by nitric acid or partially esterified by nitric acid or sulfuric acid in the acidic NC.

The side reaction of the cellulose nitration process increases with the increase of the moisture and sulfuric acid content in the mixed acid. In other words, the side reaction intensifies with decreasing concentration of nitric acid. When the refined cellulose has poor water absorption capacity, it is not easy to be mixed with nitric acid, which leads to increased side reactions; increased nitration temperature also results in increased side reactions.

(2) Nitrating agents for cellulose

The nitration of cellulose can be carried out directly with nitric acid, nitric acid vapor, or nitric anhydride (N_2O_6) without any catalyst. The addition of other agents besides nitric acid to the nitration mixture is for technical and economic reasons only. For example, industrial nitration agents are often added with sulfuric acid, which can facilitate the nitration reaction and reduce the cost of nitration agent.

In addition, mixtures of nitric acid and dehydrating agents such as nitric acid and acetic anhydride, mixtures of nitric acid, and organic inert substances such as carbon tetrachloride or inorganic salts such as magnesium nitrate can be used as nitrating agents. However, similar to nitric acid, nitric acid vapor, or nitric anhydride, these nitrating agents suffer from drawbacks, such as poor esterification capacity, significant side reactions, heterogeneous products, high cost, lack of raw material sources, and corrosion of equipment; Besides nitric acid and magnesium nitrate as a dehydrating agent are used in the factory trial, nitro-sulfur mixed acid (composed of nitric acid, sulfuric acid, and water) nitration agent is still widely used as industrial nitration agent.

(3) Nitration of refined cellulose

1. Description of nitration process

When the nitration reagent contacts the refined cellulose, the liquid first infiltrates the solid surface while wetting the entire cellulose along the capillary to the interior, thereby diffusing into the interior of the refined cellulose and causing swelling. The reaction may take place after the infiltration process or diffusion into the refined cellulose to a certain extent, depending on the relationship between the rate of chemical reaction and the rate of diffusion of the nitration reagent.

If the reagent can enter the oriented zone, V_R, V_{D1}, V_{D2}, and V_{D3} represent the chemical reaction speed, the diffusion velocity of reaction reagent in the fiber, the diffusion velocity in the non-oriented region, and the diffusion velocity in the orientation zone, respectively. If $V_R \gg V_{D1} > V_{D2} > V_{D3}$, the reaction is the most heterogeneous and shows the transition to the fiber surface reaction; if $V_{D1} > V_R \gg V_{D2} > V_{D3}$, the reagent can only enter the fiber surface and regions around large capillaries, and then gradually enter into the inside of fiber. The initial phase of the reaction takes place on the surface of fiber, which is called fiber heterogeneous reaction; if $V_{D1} > V_{D2} \gg V_R \gg V_{D3}$, the reagent first reaches the non-oriented region within the fibril and reacts with fibers, and then gradually penetrates and occurs nitration in the oriented zone. At this time, the reaction may be heterogeneous. It may be homogeneous if the reaction ends as a homogeneous phase. The heterogeneous reaction at this time transfers to the heterogeneous reaction of fibril; if $V_{D1} > V_{D2} \gg V_{D3} \gg V_R$, the reagent can contact with the macromolecule of the cellulose freely, and all the stages of the reaction can be a completely homogeneous phase. Therefore, there are two basic types of nitration of refined cellulose, namely, homogeneous and heterogeneous reactions. The rate of nitration is closely related to the diffusion velocity of the reagent and is determined by the slow step reaction.

Whether the nitration reagent can diffuse into the oriented zone, as well as the diffusion velocity, it depends mainly on the processes of cellulose swelling, dissolution, and hydrogen bond damage; it also correlates with the affinity between nitration reagent cation and nitration agent with cellulose. If the reaction medium has a strong swelling effect and even dissolving capacity, and the structure of the reagent is conducive to the diffusion and infiltration into the fiber and has a strong affinity with the hydroxyl group, the oriented zone, and hydrogen bonding may be damaged. Moreover, it will also cause the following results in the reaction process, including fast diffusion rate of nitration reagents, strong ability to penetrate into the cellulose, consistent between the amount of reaction and the number of activated hydroxyl groups, and the possibly complete reaction of all hydroxyl groups. Conversely, if the reaction terminates when the nitration reagent diffuses to a certain extent, the reaction only takes place in the region where the reagent is present. Only the activated hydroxyl group in the free hydroxyl groups participates in the nitration reaction, thus exhibiting a low reaction capacity.

The structure of the refined cellulose has a direct effect on the diffusion rate, infiltration capacity, and reactivity of the reagent. In addition, external conditions, such as temperature, stirring, refined cellulose morphology and dispersion state, will also cause some impact. The nitration reaction process also affects the degree of esterification and quality uniformity of product.

The rate of homogeneous reaction is faster than the reaction that starts from the non-oriented zone and then gradually diffuses into the oriented zone, which is estimated to be about a thousand times faster. If the oriented zone can be destroyed and the hydroxyl group is free, the nitration rate can be the same as that of the small molecule, that is, the chemical reaction rate V_R. Therefore, the diffusion rate of the reagent has a considerable effect on nitration reaction of the purified cellulose. There are some necessary measures to improve the nitration reactivity, such as improving the wetting performance of the purified cellulose, destroying the oriented zone, increasing the diffusion rate of the reagents, and changing the reaction medium and nitration reagent.

Since the reaction reagent and the medium do not dissolve the cellulose and the corresponding product, the reaction proceeds in a multiphase manner. The resulting nitrocellulose is heterogeneous in the degree of esterification and degree of polymerization. The time required for the reaction depends mainly on the diffusion rate of the nitrating agent, and the reaction product still retains the aggregation structure of the original cellulose material. The nitration of refined cellulose by nitro-sulfuric mixed acid is a typical example.

The nitration reaction of cellulose is an incomplete reversible reaction. In the process of heterogeneous reaction, with the equilibrium of nitration reagent diffusion, there is the existence of reverse reaction of the reagent, which can improve the uniformity of esterification degree of product. On the other hand, the denitration rate is slow, which results in incomplete reaction and compromised uniformity of product.

2. Process for the nitration of refining cellulose with nitro-sulfuric mixed acid

In the previous chapters, several possibilities of nitration were studied. Typically, nitro-sulfuric acid is used as a nitrating agent in the industry. From the analysis and discussion in the second chapter, the effect of the mixed acid in the nitration reaction has been summarized. Through the investigation of macromolecular packing regularity of in the process of nitration of cellulose, the following conclusion has been drawn. ① when the nitrogen content is less than 7.5%, the NC shows the characteristics of swelling of cellulose; ② when the nitrogen content is in the range of 7.5–10.5%, NC is a swelling mixture of cellulose and cellulose trinitrate, but regular arrangement still exists among macromolecules; and ③ when the nitrogen content is 12.73% or more, it exhibits complete characteristics of cellulose trinitrate after a stability treatment. According to this view, it is apparent that in the process of nitration of cellulose, nitric acid infiltrates the cellulose structure first, resulting in swelling and destroy of structural regularity, and nitration only occurs in the non-oriented zone; when the acid concentration continues to increase, the swelling of cellulose reaches the maximum, and the oriented zone is destroyed completely. As a result, the swelling and destruction of the oriented zone tend to be constant. The number of nitrates increases with increasing nitration ability, but the distribution of the nitrate groups is irregular. When the nitrate group increases to the nitrogen content of 12.73% or more, it is possible to arrange the macromolecule chain after stability treatment, thus exhibiting the structure of the trinitrate. In the case of the residual nitrating acid is present, the swelling fiber leads to the indistinct of nitrocellulose. The reactivity of the hydroxyl groups in the cellulose macromolecules is not the same. Specifically, the reactivity of the hydroxyl groups is high while the reactivity of hydroxyl having hydrogen bonding is low. Figure 10.15 depicts the infiltration process of mixed acid into the oriented zone.

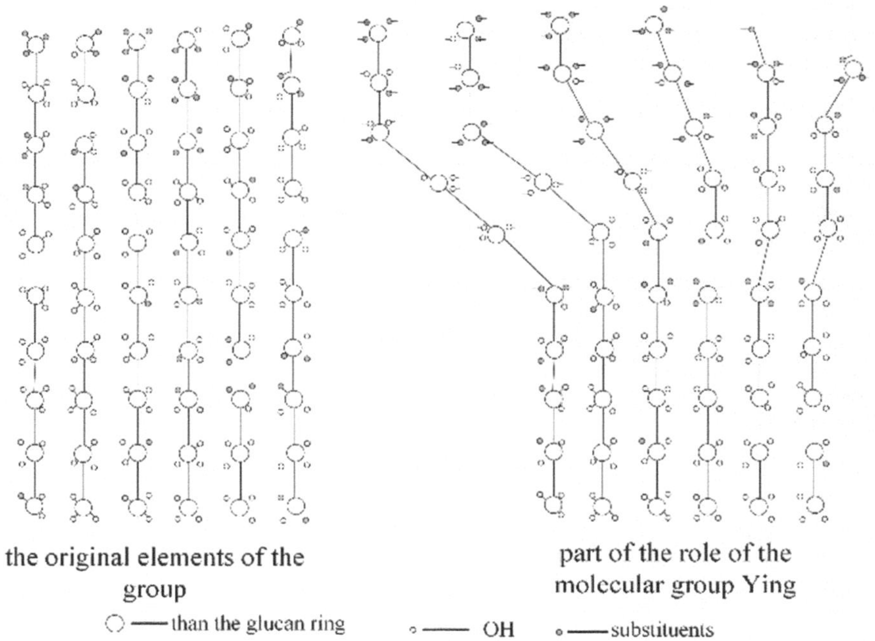

Fig. 10.15 Nitrate reaction in cellulose oriented zone [3]

Figure 10.15 reflects the nitration reaction proceeds from one end of the oriented zone gradually into the oriented zone, followed by the gradual slip of the chain.

The mixed acid used in the gunpowder industry is the preparation of the corresponding NC with the composition of 11.75% N (188 ml/g). It is expected that the cellulose has a more intense swelling during the nitration process, in which the oriented zone of the macromolecule is almost destroyed. In the preparation of NC with 12.8% N (204.8 ml/g) or more, the degree of swelling is even greater. In particular, the reaction may proceed in the form of exchange type in the latter stage of the reaction. In other words, almost all of the hydroxyl groups can participate in the reaction, which can be regarded as seemingly complete uniform. On the other hand, the nitration reaction occurs between the solid and liquid phases, which should be considered as a multiphase reaction.

If all of the hydroxyl groups have the same reactivity and can take place the transesterification reaction, the average nitration degree will be increased despite the heterogeneous quality of the nitration product. Figure 10.16 shows the distribution of substituents on different glucose units from a statistical point of view, reflecting the heterogeneity of the quality of nitration products when the average degree of substitution of cellulose nitration increases.

The swelling ability of the mixed acid is related to the composition of the mixed acid. When water or nitric acid in the mixed acid increases, the swelling rate of the cellulose increases extremely rapidly. When the water content in the mixed acid is low, the sulfuric acid increases the concentration of nitronium ion, thus changing the swelling condition of the cellulose. The increased average substitution degree of the nitrocellulose favors the preparation of NC with high nitrogen content.

3. The effect of refined cellulose structure on the nitration reaction

From the previous discussion, there are a series of correlations in the nitration of refined cellulose, such as the structure and reaction rate, the degree of cracking reaction, and the degree of product uniformity. The effects of the specificity of the refined cellulose structure on the nitration reaction are as follows: ① The reactivity of hydroxyl groups in cellulose macromolecules are different. In general, the

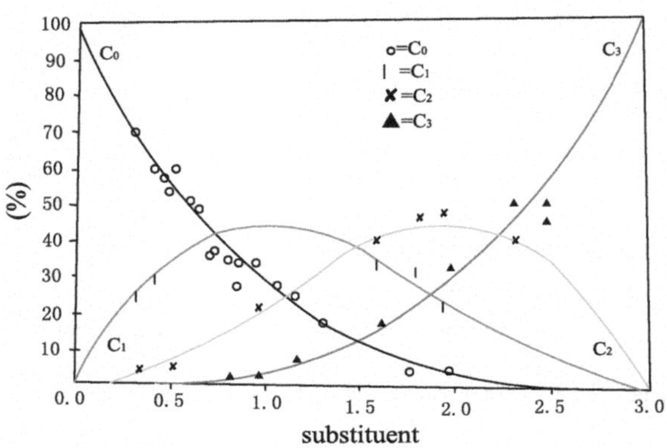

Fig. 10.16 Distribution of substituents on different glucose units [3]

reactivity of hydroxyl group in the oriented zone is lower than that in the non-oriented zone; the reactivity of hydroxyl group forming the hydrogen bond is not as active as the free hydroxyl group, and the activity of forming the hydrogen bond is the lowest. Among free hydroxyl groups, the reactivity of primary hydroxyl is the highest in the acidic medium. The reactivity of secondary hydroxyl on the second carbon atom is often higher than that of the primary hydroxyl group and the secondary hydroxyl group at the 3-position because of the large spatial barrier of the hydrogen bonding. ② In the reaction process, the formation of mononitrate is relatively easier while the formation of trinitrate is the most difficult for each glucose residue. The formation of trinitrate requires a nitration agent having a strong electrophilic capability. ③ Due to different packing regularities of macromolecules or their basic units in cellulose macromolecule aggregation state, the interactions between the macromolecules are also different, which results in different diffusion rate and penetration depth of nitration reagents in cellulose. As a result, the reaction proceeds in a multiphase manner. ④ The penetration degree and reaction rate of the nitration reagent are affected by the different morphological structure of the refined cellulose, the destruction degree of the morphological structure during refining, and the removing degree of the side products; the expansion degree of the cellulose capillary, especially the air content in the fine capillaries, has a certain effect on the nitration reaction.

4. The yield of nitrocellulose

According to the equation of cellulose nitration reaction, the relationship between theoretical yield and esterification degree of NC can be obtained. When a hydroxyl group in one glucose residue of cellulose is replaced with a nitrate group, the molecular weight of the cellulose macromolecule is increased by 45. When the degree of esterification is γ, the "molecular weight" of a basic link is 162 + 45γ. Therefore, the theoretical yield of NC is

$$\text{Theoretical yield} = 162 + 45\gamma 5\gamma / 1100\,(\%)$$

where 162 is the "molecular weight" of a basic link of cellulose macromolecules.

Actual yield = The mass of final nitrocellulose/The mass of refined cellulose 100%.

Since the chemical reactions and mechanical transportation process in the production of NC cause a certain loss of product, the actual yield in production is always less than the theoretical yield. The yield in production is an important indicator of the rationality of the production process.

5. Nitrocellulose production process [1–3, 7]

The process of producing NC with nitric and sulfuric mixed acid is shown in Fig. 10.17.

Figure 10.17 is the layout of U-type continuous nitration production line for NC, which are composed of 22 sections, including disintegrator, air drying pipe,

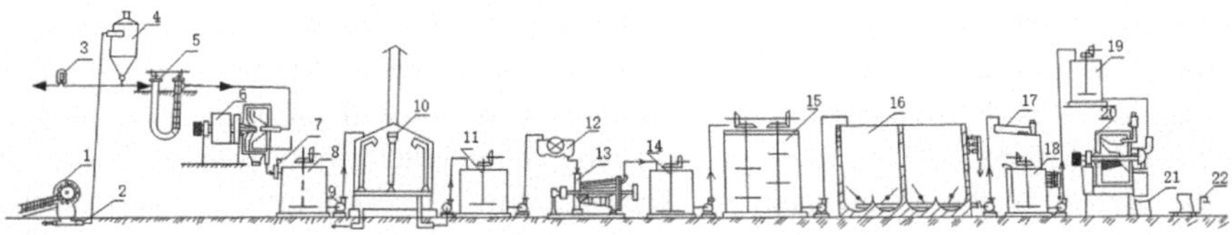

Fig. 10.17 Process flow for continuous production of NC

flowmeter, cyclone separator, U-type nitrator, horizontal acid removal machine, buffering washer, NC pump, digestion barrels, transfer tank, water filter, shredder, fine washer, blender, centricleaner, concentrator, exchange tank, water removal machine, packaging metric equipment, and so on. After nitrating mixed acid is ejected from the sprayer to impregnate the fiber material, mixed acid and cellulose move forward along the U-shaped tube and are discharged from the last U-tube into the horizontal piston discharge centrifuge. Nitrocellulose discharged from the outlet is flushed to the washer with a large amount of water, followed by the transportation of nitrocellulose by a feed pump to the digestion process for stability treatment. The exhaust gas is discharged from the top of the cyclone separator. The whole process is automatically controlled to achieve continuous nitration production.

10.3.2 Nitration, Acid Removal, and Washing Processes

Nitration is the process of nitration reaction of refined cellulose after secondary drying with temperature adjusted nitration mixed acid in the nitrator to prepare NC.

Nitration is the first stage of the preparation of NC, in which nitrating acid has a very important impact on the quality of nitration reaction. At the beginning of the nitration, freshly mixed acid is prepared by adding water to the concentrated nitric acid and sulfuric acid. When calculating acid preparation, the N_2O_4 content in the waste acid can be included in sulfuric acid, and the content of N_2O_4 in concentrated nitric acid can be included in the nitric acid.

There are two kinds of processes for acid mixing, namely, batch and continuous acid mixings. Batch acid mixing is carried out in an acid mixer. First, the required amount of waste acid for the acid mixing is collected, stirred, and sampled for analysis. According to the analysis result of raw materials, the amount of raw material acid is calculated. After the raw material acid is added and mixing evenly, the mixed acid is sampled for analysis. If the analysis result meets the requirement, the mixed acid is ready for the nitration reaction. Continuous acid mixing is the flow of several acids in the same pipeline, in which the flow rate is determined by the amount of requirement. The evenly mixed acid flows into the mixed acid tank. It is ready for nitration if the quality meets the requirement.

Since batch acid mixing has a long preparation cycle and requires more equipment, experience-based acid mixing is typically used in most plants to shorten the cycle of acid mixing. Experience-based acid mixing does not need calculation,

Table 10.12 Composition of nitrating mixed acid

NC sample lot	Composition of mixed acid (%)	
	HNO_3	H_2O
1#NC	18–23	7–14
2#NC, 3#NC, and colloidal NC	18–22	16–18
Film and paint NC	19–24	16.5–19
Plastic NC	18–24	18–21

which is achieved by adding raw material acid to the waste acid based on experience, followed by one-time mixing sampling analysis. The compositional range of commonly used nitrating mixed acid is shown in Table 10.12.

It is noteworthy that dilute acid should be used to replace water as possible in operation. The rise of acid temperature needs to be strictly controlled.

Nitration process condition is the main factor affecting NC quality. Nitration and acid removal process is one of the key processes determining NC quality.

The waste acid in NC after the nitration is driven out by centrifuge, and the separated waste acid is used for the preparation of mixed acid. After treated with acid water replacement and washing, acidic NC is processed for stability treatment by digestion.

10.3.2.1 Processes of Nitration, Acid Removal, and Washing

The cellulose nitration processes are classified into the batch and continuous nitration. The early batch nitration process is entirely manually operated kettle nitration and replacement nitration process. Later on, it is changed to mechanically operate the centrifugal nitration method. The batch nitration method is a process for the early manufacture of NC. It has the characteristics of simple operation and equipment. The outstanding drawbacks of batch nitration include poor nitration uniformity, low yield, significant acid consumption, high labor intensity, and poor operating conditions. For these reasons, batch nitration process has been replaced by stirring nitration method. Stirring nitration is a kind of nitration method with high product quality stability among batch nitration processes, which has a series of advantages, including simple equipment, high production efficiency, low acid consumption, and homogeneous quality. This method is still used in plants with low production capacity. NC continuous nitration process is the fastest growing technology in the industry, which includes four-chamber, multichamber nitrator, U-tube nitrator, multiple tandem mixing nitrators, and so on. NC continuous nitration process is characterized by continuous operation, large output, homogeneous quality, low labor intensity, and high production safety.

U-type continuous nitration process has a high degree of automation in the NC continuous production. U-type continuous nitration process increases flow resistance in the nitration of cellulose, which is favorable for the nitration, thus

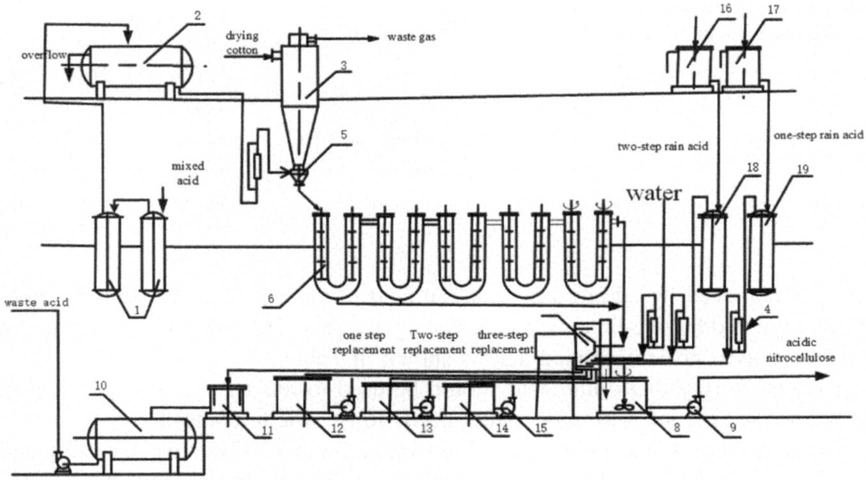

Fig. 10.18 U-type continuous nitration process

increasing the potential difference of the reaction chamber. U-type continuous nitration process is shown in Fig. 10.18.

As shown in Fig. 10.18, the qualified mixed acid is passed through a thermostat 1 and pumped into a mixed acid elevated tank 2 by a mixed acid pump. The secondary dried cellulose separated by a cyclone 3 falls into a U-type nitrator 6, in which the exhaust gas is discharged from the top of the cyclone. The mixed acid is sprayed from the sprayer 5 and impregnated with the cellulose material. The flow rate of mixed acid is adjusted by the rotor flowmeter 4. Powered by the rotation of the mixer, acid and cellulose flow along the U-tube to move forward and is discharged at the last U-tube. Subsequently, the nitration mixture enters the horizontal piston discharge centrifuge 7 for acid removal first, followed by three times of replacements. Next, it is treated with the first and second of spray acids from elevated tanks 17 and 16, respectively. The treated mixture enters into the centrifuge by the rotor flowmeter 4 through the first and second spray acid coolers 19 and 18. After the replacement, the cellulosic material is washed from the discharge port with a large amount of water to the washer 8. The washed NC is conveyed to the digestion process by the cotton pump 9 for stability treatment.

The waste acid discharged from the centrifuge flows into the waste acid collecting tank 10 through the waste acid filter 11. The acid from the first step replacement is introduced into the collecting tank 12. The collected acid in tanks 11 and 12 is used for the preparation of mixed acid. The two stages of spray acid and the replaced acid after water replacement enter the collection tanks 13 and 14, respectively, which are pumped to the elevated tanks 17 and 16 by the acid pump 15, respectively, for recycle and replacement washing.

In the continuous nitration process, the number and length of the U-tubes are determined by the time of feeding and nitration. The nitrator for the production of

Low nitrogen content NC is composed of 4–5 U-shaped tubes while the nitrator for the production of NC with high nitrogen content is composed of 5–7 U-tubes. The acid in the nitrated product is removed with a horizontal piston discharge centrifuge. To further reduce the acid consumption, acid water recovery device can be added in the acid removal equipment, which can enhance the replacement effect compared with natural filtration of acid and reduce the amount of diluted acid.

10.3.2.2 Process of Acid Removal and Washing

Acid removal and washing is one of the key processes in the manufacture of qualified nitrocellulose, which is related to equipment and involves production costs.

The acid removal in the nitrocellulose production process includes natural filtration and centrifugal removal. The waste acid after the nitration can be naturally filtered through the NC layer, by which the waste acid recovery rate is only about 60%; acid removal by centrifuge can drive out a large number of concentrated waste acid in NC so that the recovery rate of waste acid reaches more than 90%. After removing acid, NC should be quickly immersed in water to make an NC solution with a concentration of 2% (2 g/100 ml H_2O) or below, which facilitates the pumping for the digestion process.

(1) Upper suspension centrifugal acid removal equipment

Centrifugal acid removal equipment has a structure of upper suspension and bottom discharge, of which centrifuge is made of acid-resistant stainless steel. Figure 10.19 shows the structure of the upper suspension and the lower discharge centrifugal acid removal equipment. Before feeding of NC, the front cover and cone cover 5 of centrifugal acid removal equipment are closed. The rotation rate of the centrifuge is adjusted to 200–350 r/min while feeding; after loading, the centrifuge is set as 700–1000 r/min to drive acid. After a certain time, stop the centrifuge and lift the cover. The NC on the screen is divided into small pieces manually by aluminum fork or shovel. The broken NC naturally falls into the water flushing device and flush to the washer with a large amount of water.

The main drawbacks of centrifugal acid removal equipment include contamination of driven waste acid with NC particles, cloud in waste acid, NC decomposition accident during the acid removal process, acid leaking in the outlet, significant acid consumption, and so on. To reduce acid consumption, diluted acid and water can be used for washing and replacement.

(2) Circular equipment

The main structure of the circular equipment is shown in Fig. 10.20. The circular equipment has an annular shell of 7800 mm in diameter, and the inner annular groove is driven by an electric motor. The revolving body consists of 20 blocks of movable panels. The fan-shaped plate has some holes with 3 mm in diameter, which facilitates the filtration of acid water. The base of circular equipment is

Fig. 10.19 The structure of centrifugal acid removal equipment

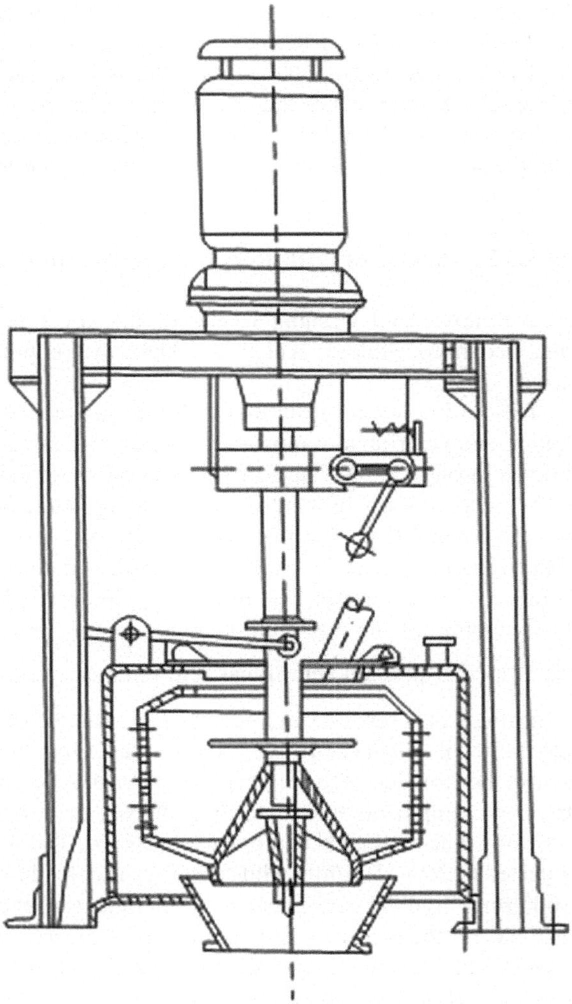

equipped with a variety of bottom tanks for the recovery of different acids, in which some baffles are used to separate the various acids, and the collected acid flow into the recycle bottom tank through bottom collecting pipe. The upper part of the tank is equipped with upper collecting pipe and spray branch pipe for the spray of various acids. The entire weight of the revolving groove is on the supported sheave. The transmission part of the equipment is the gearbox and the worm gear reducer. The motor rotates together with the rotating tank body. The rotation speed of the rotor body is different according to the time required for each turn, which can be divided into six grades. Specifically, 55 min, 50 min, 37 min, 30 min, and 19 min are classified into first, second, third, fourth, and fifth grades, respectively. The

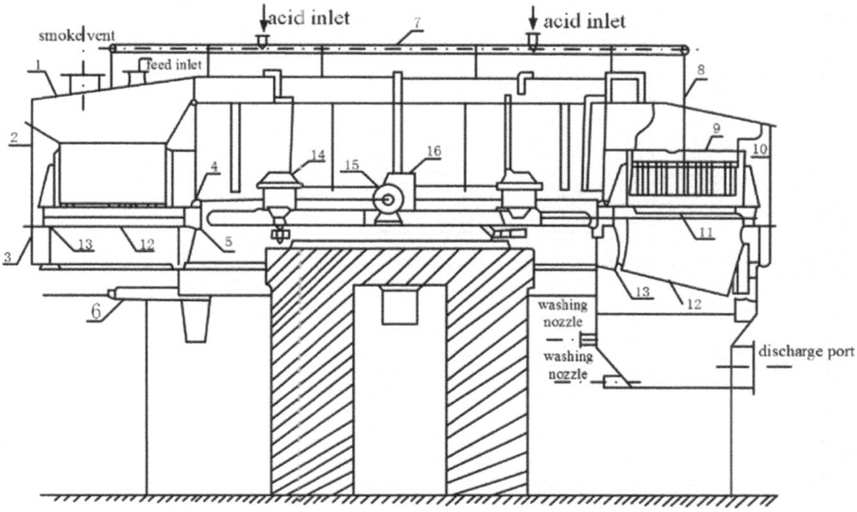

Fig. 10.20 Schematic diagram of circular equipment

speed of rotation is adjusted by the gearbox, which depends on various factors such as the difference in the raw material of the cellulose used for nitration, the difference in the grains of NC, the scale of the feeding amount, the cleaning equipment, and the handling of the accident. The power is turned on by a sliding point.

The internal rotary tank of the circular equipment can be divided into four working stages, namely, feeding, supplement nitration, replacement, and recovery, and discharge stages. According to the needs of production, the replacement and recovery stage can be further divided into three-, four-, or five-step replacement.

Circular equipment is used to replenish nitration and reduce the nitration time in the nitrator. Another function of circular equipment is to drive acid and replace and recover the waste acid. The working principle of the circular equipment is shown in Fig. 10.21.

The nitrated material is placed on the fan-shaped plate in the feeding section and evenly spread out, which rotates along with the rotary tank. The acid flowing from the orifice of the fan-shaped plate flows into the collecting tank in the bottom and is pumped to the supplement nitration section for recycling, i.e., recycled acid. The replenishment nitration is divided into two stages. The first stage is utilizing non-cooled recycled acid to continue the nitration reaction. In the second stage, cooled recycled acid with a temperature below 20 °C is employed to reduce the temperature of NC before entering the replacement reaction, thus reducing the denitration reaction of NC. The nitration degree of NC can be increased by 0.5–2.5 ml/g after the supplement nitration. After the supplement nitration, the waste acid is filtered by the fan-shaped plate and used for the mixed acid preparation. The cellulosic material enters the replacement and recovery section, in which

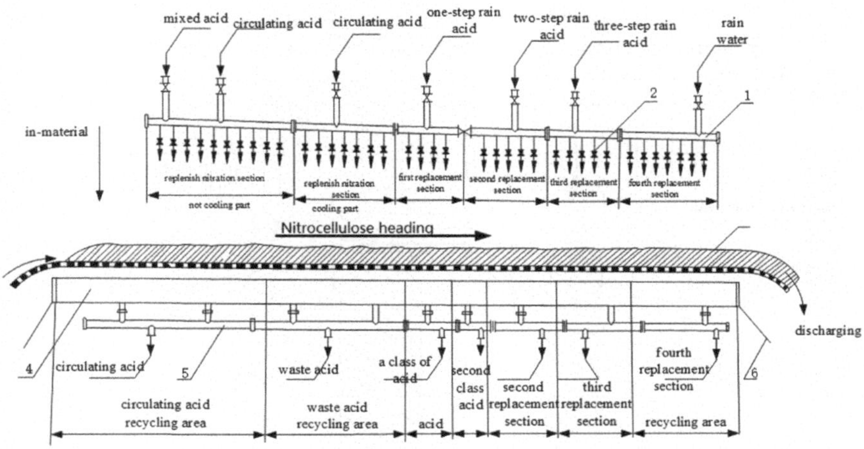

Fig. 10.21 The working principle of the circular equipment

the replacement is carried out by countercurrent washing with dilute acid and water with a gradual decrease in concentration so that a small amount of water is used to wash out the concentrated acid. With refined cellulose as raw material, four- or five-step replacement can be used. Since the temperature will rise during the replacement process, the rain acid of each section before the replacement should be cooled to below 20 °C to control the ratio of rain acid and maintain an appropriate concentration difference to reduce the dilution heat and the NC denitration value and to enhance the concentration of recovered acid. Recycled acid includes a higher concentration replacement acid and two types of dilute replacement acids with different concentrations.

(3) Horizontal piston discharge centrifugal drive acid equipment

Continuous drive acid equipment can be classified into primary and secondary according to their specifications. The main technical specifications of the horizontal piston discharge centrifuge are shown in Table 10.13.

The working process of primary horizontal piston discharge centrifuge is as follows. The nitrated material from the feeding inlet of the centrifuge is added to the conical cloth hopper. Under the action of centrifugal force, NC is swung onto the screen through the gap between the push plate and the hopper. In the meantime, the waste acid is gradually squeezed out, and NC on the screen forms a fibrous layer. Because of the continuing and reciprocating motion of driving disk, the fabric layer moves along the axial movement to the front of the drum and is swung away from the centrifuge. Under the action of gravity, the NC fabric layer falls into the washing device. To reduce acid consumption, three-stage acid replacement can be carried out in the front of the centrifuge with a certain distance to the outlet.

Table 10.13 The main technical specifications of the horizontal piston discharge centrifuge

No.	Main technical specifications	Centrifuge model no.		
		WH_1-1200	WH_1-800	WH_2-800
1	Rotary drum inner sieve diameter (mm)	1200	800	Primary 800 Secondary 880
2	Rotary drum filter area length (mm)	450	400	Primary 300 Secondary 230
3	Screen gap (mm)	0.30–0.35	0.30–0.35	0.30
4	Puncher travel (mm)	50	40	40
5	Maximum frequency of puncher travel (times/min)	20	25	35
6	Rotary drum speed (r/min)	450–550	650–850	700–900
7	The separation factor	136–205	189–323	242–398
8	Main motor power	28	17	22
9	Working pressure of hydraulic system (kg/cm^2)	25	20	18
10	Pump motor power (kW)	22	7.5	10

(4) Washer

After acid removal, NC still contains a significant amount of acid water. Therefore, it must be quickly immersed in a lot of water to reduce acidity by diluting the NC material concentration to 2% or less. The washer is a cylindrical container made of stainless steel and composed of a frame (13 rpm), agitated propeller (280 rpm) or paddle (60 rpm). The washed NC is conveyed to the digestion process by an NC feed pump.

(5) Nitration and acid removal process conditions

The process conditions for batch agitation nitration are shown in Table 10.14.

The conditions for continuous stirring nitration are shown in Table 10.15.

The process conditions for four-chamber continuous nitration are shown in Table 10.16.

The stirring speeds of the first, second, and third chambers are in the range of 150–200 rpm, and the stirring speed of the fourth chamber is 70–120 rpm. The height difference between the chambers is less than 40 mm. The process conditions for U-tube nitration are shown in Table 10.17.

Table 10.14 The process conditions for batch agitation nitration (3#NC)

Mixed acid composition (%)			Loading amount (kg) (dry weight)	Nitration coefficient	Nitration temperature (°C)	Nitration time (min)	
H_2O	HNO_3	H_2SO_4				Feeding	Nitration
16–18	18–22	60–66	24–27	36–42	28–38	2–6	25–35

Table 10.15 The process conditions for continuous stirring nitration (3#NC)

Mixed acid composition (%)			Loading amount (kg) (dry weight)	Nitration coefficient	Nitration temperature (°C)	Nitration time (min)
16–18	18–22	60–66	12–18	40–50	30–40	15

Table 10.16 The process conditions for four-chamber continuous nitration (3#NC)

Mixed acid composition (%)			Loading amount (kg) (dry weight)	Nitration coefficient	Nitration temperature (°C)	Nitration time (min)
16–18	18–22	60–66	3–7	40–50	28–38	25

Table 10.17 U-tube nitration process conditions (3#NC)

No.	Item						
	Mixed acid composition (%)			Loading amount (kg) (dry weight)	Nitration coefficient	Nitration temperature (°C)	Nitration time (min)
1	8–13	18–22	65–74	7–10	50–60	20–30	<20
2	10–18	18–22	60–66	7–10	50–65	20–30	<25
3	16–18	18–22	60–66	7–15	50–65	30–35	25

Table 10.18 Acid displacement process conditions of suspended centrifuge

Slow feeding time (min)	Fast acid displacement time (min)	Residual acid concentration (%) (based on H_2SO_4)	Product discharging time (min)
2–4	2–4	$\gg$45	2–3

The stirring speed of u-tube is in the range of 100–150 r/min, and the stirring speed of the last U-shaped tube is 80–100 r/min. The acid displacement process conditions of the suspended centrifuge are shown in Table 10.18.

The concentration of recovered dilute acid by displacement should not be too low. The relatively high concentration of recovered acid facilitates the denitration treatment and improves the recovery efficiency of sulfuric acid and nitric acid. Therefore, the amount of added water in the replacement should not be too large. The concentration difference in different time should be carefully controlled to reduce the dilution heat and the degree of denitration. The process conditions for the first stage horizontal piston discharge centrifuge are shown in Table 10.19.

Table 10.19 The acid displacement process of the primary horizontal piston discharge centrifuge (WH1-1200)

Pushing times (times/min)	Working oil pressure (kg/cm^2)	Pusher travel (mm)	Rotational speed (r/min)	Residual acid concentration (%) (based on H_2SO_4)
20	<27	<50	500–600	$\gg$45

10.3.3 Issues in the Cellulose Nitration Process

To obtain NC with different nitrogen contents, when the weight ratio of sulfuric acid to nitric acid is about 3: 1 and the moisture content of the mixed acid is within 7–21%, the nitration degree of NC can be accurately regulated by strictly control the deviation of mixed nitrating acid. Table 10.20 illustrates the effect of nitration of 0.1% change in moisture content in mixed acid on the nitration degree of NC, especially for Low nitrogen content NC.

According to numerous nitrocellulose production practices, factors affecting nitration are as follows:

(1) The weight ratio of sulfuric acid to nitric acid in the mixed nitrating acid

Typically, a higher content of sulfuric acid in the mixed acid results in a slower nitration rate, a more amount of formed sulfate, and a more severe nitration degradation. When the content of sulfuric acid in the mixed acid is considerably high, the fiber structure is damaged significantly, which further reduces the viscosity and yield of the resulting NC. The correlation between the ratio of sulfuric acid and the properties of NC are shown in Fig. 10.22 and Table 10.21.

In the blending formula of mixed acid, the contents of nitric acid and cellulose are kept constant at 30 g and 2.5 g, respectively. The amount of sulfuric acid is varied to investigate the nitration results at 19 °C.

Figure 10.22 and Table 10.21 show that when the ratio of sulfuric acid to nitric acid is in the range of 0.25–1, the rate of esterification is the maximum and the high nitrogen content can be obtained in a short time. When the ratio of sulfuric acid to

Table 10.20 The effect of 0.1% change in water content in mixed nitrating acid on the nitration degree in nitric acid

Controlled nitration range (ml/g)	The effect of water change 0.1% on the nitration (ml/g)
215.0–221.0	0.06–0.09
213.0–215.0	0.1–0.2
211–213.0	0.2–0.3
194–198.0	0.4–0.6
188–194.0	0.6–0.9
180–188	0.8–1.1
172–180	1.0–1.5

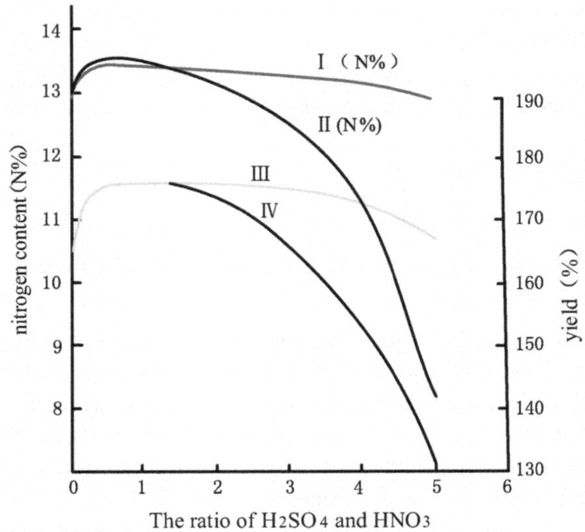

Fig. 10.22 The correlation between the ratios of H_2SO_4 to HNO_3 in the mixed acid with decreasing yield of NC: I, III—the correlation between nitrogen content and NC yield after 24 h nitration; II, IV—the correlation between nitrogen content and NC yield after 30 min nitration

Table 10.21 The effect of sulfur acid ratio in mixed acid on nitration

No.	The ratio of sulfuric acid to nitric acid	After nitration 30 min		After nitration 24 h	
		Nitrogen content (%)	Yield (%)	Nitrogen content (%)	Yield (%)
1	0	12.51	162.7	12.60	163.3
2	0.1	–	–	12.73	165.0
3	0.25	13.42	175.7	13.41	175.7
4	0.50	–	–	13.40	175.2
5	1	13.33	174.5	13.35	174.7
6	2	13.20	174.1	13.29	175.9
7	3	12.69	166.1	13.37	176.4
8	4	–	–	13.16	175.1
9	5	8.12	130.9	13.06	166.6

nitric acid is less than 0.25, the degree of nitration increases with increasing ratio of sulfuric acid to nitric acid. When the ratio of sulfuric acid to nitric acid is greater than 1, the reaction rate decreases with decreasing ratio of sulfuric acid to nitric acid. When the ratio of sulfuric acid to nitric acid is greater than 3, the reaction rate of the cellulose decreases significantly. Simultaneously, cellulose is hydrolyzed partially, resulting in reduced yield of NC. When the ratio of sulfuric acid to nitric acid is greater than 7, the fiber structure in the nitration is severely damaged and becomes powder. When the ratio of sulfuric acid to nitric acid is in the range of 12–38, the cellulose dissolves in the mixed acid completely.

Table 10.22 Composition of the mixed acid used for different types of NC

Required nitration degree (ml/g)	Composition of mixed acid	
	HNO_3 (%)	Water content (%)
>214.0	20–23	8–10
213–214	20–22	9–10.5
212–213	19–21	10–12
211–212	18–21	12–13.5
193–198	18–20	16–17.5
188–193	18–20	16.5–18.0
180–188	20–24	17.5–19.0
172–180	20–24	18.0–20.0

To reduce the consumption amount of nitric acid and ensure the quality of the product, the ratio of sulfuric acid to nitric acid is controlled in the range of 2.5–3.5, and the water content in the mixture is adjusted to control the nitrogen content of NC in the industrial production. When the water content in the mixed acid is almost constant, the change of nitric acid content affects the nitrogen content of NC to some extent. In the case of NC with high nitrogen content, the fluctuation of the nitric acid content of 1% results in the nitration degree difference by 0.1–0.6 ml/g, which is equivalent to the influence of 0.1% water fluctuation. For low nitrogen NC, the fluctuation of the nitric acid content of 1% results in the nitration degree difference by 0.5–1 ml/g.

Table 10.22 shows of mixed acid composition used for the production of NC with different grains in a plant. Under normal circumstances, the ratios of nitric acid and sulfuric acid are in the ranges of 18–25% and 60–70%, respectively. To prepared NC with nitration degree more than 212.5 ml/g, the water content in the mixed acid should be controlled at ±0.3–0.5%. For the preparation of NC with other different range of nitrogen contents, the water content should be controlled within ±0.1–0.3%. In the production of low nitrogen content NC, the contents of water and nitric acid should be controlled in the ranges of ±0.12% and ±0.5%, respectively.

The effects of water content and the ratio of sulfuric acid to nitric acid on the nitration process were studied. The correlation between nitric acid composition and nitration is shown in Fig. 10.23.

The numbers on each line in Fig. 10.23 indicate the nitrogen oxides content.

(1) The effect of nitrogen oxides content

In the industrial nitric acid and mixed acid, the decomposition of nitric acid will produce certain nitrogen oxides (such as NO_2, N_2O_4, and N_2O_3).

$$(4HNO_3 = 4NO_2 + 2H_2O + O_2)$$

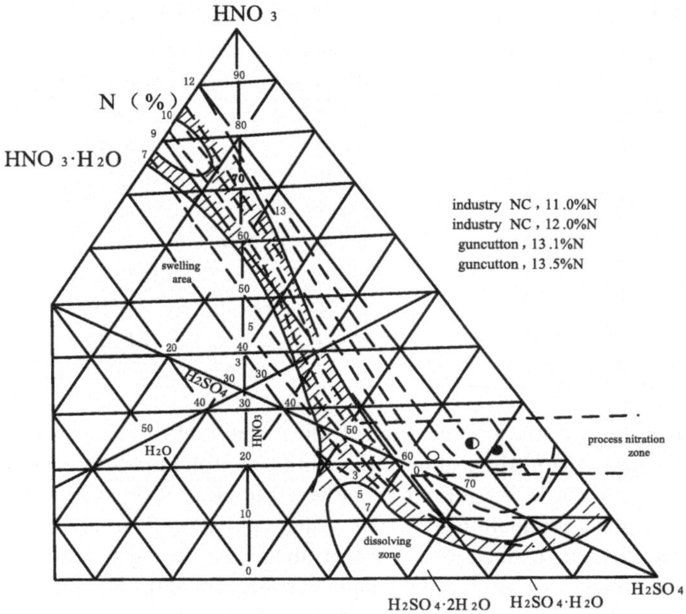

Fig. 10.23 Relationship between the acid content in the mixed acid (% by weight) and the nitrogen content of NC

$$2NO_2 = N_2O_4$$

When the water content in the mixed acid is more than 10%, the nitrogen oxides react with water to form nitric acid and nitrous acid:

$$N_2O_4 + H_2O = HNO_3 + HNO_2$$

At this point, the nitric acid content decreases relatively while the oxidation capability increases. Therefore, strict control of N_2O_4 content in the production is conducive to the nitration of nitrocellulose by mixed acid.

When the water content in the mixed acid is less than 10%, most of the nitrogen oxides are converted to nitroso sulfuric acid, which does not damage the cellulose.

The content of N_2O_4 in the mixed nitrating acid is low, about 1.0%. When the temperature exceeds 30 °C, the content of N_2O_4 content is slightly higher. After long-term recycle mixed acid, the content of N_2O_4 in the mixed acid gradually increases, and the mixed acid dissolves more low-degree nitration products increasingly, thus causing the mixed acid unclean. When the content of N_2O_4 in the mixed acid exceeds 2.5%, this mixed acid cannot be used.

In the production, the content of nitric oxide is relatively low, thus causing a negligible impact on the quality of NC. The total acid and sulfuric acid contents are analyzed, and then the contents nitric acid and water can be calculated by the difference method. However, this approach likely to cause the high value of nitric acid and low value of water. In the determination of the nitric acid composition, this factor should be taken into account to eliminate the impact on the NC quality.

(2) The relationship between the composition of mixed acid and NC solubility

The effect of the mixed acid composition in the nitration process on the solubility of the NC in alcohol and ether solvents depends on the difference in the nitrogen content of the NC and improved solubility caused by the degradation and hydrolysis of the cellulose polymer in the nitration process. Among them, the change of nitrogen content is the primary factor that causes the change of NC solubility. With increasing hydrolysis side reactions, other side reactions increase correspondingly, thus diminishing the yield of NC. The influence and patterns of water content and the ratio of sulfuric acid to nitric acid in the mixed acid on the NC solubility are shown in Table 10.23 and Fig. 10.24.

Since the area corresponding to the nitration degree and alcohol ether solubility of mixed NC is in the abrupt region, the mixed cotton blended with 1#NC and 2#NC in different ratios is used rather than a single grade of NC. The nitrogen contents of some 1#NC with alcohol ether solubility that is close to the abrupt region should be carefully controlled to prevent alcohol ether solubility from exceeding the standard. The percentage on the curve in Fig. 10.24 is the relationship between the water content and the solubility of the mixed acid.

It can be seen from Fig. 10.24 that the water content in the mixed acid is 7–14%, the NC is in the gradient region of alcohol ether solubility. The solubility is less than 15%, which belongs to the #1NC production area. When the water content is in the range of 16–21%, the solubility is greater than 95% although the nitration degree gradually decreases, which belongs to the stable solubility region. In the water content range of 14–16%, the solubility is in the abrupt region. When the nitrogen content of NC slightly fluctuates, it will cause a great change in solubility,

Table 10.23 The effect of the ratio of sulfuric acid to nitric acid in the mixed acid on the NC solubility in alcohol and ether

Ratio H_2SO_4/HNO_3 nitric acid	Solubility %						
	The water content in the mixed acid (%)						
	5	10	12	14	16	18	20
0.25	3.0	7.5	92.0	96.0	76.0	23.0	6.0
0.50	3.0	3.5	36.0	100.0	100.0	94.5	23.0
1.0	3.0	3.5	10.0	100.0	100.0	98.3	86.0
2.0	1.5	1.8	3.0	15.0	100.0	100.0	96.0
3.0	1.2	1.2	1.5	12.0	100.0	99.5	99.0
4.0	3.0	3.0	3.0	18.0	99.0	99.0	98.0
5.0	4.5	4.5	4.5	24.0	96.0	97.0	97.0
6.0	5.0	7.5	9.0	51.0	96.0	96.0	89.0
10.0	–	27.0	55.5	77.0	74.0	22.0	–

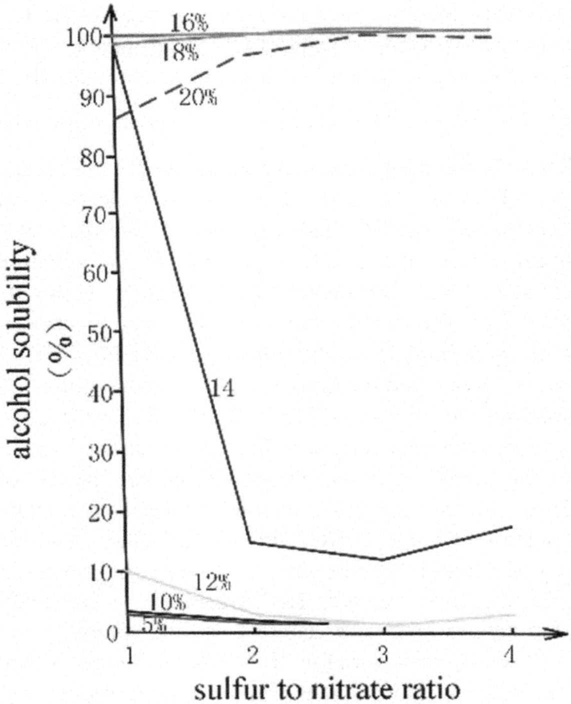

Fig. 10.24 The effect of water content and the ratio of sulfuric acid to nitric acid in the mixed acid on the solubility of NC

as great as more than 15% fluctuation. Table 10.24 illustrates the relationship between the water content in the mixed acid and the 1#NC solubility in alcohol ether.

As shown in Table 10.24, when the water content is in the range of 12.5–13.5%, the corresponding solubility is close to the abrupt region. When the water content in the mixed acid changes 0.1%, alcohol ether solubility changes 1.0–3.0%, and the NC solubility in alcohol ether is more than 15%. The composition change of the

Table 10.24 The relationship between the water content in the mixed acid and the alcohol ether solubility of NC

Water content in the mixed acid (%)	NC solubility in alcohol ether (%)	
	General range	Average range
8–10	3–7	4–6
10–11	4–9	5–7
11–12	5–10	6–8
12–12.5	6–11	7–9
12.5–13.0	7–16	8–12
13.0–13.5	8–20	10–15

mixed nitrating acid promotes the side reactions in the nitration process, which in turn increases the number of compositions with low nitrogen content and low degree of polymers, thus enhancing the solubility in ethanol.

Nitration capacity of mixed acid is a synergistic effect of nitric acid, sulfuric acid, and water. The composition of mixed acid not only affect the nitration rate but also directly influence the nitration degree of cellulose, the solubility in alcohol ether, the viscosity of NC, and the yield of NC. To ensure the quality of NC, the composition of mixed acid should be strictly controlled according to different nitration process, different product specification, and different production season. The influence of the production season is essentially a change in the temperature and relative humidity of the air, which leads to changes in the nitration and side reactions of the NC. Ultimately, it reflects the change of NC quality. According to the production experience, the high temperature and high humidity in the summer typically result in higher water content in the dried cellulose and nitric acid, which further accelerates the denitration of NC, thus reducing the nitrogen content of NC.

10.3.4 Effect of Nitration Coefficient and Quality of Refined Cellulose on the Nitration Reaction

(1) The effect of nitration coefficient

Since the nitration reaction is an incomplete reversible and heterogeneous chemical reaction, the key factor determining the nitrogen content of NC is not entirely at the beginning of the nitration reaction but in the final acid composition in the NC. Therefore, nitration coefficient between the nitration acid and refined cellulose impacts the nitration reaction significantly.

Nitration coefficient mainly affects nitrogen content of NC, uniformity, NC solubility, the final cost of production, the treatment amount of waste acid, nitration equipment, acid preparation equipment, and so on.

According to the production experience in a plant, when the nitration coefficient is in the range of 47–60, the nitration degree is insignificant; when the nitration coefficient is less than 40, the impact of nitration should be considered, especially for the preparation of Low nitrogen content NC. With increasing nitration coefficient, the nitrogen content of NC increases. The effect of nitration coefficient on nitrogen content is shown in Table 10.25.

The selection of nitration coefficient is closely related to the mode of production and the morphology of cellulose. Table 10.26 lists the nitration factor used for different nitration plants.

(2) The effect of quality of refined cellulose

Refined cellulose should have good surface properties to ensure easy acid infiltration and efficient absorption of mixed acid. These optimal surface properties depend largely on the purity and physical form of the refined cellulose.

Table 10.25 The effect of nitration coefficient on nitrogen content of NC

No.	Nitration coefficient	Loading amount of nitrator (kg)	1#NC quality		2#NC quality	
			Nitration degree (ml/g)	Solubility in alcohol ether (%)	Nitration degree (ml/g)	Solubility in alcohol ether (%)
1	50	30	213.9	5.5	195.7	99.7
2	45	35	213.8	6.2	195.4	99.7
3	40	40	213.6	6.5	195.2	99.6

Table 10.26 Variation range of nitration coefficients for different nitration equipment

Nitration coefficient	Stirring nitration	Four-chamber continuous nitration	U-type continuous nitration
Nitration coefficient	Approx. 40	40–50	50–65

The purity of the purified cellulose refers to the content of the cellulose therein. When the content of impurities in cellulose is high, the hydrophilicity is poor. As a result, the acid infiltration is slow and less acid is absorbed, which cause prolonged nitration time, enhanced side reactions, the low nitrogen content of the prepared product, increased solubility, and decreased yield and stability. In the case of the contents of fat, wax and lignin are too high, the nitration process involves severe oxidation, resulting in local overheating and NC decomposition or fire. Low molecular nitrated products dissolve in acid readily, which makes acid turbid, affecting the recycling of acid.

The morphology of the raw material affects the quality of the nitration product because the morphology of the cellulose is related to its loading density and nitration coefficient. When the absorption performance is good, a certain volume of cellulose can absorb a certain amount of mixed acid. As the loading density increases, the nitration coefficient is reduced in a certain volume of nitrator. To ensure the quality of nitration products, it is necessary to change the composition of mixed acid. Increase in the content of nitric acid in the mixed acid and reduction in the water content enables the waste acid inside the NC has the same composition as the relatively porous nitrocellulose.

The water content of the refined cellulose also affects the nitration degree. If other conditions are identical, the fluctuation of dry, refined cellulose moisture by 1% causes the degree of nitration of nitrated product to vary 0.2–1 ml/g. The lower the nitration degree, the greater the impact. The primary reason is water inside the cellulose can quickly spread to the mixed acid, resulting in increased water content in the mixed acid. Simultaneously, when the mixed acid is dispersed into the cellulose, it combines with the internal water of cellulose and generates local dilution, thus resulting in a low degree of nitration locally and enhanced side reactions. Therefore, continuous nitration requires that the water content of refined cellulose should be below 5%.

The viscosity of refined cellulose is one of the main factors determining the viscosity and solubility of NC. Different uses of NC have certain viscosity requirements for refined cellulose. After the semi-high-pressure stability treatment process is adopted for the NC production, the requirement of refined cellulose viscosity can be appropriately slacked.

10.3.5 The Effect of Nitration Temperature and Nitration Time on the Nitration Reaction

Increase in the nitration temperature not only speed up the chemical reaction but also can considerably affect the diffusion of mixed acid to internal of refined cellulose, which significantly affect the rate of nitration reaction. However, it does not significantly affect the nitrogen content when the reaction reaches equilibrium. According to the experience of factory production, if the nitration temperature of the cellulose is increased by 10 °C, the reaction rate is doubled. On the other hand, when the reaction temperature increases, the degradation of the cellulose is aggravated, and the side reactions such as hydrolysis and oxidation of the product are enhanced. Therefore, the viscosity, yield, stability of NC all decline. The relationship between nitration temperature and nitration rate and NC mass is shown in Fig. 10.25. This effect increases with increasing water content in the mixed acid.

The nitration temperature in production is not more than 40 °C. If the nitration temperature rises to 50 °C, the resulting NC is brittle. When the nitration temperature is in the range of 60–80 °C, the cellulose dissolves in the mixed acid partially. When the viscosity of the refined cellulose is too high, and NC with a lower viscosity is desired, the nitration process can be performed at elevated temperatures, but the nitration temperature should not exceed 45 °C.

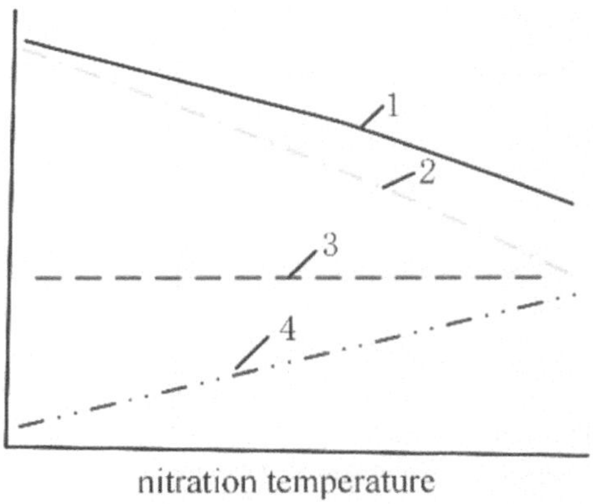

Fig. 10.25 The effect of nitration temperature on the nitration rate and NC mass: 1—nitration time, 2—solution viscosity, 3—esterification degree, and 4—solubility

Table 10.27 Nitration temperature and temperature of nitric acid in the preparation of different grades of NC

Grade	Temperature of mixed acid (°C)	Nitration temperature (°C)	Increased temperature (°C)
1#NC	18–25	23–30	5–10
2#NC	20–30	25–35	4–8
3#NC	25–35	28–38	4–7
Colloidal NC	1–5	6–15	5–10

Since the nitration of cellulose is an exothermic reaction, the temperature of mixed acid for nitration should be lower than the nitration temperature. Table 10.27 is the temperature range of nitric acid in the production plant.

According to the production experience of a plant, the increase in nitration temperature by 3 °C results in the increase in nitration degree 0.3–0.5 ml/g. However, the increase in nitration temperature slightly affects the preparation of NC with low nitrogen. The fluctuation range of mixed nitrating acid should be controlled ±2.0 °C.

The nitration time has a certain effect on the quality, uniformity, and viscosity of the NC. In the initial first 5–10 min of nitration, the nitrogen content of NC is close to equilibrium; when the nitration reaction proceeds 30–40 min or more, the impact on the nitrogen content is negligible; when the nitration reaction is less than 30 min, the impact of nitration reaction time on the NC nitrogen content should be considered. The relationship between the agitated nitration time and the nitrogen content of the refined cellulose is shown in Table 10.28.

According to the data in Table 10.28, the relationship in Fig. 10.26 is obtained.

Extended nitration time enhances side reactions, which is confirmed by the decreased NC viscosity. The relationship between nitration time and NC mass is shown in Table 10.29.

Table 10.28 The relationship between the agitated nitration time and the nitrogen content of NC

Nitration time (min)	Nitration degree (ml/g)	Solubility in alcohol ether (%)	Nitration time (min)	Nitration degree (ml/g)	Solubility in alcohol ether (%)
2	181.1	–	16	190.5	98.1
4	187.1	98.7	18	189.9	97.6
6	187.6	96.9	20	190.1	95.4
8	188.7	97.0	22	191.8	99.6
10	189.3	99.4	24	190.9	98.3
12	189.6	97.4	26	192.0	99.0
14	190.8	96.3	30	192.2	99.4

Experimental conditions: H_2SO_4, 62.51%, HNO_3, 19.92%, H_2O, 16.93%, N_2O_4, 0.64%; nitration coefficient 40; nitration temperature 28–33 °C

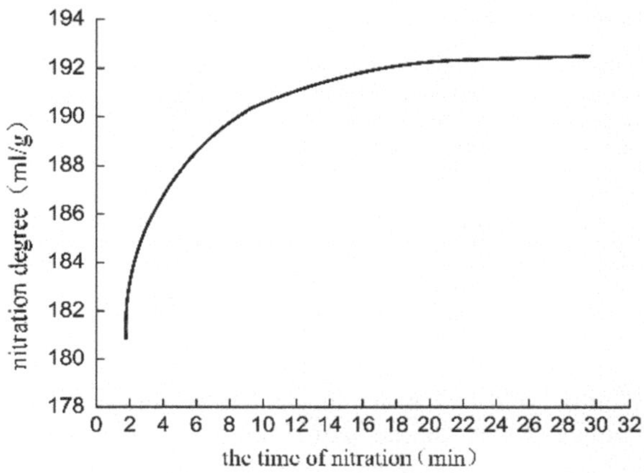

Fig. 10.26 Relationship between nitration time and the nitrogen content of NC

Table 10.29 Relationship between nitration time and NC mass

No.	Nitration time (min)	NC mass		
		Nitrogen content (%)	Viscosity (Engler degree)	Solubility in alcohol ether (%)
1	30	11.8	180	99.3
2	35	12.0	160	99.5
3	40	12.0	138	99.7
4	45	11.9	102	99.3
5	50	11.8	93	99.5

When the stirring nitration time is reduced by ten minutes, the viscosity of low nitrogen content NC increases by about 0.5 Engler degrees.

10.3.6 The Composition of Adsorbed Acid and Waste Acid of Nitrocellulose [3, 7]

The obtained NC at the end of the nitration reaction and the mixed acid with the changed composition are balanced. This mixed acid is referred to as "waste acid". If the remaining waste acid has been removed from the NC, such as the acid removal, the NC still contains a large amount of acid. Table 10.30 is some data of the acid components contained in the waste acid and NC.

It can be seen from Table 10.30 that if the prepared NC has a low nitrogen content, the absolute value of the acid content in NC is high; in any case, the composition of acid contained in NC is different from that of waste acid; the amount

Table 10.30 The composition of waste acid and the remaining acid contained in the NC

Nitrogen content of prepared NC (%)	Weight ratio of the remaining acid contained in the NC to the NC	Composition of waste acid (%)			Composition of remaining acid in NC (%)		
		H_2O	HNO_3	H_2SO_4	H_2O	HNO_3	H_2SO_4
13.43	0.73	8.3	46.3	45.4	19.5	50.5	30.0
13.28	0.79	12.4	59.8	27.8	17.1	64.1	18.8
12.20	0.74	19.1	18.1	62.8	21.2	30.3	48.5
11.75	0.92	18.7	19.4	61.9	20.1	33.6	46.3

of nitric acid and water in the NC is always higher than that in the waste acid. Compared with the ratios of sulfur acid to nitric acid in the mixed acid, waste acid, and acid contained in NC, the ratio in the waste acid increases, and the ratio of acid in NC is lower than that of the mixed nitrating acid.

From the above discussion, the acid contained in the NC consists of two parts, namely, the mixed acid filled in the fiber capillary that has the component as the waste acid, and the acid adsorbed by the NC. The acid adsorbed by NC has a high content of water and nitric acid, which is different from the waste acid because NC tends to adsorb nitric acid and water from the mixed acid.

According to experimental evidence, the adsorbed acid by the NC at the end of the nitration reaction is nitric acid and water rather than sulfuric acid, which is consistent with most of the conclusions in the literature. The adsorbed acid of NC in the mixed acid at the end of the nitration reaction is shown in Fig. 10.27.

The two sets of curves in Fig. 10.27 represent the amount of acid adsorbed by NC and the concentration of nitric acid in the adsorbed acid. The mixed acid consisting of 17% H_2O and 25% HNO_3 produces NC with 12% nitrogen content. From Fig. 10.27, 1 g of cellulose can adsorb 0.37 g (37%) of acid (HNO_3 + H_2O), of which the nitric acid content is 90%, corresponding to the adsorption of 0.333 g of anhydrous nitric acid per gram of cellulose.

The waste acid composition and the adsorbed acid in the NC are related to the residual acid (30–40%) in the NC, which can guide the recovery of the waste acid and the stability treatment to reduce the production cost of NC. A WH1-1200 pusher centrifuge is used for recovery of residual acid by water mist replacement treatment of 3#NC in one production. The composition change of adsorbed acid is shown in Fig. 10.28.

Figure 10.28 shows that sulfuric acid is readily recovered, but nitric acid is difficult to remove. If the residual acid filled in the capillaries is removed, the nitric acid content of the adsorbed acid will be higher. In other words, NC still contains a large number of residual acid, approximately 422 kg of sulfuric acid and 377 kg of nitric acid per ton of NC.

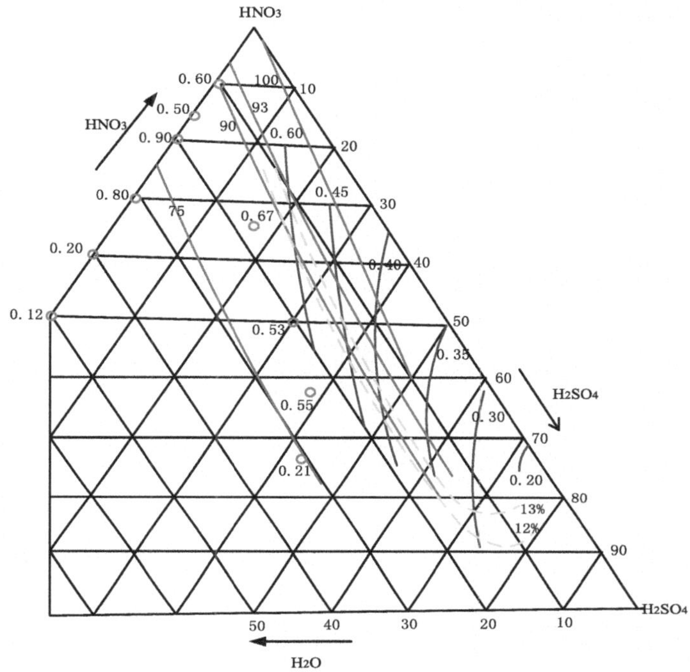

Fig. 10.27 Adsorbed acid in NC at the end of the nitration reaction: (1)—adsorbed nitric acid by 1 g of original cellulose ($HNO_3 + H_2O$); (2)—the effective concentration of nitric acid in the adsorbed acid $\left(\frac{HNO_3}{HNO_3 + \text{Free } H_2O}\right)$; O—adsorption rate of HNO_3 measured at Adil plant; and -------- isonitrogenous line. Free H_2O

According to the composition change in the residual acid within the NC in the process of acid removal, with increasing nitric acid concentration, there is supplement nitration, which is also called "post-nitration". It is also believed that the generated heat during the post-nitration may lead to the hydrolysis of NC. At the end of the nitration reaction, the acid inside the fiber is equilibrated, and the adsorbed acid is formed during the digestion process. If the acid removal process is mainly to remove the free waste acid, the acid removal process in a short time does not cause a significant change in adsorbed acid to reduce post-nitration, unless the nitration reaction has not yet reached the equilibrium before the acid removal, and acid removal is undergoing, which is different "post-nitration".

Fig. 10.28 Composition of adsorbed acid on the NC (3#NC) sieve: 1—sulfuric acid, 2—water, and 3—nitric acid

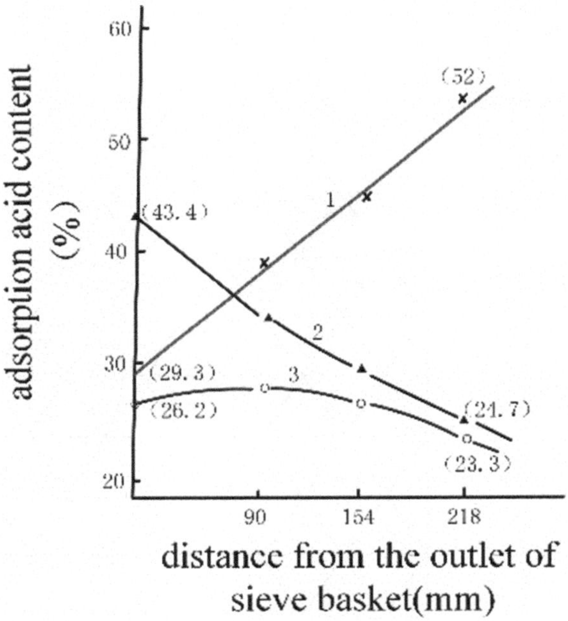

10.4 Stability Treatment of Nitrocellulose

Acid is the cause of NC instability. Degradation and denitration are the side reactions in the process of stability treatment. The basic principle of NC stability treatment is to focus on the residual acid in NC molecules. The stability treatment of nitrocellulose is a critical process in the industrial production of NC.

10.4.1 Principle of Nitrocellulose Stability Treatment

(1) The reasons for the immobilization of nitrocellulose after acid repellent

After acid removal and washing, NC contains many impurities, majorly including the residual, cellulose sulfate, nitro-sulfuric mixed ester, a variety of nitrated sugars acid attached to the NC. Cellulose sulfates, nitro-sulfuric mixed esters, and various nitrated sugars are the products of side reactions during nitration.

Cellulose sulfate ($ROSO_3H$) is acidic and readily hydrolyzed to H_2SO_4 in the presence of acid at the high temperature and. In the meantime, the ($-OSO_3H$) group has a higher electronegativity than ($-ONO_2$). Therefore, the presence of sulfuric acid ester and nitro-sulfuric ester deteriorates the stability of NC.

Low molecular carbohydrate nitrates are the nitration products of oxidized cellulose and hydrolyzed cellulose which are derived from the oxidation and hydrolysis side reactions of associated products in the cellulose and hemicellulose with a low degree of polymerization. However, low molecular weight carbohydrate nitrates only can be formed in the nitric acid with high concentration and low content of water and are stored in the NC under this condition. A higher degree of nitration is accomplished primarily because of the strong nitration capability of the mixed acid. Although the polymerization degree of low molecular weight carbohydrate nitrates is low, they cannot dissolve in the mixed acid but remain in the NC. For this reason, high nitrogen content NC contains more this type of low molecular weight carbohydrate nitrates.

Because of the short molecular chain of low molecular weight carbohydrates, their poor stability of the terminal group significantly affects the stability of the molecular chain. This effect increases with increasing degree of hydrolysis (reduced molecular chain) and oxidation (such as the formation of –COOH functional group). Their decomposition products will further, in turn, catalyze the decomposition of NC.

In the nitration process, the refined cellulose can swell in the mixed acid to expand the volume. However, because of the decline in NC hydrophilicity after the nitration, macromolecular chains shrink and move closer to each other after a large number of washing cycles. As a result, the acid mixed in the NC is difficult to diffuse out, especially sulfuric acid with a large volume, which is adsorbed by NC to form the residual acid in the NC. Although the number of residual acids contained in the NC is small, the acid concentration gradually increases with the removal of water (such as natural evaporation), which typically reaches a sufficient acid concentration to decompose NC. NCs with different nitrogen contents are washed with sufficient water until the water reaches neutral, and then the total acid content and sulfate content in the NC are determined. The results are shown in Table 10.31.

From analysis of the amount of total sulfuric acid and the amount of free sulfuric acid in the normally produced cellulose after water washing, the total sulfate content and free sulfate content of 1#NC, 2#NC, and 3#NC are in the ranges of 0.35–1.1% and 0.3–1.0% (weight percentage), respectively.

As can be seen from Table 10.31, the higher the nitrogen content of the NC, the higher the content of free sulfuric acid. On the other hand, a high amount of hydrated sulfuric acid is observed in Low nitrogen content NC. To confirm the

Table 10.31 Free sulfuric acid and hydrated sulfuric acid (sulfate) in NC

NC sample	Nitrogen content (%)	Free sulfuric acid (%)	Hydrated sulfuric acid (based on H_2SO_4) (%)
1#NC	13.4	0.69	0.08
Collodion	12.6	0.44	0.16
High viscosity NC	12.1	0.17	0.30
Low viscosity NC	10.7	Trace	0.52

major impurities in NC with high nitrogen amount are free sulfuric acid, the following experiments were carried out. To the sulfuric acid-free and stability treated NC with high nitrogen content was added 80% aqueous acetone solution containing 3–5% sulfuric acid. The treated NC swelled and absorbed the acid containing acetone. After removed the acetone, a large amount of water was added to shrink the NC capillary suddenly. Because of the large volume of sulfuric acid, it cannot completely diffuse into the surrounding medium and a part of the sulfuric acid residue in the fiber capillary. As a result, the original stable NC became unstable, even after a long period of washing with 0.1% sodium bicarbonate solution washing, the treated NC cannot reach a stable state. In Table 10.31, 1#NC was dissolved in acetone and then precipitated with water to obtain a well-characterized NC, which indicated that free sulfuric acid in NC with high nitrogen content is one of the main factors affecting the stability.

As shown in Table 10.31, low viscosity NC is also dissolved in acetone. Although the precipitated NC with water almost retains all of the acids before the purification, it is still as unstable as before the refining, thus confirming the presence of sulfate in the low nitration degree of NC is the main factor affecting the stability. The following experiments can be performed to confirm that the major impurity of low nitrogen content NC is the hydrated sulfuric acid.

1. Low nitrogen content NC is prepared with nitric acid without sulfuric acid. After soaked with 2% sodium bicarbonate solution for 30 h to neutralize the residual acid in NC, the prepared NC reaches a stable state. This shows that only neutralization of the residual acid can reach a stable state in the absence of sulfuric acid in the NC.

2. Low nitrogen content NC prepared with nitric and sulfuric mixed acid is dissolved in acetone and then immersed in sodium bicarbonate solution. The precipitated NC cannot reach a stable state. However, stable NC can be obtained by refluxing in dilute acid solution (0.5% sulfuric acid).

(2) Removal of unstable impurities in nitrocellulose

Treatment of the NC with 0.1–0.2% dilute sulfuric acid solution (weight ratio) at high temperature removes the sulfate and sulfate of the sulfate and nitrate mixed ester. Refluxing with 0.5% sulfuric acid solution for 2 h can decompose the sulfate. Treatment with other inorganic acid affords the same effect. The capability of inorganic acid to stabilize NC is in the order of $HNO_3 > HCl > H_2SO_4$. If treated with 1% sulfuric acid, the sulfate can be effectively damaged. The treated sulfuric acid in the NC exists in two different forms, namely, inorganic salts (mostly $CaSO_4$) and cellulose-combined sulfonic groups ($-OSO_3H$). The sulfuric acid content in NC after stability treatment is shown in Table 10.32.

After the refluxing of NC in acid, the total sulfate is 0.06–0.6%, of which free sulfate is 0.05–0.5%.

If treated with lye, the sulfate is difficult to remove because the hydrogen sulfate ($-SO_3H$) will react with the base to produce a stable salt ($-SO_3Na$), which is difficult to be removed.

Table 10.32 Sulfuric acid content in NC after the stability treatment

Nitrogen content of NC (%)	Sulfuric acid content		The ratio of total sulfuric acid to sulfate inorganic salt
	Total amount	Content inorganic sulfate	
11.20	0.24	0.12	2:1
11.88	0.16	0.12	1.3:1
12.45	0.16	0.10	1.7:1
13.00	0.27	0.12	2.3:1
13.19	0.25	0.10	2.4:1
13.25	0.27	0.13	2.1:1

Nitrate carbohydrates can also be destroyed in acidic solutions. If this process is carried out in alkaline solutions, the alkaline solution can neutralize a portion of the free acid and the acidic products produced by the decomposition of the impurities. Therefore, to accelerate the removal of these unstable impurities, sodium carbonate solution with a concentration of 0.01–0.03% is commonly used to digest NC. Excessive high concentration will cause NC significant denitration and decomposition of NC, which is undesirable. The use of strong bases is inappropriate because the destruction of NC by the strong base is too severe.

Most of the free acid and the generated acid by the decomposition of impurities in the NC can be washed with hot water but cannot be completely removed. If the NC is in a swollen state, the acid can be removed quickly and easily. This is because in the aqueous medium or the dilute aqueous solution medium, the acid in the NC is diffused out by the concentration difference. When the NC is in the swollen state, the distance between the macromolecules increases, which reduces the diffusion resistance and facilitates the removal of residual acid, especially the diffusion resistance of sulfuric acid. The content of acid in the medium will increase during the digestion process. The NC with 13.2% nitrogen content is prepared with the mixed acid containing 22% nitric acid and then boiled with water. The acid and composition changes into the water are shown in Fig. 10.29.

The changes in acid digestion (based on H_2SO_4) are shown in Table 10.33. The commonly used starting acidity in the production is 0.3–0.4%. After the acid digestion, the general acidity increases by about 0.1–0.2%, whereas the acidity of high nitrogen content NC increases more significantly.

As can be seen from Fig. 10.29, the concentrations of nitric acid and sulfuric acid in water increase rapidly after 90 h of boiling. The ordinal value between I and the reference line III represents the released nitric acid, which is a constant (about 2.48) at any time during the second stage of the hydrolysis. In the course of 90 h, the loss of nitric acid is 2.80%, equivalent to 0.62% loss of NC nitrogen.

It is difficult to completely remove the residual acid in the NC, but neutralize with the weak base solution can remove the residual acid completely. That is, first NC is boiled in acidic aqueous solution, followed by boiling in alkaline solution. The neutralization of acid and alkali can be utilized to remove the residual acid.

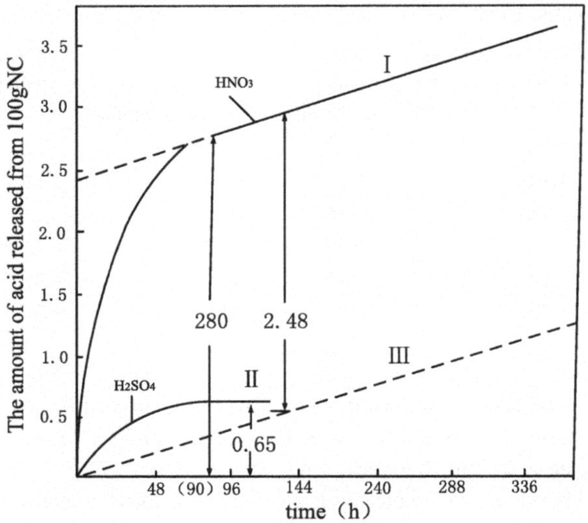

Fig. 10.29 Changes in acidity during digestion of NC with a nitrogen content of 13.2%

Table 10.33 Changes in acidity during acid digestion

No.	Starting acidity (%)	Acidity change during digestion (%) (based on H_2SO_4)			Digestion temperature (°C)
		2 h	4 h	6 h	
1	0.21	0.34	0.36	0.39	115
2	0.27	0.35	0.40	0.45	110
3	0.33	0.37	0.41	0.46	115
4	0.39	0.47	0.50	0.56	116

According to the electrolyte film theory, the equilibrated electrolyte solution on both sides of the membrane (semipermeable membrane) exhibits the following relationship.

$$X = \frac{C_2^2}{C_1 + 2C_2}$$

where

C_1 represents the initial concentration of dielectric that cannot pass through the membrane, namely, H_2SO_4 concentration;

C_2 represents the initial concentration of dielectric that can pass through the membrane, namely, Na_2CO_3 concentration; and

X represents the dielectric concentration that passes the membrane from the side of the dielectric solution which cannot diffuse at the moment of equilibrium is established, namely, the concentration of Na_2CO_3.

Typically, C_1, C_2, and X are 2–3%, 0.02–0.03%, and 2.62×10^{-5}, respectively, of which $C2 \ll C1$. Although it is assumed that H_2SO_4 cannot be diffused out of the NC, it is still possible to demonstrate that the amount of the lye diffused into the interior of the NC is insufficient to neutralize the acid. In fact, the NC, after the acidic and alkali cooking, NC is still unstable (especially high nitrogen NC), except the NC has a nitrogen content below 12%. Under the normal production conditions, the total sulfate and the free sulfate in 1#, 2#, and 3#NC are 0.06–0.3% and 0.0–0.2%, respectively. To remove the acid completely, the fine-breaking technology used in papermaking industry is adapted to chop, flatten, fibrillation (that is divided into many filaments like a broom) to increase the surface area and capillary effect, thus enhancing the acid contact area and diffusion capacity. Subsequently, the treated NC is washed in the medium with alkalinity of 0.01–0.03% (i.e., fine washing). After abovementioned treatment, a stable NC is obtained. After the fine-breaking treatment, the total sulfate is 0.04–0.08%, and the free sulfate is 0–0.05%. The total sulfate and the free sulfate are reduced to 0.03–0.07% and 0%, respectively, after fine washing.

The free acid and the nitrated sugar are removed during the boiling process of the high nitrogen content NC, in which acid is mainly removed. As the nitrogen content is high, NC is extremely difficult to swell in water; it is difficult to remove the residual acid in the fiber capillaries, thus requiring extended boiling time in acidic or basic media. For the low nitrogen content NC, the stability treatment is mainly to remove the free acid and sulfate, in which the removal of sulfate is the main process. The amount of free acid is less and easy to remove, so the boiling time is short. If the viscosity of the nitrated product is required to be reduced, the boiling time needs to increase. With increasing boiling temperature and pressure, the corresponding boiling time can be reduced. On the other hand, a decrease in viscosity and increase in denitration of NC will also be intensified, thus resulting in declined product yield. During the acid boiling process, the acidity of the solution increases with the release of residual acid. In contrast, the alkalinity of a boiling solution is decreasing in the case of alkali boiling. The alkalinity change during alkaline boiling is shown in Table 10.34.

At present, the stability treatment of NC includes boiling, fine-breaking, and fine washing, of which boiling determines the quality of NC stability treatment and the length of product preservation time.

Table 10.34 The alkalinity change during alkaline boiling

No.	The alkalinity during alkaline boiling (based on Na_2CO_3)					
	Initial	2 h	4 h	6 h	8 h	9 h
1	0.078	0.027	0.005	Weak acidic	Weak acidic	Acidic
2	0.077	0.034	0.007	Weak alkaline	Weak acidic	Weak acidic
3	0.086	0.040	0.008	Weak alkaline	Weak acidic	Weak acidic
4	0.096	0.034	0.006	Weak alkaline	Weak acidic	Weak acidic

Boiling production has two types of processes, namely, in-batch normal pressure boiling and pressure boiling. Pressure boiling is divided into pressurized tubing and in-batch high-pressure boiling. The in-batch semi-high pressure adopts the advantage of pressure boiling, thus overcoming the drawbacks of normal pressure boiling.

(3) Denitration and degradation of NC during settling

In the process of the stability treatment, nitrocellulose inevitably occurs denitration and degradation reaction because of the long time contact with dilute acid and weak alkali at the high temperature and the impact of mechanical force. In particular, the chemical process of the NC boiling in the stability treatment involves the hydrolysis of the nitrates, the cleavage of the long chain or oxygen bridges, and the thermal decomposition of the nitrates. Their synergistic effect causes the decline in the nitrogen content and viscosity of NC. The reduction of nitrogen content in the boiling process is mainly determined by the saponification and thermal decomposition, while the decrease of viscosity mainly indicates that the macromolecule is partially depolymerized. The changes in nitrogen content and viscosity during the high-pressure boil process (132 ± 2 °C) of NC with different nitrogen contents in water medium are shown in Fig. 10.30. Figure 10.31 shows the conversion of the change in the viscosity of the high nitrogen amount NC into the degree of polymerization. Figure 10.32 shows the change in viscosity of low viscosity NC at 132 °C.

As shown in Fig. 10.32, the viscosity change in the initial 1 h of oiling is very significant, which is mainly caused by degradation. In the boiling process, the macromolecule degradation of the cellulose is the major reaction while the denitration caused by saponification and thermal decomposition is the minor reaction. With the decreasing degree of polymerization, the saponification reaction becomes the major reaction. When the degree of polymerization reaches a severe reduction stage, the denitration reaction caused by the thermal decomposition becomes the main reaction. In other words, NC with low polymerization degree and short-chain

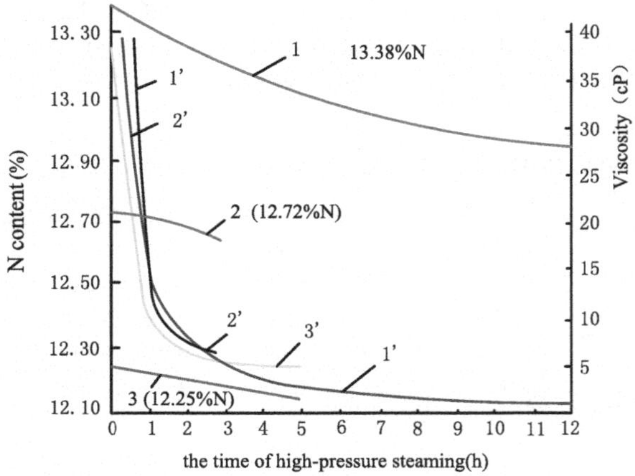

Fig. 10.30 The change in viscosity and nitrogen content of NC at 132 ± 2 °C in the boiling process

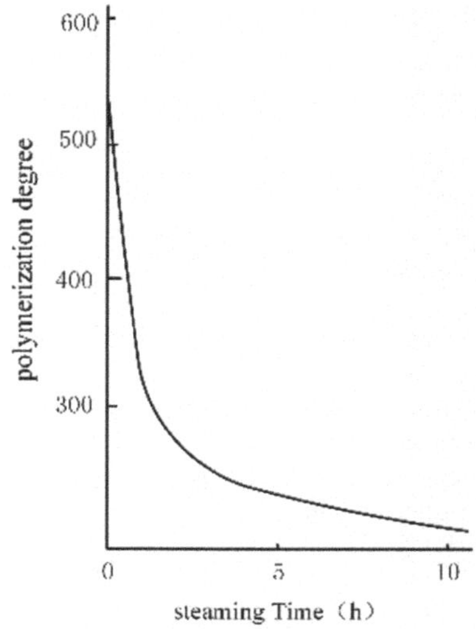

Fig. 10.31 The change in viscosity of NC with nitrogen content of 13.5% at 132 °C in the boiling process

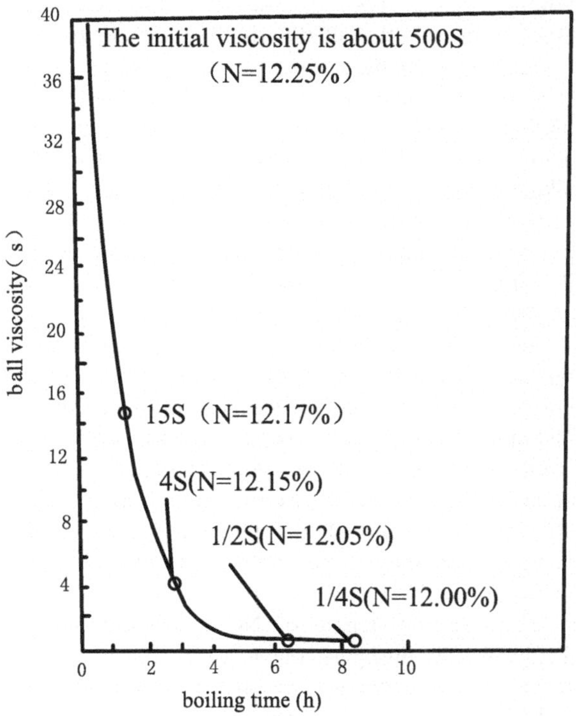

Fig. 10.32 Viscosity change of low viscosity NC at 132 °C boiling process

Table 10.35 Changes in NC after alkali boiling

Treatment method for NC	Nitrogen content (%)	Viscosity in ether–alcohol solution (S)	Solubility in 95% ethanol (%)	The copper content after denitration
Without alkali treatment	12.22	88	5.5	
Slight alkali treatment	12.18	64	5.4	
Secondary treatment with slight alkali	12.16	29	9.2	
Treatment with strong alkali	12.08	11.4	13.2	
Secondary treatment with strong alkali	11.91	8	20.0	

NC with more hydroxyl groups tend to occur thermal decomposition. In the alkali boiling process, both viscosity and nitrogen content of NC are reduced. The changes in NC after alkaline boiling are shown in Table 10.35.

The change in the viscosity of the NC caused by the late stage boiling in the acid is less than that in the late stage of alkaline boiling. Under the normal pressure boiling of 3#NC, the viscosity decreases by 0.05–0.02 Enshi degrees per hour in acidic boiling while the viscosity decreases by 0.08–0.1 Enshi degrees per hour in alkaline boiling. Similarly, the effect of acidic boiling on the viscosity of 1#NC is insignificant, whereas alkaline boiling reduces the viscosity by 0.2–0.8 En degrees per hour. The increase in the boiling temperature facilitates the process of removing the impurity, but the side reactions such as degradation are also increased accordingly.

Increasing the boiling temperature significantly accelerates the degradation reaction rate. Increased acidity or alkalinity, especially increased alkalinity, accelerates the degradation reaction. The nitrogen content loss of high nitrogen content NC in the boiling process is faster than that of low nitrogen content NC, especially at the high-temperature treatment. The decrease in nitrogen content has a significant effect on the decrease in viscosity at the end of the wash. Both acidity and alkalinity affect the reduction of nitrogen, but the effect of alkalinity is significant. With increasing boiling time, both molecular degradation and nitrogen reduction increase.

The nitration degree of NC is usually reduced by 1–3 ml/g after boiling. Typically, the nitrogen content of 1#NC decreases while the nitrogen content of 2#NC slightly increases, because the increase in nitrogen content caused by the removal of low molecular nitrate is greater than the increase in the nitrogen content caused by the denitration reaction. After the normal pressure boiling process, the nitrogen content of 3#NC nitrogen also decreases. The change in NC viscosity in each stage of the treatment is shown in Table 10.36.

Table 10.36 The change in NC viscosity in each stage of the treatment

Grade and boiling method of NC	Viscosity of refined cellulose (centipoise)	Viscosity of NC (Engler degree)				
		After washing	After acidic boiling	After alkaline boiling	After fine-breaking	After fine washing
1#NC (normal pressure)	31–45	234	28	11	10	9
2#NC (normal pressure)	31–45	44	28	10	9	9
3#NC (pressurized tubing)	10–20	6–7	2.5–3.0	1.9–2.6	–	1.9–2.6

Table 10.37 The change of NC in the process of fine-breaking

Fine-breaking time (h)	Nitrogen content of NC (%)	The polymerization degree of NC
0	12.95	1540
4	12.80	445
8	12.46	280
12	11.75	165

The changes in nitrogen content and polymerization degree in the process of fine-breaking are shown in Table 10.37.

Normal pressure fine washing process causes the following NC viscosity changes. The viscosities of 1#NC, 2#NC, and 3#NC reduces 0.1–0.3, 0.2–0.4, and 0.03–0.04 Enshi degrees per hour, respectively. Increasing temperature, extended time, and increased alkalinity of fine washing result in decreased viscosity and increased denitration.

Stability treatment of NC needs to control the stability of finished NC and the viscosity. For example, in the production of low viscosity and low nitrogen content NC, since the raw material refined cellulose is not easy to meet the requirements of low viscosity, it is typically achieved by extending the time of stability treatment or by intensifying the process conditions to reduce the viscosity. Semi-high-pressure boiling allows the refined cellulose to have a higher viscosity. Although the viscosity of the product is acceptable, the loss of NC will increase during the boiling process, thus increasing the consumption of raw materials. Normal pressure boil production consumes cellulose 700–715 kg per ton of NC while semi-high-pressure boiling process requires cellulose 730 kg per ton of NC.

After nitration, the nitrogen content and viscosity of NC fluctuate. Therefore, their nitrogen content and viscosity need to be adjusted to meet the quality indicators and achieve uniform product quality in the stability treatment process. Since this method is to some extent passive, only the strict control of each process and operation is a better option.

Finally, it is worth noting that nitrogen content changes and macromolecule degradation of NC in the process of stability treatment affect other related properties of NC because of the removal of impurities.

10.4.2 Boiling

Boiling is an important process for the stability treatment of NC stabilization. The impurity in NC is removed via decomposition or destroyed in the acidic or alkaline boiling process, thereby improving the stability of the NC. If boiling is not complete, especially for high nitrogen content NC, the improvement of NC stability is still not significant even the final fine washing treatment is intensified. For example, if the high nitrogen content NC is treated with boiling conditions for low nitrogen content NC, the finished product is very unstable, and the stability is in the range of 3.90–4.97 ml/g in general. Even if the final washing for up to seven times and a total of 30 h, the stability is still above 3.30 ml/g.

During the boiling process, the viscosity of the NC is significantly reduced because of the decomposition of macromolecules. After boiling treatment, the viscosity of NC is too high. After extended fine washing time, the degree of viscosity reduction varies. Correct selection and strict control of the boiling process conditions are critical to determining the stability and viscosity of NC.

(1) Boiling process and equipment

Boiling process includes normal pressure and pressure boiling, of which pressure boiling can be classified into in-batch pressure and continuous pressure boiling.

1. Normal pressure boiling

Normal pressure boiling is in-batch production. The drawbacks of this method include long production cycle and small production capacity. For example, the production cycle of 1#NC is up to 40–50 h. The loading amount per barrel is 1000–1500 kg NC (a small scale; 100–120 kg/m^3). The boiling barrel is made of stainless steel with a diameter of 3–4 m and height of approximately 2.15–2.75 m. The structure of the barrel is shown in Fig. 10.33.

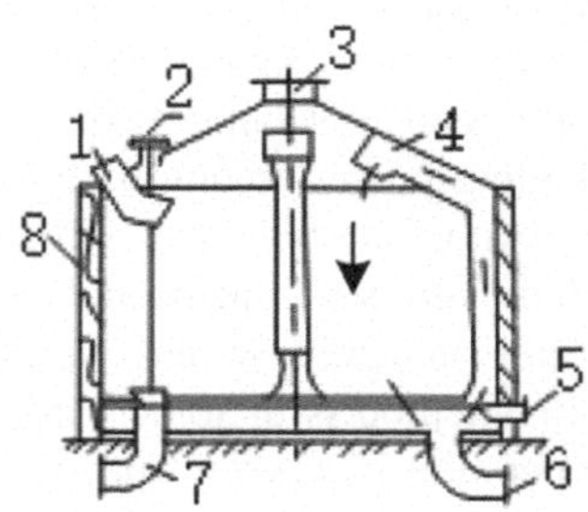

Fig. 10.33 The structure of the boiling barrel

The boiling barrel consists of inlet (1), switch (2), exhaust port (3), material circulation pipe (4), steam heating pipe (5), wastewater discharge port (6), outlet (7), and insulation baffle (8). The exterior of the barrel is equipped with 60–100 mm thick insulation layer. The bottom of the boiling barrel is a false bottom, where there are many staggering circular holes with a diameter of 5–6 mm for the filtration of wastewater in NC. To facilitate the circulation of boiling water in the boiling process, in the inside of the barrel four vertical tubes along the vertical interior wall of the boiling barrel are installed. Steam flows into four circulation tubes through the space between the bottom and false bottom from the exterior of the barrel. By the injection of steam generated through the circulation on the thin neck of the tube the water along the circulating pipe is poured on the NC when water is heated, thus forming the circulation of boiling water. Under the bottom of the false bottom, wastewater pipe and sedimentation tank are connected. Also, there is a discharging pipe on the false bottom. To ensure a successful discharge and watertight of boiling process, discharge pipe and wastewater pipes are equipped with a safety bonnet. The lid of the boiling barrel is in a shape of conical, in which the spire is connected with the ventilation pipe. The exhaust escapes from the tube to enter into the steam condensing tower for recycling exhaust. When the boiling barrel is discharging the product, NC is spread out by using the high-pressure water and pumped to the fine-breaking process.

Boiling process conditions vary with different grades of NC. For the same grade of NC, the variation in equipment and refined cellulose viscosity in different plants lead to different process conditions. The ratio of water to NC in the in-batch normal pressure boiling is approximately 13 to 9: 1. The typical process conditions are shown in Table 10.38.

2. Pressure boiling

In the process of boiling, the boiling temperature is increased by improving the boiling pressure, so that the swelling of NC is accelerated and the catalysis of acid and alkali ions is improved. In the meantime, the diffusion rate and reaction speed are enhanced, and the decomposition and destruction of unstable impurities are accelerated, thus accelerating and reducing the boiling and treatment time.

Table 10.38 The typical process conditions of in-batch normal pressure boiling

		1#NC	2#NC	3#NC
Cold water rinsing (h)		Depends on the acidity of NC		
Acidic boiling	Acidity (%) (based on H_2SO_4)	0.2–0.5	0.2–0.5	0.2–0.5
	Time (h)	15–20	5–10	10–20
Cold water rinsing (h)		0.5	0.5	0.5
Alkaline boiling	Alkalinity (%) (based on Na_2CO_3)	0.02–0.06	0.02–0.06	0.003–0.025
	Time (h)	20–25	10–15	10–25

In-batch pressure boiling equipment is a semi-pressure digester or a reinforced normal pressure digester. The boiling pressure is below 0.1 MPa (gauge pressure). Compared with normal pressure boiling, the cycle time of boiling can be reduced by half. Because of in-batch production, the main process conditions (temperature and time) are easy to adjust. When the viscosity fluctuates significantly, the refined cellulose can also be adapted. However, the production capacity and quality uniformity are not as good as the continuous boiling.

The structure and shape of refitted digester are similar to the normal pressure digester, except for the necessary sealing and reinforcement. Due to the need for pressure process, the installed structure includes explosion-proof port, safety valve, sampling tube, level gauge, and so on. At the same time, to facilitate the control of temperature, pressure, automatic instrumentation is installed for remote control. The new semi-pressure digester is a cylindrical tank with sphere crown shape in the upper and bottom to increase the pressure-resistant capacity. There is a layer of the false bottom on the top of the bottom. The false bottom has a series of holes with a diameter of 2–2.5 mm. The interior wall is equipped with four circulation pipes. A discharging pipe is connected to the top of the false bottom while a wastewater pipe is connected to the bottom of the false bottom. Both the discharge pipe and drainage pipe are equipped with a safety valve. Feeding tubes for raw material and alkali are installed on the top of the tank.

The pressure at the time of boiling is determined by the requirements of NC viscosity and the pressure resistant of the equipment. The general gauge pressure is in the range of 0.02–0.07 MPa, which is equal to 101.2–118 °C. With increasing temperature, NC viscosity diminishes dramatically. To ensure alkaline alkalinity is not too low during the digestion process, additional alkali should be added in the middle of the digestion process by using steam pressure. The steam pressure for the addition of alkali is higher than the digestion pressure by 0.05–0.10 MPa. The process conditions for in-batch pressure boiling are shown in Table 10.39.

Table 10.39 The process conditions for in-batch pressure boiling

		1#NC	2#NC	3#NC
Cold water rinsing (h)		Depends on the acidity of NC		
Acidic boiling	Acidity (%) (based on H_2SO_4)	0.2–0.5	0.2–0.5	0.2–0.5
	Temperature (°C)	Below 120	Below 120	Below 120
	Pressure (gauge pressure) (MPa) $\ngtr$	0.1	0.1	0.1
	Time (h)>	5	5	5
Cold water rinsing (h)		0.5	0.5	0.5
Alkaline boiling	Alkalinity (%) (based on Na_2CO_3)	0.02–0.06	0.02–0.06	0.02–0.06
	Temperature (°C)<	120	120	120
	Pressure (gauge pressure) (MPa)	0.1	0.1	0.1
	Time (h)>	7	5	7
Cold water rinsing (h)		0.5	0.5	0.5

Continuous pressure boiling includes long pipe and segment pipe continuous boiling processes. The long pipe employs a high-pressure tuber, which is made up of a casing series. The interior of the tube is loaded with NC material, while the exterior of the tube is the heating steam. Due to its rapid flow (about 1–1.5 m/s) and short residence time (12 min or so), the pipeline is long and requires more auxiliary equipment.

The length of pipe for acidic boiling is 1190 m. The inner pipe has a diameter of 100 mm and a thickness of 4 mm, while the exterior pipe has a diameter of 150 mm and thick 4.5 mm. The heating section of pipe consists of 6–8 tubes, which heats the NC material to the required temperature of the boiling. The insulation section has 8–10 tubes and maintains the NC material at the specified temperature for a certain time. The cooling section consists of two tubes and ensures the temperature of NC material below 100 °C. The length of each tube in cooling section is 60 m, and the external jacket is made of stainless steel. The other sections of the casing are made of steel pipe.

The slurry flow rate, boiling temperature, and pressure of NC are measured and controlled automatically. The process conditions for the production of 1#NC, 2#NC, and 3#NC are shown in Table 10.40.

In general, since the viscosity requirement of 3#NC is low, the temperature of pipe boiling is controlled in the upper limit.

Steam direct heating is used for the short pipe boiling, which results in rapid heating up and less stable control on the temperature and pressure of the boiling process than the indirect heating, thus affecting the uniformity of NC quality. In addition, because of the slow flow rate of NC in the short pipe, it requires stirring to meet the uniformity and consistency of the product. This brings difficulty to the sealing of the equipment and compromises the pressure resistance. As a result, settlement frequently occurs in the elbow and other dead ends of the pipe.

Table 10.40 The process conditions for pipe boiling

No.	Process	Process conditions
1	Preheat	NC concentration of NC 2.5–4.0% (g/100 ml)
		Temperature of NC >90 °C
		Preheating time <2.5 h
2	Feeding	Temperature of NC >95 °C
		Raw material solution level 1–3 m
3	Pipe boiling	Heating section temperature 132–145 °C
		Insulation section temperature 130–143 °C
		Cooling section temperature <95 °C
		Flow rate of NC 0.9–1.5 m/s
4	Pressure	Steam pressure 0.7–1.0 MPa
		Feeding pump pressure 0.7–0.9 MPa
		Brake pump pressure 0.2–0.45 MPa
		NC pump pressure 1.0–1.2 MPa

In addition to the abovementioned boiling process, there are other methods to accelerate the stability of the process, such as acetone, ethanol and other organic solvent treatment and ammonia treatment.

3. Recovery of boiling waste steam

The amount of steam used for the boiling process is very significant. Since a large amount of waste steam is discharged in the boiling process, the recovery of waste steam is very necessary.

(2) Factors that affect the quality of the boiling

The changes in NC stability and viscosity during the boiling process are quite complicated, which are related to the property of the feedstock, the type of boiling, the method of operation, and the conditions of the cooking process. Correct and reasonable choice of process conditions is critical to ensure the quality of boiling, reduce the production cycle, and improve labor productivity.

1. The effect of the boiling temperature

The boiling temperature is the main factor affecting the quality of the boiling. In particular, for the pressure boiling at the temperature above 100 °C, the temperature has a significant impact on the NC stability and viscosity. When the temperature is in the range of 100–130 °C, the stability treatment time is reduced by about twice if the temperature increase by 10 °C. Simultaneously, the corresponding degradation rate also increases. For the production of low viscosity NC, it is optimal to use high-temperature boiling. Since the temperature of normal pressure boiling is not high, the effect of temperature change on the stability and viscosity of NC is less significant than pressure boiling.

2. The effect of acidity and alkalinity

① Acidity

The control of acidity is very important during the acidic boiling, which is highly related to the removal of unstable impurity. The follows are the factors that affect the acidity. The fully swollen of NC by acid facilitates the diffusion of acid from the NC to the medium; acid promotes the decomposition and destruction of sulfuric acid; side reactions should controlled, especially denitration reaction cannot be too much. Experience suggests that the initial acidity range of 0.2–0.5% is more appropriate. Acidity during acid boiling should not fluctuate significantly. Since high nitrogen content NC releases more acid, the initial acidity should be controlled in the lower limit to avoid a high acidity in the late stage.

② Alkalinity

Since NC is very sensitive to alkali by causing denitration and degradation reaction, the alkalinity in the boiling process cannot be too high and boiling time should also be reduced. The general alkalinity should be less than 0.1%, and the initial alkalinity is mostly used 0.03–0.05%. Additional alkali should be added during the alkaline boiling process. NC after alkali boiling becomes brittle, which is conducive to the following fine-breaking process. NC is difficult for the fine-breaking process without alkali boiling treatment. Plastic and paint NC require boiling in water rather than in alkali. In addition to the use of bleached and refined cellulose, the prepared NC also needs bleaching treatment to ensure the final products have higher transparency and whiteness.

③ Effect of boiling time

The boil time is determined by the removal of impurities and viscosity requirement. Because the removal of impurities is more difficult in 1#NC, its acidic and alkali boiling time are long. In contrast, the removal of impurities in low nitrogen content 3#NC is easy, so its acidic boiling time is short. The long alkali boiling time is only to reduce the viscosity of the product.

④ The effect of nitration conditions

The impurity in the NC is mainly generated during the nitration reaction. The types and quantities of impurity generated are closely related to the quality of the refined cellulose used and the conditions of the nitration process. Therefore, the nitration process conditions have a direct impact on the stability of treatment.

(3) The effect of refined cellulose quality

Poorly refined cellulose exhibits poor quality, such as the low content of α-cellulose, high content of impurity, and low degree of polymerization, which increases the quantity of unstable impurities in the nitration reaction. Therefore, refined cellulose that does not meet the quality standards cannot be used for nitration reaction. In the NC production, in addition to the quality requirements of the refined cellulose to meet the criteria, its viscosity must also be strictly selected. The grades of refined cellulose for the production of different grades of NC are different, which are shown in Table 10.41.

The viscosity of the refined cellulose after nitration should be equal to the sum of the viscosity of the finished product of NC and the decrease in viscosity during the stability treatment. The viscosity of the purified cellulose after nitration depends on

Table 10.41 The refined cellulose viscosity range for production of different grades of NC

No.	NC grade	Viscosity requirement of NC (Engler degree)	Selected viscosity of NC (centipoise)
1	Antifreeze colloidal NC	>160	>1500
2	Regular colloidal NC	>50	>1000
3	Mixed NC	3–20	20–60
4	Plastic NC	2.5–4.5	20–30
5	3#NC	1.9–2.5	10–20
6	Paint NC	Falling ball viscosity 1/4–40 s	<20

the viscosity of the refined cellulose, and the nitrogen content of the NC, and the side reactions in the nitration process. The decrease in viscosity during the stability treatment depends on the mode of production and the process conditions used. Only the consideration of all these factors can obtain the required viscosity value.

(4) The effect of nitration conditions

The acid composition is the primary condition for determining the nitrogen content of NC and the generation of a specific unstable impurity. Since the mixed acid for 1#NC contains less water and high content of nitric acid, the free acid and low-degree nitrates are mainly unstable impurities. On the other hand, the mixed acid for 2#NC and 3#NC contains a high content of water and less amount of nitric acid compared with a mixed acid of 1#NC. As a result, the sulfate and nitro-sulfur mixed ester are the major unstable impurities.

High nitration temperature and extended reaction time lead to increased side reaction products. To ensure the uniform quality of the product, the nitration time cannot be too short.

(5) Other effects

When boiling with the barrel, the amount of water, the circulation of barrel interior, and boiling water quality affect the stability of the NC. ① The quality of NC in a different location of the barrel are variable, such as the upper and lower, center and edge for many reasons, including heterogeneous pH and circulation. Therefore, the loading volume of NC cannot be too much, and the steam pressure should be stable; ② hard water mainly contains calcium carbonate, magnesium carbonate, calcium bicarbonate, magnesium bicarbonate, and other impurities. Since all these impurities are weakly alkaline, they can accelerate the decomposition of impurity and neutralization of residual acid in the boiling process, which is conducive to stability treatment. However, the calcium carbonate adheres to the NC fiber as a solid powder, thus increasing the alkalinity and ash content of the NC. If the alkalinity of NC is too low, the alkalinity can be improved by adding calcium carbonate powder in the fine-breaking process; because of adsorption capability of NC, if the boiling water contains high organic matter and suspended solids, it will affect the transparency and whiteness of NC products. If the iron content in water is too high, it will affect the stability of NC. If the water quality does not meet the requirements, it should be purified; ③ stability analysis is also affected by climatic conditions. High temperature and humidity in summer cause the fluctuation of the water content of dried NC and the temperature of the cabinet, thus leading to slightly higher stability results. These factors must be taken into account in the actual production.

10.5 Fine-Breaking

The technological progress of the papermaking industry has brought vitality to the NC manufacturing. After the introduction of fine-breaking technology to the NC manufacturing, the stability issue of NC has been solved, which not only achieves a new milestone of NC production history but also realizes the wide application of NC [2, 7].

10.5.1 The Basic Principle of Fine-Breaking

Fine-breaking is the cutting and grinding of the long fiber of NC after boiling by the physical and mechanical action of the chopper to increase the surface area of NC, which facilitates the diffusion of residual acid and other unstable materials in the capillary, thus accelerating NC stability treatment.

The flow of the NC slurry between the rolling cutter and the bed knife is shown in Fig. 10.34 in the process of NC fine-breaking. Affected by centrifugal force, friction resistance, and other forces, NC slurry is strongly impacted and mixed.

The flow of the NC slurry is divided into two parts by the rolling cutter. One part flows upward along the front wall of the rolling cutter, whereas another part flows forward through the bottom of rolling cutter. The NC slurry flowing into the blade flute forms eddy motion within the groove under the effect of centrifugal force and friction, which is illustrated by the formation of a groove on the front edge of the knife after a long time use. The NC film passing from the bottom of the knife forms an NC film on the edge of the blade. The thickness of this layer of NC film and the thickness of the blade will directly affect the quality of the fine-breaking NC.

When the thickness of NC film and blade is thin, the NC slurry is mainly sheared between the rolling cutter and the bed knife, so that the fibers are cut in the transverse direction but still remain the original fibrous structure. When the thickness of NC film and blade is thick, NC slurry between rolling cutter and the bed knife is

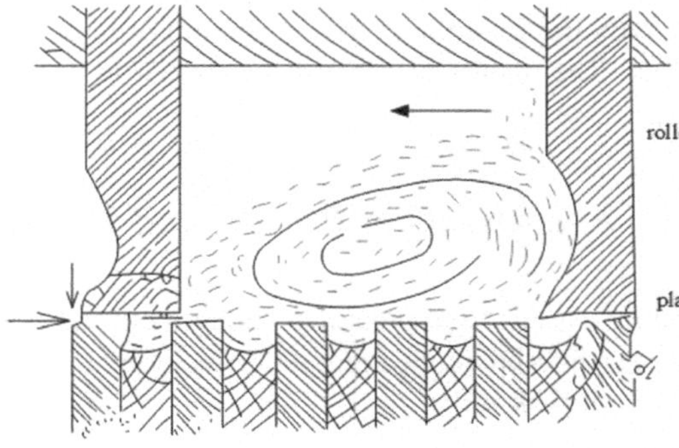

Fig. 10.34 The flow of the NC slurry in the chopper

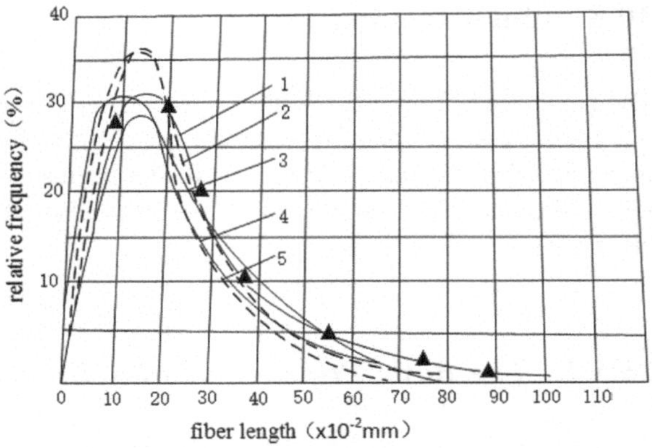

Fig. 10.35 The fiber length distribution after the fine-breaking: 1—cone-shaped chopper (2#NC, fineness 78 ml), 2—disk mill (fineness: 68 ml), 3—cone-shaped chopper (3#NC, fineness 67 ml), 4—disk mill (fineness: 58 ml), and 5—Hao chopper (fineness: 52 ml)

impacted mainly by crushing, friction and extrusion, which grinds the fiber in the longitudinal direction, thus significantly changing the structure of the fiber.

There are many factors that affect the thickness of the NC film. The high concentration of NC pulp, long fiber, high flow rate, significant intertwined fiber on the knife-edge result in a thick NC film, and vice versa. These factors in the process of fine-breaking vary frequently, and the interaction on the fiber is also complex, both the role of shear and grinding effect. The stable process condition is essential to ensure uniform quality fine-breaking of NC.

After fine-breaking, the NC is loose and easy to filter. In contrast, the grinding of NC is not easy to precipitate or filter. The currently used fine-breaking equipment in production possesses both chopping and grinding effect. According to the preliminary study on the finished product NC, it is believed that the current fine-breaking equipment has a more significant chopping effect than the grinding effect. The fiber length distribution after the chopping by the common fine-breaking equipment is shown in Fig. 10.35.

From Fig. 10.35, it is obvious that the length distribution of the fiber after the chopping with the Hao fine-breaking machine is relatively uniform and narrowly dispersed, and the 90% or more of fiber is in the range of 5–35 × mm. The chopping result by dick mill is next to the result of Hao fine-breaking machine. Increase the chopping time by disk mill form 9 times (curve 2) to 12 times (curve 4) significantly improve the even distribution of fiber length and reduce the content of long fiber. The results of the test are shown in Table 10.42.

10.5.2 Fine-Breaking Process and Equipment

Commonly used fine-breaking equipment is Hao fine-breaking machine. Continuous fine-breaking equipment includes a cone-shaped chopper, cylindrical chopper, and disk mill.

Table 10.42 Fiber length distribution after chopping by disk mill

Chopping time	1	2	4	6	8	10	Hao fine-breaking machine
Fineness (ml)	176	155	115	92	72	67	54
Average length (mm)	0.64	0.68	0.47	0.36	0.33	0.29	0.27
Length distribution (mm)	Content of length distribution (%)						
0.1–0.2	8	4	11	33.3	28	48	38
0.2–0.3	11	13	24	25.3	35	2	32
0.3–0.4	13	14	23	16.2	14	14	17
0.4–0.5	14	19	10	9.1	7	7	8
0.5–0.6	11	7	10	6.1	7	2	2
0.6–0.7	14	9	5	3	3	4	3
0.7–0.8	7	7	4	3	2	0	0
0.8–0.9	2	5	3	1	2	1	
0.9–1.0	9	5	3	2	2	1	
1.0–1.1	1	4	2	0	0	0	
1.1–1.2	1	0	5				
1.2–1.3	1	4	0				
1.3–1.4	1	1					
1.4–1.5	1	1					
1.5–1.6	3	1					
1.6–1.7	0	0					
1.7–1.8	0	1					
1.8–1.9	0	1					
1.9–2.0	1	2					
2.0–2.1	1	0					
2.1–2.2	0	1					
2.2–2.3	1	0					
2.3–2.4	0	0					
2.4–2.5	0	1					

(1) Hao fine-breaking machine

The structure of the Hao fine-breaking machine is shown in Fig. 10.36.

Hao fine-breaking machine is characterized by easy control of the process conditions, which is suitable for a variety of different requirements of the fine pulp. However, it has some drawbacks, such as low production capacity (volume of 400–500 kg), difficulty in control the quality of each batch, and high labor intensity.

In the process of fine-breaking, turn the hand wheel of the butted-knife installation (7) to adjust the height of the beater roll (1), and the knife distance is based on the loaded current of the motor. At the beginning of a feeding, the knife distance is

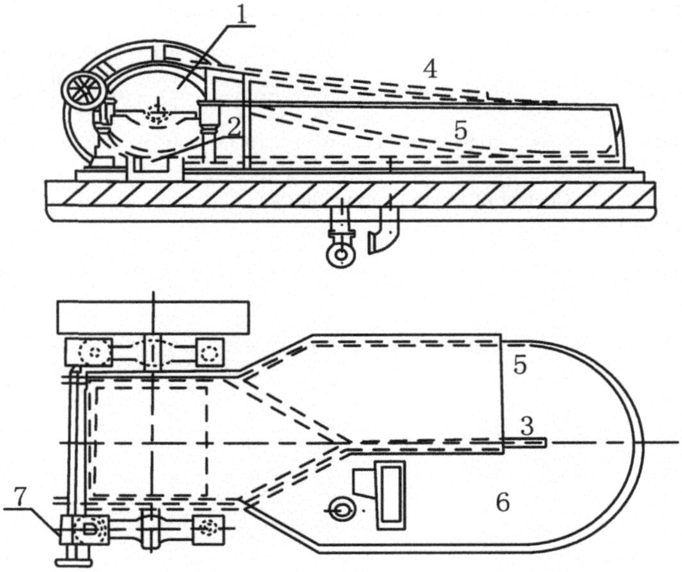

Fig. 10.36 The structure of the Hao fine-breaking machine [2]: 1—beater roll, 2—blade bottom, 3—baffle, 4—top cover, 5, 6—circulation groove, and 7—butted-knife installation

relatively wide. With the progress of fine-breaking, the knife distance gradually reduces while the current value gradually increases. The chopping should not be too fast or intense. After chopping of NC, the current value of chopper is gradually reduced, but a dramatic drop in the current value should be avoided. Normally, the NC pulp concentration is about 9–12% in the fine-breaking process with a breaking time of about 4 h temperature of 60–63 °C. During the process of fine-breaking, the addition of a certain amount of Na_2CO_3 is required to neutralize the acid.

(2) Continuous fine-breaking machine

1. Cone-shaped chopper

Cone-shaped chopper is composed of some units, which has a series of advantages, such as large production capacity, relatively stable quality, high efficiency, small site area, and low labor intensity. On the other hand, it has some shortcoming, including complex equipment structure, expensive maintenance, high motor power, and so on. The structure of the cone-shaped chopper is shown in Fig. 10.37.

The cone-shaped chopper consists of a rotor and a stator. The rotor is a cone shape, in which the top is installed with blades along the axial. There are several groups of blades with six blades in each group with a knife thickness of 3–5 mm. Each blade has a certain gap between blades while the gap between groups is wider. The stator is a sleeve located outside the rotor, fixed to the case. There are equipped with two groups of the A-shaped knife on the stator along the axial. The taper of the

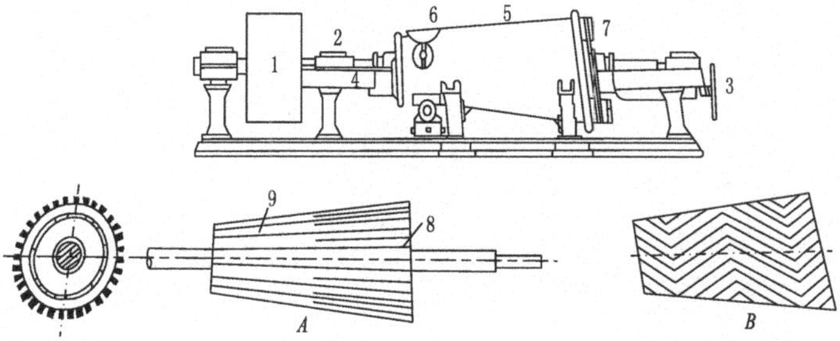

Fig. 10.37 The structure of cone-shaped chopper [2]: 1, 2—the traditional system, 3—hand wheel regulator, 4—track, 5—shell of chopper, 6—nitrocellulose slurry inlet, 7—NC slurry outlet, 8—blade, and 9—blade holder

vertebral body is 16° 32′. The rotor is supported by two rolling bearings, which can be used to move the rotor axially through the infeed device, and the rotor speed is 365 rpm. The small end of the vertebral body has a feed tube with a paddle propeller at the feed. The current level is used to measure the gap between the blades indirectly.

2. Cylindrical chopper

Cylindrical chopper can be classified into model 50 and model 100. The drive power of the model is in the range of 50–55 kW. Compared with the cone-shaped chopper, it has the advantages of compact equipment, small site area, small motor power, long residence time, high stability, and so on. However, the wear of blade is fast, and the change of blade is very complicated. In the case of eight sets of 50-model cylinder chopper, the knife is made of No. 20 chromium molybdenum steel and carburized, which can be only used for the production about 300 tons of NC. The structure of the cylindrical chopper is shown in Fig. 10.38.

The diameter of the rolling cutter is 280–400 mm, which is equipped with 38–50 pieces of steel knives with the blade thickness of 4–6 mm. The roll knife has a certain angle with the axis (such as 10°–8°), which usually operates at high speed, 900–1000 r/min. There are four sets of stator knives located evenly around the roll

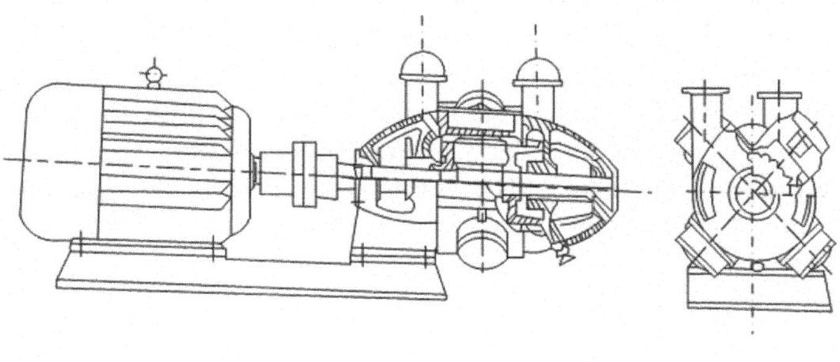

Fig. 10.38 The structure of the cylindrical chopper

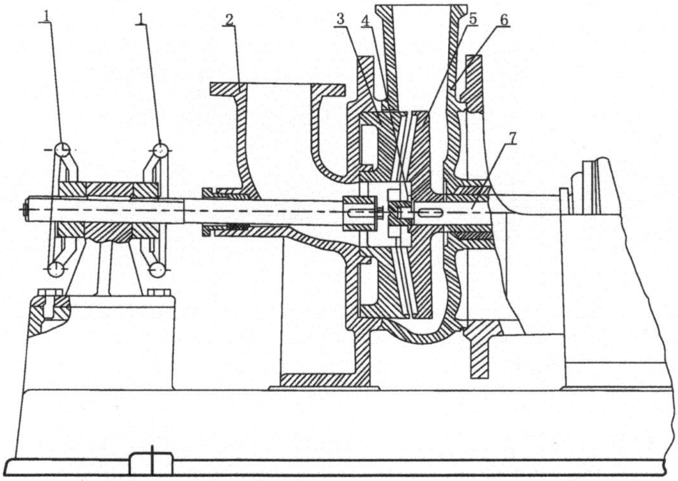

Fig. 10.39 The structure of disk mill: 1—adjustment wheel, 2—feed port, 3—fixed disk, 4—transmission turbine, 5—rotating grinding, 6—discharge port, and 7—drive shaft

knife, in which the blade is parallel to the axis. Compressed air or high-pressure water is employed for adjusting the knife. The feeding inlet of the chopper is equipped with a feeding impeller, by which the NC slurry is continuously fed evenly between the rolling cutter and the bed knife. The chopped NC pulp is discharged into the second chopper.

3. Disk grinding chopper

Disk mill is mainly composed of shell, spindle, fixed disk, and rotating disk. The structure is shown in Fig. 10.39.

The fixed disk can adjust the distance between the disks by adjusting the device. The diameter of the rotary disk is 350 mm with an inner diameter of 100 mm. There are 104 knives (teeth) between rotation mill and fixed mill with a width of about 4 mm, space of 4 mm, and height of 10 mm. There are 8 feed tracks on the ground, which divides the disk into 8 equal parts. The depth and width of the track are 6–8 mm and 15 mm, respectively. Teeth and track have a certain angle. The disk material is made of 45# steel. Connected directly to a motor with 40 kilowatts power, the rotation mill has a rotation speed of 1440 r/min. NC slurry flows from the feed pipe and passes through the center of the fixed disk into the two disks. Pushed by the feed impeller, NC slurry is transferred into the feeding track and then dispersed into the teeth. Finally, NC slurry is discharged from the outlet by the centrifugal force for the chopping in the next disk mill.

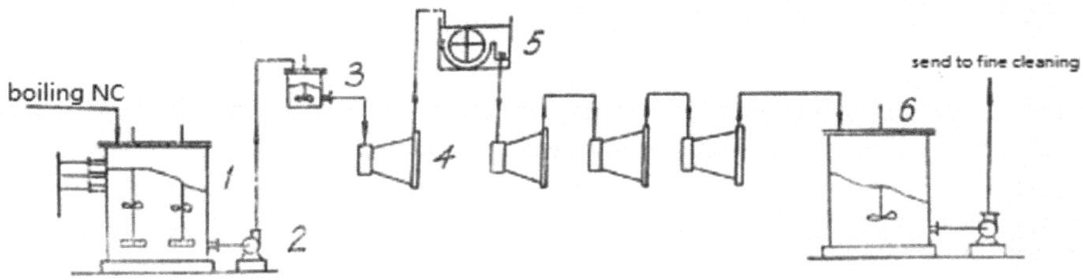

Fig. 10.40 Continuous chopping process and operation: 1—feed concentrator, 2—pulp pump, 3—high tank, 4—chopper, 5—semifinished product concentrator, and 6—tank

(3) Continuous chopping process and operation

Continuous chopping process consists of 4–5 sets of cone-shaped choppers while the cylinder chopper is composed of 6–8 units. The production flow of the cone-shaped chopper is shown in Fig. 10.40.

The NC material from the digester is fed into the feed concentrator 1 to adjust the concentration to 2–4% (g/100 ml). In the meantime, a certain amount of base is added to neutralize the free acid by adjusting the NC material to weak alkaline. Subsequently, the NC material is driven into the feed high tank 3 by the pump 2 so that the NC material flows evenly into the chopper 4. To enhance the loading amount of chopper and to ensure chopping efficiency, the first and second units of chopper can be connected in parallel. After the first time chopping, the NC slurry flows into the roller concentrator 5 to adjust the concentration to 6–10% to improve the efficiency of chopping, and then enters the subsequent chopper. After completion of the chopping, it enters the NC collection tank 6. With the rapid analysis method to determine the degree of fineness, the qualified NC is pumped to the fine washing process. Otherwise, the unqualified NC is sent to the feeder for rework. The rework does not go through the roller concentrator to reduce the loss of NC.

10.6 Refining of Nitrocellulose

Fine washing is the final process of stability treatment for NC. The chopped NC is washed in hot and cold alkaline medium to neutralize the residual acid and remove the unstable impurities. After the fine washing, NC is a single batch (small batch) of the finished product, which must meet all the required quality standards. If the product fails one of the following standards, including stability, alkalinity, viscosity, and ash, the product must be reworked. If the stability of NC is not qualified, it cannot be transferred to the next process.

10.6.1 Fine Washing Process and Equipment

Fine washing process commonly employs normal pressure fine washing at the temperature of 90–98 °C. Pressure fine washing is also utilized at the temperature above 100 °C and pressure of approximately 0.05 MPa (gauge pressure).

(1) Fine washer for NC

The washed for NC is made of steel cylindrical groove with a diameter of 3.5 m and height of 3 m, equipped with two paddle stirrer with opposite rotation direction and the rotation speed of 18–25 r/min. A small volume of fine washed is equipped with a flat paddle stirrer with an adjustable speed of 48–52 r/min. There is a hole in the upper part of the iron cover of the fine washer, in which the lid can be opened and closed. In addition, the cover of the washer is equipped with NC slurry feed tube, water pipe, and lye pipe. The bottom of the washer is equipped with an outlet with a lock valve. A thermometer hole is installed on the side wall of the washer. With direct steam heating, the steam enters the lower part of the tank through a silent nozzle, reducing vibration and noise. The side of the fine washer is equipped with wastewater switch, through which the wastewater flows into the NC recovery tank. The structure of the fine washer is shown in Fig. 10.41.

(2) Fine washing process conditions

Fine washing temperature, time, and the number of fine washing depend on the grade and quality of NC. Typical fine washing conditions are shown in Table 10.43. After the feeding for fine washing, the mixture precipitates for about 30 min and drains the wastewater. After adjusting the concentration, the stirring can be started. The charge cannot be overfilled. Slowly warming up to reach the specified temperature, then start the calculation of time. When stopping the fine washing, after turn off the steam pipe, the water pipe should be turned on to flush the connected steam pipe to prevent NC entering into the steam pipe, which could be a risk factor.

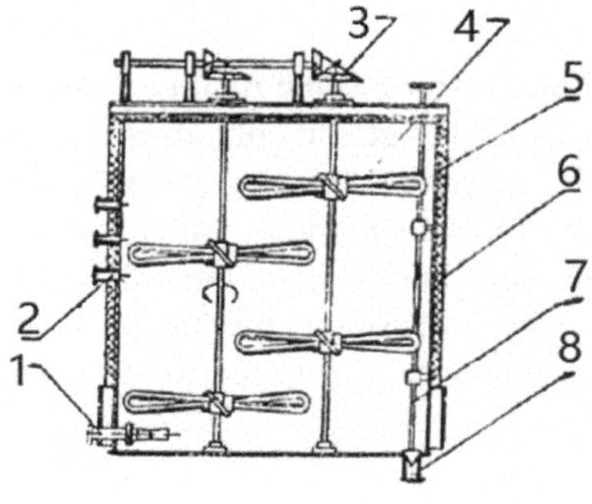

Fig. 10.41 The structure of the fine washer: 1—nozzle, 2—waste pipe, 3—drive shaft, 4—mixing, 5—tank, 6—insulation layer, 7—discharging, and 8—discharge tube

Table 10.43 Fine washing process conditions

Item	Grade		
	1#NC	2#NC	3#NC
Pulp concentration (%) (g/100 ml)	8–11	8–11	10–13
Washing time	1–2	1	1–2
Each washing time (h)	6–12	2–6	2–14
Washing temperature (°C)	>98	>98	60–98
Washing alkalinity (%, based on Na_2CO_3)	0.05	0.05	0.003–0.025

Table 10.44 Stability treatment conditions for recycling 3#NC

Item	Condition
Concentration (%)	15%
Temperature (°C)	60 ± 5
Fine washing time	No less than two times
Each fine washing time	No less than 2 h

For 2#NC and 3#NC, if the stability of NC is optimal and viscosity and other indicators meet the requirements, only room temperature washing is required rather than fine washing at elevated temperature.

If the ash content is slightly large, but no more than 1.0% and the stability is less than 3.0 ml/g, and the other parameters are in line with the requirements, the NC can be added as a single batch to the large batch but the added amount should be less than 30% of the total volume for recycling. If the NC cannot be directly added for the recovery, it should be handled separately or added to the normal batch for stability treatment. The stability treatment conditions for recycling 3#NC are shown in Table 10.44.

The complete NC analysis indicators include nitration degree, alcohol ether solubility, ethanol solubility, viscosity, fineness, stability, alkalinity (to $CaCO_3$), and ash content, which is also known as eight major indicators. In addition, 1#NC and 2#NC also need to analyze the NC pulp concentration. Qualified NC can be transferred into the mixing process. According to the requirements of the total batch mixing, 1#NC and 2#NC are transferred into the mixing machine by NC pumper to blend into a large batch of mixed NC; 3#NC can be directly mixed into a large batch.

Occasionally, if the stability alkalinity, viscosity, and nitration degree of the NC are not ideal, it needs reprocessing in the fine washing process. Generally, the decreases in viscosity per hour at 98 °C of 1#NC, 2#NC, and 3#NC are approximately 0.1–0.3 °C, 0.2–0.4 °C, and 0.03–0.04 °C, respectively. During the fine washing process, the nitration degree of 1#NC decreases considerably. According to the production experience, the nitrogen content fluctuates 0.2–0.3 ml/g after the fine washing for 3–8 h.

10.6.2 *Factors That Affect the Quality of the Fine Wash*

(1) Alkalinity

Maintain the fine washing in alkaline solution is necessary. To prevent side reactions are too intense, the alkalinity is maintained in the range of 0.01–0.03%. After the fine washing, NC should maintain a certain degree of alkalinity.

(2) Temperature

The temperature is usually in the range of 90–98 °C. With increasing temperature, the viscosity decreases.

(3) The time and number of fine washing are determined by the NC viscosity and stability. When the stability is not good, and the viscosity is high, it needs long fine washing time and more washing number, and vice versa.

(4) The loading amount for fine washing. The increase in loading amount improves the productivity. However, overloading causes poor quality uniformity. During the fine washing, the appropriate concentration of NC pulp is 10–13%.

(5) Water quality also affects the fine washing.

10.7 Blending of Refined Nitrocellulose

The blending of NC is a process of homogeneous mixing 1#NC and 2#NC with different ratios according to the requirements of the nitration degree and solubility in alcohol ether of mixed NC.

The blending of 3#NC is to mix a large batch of NC with uniform quality.

(1) Mixer

The mixer is a vertical cylindrical container, whose volume depends on the production capacity and the amount of the total batch. Typically, a total batch of NC is 20–30 tons (dry volume). The interior of the mixer is equipped with a mixing device. Previously, frame stirrer is frequently used, which is replaced by propeller-type stirrer nowadays. Propeller-type stirring includes vertical and horizontal stirring.

The capacity of the vertical mixer is small, usually about 20 tons. The tank is equipped with 4–5 propeller stirrer with a speed of 280 r/min. NC slurry is stirred to achieve uniform mixing. The structure of the vertical mixer is shown in Fig. 10.42.

The horizontal mixer has a large capacity of approximately 30 tons. There are propeller stirrers on the bottom with a stirring speed of 735 r/min. There four stirrers have an angle of 28° with the central axis. Four NC slurry draft tubes with seven holes are installed on the wall. Under the role of stirring, NC slurry flows upward from the center of the stirrer and then goes down from the guide tube to achieve the purpose of mixing. The structure of the horizontal mixer is shown in Fig. 10.43.

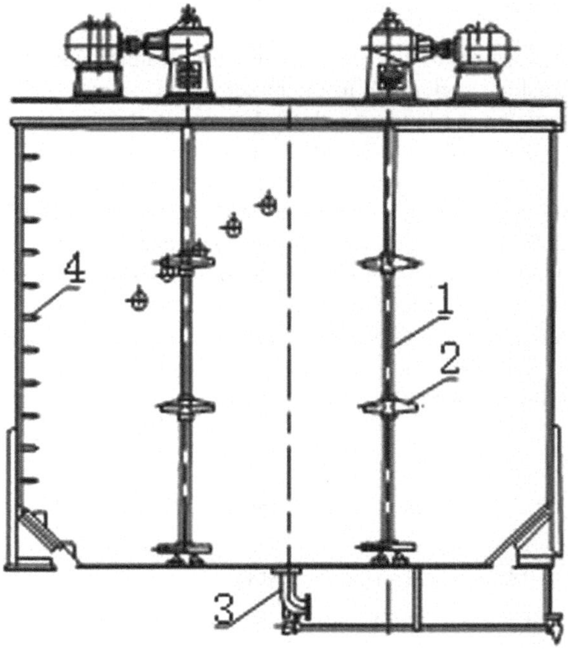

Fig. 10.42 The structure of vertical mixer: 1—stirring shaft, 2—stirring blade, 3—outlet, and 4—brake handle

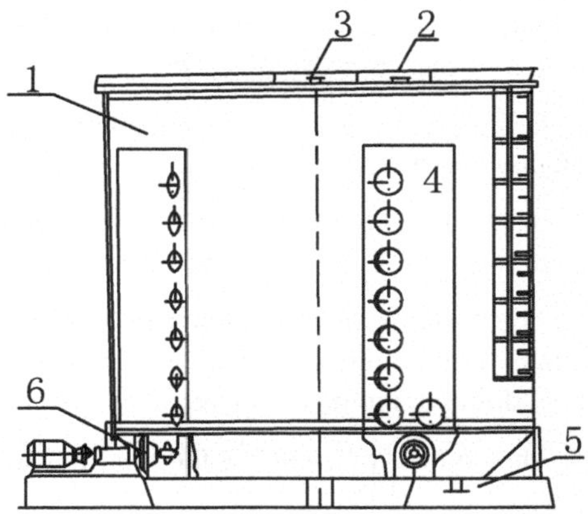

Fig. 10.43 The structure of horizontal mixer: 1—mixer body, 2—feed port, 3—vent, 4—draft-tube, 5—discharge, and 6—mixing drive system

Wastewater outlets are equipped on the side of the tank. Since 3#NC is a single product blending, sometimes fine washing and blending are performed in the same equipment equipped with a steam heating device.

In addition, there is a horizontal mixer with a semicircular bottom and stirring speed of 20–30 r/min to reduce friction. This type of equipment is not frequently used.

(2) Blending process

According to the total calculation results, qualified single batch NC after fine washing is pumped to the mixer by NC pump. If the concentration of NC slurry is low and the loading volume is too full, the feeding can be paused in the middle of feeding. After precipitate for more than an hour, the wastewater is drained and then the feeding is continued. Normally, the distance of solution level to the top of the mixer is 500 mm. After the feeding, the concentration is adjusted to the range of 10% to 12%, and then the slurry is evenly stirred for 2–4 h, followed by the full analysis of the sampling. If the nitration degree or solubility is not qualified, 1#NC or 2#NC is added to blend again until qualified. If the quality meets the requirement, the NC is processed for the removal of iron removal and dehydration.

The NC pulp concentration and the agitator speed are the main factors influencing the blending quality.

10.7.1 Group Batching of Blended NC

(1) Principle of group batching

1. According to the nitration degree and alcohol ether solubility requirements of blending NC, the nitration degree range of 1#NC and 2#NC is determined.

2. The small difference in the viscosity of each batch of NC for blending NC is preferred. To facilitate the subsequent gunpowder processing, the general requirements of the viscosity difference are less than 5 Engler degree.

3. Other quality indicators of a single batch of NC should be within the scope of quality standards.

4. The alkalinity of total NC batch should be controlled to a certain degree. Generally 0.1–0.2%, to ensure the stability of NC in storage.

5. The addition of the amount of recycled NC in the total batch depends on the quality requirements of total NC batch and the quality of recovered NC. In general, the adding amount of recovered NC is less than 25%.

6. The content of 2#NC in the total batch should meet the requirements.

(2) The quality criteria for blending NC

Different grades of single-base propellant have various nitration degrees of blending NC, alcohol ether solubility, and 2#NC content, as shown in Table 10.45. The actual quality control of the specifications in production is often stricter than the scope of the standard to ensure the optimal quality of products. The other quality indicators for blending NC are shown in Table 10.46.

Different grades of single-base propellant have different quality requirements for 1#NC and 2#NC, which not only can ensure the optimal NC processing

Table 10.45 Quality requirements of blending NC for several types of single-base propellants

Grade	Nitration degree (ml/g)	Alcohol ether solubility (%)	2#NC content (%)
14/7 propellant	206–207	36–42	33–37
5/7 propellant	207.5–208	25–34	20.5–23.5
7/14 propellant	208–209	28–40	>27
2/1, 4/7 propellant	>209.5	28–40	20.5–23.5

Table 10.46 Criteria after blending

Criteria	Criteria after blending
Alcohol ether solubility	<5%
Stability (132 °C)	≯3.0 ml/g
Fineness (sedimentation method)	≯85 ml
Alkalinity (based on $CaCO_3$)	≯0.25%
Ash content	≯0.5%
Viscosity (Engler degree)	4–15

performance but also can guarantee the ballistic performance of finished gunpowder to meet the requirements. These processing and ballistic performances are determined by the blending ratio of 1#NC and 2#NC. The ratios of 1#NC to 2#NC for preparation of several grades of propellants are shown in Table 10.47.

(3) Calculation for group batching of blended NC

The principle of the calculation is the material balance and the weighted average method.

The final check is required to ensure all the results meet the requirements. If the analysis results fail, recalculation is required.

Table 10.47 Quality indicators for 1#NC and 2#NC

Grade of propellant		Nitration degree (ml/g)	Alcohol ether solubility (%)	Alcohol solubility (%)	Viscosity (Engler degree)
14/7	1#NC	212.3–214.2	<15	<4	5–9
	2#NC	193.0–195.5	>98	<6	5–9
5/7	1#NC	210.5–212.2	<18	<4	5–9
	2#NC	192.5–195.0	>98	<6	5–9
7/14	1#NC	>213	<15	<6	5–9
	2#NC	194.0–197.0	>98	<6	5–9
2/1	1#NC	>213.5	<15	<6	5–9
	2#NC	194–197	>98	<6	5–9

10.8 Separation of Mechanical Impurities and Concentration of Pulp

10.8.1 Separation of Mechanical Impurities and Concentration of Pulp

(1) Separation of mechanical impurities

Throughout the NC production process, some mechanical impurities are inevitably mixed with NC, such as sand, iron, dust, and non-chopped long fiber. These impurities not only affect the quality of NC and finished powder but also affect the safety of gunpowder manufacture. Therefore, they must be removed.

1. Separation of mechanical impurities

The blended NC pulp is first passed through the cleaner. The cleaner is an aluminum or wooden sawtooth baffle placed in the wooden tank that cotton pulp flows. When the diluted pulp (concentration of 5–8%) flows through the baffle, the heavier debris settles off the toothed recesses in, the lower part of the baffle. Therefore, the slag removal process is a settling process. Factors affecting this precipitation process include pulp concentration, flow rate, and flow volume. To thoroughly remove the impurities, the wood groove cannot be too short. According to the production experience, the optimal conditions for the settling process are groove slope of one percent, minimum groove length of 30 m, the flow rate of 0.5–1 m/s, and pulp concentration of 6–8%. Hydrocyclone is also used as the first step for slag removal in the production.

2. Water screen machine

NC pulp flows through a screen that is vibrated vertically by an eccentric wheel, by which long fiber and mechanical impurities stayed on the screen, and qualified NC slurry passes through the screen. The water screener can be placed in front of the cleaner or the back.

3. Iron remover

Iron chip is mainly from the wear of blade. The iron chip can deteriorate the stability of NC and negatively affect the subsequent gunpowder production. Therefore, the iron chip should be removed. Iron removers are usually electromagnets that installed in the tank where pulp flows. They are made of some soft iron bars. The shape of the iron remover is shown in Fig. 10.44.

Fig. 10.44 Structure of mechanical impurity separator

One end of the iron bar is located in the slurry tank, and one end is fixed. When the direct current flows through the copper wire around the coil on the fixed part of the iron bar, the iron bar generates magnetism. The terminal of the iron bar has the strongest magnetism, which can attract and remove the iron chip in the pulp.

In the past, DC power is supplied by DC generator. Currently, Dc power is obtained by the conversion of AC to DC with selenium rectifier, which is convenient and straightforward. Cleaners, water sieve, and iron remover need regular cleaning. When cleaning the iron remover, the power should be turned off to degauss.

(2) The concentration of NC slurry

To improve the efficiency of centrifugal dehydration, it is necessary to remove some of the water in the NC pulp to increase the concentration of NC pulp.

The structures of the concentrator and fine washer are the same, both of which utilize precipitation method. After a period (1.5–5 h) settling, the clear top layer is drained. The loading process usually needs to drain 1 or 2 times to increase the concentration to 11–14%.

Concentrated NC slurry is directly heated with steam to reach the temperature of 60–75 °C, which is favorable for centrifugal dehydration. After reaching the desired temperature, NC slurry is pumped to the elevated tank and then transferred to the centrifuge for water removal.

10.8.2 Dehydration and Packaging of Nitrocellulose

NC for different applications has different water content. The water content of blended NC is 24–30%. The water content of 3#NC is typically 35–40%, which is slightly higher. If it is followed by the absorption process, 3#NC does not cause dehydration and can be used directly for the absorption process. In the case of NC needs transportation, the water content should be controlled in the range of 32–35%.

(1) Dehydration principle

NC dehydration is carried out in a centrifuge. With the high-speed rotation of centrifuge, water is separated from NC through the screen under the action of centrifugal force. NC remains on the screen to form the NC material layer. Initially, centrifugal dehydration is fast as NC contains more water. With decreasing water content and increasing resistance of NC layer, dehydration efficiency gradually reduces.

The dehydration capacity of the centrifuge is mainly related to the separation factor of the centrifuge. The separation factor f is the ratio of centrifugal acceleration to gravitational acceleration:

Table 10.48 Separation factors for centrifuge

Characteristic	Model					
	Schalzentrifuge				WH1-1200	WH2-800
Diameter of screen (mm)	1800				1200	800
Rotation speed of screen (r/min)	720	600	500	400	650–850	700–900
Separation factor	520	360	250	160	189–323	242–398

$$f = \frac{\text{centrifugal acceleration}}{\text{gravitational acceleration}} = \frac{W^2}{gR}$$

where W/R is the centrifugal acceleration and g is the gravitational acceleration. According to the relationship centrifugal acceleration, angular velocity, and rotation speed, the following equation can be obtained:

$$f = \frac{n^2 R}{900}$$

where N, R, and f represent the drum speed (r/min), the drum radius, and the dimensionless quantity, respectively. The higher the separation factor, the greater the dehydration capacity. The separation factors for the common dehydrator are shown in Table 10.48.

(2) Dehydration process and equipment

1. Dehydration process

The dehydration process is shown in Fig. 10.45. The NC pulp in the storage tank 1 is conveyed into the flow tank 4 by the pump 2. After flow through the cleaner 3 and iron remover 5, NC is transferred to concentrator 6. The concentrated and heated NC slurry is pumped to the elevated tank 7, followed by the dehydration in the centrifugal dehydrator 8. After the dehydration, NC is packaged in bags for storage [2].

2. The structure of centrifuge and dehydration process conditions

① Scraper discharge centrifuge

The structure of scraper discharge centrifuge is shown in Fig. 10.46. This equipment is continuously operated with automatic discharge, in-batch operation, high

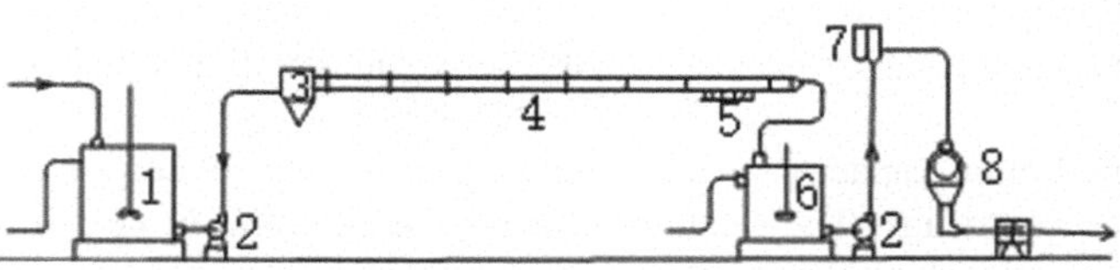

Fig. 10.45 The structure of centrifuge: 1—NC storage tank, 2—cotton pulp pump, 3—cleaner, 4—flow tank, 5—iron remover, 6—concentrator, 7—elevated tank, and 8—dehydrator

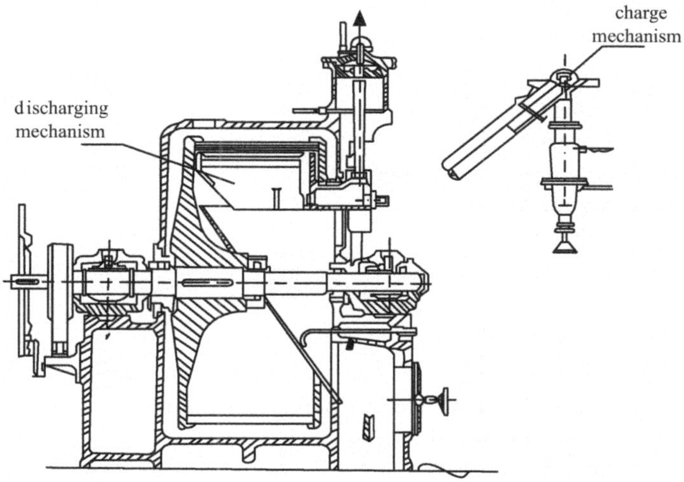

Fig. 10.46 The structure of scraper discharge centrifuge [2]

production capacity, and automatic feeding, dehydration, and discharge. However, it suffers some drawbacks, including complicated equipment structure, difficulty in controlling the operation, periodical cleaning maintenance of screen frame for improving efficiency, high labor intensity, and poor working conditions.

Scraper discharge centrifuge consists of centrifuge spindle, bearings, a discharge device, loading device, sieve, chassis, filling, moving wheel, brake, and other components.

② The process conditions for scraper discharge centrifuge dehydration are shown in Table 10.49.

Table 10.49 Dehydration process of scraper discharge centrifuge

Parameter	Specification
Thickness of NC (mm)	140–160 (gap between scraper and sieve 10–12 mm)
Loading amount (kg)	200 dry weight
Oil pressure of automatic control system (MPa)	0.6
Oil pressure of lubrication system (MPa)	0.03–0.08
Temperature of NC pulp (°C)	60–75
Concentration of NC (%)	11–14
Sieve cleaning cycle (Sieve)	10–20

Table 10.50 Dehydration process conditions for the centrifuge. Number of times the centrifuge pushes (times/min). Centrifuge pusher travel (mm)

Process	Specification
Temperature of NC pulp (°C)	60–75
Concentration of NC pulp (%)	<15
Number of the centrifuge push times (times/min)	18–26
Centrifuge pusher moving distance (mm)	<40
Oil pressure of centrifuge (MPa)	<2.0

3. Piston discharge centrifuge

Piston discharge centrifuge has a continuous feeding with complicated equipment structure and requires highly demanding maintenance. The secondary piston discharge centrifuge has high dehydration efficiency.

The dehydration process conditions for the primary piston discharge centrifuge are shown in Table 10.50.

(3) Factors that affect the efficiency of dehydration

The effect of temperature and dehydration time of NC pulp.

Increase in the temperature of NC pulp accelerates the dehydration process and reduces the dehydration time, which is attributed to multiple reasons. The increase in temperature leads to decreased interaction between NC and water and a reduced viscosity of water, thus facilitating the dehydration process.

When the temperature is high, the equilibrated water content of the NC is lower at the same relative humidity, which is favorable for the dehydration. According to the production experience, dehydration is easier and more complete in summer than in winter, which is also related to the fact that water is frozen inside the NC in the winter. The effect of temperature and time on the dehydration process is shown in Table 10.51.

Table 10.51 The effect of NC pulp temperature and dehydration time on dehydration

Dehydration time (min)	The water content of NC after dehydration (%)	
	Temperature 30 °C	Temperature 40 °C
10	35.40	33.10
15	32.92	31.18
20	30.71	28.00
25	30.00	28.20
30	30.10	28.09

10.9 Manufacture and Properties of Nanometer Nitrocellulose

The purpose of nitrating cellulose nanotechnology is to provide a new theoretical basis and technical means to improve the combustion, explosion and energy properties of energetic materials. Nano-nitrocellulose is one of the modern technologies of nitrocellulose types and preparation. It is also one of the innovative technologies of propellant. Nano-nitrocellulose is at the transitional area of atomic clusters and micro-materials (superfine powder of chopped nitrocellulose). From the general micro and macro perspective, nano-NC is a typical mesoscopic system, which derived from the traditional nitrocellulose without many special properties. Nano-nitrocellulose has the following reactions [14, 15].

10.9.1 Special Effects of Nano-nitrocellulose

(1) Surface effect of Nano-nitrocellulose

Reduction the particle size of Nano-nitrocellulose to below 200 mm not only causes the rapid increase in the number of atoms on the nitrocellulose surface bus also rapidly enhances the surface area and the surface energy. This is primarily because the molecules on the NC surface have different crystal field environment and the combination of energy than the internal molecules.

When the size of the nitrocellulose particle is much larger than the molecular diameter, the molecule on the surface cannot be neglected. When the particle size of the nitrocellulose gradually approaches the molecular diameter, the number of surface molecules and their effect cannot be neglected, which induces considerable changes in the properties of NC, such as surface area, surface energy, and surface binding energy. These various specific effects caused by nanofiber are collectively referred to as the surface effect of nitrocellulose. In other words, when the particle size of NC is 10 nm, the number of NC molecules on the surface is 20% of the total number of molecules. When the particle size is 1 nm, the percentage of the surface molecule increases to 99%. At this time, almost all the 30 molecules of NC nanocrystal are enriched on the surface.

(2) The volume effect of nanometer nitrocellulose

The reduced particle size of nitrocellulose results in two different consequence. First, the property of nitrocellulose does not change, but the combustion performance and combustion release heat and their related properties change. Under this circumstance, it becomes a semiconductor with a reduction in electronic freedom degree, electrostatic effect, and magnetic area. Alternatively, the property of nitrocellulose has changed with the decrease of particle size, thus changing the collective properties of nitrocellulose that composed of numerous atoms or molecules. When nitrocellulose is nanometer size, the size of the nitrocellulose is equal to or smaller than de Broglie wavelength that conducts electrons. The periodic boundary conditions will be destroyed, which results in a considerable change in various properties, such as magnetism, internal pressure, light absorption, thermal resistance, decomposition performance, chemical activity, sensitivity, chemical stability, and melting point, compared with the ordinary grain NC, which is the volume effect of Nano-nitrocellulose.

(3) Quantum size effect of nanometer nitrocellulose

When the size of the nitrocellulose particles decreases to a certain level, the electron energy level near the Fermi level will be changed from continuous energy level to discrete energy level, which is referred to the quantum size effect.

The most direct effect of quantum size effect is the blue boundary shift of nano-sized nitrocellulose in absorption spectra. This is because the electrons and holes generated by the light in the nanocrystals are no longer free. In other words, there is a coulomb effect, which is an exciton in a macro-like crystal material. The smaller the effective mass of the electron–hole pair, the more obvious the influence of electrons and holes, and the higher the photon energy shift and the quantum size effect.

(4) Macroscopic quantum tunneling effect of nanometer nitrocellulose

The effect of nitrocellulose microparticles to penetrate the barrier is called tunneling effect. In recent years, some macroscopic quantity has been discovered in the field of energetic materials, such as the magnetic strength of typical explosives or nitrates. The magnetic flux in the quantum coherence device exhibits a tunneling effect, which can penetrate the macroscopic barrier and change. This phenomenon is called the macro-quantum tunneling effect. This effect, along with the quantum size effect, defines the limits of further miniaturization of microelectronic devices or radio components prepared with nitrocellulose particles, as well as the shortest time for information storage and conversion of nitrocellulose products using 3D printing. This concept can be used to qualitatively explain the energy released instantaneously by the nano-sized nitrocellulose particles exceeds the energy of normal nitrocellulose and their rapid combustion.

The above four effects are the basic properties of nano-nitrocellulose, which induces nitrocellulose particles and propellants to exhibit many strange physical properties, chemical properties, and abnormal phenomena, such as decreased sensitivity, increased density, extended flame, weakened infrared characteristic. In addition, the electromagnetic, optical, and thermal properties of nano-nitrocellulose might also change.

10.9.2 Preparation of Nanometer Nitrocellulose

The nano-sized energetic materials can improve their thermodynamic properties, such as melting point and decomposition temperature, which is favorable for the rapid decomposition and complete combustion (or explosion), thereby enhancing their energy performance. Studies have shown that Nano-energetic materials will provide various potential performance advantages, including high energy release rates, extraordinary combustion (energy conversion) efficiency, high energy release, reduced sensitivity, enhanced mechanical properties of explosives [16]. As an important raw material, nitrocellulose with a nitrogen content of 11–13.5% is widely used in propellant and explosives. Traditional NC is composed of NC balls with a fiber diameter of 40–50 μm cotton balls. Because it belongs to a typical semirigid chain polymer material, it exhibits a limited thermoplastic as a propellant adhesive. In particularly for high nitrogen content NC, it is more difficult to be absorbed by the nitrate plasticizer, thus causing the propellant molding process more difficult and resulting in poor mechanical properties of the product [17]. If the traditional NC is converted to fiber nanometer, nano-sized NC fiber can have a greater specific surface area, which is conducive to be absorbed by nitrate plasticizer. In the meantime, nano-sized NC fiber can improve the NC combustion performance and energy conversion efficiency, thereby improving the performance of NC-based explosives.

(1) Preparation of nanometer cellulose suspension

The cellulose raw material is immersed in a sulfuric acid solution having a mass fraction of 64%. The reaction is stirred for 10 h in a water bath at a set temperature, and the reaction is quenched by adding a 10-time volume of water. After centrifuged for 2–3 times, the suspension is dialyzed in deionized water until the pH is constant to afford a suspension of nanofibers. The suspension is filtered with a pore size of 0.22 μm filter. The resulting nanofiber is dried in a vacuum oven at 50 °C until constant weight is reached.

(2) Nitration of nanofiber suspensions [16, 17]

The nanometer cellulose suspension is added to the fuming nitric acid. After the reaction proceeds with an ice bath for a certain period, and the reaction solution is poured into a large volume of deionized water to dilute the reaction. The collected nitrated fiber is repeated centrifuged and washed until the solution reaches a constant pH.

The prepared nano-NC is shown in Figs. 10.47 and 10.48.

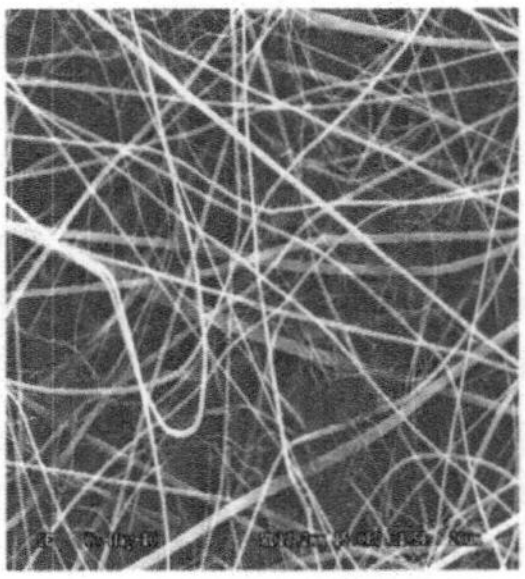

Fig. 10.47 SEM image of NC dissolved in acetone with concentration of 8% (700 nm of average diameter)

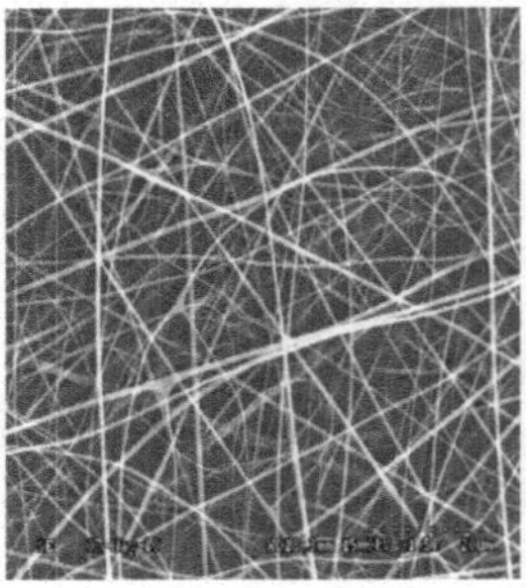

Fig. 10.48 SEM image of NC dissolved in acetone/ ethanol mixture with concentration of 10% (500 nm of average diameter)

10.9.3 Structure and Characterization of Nanometer Nitrocellulose

(1) Characterization of FTIR

As shown in Fig. 10.49, the absorption peak at 3400–3200 cm^{-1} corresponds to the stretching vibration peak of OH bond; the two absorption peaks at 1560 cm^{-1} and 1370 cm^{-1} are assigned to the stretching vibration peaks of $–NO_2$. The antisymmetric stretching vibration of the –C–O–C bond at 1123 cm indicates the presence of the C–O–C bond in the ring. The broad peak at 650 cm^{-1} is assigned to the outward bending vibration of OH bond, whereas the peak at 1400 cm corresponds to in-plane bending vibration of O–H bond. It is obvious that the infrared spectra of NC before and after spinning do not change, which indicates that the high voltage electrospinning process does not lead to the change the chain structure of NC molecule.

It can be seen from Fig. 10.50 that the thermal decomposition of the NC ball can be divided into two stages. The first stage is the initial decomposition temperature of about 120 °C, corresponding to the NO_2 elimination reaction of NC molecular chain. The second stage is the initial decomposition temperature of about 190 °C, which corresponds to the condensed phase of small molecules generated from the C–O–C breakage of the NC macromolecule chains caused by the released NO_2, which leads to the NC autocatalytic reaction to accelerate the degradation of the

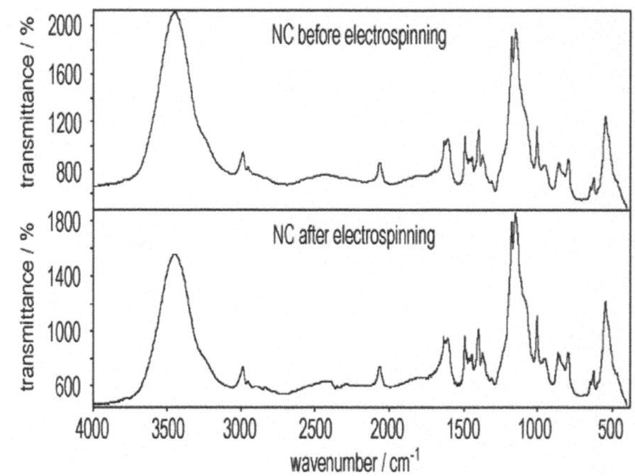

Fig. 10.49 FTIR spectra of NC tampons and NC nanofibers [16]

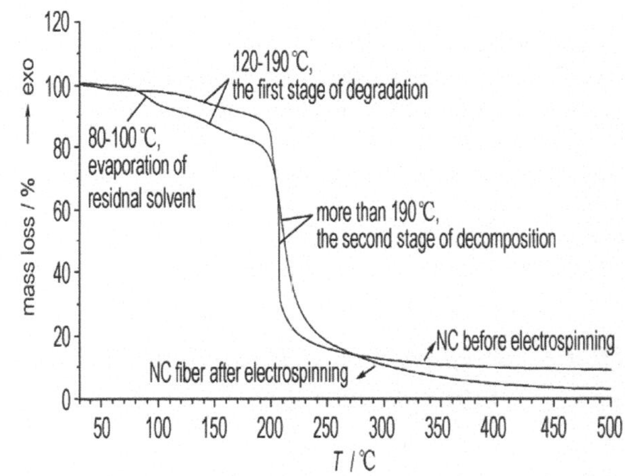

Fig. 10.50 TG curves of NC tampons and NC nanofibers [16]

denitration substrate [18]. The thermogravimetric curve of nano-NC has a weight loss step in the temperature range of 80–100 °C, which should be caused by the volatile residual solvent in nanometer NC. The thermogravimetric curve is divided into two stages, which correspond to denitration reaction and autocatalytic reactions. The weight loss curve of nano-NC corresponding to autocatalytic reaction phase is relatively gentle, and the residual rate of thermal weight loss is only 3%, which is lower than that of NC ball after thermal weight loss (about 9%). This result indicates that nano-sized NC fiber is beneficial for complete degradation of NC. The decomposition heat of NC and nano-NC is characterized by DSC. The results are shown in Fig. 10.51.

The results show that the decomposition heat of nano-NC is about 1835.80 J/g, which is significantly greater than that of NC ball at 1320.06 J/g. This difference

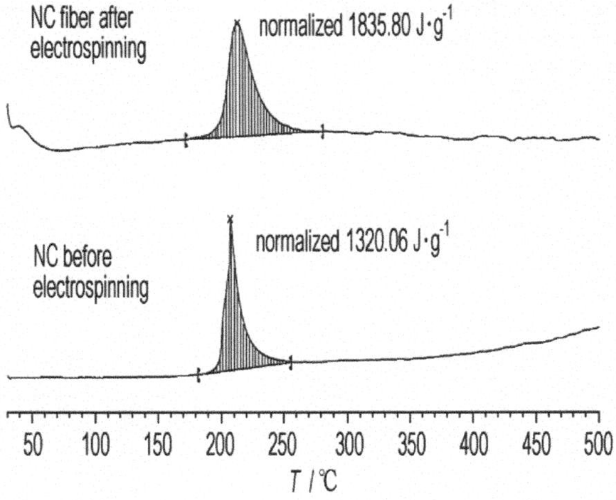

Fig. 10.51 DSC curves of NC tampons and NC nanofibers [16]

should be closely related to the size effect of nano-NC. Nano-NC has a large specific surface area, which is favorable for each fiber to absorb the external heat efficiently, thus fully trigger the denitration reaction of the NC molecular chain. Simultaneously, it produces a large number of NO_2 gas that can contact with the denitrified fiber surface tightly to trigger autocatalytic degradation reaction efficiently, which facilitates nano-NC thermal degradation and the complete heat release from the decomposition.

10.9.4 Preparation of Propellant by Nano-sized Nitrocellulose

The preparation process of nitramine propellant is the traditional semi-solvent process, consisting of absorption, tabletting gelation, hydraulic molding, cutting, posttreatment, etc. The preparation processes for the nanocellulose modified propellant and blank propellant are identical. The primary difference is that the nanocellulose suspension is added to the absorption link.

First, water is added to the absorption kettle, followed by the addition of raw material (NC + NG). Then, the ultrasonically treated homogeneous nanocellulose suspension is added and mixed with stirring for 30 min. RDX is added and heated to 60 °C. In the meantime, the medium and DEGN mixture in solution is slowly added to the reaction mixture in dropwise for at least 10 min. After the addition, the absorption is allowed for an hour. Finally, the mixture is cooled to 30 °C. The product is collected and filtrated to remove water. After the product is kept at room temperature for 1–2 days, it is dehydrated by centrifuge and goes to the rolling process. The roller temperature is controlled at 90 °C, and the distance between rollers is 0.03 mm. The product is pressed for 2–3 times by the rollers to form

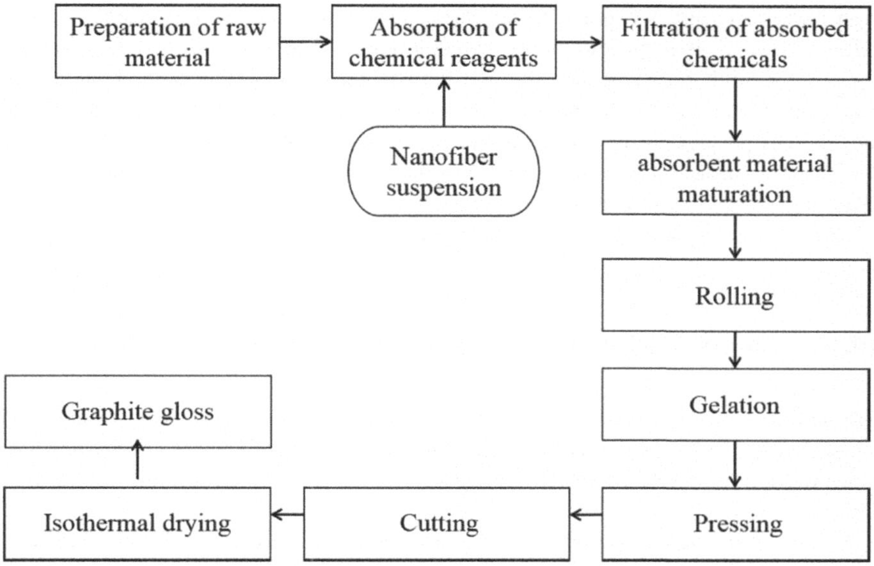

Fig. 10.52 Preparation process for nano-nitrocellulose propellant [19]

sheets. The sheet is pressed twice in a plasticizing machine and subjected to a 2L gelling press according to the ZTH-11 process conditions. Hydraulic filter molding is used for filtration. After selected a suitable mold, NC is charged densely, and the discharging speed is carefully controlled. The NC pellets are cut according to the test requirements. After air-dried for 72 h, it is dried at the temperature of 55 °C. Finally, the mechanical properties and combustion performance test are performed. The nanometer modified propellant preparation process is shown in Fig. 10.52.

Nanocellulose whiskers have various advantages, such as excellent mechanical properties, lightweight, optimal light transmission, small thermal expansion coefficient, and environment-friendly. The high crystallinity of nanocellulose whiskers endows their excellent mechanical properties and strength. The Young's modulus strength of nanocellulose whiskers is comparable with the ceramic and metal. The Young's modulus and tensile strength of pure nanocellulose are about 150 and 10 GP, respectively, which is an excellent Nano-reinforced material. Since the nitrocellulose is the main binder in the propellant, the nitrate optimization of nanocellulose whiskers not only can ensure a good interface between the nitrocellulose matrix and the nanocellulose whiskers but also increase the energy of the propellant. In recent years, as the weapon system has been advancing rapidly in the direction of precision strike and high-efficiency damage, the basic requirements for high energy and high strength have been put forward for the propellant. On the one hand, to meet the high energy requirements, the formula of propellant is added solid high-energy explosives, which has an adverse impact on the mechanical properties of the propellant. As a result, it is difficult to meet the application requirements of high barrel pressure weapons. In addition, the main binder of propellant is nitrocellulose, which is semirigid chain molecule tends to be brittle at low temperature and soft at high temperature. The mechanical strength of the propellant has an important effect on the burning of the propellant in the weapon system. If the mechanical properties do not meet the requirements of the propellant, it will be deformed or broken under the shooting conditions, which leads to abnormal ballistic performance, and even abrupt pressure increase or breech blows. Therefore, it is very significant to improve the mechanical strength of the propellant.

10.10 Safety Technology for Production of Nitrocellulose

Nitrocellulose is a flammable and explosive product. In the production, storage, and transportation of NC, we must be vigilant to overcome carelessness. The concept of safe production should be firmly established to prevent the occurrence of burning and explosion.

In the nitrocellulose production process, there are a lot of acid and alkali and other corrosive liquids, gases, and vapors. Some equipment and pipelines might be corrosive and leaking under the long-term influence of corrosive agents, which may also cause poisoning and chemical burns. In addition, it is very important to prevent electric shock and mechanical injury accidents due to the operation of electrical and operation equipment.

It is of great significance that all personnel engaged in NC production must understand the safety knowledge associated with the production process and consciously implement the technical safety procedures at work to ensure the health and safety of operators.

10.10.1 Common Incidents and Prevention

(1) Combustion and explosion NC

In particularly the NC with a nitrogen content of more than 13% can lead to NC combustion or explosion with significant accumulation or in closed conditions under the role of external energy. Under the normal conditions, NC has a characteristic of slow decomposition. Under certain conditions, the slow decomposition can be transformed into a burning or explosion. Such accidents must be prevented in the production process.

According to many bloody lessons in the NC production process, spark, local overheating of equipment, and automatic decomposition of NC can cause fire and explosion. As long as the safety regulation is strictly implemented, NC does not have the necessary conditions for combustion and explosion, and the accident will be avoided.

To this end, the equipment should always be cleaned to prevent the accumulation of NC dust and self-decomposition. Production sites must not be brought into the fireworks. Some flammable materials, such as oil, rags should be stored by the provisions of timely cleaning. The direct sunlight on the NC should be avoided. The operation to prevent the collision of metal objects should be prevented. Various measures should be implemented to eliminate the spark. When welding, the interior and surface of the equipment must be cleaned thoroughly to eliminate the residual NC. Fire prevention measures should also be well prepared. The welding operation can be only done with special approval. In summary, the relevant rules and regulations should be strictly enforced, and the objective laws should be violated. In the event of an accident, it should be reported and properly handled and controlled.

In addition, it is necessary to take various effective measures to prevent accidents caused by the overpressure of equipment.

(2) Poisoning and chemical burns

The direct splash of sulfuric acid or nitric acid on the human body will cause chemical burns. The inhalation of sulfur dioxide, nitric acid vapor, and other harmful gases, especially nitric acid vapor, will cause varying degrees of poisoning depending on the amount of inhalation. The inhalation of nitric acid vapor can cause pulmonary edema.

To avoid the occurrence of chemical burns and poisoning accidents, the operating procedures should be strictly followed in the acid preparation, nitration, and acid removal operation. The required labor protection gear should be worn. Equipment should be maintained and repaired in time. The workshop should have safety facilities, such as the accident exhaust and sodium carbonate solution (5% concentration) for flushing splashed acid on the body. The harmful gases and dust in the workplace should be below the maximum allowable concentration.

Once the accident occurs, the injured person after the necessary emergency treatment should be immediately sent to the hospital for treatment.

(3) Electric shock and electrical injury accident

When using electrical equipment, electric shock and electric injury should be prevented. Electric shock refers to the damage to the human body caused by current flowing through the body, while electric injury refers to local injuries caused by electricity. Both electric shock and electric injury on the human body are harmful. Severe electric shock is deadly.

Safety rules should be strictly enforced in the production. Electrical equipment should be grounded, and the insulation should be in good conditions. Be sure to disconnect the power supply when repairing the equipment. Long-term unused electrical equipment should be inspected before using. If it is wet, it should be dried before use.

In the event of an electric shock, turn off the power immediately so that the electric shock is disconnected from the power supply. If the shocked patient stops breathing, artificial respiration and first aid should be performed immediately.

Mechanical injury accidents are mostly caused by improper operating methods, such as improper safety procedure and violation of operating procedures. To this end, the equipment should always be inspected and repaired. The transmission part must have good protective equipment. The operator should wear protective equipment and follow the relevant safety procedure.

References

1. Liu JP, Qian HJ (1993) Production and manufacture of nitrocellulose. Air Force Institute of Aeronautics and Ammunition, Nanjing
2. 301 Teaching and Research Department (1976) Nitrocellulose process. Huazhong University of Science and Technology, Nanjing
3. Liu JP (1987) Basic chemical of propellants and explosives. GuiZhou Press, Guiyang
4. Liu JP (1991) Method of making nitrocellulose from eulaliopsis binata as raw material. China patent CN1011970
5. Liming W (1992) Study on the properties of cellulose fiber cooling and swelling (first report): cooling with cooling. Artif Fiber (5):1–6
6. Nicky (1960) Wood and its cellulose chemistry. People's Education Press
7. Urbanski T (1965) Chemistry and technology of propellants & explosives, vol II. Pergamon
8. Kast H (1921) Spreng-und Zündstoffe. Vieweg&Sohn, Braunschweig
9. Milus PR (1937) Ind Eng Chem 29:429
10. Dai AQ (1994) Analysis on nitrogen content of nitrate cotton by the five-tube method. Chin J Explos Propellant 02
11. WJ 36-2005. Test method of nitrocellulose-nitrogen content-Devarda alloy reduction method, 2005.12.12
12. Liu JP (1990) Method of making nitrocellulose from cotton stalk as raw material. China patent CN1052484
13. Liu JP (1990) Method of making nitrocellulose from bagasse as raw material. China patent CN1052489
14. Liu JP, Hao XY (2002) Nanoscience and technology. Science Press
15. Liu JP, Hao XY (2009) Polymer-based nano-modified materials. Science Press
16. Xia M (2012) Preparation and characterization of nitrocellulose nano-fibers. Chin J Energ Mater (2):167–171
17. Yang ZY (2013) Surface acetylation of cellulose nanowhiskers and its reinforcing function in cellulose acetate. Chem J Chin Univ 4(34):1021–1026
18. Wang Y, Liu R et al (1998) A study of thermal decomposition mechanism of nitrocellulose. Chin J Explos Propellant (6):157–168
19. Wang WJ, Shao ZQ, Zhang FX et al (2011) Preparation of nano-sized nitrocellulose by nanometer cellulose whisker suspension as raw material. Chin J Explos Propellant 34(2): 73–76

Chapter 11
Nitrostarch

11.1 Introduction

Nitrostarch is a light yellow amorphous powder made by the nitration of starch. Starch is a high molecular weight polymer of glucose with a chemical formula of $(C_6H_{10}O_5)_n$, which is also known as grapefruit powder in the catering industry. Maltose with a chemical formula of $C_{12}H_{22}O_{11}$ is the hydrolysis product of starch to the disaccharide stage, whereas glucose with a chemical formula of $C_6H_{12}O_6$ is the fully hydrolyzed product of starch. Starch particles contain trace amounts of noncarbohydrates, such as proteins, fat acids, and inorganic salts. Except for fatty acids adsorbed by amylose molecules and phosphates bonded with amylopectin molecules via esterification, the other substances are physically mixed with starch. The starch molecule is a polymer compound that is linked by some α-D-glucopyranosyl units via glycosidic bonds. The basic constituent unit of starch is α-D-glucopyranose (a stable chair conformation). Starches consist of amylose and amylopectin, as well as the third component in many starches, which is a transition state between amylose and amylopectin with a slightly branched amylose structure.

The reaction of starch with nitric acid or mixed nitric and sulfuric acid forms the nitrostarch precipitate as an amorphous powder that maintains the basic form of starch. Nitrostarch has a certain explosive property that can be used in the military industry as smokeless gunpowder [1]. Typically, it is used as an explosive alone or mixed with nitroglycerin to fill grenades or torpedoes.

J. Liu, *Nitrate Esters Chemistry and Technology*,
https://doi.org/10.1007/978-981-13-6647-5_11

11.1.1 Physical and Chemical Properties of Nitrostarch

11.1.1.1 Chemical Properties of Nitrostarch

The chemical formula of nitrostarch: $(C_6H_6N_4O_{13})_n$
The structural formula of nitrostarch

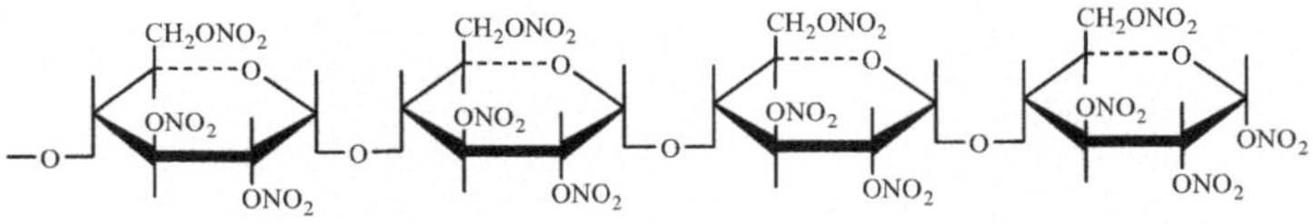

Relative molecular weight: 342.11
Oxygen balance: −9.35% (by the amount of CO_2 generation)
Nitrogen content: 16.37%
Oxygen coefficient: 86.67%

The early process for the preparation of nitrostarch consists of the dissolution of potato pound in concentrated nitric acid and the addition of water to precipitate the nitration products. [2]

With the development of chemistry and chemical engineering technology, especially the establishment of cellulose nitration theory [3–5], the process for the starch nitration is the same as the cellulose nitration method nowadays. Typically, a mixed nitric and sulfuric acid is used for the nitration. Studies have shown that [6, 7] the reduction of nitrostarch can restore the physical and chemical properties of carbohydrate. Interestingly, starch does not dissolve in water, but the nitrostarch dissolves in water well. A variety of methods [8] has been investigated to isolate nitrostarch from nitric acid. Nitrostarches with different nitrogen contents were obtained by precipitation using water or sulfuric acid. It is noteworthy that the products precipitated with sulfuric acid exhibits poor stability and tends to decompose under ambient storage conditions [9]. The stability of the nitrostarch that precipitated with sulfuric acid is low due to the partial sulfonation of nitrostarch.

The starch is dissolved into 12 parts of nitric acid (specific gravity 1.50), followed by the addition of 30 parts of sulfuric acid to precipitate the product. The nitrogen content of nitrostarch obtained by this method is up to 13.5%. However, the stability of the nitration product is very low.

When starch is nitrated with high-concentration nitric acid (specific gravity 1.52), starch nitrate is obtained with a nitrogen content of 14.04%. After the stability treatment in refluxing ethanol, the stability of nitrostarch could be dramatically improved, resulting in excellent stability.

The starch can be prepared by nitration in a manner similar to that of cellulose nitration, and the nitrostarch containing 13.9% N can be prepared with a mixed nitric and sulfuric acid.

Early process for the preparation of nitrostarch consists of the dissolution of starch dissolved in nitric acid, followed by the addition of sulfuric acid to precipitate the product and washing with diethyl ether and chloroform for stability treatment.

Nitrostarch can be mixed with KNO_3 or $KClO_3$ to make smokeless gunpowder [10]. Starch is dissolved in 10 parts of nitric acid, followed by the addition of waste acid from the production of nitroglycerin to precipitate the nitrostarch [11]. After the stability treatment by washing with water, aqueous sodium carbonate, and aqueous aniline successively, the resulting nitrostarch can be used to make smokeless gunpowder. Volney's Powder is manufactured by this process.

Since the chemical stability of nitrostarch is poor, the nitrostarch is considered to have no practical value until the discovery of stability treatment at the beginning of the past century. At that time, by licensing US patent, a British chemist [12] built and operated two nitrostarch plants to manufacture pure nitrostarch as a replacement of dynamite in the United States. In the manufacturing process, a mixed anhydrous acid consisting of nitric acid, sulfuric acid, and fuming sulfuric acid was used to nitrate starch to afford nitration products with 13.3–13.5% N. Later on, owing to an explosion accident during production, the production of nitrostarch was discontinued. For this reason, the manufacture and related applied research on nitrostarch [13] were also terminated. Nevertheless, low-freezing-point explosives and "Arctic explosives" contain nitrostarch have been used in the United States.

At the end of the First World War, the United States produced about 25,000 tons of nitrostarch per year, which were mainly used for two types of explosives (Grenite and Trojan Explosive).

In the two World wars, nitrostarch was used to fill bombs to replace TNT. During the Second World War, nitrostarch was used to fill grenades. At the peaceful time after the Second World War, the production and application of nitrostarch gradually ceased. However, nitrostarch is used as a double-based propellant or propellant again this year because of the rising price of nitroglycerin.

11.1.1.2 Physical Properties of Nitrostarch

Nitrostarch having a nitrogen content of less than 8.0% N is almost insoluble in ether or alcohol. For example, the solubility of nitrostarch with 6.4% N is only 8.4%. In general, nitrostarch is more soluble than nitrocellulose in ethanol. Nitrostarch with a nitrogen content of 10.0–11.5% can be fully dissolved in ethanol. Incomplete dissolution of nitrostarch indicates that the content of nitrogen in the nitrostarch is beyond this range. The product with a nitrogen content of 12.1% N has a solubility of 32.5%.

Table 11.1 Solubility of nitrostarch in different solvents

Solvent	Solubility (%)			
	10.0–11.5% N	12.0% N	6.5% N	8–9.5% N
Ethanol	100	100	8.0	32.7
Diethyl ether	100	100	7.3	46.6
Acetone	100	100	87.5	100
Chloroform	100	100	9.6	29.8
THF				
Ethyl acetate	100	100	58.4	92.6
Water	0	1	0	0

The solubility of nitrostarch is related to its nitrogen content. Nitrostarch can be dissolved in polar solvents, and the solubility varies with the polarity of the solvent. Nitrostarch with a nitrogen content of 6.4% or greater can be completely dissolved in acetone. Nitrostarch containing 8.0–11.5% N can be dissolved in the mixture of ether and alcohol. Specifically, the solubility of nitrostarch in the ether–alcohol mixed solvent decreases with increasing nitrogen content. The solubility of nitrostarch in different solvents is shown in Table 11.1.

Through the careful study of the morphology of coarse nitrostarch, it was found that nitration deteriorated the structural homogeneity and integrity of starch. In particular, the decrease in viscosity is a prominent performance of nitration. In contrast, the structure of amorphous starch after nitration is better than that of before nitration. During the nitration of starch, the first stage is the hydrolysis of starch to form soluble starch, and the second stage is the actual nitration reaction.

The result of the decreased viscosity of nitrostarch indicates that the starch molecules significantly degraded and turned to liquid nitrates during the starch nitration process.

The viscosity of the nitrostarch solution is significantly lower than that of nitrocellulose solution with the same degree of nitration. If the viscosity of water is regarded as 100, the relative viscosity of 1% nitrostarch varies between 45 and 70. Due to the low viscosity and amorphous–crystalline structure of nitrostarch, the film formed by the evaporation of nitrostarch is very hard, brittle, and delicate. The results of this macroscopic test show that the structure of nitrostarches is amorphous–crystalline. Therefore, nitrostarches cannot be used for coatings, plastics, and celluloid manufacturing.

There is an opinion that the structures of amylose and amylopectin are essentially the same [14], in which the oxidation capacity of mixed nitric acid on amylose with loose structure is strong while the oxidative effect on the amylopectin with compressed and cluster structure is significantly weak.

The solubility of nitrated amylose in the ether–alcohol mixture is 20% [15], while the solubility of nitrated amylopectin ranges from 86 to 87%. Their difference in solubility is attributed to the branched structure of amylopectin. In theory, the solubility of all branched polymers (except for stereospecific polymers) is higher

Table 11.2 Melting points of different nitrostarches

Names of nitrostarches	Molecular formula	Melting point (°C)	Recrystallization solvent
α-Amylose dimmer hexanitrate	$[C_6H_7O_2(ONO_2)_3]_2$	206	Acetic acid
β-Amylose trimmer hexanitrate	$[C_6H_7O_2(ONO_2)_2]_3$	203	Ethanol
β-Amylose trimmer nonanitrate	$[C_6H_7O_2(ONO_2)_3]_3$	198	Acetic acid
α-Amylose tetramer octanitrate	$[C_6H_7O_2(ONO_2)_2]_4$	204	Acetic acid

than those of linear polymers. Regarding the stability, nitrated amylose is more stable than the corresponding nitrated amylopectin.

The nitration of amylose degradation products [10], namely dimerization, trimerization, and tetramerization amylose, generates a series of nitrate compounds. The melting points of the product crystallized in different solvents are shown in Table 11.2.

The products listed in Table 11.2 are usually unstable and stored in a desiccator. After 6–8 months, their weight loss is about 75%.

11.1.1.3 Thermochemical Properties and Explosive Properties of Nitrostarch

(1) *Thermochemical properties of nitrostarch*

The formation heats of nitrostarch vary with different nitration reaction conditions [16]. The formation heats of nitrostarch with different nitrogen contents and prepared by mixed acid are listed in Table 11.3. The results in Table 11.4 indicates the combustion heats and the formation heats of nitrostarches prepared with mixed nitric and sulfuric acid are slightly different than those of nitrostarches prepared with different nitration reagent, although those nitrostarches have the same nitrogen content.

The combustion heats and formation heats of the two types of nitrostarches listed in Tables 11.3 and 11.4 are different due to the presence of sulfonates in the

Table 11.3 Nitrostarches prepared by nitration with mixed nitric and sulfuric acid

Nitrogen content (%)	Combustion heat (J/g)	Formation heat (J/g)
10.52	10777.15	2938.84
11.02	10504.35	2796.17
12.87	9525.71	2324.21
13.24	9279.69	2263.54

Table 11.4 Nitrostarches prepared by direct nitration

Nitrogen content (%)	Combustion heat (J/g)	Formation heat (J/g)
10.52	10797.65	2914.16
12.20	9952.90	2411.24
12.87	9660.86	2189.49
13.34	9511.49	2110.83

nitrostarches prepared by the mixed acid. Nitrostarch is easy to burn when ignited and burned in the same way as nitrocellulose.

(2) *Explosive properties of nitrostarch*

The explosive properties of nitrostarch are similar to those of other nitrogen-containing compounds. With the increase of nitrogen content, the detonation velocity, and lead body increase. According to the results of explosive experiments, the detonation rate was 4970 m/s in diameter 30 mm thick paper tube with a charge density of 0.90 for 13.52% N nitrostarch. The relative lead block expansion value was 117 (picric acid = 100), which is the same as the value of 13.4% N nitrocellulose.

Table 11.5 Detonation velocity and explosive power of nitrostarches

No.	Nitrogen content (%)	Detonation velocity (in the tube of 34/28 mm in diameter) (m/s)		Lead block testing (cm^3)	Shock sensitivity, the height of 5 kg weight drops onto the material to cause 80% explosion (cm)
		Detonation with No. 8 detonator	Detonation with 10 grams of picric acid		
1	13.4	6190	4790	430	21
2	13.1	5740	4340	395	23
3	12.8	5300	3950	375	28
4	12.4	4950	3520	360	36
5	12.1	4480	3480	320	38
6	11.6	3880	3460	300	70
7	11.0	3020	3460	260	100
8	10.5	2880	3380	230	100
9	9.4	1500	3200	160	100
10	8.9	1000	300	125	100
11	7.7	No explosion	No detonation	35	No explosion

Table 11.5 shows the data on the explosive power of nitrostarches (nitrogen content in the range of 7.7–13.4%).

11.1.2 Starch

The maturity and quality of starch significantly affect the properties of nitrostarch, such as nitrogen content, viscosity, and product stability treatment procedure. Thus, it is necessary to study the properties of starch when investigating nitrostarch [17–19].

Starch is a natural product of polysaccharide with a general formula of $(C_6H_{10}O_5)_n$. Usually, starches are present in the form of granules in plants, and their shape and size vary in different plants. Although starches with different origins have morphological and physical differences, their chemical properties are the same. Since the hydrolysis performance is the common basic characteristics of starches, they are important raw materials for the food industry.

Since phosphorus compounds catalyze the enzymatic processes of starch formation, a small amount of phosphorus is contained in various starches. With few exceptions, the starch has two different structures, namely amylose, and amylopectin. The hydrolysis of starch produces maltose (produced by the decomposition of enzyme-catalyzed molecules) or glucose. Amylose and amylopectin structures can be differentiated with iodine. The interaction of iodine and amylose produces a dark blue color while the interaction of iodine with amylopectin gives purple or red-purple color. Compared with amylose, amylopectin is more resistant to enzymes. For example, amylose is completely hydrolyzed to maltose in the presence of amylase, whereas only about 60% of amylopectin is hydrolyzed.

Amylose has a molecular weight of about 10,000–50,000 (measured by osmotic pressure), which is consistent with the value calculated by the method of terminal functional group measurement. The amylopectin has a molecular weight of 50,000–1,000,000. Amylose and amylopectin can be primarily separated by hot water (70 °C). Hot water can dissolve amylose but cannot dissolve amylopectin (80–90%). The basic structure of starch is shown in Fig. 11.1.

As can be seen from Fig. 11.1, the structure of amylose (a) is significantly smaller than that of amylopectin (Fig. 11.1b). According to the result of terminal functional group measurement, the long chain of amylopectin molecules likely consists of only 24–30 glucose units, which can be used to determine amylopectin.

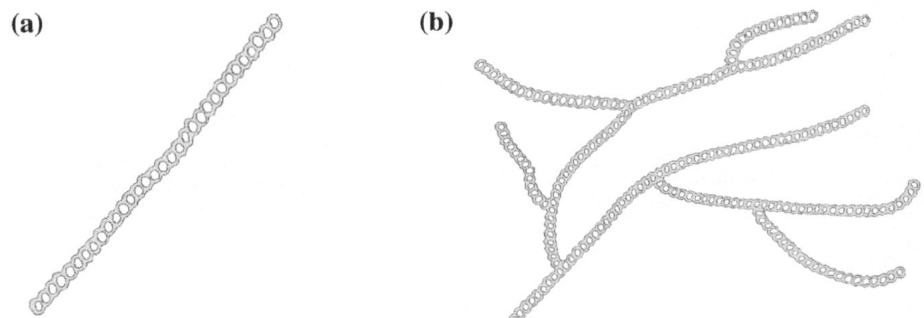

Fig. 11.1 Structures of amylose and amylopectin

Studies on the partial or complete hydrolysis of methylated starch have led to the fact that the major units of amylose and amylopectin are composed of the α-D-dehydrated glucose chain. Similar to maltose, starch units are linked through the glycosidic bond between the first and fourth carbon atoms of glucose.

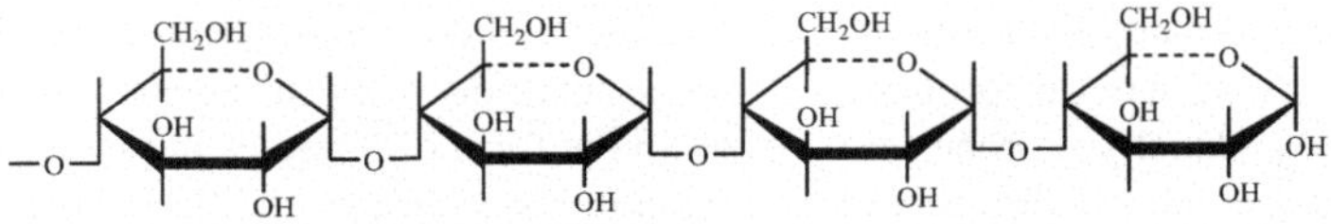

2,3,6-Trimethylglucose is the main product of the methylated starch hydrolysis. In addition, there is a small amount of 2,3,4,6-tetramethyl glucose, which is derived from the terminal cyclic structure.

After the hydrolysis of methylated amylopectin, the minor products were 2,3-dimethyl glucose and 2,6-dimethyl glucose. The presence of these two compounds confirmed that the branched chain was linked to the sixth carbon or third carbon atom of the glucose. Accordingly, the molecular structure of amylopectin is proposed as follows [20].

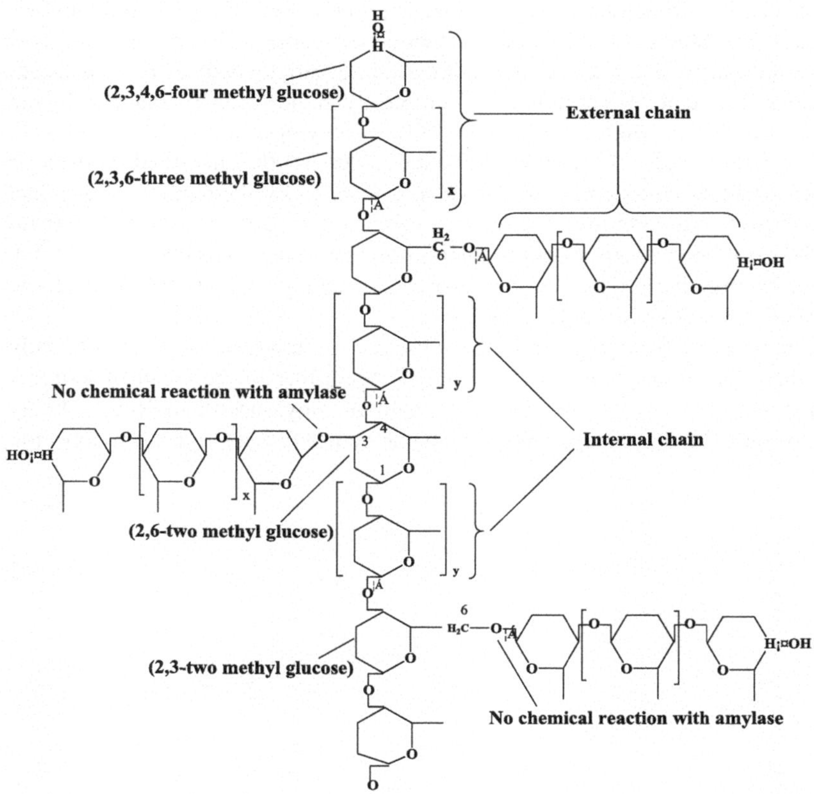

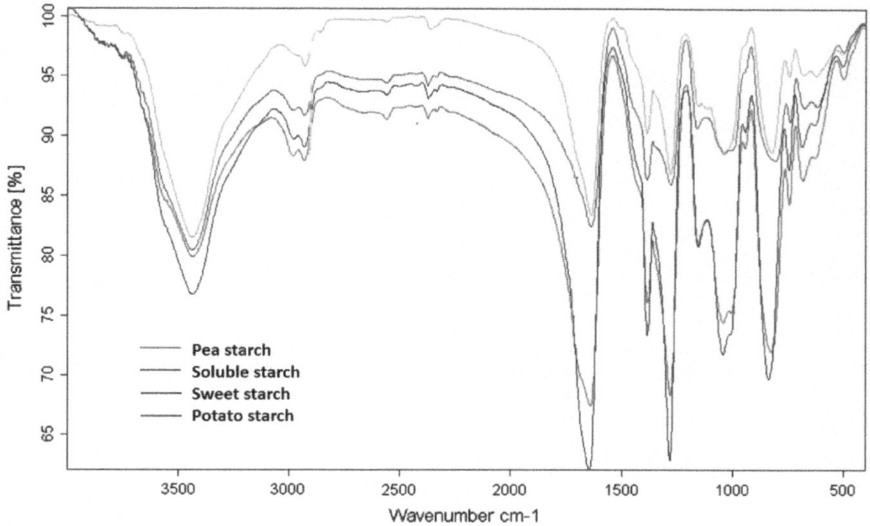

Fig. 11.2 Infrared spectra of several different starches

The purity of starch has a substantial effect on the nitration products, while the purity of starch is related to maturity and species of plants, such as potato starch, sweet potato starch, pea starch, lotus root starch, puerarin starch, and cornstarch. Starches with various sources have different starch growth periods, which results in different starch molecular weights despite their identical amylose and amylopectin structures. The infrared spectra of starches from different plants are shown in Fig. 11.2.

The results in Fig. 11.2 unambiguously show that the hydroxyl positions of different starches are the same despite their different starch sources, indicating that the starch structure is independent of the plant species.

For the production of starch with excellent performance and stable quality, it is necessary to refine the starch from different sources before nitration to ensure the product quality and homogeneity.

11.1.3 Preparation of Nitrostarch

The process for the nitration of starch is similar to that of nitrocellulose. The main chemical reaction of nitration is

$$C_6H_{12}O_6 + 4HNO_3 \rightleftharpoons 4C_6H_6N_4O_{13} + 4H_2O$$

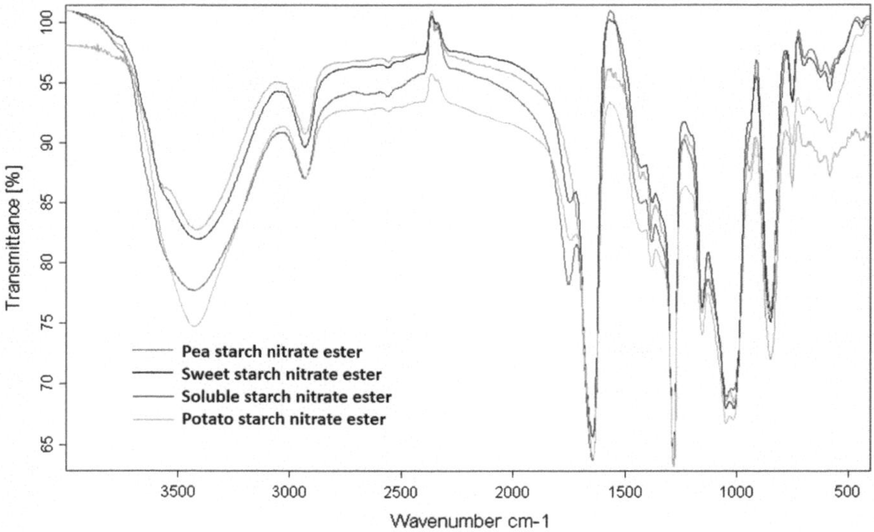

Fig. 11.3 The infrared spectra of several nitrostarches

Nitration of starch with fuming nitric acid produces nitrostarch having a nitrogen content of 14.04%. After the stability treatment of nitrostarch in refluxing ethanol, the obtained nitrostarch possesses excellent stability. A product containing 13.9% N can be prepared from a mixed nitric and sulfuric acid. The infrared spectra of several nitrostarches are shown in Fig. 11.3.

11.1.3.1 Nitration Methods and Conditions for Starch

When the nitric acid solution of the nitrostarch is poured into water, sulfuric acid, or phosphoric acid, the nitrostarch precipitated as an amorphous powder. Starch and cellulose differ not only in shape but also in structure. Their structural differences result in their different basic properties. When starch reacts with nitric acid, it only starts to swell and dissolve in the presence of nitric acid with a concentration of 76% or greater. In contrast, cellulose only slightly swells and is typically insoluble in 76% nitric acid, and it only dissolves in 80% nitric acid.

The starch was dissolved in 10 times of nitric acid (97.3%) in mass ratio. After the reaction mixture was continued at 20 °C for 6 h, the reaction mixture was precipitated and washed with a 20-time volume of water to afford nitrostarch with nitrogen content ranging from 11.7 to 11.9%, which is soluble in ethanol and ether–alcohol mixture.

The starch was dissolved in 10-time nitric acid in mass ratio. After the reaction was continued at 20 °C for 2 h, the reaction mixture was poured into a 40-time volume of 94% sulfuric acid to precipitate the nitrostarch. The resulting nitrostarch contained 12.6% N, which exhibited a solubility of 11.4% in ethanol but reached 95.7% in the ether–alcohol mixture.

When the reaction time of starch with nitric acid is less than 6 h, such as 4 h, the nitration generates nitrostarch with a low-nitrogen content (12.2% N). The process employing nitric acid as nitration agent and utilizing sulfuric acid to precipitate nitrostarch yields a paste-like intermediate product, which is difficult for workup and stability treatment involving stirring and washing. The product cannot be obtained or used. As a result, local overheating occurs during the nitration reaction, and partial denitration of the nitrostarch occurs. In general, the nitrogen content of the product prepared by this process is much lower than that of nitrostarch synthesized by direct nitration using mixed nitric and sulfuric acid.

Nitration using mixed nitric and sulfuric acid can produce a more homogeneous product. The maximum nitrogen content is reached after the reaction time of 1–2 h. This reaction is basically the same as the nitration of nitrocellulose. The nitrogen content of nitrostarch depends on the final composition and nitration coefficient of the mixed nitric acid.

To identify the correlation between the nitrogen content of nitrostarch and the composition of mixed nitric and sulfuric acid, we studied the nitration reactions of starch with mixed nitric and sulfuric acid with more than 100 different ratios based on previous studies [21, 22]. After the complete analysis of the experimental results, we designed a new nitration procedure. The nitration reactions were performed by employing a mixed nitric and sulfuric acid at 10–12 °C for 1 h with a nitration coefficient of 50, and the results are plotted in an HNO_3–H_2SO_4–H_2O three-phase diagram in Fig. 11.4.

In Fig. 11.4, figure (a) and (b) are plotted in weight percent and molar percent, respectively. The area between the curve AAA′A′ and the axis HNO_3–H_2SO_4 represents the nitration of starch with a nitrogen content of more than 12.75% by nitric acid. The area enclosed by curves AA-BB and A′A′-B′B′ corresponds to

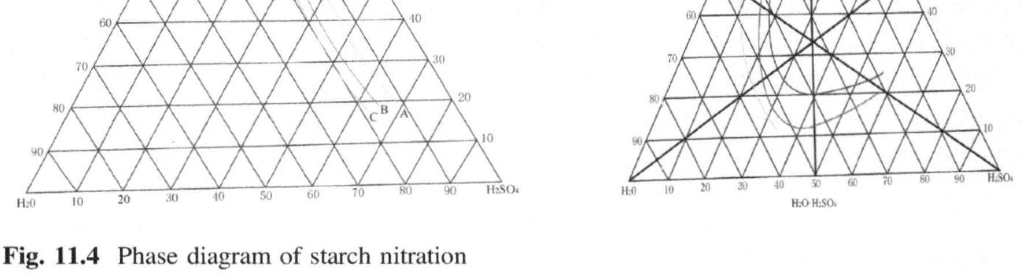

Fig. 11.4 Phase diagram of starch nitration

nitration with nitrogen contents of 11.1–12.75%. Areas under the curves of BB-CC and B′B′-C′C′ mixed acid produce nitrostarches with nitrogen contents of 9.15–11.1%. The left side of the curve CC indicates that the nitration reaction with mixed acid can only generate an oxidative or hydrolyzed heterogeneous product with a low degree of nitration. The profiles of AA, BB, and CC are similar to those of nitrocellulose (Chaps. 2 and 10 nitration theory), which are similar to nitronium NO_2 concentration curves of nitric acid in the mixed acid.

Compared with the three-phase diagrams of the nitrocellulose and the nitrostarch, there is a minimum value in the three-phase diagram of nitrostarch, which is approximately the same to the weight ratio of nitric acid to sulfuric acid of 6 (molar ratio of 9) in mixed acid.

Many studies have shown that the sulfuric acid weight ratio of more than 6 (that means less sulfuric acid) in the mixed acid is favorable for the preparation of nitrostarch with a high content of nitrogen. This result can be explained by the slopes of the curves A′A′, B′B′, and C′C′. In addition, it can also be explained by the fact that the mixed acid in this region dissolves the starch during nitration and conforms to the law of nitration in the single phase. According to the results in Fig. 11.4, the nitration effect of the mixed acid is very strong. As shown in Fig. 11.4, only at the end of the nitration process, some mixed acid systems can precipitate a certain amount of nitrostarch.

The hydrolytic degree of nitrostarch in the acid-containing trance or without sulfuric acid is not as significant as the mixed acid that usually contains significant amounts of sulfuric acid.

By using the mixed nitric and sulfuric acids for the nitration of various types of starches [23], the nitrogen contents of resulting nitration products are all 12.5% N or greater. There is an opinion that nitrostarch prepared with mixed nitric and sulfuric acid [24] decomposes considerably, but the decomposition mode is different than that of the general nitrate.

Phasing nitration can improve the chemical stability of nitrostarch, which is a notable nitration method [25]. For example, starch is nitrated with the mixed acid with 75–90% nitric acid at 5 °C for 30 min in two phases. After a series of stability treatment, the yield of nitrostarch is better than that of the general nitration with mixed nitric and sulfuric acid. The time in Abel test is up to 40 min.

On the basis of extensive studies on the nitration of starch with mixed nitric and sulfuric acid [26], the main conclusions are as follows: (1) The effects of the composition of mixed nitric and sulfuric acid and the nitric acid activity in the mixed acid on the nitrogen content and esterification heat of the nitrostarch are determined. (2) The correlation between the nitration temperature and the viscosity of nitrostarch is identified. (3) The linear relationship between the reaction time and the nitration temperature is determined. Figure 11.5 illustrates the relationship between nitric acid and sulfuric acid in the mixed acid system. The mixed acids contain the following compositions: (I) 35.0% HNO_3, 57.5% H_2SO_4, 7.5% H_2O; (II) 33.4% HNO_3, 60.75% H_2SO_4, 5.81% H_2O; (III) 30.5% HNO_3, 63.5% H_2SO_4, 6% H_2O. When the nitration temperature is above +10 °C, the nitration products with the highest nitrogen contents can be obtained almost in a few minutes.

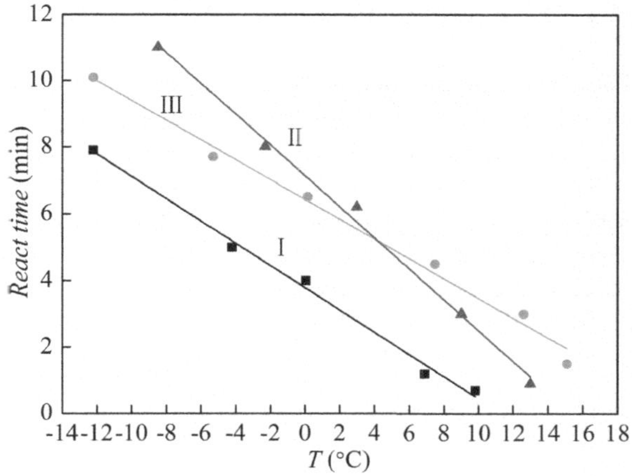

Fig. 11.5 The relationship between the reaction time of starch nitration and the nitration temperature

When the starch is nitrated with mixed nitric and phosphoric acid, and the water content in the mixed acid is 10% or more, it will cause the swelling of the starch [27] and a poor homogeneity of the nitration product. Nitrostarches with nitrogen contents ranging from 13.0 to 13.1% can be obtained by nitrating with a mixed acid containing 20% HNO_3 and 80% H_3PO_4.

When nitric acid vapor or nitrous oxide dissolved in nitric acid is employed for the nitration of starch [28], the nitrogen content of nitrostarch is up to 13.86% by using two-phase nitration, initial nitration temperature of +5 °C for 30 min, and then the reaction temperature of 0–16 °C for 30 min. The characteristic of the nitrostarch prepared with this method has high chemical stability.

During the Second World War, N_2O_6 was extensively studied for the nitration of starch [29–32]. N_2O_6 dissolved in chloroform for the nitration of starch produced nitric acid as a byproduct of the reaction, which partially dissolved nitrostarch. To suppress this side effect, sodium fluoride is added to the reaction mixture to form complexes Na–F … HO–NO_2 via hydrogen bonding with nitric acid. This method is successfully used to nitrate all of three hydroxyl groups in starch completely, resulting in starch trinitrate almost without degradation with nitrogen content ranging from 13.92 to 14.03%. The stability treatment of nitrostarch without sulfonate is straightforward without any challenge [33].

The high concentration of nitronium $NO_2^{\oplus}$ in the nitrating agent [34] is favorable for producing nitrostarch with high nitrogen content.

The reactions of mixed nitric and sulfuric acid with starches from different plants (corn, wheat, rice, cassava, and potato) were studied previously [27]. If the purities of the starches are the same, the properties of the prepared nitrostarches are almost identical (i.e., nitrogen, solubility, and viscosity).

11.1.3.2 Industrial Production Method of Nitrostarch

The industrial production of nitrostarch is basically the same as that of other nitrates. Because of its different properties, the preparation process of nitrostarch has some differences. The production process of nitrostarch is shown in Fig. 11.6.

The industrial process for the production of nitrostarch includes the initial washing with a diluted solution of sodium hydroxide or ammonium hydroxide, removing fat acids and pectic acid, and rinsing with water [35]. After drying the starch to the moisture content of 0.5% or less, it is ready for the following nitration reaction. The mixed acid composition of nitrated starch is

HNO_3	38%
H_2SO_4	62%

To a reactor containing 369 kg of mixed nitric and sulfuric acid was added 90 kg of starch at 32 °C with stirring. The temperature of nitration was maintained in the range of 38–40 °C. After the completeness of the reaction at 40 min, the resulting reaction mixture was poured into a water tank, washed to neutral, and filtered to yield the product.

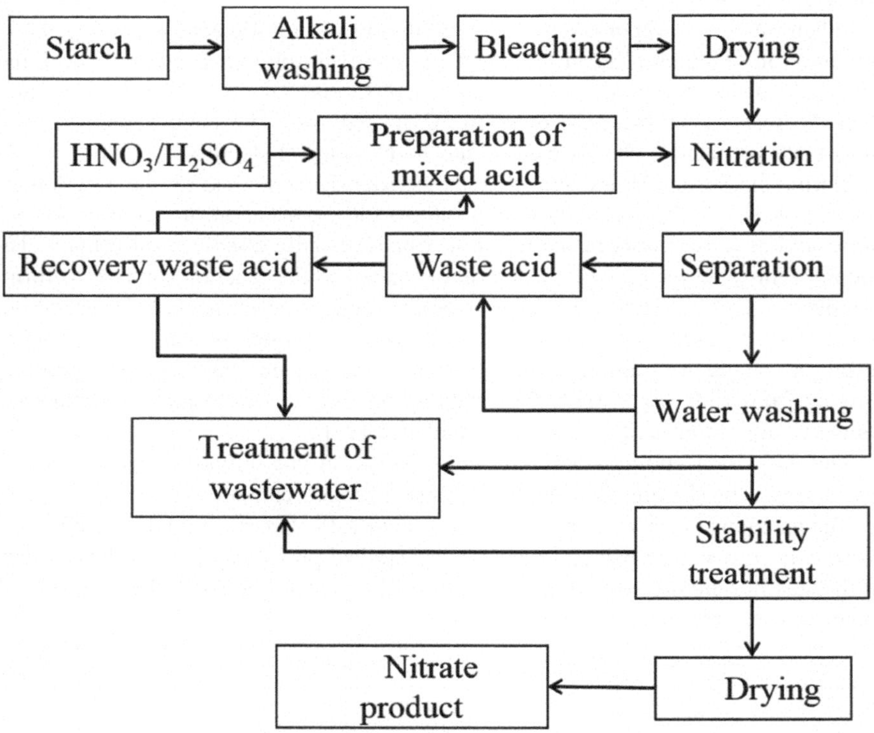

Fig. 11.6 The production process for the nitrostarch from starch

The nitrated product was refined by washing with cold water without refluxing. To neutralize the residual acid, a certain amount of ammonia was added to the water during the washing. After washed to neutral, the product was dried at the temperature of 35–40 °C to afford the product with the nitrogen amount of 12.75%.

11.1.3.3 Factors Affecting the Viscosity of Nitrostarch

The factors affecting the variations in viscosities of the nitrostarch and nitrocellulose are identical. In general, the high nitration temperature leads to the higher nitrogen content of nitrostarch. The viscosity of nitrostarch having a strong reaction during stability treatment in refluxing ethanol is higher that of nitrostarch exhibiting a mild reaction in stability treatment at low temperature. In contrast, the variation in viscosity of nitrostarch is hardly affected by the nitration reaction temperature, nitrogen content, amount of alkali and temperature during safety treatment, and other factors. The properties of the nitrostarches prepared with nitric acid as a nitrating agent are shown in Table 11.6.

There is no doubt that starch degrades vigorously in the nitration process, because of the strong oxidation and hydrolysis ability of nitric acid [18]. Table 11.7 shows the viscosity change of a 10% starch solution in the presence of 97.3% nitric acid (at 20 °C).

The viscosity of prepared nitrostarch is significantly reduced no matter the preparation method of nitrostarch. Table 11.8 illustrates the change in the viscosity of the starch before and after nitration.

If nitric acid is used to nitrate starch, the degree of degradation of nitrostarch is very significant. In contrast, the viscosity of nitrostarch prepared from nitration using mixed nitric and phosphoric acid is twice as much as that of nitrostarch prepared from mixed nitric and sulfuric acid [27]. This result demonstrates that the degradation during nitration using mixed nitric and phosphoric acid is significantly

Table 11.6 Relationship between nitration conditions and properties of nitrostarches

Nitration temperature (°C)	Nitrogen content of nitrostarch (%)	Relative viscosity (water = 100)	Solubility in water (%)
20	12.13	58.4	32.5
	11.09	52.7	100
	10.28	50.7	78.2
	9.51	50.3	41.8
0	13.19	71.6	
10	13.08	68.3	
20	12.76	61.7	
30	12.57	57.3	

Table 11.7 Relationship between reaction time and viscosity change in 10% starch dissolved in 97.3% nitric acid

Reaction time (min)	Relative viscosity (water = 100)
45	4730
75	3530
120	2720
180	1925
375	1040
720	625
1440	415

Table 11.8 Comparison of viscosity changes before and after the nitration of starch [36]

Type	Starch		Nitrostarch	
	Viscosity (Pa S)	Relative viscosity (water = 100)	Nitrogen content (%)	Relative viscosity (water = 100)
1	224	1415	13.3	138
2	165	707	13.2	136
3	151	560	13.1	134
4	144	471	13.3	131
5	122	319	13.2	132
6	99	232	13.2	131
Solubility	14	103	13.3	124

less that of using nitric acid only or mixed nitric and sulfuric acid. To further confirm this conclusion, the relative molecular weights of the nitrostarches prepared with nitric acid–sulfuric acid–water (64/26/10) and nitric acid–sulfuric acid–water (70/20/10) were determined to be 100 and 21–43, respectively.

Nitrostarch is a highly heterogeneous mixture [37], which can be graded by the addition of a certain amount of water, followed by the addition of acetone. The nitrogen content of each grade is slightly different, but their viscosities are almost identical.

Treatment of nitrostarch in refluxing ethanol isolates the low-nitrogen soluble fraction and insoluble fraction (nitrogen up to 13.9%). Another factor that causes poor homogeneity of nitrostarch is the inclusion of non-isolated amylose and amylopectin in the starch. When both the two isomers are nitrated, the resulting nitrates are different from each other. The viscosity of the nitrostarch from the nitration of amylopectin is much higher than that of amylose nitrate [37], which is primarily attributed to the fact that the molecular weight of amylopectin is higher than that of amylose.

11.1.3.4 Stability Treatment of Nitrostarch

The results of a vast number of nitration studies have shown that the nitrostarch prepared by nitration with nitric acid is safer than the product obtained by nitration with mixed nitric and sulfuric acid, which is consistent with the conclusion of other nitrated products. The reason is that the formation of sulfate-coated gels in starch is difficult to be dispersed in water, thus increasing the difficulty of stability treatment.

To address the issues of stability treatment of nitrostarch, nitrostarch is dispersed in water into fine particles so that the water can be immersed in the interior of the nitrostarch particles to replace the residual acid gradually. In principle, this substitution process is highly similar to the protocol of stability treatment of nitrocellulose.

Washing nitrostarch in refluxing solvents is an important method of stability treatment. First, the nitrostarch is washed with refluxing acidic water (washed with refluxing acidic water, and the acidity of water is gradually decreasing), followed by washing with refluxing alkaline water. If the remaining acid in the nitrostarch is not replaced with diluted acid until neutral or prematurely treated with base, which will prevent the sulfate from hydrolyzing, thus leading to the sulfate remaining in the starch nitrate will be present in the form of salt and condense in the interior of nitrostarch. In the process of stability treatment of nitrostarch, even if the nitrostarch is not washed to neutral, it is favorable for the production of a stable product by introducing ammonia. However, this method can increase the color of nitrostarch.

The method used in the Nitrokemia plant in Hungary [38, 39] is the addition of a substance that swells the nitrostarch during the stabilization treatment to facilitate the stability treatment solution to enter into the swelling of the nitrostarch and to replace the residual acid. In this process, the additive plays the role of a swelling agent.

Washing nitrostarch in refluxing ethanol is used in many laboratories for stability treatment [40]. This method can efficiently and quickly stabilize nitrostarch. The stability of nitrostarch containing 13.4–13.9% N after stability treatment is much higher than that of the crude product before the treatment. In general, the higher the nitrogen content, the lower the amount of loss of nitrostarch in the treatment.

References

1. Chatius FU, Dinglers Polyt J 132, 37, (1854); 161, 146 (1861)
2. Braconnot H (1833) Ann Chim phys 52:29; Ann 7:249 (1833)
3. Liu JP (1987) Basic chemical of propellants and explosives. GuiZhou Press, Guiyang
4. Liu JP, Qian HJ (1993) Production and manufacture of nitrocellulose. Air Force Institute of Aeronautics and Ammunition, Nanjing
5. Urbanski T(1965) Chemistry and technology of explosives. Pergaon
6. Béchamp A (1856) Compt Rend 41:817; (1856) J Prakt Chim 68:51
7. Béchamp A (1856) Ann Chim Phys 3(46):349; 48:458
8. Béchamp A (1862) Ann Chim Phys 3(64):311; Bull Soc Chim France 4:358

9. Mühlhaauser O (1892) Dinglers Polyt J 284:140
10. Holmes FB (1905) U.S. Pat. 779421, 779422; (1907) 875913, 875928
11. Dynamit AG, Nobel AG (1891) Ger. Pat. 57711
12. Hugh A (1904) U.S. Pat. 751076; Ger. Pat. 172549, 1903; Brit. Pat. 12627, 1904
13. According to CAESAR, starch nitrate. In: Advances in carbohydrate chemistry, vol 13. Academic Press, New York (1958)
14. Urbański T, Erlowski SB (1952) unpublished
15. Ashford WR, Cook LM, Hibbert H (1946) Can. J. Res 24(B):238
16. Smoleński D, Strzondala J (1954) Zeszyty Nauk Politechn. Wroclaw, no. 4, Chem. no. 1:49
17. Estrada-Leon RJ, Moo-Huchin VM, Ríos-Soberanis CR et al (2016) The effect of isolation method on properties of parota (Enterolobium cyclocarpum) starch. Food Hydrocoll 57:1–9
18. Blanco M, Gardner C et al (2016) Dosage effects of waxy gene on the structures and properties of cornstarch. Carbohyd Polym 149:282–288
19. Phuong H, Vu N, Lumdubwong N (2016) Starch behaviors and mechanical properties of starch blend films with different plasticizers. Carbohyd Polym 154:112–120
20. Staudinger H, Husemann E (1938) Ber. 71:1057
21. Hackel J, Urbański T (1932) Roczniki Chem. 12:276
22. Z.ges.Schiess-u.Sprengstoffw. 28:35 (1933)
23. Staudinger H, Eilers H (1936) Ber. 69:819
24. Staudinger H, Hsemann E (1937) Ann 527:195; (1938) Ber 71:1057
25. "Nitrokemia" Ipartelepek, Hungar. at. 118102 (1938)
26. Kunz A, Tóth I (1938) Acta Chim Hungar 13:385
27. Urbański T, Kapuscińsk Z, in Urbanski and Hackel, ref [35], p 769
28. Urbański T, Kapuscińsk Z (1937) Roczniki Chem 17:384
29. Caesar GV, Goldfrank M (1946) J Am Chem Soc 68:372
30. Caesar GV, Gruenhut NS, Cushing ML (1947) ibid. 69:617
31. Gruenhut NS, Cushing ML, Caesar GV (1948) ibid. 70:424
32. Caesar GV (1946) U.S. Pat. 2400287
33. Vollmert B (1951) Makromol Chem 6:78
34. Izraelashvili SH (1950) Nature 165:686
35. Davis TL (1943) The chemistry of powder and explosives. Wiley, New York
36. Urbański T, Golofit J in Urbański T, Hackel ref [35], p 771
37. Urbański T, Cieslak I in Urbański T, Hackel ref [35], p 772
38. "Nitrokemia" Ipartelcpek, (1939) Hungar. Pat. 121180
39. Kunz A (1940) Ger. Pat. 691154; Olivér S (1940) Hungar. Pat. 123989
40. Yang T, Chen LP, Chen WH et al (2005) Thermal stability of 2-ethylhexyl nitrate with acid. J Therm Anal Calorim 119:205

Chapter 12
Chitin and Algae Nitrates

The object discussed in this chapter is different from the previous chapters. It mainly studies three kinds of compounds, namely chitin nitrates, pectin nitrates, and algae nitrates in this chapter. Since the content of each section of this study is rarely or hardly reported in the previous literature, the content included in this chapter is likely incomplete and might require scientific researchers in this field continue to explore the study further [1, 2].

12.1 Chitin Nitrate

Chitin nitrate is prepared by the reaction of N-acetylglucosamine-based Chitin material with nitric acid or nitro-sulfuric acid to form a nitrate-containing compound with blasting ability.

12.1.1 Properties of Chitin Nitrate

Chitin nitrate has a low nitrate content but exhibits strong combustion and moderate explosion capability.

J. Liu, *Nitrate Esters Chemistry and Technology*,
https://doi.org/10.1007/978-981-13-6647-5_12

Molecular formula: $C_8H_{11}O_9N_3$
Structure:

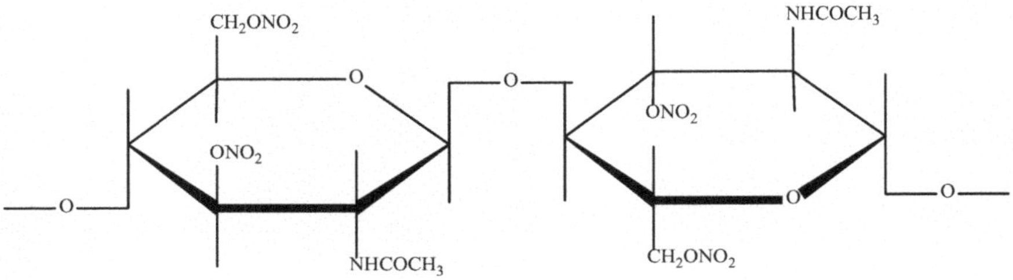

Relative molecular weight: 293.21
Nitrogen content: 15%
Oxygen balance: −68.2% (by the generation amount of CO_2)
Oxygen coefficient: 41.86%

Chitin nitrate material is easy to burn when ignited with an ignition point of about 163 °C. It can tolerate Abel heat test (65 °C) for 3–8 h without changing color using starch-potassium iodide test paper employed as an indicator.

Since chitin nitrate materials can be partially dissolved in formic acid, it can be divided into two grades by employing formic acid. Chitin nitrate solution in formic acid can be precipitated by the addition of water. By using this method, it is feasible to distinguish the nitrogen content of the chitin nitrate materials and their differences in molecular weight. Chitin nitrate materials can only swell in benzene, naphthalene ($C_{10}H_{12}$), nitrobenzene, aniline, phenol, pyridine, and furfural. With this swellability, chitin nitrate materials can be used as energetic materials or directly in certain materials with lower energy and higher gas production.

12.1.2 Refining of Chitin Material

Chitin is a long-chain polymer (similar to cellulose) consisting of β-N-acetyl-glucosamine units. Each unit contains two OH groups and one NH-$COCH_3$ group [3].

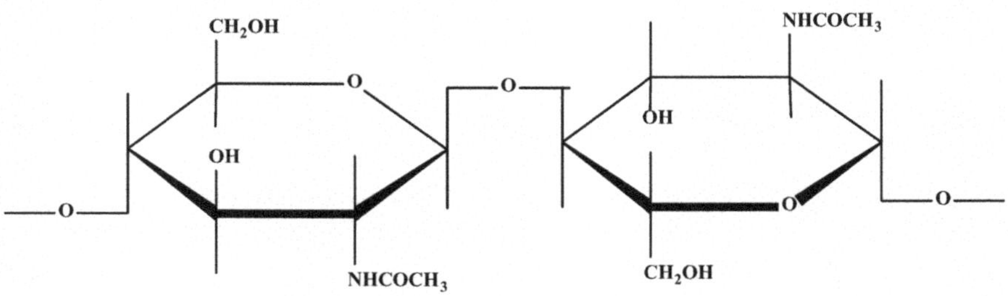

Since the surface of raw chitin is wrapped with lipids, general chemical reagents cannot penetrate the surface of chitin. It is required to remove the surface materials such as pectin, carbohydrates, and lignin via chemical treatment before the preparation of chitin nitrates.

There are two basic methods for refining of the raw chitin. Firstly, mechanical separation method. Owing to the molecular weight of the pure chitin material is high, it is easy to form cellulose-like material and to be separated by the break of shell surface pectin via mechanical effect. However, this separation and refining method only affords 75–85% purity of chitin. To prepare chitin nitrate with high nitrogen content, it is still required to use a certain amount of alkali or solvent treatment. Only the refined material with a purity of 95% or greater can be used to prepare high-quality chitin nitrates. The second method is chemical refining process. The impurities in raw chitin are removed mechanically, and the crude materials are processed into 20 × 30 mm lumps. The processed materials are fed into the reactor, followed by the addition of 5–9 times of H_2O in volume and a small amount of HCl, and then heated to about 130 °C. After the pre-hydrolysis treatment for 2–3 h, the reaction mixture is washed with water and heated with alkali for 5–6 h at the reaction temperature below 130 °C. An appropriate amount of saponification agent should be added to the alkali treatment to ensure the surface of refined chitin material is loose and has a certain water absorption function, which is favorable for the nitration. After the above treatment, the purity of the chitin can reach more than 98%, and the nitrogen content of chitin is up to 14.8% with the maximum nitrogen content of 15%.

12.1.3 Preparation of Chitin Nitrates

The nitration of chitin materials has been reported in the literature [4]. The reaction of nitric acid (density 1.525) with chitin was attempted to accomplish both the nitration and degradation of chitin, which yielded two types of products. One type is insoluble in common organic solvents while another is soluble in polar solvents, which are likely chitin nitrates and nitrate-containing oxidation products, respectively. Since chitin nitrate and nitrocellulose have a certain similarity, the studies on the birefringence properties of chitin and chitin nitrate have been reported in the literature [5].

Based on the studies on the most extensive preparation and properties of chitin nitrates [6], the authors found that it was impossible to nitrate the chitin with a commonly mixed acid containing sulfuric acid because the chitin was decomposed by the strong oxidation capability of the sulfuric acid. The method for the preparation the chitin nitrate is to dissolve the chitin in concentrated nitric acid (d = 1.50) for 1–2 h and then transfer to a large amount of water. After the nitration, the precipitated chitin nitrate is a white flake. The content of the–the ONO_2 group is 7.5% N, which is equivalent to 1.5—ONO_2 groups per glucoside unit.

According to our experimental results, the nitration of purified chitin with a mixed acid consisting of HNO_3 30%, H_3PO_4 64.3% and H_2O 5.7% afforded the nitrate with a nitrogen content of 14.6%. In another experiment, the nitration reaction was carried out with the addition of 5% magnesium nitrate to the above-mixed acid at nitration temperature of 8–12 °C for 20 min, affording the nitrate material with the nitrogen content of 15.1%. Based on a large number of experiments, the authors found that the quality of refined chitin significantly affected the nitration products. For example, the nitration of unrefined chitin in the same mixed nitric acid resulted in nitrate products with nitrogen content of 7.6–8.9%, 5 times increase in the viscosity of waste acid compared with that of before the nitration, mass loss rate of 8–12% after the stability treatment, and the product yield of 58%. In contrast, the nitration of refined chitin produced nitrate products with a nitrogen content of 14.8–15.1%, product yield of 98.9%, slightly darker color of nitration waste acid, 1–2 times increase in the viscosity of mixed acid after nitration, and less than 2% mass loss after the stability treatment.

12.2 Pectin Nitrate

Pectin nitrate is also known as nitrated pectin, which is prepared from the nitration of pectin that is extracted from natural plants and a milky white viscous colloidal solution similar to natural polymer compound.

12.2.1 Properties of Pectin Nitrate

Pectin nitrate is a colloidal or powder with a yellowish appearance. It typically exhibits a certain degree of flammability and a weak explosive performance.

Molecular formula: $C_6H_6O_{10}N_2$
Structure:

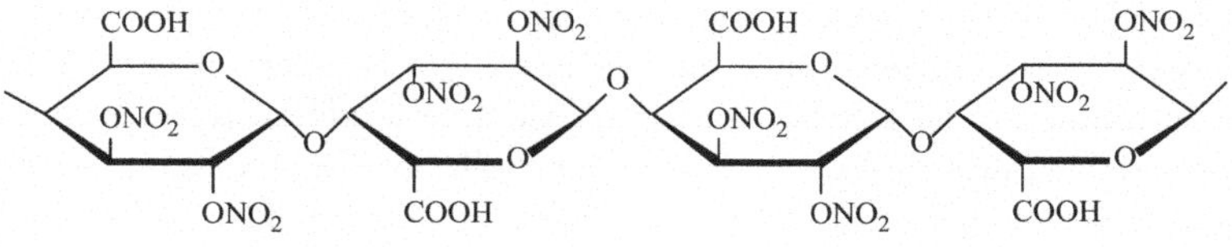

Relative molecular weight: 266.12
Nitrogen content: 10.53%
Oxygen balance: −30% (by the generation amount of CO_2)
Oxygen coefficient: 66.67%

Pectin nitrate has flammable characteristics. The flame color of pectin nitrate is bright red to bright white, and flammable speed of pectin nitrate is quick with barely burning residue after burning. The emission from the combustion is fragrant. Pectin nitrate is soluble in benzene, nitrobenzene, ethyl acetate, acetone, and other organic solvents but insoluble in ethanol.

12.2.2 Extraction and Purification of Pectin

Pectin is a kind of natural macromolecule and water-soluble compound with a molecular weight of about 500,00–300,000, which mainly exists in all advanced plants and is a major component of plant intercellular substance. Currently, pectin can be isolated in industrial scale. Pectin is a polysaccharide mixture consisting of a polygalacturonic acid and galacturonic acid, which contains a significant amount of methylated pectic acid. Under the appropriate conditions, the solution of pectin can form a gel and takes place partial methoxylation (methyl esterification, namely the formation of methyl ester). The main component of pectin is partial methyl esterification of α (l, 4)-D-polygalacturonic acid [7]. Pectin is widely used in the food industry and as a gelling agent, thickener, emulsifier, stabilizer, and so on. In addition, pectin is a natural food additive and raw material for the manufacturing of foods, such as jam and jelly.

The stable structure of pectin is as the following:

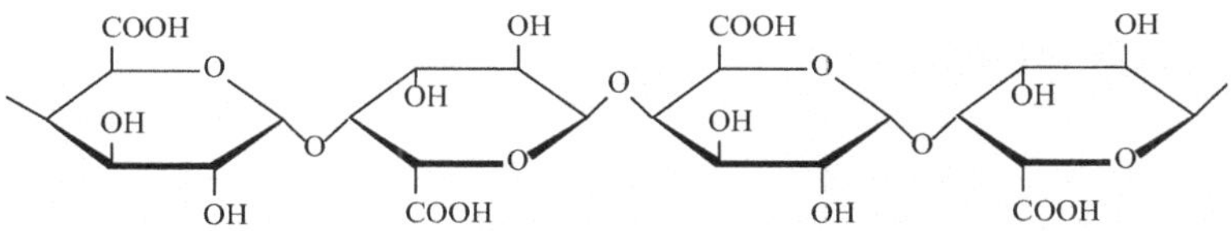

According to the structure of pectin, it can be seen that the pectin can be oxidized easily. In the presence of oxidative acid, the glycoside bonds in pectin are oxidized and decomposed to small molecule compounds. The hydroxyl group of the pectin is readily replaced by the active group to form an ester, and the hydrogen atom of the ester group replaced by the electron-withdrawing group still exists in the form of an ester. Pectin nitrate typically does not have a very high nitrogen content but exhibits poor stability, which is likely attributed to ready deformation by the formation of jelly or liable structure via oxidation.

12.2.3 Preparation of Pectin Nitrate

Pectin nitrate was invented before the Second World War. At that time, German scientists investigated the nitration of pectin extracted from chopped beets. However, the property of pectin nitrate has not been received considerable research attention. As a result, the studies on pectin nitrate has not yielded any practical application.

In some studies, the intermediates of pectin nitrate were prepared [8], followed by the esterification of the intermediates to form the corresponding nitrates via reactions with acetic anhydride.

The pectin nitrates were prepared in 1949 for the first time [9], and the properties of pectin nitrates were studied. The pectin was nitrated for 4 h at 20 °C using a mixed acid consisting of 48% HNO_3, 50% H_3PO_4 and 2% P_2O_5, and the resulting product contained 9.2% nitrogen and 7.5% carbonyl.

The results of various nitration process studies show that the number of COOH groups remains constant during the nitration process, whereas the properties of the pectin nitrate and cellouronic acid nitrate are completely different. Pectin nitrate is completely soluble in acetone, but nitrated cellouronic acid only swells or partially dissolves in acetone. This is primarily because the carboxyl group in pectic acid is partially esterified so that it cannot participate in the cross-linking between the carboxyl and hydroxyl groups. In this paper, a series of experiments were carried out to investigate the nitration of pectin. Various factors, such as the purity of pectin and nitration conditions (nitration agent, nitration temperature, nitration time, etc.) were studied systematically. Pectin with a purity of 98% was nitrated with nitrating agent consisting of acetic acid 68.2%, nitric acid 26.8%, H_2O 5.0%, magnesium nitrate 2.0%, and sodium acetate 1.5% for 12–18 min to afford pectin nitrate with a nitrogen content of 12.85%. After the stability treatment, the pectin nitrate exhibited excellent stability in the application of propellant. Specifically, it did not change color after Abel test for 200 min.

12.3 Alginate Nitrate

The alginate nitrate is similar to that of pectin nitrate. The nitrogen content of alginate nitrate is more than 10.5% with the highest nitrogen content of 14.83%, and the solubility in acetone is 96.4%. However, it cannot form the linkage that is similar to those of cellouronic acid and its nitrate. The mechanical properties of alginate nitrate are superior to nitrocellulose products in the application of propellant.

12.3.1 Properties of Alginate Nitrate

Alginate nitrate is a kind of yellowish to yellow cellulose, colloidal or powder. Typically, it has a certain burning appearance.

Molecular formula: $C_6H_6O_{10}N_2$
Structure:

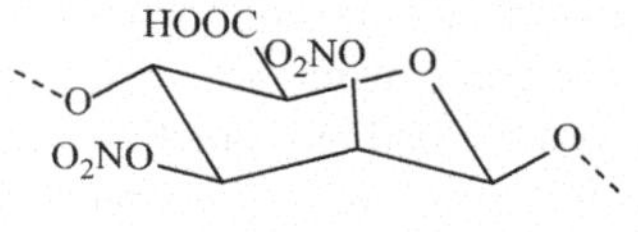

β —D —mannitol nitrate

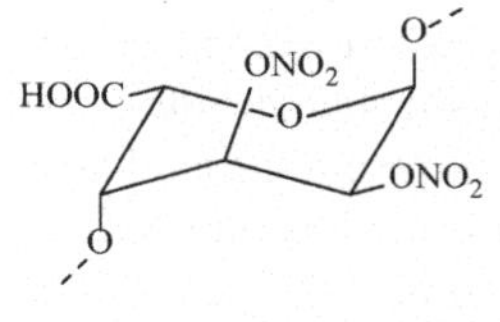

α —L —polyguluronic nitrate

Relative molecular weight: 266.12
Nitrogen content: 10.53%
Oxygen balance: −30.07% (by the generation amount of CO_2)
Oxygen coefficient: 66.67%

12.3.2 Extraction and Purification of Alginate

Algae are the oldest among all plants on earth. Most algae live in water, and their structures are very basic. Each alga is a leafy body without difference among root, stem, or leave. Algae vary considerably in body shapes. For example, diatom living in the ocean is very small and is phytoplankton in plankton. On the other hand, kelp is a type of seaweed with large size. This brown seaweed can be up to four meters in length, and Sargasso can be up to tens of meters in length. Algae also have different shapes. Some are simply linear (straight or branched), whereas others are flat or spherical with uneven edges.

According to the color of the pigments, algae fall into three categories: green algae, brown algae, and red algae. Green algae (such as sea lettuce and Spirogyra) only contain green pigment-chlorophyll; brown algae (such as the fucus algae genus) only contain brown and yellow pigments; red algae contain red and blue pigments. Algae use pigments to acquire energy, and their growth and survive also require water and light. Brown algae can only grow in seawater while both green algae and red algae can also grow in fresh water. Some algae can grow out of the water. For example, green algae can live on bark or damp old walls.

The reproduction methods of algae are various and very complicated. Fucus Algae, a type of brown algae growing on the rock, is a typical example. Fucus Algae can be male, female, or hermaphrodite. In a particular time, the edges of the foliage are swollen and wrap the reproductive organs used to produce the reproductive cells (or gametes), which are released into the water, fertilized in water, that is, the fusion of male and female gametes. Eventually, the formed single cell-fertilized egg grows into a new plant.

(1) General properties and extraction of alginic acid

Chinese scholars classify algae into eleven types [10, 11], including blue-green algae, red algae, cryptic algae, dinoflagellates, green algae, yellow algae, diatom, brown algae, euglena, green algae, and chara.

Algae collectively refer to a type of simple plants, which usually fix on the seabed or a solid structure and are single plants or a long string of plants consisting of cells. A large number of algae are aquatic plants that do not exhibit the difference between stem and leave. Algae include many kinds of species, which vary considerably in shape and across a variety of living bodies. However, their common grounds are living in the sea and photosynthesis of organic matters by utilizing their internal pigments. The structures of the alginate have the following two types:

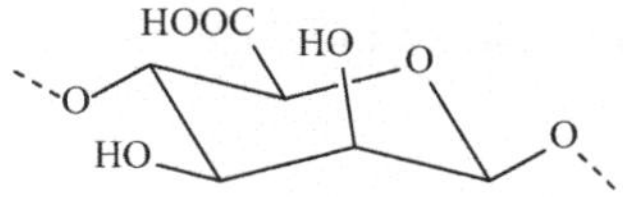

β−D−Mannuronate

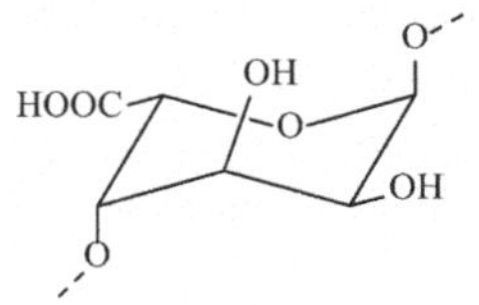

α−L−guluronate

1. Sargassum fusiforme contains alginic acid 15.32−32.18%, mannitol 2.21–7.87%, iodine 32.2–84.2%, potassium oxide 3.23–11.67%, total ash content 19.72–37.53%, Sargassum fusiforme polysaccharide A (SFPP), Sargassum fusiforme polysaccharide B (SFPPR), Sargassum fusiforme polysaccharide C (SFPPRR), and brown algae starch (laminarin).
2. Sargassum pallidum includes alginic acid, mannitol, iodine, potassium, crude protein, ash, sargassum polysaccharide (sargassum), as well as phospholipids dominated by cephalin.

   ① Sargassum fusiforme contains alginic acid (also known as seaweed acid, brown seaweed acid, alginic acid) 20.8%, crude protein 7.95%, mannitol 10.25%, ash 37.19%, potassium 12.82%, and iodine 0.03%.

   Sargassum pallidum contains 19.0% of algae acid, 9.69% of crude protein, 9.07% of mannitol, 30.65% of ash, 5.99% of potassium, and 0.017% of iodine. It is also composed of sargassum, consisting of D-galactose, D-mannose, D-xylose, L-fucose, D-glucuronic acid, and polypeptides.

(2) Purification of alginic acid

Among the raw materials that can be used for the preparation of seaweed fiber, the most commonly used is soluble sodium alginate. The treatment of seaweed with dilute acid converts insoluble alginate into alginic acid, followed by the extraction with the addition of alkali and heating. The resulting sodium alginate is soluble in water. After filtration, calcium alginate is precipitated by the addition of calcium salt. Treatment the collected calcium alginate with acidic solution affords insoluble alginic acid, which is further treated with dehydration, alkalization, and drying to afford the sodium alginate [12]. The alginate fiber is usually prepared by the wet spinning method. First, the soluble sodium alginate is dissolved in water to form a viscous spinning solution. After filtered and deaerated, the alginate solution is extruded through a spinneret hole into a coagulation bath containing divalent metal cations to form a solid and insoluble alginate filament. The filament is stretched, washed, dried and crimped to form fibers (the wet spinning process is as follows: alginate → dissolve → filter → deaeration → spinning → stretching → washing → drying → winding) [13]. The preparation and morphology of the various alginates are shown in Fig. 12.1.

Fig. 12.1 The structure and morphology of alginic acid. **a** Alginic acid in the state of seaweed on the seabed; **b** dried but unpurified alginic acid; **c** micro-scale processed alginic acid fiber; **d** prepared alginic acid powder

Alginic acid containing greater than 98% fiber can be obtained through the above-mentioned various processing methods. The prepared alginic acid fiber can be used for a variety of industrial applications.

12.4 Alginate Nitrates

The nitration of alginic acid has been studied in the early of last century [9]. The nitration with a mixed acid containing 48% HNO_3, 50% H_3PO_4 and 2% P_2O_5 at 20 °C for 4 h afforded alginate nitrate that contains 9.3% N and 16% COOH. The number of carboxyl groups did not increase during the nitration process, and the product was almost completely soluble in acetone (96.4%).

It is believed that the intramolecular esterification between the carboxyl group and hydroxyl group accounts for the solubility of the alginic acid. Thus, similar to pectin nitrate, alginate nitrate does not form the linkage that is same to cellouronic acid and cellouronic acid nitrate.

The authors also studied the purification and nitration mechanism of the alginic acid. The alginate fiber supplied by Xia Yanzhi [14] was refluxed in a saponification solution composed of sodium sulfide 1.5%, NaOH 1.2%, rosin 0.8% and castor oil 0.2% for 45 min. After washing and drying, the water absorption ratio of prepared alginate fiber reached 193.5 g, and the content of fiber was more than 98%. The alginate fiber was nitrated with a mixed acid composed of 68.2% acetic acid, 26.8% nitric acid, H_2O 5.0% at 20 °C for 22 min to afford alginate nitrate containing 14.83% nitrogen. The prepared alginate nitrate has been applied as propellant and has exhibited superior mechanical properties to those of nitrocellulose products, thus demonstrating promising prospects in the application of propellant.

References

1. He FH, Ning SZ et al (1995) Exploitation and application of chitin, chitosan. Chin J Qinghai Anim Husb Veterin Med Couege 12(1):48–50
2. Influence of nitrogen concentration on algae (2006) J Jinzhong Univ. (06):41–43
3. Meyer KH, Mark H (1951) Biological significance of folding and unfolding of protein molecules. Nature 167(4253):736
4. Fürth O, Scholl R (1933) Über Kolloidale Veränderungen, die durch Manche Antipyretica und Analgetica Hervorgerufen Werden, und Deren Angebliche Beziehung zur Erregbarkeit Nervöser Zentren und zur Leberschädigung. Klinische Wochenschrift 12(13):511–512
5. Schmidt WJ (1934) Z Wiss. Mikroskopie, mikrosk. Technik 50:296; Chem Zentr 1934, I, 3242
6. Pshorygin P, Khait EV (1934) Ber 67:1712
7. Zhou S (1995) Pectin comprehensive discussion. J Anshun Coll Chin 4:76–80
8. Schneider G, Ziervogel M (1936) Ber 69:2530
9. Rogovin ZA, Treyvas MG, Shorygina NN (1949) Zh Prikl Khim 22:1244
10. Deng J (2012) Common algae map in Chinese inland waters. Chang Jiang Press
11. Zhao W (2016) Aquatic biology. China Agriculture Press
12. Zhao X, He JX et al (2018) The performance and resent research advances of alginate fiber. Melliand China 11
13. Chen XL, Song R, et al (2012) Advances in application of Eco-environmental fiber. Synth Fiber China 12
14. Guo CX, Kong QS, Gao JX et al (2011) Dynamic adsorption of methylene blue dye from simulated waste water by alginic acid fiber. Chin J Environ Eng 5(2):297–300

Chapter 13
Sugar Nitrates

Sugar nitrate is an explosive colloidal or solid compound made by the nitration of sugar [1].

Sugar is a necessity for human living, such as glucose, fructose, sugar, lactose, galactose, and sucrose, which have attracted the considerable research attention because these substances are not only edible but also can be converted into powerful liquid explosives when treated with nitric acid or mixed nitric and sulfuric acid [2].

13.1 Physical and Chemical Properties of Sugar Nitrates

Sugar nitrates usually exist in the form of resinous colloidal or powder form with light yellow to yellow appearance.

Molecular formula: $C_6H_7O_{16}N_5$ (glucose, galactose, fructose); $C_{12}H_{14}O_{27}N_8$ (sucrose, lactose, maltose).

Structural formula: sugar has isomers [3].

The structures of different sugar nitrates are shown in Table 13.1.

Relative molecular weight: 405.12 (glucose, galactose, fructose); 702.24 (sucrose, lactose, maltose).

Oxygen balance (%): 1.975; −9.11 (by the generation amount of CO_2).

Nitrogen content (%): 17.28 (glucose, galactose, fructose); 15.95 (sucrose, lactose, maltose).

Oxygen coefficient (%): 103.23 (glucose, galactose, fructose); 87.09 (sucrose, lactose, maltose).

J. Liu, *Nitrate Esters Chemistry and Technology*,
https://doi.org/10.1007/978-981-13-6647-5_13

Table 13.1 The structures of different sugar nitrates

Name	Structure of sugar nitrate
Glucose nitrate	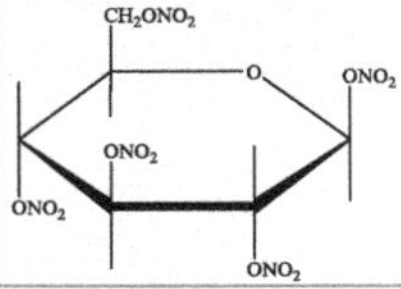
Galactose nitrate	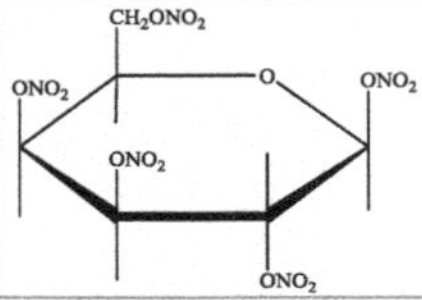
Fructose nitrate	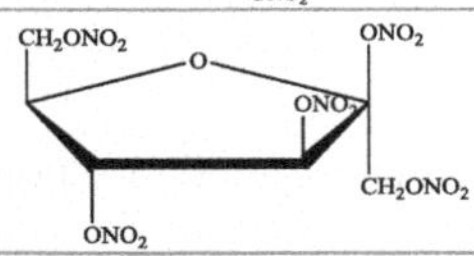
Sucrose nitrate	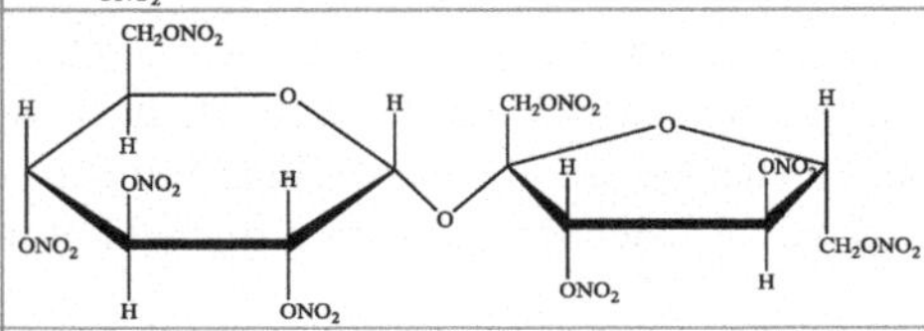
Lactose nitrate	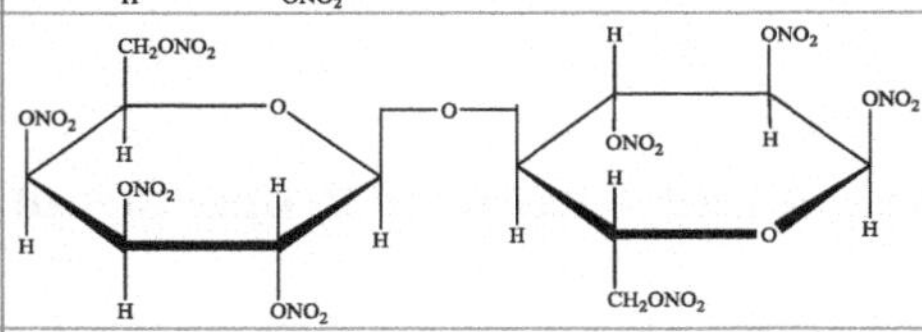
Maltose nitrate	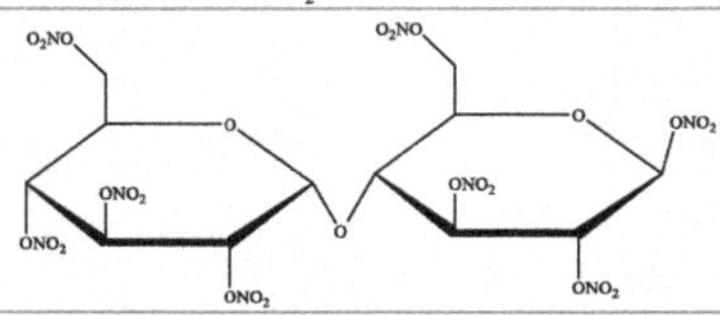

13.1.1 Physical Properties of Sugar Nitrates

The nitration process of sugar is quite straightforward in principle but complicated and challenging in practical terms. As the sugar nitration tends to generate colloid and form resinous material simultaneously, the product exists in the form of colloidal. As a result, the residual nitric acid mixed in the product is difficult to be removed. In addition, the product is very liable because it is difficult to improve the stability to a relatively stable degree through stability treatment. Most of the sugar nitrates can be prepared as crystalline products.

The nitration of sugar is usually carried out below 0° C. The early process for sugar nitration is the precipitation method employing a mixed nitric and sulfuric acid. First, sugar was dissolved in nitric acid and then was poured into the concentrated sulfuric acid at such a rate that the temperature of the reaction mixture did not exceed 0 °C in the entire process. After the addition, the reaction was continued for 20 min. The corresponding crude sugar nitrate was precipitated, which was separated by filtration. The pure product was obtained by washing with ice water and recrystallization. The melting points of different sugar nitrates are shown in Table 13.2.

The nitration of α-methyl-D-glucoside affords α-methyl-D-glucoside tetranitrate $C_7H_{10}O_2(ONO_2)_4$ with a melting point of 40–50 C, the weight loss of 0.7% at 50 C for 120 h, and decomposition temperature of 135 °C, which is much more stable than that of average glucose nitrate. The appearance of corresponding α-methyl-D-mannoside tetranitrate is needle-like with a melting point of 36 °C. α-Methyl-D-mannoside tetranitrate emits a small amount of gas at 50 °C, and the mass of α-methyl-D-mannoside tetranitrate increases with increasing heating temperature. The nitration of α-alkyl-D-glucoside with a nitrating reagent consisting of nitric acid, acetic anhydride, and acetic acid affords the corresponding nitrate. The melting points of different structures of glucose nitrates are shown in Table 13.3.

The weight loss of the above compounds at 100 °C is shown in Table 13.4.

In addition to the identification of preparation method and explosive characteristics of sugar nitrates, the authors found that sugar nitrates can be reduced by the complete denitration, but the weight of the reductive product decreased. These findings lay the foundation for the synthesis of a series of new sugars, such as D-glucose, D-galactose, and D-fructose methyl ether.

In the presence of iodomethane and silver oxide, methyl-β-D-glucopyranoside 2,3-dinitrate could be converted into the corresponding 4,6-dimethyl ether nitrate, which was treated with trityl chloride and acetic anhydride in pyridine to yield the final product methyl-4-D-acetyl-6-D-trityl-β-D-glucoside 2,3-dinitrate.

With nitrating agent that is milder than a mixed nitric and sulfuric or a mixed nitric and phosphoric acid, the nitration of sugar is not only easy to control but also generates fewer impurities. For example, the nitration of β-D-glucopyranose with nitric acid in chloroform produced 2,3,4,6-tetra-D-acetyl-α-D-glucosyl nitrate and impurities in which acetyl groups attached to the carbon atoms are replaced with nitrates [4].

The crystalline product, 2,3,5-trimethyl-α-D-glucopyranoside with the melting point 132–135 C was prepared by the nitration of trimethyl glucose ether and triacetyl-1,6-anhydroglucopyranose using nitric acid chloroform solution containing nitrous oxide [5]. Many experiments have demonstrated that nitrous oxide dissolved in chloroform is an excellent oxidizing and nitrating agent for the sugar nitration [5], which can be used for the preparation of 1,2,4,5-di-D-isopropylidene-D-fructose-3-nitrate [6]. With this nitrating agent, most of methyl-4,6-D-ethylidene-β-D-glucoside can be converted to 2,3-dinitrate at the temperature of below 0 °C for about 5 min [7].

Table 13.2 Melting points of different sugar nitrates

Sugar type	Chemical formula and nitration degree	Recrystallization solvent	M.P. (°C)	Stability	
				Mass loss at 50 °C	Decomposition temperature (°C)
Monosaccharide					
L-Arabinose	$C_5H_6O(ONO_2)_4$	–	85	960 h ca. 40%	120
D-Xylopyranose	$C_5H_6O(ONO_2)_4$	–	liquid	–	–
D-Xylopyranose	$C_5H_7O_2(ONO_2)_3$	–	141	600 h ca, 1.1%	–
L-Rhamnose	$C_6H_8O_2(ONO_2)_4$	–	135	720 h ca. 1.2%	–
L-Rhamnose	$C_6H_9O_3(ONO_2)_3$	–	ca.100	500 h ca. 6.0%	–
D-Galactose	$C_6H_7O(ONO_2)_5$	Ethanol	α 115–116	24 h ca. 42%	125
			β 72–73	24 h ca. 46%	125
D-Glucose	$C_6H_7O(ONO_2)_5$	Ethanol	liquid	1 day ca. 38%	135
D-Mannose	$C_6H_7O(ONO_2)_5$	–	81–82	24 h ca. 46%	124
α-D-Glucose	$C_7H_8O(ONO_2)_6$	–	100	–	–
Disaccharide					
Maltose	$C_{12}H_{14}O_3(ONO_2)_8$	Methanol	164–165	1080 h ca. 43%	170–180
Lactose	$C_{12}H_{14}O_3(ONO_2)_8$	Methanol or ethanol	145–146	960 h ca. 40%	135
Lactose	$C_{12}H_{16}O_5(ONO_2)_6$	Methanol or ethanol	70	–	–
Sucrose	$C_{12}H_{14}O_3(ONO_2)_8$	Methanol or ethanol	85	72 h ca. 11%	135
Trehalose	$C_{12}H_{14}O_3(ONO_2)_8$	Methanol or ethanol	124	–	136
Trisaccharide					
Raffinose	$C_{18}H_{21}O_5(ONO_2)_{11}$	–	55–65	–	136

Table 13.3 The melting points of different sugar nitrates

Sugar nitrate	Melting point (°C)
Methyl-α-D-glucopyranoside tetranitrate	48
Methyl-β-D-glucopyranoside nitrate	116.5
Ethyl-D-xylopyranoside trinitrate	95.5
Ethyl xyloside trinitrate	95.5
Methyl riboglycoside heptanitrate	134
Methyl-α-D-mannoside tetranitrate	36

Table 13.4 The weight loss of different sugar nitrates at 100 °C

Sugar nitrates	Heating time (h)	Weight loss (%)
Methyl-α-D-glucopyranoside tetranitrate	20.0	2.87
Methyl-β-D-glucopyranoside nitrate	20.0	4.69
Ethyl-D-xylopyranoside trinitrate	20.0	3.0
Ethyl xyloside trinitrate	20.0	2.66
Methyl riboglycoside heptanitrate	20.0	0.87
Methyl-α-D-mannoside tetranitrate	20.2	2.92

13.1.2 Chemical Properties of Sugar Nitrates

Sugar nitrates, including disaccharide and trisaccharide nitrates, can reduce the Fehling's reagent at the elevated temperature. In addition, sugar nitrates can occur reductive denitration reaction and other typical chemical reaction of nitrates.

One of these reactions is iodination reaction of the primary carbon atom linked to the nitrate group under the conditions of heating sugar nitrate with the Finkelstein reagent [8] (sodium iodide solution in acetone) for a long time. In this reaction, it seems that the CO bond is cleaved.

$$-CH_2-ONO_2 \quad + NaI \longrightarrow CH_2I + NaNO_3$$

In the presence of Finkelstein reagent, the nitrate group linked to the secondary carbon atom is hydrolyzed to produce an OH group rather than proceeding an iodination reaction [9].

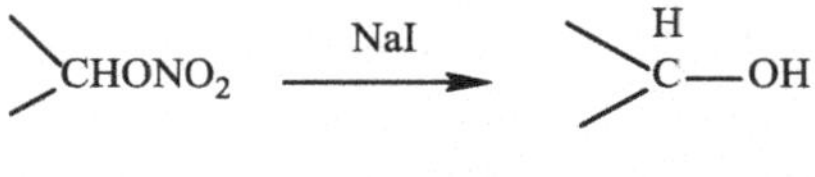

The difficult degree of this reaction depends mainly on the position of the carbon atom linked to the hydroxyl group.

The reaction of methyl-2,3-di-O-methyl-β-D-glucose with sodium iodide in acetone at 100 °C for 20 h yielded 6-deoxy-6-iodo-2,3-di-O-methyl-β-D-glucoside, as well as a small amount of tetranitrate [10]. As for the β-D-glucopyranoside tetranitrate, the nitrate group attached to the C4 carbon atom is more reactive than that attached to C3-nitrate group.

Methyl-4,6-O-ethylidene-β-D-glucoside 2,3-dinitrate was heated with NaI at 100 °C for 20 h to afford 3-nitrate [11] in good yield.

The hydrolysis mechanism of secondary nitrate group with sodium iodide seems to be different, which is likely the cleavage of the O–N bond:

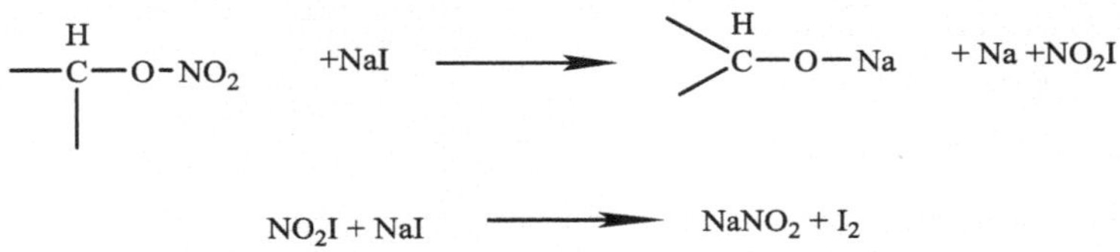

Among other characteristic chemical reactions of sugar nitrate, the reaction of sodium nitrite solution with sugar nitrate is noteworthy, which is usually used to remove the NO_2 group attached to the secondary carbon atom. For example, methyl-4,6-O-ethylene-α-D-glucoside and 4,6-O-benzylidene-α-D-glucosyl 2,3-dinitrate were converted into their corresponding 3-nitrate sugars, respectively, after refluxing with sodium nitrite for 12 h. Only under specific circumstances, the primary nitrate group could occur such as exchange reaction.

The reaction of highly nitrated sugar with pyridine or hydroxylamine alone is very mild [12]. In contrast, the reaction of sugar nitrate with a mixed pyridine and hydroxylamine solution is very intense. By using this method, methyl-α-D-glucopyranoside tetranitrate was converted to 2,3,6-trinitrate (28% yield), 3,6-dinitrate (17% yield), and other sugar nitrates with lower degree nitration. This is also the best way to remove the nitrate that is attached to the C4 carbon atom in sugar nitrate.

Sugar nitrates are hydrolyzed in alkaline media to produce complex mixtures [13].

Since base reacts with carbonyl compounds vigorously, the hydrolytic products of carbonyl in alkaline media are more complicated than those of hydrolysis of simple nitrate compound [14]. It should be noted that relatively mild basic reagents, such as barium carbonate in refluxing methanol solution, can also be used to hydrolyze nitrate compounds. For example, 2,3,4-tri-O-methyl-α-D-glucose 1,6-dinitrate was hydrolyzed to the corresponding tri-O-methyl-α-D-glucopyranoside under the conditions of barium carbonate in refluxing methanol.

When methyl-α-D-glucopyranoside 6-nitrate was treated with a base, methyl-3,6-dehydrated-α-D-glucopyranoside [15] can be prepared in high yield.

Numerous studies have demonstrated that 4,6-O-benzylidene-α-D-glucosyl 2,3-dinitrate occurred two forms of heterolytic fission reactions in the presence of sodium methoxide in methanol.

2,3-Dehydro-α-D-allopyranoside-3-nitrate and methyl 4,6-O-benzylidene-α-D-glucoside were obtained in a low yield of 7.5%. The mechanism of the formation of these two products is as follows:

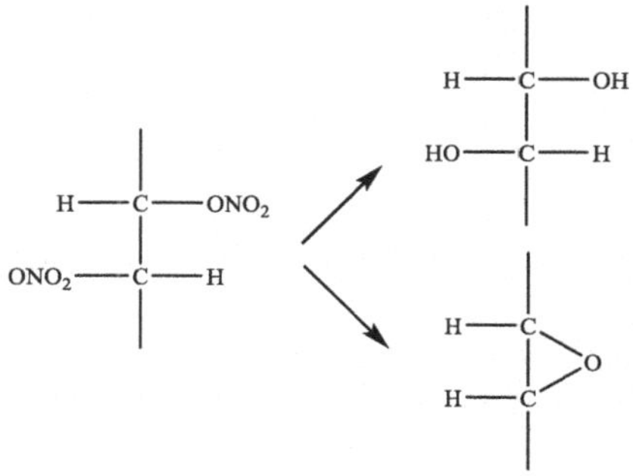

These two decomposition pathways can be confirmed by using a basic aqueous solution containing ^{18}O isotope [16].

13.1.3 Explosive Properties of Sugar Nitrates

The authors have studied the explosive properties of sugar nitrates [17]. The explosive properties of sucrose octanitrate were determined as explosion heat of 3974.8 J/g, lead block expansion value of approximately 300 cm^3. In the drop-weight sensitivity experiment, 2 kg of weight from the height of 20 cm dropped onto the material to cause 40% explosion. The maltose octanitrate exhibited a lead block expansion value of 260 cm^3.

In the case of the explosive properties of methyl-α-D-glucopyranoside tetranitrate [16], it is sufficient to detonate the sample with a density of 1.10 by using a relatively weak detonating agent, such as 0.25 g of mercury fulminate. Only the casting product (density 1.65) needs a strong detonator (1.5 g or more of mercury fulminate).

The results of the drop-weight sensitivity experiment of methyl-α-D-glucopyranoside tetranitrate showed that 1 kg of weight from the height of 20 cm dropped onto the material to cause 50% explosion, which was similar to the drop-weight sensitivity of Hexogen.

The relative lead block expansion value of methyl-α-D-glucopyranoside tetranitrate was 130 cm^3, whereas the lead block expansion value of picric acid was 100 cm^3. The detonation velocity of methyl-α-D-glucopyranoside tetranitrate increased with the increase of the charge density. The relationship between the change in charge density and the detonation velocity in a 30 mm thick paper tube is shown in Table 13.5.

Table 13.5 The relationship between the change in charge density and the detonation velocity

Density (g/cm^3)	Detonation velocity (m/s)
1.10	5630
1.50	6970
1.60	7560
1.62	7820
1.70	8340

Fused cast charge in 20 mm diameter (density 1.65) reached detonation velocity of 7830 m/s.

The explosion strength of the methyl-α-D-glucopyranoside tetranitrate is only slightly lower than that of hexogen. The mixture of methyl-α-D-glucopyranoside tetranitrate with dinitrotoluene and ammonium nitrate was also investigated as a mixed explosive.

The mixture of sugar nitrate and nitroglycerin has been only utilized in practical applications in the United States. For the preparation of such explosive, a mixture containing 20–25% sucrose and 75–80% glycerol is nitrated to produce an oily nitrate compound [18], which is used to replace pure nitroglycerin for the production of dynamite. The manufacture of this type of mixed explosive is due to the economic consideration in the United States, where sucrose is cheaper than glycerol [19].

The commercial name of mixed sucrose and glycerol nitrate product is Nitrohyrene. Since the viscosity of Nitrohyrene is much higher than that of nitroglycerin, the stability treatment and washing of Nitrohyrene are much more difficult. In particularly, Nitrohyrene tends to form aqueous emulsions. After washed with sodium carbonate solution, Nitrohyrene was treated with a sodium chloride solution to facilitate its complete separation from water. The composition of Nitrohyrene is slightly different than the original composition of ingredients. For example, the nitration of a mixture containing 80% glycerol and 20% sucrose produced Nitrohyrene 80/20 having 86% nitroglycerin and 14% nitrated sucrose. Nitrohyrene 75/25 contained 80% nitroglycerin and 20% nitrated sucrose.

The stability of Nitrohyrene is much worse than that of nitroglycerin. In the mixture of sucrose and glycerol nitrate, the higher the percentage of nitrated sucrose, the lower the stability. The combination of wood flour and dynamite can act as a stabilizer [19], which can remove impurities and decomposition products by adsorption and reaction with them. In order to achieve prolonged shelf life, 0.1–0.2% of the aniline is added to the Nitrohyrene as a stabilizer. The mixture produced by this method can be stored at 55 °C for 75 days. The lead block expansion values of Nitrohyrene are shown in Table 13.6 [19]:

Table 13.6 The lead block expansion values of Nitrohyrene

Name	Lead block expansion value (cm^3)
Nitroglycerin	550
Nitrohyrene	(80/20)530
Nitrohyrene	(75/25)515

13.2 Preparation Process of Sugar Nitrate

The industrial preparation process of sugar nitrate is the same as that of other nitrates. Because of its different properties, the preparation process of nitrate has some differences. The production process of sugar nitrate is shown in Fig. 13.1.

According to recent experimental results [12], the nitration with a mixed acid containing 2 parts of anhydrous nitric acid, 1 part of acetic anhydride, and 1 part of acetic acid could afford a quantitative yield of sugar nitrate. Moreover, the resulting product is easier to be crystallized than the product obtained by using a mixed nitric and sulfuric acid. The properties of the sugar nitrate are shown in Table 13.7.

Most of these nitrates can be crystallized from aqueous or anhydrous alcohols. These compounds are unstable. When heated at 100 °C, the weight losses of these compounds are shown in Table 13.8:

Sucrose nitrate can be prepared by using a mixed acid containing anhydrous nitric acid and 17% fuming sulfuric acid in the presence of urea nitrate at the

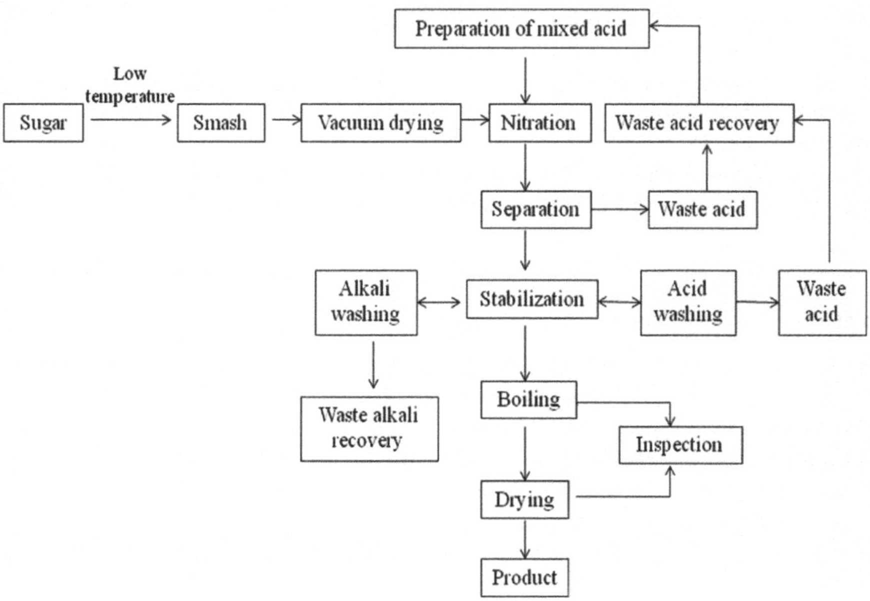

Fig. 13.1 Basic manufacturing procedure of sugar nitrate

Table 13.7 Properties of sugar nitrate

Sugar nitrate	Melting point (°C)	Ignition point (°C)
D-α-Dehydro-Glucose pentanitrate	53.5	200
D-β-Dehydro-Glucose pentanitrate	111	257
D-α-Glucopyranose tetranitrate	73.5	250
β-Cellobiose octanitrate	140	287

Table 13.8 The decomposition of several sugar nitrates

Sugar nitrates	
Compound (1)	60% within 10 h
Compound (2)	60% within 10 h
Compound (3)	60% within 10 h
Compound (4)	35% within 20 h

reaction temperature of below +2 C. The resulting sucrose nitrate was washed with water to neutral, then washed with a low concentration of sodium carbonate solution and water, and crystallized from ether–alcohol mixture. Maltose nitrate can also be prepared in the same procedure.

Monosaccharide reacted with a mixed acid in 0–15 °C for 48 h yielded highly condensed pentose and hexose nitrates as the major products. The molecular formula and melting point of D-sugar are shown in Table 13.9.

When nitric acid is used as the nitrating agent for D-glucose, the main product is methyl-β-D-glucoside 2,3,4,6-tetranitrate, which could be attributed to the strong hydrolytic ability of nitric acid that cleaves acetal in sugar.

The conversion of methyl 4,4-O-benzylidene-α-D-glucoside into the corresponding 2,3-dinitrate [2, 20] was accomplished without affecting the acetal and aromatic rings.

2,3,4,6-Tetra-O-acetyl-α-D-glucosyl chloride [21] was treated with silver nitrate in diethyl ether to afford 2,3,4,6-O-acetyl-α-D-glucose nitrate.

Table 13.9 Melting points of different D-sugar

Name	Sugar nitrate	Formula	Melting point (°C)
D-Xylose	Xylan dinitrate	$C_5H_6O_2(ONO_2)_2$	75–80
D-Galactose	Galactan trinitrate	$C_6H_7O_2(ONO_2)_3$	Ca. 76
D-Glucose	Glucose trinitrate	$C_6H_7O_2(ONO_2)_3$	82
D-Fructose(α isomer)	Fructan trinitrate	$C_6H_7O_2(ONO_2)_3$	139–140
D-Fructose(β isomer)	Fructan trinitrate	$C_6H_7O_2(ONO_2)_3$	48–52
D-Sorbose	Sorbitan trinitrate	$C_6H_7O_2(ONO_2)_3$	ca. 40–54

References

1. Liu JP (1987) Explosive and its chemical basis. Guizhou People's Publishing House
2. Liu JP (2015) Liquid explosive. Springer
3. Li WH. Yang GP, et al (1994) Studies on some characteristics of nitrate reductase from sugar beet(BetavulgarisL)Leaves. J Northeast Agric Univ (1):20–25
4. Kónigs W, Knorr E (1901) Ber. 34(957):4343
5. Holdham JW (1925) J Chem Soc 127:2840
6. Montgomery TN (1934) J Am Chem Soc 56:419
7. Synge RKM (1937) J Chem Soc 1711(1938):836
8. Finkelstein H (1910) Ber 43:1528
9. Ostrinskaya A et al (2016) Characterization of nitrate sugar alcohols by atmospheric-pressure chemical-ionization mass spectrometry. Rapid Commun Mass Spectrom
10. Oldham JWH, Rutherforo JK (1932) J Am Chem Soc 54:366
11. Ansell EG, Honeyman J (1952) J Chem Soc 2778
12. Hayward LD, Purves CB (1954) J Can Chem. 32:19
13. Su Y, Yang L (2003) Sugar complexes with neodymium nitrate. Spectroscopic and structural studies of two ND(NO3) 3-galactitol complexes. Carbohydr Reseach 338(19):2029–2034
14. Honeyman J, Morgan JWH (1953) J Chem Ind 1035
15. Gladding EK, Purves CB (1944) J Am Chem Soc 66:79
16. Anbar M, Dostrovsky I, Amuel DS, Yoffe AD (1944) J Chem Soc 66:76
17. Monasterski S (1933) Z.ges.Schiess-u.Sprengstoffw. 28:349
18. Bjórkman, Brit. Pat. 2483, 1880; Cocking and Kynoch, Brit. Pat. 17221, 1911
19. Naoüm PH (1926) Nitroglycerin u.Nitroglycerinsprengstoffe. Springer, Berlin
20. Caslin BD et al (1970) Rapid determination of nitrate-nitrogen in sugarbeets with the specific icon electrode. Am Soc Sugar Beet Technolo J 16(1):64–70
21. Schlubach HH, Stadtler, WI (1928) Ber 61:287

Chapter 14
Nitrate Esters of Lignin and Rosin Acid

On Earth, except mosses and fungi, all plants contain a large amount of lignin, which is a renewable organic resource. After a few decades or a longer period, fossil resources will run out and it will be one of the main sources of organic compounds, especially aromatic compounds [1].

Lignin has been used in dozens of departments in foreign countries to produce many products; while in Asia it is only at the beginning of lignin development and utilization with some limited products [2].

The complex structure of lignin cannot be expressed in simple language, it only can be said that lignin is a characteristically aromatic and noncrystalline three-dimensional polymer network with a structural unit of phenylpropanoid. In accordance with the different types of plants, lignin can be classified into three lignin categories, including softwood, hardwood, and herb lignins [3]. Coniferous wood lignin is composed mainly of pockwood-based propane unit, hardwood lignin is primarily from pockwood-based and lilac-based propane units, and herb lignin includes majorly pockwood-based propane unit, lilac-based propane unit, and p-hydroxy benzopropane unit [1, 2]. The structure of these units is as follows:

J. Liu, *Nitrate Esters Chemistry and Technology*,
https://doi.org/10.1007/978-981-13-6647-5_14

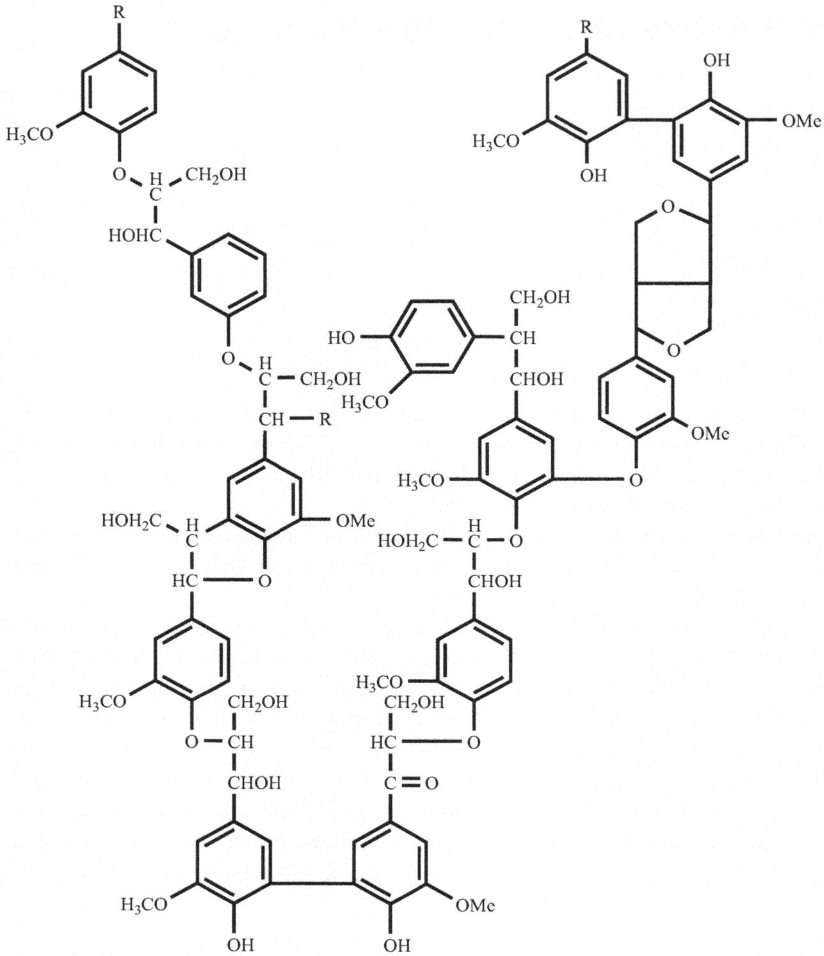

14.1 Characteristics of Hemicellulose and Pentose Nitrate

14.1.1 Nitrate Esters of Hemicellulose and Pentosane

Some scientists had run nitration experiments [4] of pentosane-containing hemicellulose from concentrated nitric-acid-treated cellulose material, in which water or sulfuric acid were used to precipitate and separate the nitrated product. And the obtained product was insoluble in ether–alcohol.

Table 14.1 Nitration of hemicellulose

Temperature (°C)	Nitrogen content of product (%)	Nitrated hemicellulose from 100 g of hemicellulose (g)
20	7.24	50.75
10	8.30	107.5
0	8.88	113.2

Nitration of hemicellulose is performed with mixed acid used in cellulose nitration. For example, the used mixed acid (used in the production of cellulose nitrate with nitrogen content of over 13%) contains the following:

HNO_3	25.10%
H_2SO_4	66.85%
H_2O	8.05%

Although the yield of lighted nitrated product from the hemicellulose is very low, it has strong explosive properties as listed in Table 14.1.

Some of these as-made nitrated products are soluble in ether–alcohol, and all can be dissolved in ethyl acetate.

With a diluted acid (16.5–21.0% in H_2O) at 20 °C in the nitration of hemicellulose, the nitrogen content of the product was very low (5.8–7.2%), and the product was almost insoluble in the mixture of ether and alcohol.

The pure Xylan isolated from straw cellulose was also nitrated with concentrated mixed nitration acid with the abovementioned proportion, and nitrated Xylan as a white powder was produced with 8.76% of nitrogen and a yield of 115.7%.

Only 61% of the as-produced was soluble in ether–alcohol and the dissolved contained 7.13% of N, and the insoluble part had 9.22% of N. Many other studies in this area had also been reported in the literature.

14.1.2 Nitrate of Fiber Uronic Acid (CMC)

After the treatment of cellulose with nitric tetroxide, the fiber uronic acid is generated. Then, it can be used to make nitrates. Compared with the cellulose, the nature of nitrate is changed significantly, even with only a few of carboxyl groups.

In order to show characteristics of the nitrated ester product, the published information [5] was collected and listed in Table 14.2.

From the above values, it can be clearly seen that, because of 2–3% COOH (a carboxyl group in every 8–10 dehydrated glucose rings), the prepared product is difficult to dissolve even in acetone, the excellent solvent of cellulose.

These authors believed that the solubility of nitrate of fiber uronic acid was very low due to the cross-linking between the adjacent carboxyl of oxidized cellulose long-chain molecules and free hydroxyl groups. Because part of the primary

Table 14.2 Nitration results of fiber uronic acid [5]

COOH content in raw material (%)	Content in product (%)	Solubility in acetone (%)
1.56	13.5	18.2
3.6	13.0	9.7
8.0	11.5	13.2
18.5	7.0	2.5

alcohol groups was replaced by carboxyl groups, the strength of the glycoside bond, linking dehydrated glucose units together, was significantly weakened. Therefore, CMC with 2–3% of COOH could start to depolymerize even at room temperature when there was a diluted base.

14.1.3 Nitrate of Dialdehyde Cellulose

As reported in [6], in the dehydrated glucose units of periodate-oxidized cellulose, because of the oxidation of the adjacent secondary hydroxyl groups (α-diol), two aldehyde groups could be formed as the following,

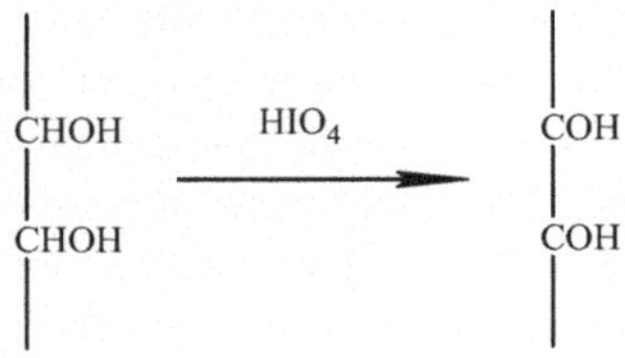

As reported in [7], the nitrate of dialdehyde cellulose was from raw materials with a different number of aldehyde groups and a mixed acid (including 48% of HNO_3, 50% of H_3PO_4, and 2% of P_2O_5) at 0 °C.

The difference between this product and nitrocellulose is its poor high hygroscopicity and different solubility in organic solvents. All data were listed in Tables 14.3 and 14.4.

Table 14.3 Hygroscopicity of nitrate of formaldehyde-contained cellulose esters

Oxidation degree (number of formaldehyde groups per 100 dehydrated glucose base units)	Nitrogen content in product (%)	Hygroscopicity (water absorbing) (%)
0 (cellulose)	13.85	0.82
7.0	13.66	1.02
43.8	12.69	2.64
72.0	11.91	3.50
144.6	9.42	11.02

Table 14.4 Hygroscopicity of formaldehyde-contained cellulose nitrate

Oxidation degree	Nitrogen content in product (%)	Hygroscopicity (water absorbing) (%)
0 (cellulose)	13.85	100
5.2	13.75	44.4
14.0	13.11	27.6
43.8	12.7	3.4
73.3	11.8	3.8
117.4	10.3	3.3
114.6	9.4	1.6

14.1.4 Copolymer of Cellulose Nitrate

Recently [8], vinyl chloride or vinylidene chloride was used to prepare graft copolymer of cellulose nitrate, in which nitrocellulose-free groups was generated first (it is more effective through γ-irradiation or ozone oxidization of nitrocellulose).

Then, when the oxidization catalyst existed, the monomer of vinylidene chloride or vinyl chloride reacted with a large group, and the following structure of copolymer was generated,

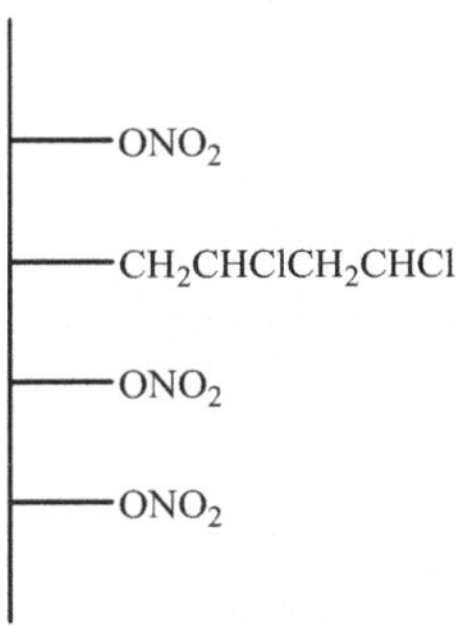

Compared with nitrocellulose, the main characteristics of this copolymers were its much lower ignition and flammability performance, which was very valuable to the manufacture of heat-resistant and nonflammable film, paint, and celluloid.

14.2 Nitrate Ester Derivatives of Lignin

Wood contains complicated complexes from different chemicals, in which the complex, that can be destroyed with the mixture containing potassium chlorate and nitric acid, is known as lignin and hemicellulose. A large part of the lignin molecule has the aromatic nature [1, 2], so, as you can imagine, it should be easy to participate in the nitration reaction. However, it was found that, when there was diluted nitric acid, lignin oxidation ability was specially dominating, which could be explained with that the aromatic part of lignin molecule was derived from catechol. The reaction of lignin with nitric acid forms the basis to separate cellulose from wood pulp, in which diluted nitric acid (3–10%) was used to treat wood pulp [1, 9].

The reaction can be used as an analysis method or an industrial method. According to the reported method in reference [10], 3–6% of nitric acid was used to produce cellulose from wood pulp.

In the reported nitration of lignin (isolate from wood pulp with hydrochloric acid in advance) [11], the mixture of fuming nitric acid and lignin was heated in a water bath until lignin was completely dissolved. The solution was diluted with water to precipitate the nitration product. The main product in the aqueous solution was oxalic acid. These authors did not study more closely the method to generate nitrated lignin.

The authors studied the influence of nitric acid or mixed nitric acid and sulfonic acid on lignin and found that the base ring of hardwood lignin contained 4-hydroxy-3,5-dimethoxy,

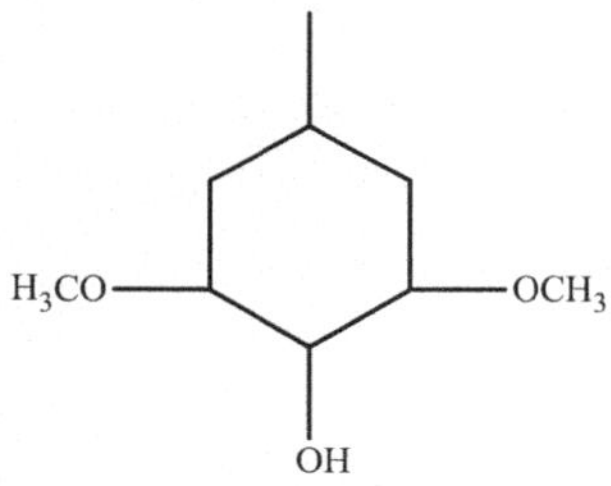

While the lignin of the tree did not contain this mentioned group.

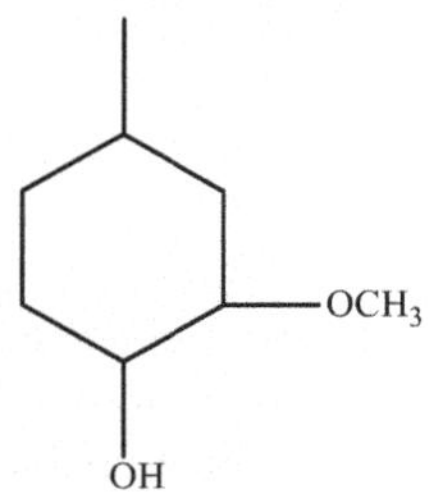

Therefore, different lignins with different structures certainly has different products of nitration.

As in the reported systematical studies [12], 31.5% of nitric acid was suited to treat the lignin at room temperature and nitrate the lignin under slow heating. It was found that the nitrogen content of the product was only 4.3% with the chemical formula of $C_{39}H_{29}O_{21}N_3(OCH_3)_3$ according to the elemental analysis. Others [13] reported that product with 3% of nitrogen was obtained in the nitration of lignin with nitric acid at a low temperature.

According to Ref. [14], in the lignin nitration with nitric and sulfonic mixed acid, water, and ice were used to precipitate and isolate the nitrated product.

These as-prepared nitrated lignins were a red powder with the following formula:

$$C_{42}H_{39}O_{13}(NO_2)_3$$

In the direct nitration of lignin [15] from pulp, nitration with an alcoholic solution of nitric acid (density of 1.40) was done at the boiling point of acetic acid. Under these conditions, cellulose remained unchanged. However, lignin was nitrated and dissolved in alcohol, and then nitrated lignin was separated from the solution with water (yellow amorphous powder). The yield of nitration of lignin was 15–17% of pulp weight. According to the different types of wood used to produce lignin, the nitrogen content of fully nitrated product was in the range of 3.0–4.5%.

According to the experimental results, the composition of the functional group in nitrated lignin was proposed as

$$C_{18}H_{13}O_2(COOH)(OH)_2(OCH_3)(NO_2)$$

In order to separate lignin and sugars, some [16, 17] used strong reagents (mixed acids) in the pulp nitration experiment. With this method, nitrated lignin with high nitrogen content was prepared.

When nitric acid and sulfuric acid were mixed at a ratio of 1:2, products containing 8.45% of N were made with a yield of 55%. Oxygen-bonded nitrate groups occupied 7.1% and carbon-bonded ones only occupied 1.35%. These authors reported, with nitric sulfonic mixed acid or a mixed acid from nitric acid, acetic acid, and acetic anhydride, the proportional relationship between oxygen-bonded nitrate groups and carbon-bonded ones was roughly the same.

One notable property of nitrated lignin is that, after the reduction, nitrated lignin does not form a detectable amino group, and the reduced product cannot be diazotized, although the reason of this phenomenon is still a matter of controversy. However, in Ref. [15], it was possible to generate a lactam ring from the reaction between amino and carboxyl group.

With nitric acid, especially in the lignin treatment process with diluted nitric acid, hydrocyanic acid was released, which can be interpreted with that the generated nitrosophenol tautomerically shift into benzoquinone dioxime, which was then oxidized and hydrolyzed. In these reactions, hydrocyanic acid was decomposed through a similar way reported in [18].

Based on these experiments, although the lignin has aromatic characteristics, it was inappropriate to use lignin to make dynamite and the nitro compounds.

However, these experiments help us to fully understand the chemistry of lignin. From this point of view, reported results in [19] (nitric anhydride or nitrogen dioxide was used to treat lignin) are also important.

It is most notable that, in the experiment with a chloroform solution of nitrogen pentoxide, the as-made product contained 4.7% of nitrogen as reported, which indicated a unit contained 0.9 nitro aromatic rings. The author believed that, in the process of nitration, methoxy was removed partially. Therefore, the reaction product contained methanol and methyl nitrite. The following reaction could be used to explain this fact.

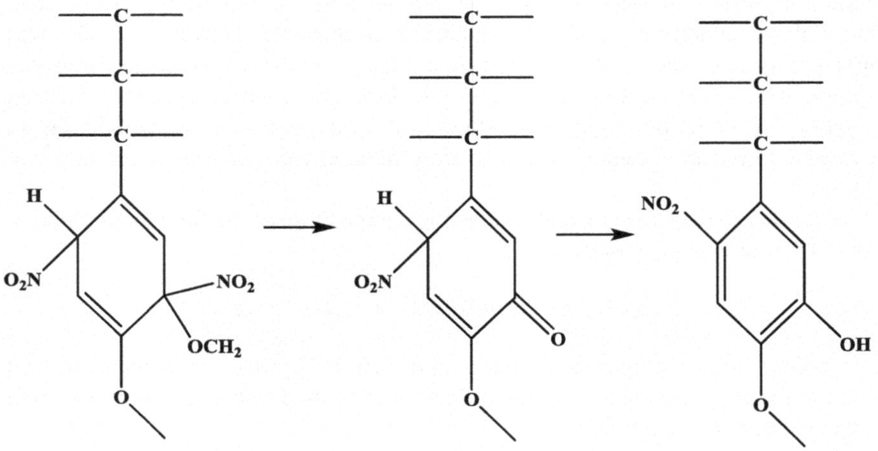

(¢ñ) £¨¢ò£© £¨¢ó£©

Molecular formula (I) represents the original nitration product of lignin, and (II) is a semi-quinone that was converted into p-nitrophenol (III).

When lignin samples from different sources were nitrated with 60% of nitric acid, nitrated derivatives with similar nature could be obtained. Lignin and nitrogen dioxide can rapidly react. The ratio of nitrogen content over methoxyl content was approximately 1:1.

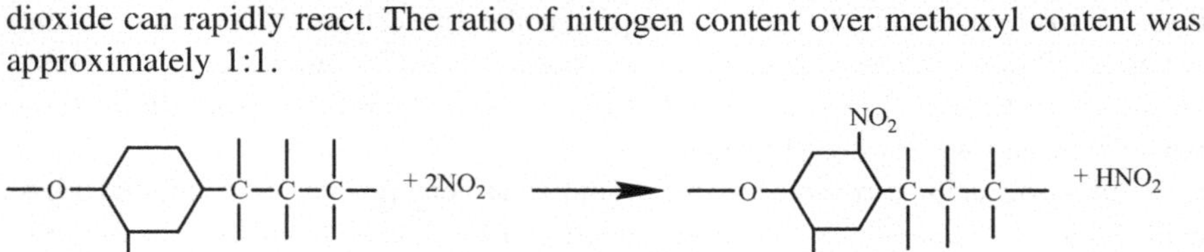

With the mixed nitration reagent from concentrated nitric acid, phosphoric acid, and acetic acid in the nitration of wood, oxidation of the nitrated product was not found. The weight of the nitrated product was increased by 50–52% from the original, the determined nitrogen content was 12.0–12.5%. In the nitration of wood, wood cellulose formed nitrocellulose and lignin was turned into nitro compounds with a content of 9–10.2%. Lignin is easier to form nitro compounds than cellulose into its nitrate.

When mixed nitration acid with 15.9% of water was used to nitrate cuprammonium lignin and hydrochloric acid lignin, the lignin content was 9.1–9.3% in the as-obtained nitrated lignin, in which the nitro nitrogen content was 4.1–5.6%. When the nitration was run at 0 °C, the recovery was 120%, and nitration yield was only 100% at 20 °C. When mixed nitration acid with 7.4% of water was used, nitrogen content could only reach to 8.4%, while the yield of nitrocellulose was 83%. When the water content in mixed acid dropped to 3% and nitration temperature was 16 °C, the recovery rate of nitrated lignin was 34% with a nitrogen content of 8.8%. These experiment results showed that lignin nitration did not exactly follow the nitration mechanism of cellulose and the reaction conditions. Therefore, it has very important significance to study the nitration mechanism of lignin, the storage and usage of the nitrated lignin products, and their explosion mechanism, as the future technology reserve.

14.3 Nitrate of Rosin Acid

The main composition of distilled solid residues–rosin or resin from coniferous tree oil (turpentine) is mainly abietic acid,

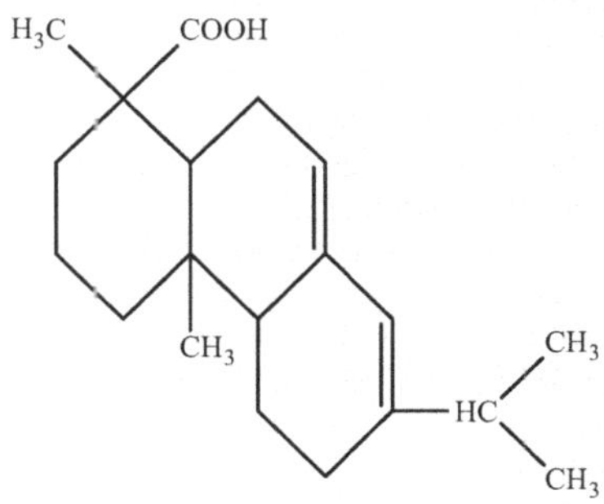

The nitration of rosin ester was first reported in 1904 [20] as a crystal nitro derivative [21], which was considered as dinitro abietic acid [22] having a chemical formula of $C_{19}H_{27}(NO_2)_2COOH$ with a melting point of 178–184 °C. Abietic acid could be nitrated in nitric acid with acetic acid (d = 1.42) to form dinitro abietic acid [23, 24] with the chemical formula of $C_{18}H_{25}(NO_2)_2COOH$, the melting point of 171 °C and yield of 33%.

With the mixed nitric and sulfonic acid, dehydroabietic acid was nitrated [25] to produce $C_{19}H_{25}(NO_2)_2COOH$, which decomposed at 178–185 °C similarly as described [21] into 6,8-nitro dehydroabietic acid. These results agreed with unidentified products containing three nitro groups as reported [23].

The author prepared trinitro rosin acid $C_{15}H_{16}(NO_2)_3COOH$ with a melting point 156–158 °C, multi-nitro rosin acids and diethyl ester of dinitro abietic acid (m. p. 157.5–185 °C) were prepared from rosin acids with hot anhydrous nitric acid [23, 24].

Although some patents were issued [26], nitro derivatives of rosin acids were still in experiments without any practical application. However, they could be reserved as emergency materials [27–29].

References

1. Роговин А, Шорыгина НН, Рогвин НН (2012) Шорыгина. ХИМИЯ ЦЕЛЛЮЛОЗЫ, И ЕЕ СПУТНИКОВ, ГОСУДАРСТВЕННОЕ НАУЧНО-ТЕХНИЧЕСКОЕ, ИЗДАТЕЛЬСТВО ХИМИЧЕСКОЙ ЛИТЕРАТУРЫ
2. Nicky (1960) Wood and its cellulose chemistry. People's Education Press
3. Liu JP (1988) Explosive and its chemical basis. Guizhou People's Publishing
4. Will W, Lenze F (2010) Nitrirung von Kohlehydraten. Eur J Inorg Chem 31(1):68–90
5. Temerdashev ZA, Konshina DN, Logacheva EY et al (2011) Sorption properties of cellulose filters with covalently immobilized thiosemicarbazide. J Anal Chem 66(10):930
6. Hudson CS, Jachson V (1937) J Am Chem Soc 59:994, 2049; (1938) 60:989; (1939) 61:1530
7. Rogovin ZA, Yashunskaya AG, Shorygina NN (1949) Zh Prikl Khim 22:865
8. Kargin VA, Usmanov KU, Aikhodzhayev BI (1959) Makromol Soyedineniya 1:157
9. Chen JC, Yang GH et al (2006) Technology and principle of fast-growing poplar pulp and papermaking. Science Press
10. Krais P (1924) Ger. Pat. 391713, 395191, 395192
11. Hagglund E (1928) Arkiv Kemi Miner Geol 7:20
12. Fisher F, Schrader H (1921) Brennstoff-Chem 2:217
13. Konlg F (1921) Cellulosechemie 2:93, 105, 117
14. Powell WG, Whitetaker H (1024) J Chem Soc 125:357, 364
15. Kurschiner K (1924) Chem Ztg 48:461; (1925) Brennstoff-Chem 6:117, 177, 188; (1928) J Parkt Chem 2:118, 238; (1931) Cellulosechemie 12:281; (1935) Zellstoff-Faster 32:17, 81, 87; (1936) Zellstoff-Faster 33:1, 49, 121; Kurschiner K, Schindler (1936) ibid. 33:121; Kurschiner K, Wittenberger K (1940) Cellulosechemie 18:21
16. Friese H, Furst H (1937) Ber Dtsch Chem Ges 70:1463; Friese H, Ludecke W (1941) Ber Dtsch Chem Ges 74:308
17. Feng H et al (2000) The main characteristics of the development of pulp and papermaking technology. Chem Ind For Prod 20(3):89–94
18. Seyewetz A (1909) Ueber die Einwirkung von Phenylhydrazin auf Phenole. Z Für Anal Chem 31(1):329–330
19. Freudenberg K, Dugrr W (1930) Ber Dtsch Chem Ges 63:2713; Freudenberg K, Lautsch W, Piazolo G (1943) Cellulosechemie 21:95
20. Easterfield TH, Bagley A (1904) J Chem Soc 85:1238
21. Johansson MD (1917) Arkiv Kemi Min Geol 6:20
22. Ashan O, Virtanen AI. Annalen 424:150
23. Dubourg J (1927) Bull Inst de Pin 41:241; (1928) Chem Abstr 22:593
24. Goldblatt A, Lowy A, Burnett WB (1930) J Am Chem Soc 52:2133
25. Fieser LF, Campbell WP (1938) J Am Chem Soc 60:159
26. Fry G (1904) U.S. Pat. 754298; Kesler JJ (1909) U.S. Pat. 922596; Arnot R (1921) Fr. Pat. 520984; (1921) Ger. Pat. 337954

27. Urbanski T (1976) Chemistry and technology of explosives (II). National Defense Industry Press, Beijing
28. Zeng ZX et al (2007) Study on modification of waterborne nitrocellulose emulsion. Coat Ind 37(2):33–35
29. Deng LR (1959) Chlorinated pulp research success. Acta Polym Sin 1959(3)

Chapter 15
Recycling and Comprehensive Utilization of Retired Nitrate Esters

In the late 1980s, with the international disarmament to promote the development of peace, security, and stability throughout the world, there will be a large amount of weapons and ammunition to destruct during the global disarmament and equipment updates, which will lead to a large amount of ammunition to retire; the decommissioning and destruction of bombs will produce a large number of decommissioned propellants and explosives; and the huge storage of expired propellants has also a become difficult problem for the military in each country. Nitrate is an important energetic material with high application importance in industry. To the drive capacity and damage effect in military technology, nitrate is especially important. However, their integrated and safe usage has not yet been resolved properly for a long time in the explosive industry of every country [1].

Up to the present, in many countries, most of retired propellants and explosives from different sources, which were so large and messy, were burnt in open air for elimination. In industrialized and developed countries, they even can only use the fluidized bed to burn out these retired propellants and explosives or sink them into the sea to destruct. Not only these destruction methods pollute the environment, but also their hazard sources are not removed.

How to utilize retired nitrate propellants and explosives? Can they be safely turned into industrial fuels or industrial raw materials to effectively prevent their industrial pollution or environmental pollution? These are the priority related to the development and process of global peace, to support global disarmament. Therefore, the recovery and comprehensive utilization of retired nitrate propellants and explosives have become a research frontier to their sustainable development. The non-destructive separation theory, proposed by us, solved three scientific problems in the separation, biodegradation, and denitrification of retired propellants to comprehensively utilize, which realized the positive production–reserve–utilization cycle of single-base propellants.

In this chapter, based on the retiring of large quantities of ammunition from international disarmament, comprehensive utilization of retired nitrate propellants and explosives will be studied.

J. Liu, *Nitrate Esters Chemistry and Technology*,
https://doi.org/10.1007/978-981-13-6647-5_15

15.1 Isolation of Retired Propellant

According to colloidal chemistry, solvent theory, and the compatibility principle, nondestructive separation theory was proposed, which solved the problem in the separation of diphenylamine, process additives, and the degraded nitrocellulose of retired single-base propellants, established the theoretical model to separate non-nitrified cellulose components quickly from retired launching propellants. A specific solvent had been found, which could make nitrocellulose in the retired launching propellants swell, dissolve diphenylamine, and process additives. The retired single-base launching propellants formed into elastic semi-colloidal, and cycling peristalsis of solvent made the semi-colloidal to form molecular channels. The dissolved molecules were spread out of the channel with the peristalsis of solvent to realize the nondestructive separation and produce pure semi-colloidal nitrocellulose. This semi-colloidal nitrocellulose was in accordance with military material requirements after testing, which could be directly used to produce different grades of single-base propellants. Other parts, which could not be used as propellants again, were converted into a civil nitrocellulose.

The studied objects were the nitrocellulose-based propellants and propellants with nitrocellulose and nitroglycerin, and also propellants with added inorganic salts.

15.1.1 Separation of Cellulose Nitrate in Retired Single-Base Propellant

Products 9/7, 11/7, 14/7, 4/7, 5/7, 7/14, 4/1, 12/1, 14/1, 18/1, 22/1, and 27/1 were the retired single-base launching propellants from the late 1950s to the early 1970s [2, 3], which were poured out and collected from the filled shells after their destruction during the disarmament, and colors of most of them had turned into black or brown. Recycling nitrocellulose of retired single-base launching propellants would follow the following steps: Through sieving with an electrostatic electricity-releasing device, after the filtration (removing non-launching ingredients from the bullet pulling out and drug pouring out), the as-obtained material was feed into a special stainless steel extraction equipment while the injection of active Solvent I with strong electron-sucking group. After 3 h of solvent cycling at 50–60 °C, diphenylamine ($C_{12}H_{11}N$), process additives, low-level nitrated compounds, and derivatives were removed by Solvent I. After the recycling and washing with Solvent II, which was safe, cheap, and miscible with Solvent I, the active strong electron group, remained in the extracted materials, was removed away to get the pure semi-colloidal nitrocellulose, the recovery ratio was 93% of the dose of retired single-base propellant. Semi-colloidal nitrocellulose through this separation technology was nondestructive based on whether its appearance or its energy state; in other words, this separation realized the nondestructive separation. In addition, performance indicators of retired single-base propellant before the nondestructive separation were listed in Table 15.1. After the nondestructive separation of retired single-base propellant, performance indicators of pure semi-colloidal nitrocellulose were listed in Table 15.2. Although this batch of propellant was already served and stored for more than 50 years, its nitration degree, viscosity, alcohol solubility, and other performance indicators basically did not change, compared with that when these retired single-base nitrocellulose propellants were manufactured. Performance indicators of these retired single-base nitrocellulose propellants when they were manufactured were listed in Table 15.3.

Table 15.1 Quality of retired single-base propellants used in experiments

Item	Results						
	1#	2#	3#	4#	5#	6#	7#
Nitration degree (mL/g)	202.33	202.29	202.16	202.42	202.23	205.11	205.16
Diphenylamine (%)	1.22	1.26	1.31	1.19	1.27	1.33	1.29
Diphenylamine and its derivatives (%)	0.35	0.33	0.22	0.32	0.26	0.19	0.23
Viscosity (Pa s)	3.5	3.8	4.9	4.8	4.8	4.8	4.7
Solubility in ethanol (%)	0.4	0.4	0.4	0.4	0.20	0.23	0.22
Dimensions method (h)	69	71	77	66	73	80	75
Methyl violet test (min)	80–70	80–70	80–70	80–70	80–70	80–70	80–70
Ash (%)	0.21	0.21	0.20	0.19	0.18	0.19	0.18

Note 1# and 2# were from 14/7-2/66; 3# and 4# were from 9/7-5/67; 5#, 6# and 7# were from 7/14-2/62 of retired single-base propellants

Table 15.2 Quality of semi-colloidal nitrocellulose produced from retired single-base propellant

	1#	2#	3#	4#	5#	6#	7#
Nitration degree (mL/g)	205.52	205.54	205.50	205.85	209.20	209.21	209.19
Solubility in ethanol–ether (%)	33.9	33.4	34.3	34.5	23.1	23.3	23.4
Solubility in ethanol (%)	0.4	0.4	0.4	0.4	0.20	0.23	0.22
Viscosity (Pa s)	3.5	3.8	4.9	4.8	4.8	4.8	4.7
Methyl violet test (min)	80–70	80–70	80–70	80–70	85–75	85–75	85–75
Alkalinity (%)	0.01	0.01	0.01	0.01	0.02	0.02	0.02
Ash (%)	0.20	0.20	0.19	0.18	0.15	0.14	0.17
Light transmittance (%)	88	87	85	85	86	85	87

Note 1# and 6# were 14/7-2/66; 3# and 4# were 9/7-5/67; 5#, 6#, and 7# were semi-colloidal NC from 7/14-2/62

Table 15.3 Quality of three different batches of nitrocellulose to manufacture retired propellants in the laboratory

	2/66-35	5/67-35	2/62-25
Nitration degree (mL/g)	205.54	205.85	209.20
Solubility in ethanol–ether (%)	33.9	34.5	23.4
Solubility in ethanol (%)	0.4	0.4	0.20
Viscosity (Pa s)	3.8	4.8	4.8
Methyl violet test (min)	80–70	80–70	80–70
Alkalinity (%)	0.01	0.01	0.01
Ash (%)	0.18	0.18	0.16

15.1.2 Recycling of Single-Base Propellant Containing Inorganic Salts

During the solvent choosing, the following factors should be considered, including dissolving performance of solvent, distribution coefficient of solvent components in the mixed solvent, operational safety, solvent toxicity, etc.

Table 15.4 lists the dissolution of propellant components in a variety of solvents. Grams of dissolved solute mass in 1.0 mL of solvent was used to evaluate the solubility. In order to study easily, solubility was divided into three levels as soluble, slightly soluble, and non-soluble. Soluble means the concentration is greater than 0.100 g/mL, concentration of slightly soluble component is 0.010–0.1 g/mL, and concentration of insoluble is less than 0.010 g/mL. In addition to distinction among these three levels of solubility, propellants were also divided into: hydrophilic, hydrophobic, insoluble, and others. In the mixed solvents, water and organic solvent were the main choices to consider.

With the component separation of single-base propellants M_1, M_6, and M_{10} as examples, separation process of the single-base propellant was described. Compared with the single-base propellant M_1, components were the same in the single-base propellant M_6 except that there was not lead carbonate, and each component had the same and similar mass fraction. Therefore, based on the separation process of M_1 components, the acidic water washing operations without the separation of nitrocellulose and lead carbonate had become the component separation process of M_6 single-base propellants. Similar to the separation of M_1 components, with limited solvent and operation times, DNT, DBP, and DPA were difficult to be separated as pure. Experiment results also confirmed that, after separation, the NC could continue to be applied in single-base propellant production.

The composition of M_{10} single-base propellant is simple, other than the nitrocellulose and diphenylamine, its additive is only potassium sulfate. In rare cases, it also contains a small amount of graphite. Figure 15.1 shows component separation of M_{10} single-base propellant. The dichloromethane extraction could remove DPA of over 90×10^{-2}. If limited water was used to extract potassium sulfate, its

Table 15.4 Solubility of propellant components

Component	Neutral water	Acidic water	Basic water	Dichloromethane	Acetone	Tetrahydrofuran	Methanol
Potassium nitrate	Soluble	Soluble	Soluble	Insoluble	Insoluble	Insoluble	Insoluble
Barium nitrate	Soluble	Soluble	Soluble	Insoluble	Insoluble	Insoluble	Insoluble
Potassium perchlorate	Slightly soluble	Insoluble	Insoluble	Insoluble	Insoluble	Insoluble	Insoluble
Lead carbonate	Insoluble	Slightly soluble	Insoluble	Insoluble	Insoluble	Insoluble	Insoluble
Potassium sulfate	Soluble	Soluble	Soluble	Insoluble	Insoluble	Insoluble	Insoluble
Graphite	Insoluble	Insoluble	Insoluble	Insoluble	Insoluble	Insoluble	Insoluble
Carbon black	Insoluble	Insoluble	Insoluble	Insoluble	Insoluble	Insoluble	Insoluble
Cryolite	Insoluble	Insoluble	Insoluble	Insoluble	Insoluble	Insoluble	Insoluble
Slightly soluble	Insoluble	Insoluble	Insoluble	Soluble	Soluble	Soluble	Slightly soluble
DBP	Insoluble	Insoluble	Insoluble	Soluble	Soluble	Soluble	Soluble
DPA	Insoluble	Insoluble	Insoluble	Soluble	Soluble	Soluble	Soluble
Binitro-diphenylamine	Insoluble	Insoluble	Insoluble	Soluble	Soluble	Soluble	Slightly soluble
Centralite	Insoluble	Insoluble	Insoluble	Soluble	Soluble	Soluble	Soluble
NC12.6%N	Insoluble	Slightly soluble	Insoluble	insoluble	Soluble	Slightly soluble	Slightly soluble
NC13.15%N	Insoluble	Slightly soluble	Insoluble	Insoluble	Soluble	Slightly soluble	Slightly soluble
NG	Insoluble	Insoluble	Insoluble	Insoluble	Insoluble	Insoluble	Insoluble
Component	Benzene	Ethyl ether	Hexane	DMSO	Isopropanol	Ethanol[a]	n-Butanol
Potassium nitrate	Insoluble	Insoluble	Insoluble	Insoluble	Insoluble	Insoluble	Insoluble
Barium nitrate	Insoluble	Insoluble	Insoluble	Insoluble	Insoluble	Slightly soluble	Insoluble
Potassium perchlorate	Insoluble	Insoluble	Insoluble	Insoluble	Insoluble	Insoluble	Insoluble
Lead carbonate	Insoluble	Insoluble	Insoluble	Insoluble	Insoluble	Insoluble	Insoluble
Potassium sulfate	Insoluble	Insoluble	Insoluble	Insoluble	Insoluble	Insoluble	Insoluble
Graphite	Insoluble	Insoluble	Insoluble	Insoluble	Insoluble	Insoluble	Insoluble
Carbon black	Insoluble	Insoluble	Insoluble	Insoluble	Insoluble	Insoluble	Insoluble

(continued)

Table 15.4 (continued)

Component	Neutral water	Acidic water	Basic water	Dichloromethane	Acetone	Tetrahydrofuran	Methanol
Cryolite	Insoluble	Insoluble	Insoluble	Insoluble	Insoluble	Insoluble	Insoluble
Slightly soluble	Soluble	Soluble	Soluble	Soluble	Soluble	Soluble	Slightly soluble
DBP	Soluble	Soluble	Soluble	Soluble	Soluble	Soluble	Soluble
DPA	Soluble	Soluble	Soluble	Soluble	Soluble	Soluble	Soluble
Binitro-diphenylamine	Soluble	Soluble	Slightly soluble	Soluble	Slightly soluble	Slightly soluble	Slightly soluble
Centralite	Soluble	Soluble	Soluble	Soluble	Soluble	Soluble	Soluble
NC12.6%N	Slightly soluble	Insoluble	Insoluble	Insoluble	Slightly soluble	Insoluble	Insoluble
NC13.15%N	Slightly soluble	Insoluble	Insoluble	Insoluble	Slightly soluble	Insoluble	Insoluble
NG	Insoluble	Insoluble	Insoluble	Insoluble	Insoluble	Insoluble	Insoluble

[a]Containing toluene with a mass fraction of 2×10^{-2}

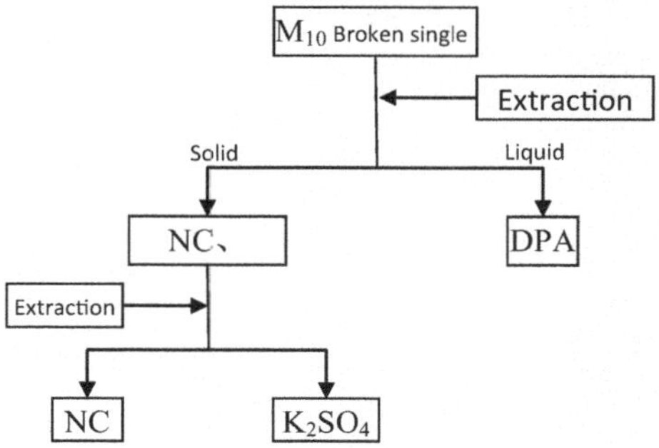

Fig. 15.1 Extraction process of M_{10} single-component propellant

recovery ratio was only up to 10×10^{-2}. However, the full recovery of NC could be achieved. Experiment results showed that the separated NC could be used as raw material to produce single-base propellant.

15.1.3 Component Separation of Two-Component Propellant [4, 5]

Double-base propellants have three categories, including double aromatic, double ethyl, and double mortar propellants. The main ingredients are nitrocellulose, nitroglycerine and/or nitrated diethylene glycol, dinitrotoluene, centralite, dibutyl phthalate, petroleum jelly, and graphite. The double-base propellants in our country generally contain or partly contain these ingredients with a small amount of metal oxides or other additives. Therefore, the double-base explosives and double-base propellants could be in one type in the separation of their components. In the components of the double-base explosives, mass fraction of nitrocellulose is 55×10^{-2} to 65×10^{-2} in which nitroglycerine or nitrated diethylene glycol account for 25×10^{-2} to 35×10^{-2}. Both of these two constitute the main subject of double-base explosives. Components of double-base propellants of M_2 and M_7 from the United States are representative, in which there are soluble organic compounds and water-soluble inorganic salts.

(1) Component separation of double-base propellant M_2

From the first extraction with dichloromethane, most of NG and centralite (EC) could be extracted, the extraction efficiency of NG was about 95×10^{-2}, and the extraction efficiency of EC was about 98×10^{-2}. The residues after the extraction were solid NC, graphite, barium nitrate, and potassium nitrate.

Through the water extraction on the solid substance, barium nitrate and potassium nitrate could be removed. Recovery ratio of potassium nitrate was around 49×10^{-2}; recovery ratio of barium nitrate was about 83×10^{-2}. Based on the analysis of extracts, NG and EC were not found.

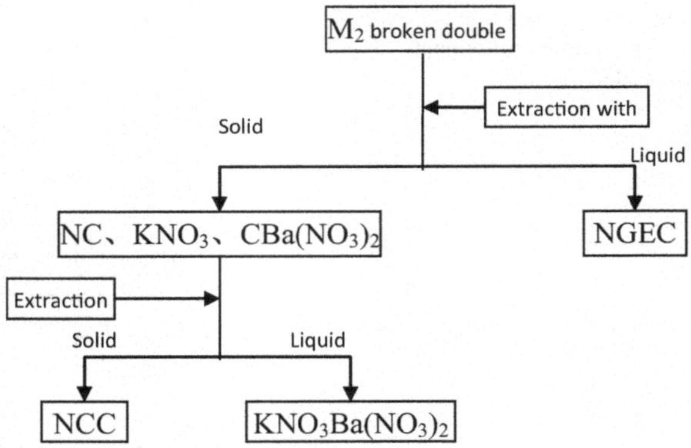

Fig. 15.2 Component separation process of double-base propellants

Residues from extraction with methylene chloride and water were NC and graphite. With NC as the base, the calculated NC recovery ratio was 101×10^{-2}. Because of the graphite and the trace amounts of NG and EC, the recovered NC could only be used as raw material to produce the double- and three-base propellants. Figure 15.2 is the component separation process of double-base propellant M_2.

(2) Component separation of double-base propellant M_7

Compared with M_2, M_7 does not contain barium nitrate and potassium nitrate, but KClO4 is added in M_7 instead. Although the components of these propellants changes, but component separation processes of M_7 and M_2 are basically the same because the potassium nitrate, barium nitrate, and potassium perchlorate are all soluble inorganic salts.

From the first extraction with dichloromethane, most of the NG and centralite (EC) could be extracted, the extraction efficiency of NG was about 85×10^{-2}, and the extraction efficiency of EC was 93×10^{-2}. After the extraction, residues were solid NC, graphite, and potassium perchlorate.

Through water extraction on the residues, 43×10^{-2} of potassium perchlorate could be extracted. Based on the analysis, water-extracted liquid did not contain NG and EC. Residue after the water extraction was black carbon and NC. With NC as the standard, the amount of recovered NC was 108×10^{-2}, which indicated that there were other components of M_7 in the recovered NC.

(3) Component separation of double-base propellant M_9

Compared with M_2, there was no barium nitrate in components of M_9, in which NG content was increased from 19.5×10^{-2} to 40.0×10^{-2}. Therefore, component separation processes of M_9 and the M_2 were the same.

From the first extraction with dichloromethane, most of the NG and centralite (EC) could be extracted, the extraction efficiency of NG was 82×10^{-2}, while the

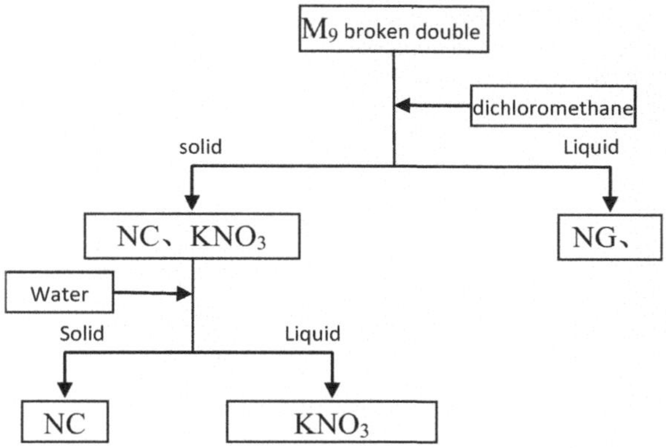

Fig. 15.3 Component separation process of double-base propellant M_9

extraction efficiency of EC was about 63×10^{-2} due to the influence of high content of NG. During the components extraction of double-base propellants with dichloromethane, the amounts of NG and EC had impacts on their extracted amounts with interdependent relations.

Residues after the dichloromethane extraction were solid NC, graphite, and potassium nitrate.

Residues were extracted with water for one time, only 58×10^{-2} of potassium nitrate could be extracted. Based on the analysis, extracted solution after water extraction did not contain NG of EC. Residue after the water extraction was carbon black and NC. With NC as the standard, the recovered amount of NC could reach 104×10^{-2}, which meant there were other components of M_9 in the recycled NC. Figure 15.3 shows the component separation process of double-base propellant M_9.

15.1.4 The Component Separation of Three-Base Propellants

Our Chinese and Russian three-base propellants belong to M_{30}-type ones. In the United States and Europe, three-base propellants are M_{30}, $M_{30}A_1$, and $M_{31}A_1$. Their main ingredients are nitrocellulose, nitroguanidine (NQ), and nitroglycerine. M_{30} has other additional ingredients as EC, cryolite, and graphite; $M_{30}A_1$ has EC, potassium sulfate, graphite; $M_{31}A_1$ contains DBP, dinitrodiphenylamine, potassium sulfonate, and graphite.

(1) The component separation of three-base propellants M_{30} and $M_{30}A_1$

From the first extraction with dichloromethane, most of NG and centralite (EC) were extracted and the extraction ratio of NG was about 96×10^{-2}. After the dichloromethane extraction, residues were solid NC, graphite, and cryolite.

Fig. 15.4 Three basic drug component separation process of three-base propellant

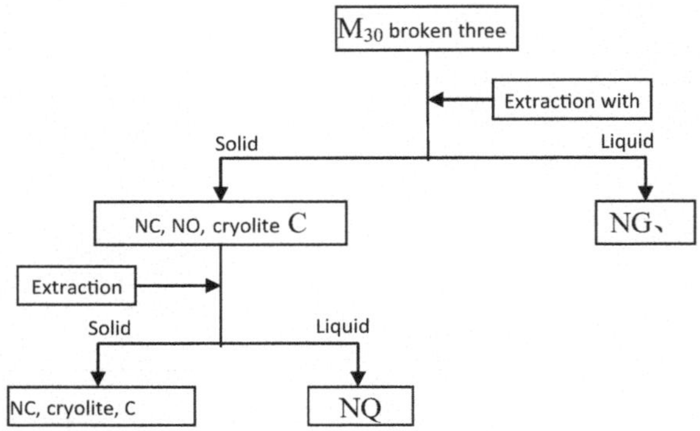

The residue was extracted with hot waters at 95 °C for the first time saturated aqueous NQ solution could be obtained. Through cooling, washing, and filtering, 88×10^{-2} of nitroguanidine could be recovered. According to the analysis of aqueous NQ solution, the extracted aqueous solution did not contain NG and EC. The residues after the water extraction were carbon black and NC. With NC as the standard, the quality content of recovered NC was 97×10^{-2}–111×10^{-2}, which meant that is there were other components of M_{30} in the recycled NC. Component separation process of three-base propellant M_{30} was shown in Fig. 15.4.

Compared with components of three-base propellant M_{30}, potassium sulfate was added in M30A1 components. In the component separation of $M_{30}A_1$, additional water washing must be added on the base of M30 component process, through which potassium sulfate will be dissolved in water to extract. This process was between the two following processes, the dichloromethane extraction and hot water extraction for NQ. Component separation process of three-base propellant $M_{30}A_1$ was shown in Fig. 15.5.

(2) Component separation process of three-base propellant $M_{30}A_1$

With dichloromethane, most of NG, dinitrodiphenylamine (2-NDPA) and DBP could be extracted. The extraction efficiency of NG was about 97×10^{-2}; that of 2-NDPA was about 91×10^{-2} to 109×10^{-2}; that of DBP reached 89×10^{-2} to 98×10^{-2}.

After the dichloromethane extraction, about 88×10^{-2} of potassium sulfate could be removed with cold water from the remained solid. According to the analysis, there were no NG and 2-NDPA in the washing liquid. Residue after the washing was extracted with hot water to get NQ, and its extraction efficiency could reach to 70×10^{-2} to 79×10^{-2}. The mass fraction of remained NC was 94×10^{-2} of the total amount. Figure 15.6 shows the component separation process of three-base propellant $M_{31}A_1$.

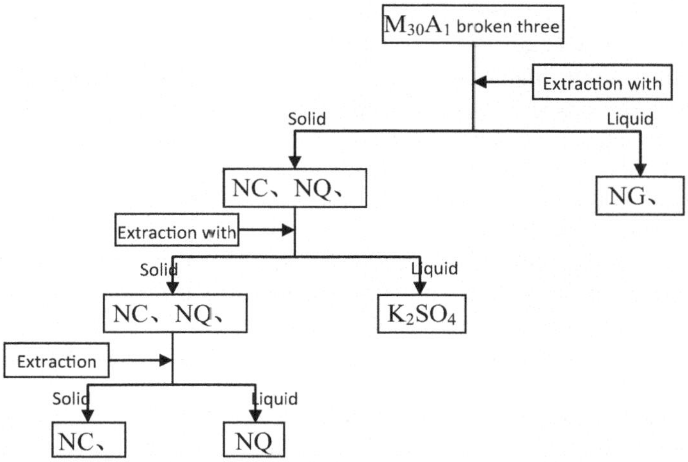

Fig. 15.5 Component separation process of three-base propellant $M_{30}A_1$

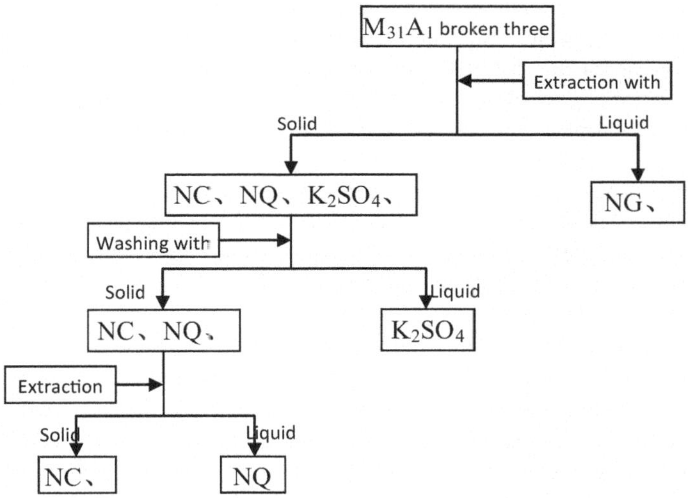

Fig. 15.6 Component separation process of three-base propellant $M_{31}A_1$

15.2 Separation of Nitrate in Composite Propellants [6, 7]

Among global equipment in service, composite propellants are different significantly in adhesives, they all have other components including (or partly) NH_4ClO_4 (AP), aluminum powder (Al), CL20, HMX, RDX, NG, NQ, and some additives, such as n-hexylcarborane (NHC). Based on the difference of adhesives and components, there are a variety of methods to separate their components and reuse them. Table 15.5 summarizes the relevant separation methods.

Up to the present, the ammonium perchlorate recycling technologies include ammonolysis during AP extraction with ammonia occurs. In this method, except that AP is recycled, aluminum powder can be recycled. In the separation process, the mainly used equipment include shredding machine, extraction tank, crystallizer, drying vessel, washing tank, and distillation reactor. After the separation, reaction wastes are adhesive residues.

Another recycling method of AP is to extract AP with water, in which only physical change exists without chemical reaction. In this process, the used equipment are shredding machine, extraction tank, drying vessel, crystallizer, filter, and centrifuge. After the separation, the wastes obtained are aluminum powder and adhesive residue.

Aluminum recycling technologies include degradation and pyrolysis of propellant and binder. In the former technology, used equipment are heating furnace and de-duster. In the pyrolysis of propellant and binder, they are drying vessel, separation reactor of solid, and solvent distillation vessel. Inverse esterification reaction is used to complete the degradation of adhesive.

In the co-recycling technology of AP and NQ, with water as solvent, propellant was crushed in the water, and intermediate products are aqueous solutions of AP and NQ. Based on the difference in solubility of AP and NQ in water, they are separated. In the separation process, used equipment are shredding machine, extraction tank, storage tank, crystallizer, filter, activated carbon adsorption column, and distillation vessel.

Ammonium perchlorate and aluminum powder can be separated and extracted from polyether-, polyester-, and butadiene-propellant. In this process, the adhesive was dissolved and degraded to lead AP aqueous solution and precipitate aluminum powder. The used equipment are shredding machine, reactor, crystallizer, distillation tank, and drying vessel.

Separation of raw materials of high-energy propellant is an important technology with practical significance. High-energy propellant contains nitrate or nitrosamine explosives, and the presence of nitrate increased the difficulty of separating. With binder dissolution and degradation method, nitrate explosive, nitramine explosive, Al, oxidant, and other products are finally obtained. In the separation process, the intermediate product is the extraction residues, which contain nitrate explosive, nitramine explosive, oxidant, solvent, and their formed sol-gel or solution. In the separation of high-energy propellant, the main used equipment are shredding machine, drying vessel, storage tank, extraction tank, filter, and crystallizer. Because of the high detonation sensitivity of nitrate and nitramine explosives (especially CL20), separation should be done with special caution.

n-Hexyl carbon borane (NHC), as a catalyst, has very high value in the propellant although its content is very low. Therefore, recovery technology of these materials with high added value is specially studied. The extraction method is to use n-pentane to extract NHC, in which intermediate products are the n-pentane

Table 15.5 Extraction and isolation of propellant components [6]

Recycling and reusing methods	Technology	Chemical reaction	Waste	Intermediate	Final Product	Special instrument
AP recycling	Extraction with NH_3 (Liquid)	Ammonolysis	Adhesive remains	AP, NH_3 solution, adhesive solution in benzene	AP, Al powder	Mill, extracting tank, crystallizer, drying vessel, washing tank, distillator
AP recycling	AP extraction with water	No	Al powder/adhesive remains	AP aqueous solution	AP	Mill, extracting tank, crystallizer, drying vessel, filter/centrifuge
Al recycling	Pyrolysis of adhesive	Pyrolysis	Product from pyrolysis	No	Al	Heater/De-duster
	Adhesive degradation	Contrary esterification	Degraded product from adhesive/remained chemicals	Ethanol ester	Al	Drying vessel, solid separation reactor, solvent distillation reactor
NQ, AP recycling	Extracting NQ and AP with water	No	Saturated carbon adhesive remains	AP aqueous solution, NQ aqueous solution	AP, NQ	Mill, extraction tank, storage tank, crystallizer, filter, carbon adsorption column, distillator
Polysther/ester propellant, AP/Al recycling	Dissolving, decomposition of adhesive, separation of solid	Ammonolysis	Adhesive remains	AP aqueous solution	AP, Al powder	Mill, extraction tank, storage tank, crystallizer, filter, carbon adsorption column, distillatory
Polybutadiene propellant, AP/Al recycling	Dissolving, decomposition of adhesive, separation of solid	Dissolving, decomposition	Adhesive remains	AP aqueous solution	AP, Al powder	Shredding machine, reaction crystallizer, distillatory, drying vessel

(continued)

Table 15.5 (continued)

Recycling and reusing methods	Technology	Chemical reaction	Waste	Intermediate	Final Product	Special instrument
HNC recycling	Extracting NHC with n-pentane	No	Adhesive remains	NHC solution in n-pentane	HNC	Shredding machine, mesh container, distillatory, extraction tank, filter, storage tank, crystallizer, condenser
Separation of raw materials of high energy propellant	Dissolving, decomposition of adhesive, separation of solid	Dissolving, decomposition	Degraded adhesive	Extraction residue, sol-gel, or solution of ammonium nitrate/oxidants/solvent	Nitrate ester, nitramine, Al, oxidant	Shredding machine, drying vessel, storage tank, extraction/boiling tank, filter, crystallizer
Extraction with chosen solvent	Extraction of main raw materials with chosen solvent	No	Al powder, Adhesive remains	Extraction residue, nitrate ester/ oxidant/solvent, sol gel, solution, mixed crystal, oxidant aqueous solution	Nitrate ester, nitramine, Al, oxidant	Shredding machine, extraction tank, crystallizer (2), filter, solvent distillatory

solution of NHC and the adhesive residue. The extraction equipment are shredding machine, mesh basket, distillation reactor, extraction tank, filter, storage tank, crystallizer, and the condenser.

15.2.1 Extraction Technology of Propellant

The basic process to extract and separate the raw material include: Grinding propellant, extracting nitrate, extracting adhesive glue, extracting ammonium nitrate and AP, and crystallizing material and other operating units. The basic extraction and separation of raw materials were shown in Fig. 15.7.

(1) Propellant crushing

The aim of crushing is to increase the "area/volume" ratio of propellant as much as possible, and this ratio is the main factor to control and decide the recycling time and recycling efficiency of the process.

(2) Separation and extraction of nitrate

In this separation unit, nitrate ester and valium were separated from the mixture, while unpolymerized or low-degree polymerized "adhesive–adhesive glue" was

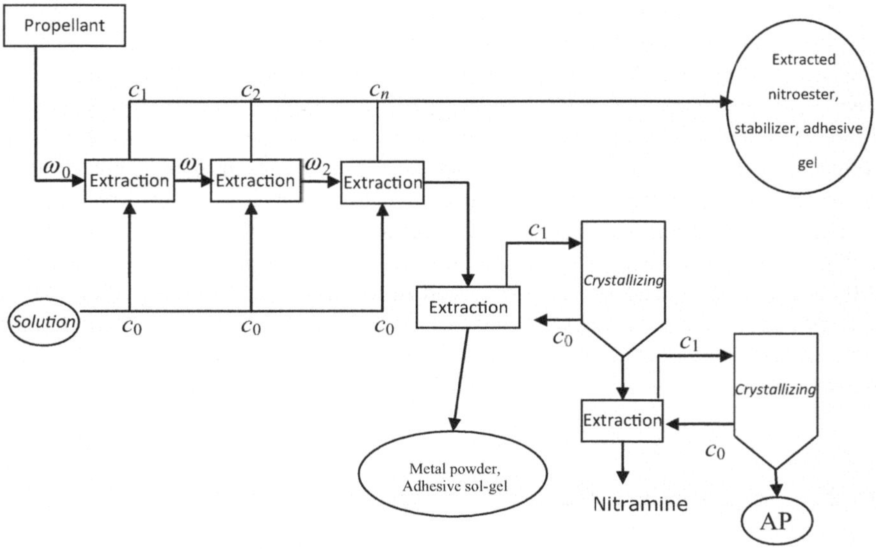

Fig. 15.7 Separation and extraction technology of propellant [6]. C_0—the mass concentration of solute in the solution, ω_0—the mass fraction of solute in propellant, ω_n—the mass concentration of solute in the liquid residue after step n, c_n—the mass concentration of solute in the extraction solution after step n

isolated. Extraction can be single-stage contact operation, parallel flow contact operation, anti-flow contact operation, and cross-flow operation. All of these are suitable to separate materials from propellants with nitrate esters. During the separation, crushed propellant is put into the sealed extraction vessel, extracted, precipitated, and separated with stirring.

In the first nitrate ester extraction, the mass fraction of solute (glue of nitrate ester, stabilizer, and adhesive) in the propellant was changed from ω_0 to ω_1 (Fig. 15.7), the mass concentration of solute in the solution was changed from C_0 to C_1. After the number n extraction of nitrate ester, the mass fraction of solute in the propellant was changed to ω_n, and the mass concentration of solute in the solution was changed to c_n. The extracted solutions from the first and (n)th were put together, separation and recovery of nitrate ester and solvent recovery operations were continued, and the recovered solvents could be recycled and reused.

Operation with some pure nitrate esters (such as nitroglycerin, 1,2,4-butanetriol trinitrate) must be strictly limited because they are very sensitive. In their unit operations, dilution method can be used. Nitroglycerin can be diluted to a mass concentration of 70×10^{-2} with the valium dissolved, which is the basic method to store and handle nitroglycerin. When the separation process was determined, the solvent that is both extracting reagent and diluting reagent is chosen. During the handling and storage, solvent losses should be strictly prevented due to evaporation or over-distillation, while long storage time of nitroglycerin should be avoided.

(3) Extraction and separation of explosive and AP

After n times of nitrate ester extraction, the remaining solid matter was sent to extraction tank to extract nitramine and AP with solvent, which could be acetone and dimethyl sulfone. The separation process of nitramine and AP consists of two steps. The first extraction was to separate nitramine and AP from the solid residue. After the extraction, solution of nitramine and AP was removed. The remaining solid materials were metal powder and adhesive gel. Then nitramines and AP were crystallized out from the solution, and the recycled solvent was recycled and reused.

The second extraction is to separate nitramine and AP, in which water are actually used to wash crystal and AP is dissolved and removed away. After the filtration, explosives are stored and treated in the form of wet block. Then, AP is crystallized out from its aqueous solution. Dry explosive (RDX, HMX, and CL20) are very sensitive to static electricity and also easy to produce dust. Thus, from a security point of view, nitramine should be treated with wet state.

15.2.2 Equilibrium During the Extraction

The extraction and separation belong to general physical process, and there is chemical reaction in the presence of adhesive glue. Effect of solvents on the extraction efficiency of propellant components was shown in Table 15.6.

Table 15.6 Effect of solvents on extraction efficiency of propellant components in the presence of adhesive glue

Solution	Recovery ratio of solvent $\times 10^2$						
	High energy explosive			Adhesive gel			
	RDX	HMX	CL20	Polyvinyl alcohol-2000	Poly adipic acid glycol acetal	Butadiene rubber-026	Poly isocyanate N-100
Toluene	2.0	1.3	0.69	12.0	102.1	99.8	100.6
Dichlorometahne	11.9	2.2	1.6	19.4	103.5	100.3	–
Diethylene glycol dimethyl ether	90.8	39.2	33.5	22.5	–	53.3	–
Acetone	100.1	93.8	91.4	9.8	100.0	6.7	100.2
Chloroform	2.2	1.4	1.6	91.7	102.3	102.9	102.6
THF	53.3	–	61.3	–	102.9	101.9	–
Hexane	1.0	0.6	0.29	0.4	0.2	0.9	0.8
Ethyl acetate	69.5	31.2	76.3	7.1	101.4	11.3	–
Acetonitrile	–	94.4	96.1	96.8	–	–	–

(1) Material balance during the extraction

F, E, R, N, O, X, B represent propellant, extracted solution, residue, high-energy explosive, oxidant, nitrate ester, and metal powder, respectively. According to the following material balance, solvent extraction process can be chosen.

1. Nitroester

$$(\omega_{FX})(q_F) = (\omega_{EX})(q_E) + (\omega_{RX})(q_R) + (\omega_{NX})(q_N) + (\omega_{OX})(q_O) \qquad (15.1)$$

2. High energy explosive

$$(\omega_{FN})(q_F) = (\omega_{EN})(q_E) + (\omega_{RN})(q_R) + (\omega_{NN})(q_N) + (\omega_{ON})(q_O) \qquad (15.2)$$

3. Oxidizing agent

$$(\omega_{FO})(q_F) = (\omega_{EO})(q_E) + (\omega_{RO})(q_R) + (\omega_{NO})(q_N) + (\omega_{OO})(q_O) \qquad (15.3)$$

4. Metal powder

$$(\omega_{FJ})(q_F) = (\omega_{EJ})(q_E) + (\omega_{RJ})(q_R) + (\omega_{NJ})(q_N) + (\omega_{OJ})(q_O) \qquad (15.4)$$

where

q_F flow rate of propellant (kg/h),
q_E flow rate of extraction solvent (kg/h),
q_R flow rate of residue (kg/h),
q_N flow rate of nitramine (kg/h),
ω_X mass fraction of nitroester ($\times 10^{-2}$),
ω_N mass fraction of nitramine ($\times 10^{-2}$),
ω_O mass fraction of oxidant ($\times 10^{-2}$),
ω_J mass fraction of metal powder ($\times 10^{-2}$).

(2) Main extracting equipment

The main equipment in the extraction and separation process of propellant components are shredding machine, extraction tank, crystallizer, filter, solvent distillator, explosion protection facilities, and manufacture safety facilities.

(3) Extraction and separation process

In order to determine the separation procedure, basic principles and basic studies on procedures should be considered first to discover the basic law in the process. With the high-energy propellant as an example, the extraction process will be further evaluated. Components of the studied objects NEPE cross-linked high energy propellants are NEPE, NG, HMX, CL20, AP, and Al. Studies are focused on the process, the evaluation on solvent, extraction rate, the purity of extracted, intermediates, and their sensitivity and phase balance, in order to get the data of engineering design and modeling of the whole process and each operating unit. Raw materials that have to be extracted are the following, ① nitrate ester, NG, trimethyl ethyl trinitrate ester, ② HMX, RDX, CL20, etc., ③ AP, ADN, etc., ④ aluminum powder and other metallic powders. Extraction process of nitrate ester, nitrosamine, and oxidant can have four types of operations to consider, including single-stage contacting operation, single-stage continuous contacting operation, multistage cross-flow contacting operation, and multistage anti-flow contacting operation.

1. Single-stage contacting operation to extract

In this process, Propellant 2 and Solvent 3 were sent into the Mixer 1 once. After the stirring, extraction, and releasing from Mixer 1, the mixture was sent into Separator 4, where solid–liquid separation was completed. If high extraction performance was needed, a lot of solvent and large extraction equipment were required. After the extraction, extracted solution and propellant residue were in equilibrium. Therefore, when the amount of used solvent is not large, propellant residue still contains a large amount of the solute. If residues are not washed with new solvent, extraction efficiency will be very low. If a large amount of new solvent is used to wash, the workload on the material de-wetting and solvent recovery will be increased. The amount of material and the used amount of solvent are related to

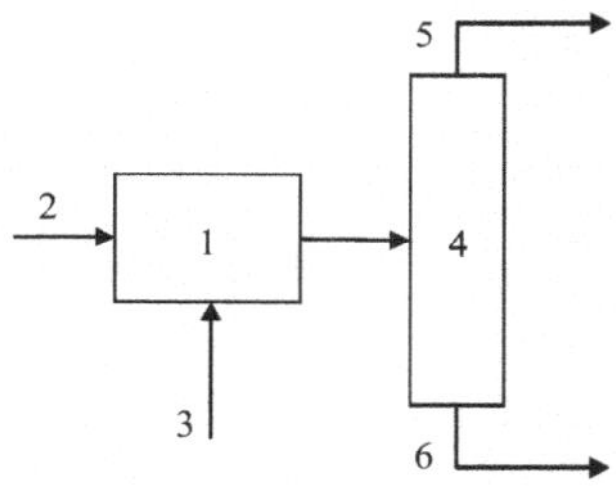

Fig. 15.8 Extraction process of single-stage contacting operation [6]. 1—mixer, 2—crushed propellant, 3—solvent, 4—separator, 5—extracted solution, 6—residue

the mixer. Single-stage contacting operations as shown in Fig. 15.8 show the extraction process.

2. Extraction of single-stage continuous contacting operation

First, crushed propellant was sent to a Mixer 1, which was followed by the extraction solvent. The extracted solution continuously flew into the Separator 6, in which recrystallization and solvent recovery were done to Recover 8. Extracted volume of the solute in the process depended on the total time of extraction (when solvent was recycling). Compared to simple noncontinuous extraction process, this process needed a small amount of solvent and small mixer volume, with low residue content and relatively simple unit operation. Single-stage continuous contacting operation of the extraction process was shown in Fig. 15.9.

3. Extraction of multistage cross-flow contacting operation

Figure 15.10 is an extraction of multistage cross-flow contacting operation, in which new solvent into the mixer 1-1 containing propellant. After the mixing and the first step extraction, the mixed materials were sent into separator 4-1, where precipitation and separation were completed. The extracted solution 5-1 was sent to recycle. The precipitated propellant 6-1 was sent to mixer 1-2 in the second step, where a new solvent 3 was added, and the first extraction process was repeated. Until it was repeated to Step n, when the quality of solute in propellant residue met the standard, the propellant residue 7 was removed. Extracted solution 5-n was vacuum distilled and crystallized to separate solute and solvent, and solvent was recycled and reused. Through the whole process, recovery ratio of solute was high. However, solute recovery ratio was not a constant, which was related to the cycling process of extraction.

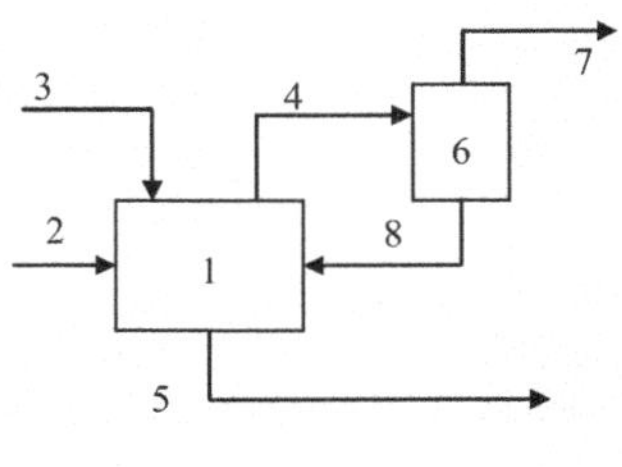

Fig. 15.9 Diagram of single-stage continuous contacting extraction process [6]. 1—mixer, 2—feeding, 3—solvent, 4—extracted solution, 5—residue, 6—separator

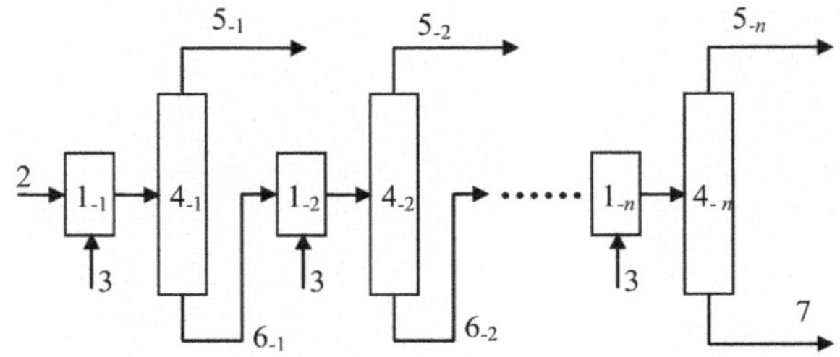

Fig. 15.10 Diagram of multistage cross-flow contacting operation [6]. 1—mixer, 2—feeding, 3—solvent, 4—separators, 5—extracted solution, 6—deposits, 7—residues

4. Multistage anti-flow contacting operations

Figure 15.11 is the multistage anti-flow contacting extraction process, in which the extracted propellant and new solvent were in multistage contact from the opposite directions. The new solvent was sent into the mixer 1-1. The propellant in mixer 1-1 was the propellant to complete n-1 extraction. After a balance of extraction was reached, the mixed material was sent into separator 4-1.

Precipitation and separation were done in the separator. After the propellant was extracted for n times, it became residues to be discharged. The extracted solution was re-sent to the extracting mixer 1-2. A second extraction was done with propellant after n-2-time extraction. When the above mentioned process was recycled to Step n, extracted solution 8 was sent to distillation and crystallization to separate solute and solvent. In this process, the recovery ration of solute was high and used solvent was small.

(4) Dissolution of the adhesive

Adhesive has an impact on extraction process of components. Decomposition or degradation of adhesive helps the extraction and for every operation of each component. It is an important step in the extraction and separation to determine the solvent to dissolve adhesive. 2-Hydroxy amine is one of the solvents to disintegrate

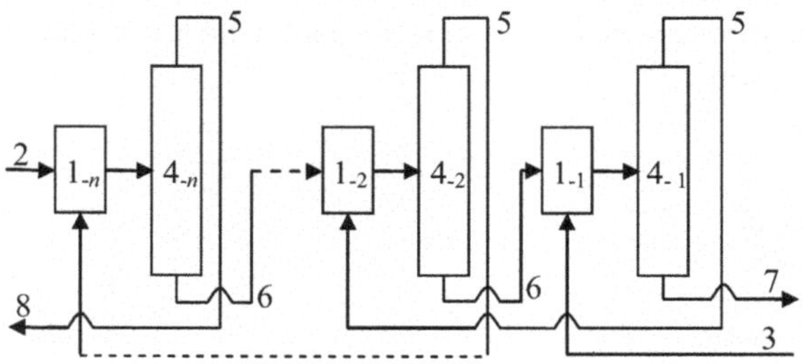

Fig. 15.11 Flowchart of extraction process of multistage anti-flow contacting operation. 1—mixer, 2—feeding, 3—new solvent, 4—separator, 5—extracted solution, 6—deposits, 7—residues, 8—final extracted solution

Table 15.7 Conditions of extraction and dissolution of propellant

Name	Parameter
Mass ration between toluene and isopropanol	1:1
Concentration of 2-hydroxyethylamine (mol/L)	2
Propellant/Solvent (g/mL)	30/100
Temperature (°C)	20
Time (h)	16

adhesives. Here, with three propellants as the studied objects, the performance of 2-hydroxy amine on adhesive disintegration was observed. These three propellants are: polyvinyl alcohol propellant (PEG) with NG, HMX, aluminum powder and AP, polyethylene glycol adipate propellant (PGA), and terminal hydroxyl polybutadiene propellants (HTPB) containing CL20, AP, and aluminum. While the solvent decomposes the adhesive, solvent also extracts related propellant components. Conditions of extraction and dissolution were shown in Table 15.7.

Results showed that PGA adhesive was completely dissolved in the process, and other materials, including NG, HMX, aluminum powder, and AP, were distributed in the free-flowing slurry. However, PEG and HTPB propellants were not dissolved and degraded, and they were only swollen. After the treatment at 80, 90, and 120 °C for 7 h, adhesives were still not up to the dissolution and degradation. Therefore, the 2-hydroxyethyl amine, as the solvent in adhesive decomposition, could not be the extraction solution in the separation of PEG and HTPB propellants.

(5) Extraction solvent of nitrate ester

1. Extraction solvent

With the presence of adhesive glue, high-energy explosive, aluminum powder, and oxidant, effects of different solvents on extraction efficiency of nitrate ester and other solutes were studied. The preferred four solvents were chloroform, toluene, methyl chloroform, and methylene chloride. In the experiment, the ratio of "solute propellant/solvent" was 1:25 (g/mL). After 4 times of extraction were done, the solute was recovered to determine the extraction efficiency of NG in the presence of adhesive glue, nitroamine, aluminum powder, and oxidant (Table 15.8), as well as the extraction efficiency of other components. It was found that dissolution abilities of various solvents on the adhesive glue were different. Especially, their difference was more significant when molecular weight of adhesive PEG had some change.

2. Dosage of Extraction solvent

Here, the whole performance of demonstrated four extraction solvents, including chloroform, toluene, methyl chloroform, and methylene chloride, was further evaluated. In the presence of other materials in propellants, solvent dissolving capability of nitrate ester and adhesive glue, their toxicity, boiling points, and flash points were investigated (Table 15.9). According to the comprehensive evaluation, dichloromethane was selected as extraction solvent of nitrate ester and adhesive glue. Methyl chloroform and toluene, not easy to dissolve polyvinyl alcohol (PEG),

Table 15.8 Solute extraction recovery ratio with different solvents [7]

Item	Solute	Solute recovery ratio $\times 10^2$			
		Methylchloroform	Chloroform	Toluene	Methylene chloride
Adhesive glue	HTPB	100	–	–	–
	CTPB	100	–	–	–
	PEG2000	–	92	12	–
	PEG4000	2	–	–	100
	PCP026	–	100	100	–
	PGA	–	100	100	100
	N-100	100	100	100	100
	BITA	100	–	–	–
Nitroamine	HMX	1.3	1.4	1.3	1.4
	RDX	–	2.2	2.0	–
	CL20	0.6	1.6	–	1.6
Nitrate ester	NG	100	100	100	100
Oxidant	AP	–	–	–	0.5

were not suitable for component separation of PEG adhesive propellant. Furthermore, toluene has a relatively high boiling point, which could lead to many difficulties in the subsequent processes. Advantages of dichloromethane as separation solvent are: it is an excellent solvent of nitrate ester and adhesive glue, and the presence of nitroamine and oxidant has no significant impact on its dissolving capability. Table 15.7 is solute extraction recovery efficiency of propellant in different solvents.

As listed in Table 15.8, the solute extraction recovery of propellant in different solvents was not only related to the solvent, but also more related to propellant components. Table 15.9 shows the relationship of solvent nature with extracting nitrate ester and adhesive glue.

3. Relationship between extraction rate and recovery ratio

With dichloromethane as extracted solvent, the extraction rate and recovery ratio of NG and stabilizers were studied in the NEPE/HMX/AP/AI/NG 1.1-stage propellant system. As shown in Figs. 15.12 and 15.13, effects of surface area, "solute/solvent" mass ratio, and extraction temperature on the extraction rate and the recovery ratio were studied.

In Figs 15.12 and 15.13, ξ represents "extracted solute/the solute in the solution" mass ratio (kg/kg) at the extraction equilibrium. The relationship between ξ and its corresponding extraction time indicates the extraction rate. Relationship between dichloromethane and its extraction rate and recovery ratio of NG and stabilizer were shown in Table 15.10.

Table 15.9 Solvent nature in the extraction of nitrate ester and adhesive glue

Solvent	Nitrate ester, adhesive	Nitroamine, oxidant	b.p. (°C)	Flashing point (°C)	Toxicity (inhaling)	Toxicity (oral)	NG using record
Methylchloroform	Soluble, except of PEG	Insoluble	74.1	No	High	Medium	No
Chloroform	Soluble	Insoluble	61.3	No	High	High	No
Toluene	Soluble, except of PEG	Insoluble	110.4	4.4	Medium	Medium	No
Methylene chloride	Soluble	Insoluble	59.8	No	Medium	Medium	Yes

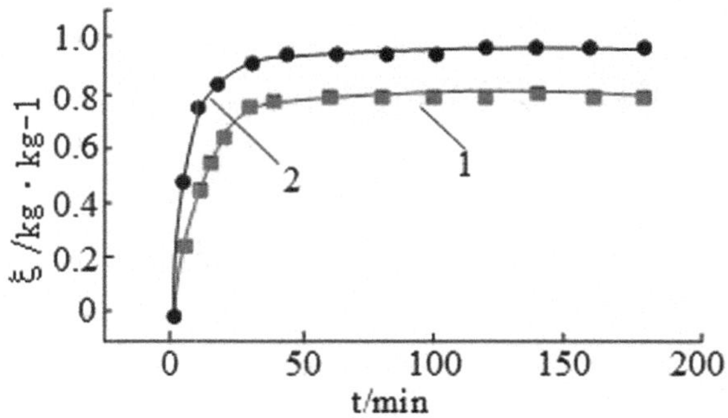

Fig. 15.12 Relationship of solute/solvent ratio, surface area, and time with extracted amount. 1—solute/solvent ratio of 0.05, 21 °C, propellant of 12.3 mm × 12.3 mm × 1.6 mm. 2—solute/solvent ratio of 0.02, 21 °C, propellant of 12.3 mm × 12.3 mm × 0.8 mm

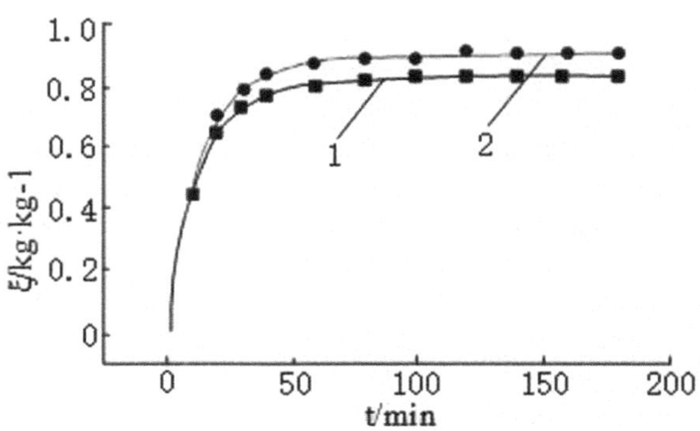

Fig. 15.13 Relationship of temperature and time with extracted amount of NG. 1—solute/solvent ratio of 0.05, 21 °C, propellant of 12.3 mm × 12.3 mm × 1.6 mm. 2—solute/solvent ratio of 0.05, 21 °C, propellant of 12.3 mm × 12.3 mm × 1.6 mm

Table 15.10 Effects of dichloromethane on extraction rates and recovery ratios of NG and stabilizers [7]

Propellant	Surface area (m^2 kg^{-1})	Temperature (°C)	Solute/ solvent ratio	Time (min)	Mass fraction of NG in the solution $\times 10^2$	Recovery ratio $\times 10^2$
NEPE /HMX /AP/Al	0.81	27	0.046	15	2.85	64.8
				30	3.71	84.4
				45	3.95	89.8
				60	4.07	92.5
				90	4.05	92.1
				120	4.02	91.4
	0.81	21	0.050	15	2.64	58
				25	3.51	76
				69	3.87	83
				128	3.67	80
				184	4.19	90
	0.81	21	0.020	16	1.25	65
				29	1.41	75
				64	1.59	82
				127	1.53	80
				215	1.61	85
	1.598	21	0.020	20	1.84	91
				30	1.83	93
				40	1.91	93
				50	1.91	95
				60	1.82	93
				120	1.88	92
				180	1.86	91
				240	1.92	96

Experiments showed that the effect of solute/solvent ratio on extraction rate was not significant. Extraction rate was increased with the increase of temperature and the surface area of material. When the surface area was 1.598 m^2/kg and the temperature was 21–27 °C, the extraction time was about 10–12 min, and the recovery ratio of solute was from 91 to 96%.

4. Equilibrium between extraction solution with residues

A—solvent (Methylene chloride),
B—insoluble solid (Adhesive, HMX, AP, Al),
C—Solute (NG, stabilizer),
R—phase of residue,

E—Liquid-extracted solute,
ζ_B—Insoluble solid/extraction solution ratio, $\zeta_B = m_B/(m_A + m_C)$,
ω_C—Mass fraction of solute in the extraction solution, $\omega_C = m_C/(m_A + m_C)$,
ω_B—Mass fraction of solute in the residue, $\omega_B = m_C/(m_A + m_C)$.

After the precipitation and separation to form two phases from the mixture of solution, insoluble solid and solute, the liquid was a solution containing solute, and the mass fraction of solute in the extraction solution was ω_{EC}. Propellant residues were formed the insoluble solid, undissolved solute, and extraction solution in the residue. "Insoluble solid/extraction solution" mass ratios in the extracted liquid phase and residue phase were ξ_{EB} and ξ_{RB}, respectively.

In order to study the equilibrium between methylene chloride extraction phase and propellant residue, recovery ratio, insoluble solid, and composition and amount of solute and solvent were studied, respectively. Results were shown in Table 15.11.

When 0.81 m^2/kg and 1. 598 m^2/kg were used as surface area and 0.25×10^{-2} to 6×10^{-2} g/mL as the solute/solvent mass ratio, the data in Table 15.11 was used to generate the equilibrium phase diagram between extraction solution and the propellant residue. Through the phase diagram, the mass fraction at the equilibrium of extraction solution and residue could be predicted.

In the multistage cross-flow contacting operation of extraction of NG and stabilizers with methylene dichloride, the calculated phase diagram of the extraction process was shown in Fig. 15.14. According to the phase diagram, the extraction level was predicted as Level 4, and the recovery rate was 95% at this time.

(4) Extraction process and simulation

1. Extraction process

After the discussion on the isolation and degradation of adhesives, and the dissolving of nitrate esters, explosives, and oxidants the whole extraction and separation process of propellant components was determined as shown in Fig. 15.15.

The surface area of crushed propellant was 1.61 m^2/kg. The extraction process of nitrate ester with dichloromethane was level-4 cross-flow process. The extraction liquid was concentrated until the nitrate ester/solvent mass ratio was 70/30. The concentrated and recovered solvent was recycled and reused. The unit operation of solvent extraction process was shown in Table 15.12.

Extraction solvent of explosives and oxidant was acetone in the single-stage continuous operation. After the extraction, the solution was concentrated to form concentrated slurry containing explosives and oxidant crystals. This concentrated slurry was filtered, washed with water to isolate solute crystal. The solvent was recycled and reused. The final product from the extraction process were: nitrate ester in inert diluent solvent, wet nitramine, and wet oxidant. Additionally, there were residues containing metal powder.

Table 15.11 Equilibrium data of NG extraction with dichloromethane

Propellant	Surface area ($m^2\ kg^{-1}$)	Temperature (°C)	Solute/solvent ratio ($kg\ kg^{-1}$)	ω_{EC}	$\zeta_{EB}\times 10^4$	ω_{RC}	ζ_{RB}
NEPE/HMX/AP/Al	0.81	21	0.0025	23	0.6	0.0100	0.640
			0.0025	25	0.6	0.0100	0.660
			0.005	49	1.4	0.0100	0.620
			0.005	48	1.0	0.0120	0.670
			0.010	93	1.6	0.0149	0.537
			0.010	93	1.2	0.0155	0.513
			0.015	138	–	0.0205	0.572
			0.020	196	0.9	0.0255	0.586
			0.020	189	2.6	0.0231	0.572
			0.025	234	1.8	0.0324	0.592
			0.030	273	6.1	0.0353	0.607
			0.035	319	8.8	0.0389	0.571
			0.040	355	4.1	0.0445	0.583
			0.045	389	14.1	0.0522	0.579
			0.050	426	18.1	0.0557	0.580
			0.055	464	24.8	0.0605	0.588
			0.060	492	21.1	0.0660	0.593
NEPE/HMX/AP/Al	1.598	21	0.010	94	2.8	0.0180	0.620
			0.015	140	1.7	0.0210	0.630
			0.020	180	9.6	0.0290	0.580
			0.020	180	8.7	0.0290	0.610
			0.025	220	7.2	0.0320	0.590
			0.030	250	24.0	0.0420	0.580

2. Process simulation

① Process simulation on nitrate ester extraction

The propellant residue (400 g) from NG extraction was extracted with the four-stage cross-flow operation, "solute/solvent" ratio of 0.04 and propellant surface area of 1.61 m^2/kg at 21 °C for 90 min. After the extraction, precipitation, and separation, its components were analyzed or calculated according to the material mass balance.

Table 15.13 shows the results from the extraction process. Here, A, B, and C represent solvent, insoluble solid, and solute. R and E represent the residue and extraction solution. Their mass fractions in the residue were m_{RA}, m_{RB}, and m_{RC}, respectively, and their mass fractions in the extraction solution were m_{EA}, m_{EB}, and m_{EC}, respectively. m_{RHMX}, m_{EHMX} represented the mass of HMX in residues and extraction solution, respectively.

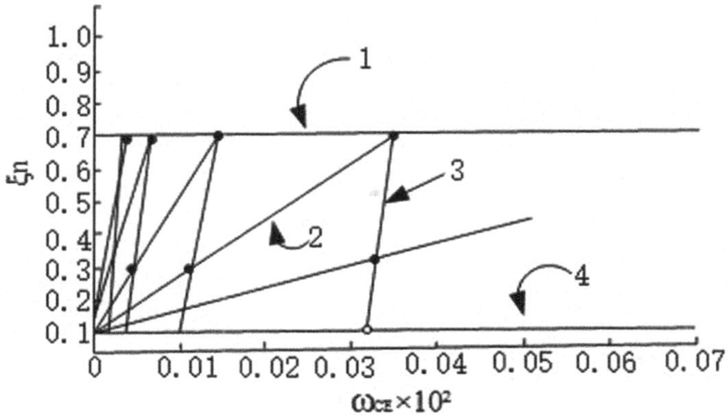

Fig. 15.14 Multistage cross-flow contacting operation chart of isolating NG from NEPE propellant [6]. 1—the propellant residue phase, 2—operating phase, 3—equilibrium phase, 4—extracted liquid phase

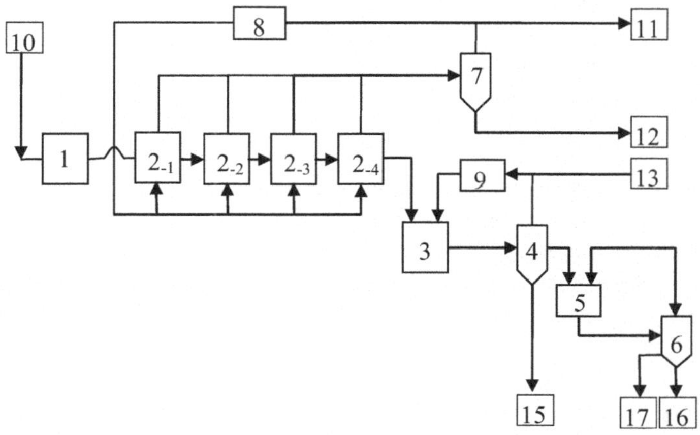

Fig. 15.15 Extraction process diagram of propellant. 1—shattered propellant powder, 2—extraction of nitrate ester, 3—extraction of nitramine, oxidizers, 4—solid–liquid separation, 5—nitramine precipitation, 6—separation of nitramine and oxidant, 7—concentrated methylene chloride and nitrate ester, 8—methylene chloride storage tank, 9—acetone storage tank, 10—propellant, 11—methylene chloride, 12—nitrate ester solution, 13—acetone, 14—water, 15—residues, 16—nitramine, 17—oxidant

In the residue, the mass ratio of insoluble over solvent and solute was $\xi_{RB} = m_{HMX}/(m_{RA} + m_{RC})$. Mass fraction of solute in the combined solvent and solute was $\omega_{RC} = m_{RC}/(m_{RA} + m_{RC})$.

Table 15.12 Choice of solvent in the whole extraction process

Item	Extraction process of nitrate ester	Extraction process of nitramine/oxidant
Process	Multistage cross-flow contacting operation	Single-stage continuous operation
Solvent	CH_2Cl_2	Acetone
Solute/solvent mass ratio	0.04	0.15–0.18
Process temperature	21 °C	21 °C
Surface area of propellant particle	1.61 m^2/kg	1.61 m^2/kg
Contacting times	4	7
Retention time	10 min every level	3 h

Table 15.13 Results of multistage cross-flow extraction of nitrate ester from NEPE/HMX/AP/Al

Stage	ω_{EC}	ζ_{EB}	ω_{RC}	ζ_{RB}	Accumulated quality of recycled nitrate ester $\times 10^2$
1	0.032	6.00	0.053	0.59	59.4
2	0.010	3.50	0.021	0.59	84.4
3	0.004	2.00	0.008	0.57	94.1
4	0.001	0.04	0.004	0.59	96.7

In extraction solution, the mass ratio of insoluble over solvent and solute was $\xi_{EB} = m_{EB}/(m_{EA} + m_{EC})$. Mass fraction of solute in the combined solvent and solute was $\omega_{EC} = m_{EC}/(m_{EA} + m_{EC})$.

The theoretical prediction and the actual results were consistent (Fig. 15.16). The total recovery rate of nitrate ester and stabilizer was calculated as 96.7%.

② Product purity

The recovery ratio of extracted NG and stabilizer was 96.4%, and the rest of NG and stabilizer was left in the adhesive glue. The recovery ratio of AP was 99.17%. Spectral analysis results showed that there were no significant traces of impurities. The recovery ratio of HMX was 96.9% and its melting point was 275 °C.

③ Detonation sensitivity

Through determination of sensitivity of intermediate and final products, their risk of explosion and burning was assessed. Their impact sensitivity, friction sensitivity, electrostatic sensitivity, and self-ignition temperature were tested, respectively. According to these tests, sealed container should be used to store NG dichloromethane solution to avoid loss of dichloromethane. In the long-time storage, diethylene glycol might be used as solvent of NG.

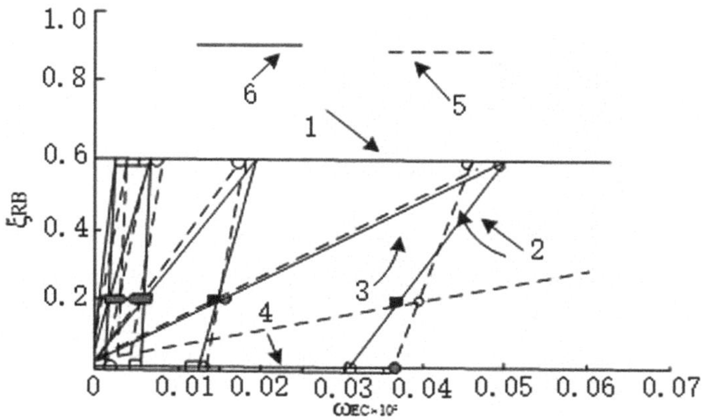

Fig. 15.16 Comparison of simulation on multistage cross-flow extraction of nitrate ester from NEPE with actual results [6]. 1—solid, 2—the equilibrium line, 3—operation line, 4—extraction solution, 5—simulated results, 6—actual results

15.2.3 Separation Engineering of the Propellant Production

Propellant extraction is a systematical engineering, including the propellant surface area, solvent permeability, effect, extraction rate, and reaction temperature to form a complete set of technological process. Solvent extraction process of propellant was shown in Fig. 15.17.

(1) The crushing step 1 was to increase the propellant surface area in contact with solvent. During the propellant crushing, size was tuned through adjusting the gap between cutters and control panels and selecting screen sieve model, then to achieve the extraction-required surface area of propellant.
(2) The NG dichloromethane extraction processes. Crushed propellant was delivered to extraction tank 3 with dichloromethane through the mesh basket. After 10 min of extraction, the solution was sent to buffering tank 6. Extraction solutions for totally 4 times were sent into continuous distillation reactor 7. After about 4–5 h of distillation, when NG/CH_2CI_2 was concentrated to 70/30, the enrichment process was completed. The concentrated product was sent into storage tank. Containers for extraction and storage of NG should be simple, less mobile, easy to clean, effective, compatible with NG, minimized transfer steps.
(3) Acetone extraction process of oxidant and explosive. The propellant residue after NG extraction was used to extract oxidant and Explosive 3 through multistage cross-flow operation. After cleaning of extraction residues with water, the residue was disposed as waste. Extraction solution was sent into the Crystallizer 10. After crystallization, crystal and liquid were separated with filters. The liquid was sent into Buffer Tank 11 to recover the solvent.

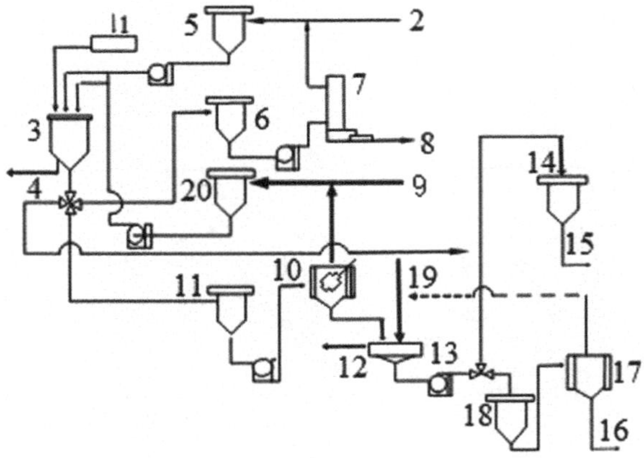

Fig. 15.17 Solvent extraction process of propellant [7]. 1—propellant shattering, 2—CH_2CI_2, 3—extractor, 4—the propellant residues after NG extraction, 5—CH_2CI_2 storage tank, 6—buffer tank of NG + CH_2CI_2, 7—distillation reactor, 8—concentrated liquid of NG + CH_2CI_2, 9—acetone, 10—crystallizer of HMX + AP, 11—buffer tank of HMX + AP, 12—HMX, 13—filter, 14—buffer tank of solvent recovery, 15—distillation tank of NG + CH_2CI_2, 16—AP, 17—AP crystal, 18—buffer tank of AP + H_2O, 19—acetone +H_2O washing solvent, 20—storage tank of acetone

(4) Separation operations of oxidant and explosive. Crystal was washed with a small amount of acetone again to remove the remnant NG, and the washing fluid was sent into the solvent recovery and buffer tank. In the isolated mixture of AP and HMX, hot water was used to dissolve AP. Through distillation, AP was recovered from the aqueous solutions of AP. Through the filtration, wet block of HMX was obtained.

15.2.4 Evaluation of Separation Extraction Technology of Propellant [8]

(1) Evaluation of the economic and technical indicators.

1. The reality of technology. To consider technology basics and the filled extent of related data.
2. Applicability. Ability to adapt to various types of propellant.
3. The complexity. To observe the complexity of operations from the whole process and the unit.
4. Raw materials. To consider the source, cost, and value of the raw material.
5. Possibilities to expand.
6. Safety. The risk and reliability of the process.
7. Environmental impact. To consider the types of wastes and their dangers.

Table 15.14 Evaluation index of extraction process

Item	Evaluated subject	Evaluation index
Reality of technology	Having pilot factory or production record	1.0
	Experiment data	0.5
	Theory	0.1
Applicability	High-energy propellant and composite propellant	1.0
	High-energy propellant and composite propellant	0.5
	Special propellant	0.1
Complexity	1–6 operation units	1.0
	7–9 operation units	0.5
	>10 operation units	0.1
Raw material	No raw material consumption	1.0
	Solvent recycling/refilling	0.5
	Consumption of recycling raw material	0.1
Possibility to expand	Standard unit operation	1.0
	Change/new applications	0.5
	New concept	0.1
Safety	Without hazardous	1.0
	High temperature, high pressure, or combustible	0.5
	Explosive or toxic	0.1
Environmental impact	No	1.0
	One or several nonhazardous effects	0.5
	One or several hazardous effects	0.1

For all of the above evaluations, the corresponding assessment index d_i was formed with the comprehensive index and could be calculated as

$$d = (d_1, d_2, \ldots, d_n)^{1/n} \tag{15.5}$$

Comprehensive evaluation index range of the extraction process is 0.25–0.82.

(2) Technology risk assessment of propellant extraction

According to Eq. (15.1) and the evaluation index in Table 15.14, the above-described various propellant extraction and recycling technologies were assessed.

① In composite propellants, liquid ammonia recovery AP technology was used. Its comprehensive evaluation index was relatively low due to the detonation risk of AP in ammonia solution.

② Comprehensive evaluation index of aluminum powder recycling was low because of high material consumption in the de-esterification and high-energy consumption in the thermal decomposition.
③ The recovery NQ-AP and the NHC's technique is only suitable for certain types of propellants and is not typical for solid propellants in service. Therefore, with the development of high-energy materials and equipment, new extraction technologies should be continuously explored to ensure the restrictions on application of these technologies after global disarmament.
④ In all recycling processes of high-energy propellant, comprehensive evaluation index is relatively low, and the whole extraction process could avoid risk factors of highly sensitive explosive materials NG, HMX, and CL20.

In addition to these mentioned technologies, other conversion technologies were taken into account, such as converting into arson, slurry explosive, booster column, and other non-incineration technologies. After a comprehensive evaluation, the following comparative economic and environmentally acceptable technologies were obtained.

15.3 Continuous Application of Extracted Nitrate Cellulose to Manufacture Propellant [1, 2, 8]

According to studies on separation of propellant components and large amount of experiment data, their recovery ratios and efficiency of separated and recovered NC, NG, NQ, HMX, RDX, and CL20 from retired propellants were very high. In addition, they could be used as raw material to produce single-base, double-base, and three-base propellants.

(1) Stabilizers in nondestructive replacement of single-base propellant to produce single-base propellant

Stabilizers in nondestructive replacement of single-base propellant, based on energy of the propellant first, were determined by nitrogen-containing semi-colloidal NC and its size after the separation of diphenylamine and process additives. The size of these launching explosives from stabilizers in nondestructive replacement of single-base propellant is usually 0.01–0.03 mm smaller than that of the same graded launching explosives from new NC. The tiny change in size was due to macromolecule loosening caused by surface-layer oxidation of retired launching explosives, which was converted into soluble small molecular low-level nitrates, and then small molecular collapsing down during the nondestructive separation expansion. Structure and mechanical performance of single-base launching explosive after nondestructive replacing stabilizer did not have a significant change. As listed in Table 15.3 of the accelerated aging, the stability and storage period of the propellant from this approach were up to 25 years [9], and its ballistic performance met the requirements of army (Table 15.15).

Table 15.15 Properties of 9/7 propellant manufactured from nondestructive replacement of stabilizer

Tested item and subject	Result
NC content (%)	95.9
Diphenylamine (%)	1.52
Inner volatilization (%)	1.36
Exo-volatilization (%)	1.22
Explosive energy (J/g)	980
Explosion heat (J/g)	3681.92
Specific capability (ml/g)	925
Ratio of specific heat	/
Average m.w. of fuel gas	/
Isovolumetric flame temperature (K)	2820
Stabilizing property test at 106.5°C (h)	7
Acceleration test (h)	79
Ballistic performance (30 mm aerogun)	
Explosive loading amount (g)	158
Average velocity (m/s)	798
The maximum velocity (m/s)	835
Average gun pressure (MPa)	295.6
The maximum gun pressure (MPa)	315.3

Note Nitrogen content of semi-colloidal NC was 205.89 ml/g, and nitrogen content of propellant after nondestructive replacing stabilizer was 204.63 ml/g

(2) Produce single-base propellant from isolated semi-colloidal NC to adjust nitrogen content

In the traditional single-base propellant production, the added recycling drug was 10–15% generally [10, 11]. Semi-colloidal NC of separated retired launching explosives with different shapes matched and adjusted the nitrogen content itself. With 100% of semi-colloidal NC and traditional single drug production process, two species of launching explosives 7/14 and 11/7 were produced. The physicochemical analysis and ballistic test results showed that these two met the application requirements. The accelerated aging test was up 112 h and they could be stored for 30 years and above (Table 15.16).

(3) Production of fast burning gunpowder from isolated semi-colloidal NC to adjust nitrogen content and added components,

Additionally, the separated semi-colloidal NC from the retired single-base propellant could be used to produce quick burning gunpowder. Retired 6/7, 7/7, 9/7, 11/7, 12/7, 13/7, 14/7, 16/7, 7/14, and 8/14 propellants, which still met the requirements of propellant, were crushed into 100 mesh particles, then they were sealed and stored for 90 min with the solvent/drug = 1/1 and alcohol/ether = 1.2/1.4. The preswollen single-base propellant was added into the kneading machine, with

Table 15.16 Properties of launching explosives 7/14 and 11/7 produced from retired 4/1, 9/7, and 5/7 launching explosives

Tested item and subject	14-Jul	7-Nov
NC content (%)	97.92	97.7
Diphenylamine (%)	1.32	1.42
Inner volatilization (%)	0.76	0.88
Exo-volatilization (%)	/	1.57
Explosive energy (J/g)	1020.57	990.39
Explosion heat (J/g)	3819.99	3623.34
Specific capability (ml/g)	978	952
Ratio of specific heat	1.2351	1.238
Average m.w. of fuel gas	24.11	23.54
Isovolumetric flame temperature (K)	2961	2805
Stabilizing property test at 106.5 °C (h)	7	7
Acceleration test (h)	102	107.5
Ballistic performance (30 mm aerogun)	37 mm anti-aircraft gun	122 mm howitzer
Explosive loading amount (g)	206	2090
Average velocity (m/s)	886	513
The maximum velocity (m/s)	905	537
Average gun pressure (MPa)	269.5	219.4
The maximum gun pressure (MPa)	283.3	232.7

Note To control nitrogen content at semi-colloidal NC matching, 208.7 ml/g in 7/14, 207.3 ml/g in 11/7

kneading and mixing, potassium nitrate powder (powder grain of >90% of potassium nitrate for 90% was through 160 mesh sieve, and water content was less than 0.1%). Then, the mixture was sealed and mixed continuously. After qualified the pressure test detection, plasticizing was stopped, the forming pressure was 32–40 MPa. After the water soaking, graphite was added for pre-glossing, and the added graphite volume was under graphite/gunpowder = 0.2/100. Production technology of quick burning gunpowder from isolated semi-colloidal NC was shown in Table 15.17 and Fig. 15.18.

(4) Producing double-base propellant from isolated semi-colloidal NC with adjusted nitrogen content

Separated semi-colloidal NC from retired single-base propellants, with self-adjusted nitrogen content, was used to produce enhanced energy double-base propellant with the isolated nitroglycerin. The operation method was: double-base propellant was crushed into particles smaller than 1 mm × 1 mm × 1 mm. After the fractional extraction, distillation was used to recover ethyl ether from extraction solution, and the left liquid is mainly NG. The isolated NC powder and recovered NG were absorbed within three times to obtain enhanced energy single-base propellant. Quality analysis on product showed that the distribution of NG was in a relatively

Table 15.17 Engineering condition control of quick burning gunpowder production from retired speed single-base propellant

Soaking times	Temperature (°C)	Time (h)
1	15–20	3
2	20–25	3
3	25–30	2
4	30–35	2
5	35–40	2
6	40–45	2
7	40–45	1–2

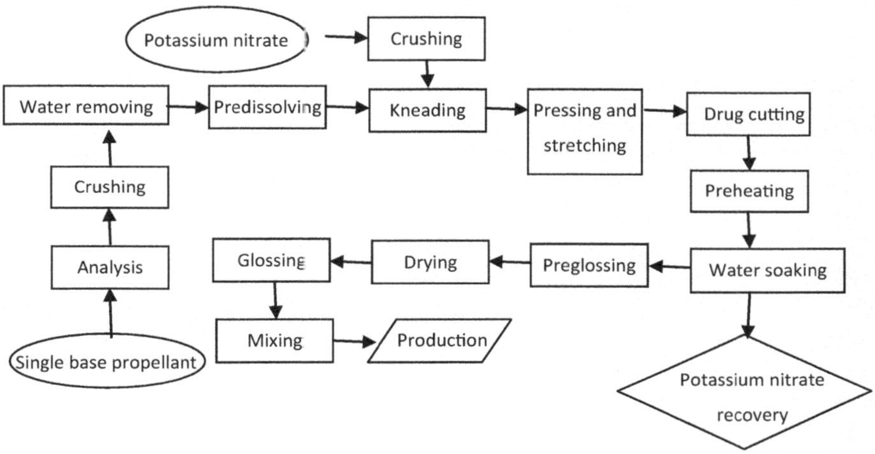

Fig. 15.18 Process of quick burning explosives from retired single-base propellant. Kneading

stable state, and its ballistic performance met weapon requirements. Accelerated aging experiments showed, its stocking period was up to 20 years. Preparation process of double-base propellant from retired launching explosives and aging mass fraction of enhanced energy double-base propellant were shown in Fig. 15.19 and Table 15.18, respectively.

(5) Mechanical properties of energy-enhanced single-base propellant

Mechanical properties of the standard single-base propellant, gunpowder M30, and energy-enhanced single-base propellant were compared. Young's modulus E_t, anti-compressing strength σ_n, and destructive compression ratio ε_n at high, low temperature, and room temperature were compared. Results showed that mechanical properties of the energy-enhanced single-base propellant were close to those of single-base propellant, but much higher than those of double- and three-base propellants. Mechanical properties at low temperature and mechanical property variation with temperature of energy-enhanced single-base propellant was similar to those of single-base propellant. But its mechanical properties at low temperature

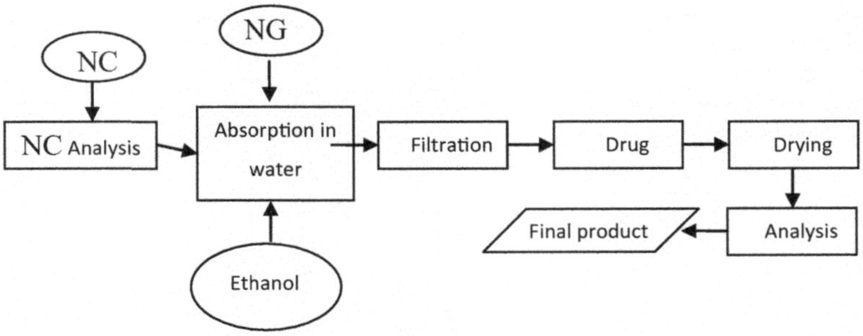

Fig. 15.19 Production process of increased energy propellant

Table 15.18 Mass fraction of NG in gunpowder with different depths and aging

Depth (mm)	NG mass percentage in gunpowder				
	No aging	10 days of aging	30 days of aging	60 days of aging	90 days of aging
0.05	32.56	30.58	27.59	23.41	19.87
0.15	28.98	27.64	25.34	21.87	18.99
0.25	24.75	25.03	22.35	20.63	17.65
0.35	19.35	20.66	19.66	18.75	16.52
0.45	13.82	14.56	15.46	16.54	16.88
0.55	6.58	7.99	10.32	13.52	14.63
0.65	2.09	3.54	7.43	9.48	12.51
0.75	0.36	0.99	3.52	6.35	11.23
0.85	3.65	3.82	4.99	6.33	7.46
0.95	9.78	10.03	9.65	8.84	8.42
1.05	13.65	12.48	11.32	10.68	10.56

was significantly higher than those of general double-base and three-base propellants. The energy-enhanced single-base propellant not only solved the problem to improve energy of single-base propellant, but also solved the poor mechanical problems of three-base and double-base propellants at low temperature. Properties of energy-enhanced propellant from retired launching explosives were shown in Table 15.19 and Figs. 15.20, 15.21 and 15.22.

Test results showed that retired propellant can be used to separate NC, NG, and other valuable raw materials through nondestructive separation recovery, and the recovery ratio and efficiency were very high. The recycled semi-colloidal NC, after self-adjusting and matching nitrogen content and adding other component materials, could be used to prepare various grades of launching propellants and their performance met weapon requirements.

Table 15.19 E_e, σ_n, and ε_n values of single-base propellant, gunpowder M30, and energy-enhanced single-base propellant

Type of gunpowder	E_e (MPa)			σ_n (MPa)			ε_n •100		
	−40 °C	15 °C	50 °C	−40 °C	15 °C	50 °C	−40 °C	15 °C	50 °C
M30	1014	382	148	140	35.1	20.4	26.4	28.6	34.0
Single base	1832	514	178	213	47.2	18.8	16.3	20.2	23.2
Energy-enhanced single base	1790	488	166	206	41.0	15.6	18.7	23.6	26.7

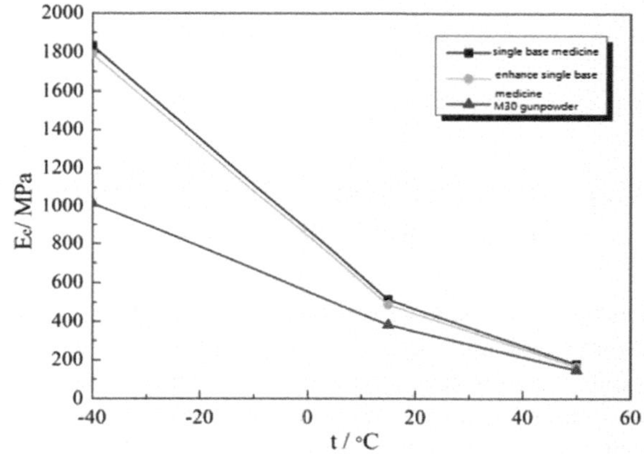

Fig. 15.20 Relationship between E_e of single-base propellant and temperature

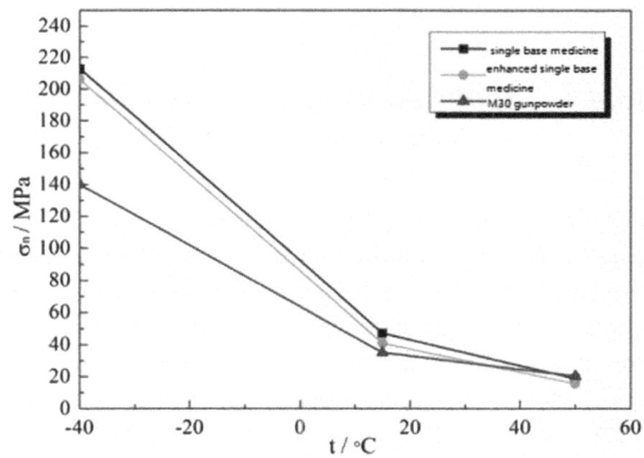

Fig. 15.21 Relationship between σ_n of energy-enhanced single-base propellant and temperature

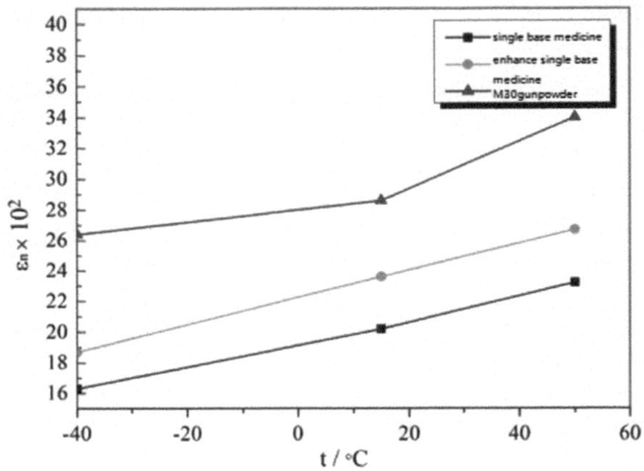

Fig. 15.22 Relationship between ε_n of energy-enhanced single-base propellant and temperature

15.4 Industrial Applications of Products Separated from Propellants

On the one hand, nitrocellulose, nitroglycerin, nitroguanidine, dinitrotoluene, dibutyl phthalate, and other raw materials could be separated from the disposed explosives. On the other hand, through chemical reactions, they could be turned into other chemicals. With the preparation of oxalic acid from single-base propellant as an example, "preparation of chemical raw materials" was shown as a way to modify and reuse abandoned explosives.

15.4.1 Applications of Semi-colloidal Nitrocellulose in Coatings Industry [1, 2, 4, 12]

In order to reuse retired propellant in the industrial products, the authors proposed nondestructive ordered degradation model of semi-colloidal nitrocellulose, which solved the problem on the control of molecular weight of semi-colloidal cellulose nitrate in the degradation process. Nitrocellulose, under ultrasonic and irradiation conditions, can be degraded with heating, which was along with hydrolysis reactions. However, degradation mechanism of semi-colloidal nitrocellulose has not been reported. In our proposed nondestructive degradation mechanism of semi-colloidal nitrated cellulose, the semi-colloid surface was nondestructive and damaged first, to cause the surface to form surface micro-holes to increase reaction contacting area. The reaction solution permeated to inside to open distribution channels of internal molecules. The unstable group was decomposed under the induced catalysis to achieve the ordered degradation of semi-colloidal nitrocellulose. The used technology was semi-colloidal nitrocellulose which was in the water interacting with the media having primary hydroxyl group. Through controlling the reaction temperature and time, ordered degradation of semi-colloidal nitrocellulose was achieved, and the viscosity could be controlled within 40–1/2 s.

The viscosity of nitrocellulose to manufacture single-base propellant was high, which was usually above 3.5°, in order to ensure a certain mechanical strength after the propellant manufacturing. Although it was stored over nearly 50 years with a slight decrease of large molecules, its viscosity was still very high with very little change. The viscosity requirements of the paint nitrocellulose was very low in order to meet the high film-forming substances in the paint, and its viscosity was calculated according to the number of dropping seconds, which was generally within 1/2–5 s. In order to make single-base nitrocellulose propellant to meet the requirements of coatings, semi-colloidal nitrocellulose–macromolecule structure must be degraded without damage on nitrocellulose properties, which only could have a relatively high application value.

Semi-colloidal nitrocellulose was added into dedicated reactors containing water with circulation before 50 mL of the solution containing solvent III was added. At a certain temperature, reaction solution was circulated continuously. By controlling the reaction time, the desired viscosity of nitrocellulose was determined. At 112 °C and under 0.05 MPa pressure, viscosity change of semi-colloidal NC with time was shown in Fig. 15.23.

After residue was washed with pumped fresh water, solution containing 100 mL solution III was added. At 120–160 °C, reaction solution was circulated. After 30–90 min of reaction to denitrify (reducing nitrogen content), the degree of reduction in nitrogen content of nitrocellulose could be accurately controlled (nitrogen content of normal civilian NC was 11.5–12.2%). According to general production process, the obtained nitrocellulose was stability treated and gradually dehydrated. Finally, all pigments on NC from the technological processes were removed with a special solvent. The obtained solid had nitrogen content of 11.5–12.2%, viscosity of 1/2–5 s, flash point of above 180 °C and Abel's test of more than 20 min, color, whose appearance and color were close to NC itself.

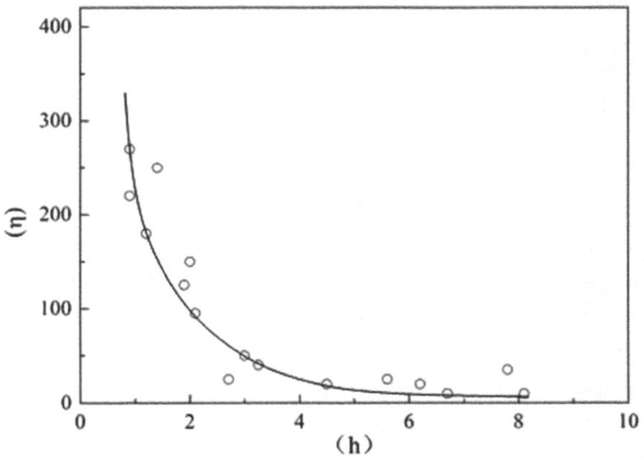

Fig. 15.23 At 112 °C and under 0.05 MPa pressure, relationship between viscosity change of semi-colloidal NC and time [2]

Table 15.20 Quality of mixed coating nitrocellulose from retired propellant mixture

Item group	Standard	1#	2#	3#
Nitration degree (mL/g)	185–195	189.32	190.27	191.29
Transmittance (%)	75–85	82	83	81
Viscosity (Pa s)	3.4–12	8.5	7.2	9.5
Exploding point (°C)	≥ 180	184	183	181
80 °C heat-resistance test (min)	≥ 15	25	25	25
Acidity (%)	≤ 0.07	0.045	0.037	0.033
Hash (%)	≤ 0.2	0.15	0.18	0.12
Content of wetting agent (%)	30 ± 2	28.5	30.2	29.7

Note The viscosity here is the actual ball dropping time

Recovery ratio was 82–84% when retired single-base propellant was used in the production of this NC. Production cost per ton of NC was about 1/7–1/6 of that of new NC. Properties of painting NC from separated retired propellant were shown in Table 15.20.

The painting NC from the abovementioned method was mechanically and evenly mixed with corresponding resin and color to obtain nitro exterior varnish, nitro exterior enamel, and acrylic resin modified external nitro enamel, respectively. Usability and aging performance showed that these products met the European quality standards. Properties of coating NC from retired and modified propellant were shown in Table 15.21. Process of comprehensive utilization of retired propellant was shown in Fig. 15.24.

The viscosity of nitrocellulose for single-base propellant manufacturing is high, usually as 3.5 Engler degrees, in order to ensure the propellant has certain mechanical strength after manufacturing. Although retired having a certain degree of molecular weight degradation after more than 20 years of storage, the viscosity of single-base propellant is still very high. The viscosity of the painting nitrocellulose requirements are very low in order to meet the high content of film-forming substances in paint, its viscosity is calculated according to the number of dropping seconds, generally 1/2–5 s is suitable. High nitrogen content seriously affected solubility and dilutability of paint. Therefore, in order to use retired propellant in the paint industry, viscosity and nitrogen content of the NC must be controlled.

(1) Relationship molecular weight and viscosity changes during the process of retired single-base propellant reforming into nitrocellulose paint.

In order to make nitrocellulose propellant to meet the requirements of paint, semi-colloidal nitrocellulose–macromolecule structure must be degraded without damage to nitrocellulose properties, thus to have a higher value. At 112° and under 0.05 MPa pressure, after 2–6 h of semi-colloidal NC degradation, the change of semi-colloidal NC viscosity was as shown in Fig. 15.25.

In Fig. 15.25, 1# was retired single-base propellant (batch number was 9/7-5/67-35). 2# was that after diphenylamine extraction and steam heat treatment.

Table 15.21 Testing results of acrylic-modified nitrate enamel from retired propellant

Name tested items results		Technical target	Testing result				
		Acrylic enamel	Pure white acrylic enamel	Red acrylic enamel	Mild yellow acrylic enamel	Mild blue acrylic enamel	Green acrylic enamel
Film color and appearance		Fitting to standard color and the range of chromatic aberration	Meet the requirements	Meet the requirements	Meet the requirements	Meet the requirements	Meet the requirements
Drying period (min)	Surface dried	<20	10	10	10	10	10
	All dried	<60	40	40	40	40	40
Viscosity (Painting —1) (s)		100–200	116	126	138	127	142
Solid content (%)		>34 (shallow), 36 (deep)	42.05	37.83	40.57	38.8	40.71
Gloss (%)		>70 (shallow), 80 (deep)	89	87	88	93	87
Hardness		>0.6	0.73	0.69	0.75	0.72	0.77
Impact strength (kg cm)		50	50	50	50	50	50
Adhesive force degree		<2	1	1	1	2	1
Flexibility (mm)		1	1	1	1	1	1
Water resistance (48 h)		Slightly whitening, loss of gloss, blistering are allowed during the 2 h of recovery	Qualified	Qualified	Qualified	Qualified	Qualified
Oil resistance (48 h)		Paint film has no blister, but gloss lose and softening are allowed. Gloss lose is allowed for aluminum and red glosses.	Qualified	Qualified	Qualified	Qualified	Qualified
Conclusion			Qualified	Qualified	Qualified	Qualified	Qualified

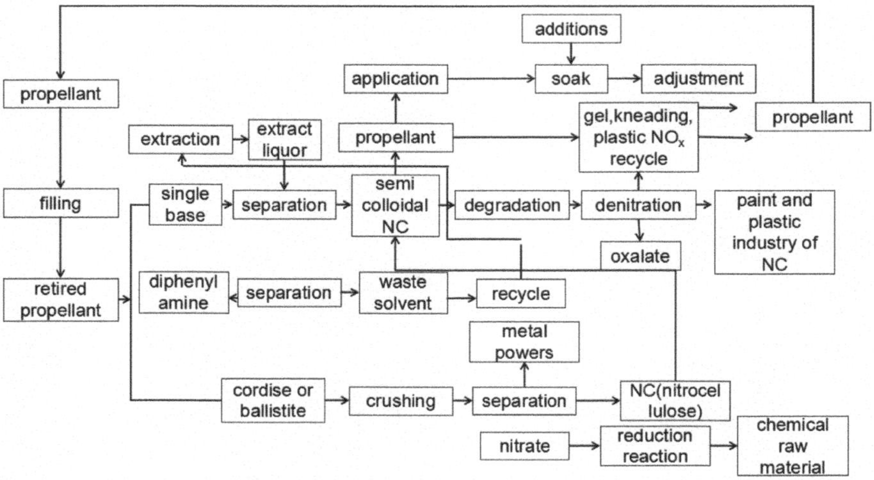

Fig. 15.24 Process of comprehensive utilization of retired propellant

Fig. 15.25 Degradation change of macromolecule NC in retired single-base propellant

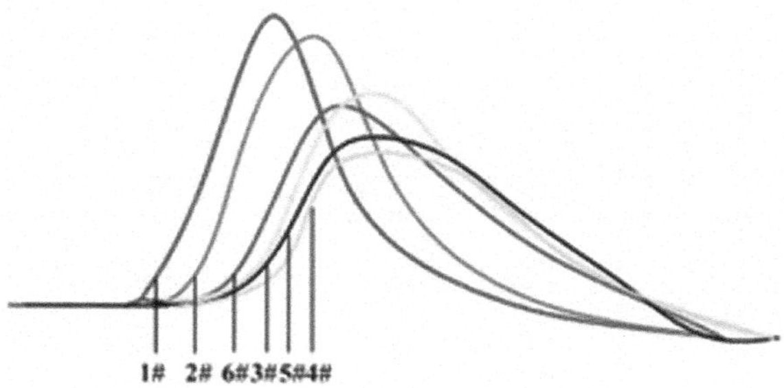

3# was that after degradation treatment (about 40 s of degradation). 4# had degradation reaction during the denitrification. 5# was nitrocellulose after stability treatment. 6# was the decolorized sample.

All NC obtained from every stage was dissolved with 85% n-butyl acetate and 15% n-butanol to prepare 20% nitrocellulose solution, and then NC was precipitated out with dropping toluene. Performance results were shown in Fig. 15.26.

As shown in Fig. 15.25 of NC degradation degree and Fig. 15.26 of NC solubility results after degradation, NC degradation was even. It also showed that heat degradation of macromolecule NC was fast after stabilizers were removed from retired single-base propellant. Especially, pure semi-colloidal NC was degraded faster.

(2) Denitrification change relationship

Nitrogen content of single-base propellant was around 12.8–13.0%, and that of painting nitrocellulose was generally in the range of 11.5–12.2%. Although there

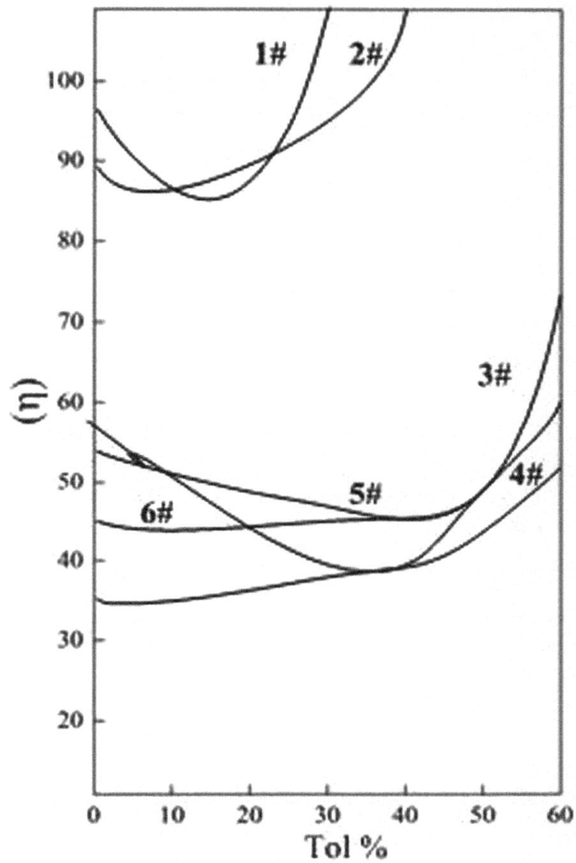

Fig. 15.26 Solubility of NC at every stage corresponding to Fig. 15.25

was little change after more than 50 years of storage, nitrogen content of propellant changed within the range of less than 10–4, which meant that nitrogen content and viscosity were little changed. When stabilizers exist in a single-base propellant, it was extremely difficult to change nitrogen content or denitrate. During the denitrification of retired propellant, stabilizers must be eliminated from the retired propellant first, to destruct the surface colloid of macromolecular, to degrade nitrocellulose–macromolecule structure. Then, denitrification could be done under the reaction with primary acid catalyst. Nobel and Russian scientist з.H and Luo Guowen thought [13] that nitrocellulose (flocculent) could be denitrated completely in sodium hydrosulfide or an alcoholic solution of sodium hydrosulfide at 80 °C within 24–48 h. In addition, the denitrated product was red-brown without any mechanical properties. Thus, it was of no value. How to denitrate granular nitrocellulose has been reported. According to principles of our proposed nondestructive controllable ordered denitrification, with water as denitrification agent, primary oxidizing acid as catalyst and reaction temperature controlled at 135 °C, ordered

Table 15.22 Control on denitrification conditions

Number	Condition factor	Condition control				
		pH of solution	Raw material: solution	Temperature (°C)	Pressure (MPa)	Time (min)
1	Water-driven feeding	4–6	1:10	Room temperature	0.25	
2	Denitrification reagent	4–6	1:10	120–150	1.5–4	90
3	Cycling					4 times/h
4	Product	>1	1:20	30–40	0.25	
5	Liquid waste	1–2		Room temperature	Disposing after treatment	

denitrification of granular nitrocellulose could be done to obtain civil nitrocellulose within 30 min.

The conditions of semi-colloidal nitrocellulose fiber denitrification were shown in Table 15.22.

Dynamics of denitrification

Assuming the denitrification reaction semi-colloidal nitrocellulose as a first-order reaction, we could know the LHHW unit model was

$$-\frac{dC_N}{d(W/F_{AO})} = \frac{K_N C_N}{1 + K_N C_N + K_C(C_{NO} - C_N)} \tag{15.6}$$

After integration, the following could be obtained:

$$(1 + K_N C_{NO})K_n C_{NO}/C_N + (K_N - K_C)(C_{NO} - C_N) = K_N(W/F_{AO})$$

When the left was set to x,W/F_{AO} was y and the obtained K_N and K_C values at various temperatures were brought into the equation, denitrification rate constants of semi-colloidal cellulose nitrate at various temperatures could be obtained after the linear regression. Furthermore, according to the Arrhenius equation, the corresponding E_n and A_n were determined and shown in Table 15.23.

According to data in Table 15.23, denitrification law of semi-colloidal law of nitrocellulose was found. Denitrification of semi-colloidal nitrocellulose was only related to temperature and time, and was not related to pressure. In addition, the yield was associated with nitrogen content. The relationship between painting

Table 15.23 Denitrification kinetics of semi-colloidal nitrocellulose

K_N	$g^{-1}h^{-1}$	mol^{-1}	E_n	A_n	
120 °C	140 °C	160 °C	180 °C	KJ/mol	
0.22	0.67	1.06	2.57	1012.99	1.93×10^8

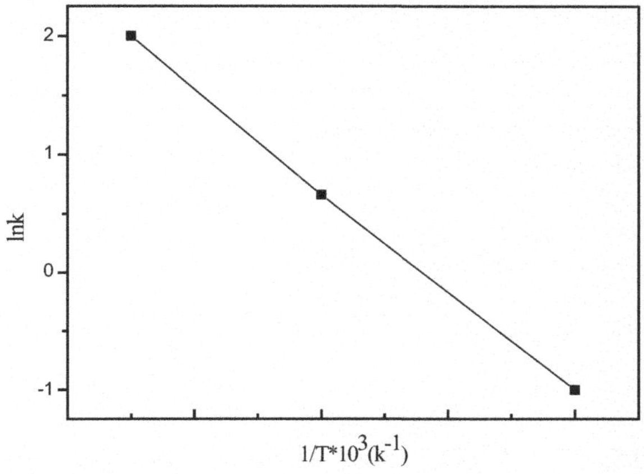

Fig. 15.27 Arrhenius of NC denitrification [2]

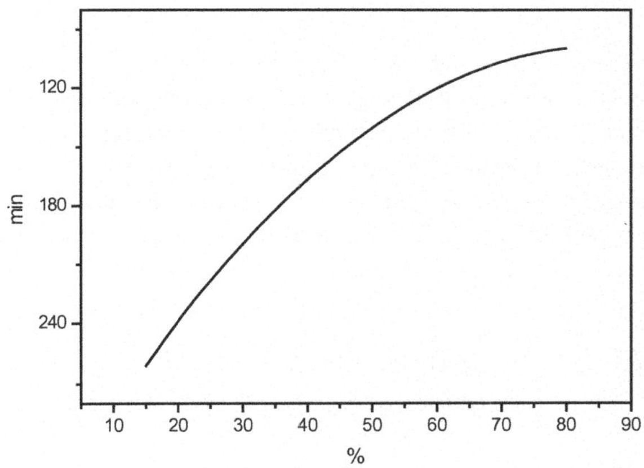

Fig. 15.28 Denitrification law of semi-colloidal NC [2]

nitrocellulose from semi-colloidal nitrocellulose and denitrification was shown in Figs. 15.27 and 15.28.

In the production of painting NC modified from retired single-base propellant, solvent consumption per ton of product was 120–160 kg. The used solvent from the first few times could be directly used without distillation, but after the filtration through a porcelain molecular sieve. After the solvent was used for several times, $C_{12}H_{11}N$, its derivatives and unstable low nitrates could be converted into dyes over Fe^{3+} catalyst.

Water from degradation, denitrification, and stabilizer treatment could be recycled and reused, and this process was waste emission free.

Table 15.24 Performance of xylonite formed from retired launching propellant

Tested subject	Performance standard of xylonite	Xylonite formed from retired launching propellant
Tensile strength (MPa)	>45.0	46.8
Elongation at break (%)	≥ 18	18.6
Solubility (h)	≤ 2.0	1.5
Water absorption (%)	≤ 3.0	1.7
Ash (%)	≤ 3.0	2.5
Thermal distortion (mm)	≤ 1.2	0.97

15.4.2 Applications of Semi-colloidal Nitrocellulose in the Plastic Industry [1, 2, 4, 14, 15]

Semi-colloidal NC was used to make chemical raw materials, such as xylonite, plastic bottle, and others. After the denitrification and dehydration of semi-colloidal NC, it was swollen with ethanol and acetone-mixed solvent. With added camphor and dibutyl phthalate esters and some plasticizer, kneading, xylonite could be made through kneading and rolling refining. The xylonite formed from retired launching propellant was yellowish, and its performance was shown in Table 15.24.

15.4.3 Industrial Oxalic Acid Produced from Separated Nitrocellulose [1, 6, 14, 15, 16]

Oxalic acid, also called as ethanedioic acid (HO_2CCO_2H), has two crystal forms: alpha and beta. Industrial oxalic acid usually contains two crystal water with a melting point of 101–102 °C, and it will lose crystal water when it is heated to 98–100 °C. Oxalic acid, used in the dyeing industry, organic synthesis, precious metals refining, and other industries, is an essential raw material in the chemical industry.

The retired launching propellant after degradation and denitrification cannot be used as industrial NC powder. NC powder was fed into the reactor, water was then added with few nitric acid, sulfuric acid, and 2–5% of V_2O_5. After reaction temperature was increased to 180 °C and then 2–6 h of reaction, six chemical materials, including oxalic acid, lysine, propanamide, ammonium oxalate, glyoxal and acetaldehyde acid, were separated and purified, respectively. Sulfonic acid was added into exhausted gases, nitrogen oxide and the reaction residues from degradation and denitrification of the recycled semi-colloidal NC. When temperature was increased to 180 °C, after 30 min. of reaction, oxalic acid and glyoxal were obtained through separation of condensation.

(1) Description of basic process

Propellants (single-base propellant, double-base propellant, and nitrocellulose) are mainly hydrolyzed in acidic aqueous solution, in which glycoside bond fracture can also occur in the molecule. The products from hydrolysis of glycoside bond cleavage were further broken down into smaller molecules. With acidic catalyst, molecule was rearranged to generate molecules with crystal water, which were dehydrated to form oxalic acid. The rate and reaction degree of hydrolysis were related to concentration of acid in acidic solution and environmental temperature.

Propellant, under heating, is prone to thermal decomposition. And, at the first phase of reaction, nitrogen dioxide was released with the formation of aldehydes. At the second stage, oxidation reactions between nitrogen dioxide and the new formed carboxyl-group compounds occurred. The final stage is bond breaking of the residual nitrate ester.

Under the catalytic conditions, the hydrolysis, decomposition, decomposition and oxidation reaction of single-base propellant would be accelerated to form alcohols, aldehydes, acids, and other products, and oxalic acid was one of the products. The preparation of oxalic acid by propellant is to carry out the hydrolysis, cracking, decomposition, oxidation and other reactions of the propellant, select and control the appropriate reaction conditions, and make the chemical reaction of the single-base propellant proceed in the direction of forming oxalic acid.

(2) Process and its control conditions

1. Process of propellant

The preparation process of oxalic acid from propellant is to hydrolyze, crack, decompose, and oxidize propellant first, and then oxalic acid is generated through molecular rearrangement. Production process of oxalic acid from propellant was shown in Fig. 15.29.

2. Conditions to control preparation process of oxalic acid from propellant

① Material preparation Single-base propellant was crushed. The product through 20-mesh sieve was collected and centrifuged to remove water. Then, mass fraction of water in single-base propellant was about 8×10^{-2} to 12×10^{-2}, and the correct values were determined.

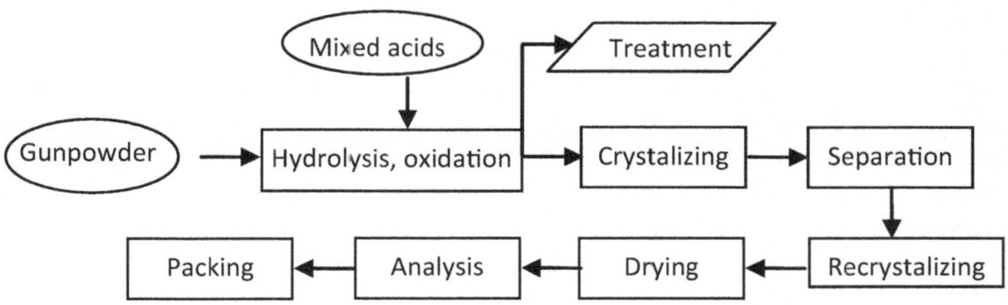

Fig. 15.29 Production process of oxalic acid through the hydrolysis of propellant

During reaction of the first time, the new mixed acid was used, which was mixed acid of sulfonic acid, nitric acid, and water according to the required preparation procedures in the reactor. During the preparation, water in the propellant powder must be taken into account. The prepared mixed acid was cooled down to 35 °C to be used. The mixed acid composition (mass fraction) was

HNO_3 30 × 10^{-2}, H_2SO_4 35 × 10^{-2}, H_2O 35 × 10^{-2};

The allowed tolerances of HNO_3, H_2SO_4 and H_2O were ± 3×10^{-2}, ± 3×10^{-2} and ± 3×10^{-2}, respectively.

Mass ratio of a catalyst over mixed acid = (0.05 − 0.5)/100.

After the first reaction, oxalic acid crystals were isolated from the reaction solution, and the remaining liquid contained HNO_3, H_2SO_4, oxalic acid, and catalysts. When the reaction occurred again, the composition and content in mother liquid should be taken into account. In the following reactions, it was simply needed to add the fraction defect to meet the required mixed acid composition. Then, the reaction liquor was stored and analyzed on its chemical composition for further recycling and reusing.

② Hydrolysis and oxidation. The reaction occurred in acid-resistant Jacket reactor with hot water and cooling water for heat exchange. First, mother liquid was injected into the reactor. Under stirring, nitric acid, sulfuric acid, catalyst, or water were added according to the component requirement and tolerance of mixed acids. When temperature of mixed acids was below 35 °C, propellant powder was added. When mass ratio of " gunpowder/mixed acid" was 1/8, hydrolysis and oxidation reactions occurred, which led the temperature of reaction solution to gradually increase. At this point, hot water was sent through the jacket to increase the temperature of reactants, where reaction temperature was controlled at 50–60 °C. Gaseous reaction product NO_x (primarily NO and NO_2), was introduced into their absorbing device through the pipeline. The speed of stirrer was 6.3 rad/s and the reaction time was 8–10 h. Before the end of reaction, compressed air was sent into the reaction solution to remove the remaining NO_x.

③ Crystallization and separation. After the reaction, reaction liquid was decanted into the crystal tank and slowly cooled down to 5 °C, and coarse crystals of oxalic acid were obtained through filtration. The reaction solution was stored for further recycling and reusing.

Isolated coarse oxalic acid was washed with ice water at near 0 °C. Mass ratio between washing water and crude oxalic acid was "water/acid" = 0.5–1.

④ Recrystallization. First, crude oxalic acid was dissolved with boiling water with water/ crude oxalic acid = 1/1.2. Then, activated carbon was added with mass ratio of activated carbon over coarse oxalic acid as 4/100–5/100. The aqueous solution was boiled for 5–10 min. Hot filtration was used to remove activated carbon and impurities. Liquid was placed in a crystal tank and cooled down to 5 °C, then oxalic acid was precipitated in crystalline form. Industrial oxalic acid was then isolated.

⑤ Drying. Recrystallized oxalic acid was dried at 50–60 °C.

⑥ Analysis and detection. Contents of oxalic acid, sulfate, burning fuel residues, heavy metals, iron, and chloride were analyzed. In the products from this method, the oxalic acid content was >99. 5 × 10^{-2}, sulfate was <0. 06 × 10^{-2}, burning residue was <0. 06 × 10^{-2}, heavy metal was < 0. 02 × 10^{-2}, iron was <0. 003 × 10^{-2}, and chloride was <0.005 × 10^{-2}.

The mass ratio of added materials: gunpowder/reaction mother solution/nitric acid (98 × 10^{-2})/sulfuric acid (98 × 10^{-2}) = 1.00/4.33/2. 3/0.4.

Output (gunpowder quality was 1): oxalic acid/reaction solution = 0.67/4.7.

(3) Factors influencing the reaction of oxalic acid

1. Effect of components of mixed acid on oxalic acid yield. In mixed acid, the mass fraction of nitric acid was 30 × 10^{-2}, mass fraction of sulfonic acid was 35 × 10^{-2}, water mass fraction was 35 × 10^{-2}, which were the appropriate proportion of the mixed acid.
2. Effect of the reaction factor mass ratio of "mixed acid/powder" on the yield of oxalic acid. The reaction factor φ was 7–8 to get a higher rate r_1.
3. Treatment of NO_x.

Hydrolysis and oxidation of single-base propellant were to release a lot of NO and NO_2, which were recycled to manufacture nitric acid. The obtained nitric acid could be reused in the hydrolysis and oxidation. Adsorption technology that could be considered was molecular sieve absorption technology and washing absorption techniques.

① Sieve adsorption

The main device in molecular sieve adsorption were double packed bed absorption tank, in which the first layer was desiccant and the second layer contained oxidation catalysts and molecular sieve to adsorb NO_x. After the gas released from hydrolysis and oxidation was sent into the double-packed bed absorption tank, NO was first oxidized to NO_2 and adsorbed. When the molecular sieve was saturated, reaction gas was sent into another back-up double-packed bed absorption tank to continue the adsorption. Adsorption-saturated double packed bed absorption tank was reactivated with cleaned exhaust gas, and NO_2 was desorbed. The desorbed NO_2 was sent into the adsorption tower to produce nitric acid again.

② NO_x absorption effect of molecular sieve

NO_x adsorption effect of molecular sieve adsorption technology was good with low energy consumption, and the content of exhausted NO met emission standards. Figure 15.30 shows the NO_x process of molecular sieve adsorption system.

③ Washing absorption process

Washing is one of the key technologies to prepare oxalic acid. Obtaining high-quality oxalic acid was related to control washing conditions and the

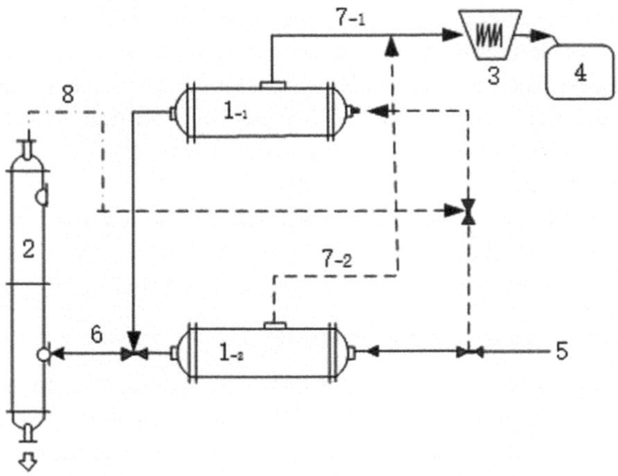

Fig. 15.30 Process of molecular sieve absorption [6]. 1—double packed bed absorption tank, 2—NO_2 adsorption tower, 3—heat exchangers, 4—heat recovery, 5—gases produced from hydrolysis and oxidation, 6—NO_2 gas desorbed from absorption tank, 7—clean exhaust gas, 8—NO_x exhaust gas

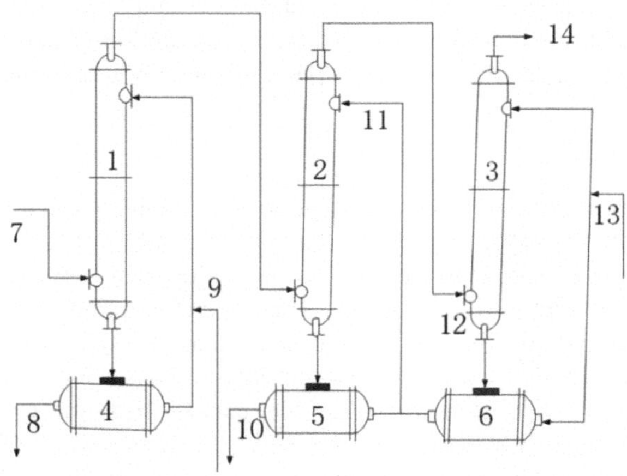

Fig. 15.31 Washing absorption process [6]. 1—foam-covered absorption tower, 2 and 3—sulfuric acid washing tower, 4—nitric acid recovery tank, 5 and 6—sulfuric acid washing and recycling tank, 7—gases from hydrolysis and oxidation, 8—diluted nitric acid, 9—water replenishing, 10—sulphuric acid feeding, 11—sulfuric acid washing solution, 12—gas washing, 13—concentrated sulfuric acid, 14—exhaust

corresponding process. Washing and absorbing controlling condition during the preparation process of oxalic acid from retired propellant was shown in Fig. 15.31.

Gases from hydrolysis and oxidation of retired propellant were sent into NO_x absorption, washing, nitric acid preparation system that was composed of three towers. First, gases went into NO_x absorbing and foam-covered NO_x absorption tower (Tower 1#), where NO_x was absorbed to generate nitric acid solution with a concentration of 55×10^{-2} to 60×10^{-2} kg/L. The generated nitric acid was sent to nitric acid production line to concentrate. Exhausted gases from foam-covered tower went through two washing towers in series (Towers 2# and 3#). In the form-covered tower (Tower 2#) sulfonic acid of 75×10^{-2} kg/L was used to wash the exhaust gases. In Tower 3#, concentration of sulfonic acid was 87×10^{-2} kg/L. After two washing steps, NO_x was basically removed from the gas. After the washing, industrial oxalic acid containing two crystal water could be obtained with melting point of 101–102 °C, which would lose crystal water when it was heated to 98–100 °C. Oxalic acid is a basic raw material for chemical industry, and widely used in dyes, organic synthesis, and precious metals refining industry.

After studying the continuous efforts of these authors, with the abovementioned various processing methods, retired propellant can be fully utilized, leading the arms industry to realize a virtuous circle of production–reserving–recycling–utilization, which has solved the long-standing problem in the military field of all countries. In the separation and degradation process, non-destruction on morphology and energy has been achieved to realize nondestructive and efficiency-enhanced comprehensive utilization of energy. These methods not only have avoided the harm from environmental pollution and arm destruction, and solved safety problems in comprehensive utilization during the elimination of weapons and ammunition after the disarmament, but also ensured the excellent ecological and environmental impacts in the process to effectively protect our resources, and created some industrial value.

References

1. Liu JP, Wang D (2016) Peace—weapons ammunition destruction and comprehensive utilization of disarmament. In: The 252nd American Chemical Society national meeting & exposition
2. Liu JP (1991) Safety utilization of out-of-service single base propellant. In: Proceedings of the 17th International pyrotechnice seminar combined with the 2nd Beijing international symposium on pyrotechnics and explosives
3. Lou JW, Long Y, Xie XB (2007) Obsolete explosives and conventional ammunition disposal of fire and destruction technology. National Defense Industry Press
4. Liu JP, Han SQ, Zhu RL (1997) Recycle of ex-service double base propellants. Energ Mater 7 (1):25–27
5. Zhu RL, Liu JP (1998) The applying study of modified plastics by ex-service double base propellants as raw materials. In: Propellant and explosive in China Ordnance Society
6. Wang Z S, Zhang L H, Cao X M(1999)Treatment and reclamation of waste propellants and explosives. Beijing National Defense Industry Press
7. Wang ZS (1996) The solution and recycle of obsolete energetic materials. Chem Ind Times 10 (6):3–6
8. Gu JL (2004) Recycle of obsolete energetic materials as resources. J Jiangsu Univ (Natural Science Edition) 25(2):180–184

9. Zhang XD, Chen HC, et al (2011) Study on the storage life of stock single-drug based on dynamic and thermal analysis. In: The 3rd national symposium on thermal analysis kinetics and thermokinetics of Chinese Chemical Society Proceedings. Nanjing, vol 10, no 1, pp 58–61
10. Lu H (1974) Single-based propellant. In: Fire and explosives series
11. Urbanski T (1976) Chemistry and technology of explosives(II). Beijing National Defense Industry Press
12. Liu JP, Zhu RL (1999) The study of utilizing ex-service double base propellants to produce coating. In: Propellant and explosive in China Ordnance Society
13. Роговин З А, Шорыгина НН (1953) А.Рогвин Н.Н.Шорыгина,ХИМИЯ ЦЕЛЛЮЛОЗЫ,И ЕЕ СПУТНИКОВ,ГОСУДАРСТВЕННОЕ НАУЧНО-ТЕХНИЧЕСКОЕ,ИЗАТЕЛЬСТВО ХИМИЧЕСКОЙ ЛИТЕРАТУРЫ
14. Tian X, Wang XF, Hang YF et al (2014) The analysis on the wasted explosive green manufacturing technology at home and abroad. In: The 9th parallel sessions of the 16th annual meeting of China association for science and technology
15. Zhang L, Li WZ, Xu LT et al (2013) Research progress of waste energy material reprocessing technologies. Equip Environ Eng 10(3):54–58
16. Wang Q (2004) New technology for preparation of oxidic acid from rice straw by hydrolysis-oxidation-hydrolysis. Jiangsu Chem Ind 32(1):37–39